D0169376

Design and Optimization of Thermal Systems

Mechanical Engineering
A Series of Textbooks and Reference Books

Founding Editor

L. L. Faulkner

Steam Generators and Waste Heat Boilers
For Process and Plant Engineers
V. Ganapathy

Heat Exchanger Design Handbook, Second Edition
Kuppan Thulukkanam

Mechanical Vibration
Analysis, Uncertainties, and Control, Fourth Edition
Haym Benaroya, Mark Nagurka, Seon Han

Blake's Design of Mechanical Joints, Second Edition
Harold Josephs, Ronald L. Huston

Logan's Turbomachinery
Flowpath Design and Performance Fundamentals, Third Edition
Bijay Sultanian

Design and Optimization of Thermal Systems, Third Edition, with MATLAB® Applications
Yogesh Jaluria

For more information about this series, please visit: https://www.crcpress.com/Mechanical-Engineering/book-series/CRCDEKMECENG

Design and Optimization of Thermal Systems

Third Edition

With MATLAB® Applications

Yogesh Jaluria

CRC Press is an imprint of the
Taylor & Francis Group, an **informa** business

CRC Press
Taylor & Francis Group
6000 Broken Sound Parkway NW, Suite 300
Boca Raton, FL 33487-2742

Printed on acid-free paper

International Standard Book Number-13: 978-1-4987-7823-7 (Hardback)

Library of Congress Cataloging-in-Publication Data

Names: Jaluria, Yogesh, author.
Title: Design and optimization of thermal systems / Yogesh Jaluria.
Description: Third edition. | Boca Raton, FL : CRC Press/Taylor & Francis Group, 2020. |
Series: Mechanical engineering | Includes bibliographical references and index.
Identifiers: LCCN 2019018480| ISBN 9781498778237 (hardback : acid-free paper) |
ISBN 9780429085789 (ebook)
Subjects: LCSH: Heat engineering. | Heat engineering—Instruments. | Engineering design.
Classification: LCC TJ260 .J353 2020 | DDC 621.402—dc23
LC record available at https://lccn.loc.gov/2019018480

Visit the Taylor & Francis Web site at
http://www.taylorandfrancis.com

and the CRC Press Web site at
http://www.crcpress.com

eResource material is available for this title at
https://www.crcpress.com/9781498778237

Contents

Preface to the First Edition

APPROACH

Design is an essential element in engineering education and practice. In recent years, there has been a growing emphasis on the design and optimization of systems because of growing worldwide competition and the development of new processes and techniques. Design has long been very important in undergraduate mechanical engineering curricula around the country. However, the effort had been largely directed at mechanical systems, dealing with areas such as transmission, vibrations, robotics, and controls, and at components such as gears, cams, springs, and linkages.

With the growth of thermal systems, such as those related to materials processing, energy conversion, pollution, aerospace, and automobiles, the need to design and optimize thermal systems has also grown. In mechanical engineering programs around the country, courses have been developed on the design of thermal systems. These are often elective courses, or, in many schools, such courses form the final capstone design course, often alternated with a design course on mechanical systems. Invariably, optimization is an important element in such courses because of the crucial need to optimize systems in practical applications.

This book is written as a textbook at the senior undergraduate or the first-year graduate level. It can also be used as a reference book for other thermal sciences courses, such as those on heat transfer, fluid mechanics, and thermodynamics, and for courses in applied areas, such as power plants, environmental control of buildings, and solar energy systems. It can be used for reference by practicing engineers as well. Although the book is written for engineering education curricula in the United States, the material and treatment can easily be used in various countries around the world. The book is largely written for mechanical engineers. However, the material is suitable for courses in other engineering disciplines, such as chemical, aerospace, industrial, and materials engineering.

The book is directed at the design of thermal systems, employing examples from diverse areas such as manufacturing, energy systems, cooling of electronic equipment, refrigeration, environmental problems, engines, and heat transfer equipment. Many such examples and an introduction to design are given in Chapter 1. Then the conceptual design and formulation of the design problem are presented in Chapter 2, along with the main constituents of design, including material selection. The design process, as predominantly based on the mathematical modeling of the system and on the results obtained from numerical simulation, is presented in the next three chapters. Analytical results, if available, are valuable for validating the numerical model as well as for providing a fundamental basis for design. The use of experimental data, mainly as correlations derived from curve fitting, is presented as an essential element in the design and optimization process.

The basic approach to the development of a suitable model is discussed in Chapter 3 in detail because this forms the most crucial step in the design of a system. Various approximations and idealizations that can be used for modeling are presented, along with examples of mathematical modeling of practical systems. This is followed in Chapter 4 by detailed discussions of numerical modeling and simulation, again linking these to the mathematical model and experimental data. These results then form the basis for creative design of thermal systems and for the evaluation of the designs obtained.

The development of a workable, or acceptable, design, one which satisfies the requirements and constraints of the problem, is discussed as a synthesis of the different design steps. Several examples are given in Chapter 5 to illustrate the overall procedure. This is followed in Chapter 6 by a discussion of economic factors in design because these often guide system design and optimization. The formulation of the optimization problem is explained in Chapter 7. The presentation on optimization includes several applicable methods such as calculus methods, search methods, and linear, dynamic, and geometric programming. These topics are covered in Chapter 8 through Chapter 10. Again, thermal systems found in several important and relevant areas are used as examples to illustrate the ideas presented. Solved examples and problems strengthen the presentation and allow the important concepts to be assimilated by the readers. Recent trends, such as knowledge-based design methodology, are included in Chapter 11. Additional practical considerations, such as economic, safety, and materials, are also discussed at various stages of the presentation.

Therefore, the book starts with the basic design concept and develops the material through modeling and simulation stages of the thermal process and system on to the creation of a workable design. The iterative process that is often used to obtain an acceptable design is presented. This logically leads to optimization procedures and the improvement of the design to obtain the best possible solution under the given constraints. The book offers a systematic approach, exploring the various considerations that lead to a workable and, finally, to an optimal design; uses up-to-date examples and problems, and presents the most current information and design tools available in this important field. Examples range from simple systems to large, complex practical systems. Synthesis of the various aspects that constitute design is discussed in detail. A few relevant computer programs are included to help with the numerical modeling and simulation. These include programs on curve fitting, solution of algebraic systems, and solution of simple differential equations. Quantitative information on materials, economics, and heat transfer correlations is also included. The mathematical modeling of systems is a particularly important aspect of design, though it is often neglected. Modeling is presented with a wide range of physical examples and a discussion of the types of approximations that can be used to simplify the problem. The inputs obtained from the model for an innovative and optimal design are outlined.

The main thrust of this book is to develop and discuss the basic considerations that arise in the design and optimization of thermal systems, as well as the appropriate methodology. Readers are encouraged to use their own backgrounds, imaginations, and available literature for designing different types of systems. Because simple closed-form, analytical solutions are rarely obtained in practical thermal systems, clear and concise answers are not readily available in many cases. A consideration of several practical systems makes this aspect of thermal system design clear.

The material included in this book has been used in courses for undergraduate seniors at Rutgers University and at the Indian Institute of Technology, Kanpur, India. The topics, examples, and problems have, therefore, been largely tested in a classroom environment. Various design projects and examples emerged from these courses, and some of them are included in the text. In keeping with the basic design process, many of the problems are open-ended and a unique solution is not obtained.

SUPPLEMENT

A solutions manual, prepared by me in order to ensure the problem-solving methodology is the same as that in the book, is available to text adopters. This manual contains possible solutions to most of the problems in this book (because many problems are open-ended and thus do not have a unique solution).

ACKNOWLEDGMENTS

Many colleagues, friends, and students have contributed in significant ways to the development of this material and to my understanding of the design process. Of particular help have been the many discussions I have had with Professors V. Sernas and N.A. Langrana of Rutgers University on the design of systems. My interactions with students over many years on design projects and on practical design problems have also provided important insights into the design process. The comments of the manuscript reviewers and of Professor J. R. Lloyd of Michigan State University, the editor of this series, have been very valuable and helpful in organizing and presenting the material. I would like to take the opportunity to thank each of the manuscript reviewers personally. The reviewers are Jamal Seyed-Yagoobi, Texas A&M University; Andrew V. Tangborn, Northeastern University; Edwin Griggs, Tennessee Technological University; Bakhtier Farouk, Drexel University; Donald Fenton, Kansas State University; John R. Lloyd, Consulting Editor, Michigan State University; Donald C. Raney, The University of Alabama; Jeffrey Hodgson, The University of Tennessee, Knoxville; Louis C. Burmeister, The University of Kansas; Edward Vendell, Utah State University; Edward Hensel, New Mexico State University; and Prasanna Kadaba, Georgia Institute of Technology. The assistance provided by the staff of McGraw-Hill has also contributed very significantly to the book.

This book is dedicated to my parents, who have been a constant source of encouragement, support, and inspiration. Certainly, the greatest contribution to this work has been the unqualified support and encouragement of my wife, Anuradha, and the patience and understanding of our children, Pratik, Aseem, and Ankur. Without their support, it would not have been possible for me to meet the strong demands placed on my time by this book.

Yogesh Jaluria

Preface to the Third Edition

The third edition of this book follows the basic approach and treatment of the previous editions. The book mainly considers thermal systems, which are largely governed by the principles of thermodynamics, fluid mechanics, and heat transfer. However, the fundamental aspects presented in the book can be extended to a much wider range of systems. The book presents a systematic approach to design and optimization. Starting with the formulation of the problem, the process involves conceptual design, modeling, simulation, acceptable design, and, finally, optimization. The inputs needed for the design and optimization process may be obtained by modeling, analysis and simulation, or by experimentation, though in many cases all three approaches are used to obtain the necessary data. Additional considerations, such as control, communication, financial aspects, safety, material selection, ethics and uncertainties, that arise in practical systems are also briefly outlined.

Examples are taken from many different areas, such as energy, environment, heating, cooling, manufacturing, aerospace, and transportation, to illustrate the basic approach and convergence to a feasible or optimal design. Solved examples and exercises are included to supplement the discussion and methods presented in the text. The material presented in the second edition has been updated to include recent books and advances on various topics. Much of the material has been expanded to include recent trends and emerging areas of interest. Additional references have been included and previous references have been updated.

Among the additional topics that were included after the first edition are artificial-intelligence-based techniques like genetic algorithms, fuzzy logic, and artificial neural networks. Response surfaces and other optimization techniques are included in the discussion, along with effective use of concurrent experimental and numerical inputs for design and optimization. Multi-objective optimization is particularly important for thermal systems, since more than one objective function is typically important in realistic systems, and a detailed treatment is included. Other strategies to optimize the system are presented. The application of these ideas to the optimization of thermal systems is reiterated with examples of actual, practical systems.

The book has been used as a textbook for engineering senior undergraduate or first-year graduate level courses in design, for capstone design courses, and for courses in thermal engineering. It can also be used as a reference in courses in relevant basic and applied areas in engineering. It is also a useful reference for practicing engineers.

This edition continues the emphasis on the use of Matlab for numerical modeling and simulation, because Matlab is often the dominant computational environment for the numerical solution of mathematical equations. Several examples and exercises are given to illustrate the use of Matlab in numerical modeling. The computer programs in Matlab have been expanded from the second edition to include several other methods of solution and also for solving additional problems such as partial differential equations. Additional exercises and examples are also included. Clarifications and simplifications are added at various places to make learning easier.

As mentioned in the earlier editions, the material presented in this textbook is the outcome of many years of teaching modeling, design, and optimization of thermal systems, in a variety of elective courses and in capstone design courses. Several colleagues and former students have been indispensable in the choice of topics and the depth and breadth of coverage. The support, assistance, and patience of the editorial staff of CRC Press, particularly Jonathan Plant, have been valuable in the development of the third edition. This book, like all my other efforts, is dedicated to my late parents, who had been a constant source of encouragement and inspiration.

Finally, I would like to acknowledge the strong and unwavering encouragement and support of my wife, Anuradha, of our children, Pratik, Aseem, and Ankur, and of their spouses, Leslie, Karishma, and Russell. My grandsons, Vyan, Nalin, Zev, and Jai, provided the incentive to complete this effort.

Yogesh Jaluria

Author

Yogesh Jaluria, MS, PhD, is currently Board of Governors Professor and Distinguished Professor at Rutgers, the State University of New Jersey, New Brunswick. He earned a BS at the Indian Institute of Technology, Delhi, India. He earned an MS and a PhD in mechanical engineering at Cornell University.

Jaluria has contributed more than 500 technical articles, including over 210 in archival journals and 18 chapters in books. He has three patents in materials processing and is the author/co-author of nine books in the areas of natural convection, computational methods, design, optimization and materials processing. Jaluria received the 2003 Robert Henry Thurston Lecture Award from the American Society of Mechanical Engineers (ASME), and the 2002 Max Jakob Memorial Award for eminent achievement in the field of heat transfer from ASME and the American Institute of Chemical Engineers (AIChE). In 2002, he was named Board of Governors Professor of Mechanical and Aerospace Engineering at Rutgers University. He was selected as the 2000 Freeman Scholar by the Fluids Engineering Division, ASME. He received the 1999 Worcester Reed Warner Medal and the 1995 Heat Transfer Memorial Award for significant research contributions to the science of heat transfer, both from ASME. He also received the 1994 Distinguished Alumni Award from the Indian Institute of Technology, Delhi.

Jaluria is an Honorary Member of ASME and a Fellow of the American Physical Society and the American Association for the Advancement of Science. He served as the chair of the Heat Transfer Division of ASME during 2002–2003 and has chaired several international conferences. He served as the editor of the ASME *Journal of Heat Transfer* during 2005–2010. He is currently the President of the American Society of Thermal and Fluids Engineers (ASTFE), an international organization that brings thermal and fluids engineering researchers together and focuses on industry, international collaboration, and young researchers entering the field.

1 Introduction

Design is generally regarded as a creative process by which new methods, devices, and techniques are developed to solve new or existing problems. Though many professions are concerned with creativity leading to new arrangements, structures, or artifacts, design is an essential element in engineering education and practice. Due to increasing worldwide competition and the need to develop new, improved, and more efficient processes and techniques, growing emphasis is being placed on design. Interest lies in producing new and higher quality products at minimal cost, while satisfying increasing concerns regarding environmental impact and safety. It is no longer adequate just to develop a system that performs the desired task to satisfy a recognized societal need. It is crucial to optimize the process so that a chosen quantity, known as the objective function, is maximized or minimized. Thus, for a given system, the output, profit, productivity, product quality, etc., may be maximized, or the cost per item, investment, energy input, etc., may be minimized.

The survival and growth of most industries today strongly depend on the design and optimization of the relevant systems. With the advent of many new materials, such as composites and ceramics, and new manufacturing processes, such as three-dimensional printing, several traditional industries, such as the steel industry, have diminished in importance in recent years, while many new fields have emerged. It is important to keep abreast of changing trends in these areas and to use new techniques for product improvement and cost reduction. Even in an expanding engineering area, such as consumer electronics, the prosperity of a given company is closely linked with the design and optimization of new processes and systems and the optimization of existing ones. Consequently, the subject of design, which had always been important, has become increasingly critical in today's world and has also become closely coupled with optimization.

In recent years, we have also seen tremendous growth in the development and use of thermal systems in which fluid flow and transport of energy play a dominant role. These systems arise in many diverse engineering fields such as those related to manufacturing, power generation, pollution, air conditioning, heating, and aerospace and automobile engineering. Therefore, it has become important to apply design and optimization methods that traditionally have been applied to mechanical systems, such as those involved with transmission, vibrations, controls, and robotics, to thermal systems and processes. In this book, we shall focus on thermal systems, considering examples from many important areas, ranging from classical and traditional fields like engines and heating/cooling to new and emerging fields like materials processing, data centers, nanomaterials, and alternative energy sources. However, many of the basic concepts presented here are also applicable to other types of systems that arise in different fields of engineering, for example, civil, chemical, electrical, and industrial engineering.

In this chapter, we shall first consider the main features of engineering design, its importance in the overall context of an engineering enterprise, and the need to optimize. We will also examine design in relation to analysis, synthesis, selection of equipment, and other important activities that support design. This discussion will be followed by a consideration of systems, components, and subsystems. The basic nature of thermal systems will be outlined, and examples of different types of systems will be presented from many diverse and important areas.

1.1 ENGINEERING DESIGN

One of the most important tasks confronted by engineers is that of design. It may be the design of an individual component, such as a thermostat, flow valve, gear, or spring, or it may be the design of a system, such as a furnace, air conditioner, or an internal combustion engine that consists of several components or constituents interacting with one another. It is, therefore, fair to ask what design is and what distinguishes it from other activities, such as analysis and synthesis, with which engineers are frequently concerned. However, design has come to mean different things to different people. The perception of design ranges from the creation of a new device or process to the routine calculation and presentation of specifications of the different items that make up a system. However, design must incorporate some element of creativity and innovation, in terms of a new, different, and better approach to the solution of an existing engineering problem that has been solved by other methods or the solution to a problem that has not been solved before. The process by which such new, different, or improved solutions are derived and applied to engineering problems is termed *design*.

1.1.1 Design versus Analysis

We are all quite familiar with the analysis of engineering problems using information derived from basic areas such as statics, dynamics, thermodynamics, fluid mechanics, and heat transfer. The problems considered are often relevant to these disciplines and little interaction between different disciplines is brought into play. In addition, all the appropriate inputs needed for the problem are usually given and the results are generally unique and well-defined, so that the solution to a given problem may be carried out to completion, yielding the final result that satisfies the various inputs and conditions provided. Such problems are usually termed as *closed-ended*.

The calculation of the velocity profile for developed, laminar fluid flow in a circular pipe to yield the well-known parabolic distribution shown in Figure 1.1(a) is an example of analysis. Similarly, the analysis of steady, one-dimensional heat conduction in a flat plate results in the linear temperature distribution shown in Figure 1.1(b). Textbooks on fluid mechanics and heat transfer, such as Pritchard and Mitchell (2015) and Incropera and Dewitt (2001), respectively, present many such analyses for a variety of physical circumstances. Many courses are directed at engineering analysis and engineering students are taught various techniques to solve simple as well as complicated problems in fundamental and applied areas. Most students thus acquire the skills and expertise to analyze well-defined and well-posed problems in different engineering disciplines.

The design process, on the other hand, is *open-ended*; that is, the results are not well-known or well-defined at the onset. The inputs may also be vague or incomplete, making it necessary to

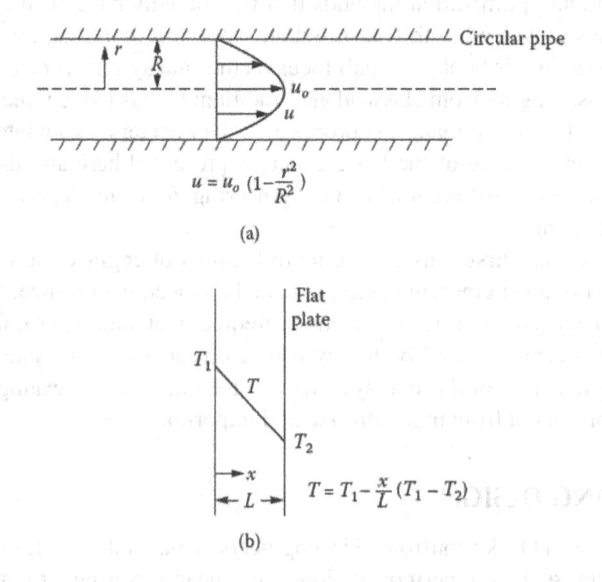

FIGURE 1.1 Analytical results for (a) developed fluid flow in a circular pipe and (b) steady-state one-dimensional heat conduction in a flat plate.

seek additional information or to employ approximations and assumptions. There is also usually considerable interaction between various disciplines, particularly between technical areas and those concerned with cost, safety, and the environment. A unique solution is generally not obtained and one may have to choose from a range of acceptable solutions. In addition, a solution that satisfies all the requirements may not be obtained and it may be necessary to relax some of the requirements to obtain an acceptable solution. Therefore, *trade-offs* generally form a necessary part of design because certain features of the system may have to be given up in order to achieve some other goals such as greater cost effectiveness or smaller environmental impact. Individual or group *judgment* based on available information is needed to decide on the final design. *Inverse problems*, in which the desired outcome is known but the conditions that would lead to this outcome are to be determined, are also commonly encountered. Again, the solution obtained is not unique and optimization methods are needed to narrow the region of uncertainty and thus achieve a physically acceptable result.

1.1.2 A Few Examples

Consider the example of an electronic component located on a board and being cooled by the flow of air driven by a fan, as shown in Figure 1.2. The energy dissipated by the component is given. If the temperature distributions in the component, the board, and other parts of the system are to be determined, analysis or numerical calculations may be used for this purpose. Even though the numerical procedure for obtaining this information may be quite involved, the solution is unique for the given geometry, material properties, and dimensions. Different methods of solution may be employed but the problem itself is well-defined, with all the input quantities specified and with no variables left to be chosen arbitrarily. No trade-offs or additional considerations need to be included.

Let us now consider the corresponding design problem of finding the appropriate materials, geometry, and dimensions so that the temperature T_c in the component remains below a certain value, T_{max}, in order to ensure satisfactory performance of the electronic circuit. This is clearly a much more involved problem. There is no unique answer because many combinations of materials, dimensions, geometry, fan capacity, etc., may be chosen to satisfy the given requirement $T_c < T_{max}$. There is obviously considerable freedom and flexibility in choosing the different variables that characterize the system. Such a problem is, thus, open-ended and many solutions may be obtained to satisfy the given need and constraints, if any, on cost, size, dimensions, etc. It is also possible that a satisfactory solution cannot be found for the given conditions and an additional cooling method such as a heat pipe, which conveys the heat dissipated at a much higher rate by means of a phase change process, may have to be included, as shown by the dotted lines in Figure 1.2. Then the design process must consider the two cooling arrangements and determine the relevant characteristic parameters

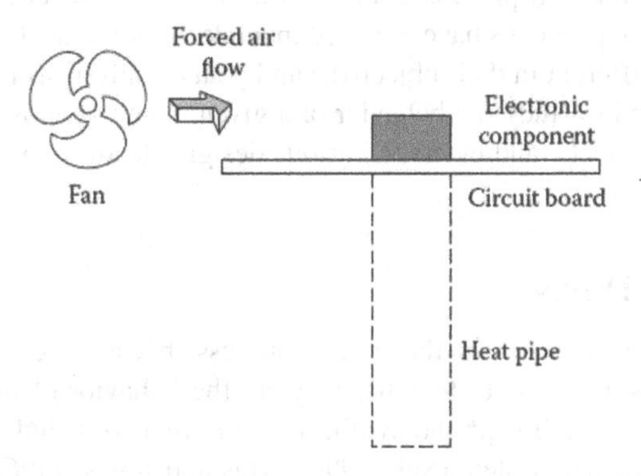

FIGURE 1.2 An electronic component being cooled by forced convection and by a heat pipe.

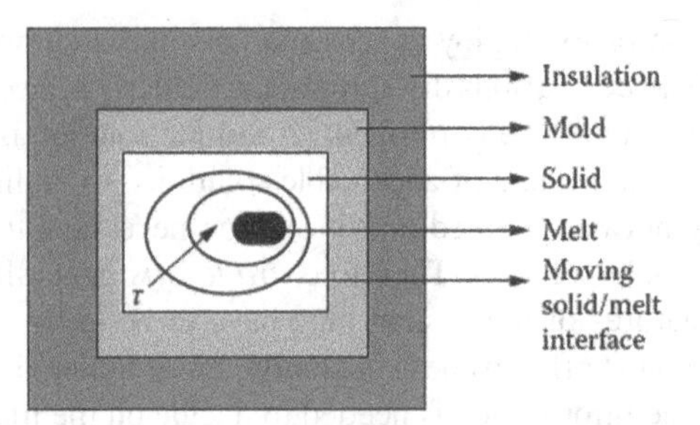

FIGURE 1.3 The casting process in an enclosed region.

for these cases. Thus, different approaches, often known as conceptual designs, may be considered for satisfying the given requirements.

Another example that illustrates the difference between analysis and design is that of a casting process, as sketched in Figure 1.3. Molten material is poured into a mold and allowed to solidify. If the properties of the material undergoing solidification and of the various parts of the system, such as the mold wall and the insulation, are given along with the relevant dimensions, the initial temperature, and the convective heat transfer coefficient h at the outer surface of the mold, the problem may be solved by analysis or numerical computation to determine the temperature distributions in the solid material, liquid, and various parts of the system, as well as the rate and total time of solidification for the casting (Flemings, 1974). The problem can often be simplified by using approximations such as constant material properties, negligible convective flow in the melt, uniform heat transfer coefficient h over the entire surface, etc. But once the problem is posed in terms of the governing equations and appropriate boundary conditions, the results are generally well-defined and unique.

We may now pose a corresponding design problem by allowing a choice of materials and dimensions for the mold wall and insulation and of the cooling conditions at the outer surface, in order to reduce the solidification time below a desired value τ_{cast}. Then, many combinations of wall material and thickness, cooling parameters, insulation parameters, etc., are possible. Again, there is no unique solution, and, indeed, there is no guarantee that a solution will be found. All that is given is the requirement regarding the solidification time and quantities that may be varied to achieve a satisfactory design. In other cases, the requirements may be specified as limitations on the temperature gradients in the casting in order to improve the quality of the product. Clearly, this is an open-ended problem without a unique solution.

It is largely because of the open-ended nature of design problems that design is often much more involved than analysis. Consequently, while extensive information is available in the literature on the analysis of various thermal processes and on the resulting effects of the governing variables, the corresponding design problems have received much less attention. However, even though design and analysis are very different in their objectives and goals, analysis usually forms the basis for the design process. It is used to study the behavior of a given system, choose the appropriate variables to achieve the desired effects, and evaluate various designs, leading to satisfactory and optimized systems.

1.1.3 Synthesis for Design

Synthesis is another key element in the design process, because several components and their corresponding analyses are brought together to yield the behavior of the overall system. Results from different areas must be linked and synthesized in order to include all of the important concerns that arise in a practical system (Suh, 1990; Ertas and Jones, 1996; Dieter, 2000; Dieter and Schmidt, 2012). We cannot consider only the heat transfer aspects in the casting problem while

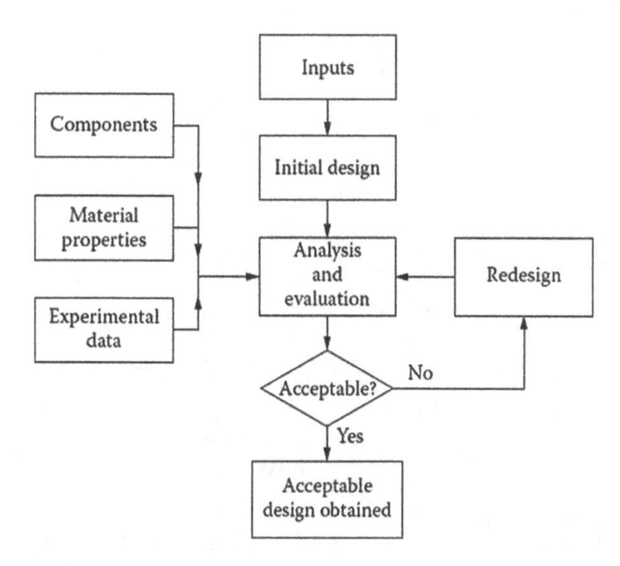

FIGURE 1.4 Schematic of a typical design procedure.

ignoring the strength of materials and manufacturing aspects. Information from different types of models, including experimental and numerical results, and from existing systems is incorporated into the design process. The cost, properties, and characteristics of various materials that may be employed must also form part of the design effort, because material selection is a very important factor in obtaining an acceptable or optimal system. Additional aspects, such as safety, legal, regulatory, and environmental considerations, are also synthesized in order to obtain a satisfactory design. Figure 1.4 shows a sketch of a typical design process for a system, involving both analysis and synthesis as part of the overall effort.

1.1.4 Selection versus Design

We are frequently faced with the task of selecting parts in order to assemble a system or a device that will perform a desired duty. In several cases, the entire equipment may be selected from what is available on the market, for instance, a heat exchanger, a pump, or a compressor. Even though selection is an important ingredient in engineering practice, it is quite different from designing a component or device and it is important to distinguish between the two. Selection largely involves determining the specifications of the item from the requirements for the given task. Based on these specifications, a choice is made from the various types of items available with different ratings or features. Design, on the other hand, involves starting with a basic concept, modeling and evaluating different designs, and obtaining a final design that meets the given requirements and constraints. The system may then be fabricated and tests carried out on a prototype before going into production. Therefore, design is directed at creating a new process or system, whereas selection is concerned with choosing the right item from those that are readily available for a given job.

Selection and design are frequently employed together in the development of a system, using selection for components that are easily available over the ranges of interest. Standard items such as valves, control sensors, heaters, flow meters, and storage tanks are usually selected from catalogs of available equipment. Similarly, pumps, compressors, fans, and condensers may be selected, rather than designed, for a given application. Obviously, design is involved in the development of these components as well; however, for a given system, the design of these individual components may be avoided in the interest of time, cost, and convenience. For instance, a company that develops and manufactures heat exchangers would generally design different types of heat exchangers for different fluids and applications, achieving different ranges in heat transfer rate, area, effectiveness,

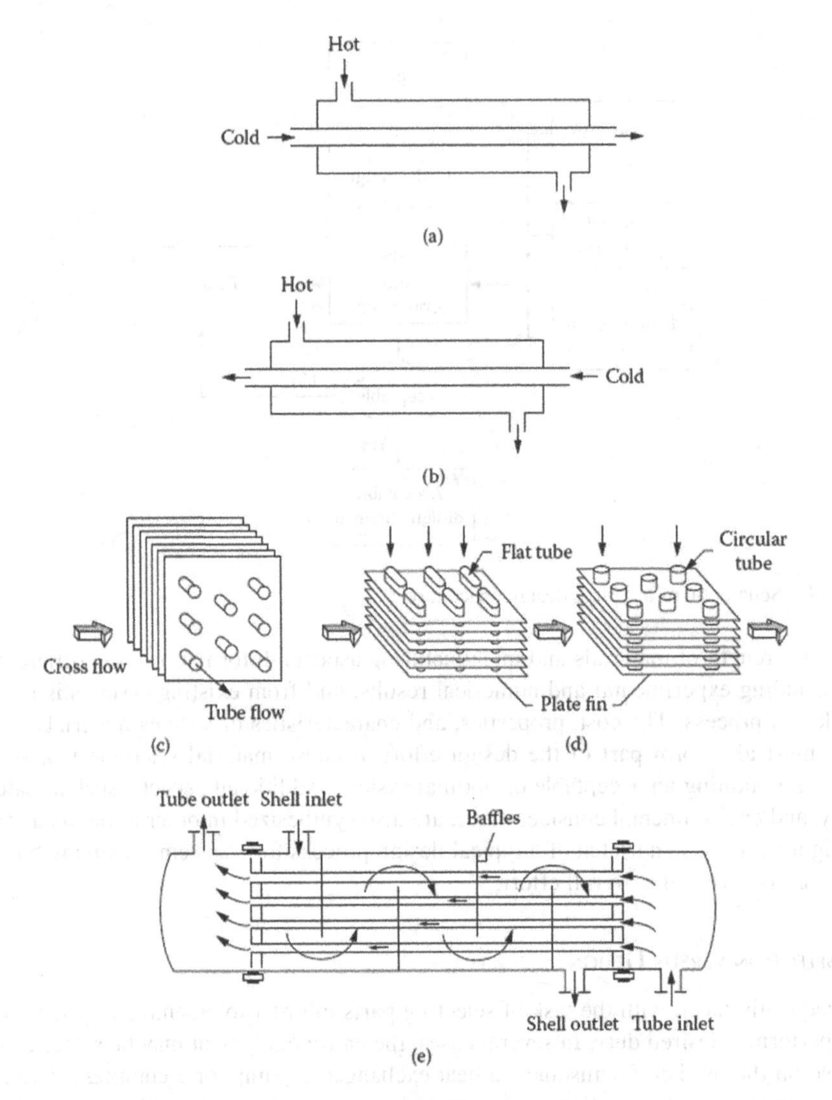

FIGURE 1.5 Common types of heat exchangers. (a) Concentric pipe parallel-flow, (b) concentric pipe counter-flow, (c) cross-flow with unmixed fluids, (d) fin-tube compact heat exchanger cores, (e) shell-and-tube. (Adapted from Incropera, F.P. and Dewitt, D.P., 2001.)

flow rate, etc. Different configurations such as counter-flow and parallel-flow heat exchangers, compact heat exchangers, shell-and-tube heat exchangers, etc., as shown in Figure 1.5, may be considered for a variety of applications. These may then be designed to obtain the desired parametric ranges of heat transfer rate, output temperature, size, and so on (Kays and London, 1984). Design engineers working on another thermal system, such as air conditioning or indoor heating, may simply select the condenser, evaporator, or other types of heat exchangers needed, rather than design these.

Selection is clearly a much less involved process, as compared to design. The requirements and specifications of the desired component or equipment are matched with whatever is available. If an item possessing the desired characteristics is not available, design is needed to obtain one that is acceptable for the given purpose. Because selection is often used as part of the overall system design, the two terms are sometimes interchanged. We are mainly concerned with the design of thermal systems and, as such, selection of components needed for a system will be considered only as a step in the design process, particularly during the synthesis of the various parts.

1.2 DESIGN AS PART OF ENGINEERING ENTERPRISE

Before proceeding to a discussion of the characteristics and types of thermal systems, it will be instructive to consider the position occupied by design and optimization in the overall scheme of an engineering undertaking. The planning and execution of such an enterprise involve many aspects that are engineering based and several that are not, for example, economic and market considerations. Engineering design is one of the key elements in the development of a product or a system and is coupled with the other considerations to obtain a successful venture. Let us follow a typical engineering undertaking from the initial recognition of need for a particular item, device, or process to its final implementation.

1.2.1 Need or Opportunity

Defining a need or opportunity is always the first step in an engineering undertaking because it provides the impetus to develop a product or system. *Need* refers to a specific requirement and implies that a suitable item is not available and must be developed for the desired purpose. The need for a given item may be felt at various levels, ranging from the consumer and the retailer to the industry itself, and may involve developing a new system or modifying and improving existing ones. *Opportunity* is the recognition of a chance to develop a new product that may be superior to existing ones or less expensive. It may also be an item for which the market is expected to develop as it becomes available.

Consumers' need for a new or improved product is often discovered through surveys conducted by the marketing and sales division and through consumer interactions with salespersons. In some cases, individual consumers and consumer groups may also provide information on their needs and requirements. The problems or limitations in existing products may become evident from such inputs, indicating the need for developing a new or improved product. The development of the hard disk in personal computers arose mainly because of consumers' need for larger data storage capacity. Similarly, CD-ROMs and memory sticks were introduced because of the need to facilitate storage and transfer of data and information. In automobiles, antilock brakes, air bags, computer-controlled fuel injection, and streamlining of the body have been introduced in response to safety and efficiency needs. The need for specific components or systems may also arise in auxiliary industrial units that are dependent on the main industry. For instance, the development of larger and improved television systems, such as the high-definition television, has generated demand for a range of electronic products and systems that will be met by other specialized industries.

The opportunity to move into a new area, develop a new product or system, substantially increase the quality of an existing item, or significantly reduce the cost of an item can also form the starting point for an engineering undertaking. This is particularly true of new materials because the substitution of materials in existing systems by new or improved materials could lead to substantial improvement in the system performance and/or reduction in cost. The replacement of metal casings in electronic equipment by plastic or ceramic ones and of metal frames in sports equipment by composites represents such changes. The personal computer is an interesting example of such an opportunity-based development. An opportunity was perceived by the industry, mainly by Apple Computers, Inc., and adequate technical expertise was available to develop a personal computer. This led to an expanding market and the use of personal computers in a variety of applications, ranging from word processing, information storage, and accounting to instruction and data acquisition. The video cassette recorder, fiber-optic cable, compact disc player, microwave oven, and the Apple iPod, iPhone, and iPad represent new products that were developed in recent years with possible opportunities and expanding markets in mind.

The industry today is very dynamic and is always on the lookout for opportunities where the available technical know-how can be used effectively to develop new ideas, leading to new products and systems. The research and development division of a given industrial concern is often the source of such opportunities because of its interest in new materials and techniques being developed in the academic, industrial, and research environments outside the firm. However, a new idea may also arise from other divisions in the company based on their involvement with various processes and products.

1.2.2 Evaluation and Market Analysis

An important consideration in the development of a new concept is its evaluation for economic viability, because profit is usually the main driving force in engineering undertakings. Even if need and opportunity have indicated that a particular product or system will be useful and will have a secure market, it is necessary to determine how big the market is, what price range it will bear, and what the possible expenses involved in taking the concept to completion are. The sales and marketing division of the company could target typical consumers, who may be individuals, organizations, or other industries. The information regarding price, consumption level, desired characteristics of the product, and nature of the intended application could be gathered through surveys, mail, telephone or individual contact, interactions with product outlets and sales organizations, and inputs from consumer groups. Earlier studies on similar products may also be used to provide the relevant information for evaluating the proposed venture. For instance, many products have recently been reduced in size and weight because of consumer demand. These include video recorders, laptop computers, digital cameras, and cars. In each case, a market analysis must have been carried out to ensure that the price and the demand were satisfactory to justify the time, money, and effort spent in developing these items. Of course, in the case of cars, the need to reduce fuel consumption was one of the main motivations for size reduction.

Once information from various sources is obtained on the product being considered, the marketing division may carry out a detailed market analysis to determine the anticipated volume of sales and the effect of the price on the sales. As the price increases, the volume of sales is expected to decrease. Consider the development of a new gas water heater for residential use. The cost increases as the capacity of the tank is increased. Similarly, a faster response to an increased demand for hot water, though desirable, would require larger heaters, leading to higher costs. Better safety and durability features will also raise the price. Clearly, additional features and higher quality make it attractive to various consumers and may open additional markets. However, as the price continues to increase, the sales volume will generally decrease, partly because of less frequent replacement, resulting from improved quality, and partly due to loss in sales to less expensive versions. Very selective models may have a small volume of sales but a large profit, or return, per unit. Figure 1.6 shows typical sales volume versus price curves. The curves are separated by differences

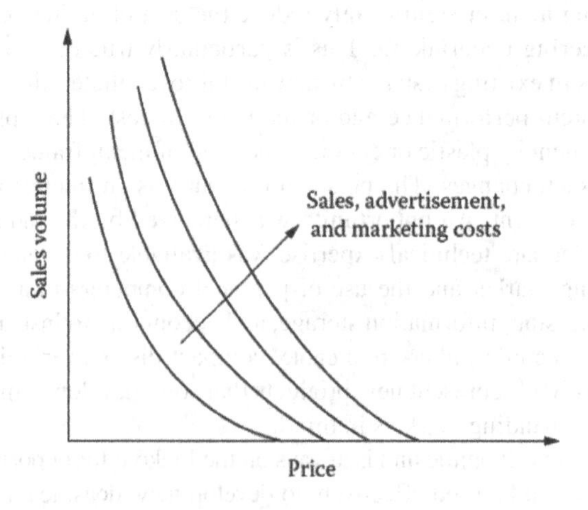

FIGURE 1.6 Typical variation of volume of sales with price.

in the expenditure involved in marketing, advertising, and sales. The profit per item is smaller at a given price if the expense in advertising is increased. However, it is expected that the total volume will increase due to better advertising, making the overall venture more profitable (Stoecker, 1989).

The evaluation of the enterprise must include all expenses that are expected to be incurred. Besides the cost of manufacture of the given item and the expense of advertising and sales, the cost of designing and developing the system, from the initial concept to the prototype, must also be considered. The cost must include both labor and the capital investment needed for equipment and supplies. Considering all the relevant costs and the anticipated sales volume (employing economic concepts as outlined in Chapter 6), the given undertaking may be evaluated to determine the profit or the percentage return on the investment. If the return is too low, the process may be terminated at this stage. Several new ideas and concepts are evaluated by typical industries, and many of these do not go much further because of an expected small volume of sales or a large investment needed for development and manufacture. In several cases, specialized companies exist in order to fabricate custom-made or one-of-a-kind products at the specific request of a client. The price may be exorbitant in these cases, but only one or two systems are made, providing a satisfactory return because of the high price rather than the large sales volume.

1.2.3 Feasibility and Chances of Success

It is important to determine if a particular enterprise is feasible. It is also necessary to evaluate the chances of success. These considerations are usually brought up early in the project, though inputs from research, development, and design may be needed to make a reliable judgment. The future of the project is strongly influenced by the results obtained from this study.

1.2.3.1 Measure of Success

The basis for evaluating success must be defined first. This would depend on the nature of the enterprise and the product under consideration. The return on investment is the criterion used by most engineering companies to determine if an undertaking is successful. The dividends paid to investors or the value in the stock market are also important measures of success of an enterprise. Sometimes, other considerations are more important than profit for a given undertaking. Pollution and environmental requirements due to government regulations may be crucial factors. For instance, the deterioration of the ozone layer has made it necessary to seek alternatives to traditional refrigerants, such as refrigerant 12 (Freon 12), which is a chlorofluorocarbon (CFC), and considerable effort has been directed at the development and testing of other fluids for this purpose. Satisfactory hazardous waste disposal similarly may be the dominant consideration in a chemical plant. Cooling towers may have to be used instead of an available lake for cooling the condensers of a power plant, again because of the undesirable environmental impact on the lake. The desire to reduce the dependence on imported oil due to national or political reasons has similarly led to work on synthetic fuels and nonconventional energy sources. Global climate change concerns have led to major investments in renewable energy sources. Safety aspects may also be used as criteria to evaluate success, particularly in nuclear reactors. National defense may require the indigenous development of certain components or systems, even though these may be procured cheaply abroad. Thus, even though profit is usually the main criterion of success, other considerations may also be used to evaluate the success of an engineering venture.

1.2.3.2 Chances of Success

Once the basis for evaluating success is chosen, the next step is to determine the chances of success. Because success depends on many events in the future that cannot be predicted with certainty, evaluation of the chances of success is based on a probabilistic analysis of the various items that are involved in the enterprise, such as financing, design, research and development, manufacturing, testing, government approvals, sales, advertising, and marketing. The probability of success must be considered over the entire duration of the project and may be expressed in terms of the probability of achieving the chosen measure of success. Suppose the rate of return r is taken as the criterion of success for a given undertaking. The probability P of achieving a return between r_1 and r_2 is given in terms of the probability function $f(r)$, which gives the probability of the return lying between r and $r + dr$ as

$$P = \int_{r_1}^{r_2} f(r)\,dr \tag{1.1}$$

with

$$\int_{-\infty}^{+\infty} f(r)\,dr = 1 \tag{1.2}$$

indicating that the probability of the return lying somewhere between $-\infty$ and $+\infty$ is 1, or 100%. The probability distribution is often a normal distribution curve given by

$$f(r) = \frac{1}{(2\pi)^{1/2}\sigma} \exp\left[-\frac{1}{2}\left(\frac{r-\mu}{\sigma}\right)^2\right] \tag{1.3}$$

This distribution has a maximum, which occurs at the mean value μ, and a standard deviation σ, which gives the spread of the curve, as shown in Figure 1.7. Thus, a larger maximum indicates a higher probability of attaining values around μ and a larger deviation σ indicates a larger spread or

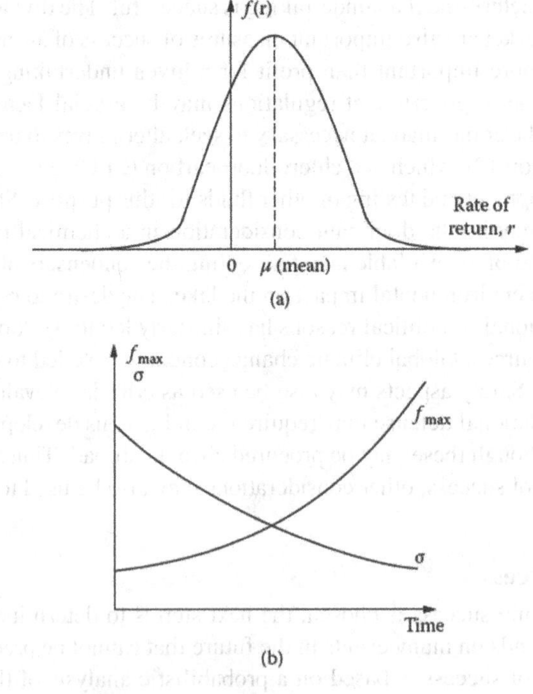

FIGURE 1.7 Probability distribution curve for the rate of return r, along with anticipated change in the maximum value f_{max} and the deviation σ with time.

uncertainty. Other distributions also arise in different cases and the corresponding characteristics may be determined. The probabilities of the occurrence of various events that make up the enterprise are considered and a statistical analysis is carried out to determine the probability function for the chosen criterion for success, such as the rate of return, margin of safety, or level of environmental pollution.

At the very beginning of the enterprise, the probability curve is expected to be spread out, indicating the large amount of uncertainty stemming from many aspects that have to be taken care of in the future. The maximum value is small, suggesting a small probability of the rate of return lying within a given range. Remember, the total area under the $f(r)$ curve must be 1 because r must have a value in the entire range, as seen from Equation (1.2). As time elapses and various concerns are resolved, the uncertainty decreases and the spread of the distribution curve reduces while the maximum value increases (Stoecker, 1989). These basic trends are also shown in Figure 1.7, indicating the increasing maximum with time and the reducing deviation. If the predicted results on the chances of success are not satisfactory, the effort may be terminated before much expense has been incurred.

1.2.3.3 Feasibility

Another important consideration is whether the enterprise is possible at all. There is no point in proceeding any further unless there is a clear indication that it is achievable. It may be infeasible because of many reasons, some of which may not be technical. We have already considered the economic viability of the project. If the rate of return on the investment is too small, or if the chances for success are not satisfactory, the enterprise may be terminated. However, even if the project is economically viable, it may not be possible technically because of constraints with respect to available materials, design, or fabrication of the system. The enterprise may also be infeasible because of lack of investment capital, industrial site and facilities, labor, transportation, waste disposal facilities, etc. It may be judged to be impractical because of safety, environmental, and other regulations. For instance, even if everything is found to be satisfactory for the establishment of a factory at a particular location, it may not be possible to proceed due to denial of the required approvals because of safety and waste disposal concerns. In recent years, the nuclear industry has run into many obstacles from regulatory bodies as well as opposition from local groups due to concerns with safety and nuclear waste disposal. Similarly, transportation facilities needed for a steel plant may not be satisfactory and the expense needed to bring these up to the desired level may be prohibitive. In such cases, where the undertaking is found to be infeasible, the effort may be terminated or alternatives to the original concept may be sought. It is important to consider all possible scenarios and difficulties that may be encountered. In some cases, the difficulties or problems may be overcome by modifications in the overall planning of the undertaking. If, despite such modifications and alternatives, the project is seen to be infeasible, the enterprise is terminated to avoid any further expense.

1.2.4 Engineering Design

Following a detailed market analysis and evaluation of the chances of success and the feasibility of the undertaking, an engineering design of the system is initiated if all of these indicators are acceptable. Design will determine the specifications of the various components of the system, often termed *system hardware*, and also the range of *operating conditions* that would yield the desired outputs for satisfying the perceived need or opportunity. Thus, design involves a consideration of the technical details of the basic concept and creation of a new or improved process or system for the specified task. The design process starts with the basic concept; then models and analyzes various constituents of the system; synthesizes information on materials, existing systems, and results from different models; evaluates the design with respect to performance; and finally communicates the design specifications for fabrication and prototype development. Numerical or computational methods are generally involved at various stages to model, study, and characterize components and the overall system. As part of the design of the system, the effort may also involve the selection of components that are easily available rather than designing these, as discussed earlier. Safety and environmental considerations usually form part of the design process. Though the focus in engineering system design is on the technical aspects of the system, the interaction with other groups and involvement with larger issues concerning the undertaking are generally unavoidable and often influence the final design.

The design phase of the enterprise is where much of the effort and time are spent and determines, to a large extent, the final outcome of the undertaking. The design process could and usually would seek technical inputs from many other groups within the company, particularly from the research and development section. Such inputs may concern information on available materials and their properties, on new techniques and processes, on the analysis and evaluation of different designs, and on possible solutions to various problems encountered during design. The design effort may be concerned with a single device such as a heater; a component or subsystem of the system, such as a pump; or the overall system itself, such as a solar energy water heating plant. Though the design of components is an important consideration in design, in this book we are mainly concerned with the design of systems consisting of several components interacting with each other. System design may be directed at different types of systems such as electronic, mechanical, thermal, or chemical. The design of thermal systems is obviously of particular interest to us.

1.2.5 Research and Development

Frequently, the information needed for design and optimization is not readily available and the research and development division of the company is employed to obtain this information from the literature on relevant processes and systems and from independent detailed investigations of the basic aspects involved. The research and development group normally interacts with most engineering activities within the company and provides inputs at various stages of product or system development. The main distinguishing feature of the research and development effort is the generally long-range interest of the various activities undertaken. Problems that arise during the normal course of operation of an establishment are brought to the research and development division only if a long-term solution is being sought or if new concepts are to be investigated for solving long-standing problems.

The research and development group also keeps track of the progress being made in research establishments around the world in academia, industry, and national and industrial laboratories. Efforts are made to store and have easy access to the literature emerging from such research efforts. Different computational software, equipment for diagnostics and measurements, and information on new materials and emerging techniques are generally housed in this division. Research activities in the group obviously focus on the processes and systems that are of particular relevance to the company. Thus, the group devotes its efforts to developing new techniques for improving existing processes and to new ideas that may be applied to develop new products. As mentioned earlier, the group may be the initiator of a given engineering enterprise by recognizing the opportunity presented by new materials or processes. In addition, a close interaction and collaboration between the engineering design team and the research and development group is generally essential to the success of the undertaking.

The lack of an established or available procedure often leads to research. For instance, safety considerations with respect to the disposal of nuclear waste have led to detailed investigations of the nature of the waste, its decay with time, effect on neighboring materials, and possible ways of neutralizing it. Similarly, a substantial amount of research has been devoted to the disposal of hazardous waste from chemical plants and other industrial sources. The accurate control of a thermal system, such as an optical fiber drawing furnace, may demand innovation, leading to research into available strategies and the development of new techniques to obtain the desired characteristics. Because the research and development effort is not involved with the routine, day-to-day activities of the company, the group is able to consider many diverse solutions to a given problem, investigate the basic characteristics of relevant processes in an attempt to improve these, consider the applicability of new techniques and developments to the company enterprises, and provide the long-term support needed by engineering design. Consequently, most big companies have well-established research and development divisions and many important and original concepts originate here, frequently leading to major changes in the company. The developments of semiconductor devices and fiber-optic cables are examples of concepts that were initiated by the research and development division of AT&T. In the absence of research and development groups within the company or if a particular expertise is lacking, external experts are often sought as consultants to supplement the effort within the company and to provide the information needed for solving short-range and long-range problems that may arise.

1.2.6 Need for Optimization

It is no longer sufficient to develop a workable or feasible system that performs the desired task while staying within the constraints imposed by safety, environmental, economic, material, and other considerations. Due to the growing worldwide competition and need to increase efficiency, it has become essential to optimize the process in order to maximize or minimize a chosen variable or criterion. This variable is generally known as the *objective function* and may be related to quantities such as profit, cost, product quality, and output. The days when a company could monopolize several products, particularly in the consumer market, are long gone. For each item, say a portable stereo system, a digital camera, or a clothes dryer, many price ranges and performance specifications are available from different manufacturers. The survival of a given product is largely a function of its performance per unit cost. Though the resulting sales are also affected by promotion and advertisement and by other factors such as durability, service, and repair, the optimization of the manufacturing process in order to obtain the best quality per unit cost is extremely important in the survival and success of the item.

Optimization of a system is often based on the profit or cost, though many other aspects such as weight, size, efficiency, reliability, and output may also be optimized, depending on the particular application. For instance, a refrigerator may be designed for a given rate of heat removal, with different temperatures being obtained in the freezer by means of a thermostat control. However, different types of refrigerator systems are possible, such as vapor compression and vapor absorption systems, sketched in Figure 1.8. If a vapor compression system is chosen, the various components, such as the compressor, condenser, evaporator, and valve, may be designed or selected for a wide range of specifications and characteristics. The control system and the operating conditions can also be varied. The inside geometry, dimensions, and materials, as well as the outside materials and appearance, are also important variables. Thus, clearly, a unique system is not obtained and the design may vary over wide ranges, given in terms of the hardware as well as the operating conditions. All these designs may be termed as *acceptable* or *workable* because they satisfy the given requirements and constraints. However, it is necessary to seek an optimal design that will, for instance, consume the least amount of energy per unit cooling effect. This measure is closely linked with the overall efficiency of the system. As we well know, the energy rating of the system, which is an indicator of the energy consumed for achieving a unit of the desired task such as cooling or heating, is an important selling point for such systems. Therefore, optimization of thermal systems is of particular interest to us, and several chapters are devoted to the basic formulation and different strategies for obtaining an optimal design.

1.2.7 Fabrication, Testing, and Production

The final stages in an engineering enterprise, before proceeding to advertising, promotion, and sales, are the fabrication and testing of a prototype of the designed system and production of the

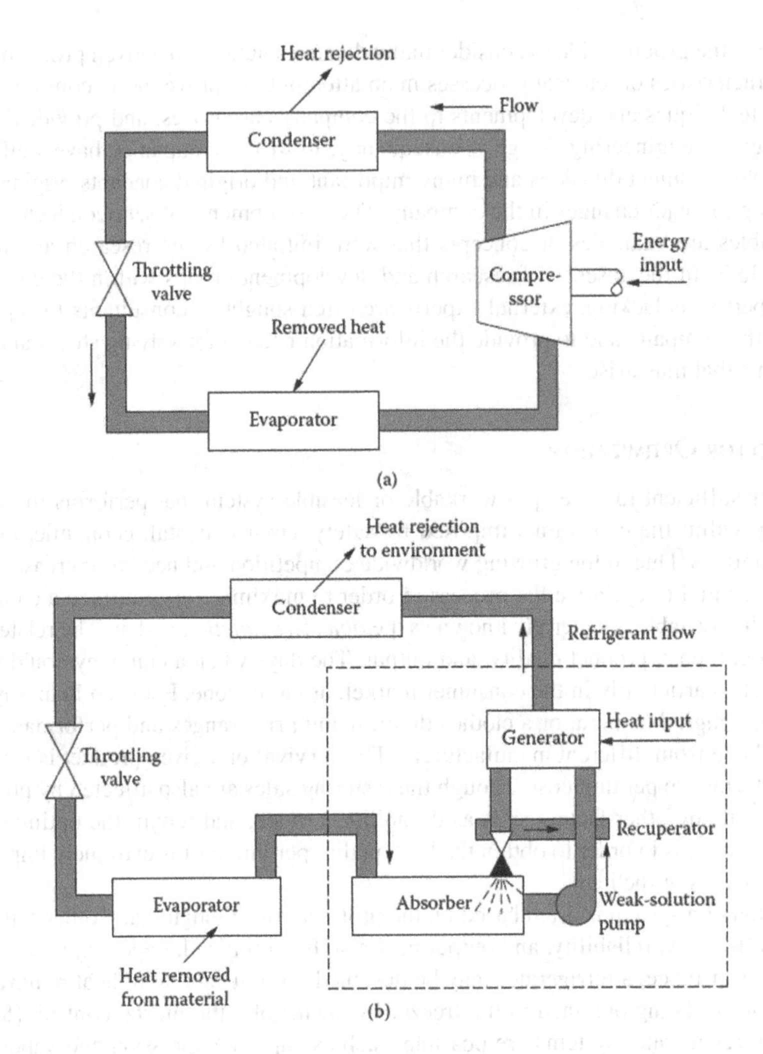

FIGURE 1.8 Vapor cooling systems. (a) Vapor compression, (b) vapor absorption. (Adapted from Howell, J.R. and Buckius, R.O., 1992.)

system in the desired quantities for sale. The outputs from the design process must be communicated to the appropriate technical facilities in order to fabricate, operate, and test the system. This communication may include many items such as engineering drawings to indicate dimensions and tolerances, design specifications, particulars of selected components, ranges of operating conditions, chosen materials, power and space requirements, details of waste and energy disposal, system control strategy, and safety measures. The information provided must be detailed enough to allow the machine shop and other relevant facilities to proceed with fabrication of the system. The overall fabrication and assembly of the system may continue to be under the control of the design group or a project manager, who coordinates the design and engineering activities, and may oversee the development of a prototype.

Once the prototype is obtained, it is subjected to extensive testing over the expected range of operating conditions. Accelerated tests may be carried out to study the reliability of the system over its expected life. Conditions much worse than expected in normal use are usually employed for such performance tests. For instance, an air conditioner or a refrigerator may be kept on for several days to test if it can survive such a punishing use. A car engine may be run at speeds higher than the recommended range to simulate variations in real life and to determine how much overload the system can safely withstand. In some cases, the temperature, speed, pressure, etc., are raised until permanent damage occurs in order to determine the maximum safe levels for the system.

The tests on the prototype are used to establish the major specifications, to ensure that the desired task is being performed satisfactorily, to validate and improve the mathematical model of the system, to establish safety levels, and to obtain the system characteristics. The prototype is also used for improvements in the design based on actual tests and measurements.

Following prototype development and testing, the system goes into production. Existing facilities are modified or new ones procured to mass produce the product or system. Economic considerations play a very important role in the development of the production facilities needed. The mass production of the product is also closely coupled with its marketing, which involves advertising, promotion, and sales.

Figure 1.9 shows the various steps discussed here for a typical engineering enterprise. The important position occupied by engineering design is evident from this sketch. However, this figure

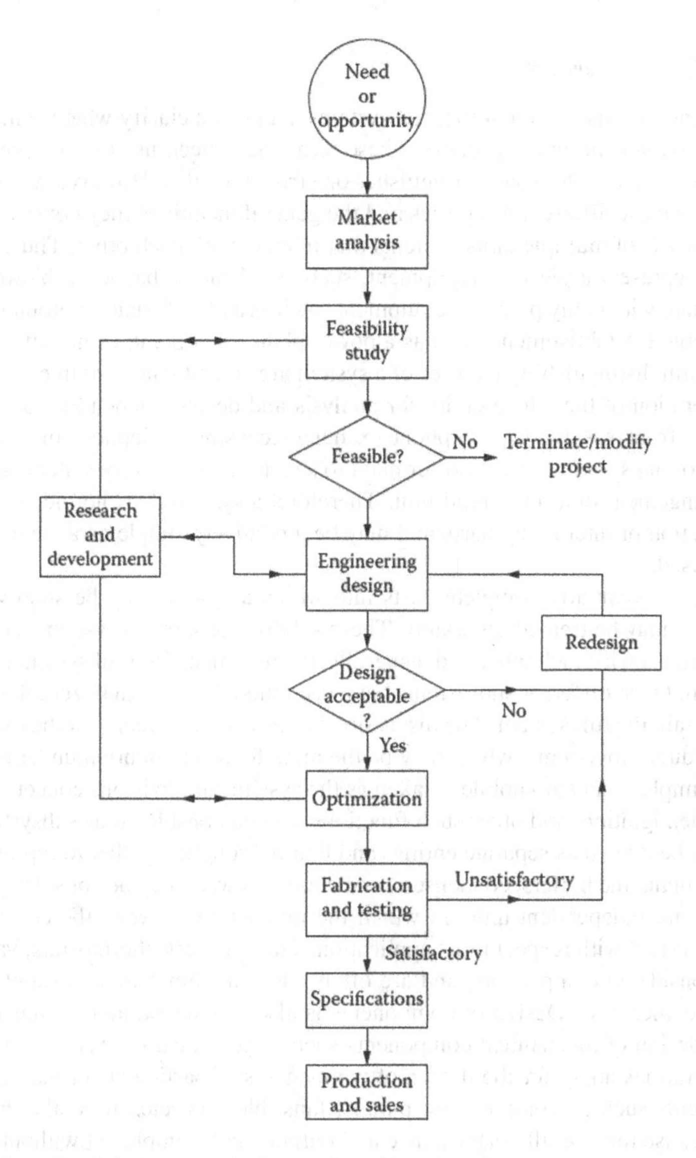

FIGURE 1.9 Schematic of design as part of an engineering enterprise.

represents just one possible sequence of events. In most cases, there is considerable interaction between various groups and there is a fair amount of overlap between the different steps. The sequence used and the importance of each step may vary depending on the product and the nature of the industry. Of course, not all design efforts end in fabrication. Several involve the selection and procurement of various components, which are then assembled. Construction of only a few select items is undertaken for custom, or one-of-a-kind, systems. However, the design steps in such cases are similar to those outlined here, though they may differ in intensity and sequence.

1.3 THERMAL SYSTEMS

Let us now turn our attention to thermal systems and consider the nature of these systems and the various types of systems that are commonly encountered in industry and in general use. As is evident from the variety of examples mentioned in the preceding sections, thermal systems are important in many different applications and occupy a very prominent place in our lives.

1.3.1 Basic Characteristics

Before proceeding to a discussion of thermal systems, let us first clarify what we mean by a *system*, a *subsystem*, a *component*, and a *process*. These terms have been used in the preceding material without much discussion as to what distinguishes one from the other. However, a few examples were given to illustrate these different categories and the general meanings they convey.

A system consists of multiple units or items that interact with each other. Thus, the term *system* can be used to represent a piece of equipment, such as a heat exchanger, a blower, or a pump; a larger arrangement with many pieces of equipment, such as a blast furnace, automobile, or a cooling tower; or a complete establishment, such as a power plant, steel plant, or manufacturing assembly line. The two main distinguishing features of a system are constituents that interact with each other and the consideration of the whole entity for analysis and design. Depending on our interest, the system may vary from, say, the full telephone exchange to a single telephone unit, from an airplane to its air conditioning system, from a power plant to a turbine, from a city water distribution system to the flow arrangement in a residential unit. Therefore, a system does not necessarily have to be a massive collection of interacting parts and may be a relatively simple arrangement on which our attention is focused.

Subsystems are essentially complete parts into which a system may be subdivided for convenience and which may be treated separately. These subdivisions, or subsystems, consist of individual parts that interact with each other and, generally, the treatment for a subsystem is quite similar to that for a system. Once different subsystems have been modeled and analyzed, they are assembled or coupled to obtain the full system. The discussion in this book is directed at the overall system and not at the individual subsystems, which may be the main focus of attention under different circumstances. For example, if an automobile is taken as the system, subdivisions concerned with cooling, transmission, fuel, ignition, and other such functions may be considered as subsystems. Then these subsystems may be treated as separate entities and finally brought together to represent the full system. In a power plant, the boilers, condensers, and cooling towers may be considered as subsystems.

Components are independent units in which the interaction between the constituents is either absent or unimportant with respect to its application. Thus, heaters, thermostats, valves, and extrusion dies are considered components, and are often selected from available supplies or fabricated according to specifications. Design of components is also of interest, and engineering courses are devoted to the design of mechanical components such as gears, cams, springs, chains, and shafts. Similar considerations apply for the design of components of particular relevance to thermal systems. Larger items such as compressors, pumps, fans, blowers, etc., may also be considered as components because the overall performance and output can be employed without considering the interaction between the various parts. These components are available as standard items and are usually selected rather than designed in the system design process. Except for some reference to component design, as needed, the discussion in this book will focus on the design of systems.

Finally, a *process* refers to the technique or methodology of achieving a desired goal. For instance, manufacturing processes such as casting, extrusion, hot rolling, and welding refer to the basic procedure and concept involved without specifying the relevant hardware. Generally, a process is used to indicate the conditions, such as the temperature and pressure, to which a material undergoing thermal processing is subjected. A system, on the other hand, is defined in terms of the hardware as well as the operating conditions.

Different types of systems arise in engineering design depending on the main features that characterize these systems. Therefore, electronic systems are concerned with electrical circuits and devices, mechanical systems with the mechanics of components such as springs and dampers, chemical systems with the chemical characteristics of mixtures and reactants, structural systems with the strength and deformation of structures, and so on. Systems that involve a consideration of thermal science and engineering to a significant extent in their analysis and characterization are termed as *thermal systems*. Thermal science and engineering, as used here, include areas such as heat transfer, thermodynamics, fluid mechanics, and mass transfer. Therefore, even though a computer is an electronic system, if one's interest lies in its cooling system in order to restrict the component temperature levels, it becomes a thermal system for this particular consideration. The focus in thermal systems is on the transport of energy, particularly thermal energy, and fluid flow and mass transport are important additional ingredients in these systems.

It is important to recognize that thermal systems arise in many diverse fields of engineering, such as aerospace engineering, manufacturing, power generation, and air conditioning. Consequently, a study of thermal systems usually brings in many additional mechanisms and considerations, making the problems much more complicated than what might be expected from a study of thermal sciences alone.

1.3.2 Analysis

As mentioned earlier, thermal systems are largely governed by transport mechanisms that arise in flow and in heat and mass transfer. Thermodynamics also plays an important role in a wide range of thermal systems. Thus, the analysis of thermal systems is often complicated because of the complex nature of fluid flow and of heat and mass transfer processes that arise. As a result, typical thermal systems have to be approximated, simplified, and idealized in order to make it possible to analyze them and thus obtain the inputs needed for design. Computational methods are generally needed to obtain the desired results on components and processes. Specialized commercial software for the given industry or generalized software such as Matlab or Ansys are commonly used.

Following are some characteristics that are commonly encountered in thermal systems and processes:

1. Time-dependent
2. Multidimensional
3. Nonlinear mechanisms
4. Complex geometries
5. Complicated boundary conditions
6. Coupled transport phenomena
7. Turbulent flow
8. Change in phase and material structure
9. Energy losses and irreversibility
10. Variable material properties
11. Influence of ambient conditions
12. Variety of energy sources

Because of the time-dependent, multidimensional nature of typical systems, the governing equations are generally a set of partial differential equations, with nonlinearity arising due to convection of momentum in the flow, variable properties, and radiative transport. However, approximations and idealizations are used to simplify these equations, resulting in algebraic and ordinary differential equations for many practical situations and relatively simpler partial differential equations for others. These considerations are discussed in Chapter 3 as part of modeling of the system. However, the equations for a few simple cases are given here to illustrate the nature of the governing equations and the effect of some of these complexities.

The simplest problems are those that assume steady-state conditions, with or without flow, while also assuming uniform conditions in each part of the system. These problems lead to algebraic equations, which are often nonlinear for thermal systems. This situation is commonly encountered in thermodynamic systems such as refrigeration, air conditioning, and energy conversion systems. Then, the governing set of algebraic equations may be written as

$$
\begin{aligned}
f_1(x_1, x_2, x_3, \ldots, x_n) &= 0 \\
f_2(x_1, x_2, x_3, \ldots, x_n) &= 0 \\
f_3(x_1, x_2, x_3, \ldots, x_n) &= 0 \\
&\vdots \\
f_n(x_1, x_2, x_3, \ldots, x_n) &= 0
\end{aligned}
\tag{1.4}
$$

where the x_i are the unknowns and the functions f_i, for $i = 1, 2, 3, \ldots, n$, may be linear or nonlinear. Such problems are generally referred to as steady, lumped circumstances and have been most extensively treated in papers and books dealing with system design, such as Hodge (1989), Stoecker (1989), and Janna (2014). However, these approximations are applicable for only a few idealized thermal systems. Additional complexities, mentioned earlier, generally demand analysis that is more accurate and may involve the solution of ordinary and partial differential equations. Nevertheless, because of the ease in analysis, steady lumped systems are effective in illustrating the basic ideas of system simulation and design. Therefore, these are used as examples throughout the book while bearing in mind that, in actual practice, more detailed analysis would generally be needed. Here, simulation refers to the process of studying the behavior and response of the system through modeling, without building an actual system.

If the time-dependent behavior of the system is sought, for a study of the dynamic characteristics of the system, the resulting governing equations are ordinary differential equations in time τ, if the assumption of uniform conditions within each part is still employed. Then the governing equations may be written as

$$
\frac{dx_i}{d\tau} = F_i\left(x_1, x_2, x_3, \ldots, x_n\right) \quad \textit{for } i = 1, 2, 3, \ldots, n \tag{1.5}
$$

where, again, the functions F_i may be linear or nonlinear. The systems for which these approximations can be made are known as dynamic, lumped systems. Such a treatment is valuable in many cases because of the resulting simplicity. In many thermodynamic systems, such as heat engines and cooling systems, the individual components are approximated as lumped and the dynamic analysis is of interest in the startup and shutdown of the systems, as well as in determining the effects of changes in operating conditions such as flow rate, pressure, and heat input.

An example of such a model is the cooling or heating of a metal piece, whose temperature T may be assumed to be uniform. Then the governing energy equation may be written as

$$
\rho C V \frac{dT}{d\tau} = A(q_{in} - q_{out}) \tag{1.6}
$$

where ρ is the material density, C the specific heat, V the volume of the item, A its surface area, q_{in} the heat flux input, and q_{out} the heat flux lost. This equation can easily be solved if the initial temperature T_i is given, along with the heat fluxes. For a heated piece cooling by convection in a fluid at temperature T_a and with a heat transfer coefficient h, the equation becomes

$$\rho C V \frac{dT}{d\tau} = -hA(T - T_a) \tag{1.7}$$

which yields an exponential decay in temperature with time, as discussed in Chapter 3.

If uniform temperature cannot be assumed in the material under consideration, for instance, in the circumstance of conduction in a stationary solid body such as the wall of a building or of a blast furnace, the energy equation for a two-dimensional constant-property circumstance becomes

$$\frac{\partial T}{\partial \tau} = \alpha \left(\frac{\partial^2 T}{\partial x^2} + \frac{\partial^2 T}{\partial y^2} \right) \tag{1.8}$$

where α is the thermal diffusivity. This equation is linear because T appears in its first power throughout the equation. It is a partial differential equation and is written for the relatively simpler constant-property, two-dimensional circumstance for the Cartesian coordinate system. In practical systems, we often encounter many additional complexities that make the analysis a very difficult and challenging affair.

Inclusion of variable properties and/or radiative transport can give rise to nonlinear mechanisms, the former due to the dependence of the properties on the dependent variable such as temperature T and the latter due to the variation of radiation heat transfer as T^4. For example, if the thermal conductivity k, density ρ, and specific heat at constant pressure C_p vary with temperature T, the energy equation becomes

$$\rho(T) C_p(T) \frac{\partial T}{\partial \tau} = \frac{\partial}{\partial x}\left[k(T) \frac{\partial T}{\partial x} \right] + \frac{\partial}{\partial y}\left[k(T) \frac{\partial T}{\partial y} \right] \tag{1.9}$$

Thus, nonlinearity arises due to nonlinear powers of T resulting from property variation with T. If radiative heat loss occurs at a surface, the corresponding energy exchange rate q, per unit area, with a black environment at temperature T_e may be written as

$$q = \varepsilon \sigma \left(T^4 - T_e^4 \right) \tag{1.10}$$

where ε is the surface emissivity and σ is the Stefan–Boltzmann constant. This equation is nonlinear in T due to the presence of temperature as T^4. Thus, nonlinear equations are frequently obtained, making the solution difficult. Iterative methods are often needed to obtain the solution. Nonlinearity also makes it difficult to scale up the results from a laboratory model to the full-size system. Many of these considerations are discussed in detail in Chapter 3.

An associated fluid flow arises in many thermal systems. Then the flow must be solved in addition to the energy equation. The flow is governed by the conservation of mass and the momentum-force balance equation, which is generally nonlinear. In some cases, such as constant-property circumstances, the flow may be determined independent of the energy equation, which is solved after the flow has been obtained. However, if property changes are important or if buoyancy effects due to temperature are significant, the flow and energy equations have to be solved simultaneously, making the solution particularly challenging. Examples in the book consider a wide range of problems with different analysis and concerns.

The various other complexities mentioned earlier also complicate the analysis and design of thermal systems. Complex geometry and boundary conditions arise in most practical systems, making it necessary to use simplifications and versatile numerical techniques such as finite element and boundary element methods. Turbulent flow is encountered in many important processes, particularly in energy systems and environmental transport. Special numerical models and experimental procedures have been developed to take turbulent transport into account. Phase change, coupling with material characteristics, time-varying ambient conditions, irreversibility, and different energy sources, such as lasers, gas, oil, electricity, and viscous heating, further complicate the analysis of thermal systems and processes. Several of these aspects will arise in examples given in later chapters.

However, our focus is not on analysis but on design, even though analysis provides many of the inputs needed for design. Therefore, only a brief outline of the basic characteristics of thermal systems is given here. Specialized books, such as Ozisik (1985), Burmeister (1993) and Incropera and Dewitt (2001) in heat transfer, Pritchard and Mitchell (2015) and Shames (1992) in fluid mechanics, Howell and Buckius (1992), Cengel and Boles (2014), and Moran and Shapiro (2014) in thermodynamics, among others, may be consulted for details on different analytical and experimental techniques as well as for results obtained on a variety of fundamental and applied problems.

1.3.3 Types and Examples

As mentioned earlier, thermal systems are important in a wide range of engineering fields, practical applications, and disciplines. Let us consider some important examples, types, and applications of these systems. Several different ways of classifying thermal systems may be employed because of their diversity. A common method is in terms of the function or application of the system. Using this approach, several important types of thermal systems, along with commonly encountered applications and examples, follow.

1.3.3.1 Manufacturing and Materials Processing Systems

Examples include processes such as casting, crystal growing, heat treatment, metal forming, drying, soldering and welding, laser and gas cutting, plastic extrusion and injection molding, powder metallurgy, optical fiber drawing, ceramics, and glass processing. Also included are food processing systems as well as common household appliances such as ovens and cooking ranges. This is an important area and many diverse thermal systems are employed for the different manufacturing processes used in practice (Jaluria, 2018). We have already discussed ingot casting, as sketched in Figure 1.3. Figure 1.10 shows the schematics for continuous casting, plastic extrusion, optical glass fiber drawing, and hot rolling processes. In all these processes, thermal aspects are critical in determining the rate of production and the characteristics of the product.

In continuous casting, molten material is allowed to solidify across an interface as the bulk material is withdrawn at a given speed through a mold, which is usually water cooled. In plastic screw extrusion, the solid plastic is fed through a hopper, melted by energy input, and conveyed downstream by the rotation of the screw, with an associated pressure and temperature rise, finally pushing the molten material through a die to obtain a desired shape. The material may also be injected into a mold and solidified in a process known as injection molding. Similarly, a specially fabricated silica glass rod, typically 5–10 cm in diameter and known as a preform, is heated in a furnace and pulled to sharply reduce the diameter to about 125 μm, yielding an optical fiber in glass fiber drawing. In hot rolling, the material is heated and reduced in thickness by pushing it through two rollers that are at a given distance apart. Several sets of rollers may be used to obtain the desired decrease in thickness or diameter. Similarly, new thermal processes have been developed for the fabrication of nanomaterials through chemical vapor deposition, thermal sprays, and other approaches.

Heat transfer is very important in these processes because the temperature determines the forces needed, the withdrawal speed, and the quality of the final product. Further details on these and other processes may be found in specialized books on manufacturing, such as Ghosh and Mallik (1986), Doyle et al. (1985), and Kalpakjian and Schmid (2013). Some of these processes will be considered again later in the book as examples. With the development of new and improved materials, the design of thermal systems for materials processing has become crucial for manufacturing new products and for meeting international competition.

1.3.3.2 Energy Systems

Examples of energy systems include power plants, solar power towers, geothermal energy systems, energy storage, solar ponds, and conventional and nonconventional energy conversion systems.

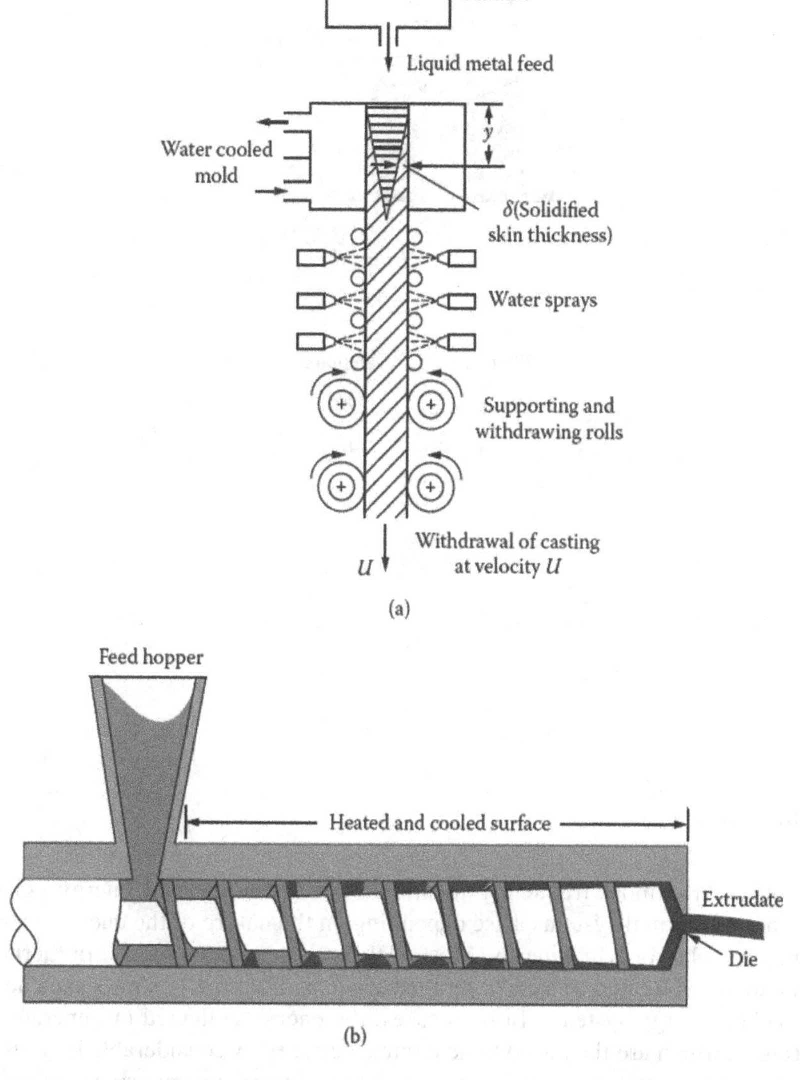

FIGURE 1.10 A few manufacturing systems. (a) Continuous casting, (b) plastic screw extrusion, (c) optical fiber drawing, (d) hot rolling. (Figure 1.10(a) adapted from Ghosh and Mallik, 1986; Figure 1.10(b) from Tadmor and Gogos, 1979.)

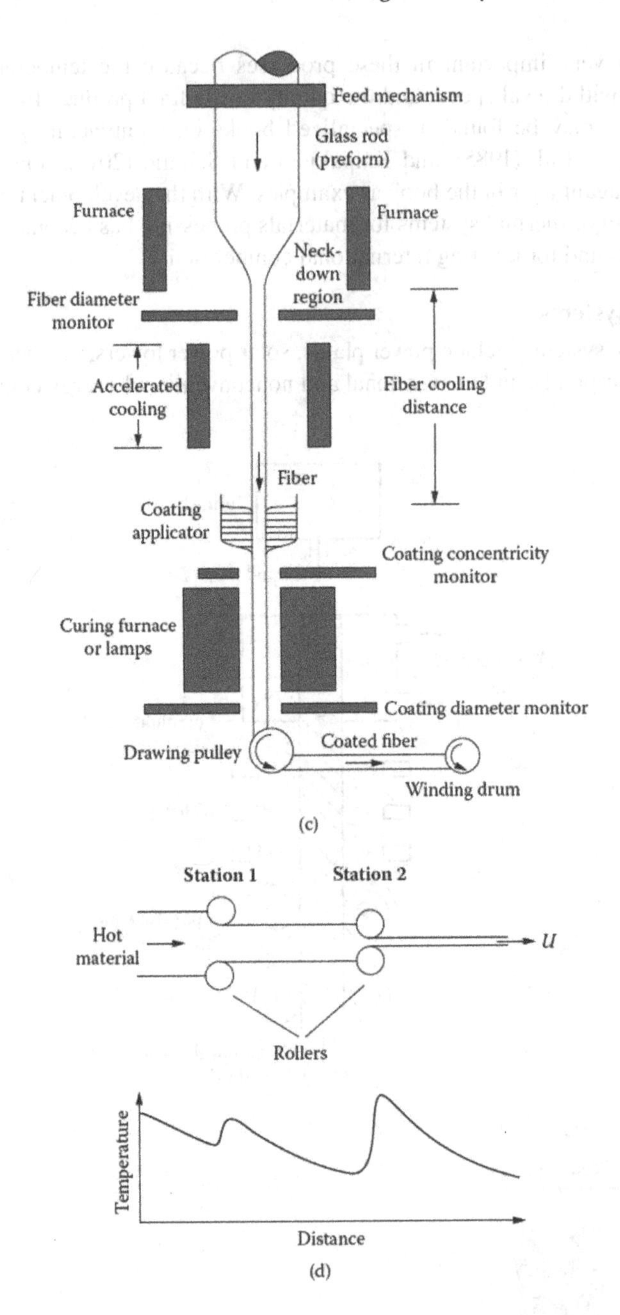

FIGURE 1.10 *(Continued).*

This is one of the most frequently mentioned areas for thermal energy considerations. Different types of thermal systems arise depending on the nature of the energy source, such as nuclear, oil, gas, solar, or wind energy. Most of these systems are covered in thermodynamics courses and are often treated as steady, lumped systems. Figure 1.11 shows sketches of typical solar and nuclear energy systems. In both cases, the energy collected or generated is used to run the turbines, which are then used to generate electricity. A considerable literature exists on thermal systems of interest in this field because of the tremendous importance of power generation in our society; see, for instance, Howell et al. (1982), Hsieh (1986), Van Wylen et al. (1994), and Duffie and Beckman (2013).

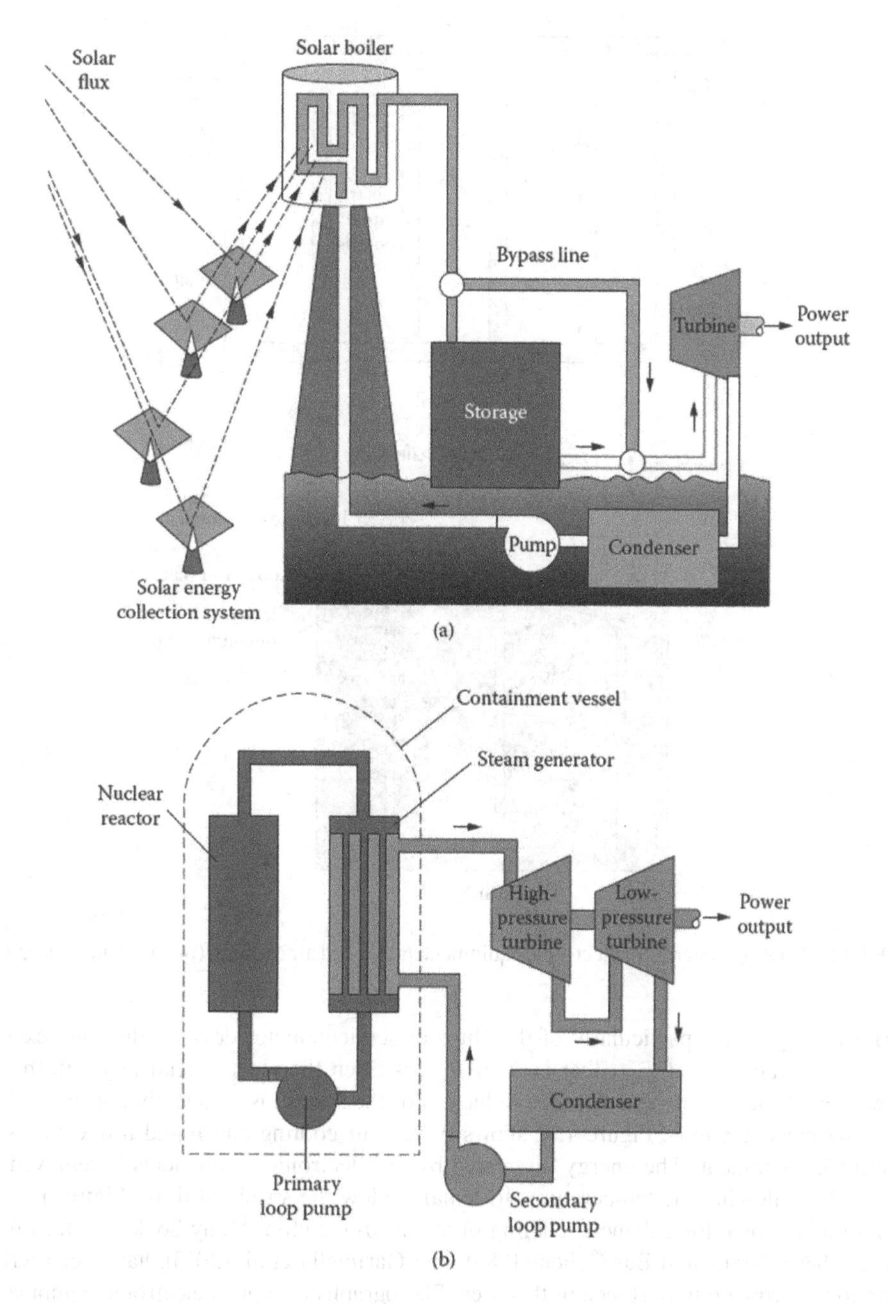

FIGURE 1.11 Power systems based on (a) solar energy and (b) nuclear energy. (Adapted from Howell and Buckius, 1992.)

1.3.3.3 Cooling Systems for Electronic Equipment

Systems that are of interest in this area include air cooling, liquid immersion, heat pipes, heat sinks, heat removal by boiling, and microchannel systems.

This is one of those areas where thermal considerations are extremely important for the satisfactory performance of the system even though the main application is in a different area. Thus, electronic systems, such as computers, televisions, digital multimeters, and signal conditioners, are used for a variety of applications, most of which are not directly connected with fluid flow and heat transfer. But the cooling of the electronic system in order to ensure that the temperature T_c of

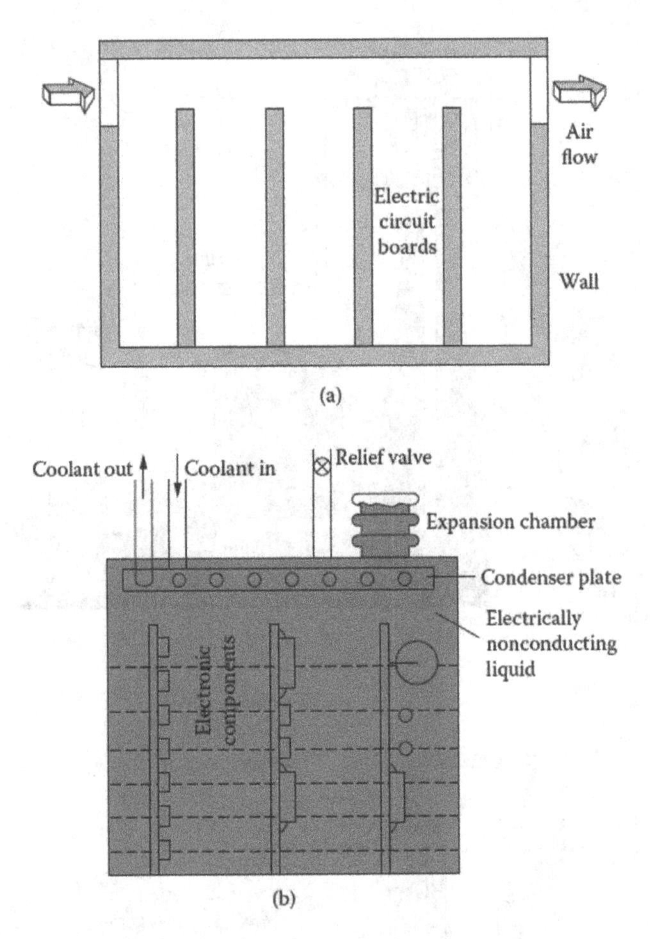

FIGURE 1.12 Cooling systems for electronic equipment. (a) Forced air cooling, (b) liquid immersion cooling.

the various components, particularly of the chips or semiconductor devices, does not exceed the allowable temperature level T_{max}, that is, $T_c \leq T_{max}$, is often the most crucial factor in the design and operation of the system. Further size reduction of the system is frequently constrained by the heat transfer considerations. Figure 1.12 shows typical air cooling and liquid immersion systems for electronic equipment. The energy dissipated by the electronic components is removed by the fluid flow, thus allowing the temperatures to remain below the specified limit. Figure 1.2 showed a sketch of a heat pipe for enhanced cooling of an electronic chip. Many books, such as those by Steinberg (1980), Kraus and Bar-Cohen (1983), and Garimella et al. (2013), have been written in response to the growing importance of this area. Photographs of typical electronic equipment with air cooling by means of a fan or a blower are shown in Figure 1.13, indicating the complexity of such systems in actual practice.

1.3.3.4 Environmental and Safety Systems

Examples of these systems include arrangements for heat rejection to ambient air and water, control of thermal and air pollution, cooling towers, incinerators, waste disposal, water treatment plants, smoke and temperature control systems, and fire extinguishing systems.

The growing concern with the environmental impact of waste and energy disposal, including global climate change and depletion of the ozone layer, has made it essential to minimize the effect on our environment by developing new and improved methods for disposal. Many thermal systems

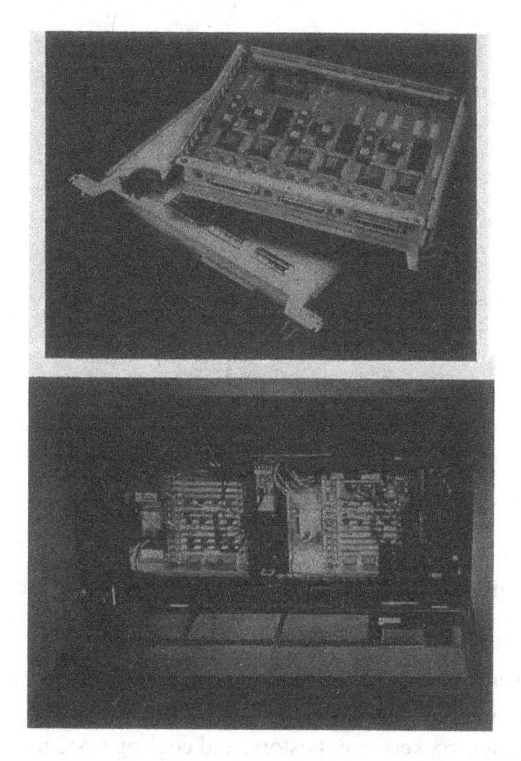

FIGURE 1.13 Typical electronic systems with air cooling by means of a fan. (From Steinberg, 1980.)

have been developed in response to this need. These include systems based on fluids that would substitute refrigerants such as CFCs that adversely affect the ozone layer, improved incineration techniques for solid waste disposal, catalytic converters in automobiles to reduce harmful emissions, and scrubbers in power plants to reduce pollutants. Figure 1.14 shows sketches of typical heat rejection systems from power plants, employing a lake as a cooling pond in the first case and a natural draft cooling tower in the second. The effect on the local environment, in terms of temperature rise, increased flow, and disturbance to natural yearly cycle, is of particular concern in these cases. Safety is also a very important consideration. Figure 1.15 shows a sketch of a room fire, indicating a hot upper layer containing the toxic and hot combustion products and a relatively cooler and less toxic lower layer that is often safe until flashover occurs when everything in the room catches fire and the room is engulfed in flames. Thus, the design of the system, which may be a building, ship,

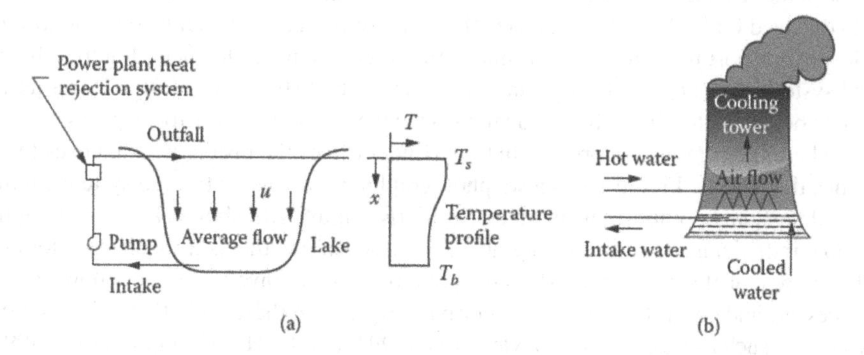

FIGURE 1.14 Systems for heat rejection from a power plant. (a) Natural lake as a cooling pond, (b) natural draft cooling tower.

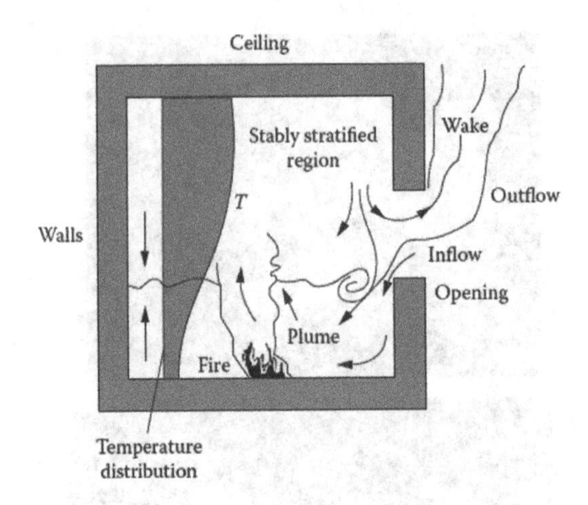

FIGURE 1.15 Flow and temperature due to fire in a room with an opening.

submarine, or airplane, for fire safety is clearly an important element in the overall construction and operation of the system.

1.3.3.5 Aerospace Systems

Many thermal systems in aerospace applications are of interest here. Some of the common ones are gas turbines, rockets, combustors, and cooling systems.

This has been a particularly important area over the last few decades because of the space program. Considerable progress has been made on the various thermal systems and subsystems that are needed. Because of the large thrust needed at rocket launch and high cooling rates during reentry, much of the effort in designing efficient systems has been directed at these two stages. However, cooling, air conditioning, and electronic and energy systems during orbit, as well as for a space station, have their own requirements and challenges. Thermal systems are also of interest in possible lunar and space settlements in the future.

1.3.3.6 Transportation Systems

Most of the relevant systems in this area are thermal in nature. These include internal combustion engines such as spark ignition and diesel engines; steam engines; fuel cells; and modern automobile, airplane, and railway train engines.

This is an extensive field, closely associated with different kinds of thermal systems. Though a traditional mechanical engineering field, this area has seen many significant changes in recent years, most of these being related to the optimization of existing systems. New systems have also evolved in response to the need for higher efficiency, size that is more compact, greater safety, and lower costs. Supersonic air transport has led to several interesting innovations in this field. Figure 1.16 shows a few typical systems that arise in transportation. Figure 1.16(a) shows two designs for a jet engine, with hot gases being ejected from the nozzle to provide the thrust. Figure 1.16(b) shows a spark ignition engine where the combustion process in the cylinder drives the piston, which moves the crankshaft and thus the wheels. Figure 1.17 shows photographic views of gas turbine systems, indicating the intake, exhaust, and combustion chamber. Similarly, Figure 1.18 shows sketches of engines for transportation, indicating a lightweight engine and a diesel engine that is turbocharged for boosting power. Clearly, practical systems are extremely complicated and involve intricate flow paths, combustion processes, and control mechanisms. For basic details on these and other systems, books on thermodynamics, such as those by Van Wylen et al. (1994) and by Howell and Buckius (1992), and more specialized books on various relevant topics, such as Ferguson and Kirkpatrick (2001) and Heywood (2018), may be consulted.

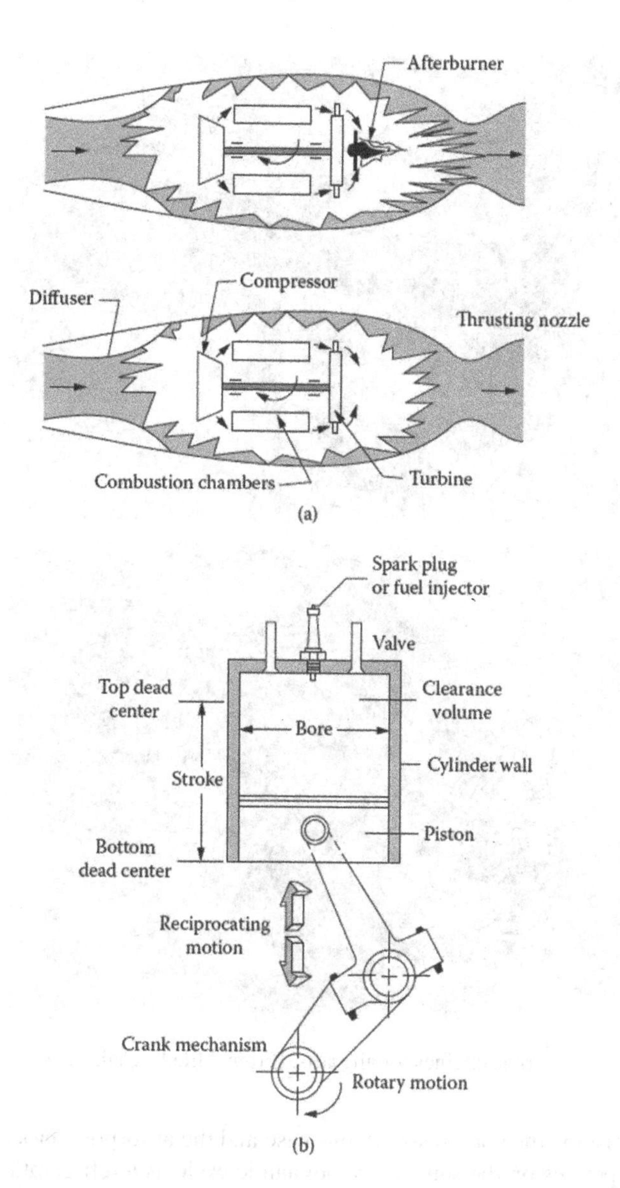

FIGURE 1.16 Thermal systems for transportation. (a) Thrusting systems for aircraft propulsion: Turbojet engine with and without afterburner, (b) reciprocating internal combustion engine. (Figure 1.16a adapted from Reynolds and Perkins, 1977, and Figure 1.16b from Moran and Shapiro, 2014.)

1.3.3.7 Air Conditioning, Refrigeration, and Heating Systems

Several different thermal systems are associated with this application, which is of considerable interest to us in our daily lives. These include vapor compression and vapor absorption cooling systems; heat pumps; ice and food freezing plants; gas, oil, and water heating systems; and refrigerators.

Even though this field has been around for a long time, the need for more efficient, dependable, and safe systems, at lower cost, has led to many improvements. In particular, better design of the main components such as the compressor and the condenser, better control of the system, and better design of the overall system to minimize losses have resulted in reduced energy consumption and lower costs. Figure 1.8 presented sketches of the vapor compression and vapor absorption systems for refrigeration. In both cases, energy is removed from a given space or material due to the evaporation of the working fluid in the evaporator and heat is rejected to the ambient in the condenser.

FIGURE 1.17 Typical gas turbine engines for aircrafts. (From AlliedSignal, Inc.)

The driving mechanism is the compressor in one case and the absorption process in the other. In a heat pump, which operates on the same thermodynamic cycle as a refrigeration system, energy is extracted from a colder environment and supplied to a warmer region, such as a house. Photographs of practical heat pumps are shown in Figure 1.19. Though these are often treated as components, they are actually thermal systems with many interacting parts. Specialized books such as those by Stoecker and Jones (1982), Cooper (1987), and Kreider et al. (2001) may be consulted for details on these systems.

1.3.3.8 Fluid Flow Systems and Equipment

These include components and fluid flow circuits such as pipe flows, hydraulics, hydrodynamics, fluidics, turbines, pumps, compressors, fans, and blowers.

Many of these are auxiliary subsystems to the main thermal systems and may be used for control; power transmission; cooling; and transport of mass, energy, and momentum. Fluid mechanics itself is closely linked with thermal energy transport in most practical processes and fluid flow systems refer to only a subset dealing with flow circuits. Figure 1.20 shows the sketches of a few typical flow distribution systems. Fluid flow equipment such as pumps, fans, and blowers are extensively used in thermal systems. Books on fluid mechanics, such as those by Pritchard and Mitchell (2015),

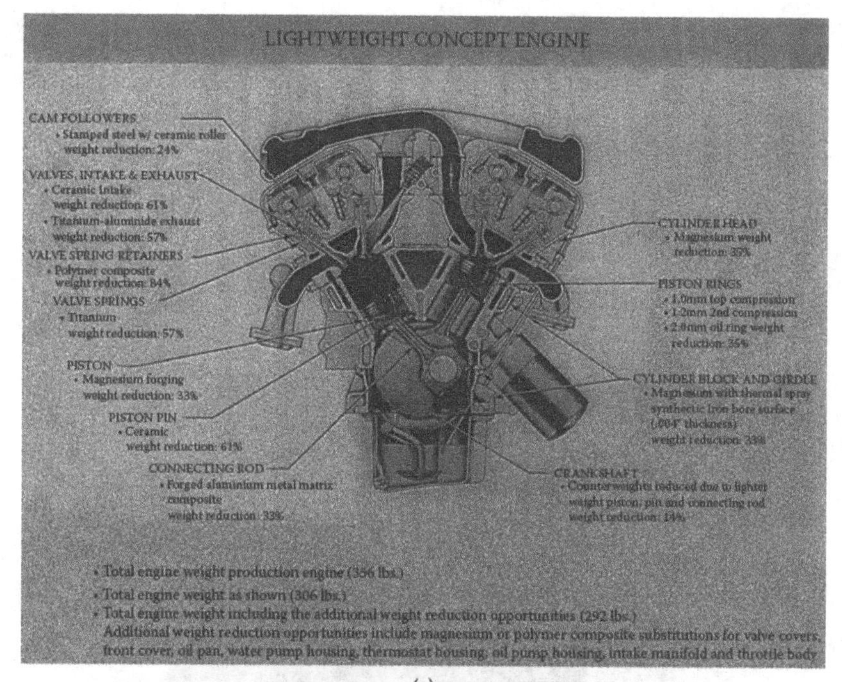

(a)

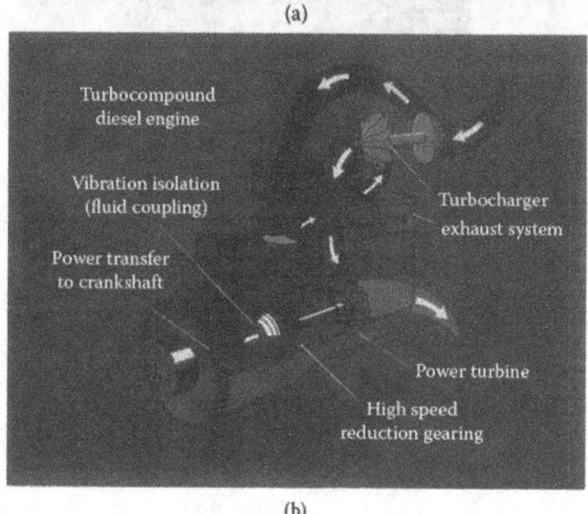

(b)

FIGURE 1.18 (a) A lightweight engine for an automobile. (From Ford Motor Co.) (b) A turbocharged diesel engine. (From Cummins Engine Co.)

White (2015), and John and Haberman (1988), contain information on the analysis of such fluid flow systems and equipment.

1.3.3.9 Heat Transfer Equipment

Such equipment includes heat exchangers, condensers, boilers, furnaces, ovens, hot water baths, and heaters.

Heat transfer equipment often forms part of the various other applications mentioned earlier. Thus, condensers and boilers may be part of a power system. Similarly, furnaces may be regarded as constituents in a heat treatment system. However, such equipment frequently can be designed without considering the application. As mentioned earlier, in the design of a thermal system some of these items may be procured through selection rather than through design. In this case, companies

FIGURE 1.19 Photographs of practical heat pumps. (From KIST, Korea.)

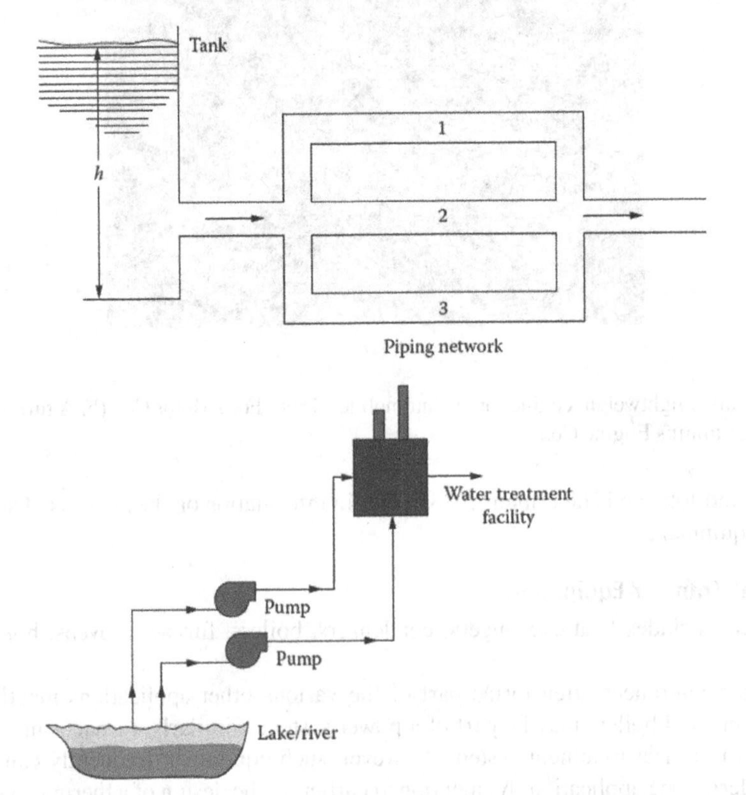

FIGURE 1.20 Examples of fluid distribution systems.

specializing in, say, heat exchangers, would design and manufacture these. Different types of heat exchangers, such as those seen in Figure 1.5, would then be produced for selected ranges of design specifications and made available for marketing (Kays and London, 1984). Similar considerations apply for drying ovens, furnaces, heated oil baths, etc., which may be designed for a specific application or for general use.

1.3.3.10 Other Systems

There are several other thermal systems that may not be as easily classified as was done here for some of the more common and practical systems. Thus, chemical reactors and systems for experimentation, space systems, construction systems, etc. also often involve thermal considerations in their design and may be treated by the techniques discussed in this book.

Other methods of classifying thermal systems can also be used. The following approach divides these systems into three types, representing the three main stages undergone by thermal energy:

1. *Generation:* Solar, geothermal, nuclear and oil-fired power systems, combustors, engines, energy conversion systems, turbines, boilers, and chemical reactors
2. *Utilization:* Manufacturing, car engines, airplanes, and rockets
3. *Rejection:* Heat removal, pollution, waste disposal, electronic systems, air conditioning, heat pumps, cooling towers, and radiators

Even though such a classification would cover most practical systems, several systems are left out and may, again, be categorized under other systems. In addition, systems such as automobiles involve all three aspects of generation, utilization, and rejection. Thermal systems may also be classified by their size, by the nature and number of the constituents, by their interaction with other systems, and so on. However, classification by the application of the system is probably the most useful and also the most frequently employed.

The preceding discussion has presented many different types of thermal systems that are of interest in a wide range of applications. Clearly, different systems have different concerns, and the design specifications and requirements are also different. However, they are all governed by basic considerations in heat and mass transfer, fluid flow, and thermodynamics. Consequently, the basic techniques for design and optimization are similar, making it possible to discuss the fundamental issues and procedures involved in their design.

1.4 OUTLINE AND SCOPE OF THE BOOK

This book focuses on the design and optimization of thermal systems, several examples of which have been given in the preceding section. The importance of thermal systems in a wide range of applications makes it essential to optimize existing and new systems and processes in order to achieve the best performance or output per unit input. In addition, the development of new techniques and materials demands new and improved systems to take advantage of such innovations. However, the design of thermal systems is usually complicated by the complexity of the underlying mechanisms and the resulting lack of adequate information on the system for obtaining a satisfactory design. Therefore, the design process often involves obtaining the relevant inputs from analysis or experiment and incorporating them into existing information on similar systems and processes to generate an acceptable design.

We shall first consider the basic features of the design process, highlighting the various steps that are involved in the design of a thermal system. These considerations are generally common to other types of systems as well. Starting with the problem statement in terms of the requirements, constraints, and other specifications, a conceptual design, which is based on creativity and existing systems, is obtained. The design variables that arise in the problem are determined and varied to obtain a variety of designs, which are evaluated through analysis to determine if an acceptable design can be chosen. Because the solution is not unique, a range or domain of acceptable designs is generally obtained. The evaluation of the designs requires detailed information on the performance of the system. Because of the complexity of most practical thermal systems, it is necessary to develop a mathematical model of the system by simplifying and idealizing the processes involved. Several types of models are discussed in Chapter 3, particularly mathematical, numerical and experimental models. Modeling of the system is one of the most important and creative elements in the design process because it allows relevant inputs to be generated.

Numerical modeling and simulation are generally needed for most practical systems, as considered in detail in Chapter 4. Numerical simulation approximates the actual system and yields quantitative information on the behavior of the system for a wide range of design and operating conditions. Therefore, the characteristics of the system can be investigated for different designs, making it possible to evaluate the design. The various methods and techniques available for modeling and simulation are discussed. The presentation of these results in a form suitable for design is also discussed. Matlab is used as the computational environment to illustrate many basic modeling and simulation ideas. Several examples of thermal systems are then taken from different application areas in Chapter 5 to discuss the synthesis of all these aspects to obtain a design that meets the given requirements and constraints. The modeling, simulation, and design of large, practical systems are also considered. Trade-offs have to be made to meet constraints due to regulations, economics, safety, and other such considerations. In most cases, a domain of acceptable designs is obtained, with the best or optimal design to be chosen from these. If an acceptable design is not obtained, the requirements may be relaxed, new concepts considered, or the effort terminated. The common challenges that arise in the design process and possible approaches to avoid these are also outlined.

This brings us to the problem of optimization. Because of growing competition in the world today, it is essential to optimize the system with respect to a chosen objective such as the output per unit cost or quality per unit energy consumption. The range of design variables over which acceptable designs are obtained may be quite large, making it necessary to narrow the domain to choose the best design that optimizes the cost or some other chosen quantity. Because economic considerations play an important role in the successful completion of the project and in the optimization effort, basic considerations in engineering economics are presented in Chapter 6. This chapter brings out the economic evaluation of an enterprise in terms of the return on investment, costs, financing, present and future worth, and depreciation.

The formulation of the basic problem for optimization is discussed in Chapter 7, indicating the need for optimization and the different approaches available for thermal systems. Optimization with respect to the hardware of the system, as well as to the operating conditions, is discussed. The next three chapters present different methods used for the optimization of thermal systems, employing examples from a variety of practical areas to illustrate the basic approaches and their limitations and advantages. Among the methods considered are calculus methods, particularly the method of Lagrange multipliers, search methods, geometric programming, and linear and dynamic programming. Different systems that are particularly suited to each of these methods are considered and the resulting optimal conditions determined. Several examples are employed to illustrate the strategies involved.

Many additional considerations are covered in Chapter 11. This chapter discusses knowledge-based design, which has become an important and valuable element in design methodology today. The use of existing knowledge, databases, heuristic arguments, and expert systems is outlined. Other techniques such as response surfaces and genetic algorithms are presented. The improvements over classical approaches for certain types of systems are presented. Other considerations such as professional ethics, sources of information, and additional constraints on the design are also discussed.

It must be noted that the book presents all the major elements needed for the design and optimization of thermal systems. However, some of these may have been covered in earlier courses at a particular college or university. The instructor could then decide to avoid covering these in a given design course. Examples of such topics are various aspects in economic considerations given in Chapter 6, physical modeling and dimensional analysis in Section 3.4, and solution procedures in Section 4.2. Though obviously needed for design and optimization, coverage of such topics may be curtailed or eliminated depending on the background and preparation of the students. Similarly, examples of thermal systems range from simple pipe and channel flows through thermodynamic systems to more involved heat transfer processes. Again, the instructor may choose to emphasize simpler thermodynamic and flow systems, rather than the more complicated systems that involve multidimensional heat transfer mechanisms, depending on the background of the students. In many curricula, heat transfer is taught much later than thermodynamics and fluid mechanics, making it easier to consider lumped, steady or transient, thermodynamics and fluid flow systems for design, rather than distributed ones. However, wide ranges of examples, problems, and exercises are presented in this book, along with all the ingredients needed for design and optimization, to make such a choice possible on the basis of the needs and preparation of the class.

1.4.1 A Note on Problems and Examples

Several examples and problems, or exercises, are given on each topic in order to strengthen the discussion and clarify the important issues involved. In many cases, the problems are reasonably straightforward and build on the material presented in the book. This is particularly true for examples and exercises in optimization and on other topics where a particular aspect of analysis or simulation is being demonstrated. However, design involves open-ended problems and synthesis of information from different sources. Many problems are given to bring out these features of the design process. A unique solution is typically not obtained in these cases and the reader may have to make certain decisions, providing appropriate information or personal choice, to solve the problem. The lack of particular information in a problem does not mean that it cannot be solved; instead, it implies that there are choices and inputs that must be provided by the reader to obtain different acceptable designs for the given requirements and constraints. Thus, many different answers may be acceptable for a given problem. Several exercises, particularly the design projects, are given with this flexibility and personal selection and input in mind.

Effort is also made to link the different topics and considerations that arise in the design process. Thus, an acceptable design of a given thermal system in an earlier chapter may be optimized in later chapters and different techniques may be employed for the same problem to demonstrate the difference. Relatively simple examples and problems are given in certain cases to illustrate the methodology. However, the overall focus of the book is on the design of thermal systems, which may range from very simple systems consisting of a small number of parts to complex systems that have a large number of components and subsystems and involve many additional considerations. The basic approach is similar in these two extreme circumstances, and, therefore, the discussion, treatment, and problems can be varied easily to consider different types of systems.

Effort was made to choose examples and problems from both traditional thermal systems as well as from new and emerging areas such as fabrication of advanced materials, cooling of electronic equipment, and new approaches in energy and environmental systems. This allows the reader to see the field as vibrant and growing, with an excitement about new technologies and important practical applications. These examples also demonstrate the critical importance of optimizing many of these systems due to the growing need to reduce cost and energy consumption while enhancing product quality and reducing the environmental impact.

1.5 SUMMARY

This chapter presents the introductory material for a study of design and optimization of thermal systems. It introduces three main topics: engineering design, thermal systems and processes, and optimization. In addition to providing definitions for the relevant terms, the discussion considers the basic characteristics and relevance of thermal systems and design to engineering enterprises. Design, which is a creative process undertaken to solve new or existing problems, is an extremely important engineering task because it leads to new and improved processes and systems. Design, which involves an open-ended solution with multiple possibilities, is contrasted against analysis, which gives rise to unique, well-defined, closed-ended results. Thus, design generally involves considering many different solutions and finding an acceptable result that satisfies the given problem. Synthesis brings together several different analyses and types of information, thus forming an important aspect in system design. In many applications, components or equipment are to be chosen from available items. This is the process of selection, rather than of design, which starts from the basic concept and develops a system for a given application. The focus in this book is on the design of systems and not on selection, although in several instances a particular component may be selected from those available in the market.

Design is also considered as part of an overall engineering enterprise. The project starts with the definition of a need or opportunity and is followed by market and feasibility analyses. Once these are established, engineering design is initiated with inputs from research and development. The design process is expected to result in a domain of acceptable designs from which the best or optimal design may be obtained. Finally, the results are communicated to other divisions of the company for fabrication, testing, and implementation. Thus, design occupies a prominent position in typical engineering enterprises. In most cases, optimization of the design is essential in order to obtain the best output/input ratio.

Processes, systems, components, and subsystems are discussed in terms of their basic features. A system consists of individual constituents that interact with each other and that must be considered as coupled for a study of the overall behavior. Thermal systems, which are governed by the principles of heat transfer, thermodynamics, fluid mechanics, and mass transfer, arise in a wide range of engineering applications. The basic characteristics of these systems are outlined, along with a few typical mathematical equations that describe them. Different types of thermal systems are considered and examples are presented from many diverse areas such as manufacturing, energy, environment, electronic, aerospace, air conditioning, and transportation systems. These examples serve to indicate the considerable importance of thermal systems in industry and in many practical applications. The diversity of thermal systems and the range of concerns in these systems are also examined. The need for design and optimization of these systems is clearly indicated.

REFERENCES

Burmeister, L.C. (1993). *Convective heat transfer* (2nd ed.). New York: Wiley.

Cengel, Y.A., & Boles, M.A. (2014). *Thermodynamics: an engineering approach* (8th ed.). New York: McGraw-Hill.

Cooper, W.B. (1987). *Commercial, industrial and institutional refrigeration: design, installation and troubleshooting.* Englewood Cliffs, NJ: Prentice-Hall.

Dieter, G.E. (2000). *Engineering design: a materials and processing approach* (3rd ed.). New York: McGraw-Hill.

Dieter, G.E., & Schmidt, L. (2012). *Engineering design* (5th ed.). New York: McGraw-Hill.

Doyle, L.E., Keyser, C.A., Leach, J.L., Schrader, G.F., & Singer, M.B. (1985). *Manufacturing processes and materials for engineers* (3rd ed.). Englewood Cliffs, NJ: Prentice-Hall.

Duffie, J.A., & Beckman, W.A. (2013). *Solar energy thermal processes* (4th ed.). New York: Wiley.

Ertas, A., & Jones, J.C. (1996). *The engineering design process* (2nd ed.). New York: Wiley.

Ferguson, C.R., & Kirkpatrick, A.T. (2001). *Internal combustion engines* (2nd ed.). New York: Wiley.

Flemings, M.C. (1974). *Solidification processing.* New York: McGraw-Hill.

Garimella, S.V., Tannaz, H., Kraus, A., Geisler, K.J.L., & Bar-Cohen, A. (2013). *Encyclopedia of thermal packaging: thermal packaging techniques: set 1.* Hackensack, NJ: World Scientific Pub.

Ghosh, A., & Mallik, A.K. (1986). *Manufacturing science.* Chichester, U.K.: Ellis Horwood.

Heywood, J.B. (2018). *Internal combustion engineering fundamentals* (2nd ed.). New York: McGraw-Hill.

Hodge, B.K. (1989). *Analysis and design of energy systems* (2nd ed.). Englewood Cliffs, NJ: Prentice-Hall.

Howell, J.R., & Buckius, R.O. (1992). *Fundamentals of engineering thermodynamics* (2nd ed.). New York: McGraw-Hill.

Howell, J.R., Vliet, G.C., & Bannerot, R.B. (1982). *Solar thermal energy systems: analysis and design*. New York: McGraw-Hill.

Hsieh, J.S. (1986). *Solar energy engineering*. Englewood Cliffs, NJ: Prentice-Hall.

Incropera, F.P., & Dewitt, D.P. (2001). *Fundamentals of heat and mass transfer* (5th ed.). New York: Wiley.

Jaluria, Y. (2018). *Advanced materials processing and manufacturing*. Cham, Switzerland: Springer.

Janna, W.S. (2014). *Design of fluid thermal systems* (4th ed.). Boston, MA: Cengage Learning.

John, J.E.A., & Haberman, W.L. (1988). *Introduction to fluid mechanics* (3rd ed.). Englewood Cliffs, NJ: Prentice-Hall.

Kalpakjian, S., & Schmid, S.R. (2013). *Manufacturing engineering and technology* (7th ed.). New York: Pearson.

Kays, W.M., & London, A.L. (1984). *Compact heat exchangers*. New York: McGraw-Hill.

Kraus, A.D., & Bar-Cohen, A. (1983). *Thermal analysis and control of electronic equipment*. Washington, D.C.: Hemisphere.

Kreider, J.F., Rabl, A., & Curtiss, P. (2001). *Heating and cooling of buildings: design for efficiency* (2nd ed.). Blacklick, OH: McGraw-Hill.

Moran, M.J., & Shapiro, H.N. (2014). *Fundamentals of engineering thermodynamics* (8th ed.). New York: Wiley.

Ozisik, M.N. (1985). *Heat transfer: a basic approach*. New York: McGraw-Hill.

Pritchard, P.J., & Mitchell, J.W. (2015). *Fox and McDonald's introduction to fluid mechanics* (9th ed.). New York: Wiley.

Reynolds, W.C., & Perkins, H.C. (1977). *Engineering thermodynamics* (2nd ed.). New York: McGraw-Hill.

Shames, I.H. (1992). *Mechanics of fluids* (3rd ed.). New York: McGraw-Hill.

Steinberg, D.S. (1980). *Cooling techniques for electronic equipment*. New York: Wiley-Interscience.

Stoecker, W.F. (1989). *Design of thermal systems* (3rd ed.). New York: McGraw-Hill.

Stoecker, W.F., & Jones, J.W. (1982). *Refrigeration and air conditioning* (2nd ed.). New York: McGraw-Hill.

Suh, N.P. (1990). *The principles of design*. New York: Oxford University Press.

Tadmor, Z., & Gogos, C.G. (1979). *Principles of polymer processing*. New York: Wiley.

Van Wylen, G.J., Sonntag, R.E., & Borgnakke, C. (1994). *Fundamentals of classical thermodynamics* (4th ed.). New York: Wiley.

White, F.M. (2015). *Fluid mechanics* (8th ed.). New York: McGraw-Hill.

2 Basic Considerations in Design

The important terms and aspects that arise in the design and optimization of thermal systems have been defined and discussed in the preceding chapter. We are interested in thermal systems that are governed by considerations of fluid flow, thermodynamics, and heat and mass transfer. The interaction between the various components and subsystems that constitute a given system is an important element in the design because the emphasis is on the overall system. Additional considerations, which may not have a thermal or even a technical basis, also have to be included in most cases for a realistic and successful design. Though selection of components or devices may be employed as part of system design, the focus is on design and not on selection. Similarly, analysis is used only as a means for obtaining the inputs needed for design and for evaluating different designs, not for providing detailed information and understanding of thermal processes and systems. The synthesis of information from a variety of sources plays an important part in the development of an acceptable design. With this background and understanding, we can now proceed to the main steps that comprise the design process.

2.1 FORMULATION OF THE DESIGN PROBLEM

A very important aspect in design, as in other engineering activities, is the formulation of the problem. We must determine what is required of the system, what is given or fixed, and what may be varied to obtain a satisfactory design. The final design obtained must meet all the requirements, while satisfying any constraints or limitations imposed due to safety, environmental, economic, material, and other considerations. The design process depends on the problem statement, as does the evaluation of the design. In addition, the formulation of the problem allows one to focus on the quantities and parameters that may be varied in the system. This gives the scope of the design problem, ranging from relatively simple cases where only a few quantities can be varied to more complicated cases where most of the parameters are variable.

2.1.1 Requirements and Specifications

Certainly the most important consideration in any design is the desired function or task to be performed by the system. This may be given in terms of requirements to be met by the system. A successful, feasible, or acceptable design must satisfy these. The requirements form the basis for the design and for the evaluation of different designs. Therefore, it is necessary to express the requirements quantitatively and to determine the permitted variation, or tolerance level, in each requirement. Suppose a water flow system is needed to obtain a specified volume flow rate R_o. Because there may be variations in the operating conditions that may result in changes in the flow rate R, it is essential to determine the possible increase or decrease in the flow rate that can be tolerated. Then the system is designed to deliver the desired flow rate R_o with a possible maximum variation of $\pm \Delta R$. This may be expressed quantitatively as

$$R_o - \Delta R \leq R \leq R_o + \Delta R \tag{2.1}$$

Similarly, if a water cooler is being designed, the flow rate R_o and the desired temperature T_o at the outflow become the requirements. The former is expressed as given in Equation (2.1) and the latter as

$$T_o - \Delta T \leq T \leq T_o + \Delta T \tag{2.2}$$

where $\pm \Delta T$ is the acceptable variation in the outflow temperature T.

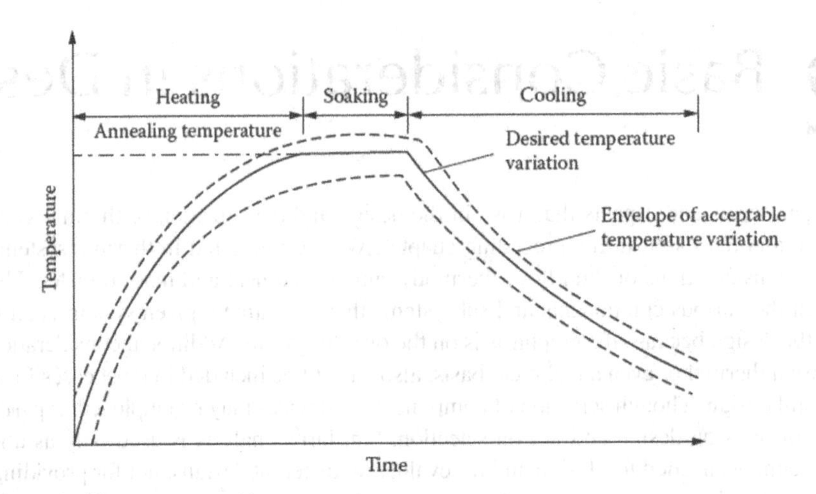

FIGURE 2.1 Required temperature variation, with an envelope of acceptable variation, for the thermal process of annealing of a given material.

In the design of thermal systems, common requirements concern temperature distributions and variations with time, heat transfer rates, temperature levels, and flow rates. Total pressure rise, time needed for a given process, total energy transfer, power delivered, rotational speed generated, etc., may also be the desired outputs from a thermal system, depending on the particular application under consideration. Consider the thermal annealing process for materials such as steel, glass, and aluminum. The material is heated to a given elevated temperature, known as the *annealing temperature*; held at this temperature level for a specified time, as obtained from metallurgical or microscale structural considerations of the chosen material; and then cooled very gradually, as shown in Figure 2.1. By heating the material beyond a particular temperature T_o, known as its *recrystallization temperature,* and maintaining it at this temperature, the internal stresses are relieved and the microstructures become relatively free to align themselves. A slow cooling allows the removal of residual and thermal stresses and refinement of the structure to restore the ductility of the material. The desired temperature cycle, including the maximum allowable temperature at which the process becomes unsatisfactory and the acceptable variation in the cycle, is shown in the figure. The duration τ_{soaking}, over which the temperature is held constant, within the two limits shown, is known as the *soaking time* and is also determined by the characteristics of the material. These requirements may, thus, be written quantitatively as

$$T_{\text{reqd}} \geq T_o, \quad \tau_{\text{soaking}} \geq \tau_o, \quad \left|\frac{\partial T}{\partial \tau}\right|_{\text{cooling}} \leq B \tag{2.3}$$

where T_o, τ_o, and B are specified constants, obtained from the basic characteristics of the given material. The acceptable variations in these constants, often given as percentages of the desired values, may also be included in these equations. Then, a thermal system is to be designed so that the given material or body is subjected to the required temperature cycle, with the allowable tolerance.

Similarly, the requirements for other thermal systems outlined in Chapter 1 may be considered. For instance, the mass flow rate, as well as the temperature and pressure at the entrance to the die in the plastic extrusion process, shown in Figure 1.10(b), are the requirements for a screw extruder. The rate of heat removal and the lowest temperature that can be obtained in the freezer could be taken as the requirements for a refrigeration system. The maximum power delivered and speed attained could be the requirements for a transportation system. The energy removal rate and the maximum allowable temperature may be the requirements for the cooling system of electronic equipment.

It is critical to determine the main requirements of the system and to focus our efforts on satisfying these. Because it is often difficult to meet all the desired features of the system, requirements that are not particularly important for the chosen application may have to be ignored. It is best to first satisfy the most essential requirements and then attempt to satisfy other less important ones by varying the design within the specified constraints and limitations. For instance, after a refrigeration system has been designed to provide the specified temperature and heat removal rate, effort may be exerted to find a satisfactory substitute for the refrigerants R-11 and R-12, both of which are chlorofluorocarbons, or CFCs; to replace the compressor with one that is more efficient; to vary the dimensions of the freezer; or to improve the temperature control arrangement. Thus, it is important to recognize the main requirements of the system and to design the system to achieve these, rather than consider every desired feature of the system at the very start of the design process.

2.1.1.1 Specifications

The system designed on the basis of the given requirements can be described in terms of its main characteristics. These form the *design specifications*, which list the requirements met by the system and the outputs that characterize the system. The final specifications of the system may include the performance characteristics; expected life of the system; recommended maintenance, weight, size, safety features; and environmental requirements. For instance, the specifications of a heat exchanger could be the overall heat transfer rate for given fluids and its dimensions. For a water chilling system, these could be the lowest attainable temperature and the corresponding flow rate and power consumption. The specifications of the system are, thus, the means of communication between the consumer and the designer/manufacturer.

2.1.2 Given Quantities

The next step in the formulation of the design problem is the determination of the quantities that are given and are, thus, fixed. These items cannot be changed and, as such, are not varied in the design process. Materials, dimensions, geometry, and the basic concept or method, particularly the type of energy source, are some of the features commonly given in the design of a thermal system. Thus, some of the materials and dimensions may be given, while others are to be determined as part of designing the system. For a particular system, if most of the parameters are fixed, the design problem becomes relatively simple because only a small number of variables are to be determined. If the basic concept is not fixed, different concepts may be considered, resulting in considerable flexibility in the design.

Let us consider the injection molding process for plastics, as shown schematically for two different machines in Figure 2.2. It is similar to the metal casting process described earlier and is thus a system dominated by heat transfer and fluid flow considerations (Tadmor and Gogos, 1979). It is an extensively used manufacturing process for a variety of parts ranging from plastic cups and toys to bathtubs, car bumpers, and molded parts made of composite materials. As shown here, the polymer is melted and injected into a mold cavity by applying force on the melt by means of a plunger or a rotating screw. As the polymer starts to solidify, additional amounts of melt may be injected to fill the gaps left due to shrinkage during solidification. The mold is held together by a clamping unit, which opens and closes the mold and also ejects the final solidified product.

For system design, the mold and the injected material may be kept fixed, while the melting and injection processes are varied. The system is a complicated one, but it can be considerably simplified by keeping several components and features fixed while a few components, such as the injection mechanism, are varied during design. In addition, the basic concept may be kept unchanged, using, for instance, either of the two schemes shown in the figure. Other approaches to melt and inject the mold, as well as to clamp and open the mold, are also possible. All these considerations substantially influence the design process.

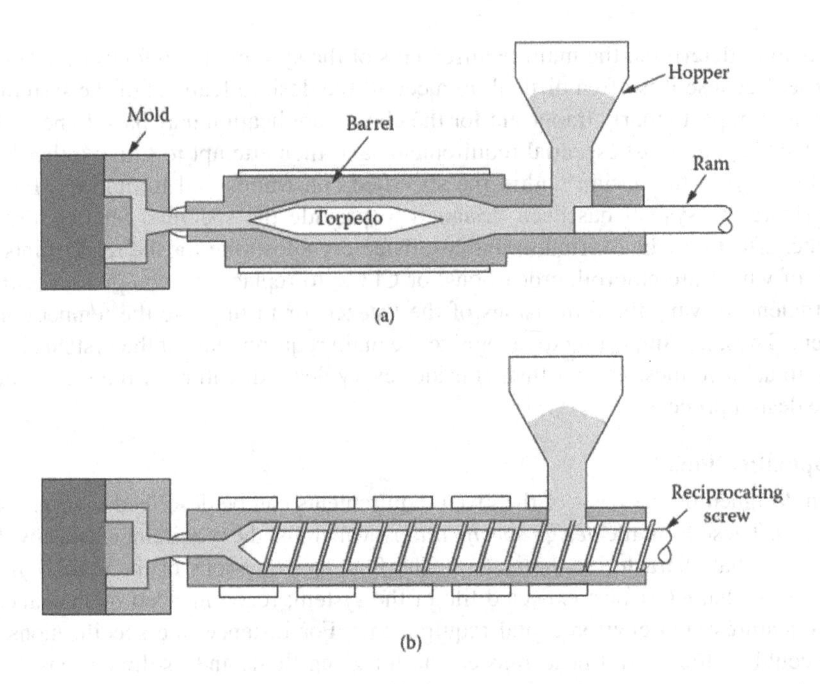

FIGURE 2.2 (a) Ram-fed injection molding machine; (b) screw-fed injection molding machine. (Adapted from Tadmor and Gogos, 1979.)

Similarly, in the design of the cooling system for an electronic system, consisting of electronic components located on circuit boards, the electronic component size, the geometry and dimensions of the board, the number of electronic components on each board, and the distance between two boards are often given. The design focuses on the cooling system, such as a fan and duct arrangement, keeping the geometry and heat dissipated by the components as fixed. A two-stroke engine may be chosen for the design of a transportation system, thus fixing the basic concept. In a solar energy system, sensible heat storage in water may be chosen as the concept, with the dimensions, geometry, and material of the tank being varied for the design. In the design of a cooling pond for cooling the condensers of a power plant, the location of the pond, which determines the local ambient conditions, is fixed and its dimensions are varied. In all of these cases, some of which are considered in later chapters, the given quantities are kept unchanged during the design process.

2.1.3 Design Variables

The design variables are the quantities that may be varied in the system in order to satisfy the given requirements and constraints. Therefore, during the design process, attention is focused on these parameters, which are varied to determine the behavior of the thermal system and are then chosen so that the system meets the given requirements. As mentioned earlier, it is important to focus on the main design variables in the problem because the complexity of the design procedure is a strong function of the number of variables.

Let us consider again the plastic injection molding system discussed in the preceding section and shown in Figure 2.2. If only the cooling of the mold is left to be designed, while the other components in the system are fixed, the problem is simplified. However, even this is an involved design problem and has generated much interest and effort over the last three decades. Cooling may be achieved by the flow of a cooling fluid through channels in the mold. Different types, configurations, and dimensions of cooling channels may be considered, obtaining the thermal characteristics of the system for each case. The solidification rate and temperature gradients in the material are usually given as the requirements that must be satisfied by using a variety of cooling channels. This leads to a domain of acceptable designs. An appropriate design may be chosen on the basis of additional considerations such as cost, power requirements, size, etc. If the other components of the system, such as geometry and dimensions of the melting and injection section, are to be varied as well, the design becomes much more involved and the domain of acceptable designs is much larger.

The design variables are usually taken to represent the hardware of the system such as the plunger, heating arrangement, mold, clamping unit, cooling channels, and so on, in the above example. However, the system performance is also affected by the operating conditions, which can be adjusted over ranges determined by the hardware. Therefore, the variables in the design problem may be classified in the following two categories.

2.1.3.1 Hardware

Hardware includes the components of the system, dimensions, materials, geometrical configuration, and other quantities that constitute the hardware of the system. Varying these parameters generally entails changes in the fabrication and assembly of the system. As such, changes in the hardware are not easy to implement if existing systems are to be modified for a new design, for a new product, or for optimization.

2.1.3.2 Operating Conditions

Operating conditions are the quantities that can often be varied relatively easily, over specified ranges, without changing the hardware of the given system, such as the settings for temperature, flow rate, pressure, speed, power input, etc. The design process would generally yield the ranges for such parameters, with optimization indicating the values at which the performance is optimal.

The design of a thermal system must include both types of variables and the final design obtained must indicate the materials, dimensions, and configurations of the various components, as well as the ranges over which the operating conditions such as pressure, temperature, and flow rate may be varied. These ranges are fixed by the hardware design; for instance, the temperature range may be determined by the heaters employed or flow rates by the pumps chosen. However, because the product obtained is a function of the operating conditions, these are often given as part of the specifications of the system. The following example illustrates the choice of variables in a practical thermal system.

Example 2.1

For the plastic screw extrusion system sketched in Figure 1.10(b), give the hardware variables and the operating conditions in the problem.

SOLUTION

The physical system under consideration consists of the following main parts: barrel, heating/cooling arrangement, screw, die, feed hopper, and the drive mechanism, which includes the motor, bearings, and gear system. Therefore, the hardware variables can be listed as

1. Geometry, material, and dimensions of the hopper
2. Geometry, material, and dimensions of the barrel
3. Dimensions, energy source, and configuration of heating/cooling arrangement
4. Diameter and material of the screw
5. Shape, height, thickness, and pitch of screw flights
6. Geometry, material, and dimensions of the die
7. Physical characteristics of the drive, motor, and gear system

Clearly, the above list includes a large number of variables. A design problem in which all of these can be varied is extremely complicated. Therefore, several of these are generally kept fixed and the ranges over which the others can be varied are determined from physical constraints, availability of parts, and information available from similar systems.

The operating conditions refer to the quantities that may be varied without changing the hardware. These may be listed as

1. Plastic flow rate or throughput
2. Speed (revolutions/minute)
3. Temperature distribution at the barrel
4. Material extruded

All of these operating conditions can be varied over ranges that are determined by the hardware design of the system. In addition, in actual practice these may not be varied completely independent of each other. For instance, the screw geometry and dimensions, along with the speed, will determine the maximum flow rate in the extruder. The heating/cooling arrangement determines the range of temperature variation. The plastic or polymer used may limit the speed or the temperature level, and so on.

2.1.4 Constraints or Limitations

The design must also satisfy various constraints or limitations in order to be acceptable. These constraints generally arise due to material, weight, cost, availability, and space limitations. The maximum pressure and temperature to which a given component may be subjected are limited by the properties of its material. For instance, a plastic or metal component may be damaged if the temperature exceeds the melting point. The performance of semiconductor devices is very sensitive to the temperature and, therefore, the temperatures in electronic equipment are constrained to values less than 80°C. The pressure rise in a thermal system is constrained by the strength of the materials at the operating temperature levels. Such constraints may be written for temperature T, pressure P, and volume flow rate R as

$$T \leq T_{max}, \quad P \leq P_{max}, \quad R \leq R_{max} \tag{2.4}$$

Generally, the maximum values, indicated here by the subscript *max*, would be considerably less than levels at which permanent damage to the component or system might occur. Therefore, T_{max} may be taken as, say, 50°C lower than the melting point of the material of which a given component is made, depending on the desired safety, accuracy of the model on which the design is based, and the material.

The choice of the material itself may be limited by cost, availability, waste disposal, and environmental impact even if a particular material has the best characteristics for a given problem. In fact, material selection is a very important element in design, as discussed later in this chapter. Volume and weight restrictions also frequently limit the domain of acceptable design. Again, these may be given as

$$W \leq W_{max}, \quad L \leq L_{max}, \quad V \leq V_{max} \tag{2.5}$$

where W, L, and V are the weight, length, and volume, respectively. Such constraints arise from the expected application of the system. For instance, weight restrictions are very important in the design of portable computers, airplanes, rocket systems, and automobiles. Similarly, volume constraints are important in room air conditioners, household refrigerators, and industrial furnaces. All such constraints and limitations determine the range of the design variables and, thus, indicate the boundaries of the domain over which an acceptable design is sought.

Constraints also arise due to conservation principles. For instance, mass conservation dictates the speed of withdrawal in a hot rolling process. For a two-dimensional flat plate being reduced in thickness from D_1 to D_2 across a set of rollers, as shown in Figure 1.10(d), mass conservation leads to the equation $U_1D_1 = U_2D_2$, where U_1 is the speed before the rollers and U_2 after, if the density of the material remains unchanged. Then this equation serves as a constraint on the speed after the rollers if the remaining quantities are specified.

Similarly, the energy rejected $Q_{rejected}$ from a power plant to a cooling pond is $\dot{m}C_p\Delta T$ where $\dot{m}$ is the mass flow rate of the cooling water, ΔT is its temperature rise in going through the condensers, and C_p is the specific heat at constant pressure. This energy must be rejected to the environment through heat loss at the water surface and to the ground. If the latter is negligible, as is often the case, the surface temperature must rise in order to lose the energy to the ambient medium. An energy balance equation may thus be written to determine the average surface temperature rise as

$$Q_{rejected} = \dot{m}C_p\Delta T = hA_{surface}\left(T_{new} - T_{old}\right) \tag{2.6}$$

where h is the overall heat transfer coefficient, $A_{surface}$ is the surface area, and $(T_{new} - T_{old})$ is the rise in the average surface temperature. A limitation of around 5°C on this temperature rise is specified by federal, state, county, or city regulations directed at minimizing the environmental effect. Therefore, the maximum amount of energy that may be rejected to the pond may be calculated. Similar considerations could lead to restrictions on temperature rise in the condensers, as well as on the total flow rate (Moore and Jaluria, 1972).

2.1.5 Additional Considerations

Several additional considerations have to be taken into account for obtaining an acceptable or workable design. These considerations may arise from safety and environmental concerns, procurement of supplies needed, availability of raw materials, national interests, import and export concerns, waste disposal problems, financial aspects, existing technology, and so on. Many of these aspects affect the overall engineering enterprise, as discussed in Chapter 1. However, the design itself may be strongly influenced by these considerations, particularly those pertaining to the environmental and safety issues. For instance, even though nuclear energy is one of the cheapest and cleanest methods of generating electricity, concerns on radioactive releases have strongly curbed the growth of nuclear power systems. Systems are designed in the steel industry to use the hot combustion products from the blast furnace in order to reduce the discharge of pollutants and thermal energy into the environment, while also decreasing the overall energy input. Thermal pollution concerns could make it undesirable to depend only on a lake or river for discharge of thermal energy from a power plant, making it necessary to design additional systems such as cooling towers for heat disposal.

Disposal of solid waste, particularly hazardous waste from chemical plants and radioactive waste from nuclear facilities, is another very important consideration that could substantially affect the design of the system. The energy source is chosen in order to meet the federal or state guidelines for solid waste disposal. Adequate arrangements have to be included in the design to satisfy waste disposal requirements.

Safety concerns, particularly with nuclear facilities, demand that adequate safety features be built into the system. For instance, if the temperature or heat flux levels exceed safe values, the system must shut down. If the fluid level were too low in a boiler, a safety feature would not allow it to be turned on, thus avoiding damage to the heaters and keeping the operation safe. Similarly, the energy source may be changed from gas to electricity because of safety concerns in an industrial system.

The formulation of the design problem is based on all of the above aspects. Therefore, before proceeding to the design of the thermal system, the problem statement is given in terms of the following:

1. Requirements
2. Given quantities
3. Design variables

4. Limitations or constraints
5. Safety, environmental, and other considerations

Because the design strategy, evaluation of the designs developed, and final design are all dependent on the problem statement, it is important to ensure that all of these aspects are considered in adequate detail and quantitative expressions are obtained to characterize these. It is worthwhile to investigate all important considerations that may affect the design and to formulate the design problem in exact terms, as far as possible, along with allowable variations or trade-offs in the various quantities and parameters of interest. Once the design problem is formulated, we can proceed to the development of the design, starting with the basic concept.

Example 2.2

An air conditioning system is to be designed for a residential building. The interior of the building is to be maintained at a temperature of 22°C ± 5°C. The ambient temperature can go as high as 38°C and the rate of heat dissipated in the house is given as 2.0 kW. The location, geometry, and dimensions of the building are given. Formulate the design problem and give the problem statement.

SOLUTION

The given quantities are the maximum ambient temperature, which is 38°C, and the rate of energy input due to activities in the house, specified as 2.0 kW. The location, geometry, and dimensions of the house are all fixed quantities. The requirements for the system to be designed are given in terms of the temperature range, 17°C – 27°C (22°C – 5°C to 22 + 5°C), that is to be maintained in the house. No constraints are given in the problem. However, typical constraints will involve limitations on the size and volume of the system, on the flow rate of air circulating in the building, and on the total cost. Use of CFCs as refrigerants will be unacceptable due to environmental considerations.

The thermal load due to heat transfer to the house from the ambient must first be determined. This load will involve absorbed solar flux, back radiation to the environment, convective transport from ambient air, evaporation or condensation of moisture, and conductive energy loss to the ground. The ambient thermal load is a function of ambient conditions, geometry of the building, its geographical location, and dimensions. It can often be modeled as $hA\Delta T$, where h is the overall heat transfer coefficient, A is the total surface area, and ΔT is the temperature difference between the ambient and the house. The overall heat transfer coefficient includes all the transport mechanisms, particularly convective denoted by h_c and radiative denoted by h_r, giving $h = h_c + h_r$. The total thermal load Q is then the ambient load plus the rate of energy dissipated inside the building. The rate of heat removal Q_r by the thermal system shown in Figure 2.3 must be greater than this total load.

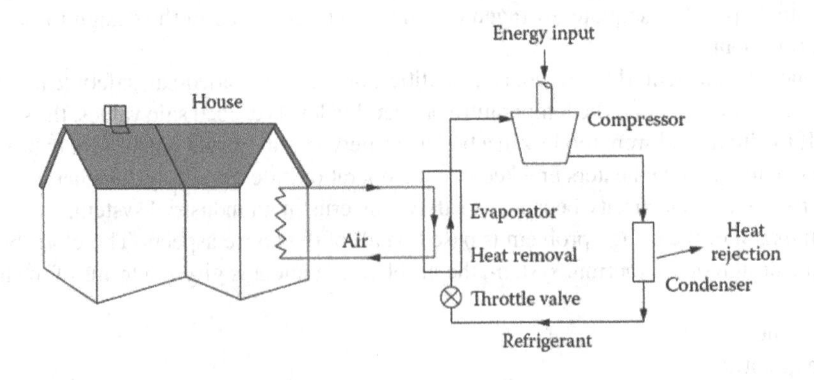

FIGURE 2.3 A thermal system for the central air conditioning of a house.

The transient cooling of the building is also an important consideration. If the total thermal capacity of the building (mass × specific heat) is estimated as *S*, then its average temperature *T* is governed by the energy balance equation

$$S\frac{dT}{d\tau} = Q - Q_r$$

From this equation, the time τ_r needed to cool the building to 1/*e* of its initial temperature difference from the ambient, i.e., the characteristic response time, may be calculated, as discussed later in Chapter 3. If this time is posed as a requirement, the heat removal rate Q_r or the capacity of the system may be appropriately determined; otherwise Q_r must simply be greater than *Q*.

The system is designed for the highest load, which arises at an ambient temperature of 38°C and inside temperature of 17°C. Simulation is used to determine the effect of ambient conditions as well as the transient response of the building. From these considerations, an acceptable design is obtained for the given design problem.

The problem statement for the given system design may, thus, be summarized as

Given: Building geometry, location, and dimensions. Maximum ambient temperature as 38°C. Rate of heat dissipated inside the house as 2.0 kW.

Requirements: Temperature inside the building must be maintained within 17°C and 27°C. In typical cases, the rate of cooling or response time τ_r is also a requirement.

Constraints: Limitations on size, volume, weight, and cost of the air conditioner. Limitation on maximum air flow rate circulating in the house may also be specified.

Design variables: Systems parts, such as condenser, evaporator, compressor, and throttling valve. Also, the refrigerant may be taken as a design variable.

Because of these requirements and constraints, the evaporator must operate at temperatures lower than 17°C to extract heat at the lowest temperature in the building. The condenser must operate at temperatures higher than 38°C in order to reject heat at the highest ambient temperature. Similarly, the total load will determine the capacity of the system. This specification is usually given in *tons*, where 1 ton is 3.52 kW and refers to the energy removal rate required to convert 1 ton (2000 lb) of water to ice in 1 day. A thermostat control with an on/off mechanism is often used with the designed thermal system to maintain the desired temperature levels.

2.2 CONCEPTUAL DESIGN

At the very core of any design activity lies the basic concept for the process or the system. The design effort starts with the selection of a conceptual design, which is initially expressed in vague terms as a method or scheme that might satisfy the given requirements and constraints. As the design proceeds, the concept becomes better defined. Conceptual design is a creative process, though it may range from something quite innovative, representing an invention or a new approach not employed before, to modifications in existing systems. Inventions may lead to patents, as discussed later. Creativity, originality, experience, knowledge of existing systems, and information on current technology play a large part in coming up with the conceptual design. For instance, microprocessors, laser-Doppler velocimeters, ultrasonic probes, composite materials, iPhone, iPad, digital cameras, light-emitting diodes (LEDs), and liquid crystals represent some of the innovative ideas introduced in recent years. Solutions based on existing and developing technology can also lead to valuable conceptual designs such as those of interest in computer workstations and laptops, automobile fuel injection systems, hybrid cars, and solar power stations. Changes can be made in existing systems to meet the given need or opportunity. In fact, much of the present design and development effort is based on improvements in current processes and systems.

For a given problem statement, several concepts or ideas may be considered and evaluated to estimate the chances of success. The ideas at this stage are necessarily fuzzy and rough estimates are carried out to determine if the concepts are feasible or if there are problems that may be difficult to overcome. Sometimes, these are simply *back-of-the-envelope* calculations that yield the overall inputs, outputs, expected ranges, etc. Such estimates allow the design group to narrow down the selection of the conceptual design to a few possible approaches. The selected conceptual designs are then subjected to the detailed design process, which would hopefully yield an acceptable design.

In order to illustrate the availability of different concepts and the choice of the most suitable one, let us consider the task of transporting coal from the loading dock to the blast furnace in a steel plant. Obviously, this can be achieved in many ways. Trucks, trains, conveyor belts, pipes, and carts are some of the methods that may be used. Each of these represents a different concept for the transportation system. The final choice is guided by the distance over which the material is to be transported, size and form in which coal is available, and rate at which the material is to be fed. For small plants, individual carts and trucks driven by workers may be adequate, whereas trains may be the most appropriate method for large distances and large plants. Clearly, there is no unique answer. In addition, within each concept, different techniques may be used to achieve the desired goals.

2.2.1 Innovative Conceptual Design

Innovative ideas can lead to major advancements in technology and must, therefore, be encouraged. Not all original concepts are earth-shattering and not all of these are practical. However, an environment conducive to the generation of creative and innovative solutions to the given problem must be maintained and various ideas brought forth must be examined before they are discarded. Such ideas may originate in different divisions within a company, such as manufacturing, research and development, and marketing. In many cases, the concept may be infeasible because of cost, technical limitations, availability of materials, and so on. But the concepts that appear to have promise must be considered further to determine if it is possible to develop a successful design based on them.

It is not easy to teach someone how to be creative and innovative. In most cases, creativity is a natural talent and some people tend to be more original than others. There are no set rules that one might follow to become creative. However, experience with current technology and knowledge of systems being used for applications similar to the one under consideration are a big help in the search for a suitable conceptual design. In addition, it is necessary to provide an environment that is open to new ideas. Creative problem solving requires imaginative thinking, persistence, acceptance of all ideas from different sources, and constructive criticism. Several such methods that may help to develop creative thinking are discussed by Alger and Hays (1964) and by Lumsdaine and Lumsdaine (1995). Techniques such as brainstorming, where a group of people collectively try to generate a variety of ideas to solve a given problem, design contests, and awards to employees with the best ideas also promote the generation of innovative solutions. Many impressive designs, such as the Vietnam Veterans Memorial in Washington, D.C., have arisen from design competitions.

2.2.1.1 An Example

In the manufacture of electronic systems, a classical process that is frequently used is that of soldering a pin to a board. Solid solder is placed around the pin in the form of a doughnut, as shown in Figure 2.4, and heated to a temperature beyond its melting point. The molten solder is driven by surface tension forces to form a joint, which solidifies on cooling to give the desired connection between the pin and the copper plated-through hole in the board. The heating had traditionally been done by radiation or by convection, using air or a liquid for immersion. Excessive and nonuniform heating of the boards was a common problem with radiation. Cleaning of the fluid and low heat transfer rates were the concerns with convection. In response to the need for an improved technique for this problem, a new and innovative method based on condensation of a vapor was proposed to yield a rapid heat transfer rate, while ensuring a clean environment with no overheating of the board. This resulted in the design of a thermal system to generate the vapor of a fluid with the appropriate boiling point. This vapor would then condense on a circuit board immersed in the condensation region, thus heating the material and forming the desired solder joint. Higher and more uniform

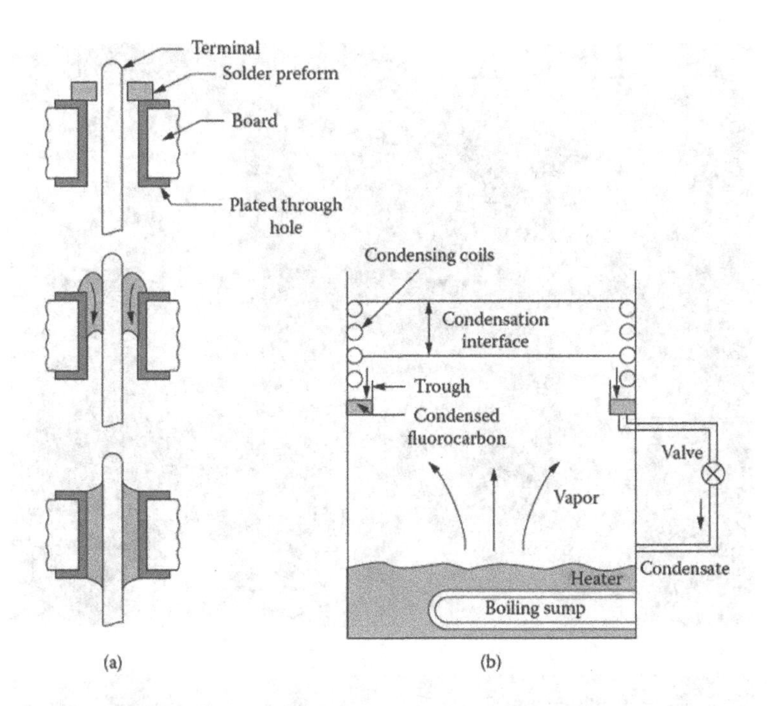

FIGURE 2.4 (a) Solder flow for forming a bond between a pin, or terminal, and a plated-through hole. (b) Schematic of a condensation soldering facility for electronic circuitry manufacture.

heat transfer rates are achieved by this method. The quality of the joint and the production rate are improved. Figure 2.4 gives the basic features of the process and of a simple condensation soldering system that can be used for such applications. Figure 2.5 shows a photograph of a condensation soldering facility, based on this concept, for large electronic components, indicating the typical scale of such practical systems. Example 2.3 discusses this process in greater detail. Figure 2.6 shows a different type of facility that uses the same basic concept and is available commercially. This system is more compact, easier to control, and has less fluid loss than the one shown in Figure 2.4(b). Dally (1990) may be consulted for further details on this and other soldering processes used in the manufacture of electronic circuitry. This system may also be used for other applications that require high rates of heating without overheating.

Many such innovative ideas have been introduced in recent years, particularly in the area of materials processing. Consequently, new materials, with a wide range of desired characteristics, and new processing techniques have been developed. Graphite tennis rackets, Teflon-coated cookware, lightweight camping equipment, lightweight laptop computers, and many such items in daily use are examples of these materials. Similarly, concerns with our environment and energy supply have resulted in many innovative systems for waste disposal, particularly for solid waste using methods such as incineration, and for unconventional energy sources such as solar, wind, and geothermal energy. Aerospace engineering is another area that has benefited from many new and original ideas that arose in response to the many challenging problems encountered due to, for example, high temperature, pressure, and velocity during rocket launching and re-entry. The space program has led to many significant advances like new alloys and composites, cellular phones, global positioning systems (GPS), and wireless accessories. Even in traditional fields, such as automobiles, many new concepts, such as microprocessor control, robotics, GPS navigation systems, and monitoring of the different subsystems, have been introduced in recent years. Therefore, creative and innovative concepts are crucial to the advancement of technology, with some of these resulting in major changes in current practice and others introducing only marginal improvements. Patents, copyrights, trademarks, and so on are needed to protect intellectual property, as discussed later.

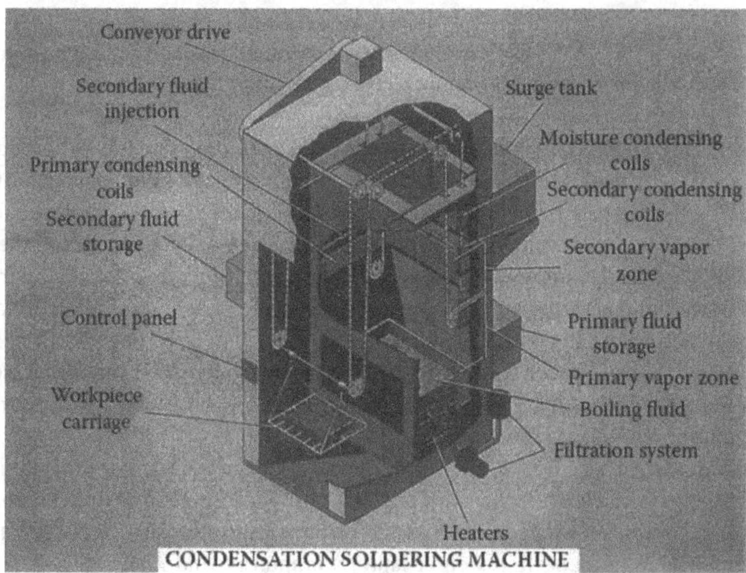

FIGURE 2.5 A practical condensation soldering facility. (From Lucent Technologies. With permission.)

2.2.2 Selection from Available Concepts

In an attempt to meet the given design requirements, concepts that have proved to be successful in the past for similar problems frequently provide a valuable source of information. With the technological advancements of recent years, a large variety of problems have been considered and many different solutions have been tried. In a given industry, the ideas that have been attempted in the past to solve problems similar to the one under consideration are well-known. Existing literature can also be used to generate additional information on various concepts and solutions that have been previously employed. The conceptual design for a given problem may then be selected from the list of earlier concepts or developed on the basis of this information. In this case, only the basic concept is similar to the earlier concepts; the system design may be quite different.

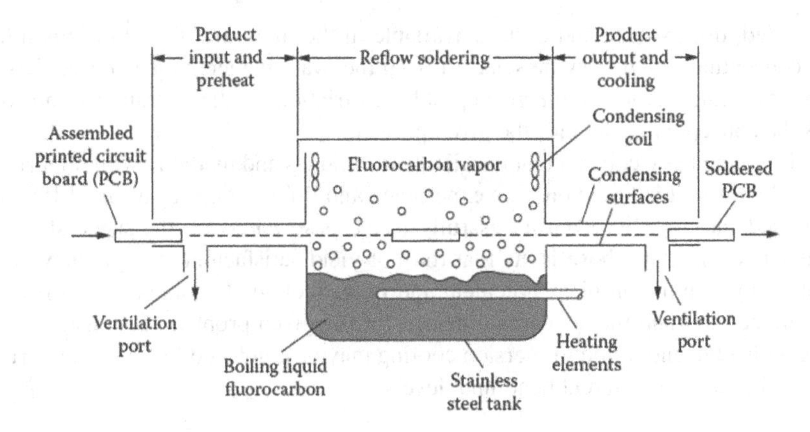

FIGURE 2.6 Condensation soldering machine for surface mounted components. (Adapted from Dally, 1990.)

Let us consider the problem of thermal management of electronic systems. If forced convective cooling is to be employed for a given electronic circuitry, the extensive information available in the literature on these cooling systems may be used to select or develop the conceptual design. Figure 2.7 shows the schematics of some of the arrangements and processes used in practice. Additional information on the characteristics of each system, for example, on the heat removal rate,

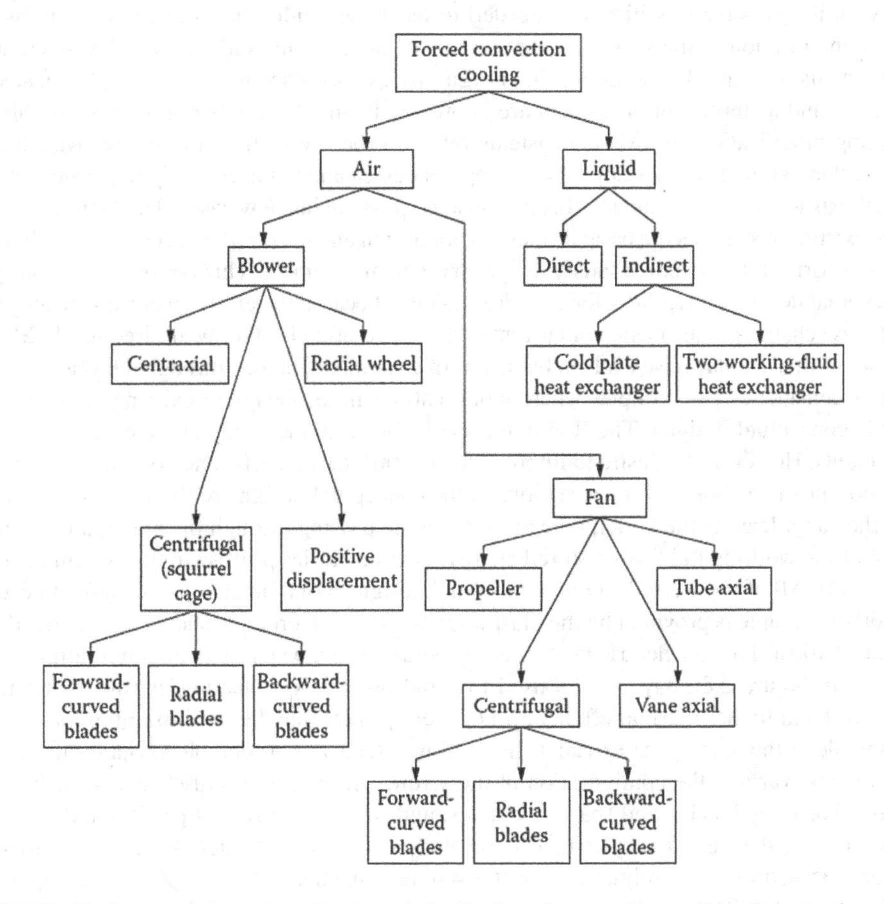

FIGURE 2.7 Various arrangements and processes for the forced convective cooling of electronic systems.

pressure needed, dimensions, and cost, is available in the literature. Based on this information, a particular conceptual design may be selected from the available techniques for cooling. If none of the approaches is satisfactory for the given problem, variations of these strategies and concepts may be used as the conceptual design for the given problem.

The choice of the basic concept from available techniques and methods is an important approach to conceptual design. It is based on both experience and information regarding different ideas that have been tried successfully or unsuccessfully in the past. Although the successful concepts are of particular interest, even those ideas that did not yield satisfactory designs may be considered because of changes in the problem statement and in technology. In some cases, different concepts may be combined to yield the conceptual design for the given problem. For instance, both forced air cooling with a fan and liquid immersion cooling may be employed for different parts of an electronic system because of different heat input levels.

2.2.3 Modifications in the Design of Existing Systems

In many cases, existing or available systems may form the basis for design of a new system to meet the requirements and constraints of a new application. This is clearly the simplest approach for obtaining a conceptual design for the given problem. However, it would work only if relatively small changes in the requirements are of interest. Improvements in the performance and characteristics of the system and in the quality of the product can also often be obtained simply by modifying the design of existing systems. Frequently, optimization of the system or of the process is achieved by such changes in the design. The conceptual design is then simply the design of the existing system, along with the possible modifications needed to meet the requirements of the new problem. The overall configuration of the system is kept largely unchanged and only a few relevant components or subsystems are varied. Therefore, the design process becomes relatively simple because many parameters and quantities in the system are known, reducing the number of design variables.

Making modifications in existing systems refers to the use of the information available on the design of these systems for developing a conceptual design and not necessarily to physical alterations in actual existing systems, although this may also be possible in a few cases. The main idea here is to employ existing systems as the basic framework for design and to consider variations in different components or parts of the system to satisfy the given problem statement. This is a very common approach in conceptual design, particularly for complex systems, because the effort involved is relatively small and because changes in the design of current systems can often lead to the desired result. Many thermal systems in use today have evolved by means of such modifications through the years.

Let us consider a few examples where modifications in the design of existing systems may lead to viable conceptual designs. The Rankine cycle is the basic thermodynamic cycle used for steam power plants. However, the desire to improve the overall thermal efficiency of the system has led to many modifications. Some of the variations in the conceptual design are those related to superheating of the vapor leaving the boiler, reheating the steam passing through the boiler, and regenerative heating of the working fluid using stored energy from an earlier process in the system (Cengel and Boles, 2014). All of these are different conceptual designs based on an existing system design.

Another example is provided by the plastic screw extrusion process, shown schematically earlier in Figure 1.10(b). Though electric heaters are generally used, water or steam circulating in jackets, as shown in Figure 2.8, may also be used to avoid possible overheating and for better temperature control and higher thermal efficiency. Different jackets may be used to impose a temperature variation along the axis of the extruder. In a screw extruder, considerable variation in the product is obtained by varying the configuration of the screw. Different types of elements, such as reverse elements, kneading blocks, and spacer elements, and screws of different profiles and pitch may be used to alter the design of the system. The die at the end of the extruder may also be varied. Thus, the overall structure and configuration of the system is unchanged and individual components are varied to achieve different characteristics and performance. Figure 2.9 shows photographs of a

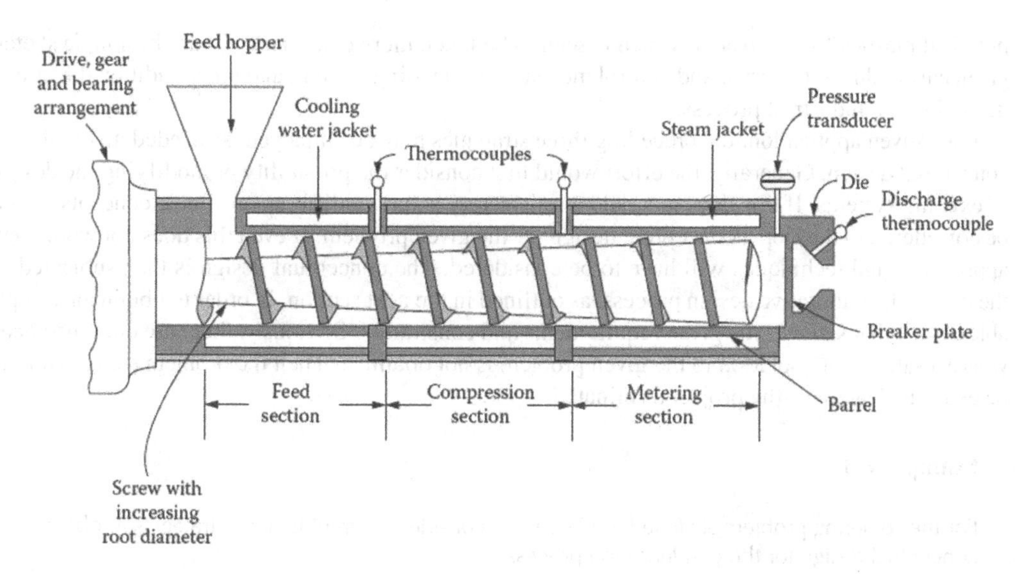

FIGURE 2.8 Schematic of a single screw extruder heated or cooled by the flow of steam or water in jackets at the extruder barrel.

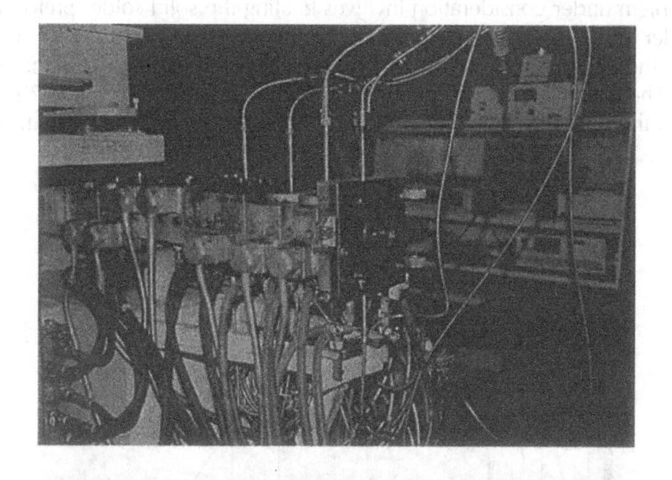

FIGURE 2.9 A practical plastics/food extrusion system. (From Center of Advanced Food Technology, Rutgers University, New Jersey. With permission.)

practical plastics/food extruder, which is seen to be much more complicated than the simple sketch given earlier due to the drive and control mechanisms, feeding system, and other additional features needed for an industrial process.

For a given application, the preceding three strategies may be employed, as needed, to obtain the conceptual design. Generally, the effort would first consider the possibility of modifying the design of existing systems. If this does not yield a satisfactory solution, different available concepts would be considered to develop a conceptual design for the given problem. If even this does not work, new approaches and techniques will have to be considered. The conceptual design is then subjected to the detailed, quantitative design process, as outlined in the next section, in order to obtain an acceptable design that satisfies the given requirements and constraints. Obviously, there are circumstances where a satisfactory solution to the given problem is not obtained. Then the problem statement may be examined again or the project terminated.

Example 2.3

For the soldering problem sketched in Figure 2.4, consider different heating strategies to obtain a conceptual design for the condensation process.

SOLUTION

The basic problem under consideration involves heating the solid solder preform so that it melts and flows under the action of surface tension, gravitational, and viscous forces to yield the solder fillet that joins the pin or terminal with the copper plated-through hole and thus with the printed circuit board. The fillet solidifies on cooling to yield the desired bond. Figure 2.10 shows the typical variation of the solder temperature with time, indicating melting and solidification at constant

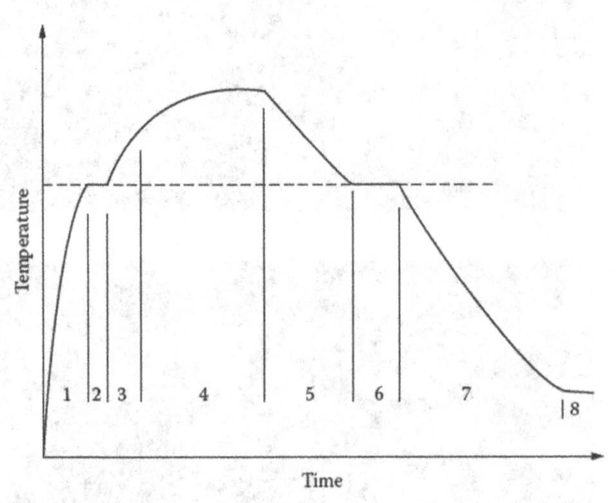

FIGURE 2.10 Typical temperature cycle undergone by a solder joint formed by melting of a solid preform, followed by solidification.

temperature. In common electrical circuitry, several such pins occur on each board and interest lies in a thermal system that achieves:

1. Rapid heating
2. Uniform heating of board materials
3. No damage to materials by overheating
4. Electrically insulating environment, so that electrical properties are not altered
5. Clean, nontoxic medium

Thermal systems may be designed for different heating mechanisms. Some of these, along with typical values of the corresponding heat transfer coefficient h for common geometries and dimensions, are estimated as (Incropera and Dewitt, 2001)

	$h(\text{W/m}^2\cdot\text{K})$
Natural convection in air and gases	5–10
Forced convection in air and gases	50–100
Natural convection in common liquids	350–550
Forced convection in common liquids	500–2,500
Radiative transport	600–10,000
Condensation	600–10,000
Fluidized bed	600–5,000

Convection has the advantage of heating the materials only up to the fluid temperature. As such, overheating can be avoided easily by choosing the fluid temperature below the temperature limitation of the materials involved. However, the heat transfer coefficient for natural convection in air or gases is extremely small. This is undesirable unless the fluid temperature is taken very large to obtain high heat transfer rates. If this is done, the materials may overheat and be damaged. Forced convection has higher heat transfer coefficients than natural convection. However, forced flow is strongly geometry-dependent and can lead to non-uniform heating due to separation and wakes, as shown in Figure 2.11(a). In addition, it will affect the shape of the solder fillet by exerting drag on the molten solder.

Natural convection using a liquid is attractive because it has reasonably high heat transfer coefficients and provides uniform heating. However, immersion in a liquid has the problem of accumulation of impurities, dust particles, and other undesirable deposits. Therefore, cleaning is a major concern in this case. Radiation provides a clean environment, but the heat flux absorbed is a strong function of the geometry and the surface properties of the material. Therefore, overheating is commonly encountered when radiation is used to heat the preform. Radiation masks, as shown in Figure 2.11(b), are generally needed to avoid overheating. Different masks are required for different geometrical configurations, making this a difficult and time-consuming effort. Fluidized bed heating has the same problems as forced convection.

The previous discussion indicates the kind of thinking that goes into the development of a conceptual design. Here, the heat transfer coefficients are obtained from the literature. Different heating mechanisms are considered and evaluated. The various strategies for heating, mentioned here, have been employed for different applications, despite their shortcomings. Finally, we come

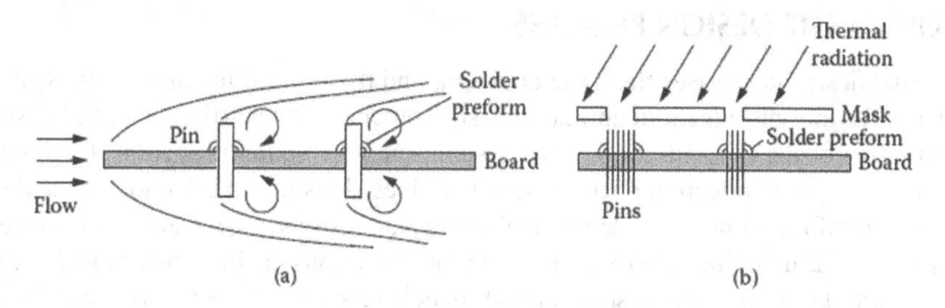

FIGURE 2.11 Heating of the solid solder preform by (a) forced convection and (b) thermal radiation.

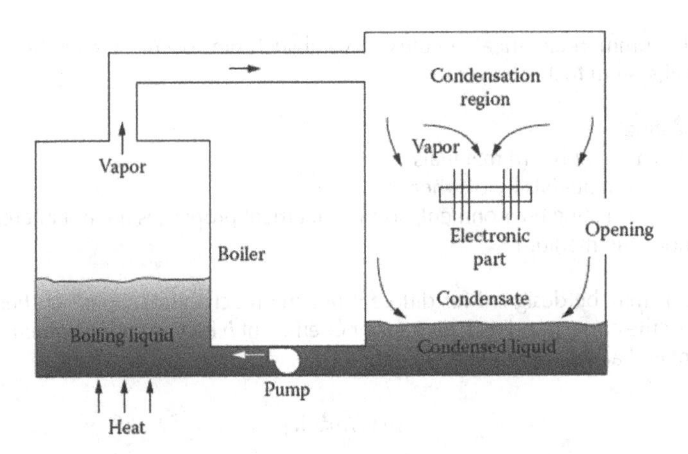

FIGURE 2.12 A possible conceptual design for a condensation soldering facility.

to condensation as a means to heat the solder preform. This process has a high heat transfer coefficient and provides uniform heating because an externally induced flow is not involved in the transport. The environment is clean because vapor obtained by boiling the liquid is used. The impurities, dust particles, and deposits are left behind in the liquid, which may be cleaned periodically. However, the success of this approach depends on the availability of a nontoxic vapor at the appropriate temperature. The melting point is around 182°C for common solders. Therefore, fluids with boiling points higher than this temperature are needed. Several high-boiling fluorocarbons are suitable for the purpose because these are nontoxic and relatively inert. However, these fluids are expensive and the system design must consider minimizing fluid losses. With all these considerations in mind, condensation heating may be chosen for the process.

Even after condensation has been selected as the method for heating and an appropriate fluid has been found, several conceptual designs for the system can be developed. We need a boiling sump where the liquid is heated to provide the vapor region where the vapor condenses on the circuitry to heat the solder preform. The condensed vapor must be returned to the sump. One possibility is to have a boiler and transport the vapor to a condensing chamber where the soldering takes place. The condensate is then pumped back to the sump. Leakage of the vapor is minimized by proper design of entry and exit ports for the electronic part. Figure 2.12 shows a sketch of such an arrangement.

The systems shown in Figure 2.4(b) and Figure 2.6 are other conceptual designs. In these cases, the boiling liquid sump and the condensing vapor region are located in the same container. Condensing coils, which are cooled by circulating cold water, condense the vapor and generate a vapor region. If a part is immersed in this region, the vapor condenses on it and thus heats it at the desirable high heat transfer rates. Though the vapor region is physically contained in Figure 2.6, it is not contained in Figure 2.4(a), resulting in greater fluid loss in this design. The part to be heated passes through the top as well. However, the interface generated at the top reduces the fluid loss. Additional mechanisms to minimize fluid loss can also be devised because the fluid is generally quite expensive. Again, the conceptual design is not unique and several other solutions and systems are possible.

2.3 STEPS IN THE DESIGN PROCESS

The conceptual design comprises the basic approach and the general features of the system. These form the basis for the subsequent quantitative design process. The starting or initial design is then specified in terms of the configuration of the system, the given quantities from the problem statement, and an appropriate selection of the design variables. This initial selection of the design variables is based on information available from other similar designs, on current engineering practice, and on experience. Employing approximations and idealizations, a simplified model is then developed for this initial design of the system so that its behavior and characteristics may be analyzed. Generally, the system behavior under a variety of conditions is investigated on the computer, by a

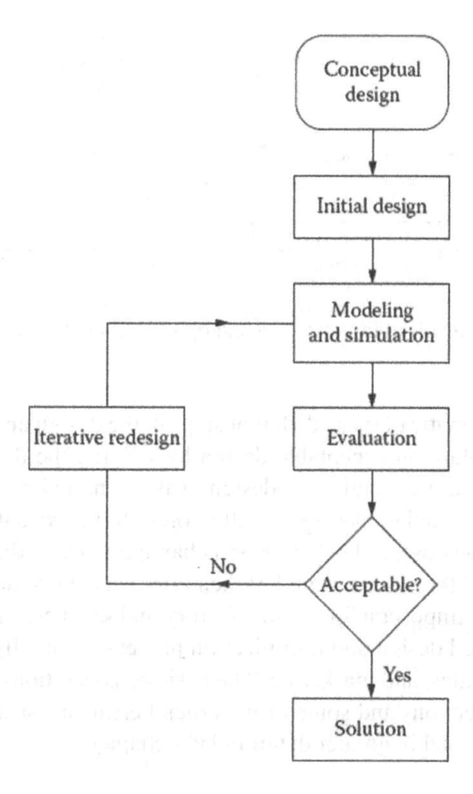

FIGURE 2.13 Iterative process to obtain an acceptable design.

process known as *simulation*, because of the complexity of the governing equations in typical thermal systems. An experimental or physical model may also be employed in some cases. The outputs from the modeling and simulation effort allow the designer to evaluate the design with respect to the requirements and constraints given in the problem statement. If an acceptable design that satisfies these requirements and constraints is obtained, the process may be terminated or other designs may be sought with a view to improve or optimize the system. If an acceptable design is not obtained, the design variables are adjusted and the processes of modeling, simulation, and design evaluation repeated. These steps are carried out until a satisfactory design is obtained. Different strategies may be adopted to improve the efficiency of this iterative procedure. Figure 2.13 shows a typical design procedure, starting with the conceptual design and indicating the steps mentioned here.

Usually, the engineering design process focuses on the quantitative design aspects after the problem statement and the conceptual design have been obtained. Then, the overall design process starts with the initial design of the physical system and ends with communication of the design to fabrication and assembly facilities involved in developing the system. The formulation of the design problem and conceptual design are precursors to this process and play a major role at various stages. Thus, the main steps that constitute the design and optimization process may be listed as:

1. Initial physical system
2. Modeling of the system
3. Simulation of the system
4. Evaluation of different designs
5. Iteration and obtaining an acceptable design
6. Optimization of the system design
7. Automation and control
8. Communicating the final design

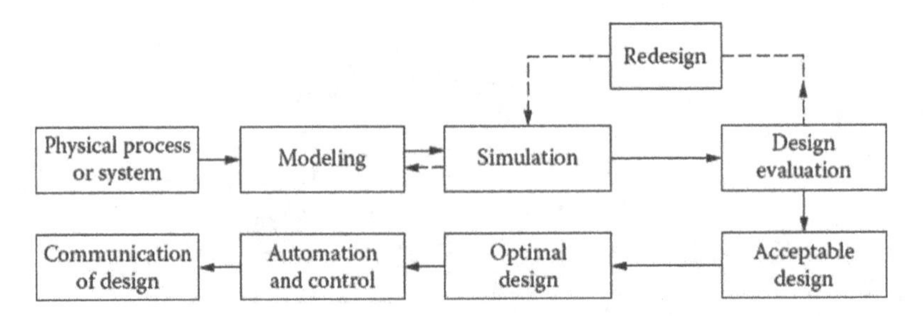

FIGURE 2.14 Various steps involved in the design and optimization of a thermal system and in the implementation of the design.

Figure 2.14 shows a schematic of these different steps in the design and optimization of a system. The iterative process to obtain an acceptable design by varying the design variables is indicated by the feedback loop connecting simulation, design evaluation, and acceptable design. There is a feedback between simulation and modeling as well in order to improve the model representation of the physical system based on observed behavior and characteristics of the system, as obtained from simulation. Optimization of the system is undertaken after acceptable designs have been obtained. Automation and control are important for the satisfactory and safe performance of the given system. The results from the detailed design and optimization process are finally communicated to groups involved with fabrication, sales, and marketing. The basic considerations involved in these steps are outlined in the following sections and some of the crucial elements, such as modeling, simulation, and optimization, are discussed in greater detail in later chapters.

2.3.1 Physical System

The starting point of the quantitative design process is the physical system obtained from conceptual design. This serves as the initial design that is modeled, simulated, and evaluated in the search for an acceptable design. Therefore, the system must be well-defined in terms of the following:

1. Overall geometry and configuration of the system
2. Different components or subsystems that constitute the system
3. Interaction between the various components
4. Given or fixed quantities in the system
5. Initial values of the design variables

A sketch may be used to represent the system configuration and the various components that interact with each other. Several of these were given in Chapter 1. For instance, Figure 1.8 presented the schematic for vapor compression and vapor absorption systems for refrigeration and air conditioning. Similarly, Figure 1.10 gave the physical representations for several manufacturing thermal systems and Figure 1.12 those for electronic equipment cooling systems. These sketches indicate the different components and subsystems that are part of the overall thermal system. The physical characteristics of these components and how they are linked with the others, particularly in terms of material, heat, and fluid flow, are also included.

In several cases, particularly for thermodynamic systems, the behavior and characteristics of the system may be represented graphically. State diagrams, which represent the equilibrium states through which a given material goes, are commonly used to indicate the thermodynamic cycle in many applications such as those related to refrigeration, power plants, and internal combustion engines. Similarly, changes in temperature, pressure, and velocity with location and time are used to indicate the basic nature of the process in many systems. Such graphical representations are largely qualitative and often idealized, thus modeling the physical system. The actual numbers and

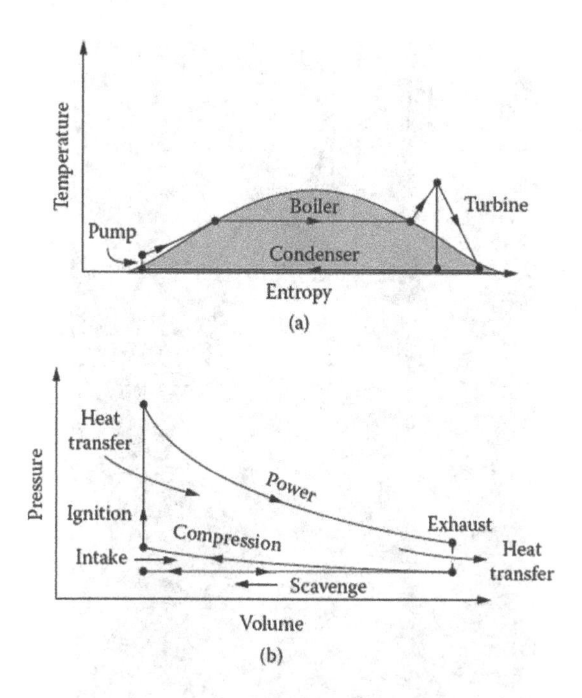

FIGURE 2.15 Thermodynamic cycles for (a) a Rankine engine with superheating of steam for power generation, and (b) an internal combustion engine based on the four-stroke Otto cycle.

other quantitative details are obtained through analytical and numerical calculations. Figure 2.15 shows qualitatively the typical thermodynamic cycles for a power plant and for a four-stroke internal combustion engine, indicating the various stages in the two processes. An analysis of these systems would then yield the actual pressures and temperatures involved (Moran and Shapiro, 2014).

The physical system must also include information on the given and, thus, fixed quantities in the problem and on the initial values of the design variables. Again, these may also be given in the form of sketches or graphs, as well as in symbolic or mathematical forms. Quantities that are often fixed are dimensions; materials and their characteristics; flow rates; and torque, pressure, or force exerted. Quantities that may be varied to obtain a satisfactory design are determined from the parameters that are not given, from operating conditions, and from the configuration of the system.

Consider the glass fiber drawing system shown in Figure 1.10(c). The basic configuration of the system is sketched in this figure. In addition, the dimensions and material of the fiber are given quantities, with specified tolerance levels. The draw speed of the fiber is a requirement in most cases for the desired productivity. The dimensions, material, and heating arrangement of the furnace could be taken as the design variables, although the given constraints will generally fix the domain of variation to fairly tight limits. The tension exerted on the fiber is to be determined for given operating conditions. Therefore, the physical system is specified in terms of these inputs. Figure 2.16 shows a photograph of the actual optical fiber drawing system, known as the *draw tower*. The simple sketch shown in Figure 1.10(c) is a schematic that gives the essential features of the system, which is much more complicated in actual practice due to power supply, control arrangement, feed mechanism, and other practical considerations.

2.3.2 Modeling

The modeling of the physical system, obtained from the conceptual design and from the formulation of the design problem, is an extremely important step in the design and optimization of the system. Because most practical thermal systems are fairly complex, it is necessary to focus on

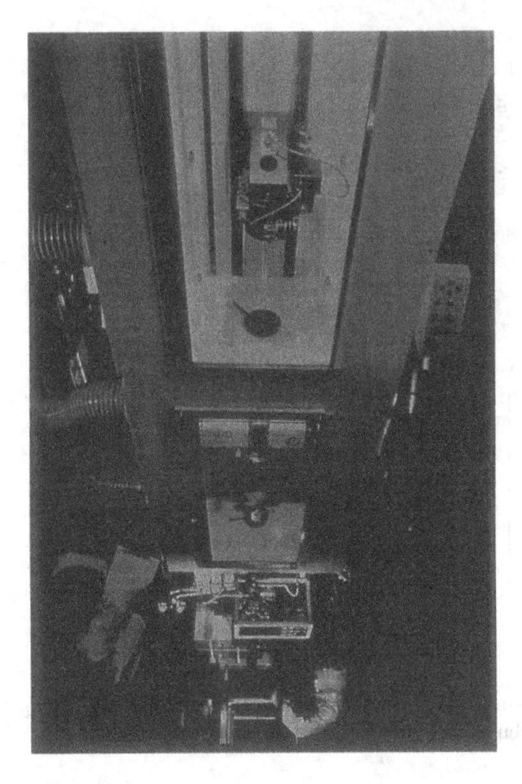

FIGURE 2.16 Draw tower for the manufacture of optical fibers. (From Fiber Optic Materials Research Program, Rutgers University, New Jersey.)

the dominant aspects of the system, neglecting relatively small effects, in order to simplify the given problem and make it possible to investigate its characteristics and behavior for a variety of conditions. Idealization and approximation of the processes that govern the system are also used to simplify the analysis. The basic conservation principles and properties of the materials involved are also important elements in modeling of thermal systems. Chapter 3 is devoted to modeling of thermal systems and only a brief outline is given here as an introduction to this process.

Both analytical and experimental procedures are employed to model the system. Because experimentation usually involves much greater time, effort, and cost, as compared to analysis, experimental methods are used sparingly and largely for the validation of the analytical/numerical model or when the inputs needed for design are not easily obtainable by analysis. Modeling of the thermal system yields a set of algebraic, differential, or integral equations, which determine the behavior of the actual system. These may be written as

$$F_i(x_1, x_2, x_3, \ldots, x_n) = 0 \quad \text{for } i = 1, 2, 3, \ldots, n \tag{2.7}$$

where x_i and F_i represent, respectively, the physical variables and the equations, algebraic or differential, that describe the problem. In most cases, numerical methods are necessary to solve these equations, particularly the nonlinear ordinary and partial differential equations often encountered in thermal systems. Discretized equations are then derived based on numerical techniques such as the finite difference and finite element methods for implementation on the computer, giving rise to a numerical model for the process or system. The analytical and/or numerical results obtained must be validated, preferably by comparisons with available experimental data, to ensure that the model is an accurate and valid representation of the physical system. The results obtained from experimental and numerical methods are frequently represented in terms of simple algebraic equations by means of curve fitting. These equations can then be used to characterize the system behavior and to optimize its performance.

Modeling of the thermal system also allows one to determine the conditions under which the results from an experimental scale model can be used to predict the behavior of an actual physical system. This involves dimensionless parameters that must be identical for the scale model and the actual system for obtaining similar distributions of flow, forces, heat transfer rates, and so on. Using the basic principles of dimensional analysis, the governing dimensionless groups are determined for a given thermal process or system. This also simplifies the experiment by reducing the number of parameters that need to be varied to characterize a given process, because most thermal systems are governed by a much smaller number of dimensionless groups as compared to the total number of physical variables in the problem.

As a result of the various simplifications and approximations, the given problem is brought to a stage where it may be solved analytically or numerically. Modeling not only simplifies the problem, but also eliminates relatively minor effects that only serve to confuse the main issues. It also provides a better understanding of the underlying mechanisms and thus allows a satisfactory inclusion of the experimental results into the overall model. Material property data and empirical results, available on the characteristics of devices and components that comprise the system, are also incorporated into the model.

Modeling is generally first applied to individual components, parts, or subsystems that make up the thermal system under consideration. Using the various experimental and analytical methods for modeling, separate models are thus developed for the constituents of the system. These individual models, or submodels, are then brought together or assembled in order to take into account the interaction between the various parts of the system. The different submodels are linked to each other through boundary conditions and the flow of mass, momentum, and energy between these. When these individual models are coupled with each other, the overall model for the thermal system is obtained. This model is subjected to a range of conditions to study the behavior of the system and thus obtain a satisfactory or optimal design.

Consider the simple power plant system sketched in Figure 2.17. The various subsystems, such as the boiler, condenser, turbine, and pump, are first considered individually and the corresponding models developed. After all these individual models, or submodels, have been developed, they must be brought together to yield the model for the complete thermal system, as shown schematically in Figure 2.18. In this particular example, the models of the individual subsystems are coupled through the fluid flow and the energy transport. Thus, the outflow from the boiler is the inflow to the turbine, whose outflow is the inflow to the condenser. Using such conditions, the different parts of the system are linked to each other through a central control unit. This then yields the model of the complete power plant. Additional subsystems such as the superheater, feedwater heater, and cooling tower may also be brought in for practical and more complicated systems. Similar considerations apply in the development of models for other thermal systems.

2.3.3 Simulation

Simulation is the process of subjecting the model for a given thermal system to various inputs, such as operating conditions, to determine how it behaves and thus predict the characteristics of the actual physical system. Though simulation may be carried out with physical scale models and prototypes, the expense and effort involved generally make it impossible to use these for design because many different designs and operating conditions need to be considered and evaluated. Prototype testing is largely used before going into production, after the system design has been completed. Therefore, simulation with mathematical models is particularly valuable in the design process because it provides information on the behavior of the given system under a range of conditions without actually constructing a prototype.

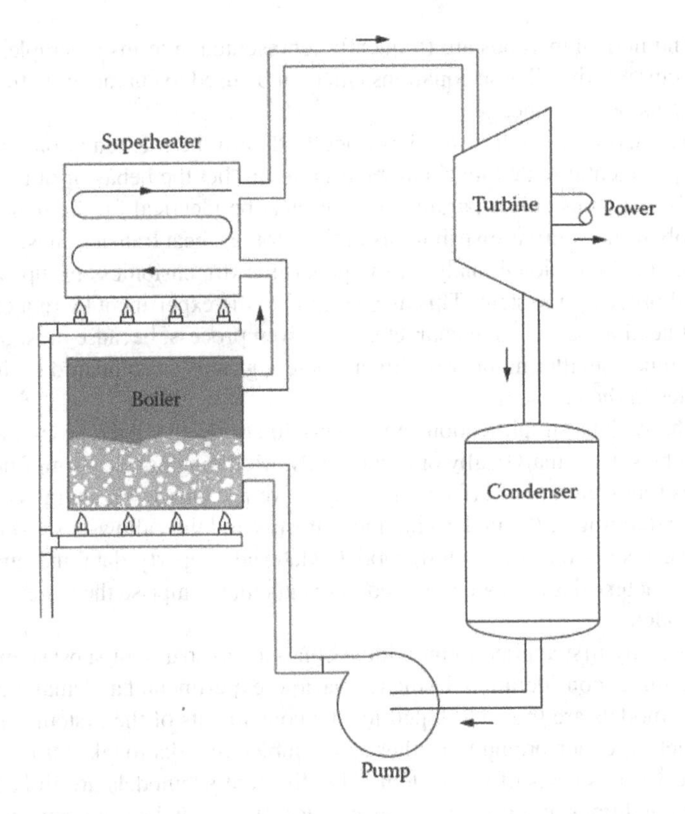

FIGURE 2.17 The physical system corresponding to the thermodynamic cycle shown in Figure 2.15(a). (Adapted from Howell and Buckius, 1992.)

The mathematical models derived for thermal systems are generally implemented on digital computers because of the complex nature of the governing equations, complicated boundary conditions, and complicated geometrical configurations that are usually encountered. The presence of several coupled submodels, representing different components of the system, and the incorporation of material properties, experimental data, and other empirical information further complicate the model. The resulting numerical model is then subjected to different values of the design variables, over the ranges determined by the constraints. Both the hardware and the operating conditions are varied to study the system characteristics. This process is known as *numerical simulation* and is an important step in the design and optimization process. Only a brief outline of numerical simulation

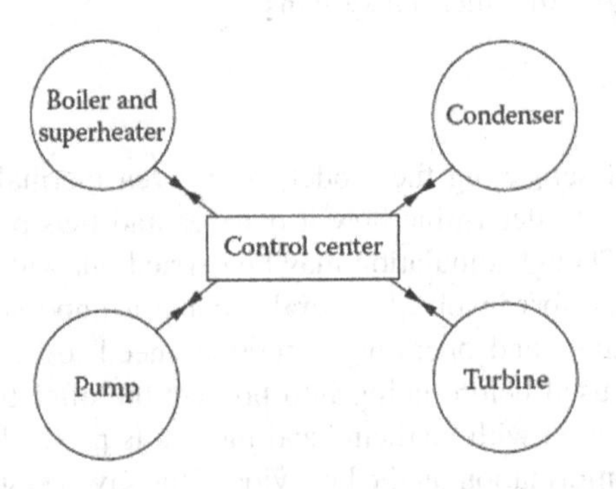

FIGURE 2.18 The main subsystems that combine to make up a power plant.

is given here, with a detailed discussion of the various procedures, types, and considerations, along with examples, given in Chapter 4.

An important question that must be answered in any numerical simulation is how closely or accurately it represents the actual, real-world, system. This involves ascertaining the validity of the various approximations made during modeling, as well as estimating the accuracy of the numerical algorithm (Roache, 2009). Certainly, if experimental data from a prototype are available, a comparison between these and the results from the simulation could be used to determine the validity and accuracy of the latter. However, such experimental data are rarely available, at least not during the design process. Consequently, the first step is to consider the simulation results in terms of the physical nature of the system and to ascertain that the observed trends agree with the expected behavior of the real system. Numerical parameters chosen by the designer or engineer, such as grid size, time step, computational domain, and so on, are then varied to ensure that the results are independent of these. Sometimes, simpler or similar systems for which experimental results are available may be simulated to validate the model. For instance, if a new system for plastic injection molding is being developed, the simulation scheme may be applied to an earlier version for which experimental data are available. Comparisons between the simulation results and experimental data could then be used to estimate the accuracy of the simulation. Therefore, considerable effort is directed at obtaining an accurate one-to-one correspondence between the model and the actual system.

After the final design is approved and a prototype is fabricated, more detailed results are obtained for the validation and improvement of the model and the simulation. In fact, results obtained over the years from systems on the market are also used to modify and improve the models and the simulation for the design and optimization of these systems in the future.

Simulation is mainly used to determine the behavior of the thermal system so that the design can be evaluated for satisfactory performance. It also provides inputs for optimization. Though there are many strategies that can be used for simulating thermal systems, as discussed in Chapter 4, a common approach is to fix the hardware and vary the operating conditions over the desired ranges. The hardware is then changed to consider a different design and the process repeated. The simulation of the system is carried out with different design variables until an acceptable design or a range of acceptable designs is obtained.

Simulation of practical thermal systems is often quite involved and can take a considerable amount of effort and time. To reduce the computational burden, approximate models, known as *surrogate* or *response surface* models, are often employed to represent the behavior of the system without extensive computational runs. The model is based on the responses at selected points and is used to represent the system over the given domain of design parameters and operating conditions. More is said on response surfaces later in the book.

2.3.3.1 An Example

Suppose a simple counterflow heat exchanger, as shown in Figure 2.19(a), is to be designed. The design variables are the two outer diameters D_1 and D_2 of the inner and outer tubes, respectively; the two wall thicknesses t_1 and t_2; and the length L of the heat exchanger. The operating conditions are the inlet temperatures $T_{1,i}$, $T_{2,i}$ and the mass flow rates $\dot{m}_1$ and $\dot{m}_2$ of the two corresponding fluid streams. Let us assume that a mathematical and numerical model has been developed for this system, allowing the calculation of the heat transfer rates and temperature distributions in the two fluid streams, as sketched in Figure 2.19(b). Let us take the heat transfer rate Q and the outlet temperature $T_{2,o}$ of the outer fluid stream as the outputs from the model and the remaining variables as inputs. Then these quantities may be given in terms of the design variables and the operating conditions, for given fluids, as

$$Q = F\left(D_1,\ D_2,\ L,\ t_1,\ t_2,\ \dot{m}_1,\ \dot{m}_2,\ T_{1,i},\ T_{2,i}\right) \tag{2.8}$$

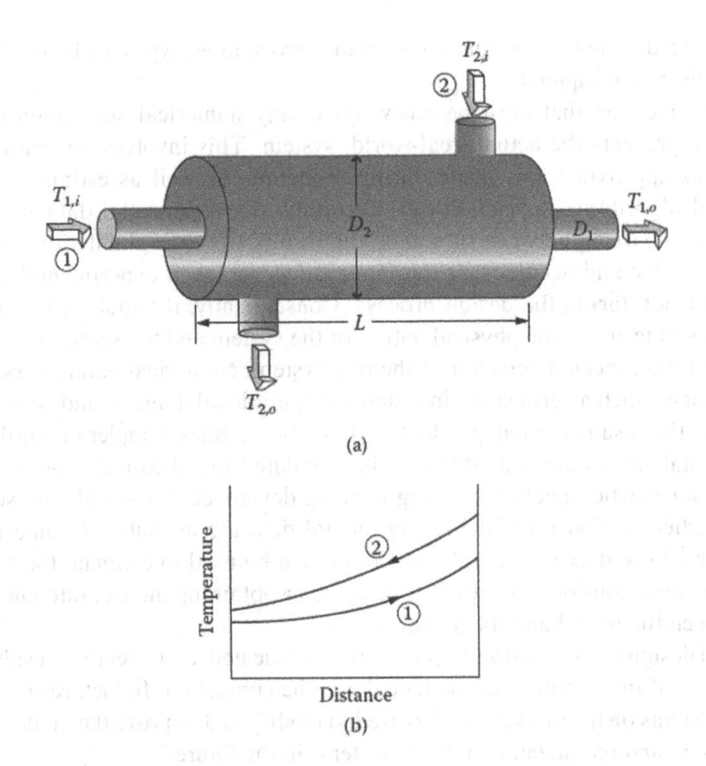

FIGURE 2.19 (a) A counterflow heat exchanger, and (b) typical temperature distributions in the two fluid streams.

$$T_{2,o} = G\left(D_1,\ D_2,\ L,\ t_1,\ t_2,\ \dot{m}_1,\ \dot{m}_2,\ T_{1,i},\ T_{2,i}\right) \tag{2.9}$$

Simple analytical expressions may be derived if the overall heat transfer coefficient is taken as a known constant (Incropera and Dewitt, 2001). The diameters and the length may be chosen so that the constraints due to size or space limitations are not violated. Tube diameter and thickness choices may be restricted to those available from the manufacturer to reduce costs. The length L and diameter D_1 may initially be held constant while different values of D_2 are considered. Then L and D_2 may be kept fixed, while D_1 is varied, and so on. Each combination of these three design variables represents a different system design that is subjected to different flow rates and temperatures, which represent the operating conditions, to study the behavior of the system in terms of outlet temperatures and overall rate of heat transfer. Thus, the model is used to consider many different designs and operating conditions in order to obtain the inputs for evaluating the design as well as for optimizing the system. Different design possibilities can be considered easily once the model and simulation scheme have been developed. Numerical simulation is, therefore, the appropriate approach even for such a simple system. Additional considerations arise in practical heat exchangers, such as different tube materials, ambient heat loss, insulation, and so on, making numerical simulation an important element in the design process. Further consideration of heat exchangers is given in Chapter 5.

The operating conditions for a particular system design are usually varied over fairly wide ranges. Certainly, the ranges expected in practice are taken into account during the simulation. But it is okay to get carried away and consider values far beyond the expected domain because these results will indicate the versatility of the system and how it would perform if the operating conditions exceeded the ranges for which the system is designed. Conditions beyond those employed for the design are often known as *off-design* conditions and simulation at these conditions provides valuable information on the operation of the system and on the model, particularly on its range of applicability. This also relates to the safety of the system because operating temperature, pressure, speed, and so on may exceed the design conditions due to a malfunction in the system or operator error. Simulation would indicate if the system would be damaged under these conditions and how its performance would be affected. In the foregoing heat exchanger example, simulation would yield the heat transfer rate and the outlet temperatures of the two fluids for different designs, given by the tube diameters D_1 and D_2 and the length L, and for different operating conditions, including off-design conditions, given by the flow rates $\dot{m}_1, \dot{m}_2$, and the inlet temperatures $T_{1,i}$ and $T_{2,i}$.

2.3.4 Evaluation: Acceptable Design

The next step in the design process is the evaluation of the various designs generated for determining if any of them are acceptable for the given design problem. As discussed earlier, an acceptable design is one that satisfies the given requirements for the system without violating the constraints. Therefore, the results from the simulation of the system are considered in terms of the problem statement to determine if a particular design is acceptable. Safety, environmental, regulatory, and financial constraints are also considered at this stage. If the design is not satisfactory because it violates the constraints or does not meet the requirements, a different design is chosen, simulated, and evaluated. This process is continued until an acceptable design is obtained. If none of the designs chosen over the given ranges of the design variables is found to be satisfactory, we may terminate the process or go back to the conceptual design stage and seek other alternatives.

If the design under consideration is found to meet all the requirements and constraints, an acceptable or workable design is obtained and the design specifications are noted. If we are only interested in obtaining a workable design, the design process may be terminated at this stage. However, in almost all practical cases, there are many possible solutions to the given design problem and the acceptable design obtained is, by no means, unique. Therefore, it is more useful to seek additional satisfactory designs by continuing the simulation with different values of the design variables. This effort would generally lead to a domain of acceptable or workable designs. From this domain, the best design may be chosen based on a given criterion such as minimum cost or highest efficiency.

Evaluating the design in terms of the results from the simulation and the given design problem statement is an important step in the design process because it involves the decision to continue or stop the process. Though several different possibilities exist, the following are some of the common ones:

1. Acceptable design obtained. Terminate iteration, communicate design.
2. Acceptable design obtained. Continue iteration to cover the given ranges of the design variables.
3. Acceptable design not obtained. Continue iteration with different design variables.
4. Acceptable design not obtained over ranges of design variables. Terminate iteration.

The first and the third conditions are the ones shown in Figure 2.13. The second one yields a region of acceptable designs from which an optimal design may be developed, as mentioned previously. The last condition indicates that a satisfactory design is not obtained over the given ranges of the design variables for the chosen concept. If additional conceptual designs are available, the design process may be reapplied to a different conceptual design; otherwise a solution to the given design problem is not obtained. All these possibilities, along with some others, do arise in actual practice because there are cases where an acceptable design is not achieved with the given requirements and constraints. In such cases, some of the requirements may be relaxed in order to obtain an acceptable design.

Considering again the simple counterflow heat exchanger discussed in the preceding section, the requirements and constraints may be written as

Requirements: $$Q = Q_o \pm \Delta Q \quad T_{2,o} = T_o \pm \Delta T \tag{2.10}$$

Constraints: $$(D_1)_{min} < D_1 < D_2 - 2t_2 \quad D_2 < (D_2)_{max} \quad L < L_{max} \tag{2.11}$$

Operating Conditions: $$\dot{m}_1 \quad T_{1,i} \tag{2.12}$$

Fixed Quantities: $$\dot{m}_2 = (\dot{m}_2)_o \pm \Delta \dot{m}_2, \; T_{2,i} = \left(T_{2,i}\right)_o \pm \Delta T_{2,i} \tag{2.13}$$

where the subscript *o* refers to specified values and *min* and *max* refer to the minimum and maximum allowable values, respectively. The minimum and maximum values are based on space limitations, manufacturing, cost, and other considerations. Specified tolerance levels or variations in the values are also given. Obviously, different requirements and constraints may be given for different applications. Here, fluid stream 2 is taken as fixed, whereas fluid stream 1 is varied. Clearly, the tube material is another important consideration that may be included in the problem. Thus, the simulation of the system may be carried out for different designs and for different operating parameters, with the preceding equations as the requirements and constraints. All the quantities are varied over the permissible ranges.

If numerical simulation is carried out with different designs, obtained by varying the design variables over the given ranges, and if all acceptable designs are collected, a region over which the design is satisfactory is obtained. This region may be represented mathematically in terms of the design variables as

$$F_b(x_1, x_2, x_3, \ldots, x_n) = 0 \tag{2.14}$$

where the function F_b indicates the boundary of the region and x_1, x_2, ..., x_n are the design variables. The boundary may also be shown graphically in terms of two variables taken at a time. For instance, the ranges of the design variables D_2 and length L for the heat exchanger problem outlined previously, along with the domain of acceptable designs, may be sketched as shown in Figure 2.20 for a particular value of D_1. Similar regions may be shown for other values of D_1, as well as for D_1 and L as the two variables, with D_2 as given. Such graphical representations are obviously difficult to obtain or use for a large number of design variables. However, this could be done easily on the computer. The main idea here is that a number of acceptable designs may be obtained on the basis

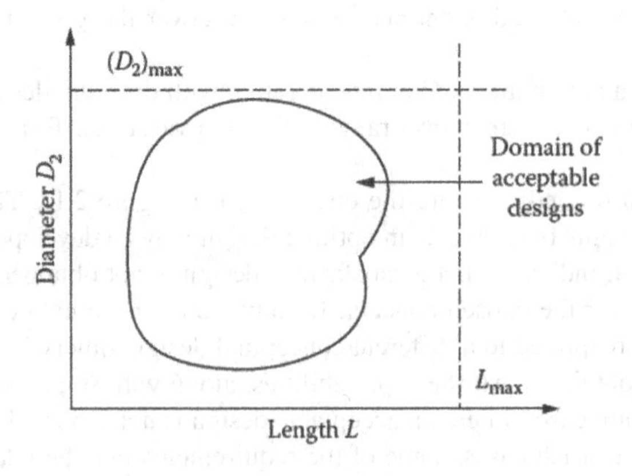

FIGURE 2.20 Domain of acceptable designs, along with the given constraints, for a heat exchanger.

of simulation. The selection for the best or optimal design may then be carried out from this region of acceptable designs.

2.3.5 Optimal Design

It is rare that the design process would be terminated as soon as an acceptable design is obtained. Only when the cost or effort involved in optimization is deemed too high would the design activity stop after an acceptable design is obtained. With growing competition in the world today, it has become necessary to reduce costs while improving product quality. Therefore, working with the first acceptable design obtained is no longer adequate. At the very least, several possible designs must be considered and the best chosen from among these, as measured in terms of an appropriate quantity such as cost, efficiency, or product characteristics. Optimization refers to a systematic approach to minimize or maximize a chosen quantity or function. The optimization process is obviously applied to acceptable designs so that the given requirements and constraints are satisfied. Then the design finally obtained is an optimal one, not just an acceptable one. Much of the latter portion of this book is devoted to optimization of thermal systems and only a brief introduction to the subject is given here to indicate its importance and position in the design process.

Optimization is of particular importance in thermal systems because of the strong dependence of cost and output on system design. Usually, the optimal design is not easily determined from available simulation or acceptable design results. A fairly elaborate effort has to be exerted in most cases to obtain the optimal design. Because simulation is generally an involved and time-consuming process for most practical thermal systems, special techniques that reduce the number of designs to be simulated are of particular interest. Frequently, there are large differences between the performance of optimized and nonoptimized systems in terms of energy consumption, product quality, overall thermal efficiency, and total costs, making it imperative to seek an optimal design.

Optimization of a thermal system can be carried out in terms of the design hardware or the operating conditions. The latter approach is particularly valuable because it allows one to operate a given thermal system under optimum conditions, thus minimizing costs and maximizing efficiency and product quality. No changes in the design hardware are needed; only the conditions such as temperature, pressure, flow rate, and speed at which the system is operated are adjusted to deliver optimum performance. The design specifications for many thermal systems thus include the information on optimal operating conditions. For instance, the best setting for the temperature in a refrigerator and the optimal speed for an engine may be given. Similarly, the design hardware for a given thermal system may be optimized in order to obtain the best performance for a desired set of operating conditions.

Generally, optimization follows the iterative design stage that yields the acceptable designs for a given application, as shown schematically in Figure 2.14. An appropriate quantity or function, known as the objective function and denoted by $U(x_1, x_2, x_3, \ldots, x_n)$ in terms of the design variables, is chosen for minimization or maximization, that is,

$$U(x_1, x_2, x_3, \ldots, x_n) \rightarrow \text{Minimum/Maximum} \tag{2.15}$$

The constraints arising from conservation principles and from physical limitations in the problem, such as those pertaining to size, weight, strength, temperature, energy input, and so on, may be equalities or inequalities, given respectively as

$$G_i(x_1, x_2, x_3, \ldots, x_n) = 0 \tag{2.16}$$

and

$$H_j(x_1, x_2, x_3, \ldots, x_n) \leq C_j \tag{2.17}$$

Here G_i represents the equality constraints, with $i = 1, 2, 3, \ldots, m$, where m is the number of equality constraints. C_j is a given quantity corresponding to the constraint H_j and $j = 1, 2, 3, \ldots, l$, where l is the number of inequality constraints.

Depending on the nature of the problem, particularly on the form in which the simulation results are available, various optimization methods may be applied to find the extremum of the chosen objective function. A sensitivity analysis may then be undertaken to determine how this function varies with the design variables and the operating conditions in order to choose the most appropriate, convenient, and cost-effective values at or near the extremum that would optimize the system or its operation. In addition, practical considerations related to safety, economic, and environmental issues have to be taken care of, resulting in trade-offs, before the final design is decided. The resulting objective function may also be compared with the values for the various acceptable designs encountered during the optimization process in order to estimate the improvements obtained as a result of optimization.

In many practical thermal systems, it is often difficult to work with a single-objective function. There may be several criteria or design objectives, making the optimization process more complicated than the approach outlined previously for a single-objective function. For instance, in the cooling of electronic equipment, the overall heat transfer rate is to be maximized. But the pressure head needed for the flow is also an important consideration because it affects the cost and operation of the system, and this quantity is to be minimized. The two objectives, heat removal rate and pressure head, oppose each other because an improvement in one objective leads to a deterioration in the other. The optimal solutions obtained by separate optimization of the two objective functions do not yield optimal solutions to the multi-objective problem. In a few cases, the different objective functions may be combined to yield a single-objective function, which tries to capture the individual behavior of the objective functions. Then a single-objective function optimization problem is solved. However, there is inherent arbitrariness in combining different objectives and a true optimal design may not be obtained. Thus, a multi-objective design optimization problem has to be solved in many cases of practical interest. The solution is obtained in a manner similar to single-objective function optimization and a trade-off between different objectives is employed, as discussed in later chapters.

2.3.5.1 Examples

Let us consider a few simple systems to illustrate these ideas. In the case of the heat exchanger discussed in the preceding sections, it is clear that there would generally be a large number of acceptable designs that would satisfy the requirements and constraints such as those given by Equation (2.10) through Equation (2.13), yielding the domain sketched in Figure 2.20. Let us assume that the cost of the equipment is to be minimized. If the material of the tubes were kept fixed, this would require minimizing the total material used. Manufacturing costs may also be included, often taken as an overhead on material cost. An expression for the total volume V of the material may be written in terms of the design variables as

$$V = \pi D_1 L t_1 + \pi D_2 L t_2 \tag{2.18}$$

These quantities are to be varied in the domain of acceptable designs in order to minimize the objective function V. The market availability of different tube sizes may also be included in this process to employ dimensions that are easily obtainable without significantly affecting the optimum, as confirmed by sensitivity studies. Once this optimal design is obtained, the operating conditions $\dot{m}_1$ and $T_{1,i}$ may also be varied to determine if the costs could be further minimized by reducing the flow rate and the inlet temperature of fluid stream 1, while the other fluid stream is fixed in the problem. Thus, the overall costs may be minimized. Additional aspects such as pressure needed for the flow may be included for a more complete design of a practical heat exchanger.

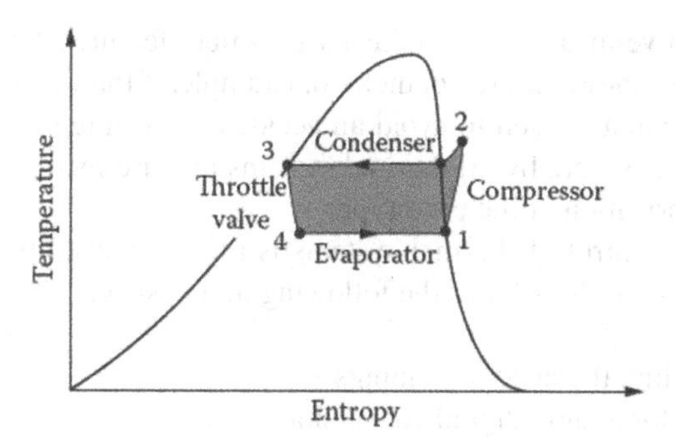

FIGURE 2.21 Thermodynamic cycle for a vapor compression cooling process, indicating the various components of the thermal system.

Similar considerations arise for other thermal systems. For instance, the desired heat removal rate Q in a refrigerator can be achieved by using a vapor compression system, as sketched in Figure 1.8. The corresponding thermodynamic cycle is shown in Figure 2.21. Because $Q = \dot{m}\Delta h$, where Δh is the change in enthalpy h per unit mass of the refrigerant in the evaporator and $\dot{m}$ is the mass flow rate of the refrigerant, the thermodynamic cycle is not unique, even if the fluid is kept fixed. Different temperatures and pressures for the condenser, evaporator, and compressor and different flow rates $\dot{m}$ can be employed to obtain an acceptable design. However, if the coefficient of performance (COP), which gives the ratio of the heat removed to the required input work, is to be maximized, the desired optimal design is more clearly defined. In terms of the state points shown in Figure 2.21, the COP is given by

$$\text{COP} = \frac{h_1 - h_4}{h_2 - h_1} \tag{2.19}$$

where the compression and throttling processes may not be at constant entropy, or isentropic. The domain of acceptable designs is thus narrowed down to obtain an optimized system. Though a unique solution is possible, generally one tends to narrow the region to a sufficiently small domain so that an appropriate optimal design may be chosen from it based on convenience and availability of parts needed for the system. Different temperature settings of the refrigerator and different heat loads may finally be considered to determine if optimum operating conditions exist at which, say, the total energy consumption is the least.

Because of the considerable improvement in most practical thermal systems through optimization, in terms of cost, efficiency, power consumption, product quality, and so on, it is now accepted that the design process would generally include optimization of the system.

2.3.6 Safety Features, Automation, and Control

An important ingredient in the successful operation of a thermal system is the control scheme, which not only ensures safety for the system and the operator but also maintains the operating conditions within specified limits. Sensors that monitor the temperatures, pressures, flow rates, and other physical quantities in the system are employed to turn off the inflow of material and energy into the system if the safety of people working on it or that of the system is threatened. For instance, if the temperature T_c of the compressor in an air conditioning system rises beyond a given safe level T_{max}, the system is turned off. Similarly, the gas flow into a furnace, power supplied to an electronic system, fuel input to an energy conversion system, and gasoline flow into an internal combustion engine may be reduced or cut off if overheating of the components occurs. In a car, sensors indicate overheating of the engine, as well as malfunction of other components, allowing the driver to turn it off or take other corrective measures. In many cases, safety features do not allow the system to be turned on if appropriate conditions are not met. For example, if the water level in a boiler is too low, a sensor-driven arrangement is used to avoid an accidental turning on of the heaters. Such safety features are employed in essentially all thermal systems and are incorporated into the final design of the system before fabrication of the prototype.

The automation and control of thermal systems is more involved than the simple inclusion of safety features in the system. It includes the following main aspects:

1. Sensors for providing the necessary inputs
2. Process interface for analog/digital conversion
3. Control strategies
4. Actuators
5. Safety features
6. Process programming

The input signals from the sensors (thermistors, thermocouples, flux meters, flow meters, pressure transducers, etc.) are electronically processed and fed into the control system, which determines the action to be taken. An appropriate signal is then given to the actuators, which make the desired changes in the system, such as reducing the flow rate, increasing the heat input, and turning off the power. Figure 2.22 shows a schematic of a typical control arrangement for a thermal system.

Most thermal systems need appropriate control arrangements for satisfactory performance. For instance, in a plastic extrusion process, the barrel is to be maintained at a specified temperature level for a given application. Temperature sensors, such as resistors or thermocouples, are located in the barrel and their output is coupled with a scheme to control the heating/cooling arrangement, for example, energy input to the heaters, in order to maintain the temperature at the given value with an acceptable tolerance. Similarly, temperature control is used in cooling and heating systems to ensure that desired conditions are maintained. Several different sensors are available for temperature, velocity, flow rate, pressure, and other variables (Figliola and Beasley, 2014). Though an on/off arrangement with inputs from a sensor is commonly used in practical thermal systems, other control strategies such as proportional, derivative, integral, and combinations of these control methods are also used. Microprocessors are used extensively to automate and control thermal systems. The inputs from the sensors, along with the desired values of the various operating parameters, are employed to maintain the appropriate levels and thus ensure automatic, safe, and satisfactory operation of the system. The subject of automation and control is a broad one, and it is not possible to discuss the extensive information available in the literature on different types of control systems; see, for instance, Palm (1986) and Raven (1994). An example on locating sensors for safety follows.

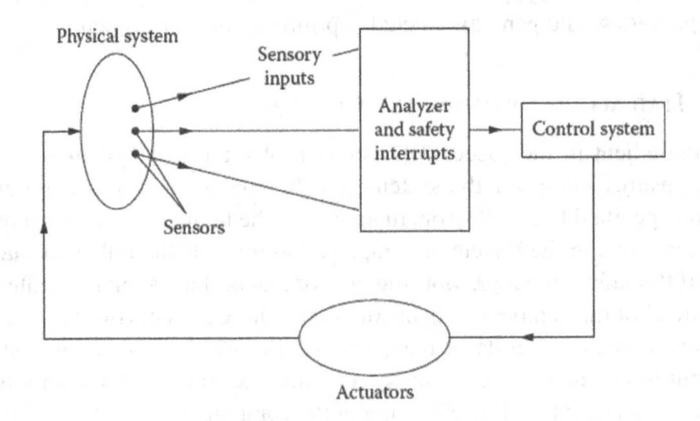

FIGURE 2.22 Schematic showing the use of sensors for control and safety.

Example 2.4

For the condensation soldering facility shown in Figure 2.4(b), give the types of sensors and locations of these that you would employ for ensuring safety of the system and the operator, as well as for control of the process.

SOLUTION

Several aspects must be considered for safety and for control. Considering first the safety issues, the important items are:

1. The heater must not overheat. A temperature sensor must be used to measure the heater temperature and turn it off if the temperature exceeds safety limits specified by the manufacturer.
2. The liquid level in the sump must be adequate to provide the vapor and cover the heater. A sensor, such as an ultrasonic sensor, may be used to ensure that the liquid level is not below a specified value. If the level is too low, it should not be possible to turn on the heater.
3. The temperature in the condensation region may be monitored to ensure that the operator does not venture into the facility or open up a side port for cleaning or maintenance unless the temperature is low enough. The temperature indicates that the concentration of vapor is low and temperature levels are safe. Though nontoxic, the vapor is heavier than air and can be dangerous because of lack of oxygen.

Considering now the sensors for control, the main items to be addressed are:

1. The vapor temperature must be high enough for reflow soldering to occur, say 183°C for the typical solder alloy. A temperature sensor in the condensing vapor region monitors this. When the temperature reaches the appropriate levels, a signal, such as a green light, may be given to indicate that the facility is ready.
2. The outlet temperature of the cooling water circulating through the condensing coils must not be too high because this would reduce the condensing effect. Generally, only a few degrees' temperature rise in the water is acceptable. Again, a temperature sensor at the outlet of water flow may be used. If the temperature is higher than the specified value, the heater input may be reduced or the water flow rate increased. A flow meter may be used to measure the water flow rate.
3. A temperature or concentration sensor above the condensation interface may be used to determine if an excessive amount of vapor is escaping from the facility. If the loss is judged to be excessive, the heater input may be reduced or the water flow rate increased. In some cases, the water temperature may also be reduced by chilling.

Clearly, the answer to this problem is not unique and several other arrangements for obtaining the required information can be devised. Temperature sensors are among the cheapest and the easiest to use, making them very attractive for thermal system control. Ultrasonic and optical sensors are useful in determining presence of fluids. Flow meters, such as rotameters and anemometers, are used for measuring flow. Figure 2.23 gives a sketch showing the locations of these sensors. The control system is based on the outputs obtained from these sensors and adjusts mainly the water flow rate and the heater input to control the process.

2.3.7 Communicating the Design

The communication of the final design to the client or customer and to those who will implement the design is an important ingredient in the overall success of the project. It is necessary to bring out the salient features of the design, particularly how it meets the requirements of the application and constraints imposed on the design. The basic approach adopted in the development of the design,

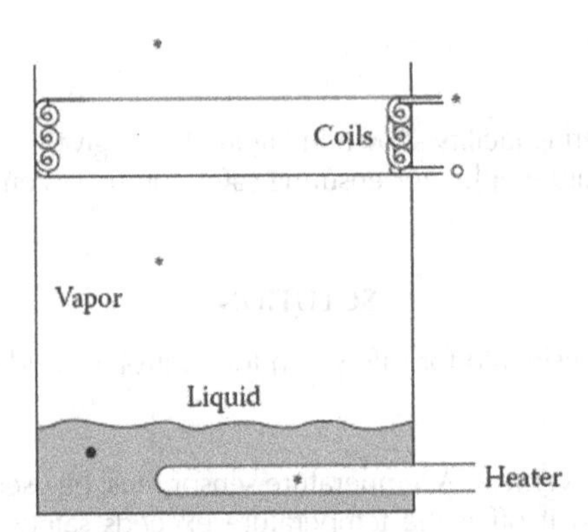

FIGURE 2.23 Locations of different sensors for safety and control in Example 2.4.

including information on modeling and simulation, must also be presented in order to stress the accuracy and validity of the results obtained. The impact and significance of the design with respect to the need or opportunity that initiated the effort (e.g., through a request for proposals [RFP] or other communication from a client) as well as to the relevant industry must be communicated.

Though it has always been important, communication has become particularly crucial these days, because team effort is quite common. Different individuals, often with different backgrounds and expertise, study different aspects of the overall problem. Different parts of the system may be designed separately and brought together during the course of the design process. An example of this approach is the design and assembly of a rocket launching system. Different groups are tasked with the design of different subsystems, such as the shields, solid propellant combustors, thrusters, landing modules, and electronic systems. All these designs and parts are brought together for the final design and assembly. The success of the entire project is strongly dependent on the interaction and communication among different groups. Therefore, good communication between the various people working on the project is very important. A project leader or head of the design group may be responsible for bringing everything together and for presenting the results to the management. Because the final decision regarding the undertaking usually rests with managers, who must take financial, personnel, and other company-related aspects into consideration, it is crucial that the design be presented in proper terms and at the appropriate level.

There are several ways to communicate the details of the final design, the chosen approach being dependent on the target audience. Detailed engineering drawings, along with a list of parts and materials selected, are needed for fabrication of the designed thermal system. Computer programs and numerical simulation results may be more appropriate for prototype testing. Working models and results on important outputs from the system under different operating conditions, often shown as charts and graphs, are useful for presentation to the customer. An outline of the final report may be sent to different levels of management to make them aware of what has been achieved. The communication between the design group and other units generally continues throughout the duration of the effort to ensure that all the important elements in the design are considered. Also, at various stages, changes in direction or inclusion of additional aspects may involve different groups. The communication at the end of the design process is simply the culmination of all these efforts and interactions.

Some of the important modes of communication are as follows:

1. *Technical reports.* These may be short memoranda or reports to communicate status or new findings to specific, interested groups. Formal detailed reports are written at the end of the project to communicate the methodology and results to diverse groups.

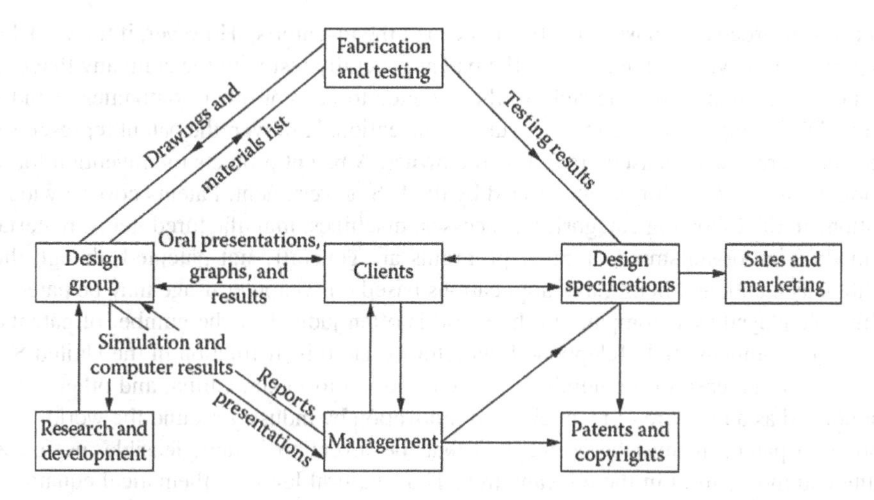

FIGURE 2.24 A possible scheme for different levels of communication at the conclusion of the design process.

2. *Oral presentations.* These may be held to give general or technical details on the design. Again, the presentation depends on the audience.
3. *Graphics and visual aids.* These would help in explaining the main ideas and the results in a presentation.
4. *Engineering drawings.* These give the detailed information on the components and subsystems of the thermal system. Materials used and important fabrication details are also given.
5. *Design specifications.* The specifications of the thermal system are given in order to indicate what might be expected of the system in terms of performance and characteristics.
6. *Computer programs and simulation results.* These are for interested technical personnel who may want to evaluate the effort, as well as for future design and optimization efforts.
7. *Working models.* Physical working models and results obtained from these are of interest to the customer as well as to technical personnel involved with the development of the system. Results from a prototype may also be included, if available.

In conclusion, a few methods that are employed to keep the appropriate people and groups informed of the progress are outlined here. Figure 2.24 shows a scheme for communication at various levels and between groups. Additional methods are available for other purposes (Dieter, 2000). Patents and copyrights communicate the important ideas to the world at large, as well as protect the invention, discovery, or creation, as discussed in the next section. When the design is finalized, the results are communicated to various groups in order to implement the design. The design then proceeds to the fabrication stage. A prototype may be developed and tested before going into production. Finally, sales and marketing personnel take over and the system goes into its intended application in the outside world.

2.3.8 Patents and Copyrights

As has been mentioned earlier, the design process involves creativity, leading to new concepts, methods, and devices. Because considerable effort is generally used in coming up with new ideas, products, and techniques, it is important to protect the investment made in such efforts and the resulting inventions. The ideas and intellectual work done in developing the relevant technology are collectively known as *intellectual property.* Patents and copyrights are the means used for the protection of intellectual property of a company or an individual.

Patents are awarded to and issued in the name(s) of the inventor(s). However, if the work has been done as part of employment, the rights of the patent generally rest with the company that employed the inventor(s). A patent gives the right to the patentee to prevent other companies or individuals from using, fabricating, or marketing the patented invention. However, the patent represents a property and can, therefore, be leased or sold by the owner. A patent protects the invention for a specified period, being 17 years for patents issued by the U.S. government. Patents cover a wide variety of inventions in the following categories: processes, machines, manufactured items, materials, and human-made microorganisms. Computer programs are generally not patented, though these are copyrighted, as mentioned below, and applications based on computer usage may be patented. The leadership role played by a company in the world is often judged by the number of patents issued annually to the company. Bell Telephone Laboratories had this distinction in the United States for many years. The increase in the number of patents issued to Japan, China, and other countries is often mentioned as a measure of technological innovation by industry around the world.

To obtain a patent, it must be established that the invention is new, feasible, useful, and not something commonly used in the relevant area. Thus, natural laws, mathematical equations, commonly used procedures, and fundamental concepts cannot be patented. A thorough search is first carried out to determine if the idea is new. If it has been published in the literature more than a year before applying for a patent, it is not treated as new. Judgment has to be made by the Patent Office whether sufficient details have been provided in an earlier public disclosure to merit rejection of the application. An invention made abroad may be patented in the United States for use and development, if it is not used or known here. To prove the authenticity of an invention, good records are essential because the patent is given to the person who can prove that he or she was the first to conceive or develop it. A bound laboratory notebook is satisfactory proof for the date of the invention if a person capable of understanding the concept witnesses the entry. Similarly, reports and other documents may be used to establish the date of the invention. The feasibility and usefulness of the invention must be demonstrated, preferably through a working model. Many inventions that violate basic laws, such as the first and second laws of thermodynamics, have been proposed in the past and turned down.

Patents form a very useful source of information. An annual index of patents is published each year and details on the invention can be obtained from the description given for the patent. A patent is a legal document and contains enough information to allow one to use the invention if the patent is licensed and after it expires. Each patent is assigned a number and the inventors are listed at the very top of the patent. The title, particulars about filing, search for patentability, and relevant references are then listed. An abstract communicates the main idea of the invention, followed by a sketch in many cases and details on the invention. Figure 2.25 shows these features on the first page of a typical patent concerning a manufacturing process. The objectives of the invention, the field in which the invention lies, and the background material are also included in the patent. The claims of the invention are the description of its legal rights, ranging from very broad and general claims, which may often be disallowed, to very specific claims. By broadening the claims, the patent seeks to cover a wide range of applications that may not even have occurred to the inventor(s). For instance, if a process is developed for heating materials for bonding purposes, the patent may take the broad claim of heating articles rather than for bonding purposes alone. Detailed discussion of the theoretical basis of the invention is generally avoided, again to avoid restrictions on the patent. Infringement of a patent usually leads to extensive litigation, depending on the financial effects of the infringement.

Copyrights are used for a variety of items that represent creative expressions in the arts and sciences. These include books, computer software, music, audio and video recordings, drawings, paintings, and so on. The term of the copyright is 50 years beyond the life of the writer or composer. For a company, it is 75 years from the publication of the material. However, copyrights do not cover ideas, only the expression of the idea. In recent years, copyrights have become important because of the substantial investments made in computer software development, in books,

United States Patent [19]
Chu et al.

[11] **4,032,033**
[45] **June 28, 1977**

[54] **METHODS AND APPARATUS FOR HEATING ARTICLES**

[75] Inventors: **Tze Yao Chu; Yogesh Jaluria,** both of Lawrence Township, Mercer County; **Peter Frederick Lilienthal, II,** Princeton; **George Michael Wenger,** Franklin Township, Somerset County, all of N.J.

[73] Assignee: **Western Electric Company, Inc.,** New York, N.Y.

[22] Filed: **Mar. 18, 1976**

[21] Appl. No.: **668,012**

[52] **U.S. Cl.** **228/200;** 228/201; 228/242; 134/11; 134/31; 134/75; 134/108
[51] **Int. Cl.**[2] **H05K 3/34; B23K 3/04**
[58] **Field of Search** 228/200, 239, 242, 201; 134/11, 26, 30, 31, 105, 107, 108, 109, 75

[56] **References Cited**

UNITED STATES PATENTS

3,001,532 9/1961 Plassmeyer 134/108
3,028,267 4/1962 Edhofer et al. .... 134/105 X
3,106,928 10/1963 Rand .... 134/30 X
3,229,702 1/1966 Murdoch, Jr. .... 134/109 X
3,375,177 3/1968 Rand .... 134/12 X
3,881,949 5/1975 Brock 134/31
3,904,102 9/1975 Chu et al. .... 228/242 X

FOREIGN PATENTS OR APPLICATIONS

2,014,177 10/1971 Germany 228/200

Primary Examiner—James L. Jones, Jr.
Assistant Examiner—K. J. Ramsey
Attorney, Agent, or Firm—D. J. Kirk

[57] **ABSTRACT**

Articles are heated in a condensation heat transfer facility to effect soldering thereof by transporting said articles into a body of hot saturated vapor within the facility. The hot saturated vapor condenses on, and gives up latent heat of vaporization to, the articles to effect solder reflow. The articles are subsequently quenched in a liquid prior to the withdrawal of the articles from the facility.

7 Claims, 5 Drawing Figures

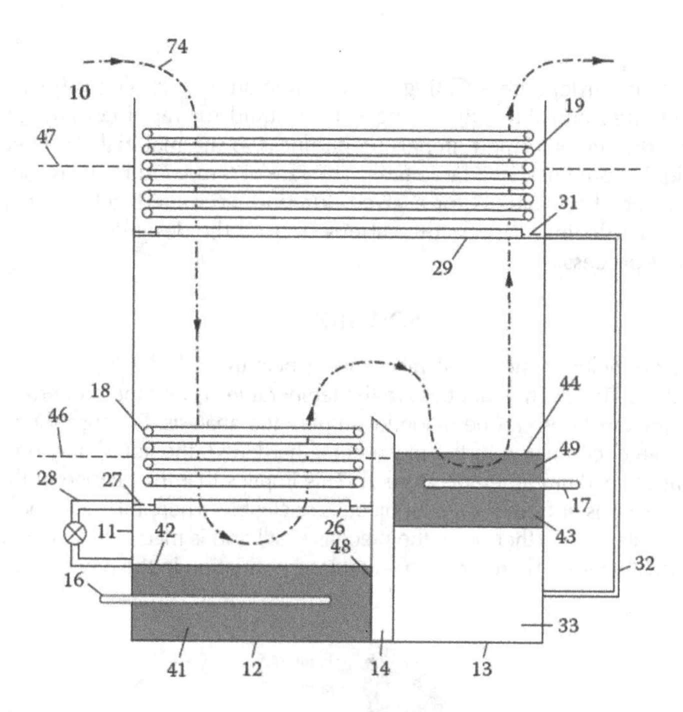

FIGURE 2.25 The first page of a typical U.S. patent showing the main features and items included in the description of the invention.

and in other creative activities in different fields. Strong legal action is often taken if infringement of copyright occurs.

Trademarks are symbols, names, words, patterns, etc., used by a company to indicate its products, and they may be used and protected indefinitely. Trademarks tend to be simple and easy to remember so that their appearance in a magazine, newspaper, or television will immediately reveal the association with a given company or product. A trademark is a company property and the symbol ® is used to indicate that it has been registered. Apple, Nike, Ford, GM, and Microsoft have well-known trademarks, as do most other prominent companies. Formulas, procedures, and information that a company wants to maintain as secret are not patented, but kept as trade secrets. The formula for Coca-Cola is a well-known example of a trade secret that is kept locked in a safe in Atlanta. There is no legal protection and the company is responsible for keeping it a secret.

Licensing of a patented invention may be undertaken by a company or an individual by giving exclusive rights to another company to manufacture, use, and sell the item over a specified region. Several companies may be licensed or a single company may be chosen. Royalties are paid, usually as a percentage of the profit, to the holders of the patent. A fixed sum of money may also be paid. Thus, patents can become a source of revenue, being quite substantial in many cases. Computer software has become very important in the last two decades, with many companies such as Microsoft Corporation engaged in developing, selling, and leasing software. Appropriate pricing and sale of the software recover the expenses borne by the company in the development of the software. General-purpose, as well as application-specific, computer programs for simulating engineering systems are extensively available and are in common use, despite large leasing and purchasing costs. For further details on patents and copyrights, books such as those by Pressman (1979) and Burge (1999) may be consulted.

The following simple example illustrates the main steps in the design of a thermal system.

Example 2.5

A small steel piece is hardened by heating it to a temperature T_o, which is beyond its recrystallization temperature, and then quenching it in a liquid for rapid cooling. A microstructure known as martensite is formed, imparting hardness to the material. The heated piece is immersed in a liquid contained in a large tank. The rate of temperature decrease $|\partial T/\partial \tau|$ must be greater than a specified value B for a given duration $\Delta\tau$ immediately after quenching to obtain satisfactory hardening. Discuss the various steps in the design of the appropriate system to achieve this process.

SOLUTION

The physical system includes the metal piece being heat-treated, the liquid, and the tank, as shown in Figure 2.26. The given quantities are the temperature T_o and the properties of the metal piece. Several approximations can be made to simplify the analysis. Because the volume of the liquid is given as large, compared to the piece being hardened, the liquid may be treated as an isothermal, extensive medium at temperature T_a. This implies that the change in the liquid temperature due to heat transfer from the metal piece is negligible. Then, the tank does not play any part in the energy balance. Furthermore, the piece is small and is made of steel, which is a good conductor of thermal energy. Therefore, the temperature variation in the piece may be neglected

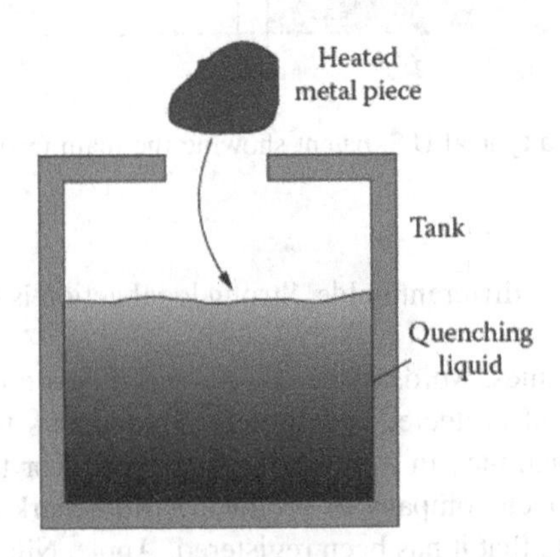

FIGURE 2.26 Thermal system for quenching of a heated steel piece for hardening.

and its temperature T is taken as uniform and as a function of time τ only, i.e., $T(\tau)$. This is the lumped mass approximation, which is discussed in greater detail in Chapter 3. Because T depends only on τ, full differentials $dT/d\tau$ may be used, instead of partial ones.

An energy balance for the steel piece gives the equation for the temperature $T(\tau)$, which decreases with time due to convective cooling, as

$$\rho CV \frac{dT}{d\tau} = -hA(T - T_a)$$

where ρ is the density of the material of the steel piece, C is its specific heat, V is the volume of the steel piece, A is its surface area, and h is the convective heat transfer coefficient at the surface of the piece. The natural convection flow generated in the fluid by the heat transfer process has to be determined to obtain the heat transfer coefficient h. However, this is an involved process and an average value of the heat transfer coefficient h may be obtained from heat transfer correlations available in the literature. Even though h depends on the temperature difference $T - T_a$ and thus on time, it is assumed that it is a constant in order to simplify the problem. The initial condition for the above equation is $T = T_o$ at $\tau = 0$. Thus, the mathematical model for the system is obtained.

Let us now consider the simulation of the system. The solution to the preceding equation is

$$T - T_a = (T_o - T_a)\exp\left(-\frac{hA}{\rho CV}\tau\right)$$

Therefore, at $\tau = 0$, $T = T_o$ and as $\tau \to \infty$, $T \to T_a$, i.e., the metal piece finally cools down to the fluid temperature. The rate of temperature decrease $dT/d\tau$ is given by

$$\frac{dT}{d\tau} = -(T_o - T_a)\left(\frac{hA}{\rho CV}\right)\exp\left(-\frac{hA}{\rho CV}\tau\right)$$

From the given requirement for hardening

$$(T_o - T_a)\left(\frac{hA}{\rho CV}\right)\exp\left(-\frac{hA}{\rho CV}\Delta\tau\right) > B$$

because the magnitude of the temperature decrease rate must exceed B for a duration of $\Delta\tau$. Obviously, this rate is greater than B for time less than $\Delta\tau$, being highest at $\tau = 0$. Therefore, the design variables may be chosen to satisfy this requirement. Because the metal piece is given, along with temperature T_o, all the quantities in the above solution are fixed except T_a and h. Therefore, these may be varied to obtain the desired rate of cooling over the given duration Δt.

The average heat transfer coefficient h is a function of the fluid, the geometry and dimensions of the piece, and the temperature difference $T_o - T_a$. Correlations for the heat transfer coefficient for natural and forced convection under various conditions and for different geometries are available in the literature (Incropera and Dewitt, 2001). The use of these correlations brings in the dependence of the cooling rate on the physical variables in the problem. The fluid is the most important parameter and may be chosen for high thermal conductivity, which yields a high heat transfer coefficient, low cost, easy availability, nontoxic behavior, and high boiling point, if boiling is to be avoided in the liquid. If boiling is allowed, the latent heat of vaporization becomes an important variable to obtain a high heat transfer coefficient. Oils with high boiling points are generally used for quenching. The temperature T_a is another variable that can be effectively used to control the cooling rate. A combination of a chiller and a hot fluid bath may be used to vary T_a over a wide range. Clearly, many solutions are possible and a unique design is not obtained. Different fluids that are easily available may be tried first to see if the requirement on the cooling rate is satisfied. If not, a variation in T_a may be considered. Optimization of the system may then be based on cost.

2.4 COMPUTER-AIDED DESIGN

An area that has generated a considerable amount of interest over the last three decades as a solution to many problems being faced by industry and as a precursor to the future trends in engineering design is that of *computer-aided design* (CAD). With the tremendous growth in the use and availability of digital computers, resulting from advancements in both the hardware and the software, the computer has become an important part of design practice. Much of engineering design today involves the use of computers, as discussed in the preceding sections and as presented in detail in later chapters. However, the term *computer-aided design*, as used in common practice, largely refers to an independent or stand-alone system, such as a computer workstation, and interactive usage of the computer to consider various design options and obtain an acceptable or optimal design, employing the software for modeling and analysis available on the system. Still, the basic ideas involved in a CAD system are general and may be extended to more involved design processes and to larger computer systems.

2.4.1 Main Features

As mentioned above, a CAD system involves several items that facilitate the iterative design process. Some of the important ones are:

1. Interactive application of the computer
2. Graphical display of results
3. Graphic input of geometry and variables
4. Available software for analysis and simulation
5. Available database for considering different options
6. Knowledge base from current engineering practice
7. Storage of information from earlier designs
8. Help in decision making

Thus, the system hardware consists of a central processing unit (CPU) for numerical analysis, hard disk for storage of data and design information, an interactive graphics terminal, and a plotter/printer for hard copy of the numerical results.

The computer software codes for analysis are often based on finite-element methods (FEMs) for differential equations because this provides the flexibility and versatility needed for design (Zienkiewicz, 1977; Reddy, 2009). Different configurations and boundary conditions can be easily considered by FEM codes without much change in the numerical procedure. Other methods, particularly the finite-volume and the finite-difference method (FDM), are also used extensively for thermal systems (Patankar, 1980). The software may also contain additional codes on curve fitting, interpolation, optimization, and solution of algebraic systems. Some of the important numerical schemes are discussed in Chapter 4. Analytical approaches may also be included. Commercially available computer software, such as Maple, Mathematica, Mathcad, COMSOL Multiphysics, and Matlab, may be used to obtain analytical as well as numerical solutions to various problems such as integration, differentiation, matrix inversion, root solving, curve fitting, and solving systems of algebraic and differential equations. The use of Matlab for these problems is discussed in detail in Appendix A.

The interactive use of the computer is extremely important for design because it allows the user or designer to try many different design possibilities by entering the inputs numerically or graphically, and to obtain the simulation results in graphical form that can be easily interpreted. Iterative procedures for design and optimization can also be employed effectively with the interactive mode. A graphics terminal is usually employed to obtain three-dimensional, oblique, cross-sectional, or other convenient views of the components.

The storage of data needed for design, such as material properties, heat transfer correlations, characteristics of devices, design problem statement, previous design information, accepted engineering practice, regulations, and safety features can also substantially help in the design process. In this connection, *knowledge-based* design procedures, which employ the available experience on the system or process, may also be incorporated in the design scheme. Besides providing important relevant information for design, the rules of thumb and heuristic arguments used for design can be built into the system. Therefore, such systems are also often known as *expert systems* because expert knowledge from earlier design experience is part of the software, providing help in the decision-making process as well. Because knowledge acquired through engineering design practice is usually an important component in the development of a successful design, knowledge-based systems have been found to be useful additions to the CAD process. Chapter 11 presents details on knowledge-based systems for design, along with several examples demonstrating concepts that can substantially aid the design process.

2.4.2 Computer-Aided Design of Thermal Systems

The main elements of a CAD system for the design of thermal processes and equipment are shown in Figure 2.27. The various features that are usually included in such CAD systems are indicated. The modeling aspect is often the most involved one when dealing with thermal systems. The remaining aspects are common to CAD systems for other engineering fields. Much of the effort in CAD has, over recent years, been largely devoted to the design of mechanical systems and components such as gears, springs, beams, vibrating devices, and structural parts, employing stress analysis, static and dynamic loading, deformation, and solid body modeling. Many CAD systems, such as AutoCAD and ProE, have been developed and are in extensive use for design and instruction.

Because of the complexity of thermal systems, it is not easy to develop similar CAD systems for thermal processes. However, the availability of numerical codes for many typical thermal components and types of equipment has made it possible to develop CAD systems for relatively simple applications such as heat exchangers, air conditioners, heating systems, and refrigerators. Even for these systems, inputs from other sources, particularly on heat transfer coefficients, are often employed to simplify the simulation. For more elaborate thermal systems, interactive design generally is not possible because numerical simulation might involve considerable CPU time and memory requirements. Supercomputers are also needed for accurate simulations of many important thermal systems, such as those in materials processing and aerospace applications. However, *parallel machines* that employ a large number of computational processors to accelerate numerical analysis are being used in powerful workstations that may be used for CAD of practical thermal processes. In addition, detailed simulation results from large machines such as supercomputers may be cast

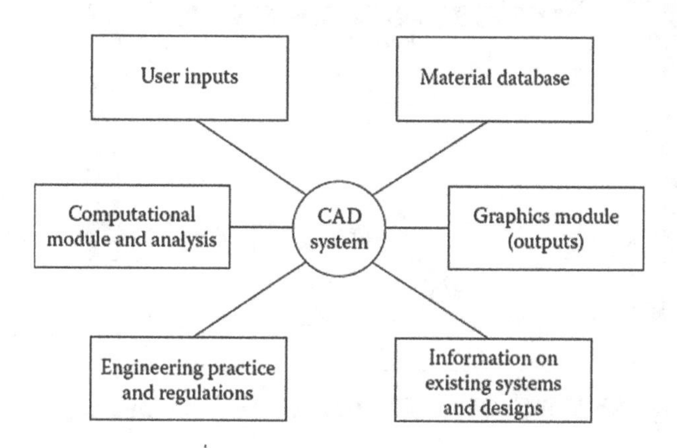

FIGURE 2.27 Various elements or modules that constitute a typical CAD system.

in the form of algebraic equations by the use of curve fitting. If a given thermal system can be represented accurately by such algebraic systems, the design process becomes considerably simplified, making it possible to develop a CAD system for the purpose. *Response surfaces* may also be obtained from the simulation to study the response of important physical quantities to the variables in the problem. This facilitates the search for an optimum, as discussed in later chapters.

Example 2.6

Discuss the development of a CAD system for the forced-air baking oven shown in Figure 2.28. The electric heater is made of 5% carbon steel, the gas inside the oven is air, the wall is brick, the insulation is fiberglass, and the material undergoing heat treatment is aluminum. The geometry and dimensions of the oven are also given, or fixed, and only the heater and the fan are the design variables.

SOLUTION

This problem is taken as an example to illustrate the basic ideas involved in the CAD of thermal systems. The main components of this thermal system are:

1. Heater
2. Fan
3. Wall
4. Insulation
5. Air
6. Material to be baked or heated

The basic thermal cycle that the material must undergo is similar to the one shown in Figure 2.1. Thus, an envelope of acceptable temperature variation, giving the maximum and minimum temperatures within which the material must be held for a specified time, provides the design requirements. The constraints are given by the temperature limitations for the various materials involved and any applicable restrictions on the airflow rate and heater input. The materials, dimensions, and geometry are given and are, thus, fixed for the design problem. Only the fan and the heater may be varied to obtain an acceptable design.

The first step is to develop a mathematical and numerical model for the physical system shown in Figure 2.28. The basic procedures for modeling are discussed in the next chapter and

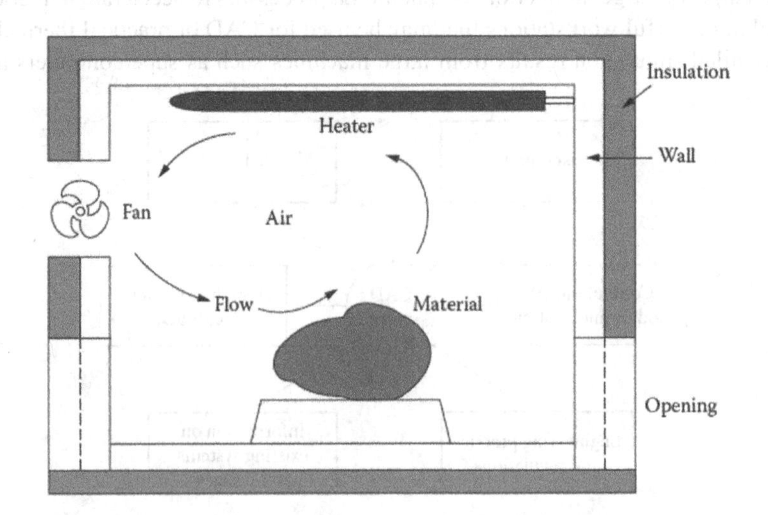

FIGURE 2.28 Forced-air oven for thermal processing of materials.

a relatively simple model to obtain the temperatures in the various parts of the system is outlined here. The simplest model for this dynamic problem is one that assumes all components have uniform temperature within the component at a given time. Thus, the material, air, heater, wall, and insulation are all treated as lumped, with their temperatures as functions of time τ only. The governing equations for these components may then be written as

$$\rho C V \frac{dT}{d\tau} = A\left(q_{in} - q_{out}\right)$$

where ρ is the density, C is the specific heat at constant pressure, V is the volume, A is the surface area, q_{in} is the input heat flux, and q_{out} is the heat flux lost at the surface. All the properties are taken as constant to simplify the analysis. Thus, a system of ordinary differential equations is obtained.

For the boundary conditions that link the energy equations for the various system parts, both convection and radiation are considered, assuming gray-diffuse transport with known surface properties. The properties for different materials are used when considering each component of the system. The conditions under which such a model is valid are discussed in detail in Chapter 3. Even though analytical solutions may be possible in a few special cases, all of these equations are coupled to each other through the boundary conditions and are best solved numerically to provide the desired flexibility and versatility in the solution procedure.

With the mathematical and numerical model defined, the fixed quantities in the problem may be entered. These include the geometry and the dimensions of the system. The size and weight of the item undergoing thermal processing are given. The relevant material properties must also be specified. Frequently a material database is built into the system for common materials, such as metals, ceramics, composite materials, and air, and may be used to obtain these properties. The requirements for the design, as well as the constraints (particularly the temperature limitations on the various materials), are also entered. All of these inputs are given interactively, so that the design variables and operating conditions can be varied and the resulting effects obtained from the CAD system. This allows the user to select the input parameters based on the computed outputs.

We are now ready for simulation and design of the given thermal system. The heater design involves its location, dimensions, and heat input. If the location is fixed at the top surface, as shown in Figure 2.28, and if the effect of dimensions is assumed to be small, which is reasonable, the heat input Q is the design variable that represents the heater. Similarly, the fan affects the flow rate $\dot{m}$ and, thus, the heat transfer coefficients at the material surface, h_m, at the heater h_h, and at the oven walls, h_w. We could solve for the flow and thermal field in the air and obtain these heat transfer coefficients from the numerical results. However, this is a more complicated problem than the one outlined here. Thus, the heat transfer coefficients may be taken from correlations available in the literature.

Simulation results are obtained by varying the heat input Q and the convective heat transfer coefficients, h_m, h_h, and h_w, all these being dependent on the flow rate, geometry, and dimensions. Figure 2.29 and Figure 2.30 show typical numerical results obtained during the heating phase, indicating the temperatures in the heater, material, gas, and wall for different parametric values. The validity of the numerical model is confirmed by ensuring that the results are independent of numerical parameters such as the time step used, studying the physical behavior of the results obtained, and comparisons with analytical and experimental results for individual parts of the system and for the entire system, if available. In most cases, results for the system are not available until a prototype is developed and tested before going into production. However, a higher Q results in higher temperatures, with the heater responding the fastest and the walls the slowest. An increase in h increases the energy removed by air and lowers the temperature levels. This is the expected physical behavior.

The next step is to consider various combinations of Q and the flow rate $\dot{m}$, which yields the convection coefficients, and to determine if the desired requirements are satisfied without violating the given constraints. The duration during which the heater or the fan is kept on can be varied. In addition, different variations of these with time can be considered to obtain the desired variation in the material temperature. Obviously, many different designs and operating conditions are possible. Again, interactive usage of the CAD system is extremely valuable in this search for an acceptable

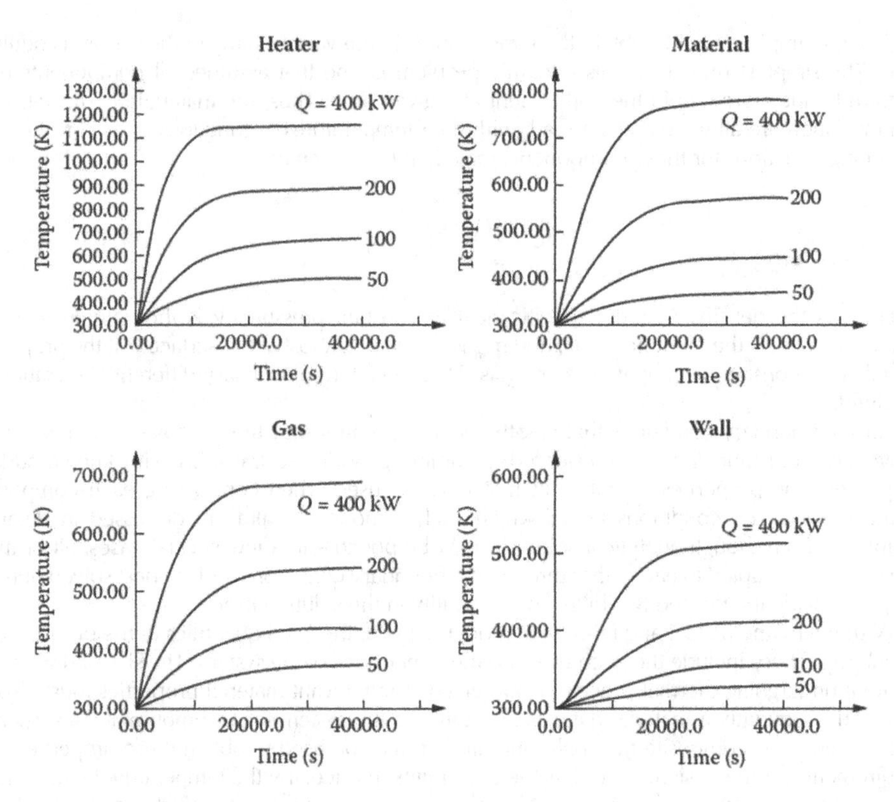

FIGURE 2.29 Variation of the heater, material, gas, and inner wall temperatures with time for different values of the energy input Q to the heater at a fixed air flow rate $\dot{m}$.

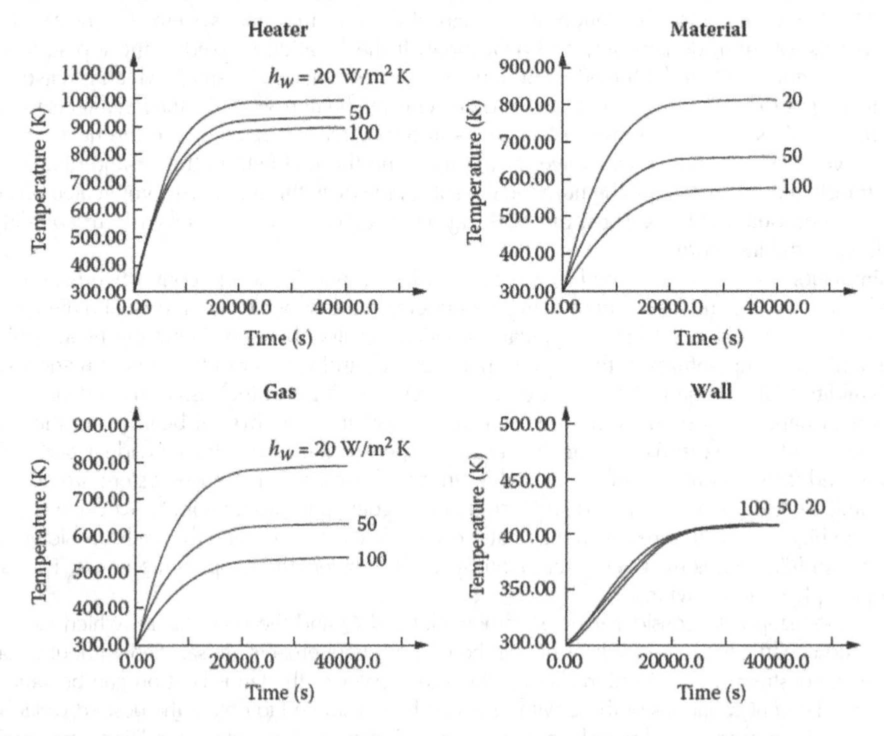

FIGURE 2.30 Results for different values of the convective heat transfer coefficient h_w, which represents the air flow rate $\dot{m}$, at a fixed Q.

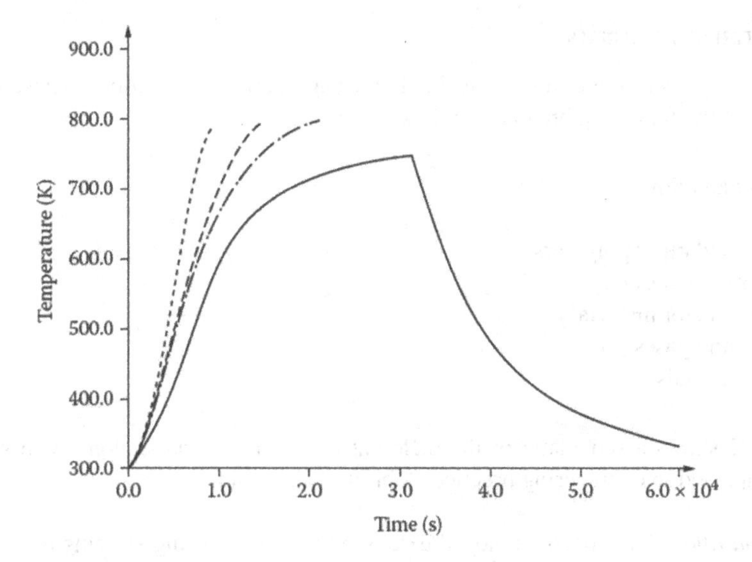

FIGURE 2.31 Results from iterative redesign to obtain an acceptable design, indicated by the solid line, that satisfies the given requirements and does not violate any constraints.

design. An acceptable design is obtained when all of the requirements and constraints are met, such as that indicated by Figure 2.31. A large number of cases are simulated even for a relatively simple problem like this one. The graphical displays help in determining if the design process is converging. The software can be used to monitor the temperatures and indicate if a violation of the constraints has occurred in any system part. In addition, the temperature of the piece being heated is checked against the envelope of acceptable variation to see if an acceptable design is obtained.

This example briefly outlines some of the main considerations in developing a CAD system for thermal processes. The model is at the very heart of a successful design process, and, therefore, it is important to develop a model that is a valid representation of the system, has the needed accuracy, and is appropriate for the given application. A knowledge-based design procedure could also be included during iterative design to accelerate convergence and to ensure that only realistic and practical systems emerge from the design (Jaluria and Lombardi, 1991). As mentioned previously, the fluid flow problem needs to be solved for a more accurate modeling of the convective heat transfer and for a proper representation of the fan as a design variable. However, the problem may then become much too complicated for a simple interactive CAD system and would probably involve detailed simulation on larger machines to obtain the inputs needed for design.

2.5 MATERIAL SELECTION

The choice of materials for the various parts of the system has become an important consideration in recent years because of the availability of a wide range of materials, because material cost is a substantial portion of the overall cost, and because the performance of the system can often be substantially improved by material substitution. Recent advancements in material science and engineering have made it possible to produce essentially custom-made, engineered materials to satisfy specific needs and requirements. In the past, the choice of material was frequently restricted to available metals, alloys, and common nonmetals. Thus, it used to be a fairly routine procedure to select a material that would satisfy the requirements of a given application. However, material selection today is a fairly sophisticated and involved process. The properties of the material, as well as its processing into a finished component, must be considered in the selection. The substitution of the material currently being used by a new or different material is also commonly employed to reduce costs and improve performance. However, material substitution should be carried out in conjunction with design in order to derive the full benefits of the new material (Budinski and Budinski, 2009).

2.5.1 Different Materials

Many different types of materials are available for engineering applications. These may be classified in terms of the following broad categories:

1. Metals and alloys
2. Ceramics
3. Plastics and other polymers
4. Composite materials
5. Semiconductor materials
6. Liquids and gases
7. Other materials

Figure 2.32 shows a schematic of the different types of materials, along with some common materials employed in engineering practice. A brief discussion follows:

Metals and alloys have been employed extensively in engineering systems because of their strength, toughness, and high electrical and thermal conductivity. Availability, cost, and ease in processing to obtain a desired finished product, through processes such as forming, casting, heat treatment, welding, and machining, have contributed to the traditional popularity of metals. A variety of metals have been employed in different applications to satisfy their special requirements. Thus, copper has been used for tubes and pipes because of its malleability, which allows easy bending, and for electrical connections because of its high electrical conductivity. Similarly, aluminum has been used for its low weight in airplanes and in other transportation systems. High thermal conductivity of both aluminum and copper make them good materials for heating and cooling systems. Gold has been used in electronic circuitry because of its resistance to corrosion. Alloys substantially expand the range of applicability of metals due to significant changes achieved in the properties. Steel, in its different compositions, is probably the most versatile and widely used material in practical systems, from automobiles and trains to turbines and furnaces. Solder, which is an alloy of tin and lead, is widely employed in electronic circuitry to make electrical connections. Changes in its composition can be used to obtain different strengths and melting points. For instance, a eutectic mixture of 63% tin and 37% lead has a melting point of 183°C and a mixture of 10% tin and 90% lead melts over the temperature range of 275°C to 302°C. Additions of silver also affect the melting point and other properties, as discussed by Dally (1990). Similarly, other alloys such as brass, Inconel, nichrome, and titanium alloys are used in different applications.

Ceramics, which are generally formed by fusing powders, such as those of aluminum oxide (Al_2O_3), beryllium oxide (BeO), and silicon carbide (SiC), under high pressure and temperature, have many characteristics that have led to their increased usage in recent years. These include high temperature resistance, low electrical conductivity, low weight, hardness, corrosion resistance, and strength, though they are generally brittle. They have a relatively low thermal resistance, as compared to other electrical insulators. Consequently, ceramics are extensively employed in electronic circuitry, particularly in circuit boards. They are also used in high temperature and corrosive environments, as tool and die materials, and in engine components. Ceramics also include glasses as a subdivision and these have their own range of applications due to transparency. The optical fiber is a recent addition to this group of materials, with applications in telecommunications, sensors, measurements, and controls. Various other optical materials used in television screens, optical networks, lasers, and biosensors are also of considerable interest to industry.

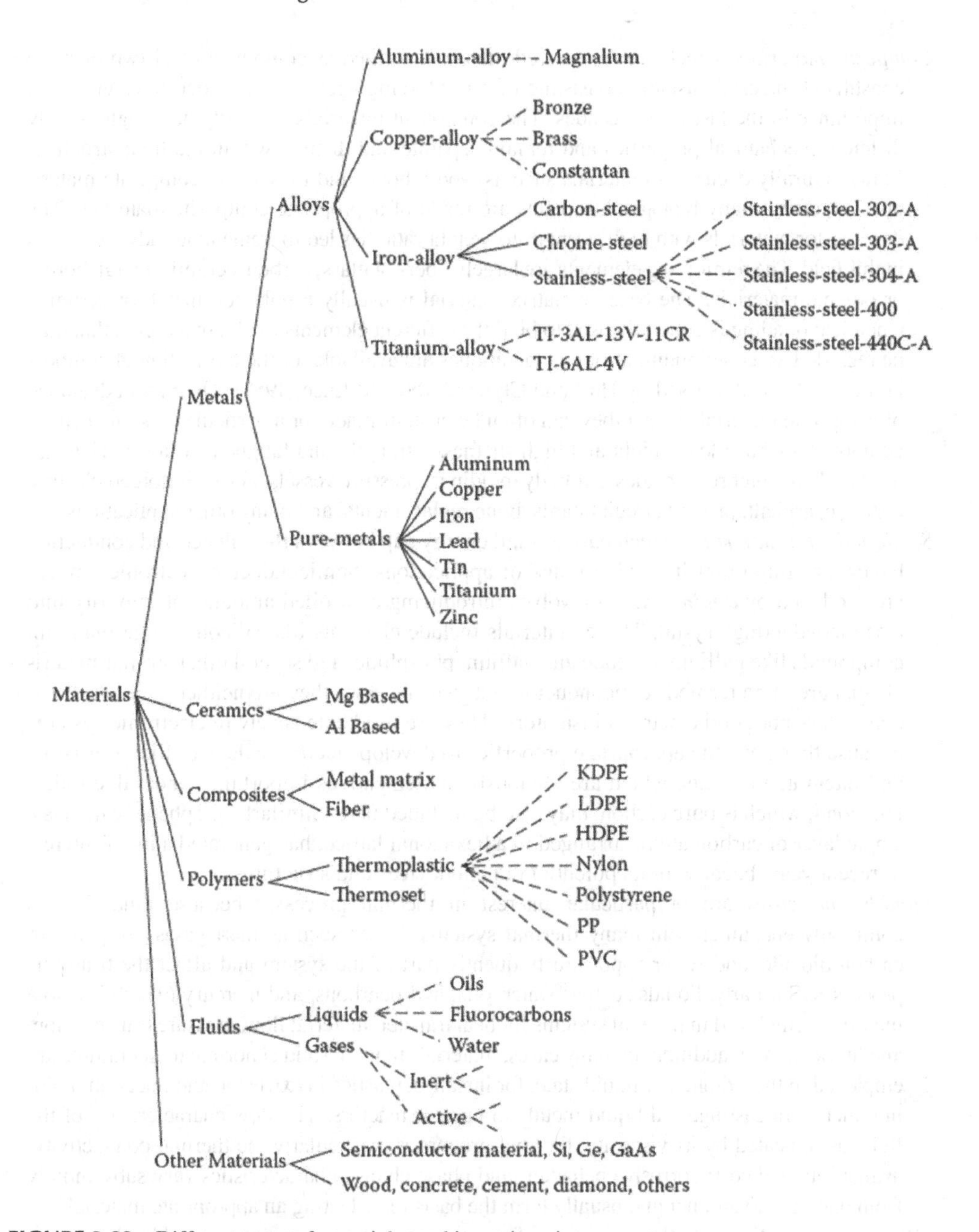

FIGURE 2.32 Different types of materials used in engineering systems.

Polymers, which include plastics, rubbers or elastomers, fibers, and coatings, have the advantages of easy fabrication, low weight, electrical insulation, resistance to corrosion, durability, low cost, and a wide range of properties with different polymers. Consequently, plastics have replaced metals and alloys in a wide range of applications. Because these materials are electrically insulating, they find use in plastic-coated cables, plastic casings for electronic equipment, and electrical components and circuitry. Similarly, the ease of forming or molding polymeric materials has led to their use in many diverse areas ranging from containers, trays, and bottles to panels, calculators, and insulation. Clearly, polymers are among the most versatile materials today, despite the temperatures that can be withstood by them without damage being limited to 200°C to 300°C in most cases.

Composite materials, which are engineered materials formed as combinations of two or more constituent materials usually consisting of a reinforcing agent and a binder, have grown in importance in the last three decades. The component materials generally have significantly different mechanical properties and remain separate and distinct within the final structure. Many naturally occurring materials such as wood, bone, and muscle are composite materials. Therefore, many biological implants are made of appropriate composite materials. The demand for materials with high strength-to-weight ratio has led to tremendous advancements in this field. The reinforcing elements are largely fibers of glass, carbon, ceramic, metal, boron, or organic materials. The base or matrix material is usually a polymer, metal, or ceramic. Chemical bonding is generally used to bind the different elements to obtain a region that may be regarded as a continuum. Different techniques are available for the fabrication of composite materials, as discussed by Hull and Clyne (2008) and Luce (1988). The main advantage of composite materials is that they can often be custom-made for a particular design need. In addition, they have low weight and high stiffness, strength, and fatigue resistance. They are used for helicopter rotor blades, car body moldings, pressure vessels, glass-reinforced plastics, concrete, asphalt, printed circuit boards, bone replacements, and many other applications.

Semiconductor materials, which have a small energy gap between the valence and conduction bands, are important in a wide range of applications. Semiconductor electronic devices are produced by *doping*, which involves introducing controlled amounts of impurity into a semiconducting crystal. These materials include elements like silicon and germanium, compounds like gallium arsenide and gallium phosphide, and several other similar materials that are often termed semiconductor materials because they are neither good electrical conductors nor good electrical insulators. They are used extensively in electronic systems because they have the appropriate properties to develop electronic devices like transistors and integrated circuits, which are obviously of tremendous importance and value today. Diamond, which is pure carbon, may also be included here. Similarly, graphene, which is a single layer of carbon atoms arranged in a hexagonal lattice, has generated a lot of interest in recent years because of its potential in nanodevices and structures.

Liquids and gases are of particular interest in thermal processes because fluid flow is commonly encountered in many thermal systems. Gases such as inert gases, oxygen, air, carbon dioxide, and water vapor are frequently part of the system and affect the transport processes. Similarly, liquids such as water, oils, hydrocarbons, and mercury (which is also a metal) are employed in thermal systems for heat transfer, material flow, pressure transmission, and lubrication. In addition, in many cases, materials that are solid at normal temperatures are employed in their molten or liquid state, for instance, plastics in extrusion and injection molding, metals in casting, and liquid metals in nuclear reactors. The flow characteristics of the fluid, as indicated by its viscosity; thermal properties, particularly the thermal conductivity; availability and cost; corrosive behavior; and phase change characteristics vary substantially from one fluid to another and usually form the basis for selecting an appropriate material.

Other materials. Several other materials of engineering interest are not covered by the groups given earlier. These include materials such as different types of wood, stone, rock, biological materials, and other naturally occurring materials that are of interest in various applications. New materials in each category are continually being developed to meet the demand for specific properties and characteristics and to improve existing materials in a variety of applications. Substantial research and development effort is directed at obtaining new and improved materials for enhancing the performance of present systems, reducing costs, and helping future technological advancements. Coming years are expected to see a variety of new and emerging materials, as well as new fabrication techniques (Jaluria, 2018).

Therefore, the main categories of materials are metals and alloys, ceramics, polymers, composite materials, fluids, and semiconductor materials. Each group has its own characteristics. Some

TABLE 2.1
Typical Characteristics of Common Materials

Metals and Alloys	Ceramics	Polymers
Strong	Strong	Weak
Tough	Brittle	Durable
Stiff	Stiff	Compliant
High electrical conductivity	Electrically insulating	Electrically insulating
High thermal conductivity	Low thermal conductivity	Low thermal conductivity
Easy processing	Difficult processing	Easy fabrication
Susceptible to corrosion	Corrosion resistance	Corrosion resistance
Easily available	Light weight	Low cost
	Temperature resistance	Temperature sensitive
Composites	**Liquids and Gases**	**Semiconductor Materials**
Strong	Material flows	Specialized characteristics
Fatigue resistant	Inert or corrosive	Not good electrical conductor
Stiff	Wide range of properties	Not good electrical insulator
Range of electrical conductivity	Low electrical conductivity	Electrical insulator at low temperatures
Range of thermal conductivity	Low thermal conductivity	Electronic properties altered by doping
Versatile	Versatile	Wide range of other properties
Low weight	Generally low weight	
Low cost	Generally low cost	

were just mentioned; see also Table 2.1. The range of application of each type of material is determined by the physical characteristics and the cost. Materials may also be categorized in terms of their applications, for instance, electronic, insulation, construction, optical, and magnetic materials. However, it is more common and useful to discuss materials in terms of their basic characteristics and to use the classes of materials outlined above.

2.5.2 Material Properties and Characteristics for Thermal Systems

We have discussed different types of materials, their general properties, and typical areas of application. Though most of the properties mentioned earlier are of interest in engineering systems, let us now focus on thermal processes and systems. Obviously, many material properties are of particular interest in thermal systems; for instance, a low thermal conductivity is desirable for insulation and a high thermal conductivity is desirable for heat removal. A large thermal capacity, which is the product of density and specific heat, is needed if a slow transient response is desired and a small thermal capacity is necessary for a fast response. The material properties that are of particular importance in thermal systems, along with their usual symbolic representation employed in this book, are:

1. Thermal conductivity, k
2. Specific heat, C
3. Density, ρ
4. Viscosity, μ
5. Latent heat during phase change, h_{sl} or h_{fg}
6. Temperature for phase change, T_{mp} or T_{bp}
7. Coefficient of volumetric thermal expansion, β
8. Mass diffusivity, D_{AB}

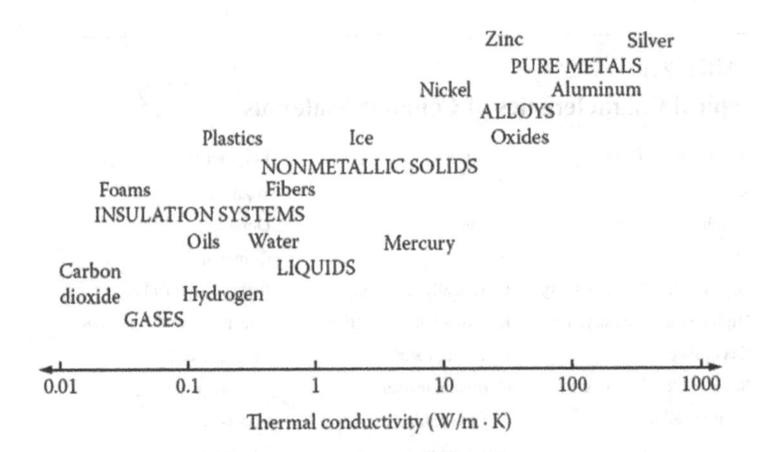

FIGURE 2.33 Range of thermal conductivity k for a variety of materials at normal temperature and pressure.

Here, the subscripts sl, fg, mp, bp, and AB refer to solid-liquid, liquid-vapor, melting point, boiling point, and species A diffusing into species B, respectively. The phase change may also occur over a range of temperatures, which is the case for an alloy or a mixture, rather than at a fixed temperature. The specific heat may be at constant pressure or at constant volume, these being essentially the same for solids and liquids, which may generally be taken as incompressible. Several other thermal properties such as the coefficient of linear thermal expansion, heat of sublimation, and thermal-shock resistance are also of interest in thermal systems.

All these properties vary tremendously among the common materials used in thermal processes. For instance, the thermal conductivity varies from around 0.026 for air to 0.61 for water to 429.0 W/mK for silver. Typical ranges are shown in Figure 2.33. Similarly, other properties are available in the literature (Touloukian and Ho, 1972; American Society of Metals, 1961; ASHRAE, 2017; Eckert and Drake, 1972; Incropera and Dewitt, 2001). In addition, properties such as thermal diffusivity α, where $\alpha = k/\rho C$, and kinematic viscosity ν, where $\nu = \mu/\rho$, are also frequently used to characterize the material. Many common materials and their properties are given in Appendix B.

In addition to the aforementioned thermal properties of the material, several characteristics discussed in the preceding section are important in the design of thermal systems. Certainly, corrosion resistance and range of temperature over which the material can be used are important considerations. Similarly, strength, toughness, stiffness, and others are important in the design because of the need to maintain the structural integrity of the system. Material cost and availability are obviously important in any design process. Manufacturability of the material is also important, as mentioned earlier. Waste disposal and environmental impact of the material are additional considerations in the characterization and evaluation of the material.

2.5.3 Selection and Substitution of Materials

In view of the material properties and characteristics discussed in the preceding section, the factors involved in the selection of a suitable material in the design of a thermal system are:

1. Satisfactory thermal properties
2. Manufacturability
3. Static, fatigue, and fracture characteristics
4. Availability
5. Cost
6. Resistance to temperature and corrosion
7. Environmental effects
8. Electric, magnetic, chemical, and other properties

Material selection is not an easy process because of the many considerations that need to be taken into account. These lead to a variety of constraints, many of which may be conflicting. Though cost is an important parameter in the selection, it is not the only one. We want to choose the best material for a given application while satisfying many constraints. However, information on material properties is often not available to the desired detail or accuracy. The range of available materials has increased tremendously in recent years, making material selection a very involved process. However, the choice of the most appropriate material for a given application is crucial to the success of the design in today's internationally competitive environment. With a proper choice of materials, the system performance can be improved and costs reduced. In several cases, material substitution is necessary because of regulations stemming from environmental or safety considerations. For example, the incentive for improvements in gasoline, including addition of ethanol, arises from pollution, availability, cost, and political considerations. Substitution of asbestos by other insulating materials is due to the health risks of asbestos. Obviously, all such considerations complicate material selection and substantial effort is generally directed at this aspect of design.

The basic procedure for material selection may be described in terms of the following steps.

1. *Determination of material requirements.* The thermal process or system being designed is considered to determine the conditions and environment that the chosen material must withstand. From this consideration, the desired properties and characteristics, along with possible constraints, are obtained. For example, the simulation of a furnace would indicate the temperatures that the materials exposed to this environment must endure. Similarly, the expected pressures in an extruder would provide the corresponding requirements for the selected material.
2. *Consideration of available materials.* Material property databases are available and may be employed to compare the material requirements with the properties of available materials. In such a search, the focus is on the desired properties and characteristics. The requirements in terms of thermal properties will be largely considered at this stage for thermal processes. Cost, environmental effects, and other considerations and constraints are not brought in. Therefore, a large number of material choices may emerge from this step. This is done mainly to avoid eliminating any material that meets the appropriate requirements.
3. *Selecting a group of possible materials.* From the materials that would satisfy the main requirements of the application, a smaller group is chosen for a more detailed consideration. At this stage, other considerations and constraints are brought in. Thus, a material that is very desirable due to its thermal properties may be eliminated because of cost or undesirable environmental impact. Gold, which is a good choice for electronic circuit elements because of its inert nature, is retained only for surface plating due to the cost. Manufacturability of the material to obtain a given part is also an important consideration at this stage. Information on previously used materials for the given problem and for similar systems may also be used to narrow the list of possible materials. Because there may be several requirements for the material properties, a weighted index that takes all of these into account, according to their relative importance, may also be employed. A short list of possible materials is thus obtained.
4. *Study of material performance.* A detailed study of the materials obtained from the preceding step is undertaken to determine their performance under the specific conditions expected to be encountered in the given application. Experimental work may also be carried out to obtain quantitative data and to characterize these materials. Available literature on these materials and information on their earlier use in similar environments are also employed. There are many standard sources for material property data (Dieter, 2000); some of them were mentioned earlier.
5. *Selection of best material.* Based on the information gathered on the short list of possible materials, the most appropriate material for the given application is chosen. The cost and availability of the material are very important considerations in the final selection.

However, in many cases, cost may have to be sacrificed in the interest of superior performance. In a few cases, the material may be developed to meet the specific needs of the problem. This is true in many electronic systems where the materials employed for the circuit board, the circuitry, and the connections are developed as variations from existing composite materials, ceramics, solder, etc. (Dally, 1990).

2.5.3.1 Final Comments

Material selection is an involved process and is somewhat similar to the iterative design process discussed earlier for thermal systems. Several options are considered and the best one is chosen based on available property data and material characteristics. Prior experience may also be used to help in this selection process by bringing in existing expert knowledge on materials and information on current practice. Then the decision-making process may be automated by using a large database on available materials and their characteristics. In many cases, an existing process or system is to be improved by substituting the current material with a different material. In several applications, plastics, ceramics, and composite materials have recently replaced metals and alloys. Plastics are now used for most containers and housings because of lower weight and cost involved. Similarly, composite materials led to improvements over metals in many of their important characteristics, while keeping the cost lower. Thus, substantial improvements in system performance and reduction in costs are obtained by material substitution. However, redesign of the component, subsystem, or system should be undertaken to obtain the maximum benefit from material substitution.

Example 2.7

a. In a food processing system, food materials are placed on flat plates that are attached to and moved continuously by a conveyor belt. The food is subjected to gas heating at the bottom of the plate for a given amount of time. Select a suitable material for the plates.
b. Select suitable materials for an electronic system, considering the board on which electronic components are located and electrical connections between these components by means of exposed circuitry on the board.

SOLUTION

a. In this problem, a high thermal conductivity material is desirable because of heat conduction through it to the food material. In addition, the material must be strong, durable, and corrosion resistant because of the application. Table 2.1 indicates that metals and alloys would satisfy these requirements. Ceramics have lower conductivity and may be too brittle for this application. Though copper and aluminum have high thermal conductivities (401 and 237 W/mK, respectively, at 300 K), alloys such as bronze and brass are easier to form into the desired shape and to bond to the conveyor. But then the conductivities are much smaller (around 50 W/mK). Steel is a better choice because of better corrosion resistance and cost. Stainless steel can be chosen due to its high corrosion resistance, but it is a difficult material to work with for fabrication, it is relatively expensive, and it has a lower thermal conductivity (approximately 15 W/mK). Carbon steels are cheaper, easier to form, and better conductors of heat (thermal conductivity around 60 W/mK).

   In view of the above considerations, carbon steel may be chosen as the appropriate material, with the exact percentage of carbon chosen based on cost and availability. Because food is involved, a nonstick surface is desirable. A Teflon coating on the surface can be used for this purpose.
b. For the electrical connections, a high electrical conductivity is needed, pointing to metals. Ceramics and polymers are electrical insulators and composites are generally not good conductors either. Silver, copper, gold, and aluminum are very good electrical conductors, with conductivities of 6.8, 6.0, 4.3, and 3.8×10^7 (ohm-m)$^{-1}$. Aluminum is useful if weight considerations are important. However, copper is a good choice because it is relatively cheap and easy to form and bond to obtain the desired configuration of

electrical circuitry. Its melting point is high (1358 K). However, it does not have good corrosion resistance and may cause problems if the system is to be used under humid conditions. Gold is excellent in corrosion resistance, is a good conductor, and has a high melting point (1336 K). However, it is much more expensive than copper and is hard to bond to other metals. Therefore, the electrical circuitry connections may be made of copper with gold plating used for corrosion resistance. Silver plating may also be used, but it is not as corrosion resistant and durable as gold.

For the board material, on the other hand, we need an electrical insulator. It must be strong enough to support the circuitry and components. Therefore, polymers, ceramics, or composites may be used. However, ceramics are brittle and relatively difficult to machine. Polymers are good for the purpose, but they may be too flexible unless thick plates are used. Composite materials are a good choice because these could be reinforced with metal or glass fibers to obtain the desired strength. The other properties could also be varied by the choice of the structure of the material. Therefore, a variety of composite materials may be chosen for the purpose.

Clearly, several other material options are possible for these applications and a unique answer is rarely obtained. However, these examples indicate the initial selection of the type of material, narrowing of the available choices, and final selection of an appropriate material.

2.6 SUMMARY

This chapter presents the basic considerations in the design of thermal systems. Several important concepts and ideas are introduced and discussed in terms of typical thermal processes.

The formulation of the design problem is the first step in design; the entire design process and the success of the final design depend on the problem statement. The formulation involves determining the requirements that must be met by the system; parameters that are given and are thus fixed; design variables that may be changed in order to seek an acceptable or workable system; constraints or limitations that must be satisfied; and any additional aspects arising from safety, environmental, financial, and other concerns. The final design must satisfy all the requirements and must not violate any of the constraints imposed on the system, its parts, or the materials involved. It is important to formulate the design problem in clear and quantitative terms, while focusing on the important features of the design and neglecting minor ones because it may be difficult or impossible to solve the problem if every possible requirement and constraint is to be satisfied.

Conceptual design is the next step in the design of a thermal system to meet a given need or opportunity. Originality and creativity are expressed in the form of the basic concept or idea for the design. The configuration and main features of the thermal system are given in general terms to indicate how the requirements and constraints of the given problem will be met. The conceptual design may range from a new, innovative idea to available concepts applied to similar problems and modifications in existing systems. Many conceptual designs are based on available designs and concepts, incorporating new materials and techniques. Knowledge of current technology, existing systems and processes, and advances in the recent past is a strong component in the development of appropriate conceptual designs. Usually, several concepts are considered and evaluated for a given application, and the one that has the best chance of success is ultimately chosen.

The selected conceptual design leads to an initial physical system that is subjected to the detailed design process, starting with the modeling and simulation of the system. Modeling involves simplifying and approximating the given process or system to allow a mathematical or numerical solution to be obtained. However, it must be an accurate and valid representation of the physical system so that the behavior of the system may be investigated under a variety of conditions by using the model. Modeling of thermal systems is an extremely important aspect in the design process because most of the inputs needed for design and optimization are obtained from a numerical simulation of the model. Experimental results, material property data, and information on the characteristics of various devices are also incorporated in the overall model to obtain realistic and practical results from the simulation.

The outputs from the simulation are used to determine if the design satisfies the requirements and constraints of the given problem. If it does, an acceptable or workable design is obtained. Several such acceptable designs may be sought to establish a domain from which the best or optimal design may be determined. Though several designs may be acceptable, the best design, optimized with respect to a chosen criterion, is essentially unique or may be selected from a narrow region of design variables. In many cases, multiple objective functions are of interest and the optimization strategy must consider these. For a chosen system hardware, the operating conditions may also be optimized to obtain the best outcome. The need for optimization of thermal systems has grown tremendously in recent years due to international competition. Additional aspects such as safety and control of the system, environmental issues, and communication of the design are also discussed.

The basic features of a CAD system are also outlined. Such a system involves interactive use of a stand-alone computer to help the design process by providing results from the simulation of the system that is being designed. Storage of relevant information, graphical display of results, and knowledge base from current engineering practice, including rules for decision-making, add to the usefulness of a CAD system. However, because of the complexity of typical thermal systems and processes, such CAD systems are often limited to the design of relatively simple systems and equipment.

Finally, the important aspect of material selection is considered in this chapter. The crucial part played by materials in the design of thermal systems cannot be exaggerated because the success of a design is strongly affected by the choice of suitable materials for the various parts of the system. With the advent of new fabrication techniques such as three-dimensional printing and new materials, particularly ceramics, semiconductor materials, and composite materials, it is essential that we seek out the most appropriate material and manufacturing process for each application. Substitution of currently used materials by new and improved ones is also undertaken to improve the system performance and reduce costs. However, redesign of the system must generally be undertaken when material substitution is considered in order to obtain maximum benefit from such a substitution. Different types of materials and the basic procedure for material selection are presented.

REFERENCES

Alger, J.R.M., & Hays, C.V. (1964). *Creative synthesis in design.* Englewood Cliffs, NJ: Prentice-Hall.

American Society of Heating, Refrigeration and Air Conditioning Engineers. (2017). *ASHRAE handbook—fundamentals.* New York: ASHRAE.

American Society of Metals. (1961). *Metals handbook: vol. 1-properties and selection of metals* (8th ed.). Metals Park, OH: American Society of Metals.

Budinski, K.G., & Budinski, M.K. (2009). *Engineering materials: properties and selection* (9th ed.). New York: Pearson.

Burge, D.A. (1999). *Patents and trademark tactics and practice* (3rd ed.). New York: Wiley.

Cengel, Y.A., & Boles, M.A. (2014). *Thermodynamics: an engineering approach* (8th ed.). New York: McGraw-Hill.

Dally, J.W. (1990). *Packaging of electronic systems: a mechanical engineering approach.* New York: McGraw-Hill.

Dieter, G.E. (2000). *Engineering design: a materials and processing approach* (3rd ed.). New York: McGraw-Hill.

Eckert, E.R.G., & Drake, R.M. (1972). *Analysis of heat and mass transfer.* New York: McGraw-Hill.

Figliola, R.S., & Beasley, D.E. (2014). *Theory and design for mechanical measurements* (6th ed.). New York: Wiley.

Howell, J.R., & Buckius, R.O. (1992). *Fundamentals of engineering thermodynamics* (2nd ed.). New York: McGraw-Hill.

Hull, D., & Clyne, T.W. (2008). *An introduction to composite materials* (2nd ed.). Cambridge, U.K.: Cambridge University Press.

Incropera, F.P., & Dewitt, D.P. (2001). *Fundamentals of heat and mass transfer* (5th ed.). New York: Wiley.

Jaluria, Y. (2018). *Advanced materials processing and manufacturing.* Cham, Switzerland: Springer.

Jaluria, Y., & Lombardi, D. (1991). Use of expert systems in the design of thermal equipment and processes, *Research in Engineering Design*, 2(4), 239–253.

Luce, S. (1988). *Introduction to composite technology.* Dearborn, MI: Society of Manufacturing Engineers.

Lumsdaine, E., & Lumsdaine, M. (1995). *Creative problem solving* (3rd ed.). New York: McGraw-Hill.
Moore, F.K., & Jaluria, Y. (1972). Thermal effects of power plants on lakes, *ASME J Heat Transfer*, 94, 163–168.
Moran, M.J., & Shapiro, H.N. (2014). *Fundamentals of engineering thermodynamics* (8th ed.). New York: Wiley.
Palm, W.J. (1986). *Control systems engineering*. New York: Wiley.
Patankar, S.V. (1980). *Numerical heat transfer and fluid flow*. Washington, D.C.: Taylor & Francis.
Pressman, D. (1979). *Patent it yourself? How to protect, patent and market your inventions*. New York: McGraw-Hill.
Raven, F.H. (1994). *Automatic control engineering* (5th ed.). New York: McGraw-Hill.
Reddy, J.N. (2009). *An introduction to the finite element method* (3rd ed.). New York: McGraw-Hill.
Roache, P.J. (2009). *Fundamentals of verification and validation*. Albuquerque, NM: Hermosa Pub.
Tadmor, Z., & Gogos, C.G. (1979). *Principles of polymer processing*. New York: Wiley.
Touloukian, Y.S., & Ho, C.Y. (Eds.). (1972). *Thermophysical properties of matter*. New York: Plenum Press.
Zienkiewicz, O. (1977). *The finite element method: its basis and fundamentals* (7th ed.). Oxford, UK: Butterworth-Heinemann.

PROBLEMS

Note: All the questions given here are open-ended. Thus, some inputs and approximations may have to be supplied by you and several acceptable solutions are possible. Appropriate literature may be consulted for these problems as well as for similar open-ended problems in later chapters.

2.1 In the Czochralski crystal growing process, a solid cylindrical crystal is grown from a rotating melt region, as shown in Figure P2.1. We are interested in obtaining a homogeneous cylinder of high purity of a given material such as silicon and with a uniform specified diameter. For this manufacturing process, list the important inputs, requirements, and specifications needed to design the system. Also, give the design variables and constraints, if any.

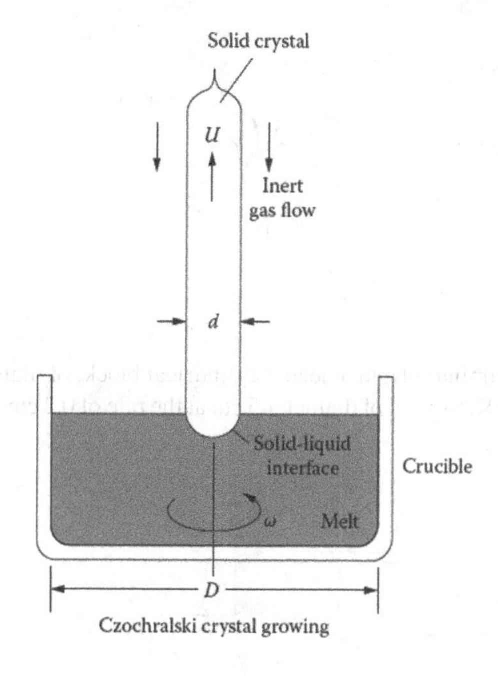

FIGURE P2.1

2.2 For a continuous casting system, shown in Figure 1.10(a), formulate the design problem in terms of given quantities, design variables, and constraints, employing symbols for the dimensions, temperatures, and other physical quantities. We wish to obtain a casting of given diameter for the chosen material.

2.3 Give the design variables and operating conditions for the following manufacturing processes:
  a. Ingot casting
  b. Plastic extrusion
  c. Hot rolling
2.4 Cooling towers, as shown in Figure 1.14(b), are to be designed for heat rejection from a power plant. The rate of heat rejection to a single tower is given as 200 MW. Ambient air at temperature 15°C and relative humidity 0.4 is to be used for removal of heat from the hot water coming from the condensers of the power plant. The temperature of the hot water is 20°C above the ambient temperature. Give the formulation of the design problem in terms of the fixed quantities, requirements, constraints, and design variables.
2.5 The condensers of a 500 MW power plant operating at a thermal efficiency of 30% are to be cooled by the water from a nearby lake, as sketched in Figure 1.14(a). If the intake water is available at 20°C and if the temperature of the water discharged back into the lake must be less than 32°C, quantify the design problem for the cooling system. How is the net energy removed from the condensers finally lost to the environment?
2.6 Formulate the design problem for the following manufacturing processes, employing symbols for appropriate physical quantities.
  a. Hot rolling of a steel plate of thickness 2 cm to reduce the thickness to 1 cm at a feed rate of 1 m/s; see Figure 1.10(d).
  b. Solder plating of a 2-mm-thick epoxy electronic circuit board by moving it across a solder wave at 350°C, the solder melting point being 275°C. See Figure P2.6(b).

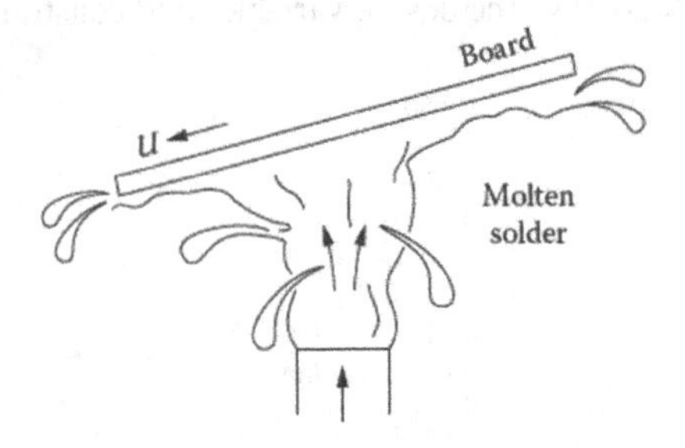

FIGURE P2.6(b)

  c. Extrusion of aluminum from a heated cylindrical block, of diameter 15 cm at a temperature of 600 K, to a rod of diameter 5 cm at the rate of 0.2 cm/s. See Figure P2.6(c).

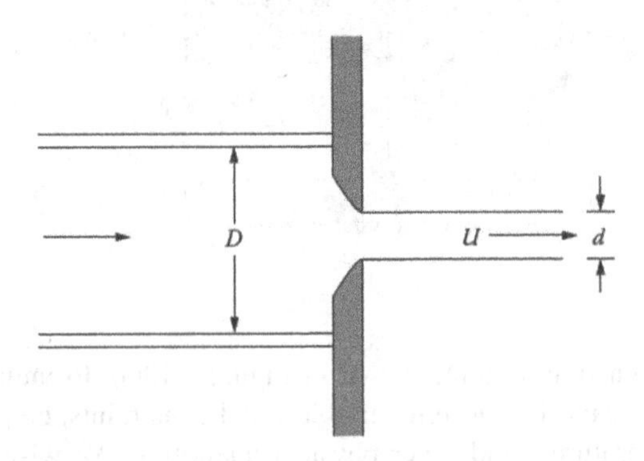

FIGURE P2.6(c)

d. Arc welding by means of an electrode moving at 5 cm/s and supplying 1000 W to join two metal plates, each of thickness 5 mm. See Figure P2.6(d).

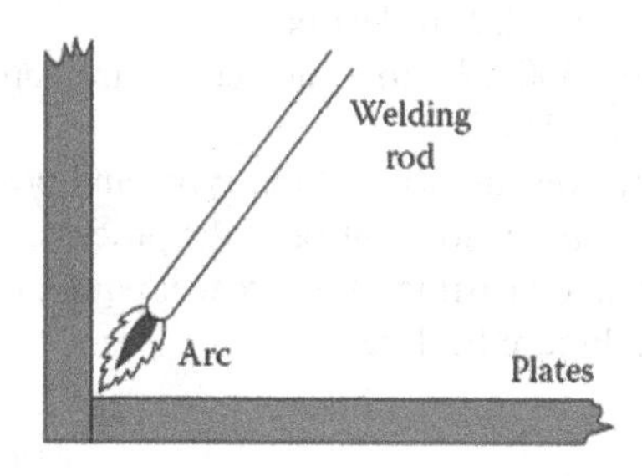

FIGURE P2.6(d)

2.7 A system for the storage of thermal energy is to be designed using an underground tank of water. The tank is buried at a depth of 3 m and is a cube of 1 m side. The water in the tank is heated by circulating it through a solar energy collection system. A given heat input to the water may be assumed due to the solar energy flux. Characterize the design problem in terms of the fixed quantities and design variables.

2.8 Consider a typical water cooler for drinking water. If the water intake on a summer day is at 40°C and the cooler must supply drinking water in the range of 14°C to 21°C at a maximum flow rate of 1 gallon/min (3.785×10^{-3} m^3/min), give the requirements for the design. Also, choose an appropriate conceptual design and suggest the relevant design variables and constraints.

2.9 For the plastic extrusion system considered in Example 2.1, formulate the design problem in terms of quantities that would generally be given, quantities that may be varied to obtain an acceptable design, and possible design requirements and constraints.

2.10 Coal for a steel plant is delivered by train at a station that is 10 km from the storage units of the plant. List different ways of transporting the coal from the station to the storage units and discuss the possible advantages and disadvantages of each approach. Choose the most appropriate system, giving reasons for your choice. Take the typical daily consumption of coal to be 10^4 kg.

2.11 Water from a purification plant is to be stored in a tank that is located at a height of 100 m and supplies the water needed by a chemical factory. Develop different conceptual designs for achieving this task and choose the most suitable one, justifying your choice. The average consumption of water by the factory may be taken as 1000 gallons/h (3.785 m^3/h).

2.12 For the following tasks, consider different design concepts that may be used to achieve the desired goals. Compare the different options in terms of their positive and negative features. Then narrow your deliberations to one concept. Sketch the conceptual design thus obtained and give qualitative reasoning for your choice. Remember that the design chosen by you may not be the only feasible one.

a. Scrap plastic pieces are to be melted and then solidified in the form of cylindrical rods at a rate of about 20 kg/h.

b. Solar energy collected by a flat plate collector system is to be stored to supply hot water at a temperature of 70°C ± 5°C to an industrial unit.

c. Water from a purification plant is to be transported to and stored in a tank at a height of 5 m above the plant. A maximum flow rate of 10 gallons/min (0.03785 m^3/min) is desired.

d. The water from a river is to be supplied at a flow rate of 50 gallons/min and a pressure of 5 atm to a water treatment plant.

e. A company wants to discharge its nontoxic chemical waste into a river, with the smallest impact on the local water region, within 25 m of the discharge point.
f. Food materials are to be frozen by reducing the temperature to below –15°C. A net energy removal rate of 100 kW is desired.
g. A building of floor area 500 m² is to be heated by circulating hot air. The temperature of the air must not exceed 90°C.

2.13 For the following systems, discuss the nature, type, and possible locations of sensors that may be used for safety as well as for control of the process.
a. A water heating system consisting of a furnace, pump, inlet/outlet ports, and piping network, as shown in Figure P2.13(a)

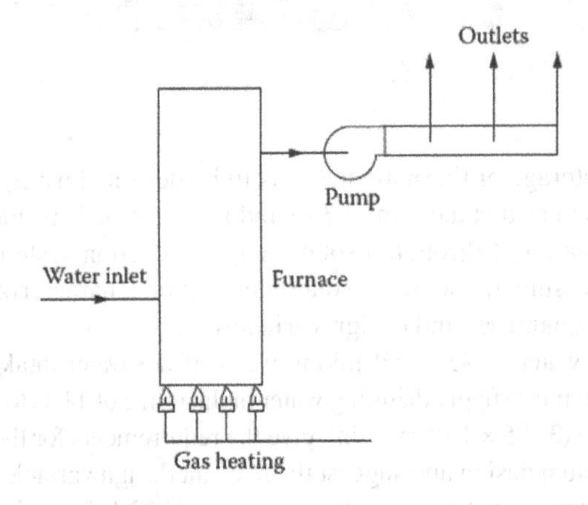

FIGURE P2.13(a)

b. A system to heat short metal rods in a gas furnace and then bend these into desired shapes in a metal-forming process
c. Electronic circuitry for a data center
d. Cooling and fuel systems of a typical car
e. A forced-hot-air-flow oven for drying paper pulp, as shown in Figure P2.13(e)

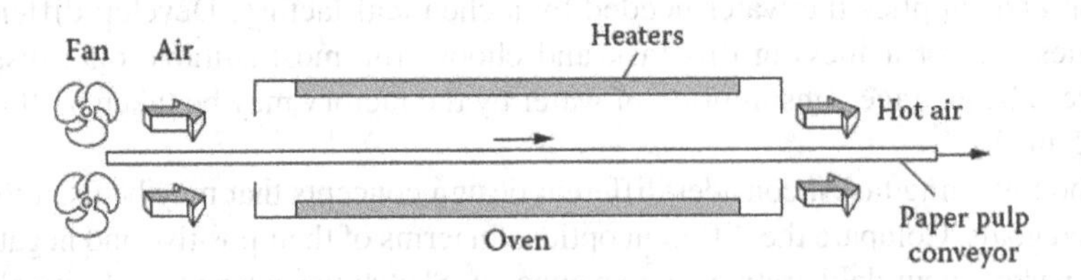

FIGURE P2.13(e)

2.14 For the air conditioning system considered in Example 2.2, discuss the types and locations of sensors that may be employed for the safety and control of the system.

2.15 Look up any patent in the literature. List the different parts of the patent and outline the information conveyed by such a document. How does one ensure that the basic concept is protected and that a slight change in the method is not treated as something new and not covered by the patent?

2.16 Copyrighting of computer software is quite prevalent today because its development is generally expensive. However, most details on the algorithm are to be provided for copyrighting.

Suggest a few approaches that may be employed to avoid duplication and use of the software by others without appropriate permission and licensing.

2.17 If a CAD system is envisaged for the design of HVAC (heating, ventilation, and air conditioning) systems, what relevant characteristics would be desirable? What should the different parts of the CAD system contain? Are there some features that are crucial to the successful use of the CAD approach for this problem?

2.18 Repeat the preceding problem for a power plant heat rejection system consisting of condensers, circulating water, and cooling towers.

2.19 In view of the increasing speed and storage capacity of computer workstations, discuss what additional features could be included in the CAD system outlined in Example 2.6 to make the system more versatile and useful for practical processes.

2.20 Consider different materials that may be used for the following applications. Using the general characteristics of these materials, choose the most appropriate one, giving reasons for the choice. The final material selected is not unique and several options may be possible. Discuss your selection criteria. Remember to include cost, availability, and safety issues in your considerations of different material choices.

a. Outer casing for a personal computer
b. Material for the boards used in the electronic circuitry of a television
c. Materials for the tube and shell of a heat exchanger
d. The mold material for the casting of aluminum, as shown in Figure 1.3. How will the material differ if steel were being cast instead?
e. Materials for the seats in an airplane. Are any thermal considerations involved in the material selection?
f. Electronic circuitry used in an airplane
g. Materials for the wall and the insulation of a gas furnace used for melting scrap steel pieces
h. Liquid that may be used for immersion cooling of an electronic system

2.21 Consider the cooling systems for an automobile and for a personal computer. Suggest various materials that may be employed, discussing the differences between the two applications. Narrow your choices to the best one or two candidates, giving reasons for this selection.

2.22 There are several subsystems in an automobile. List a few of these. Pick any one thermal subsystem and, using your imagination and experience, give a set of requirements and constraints that must be satisfied for a workable design. Also, give the design variables that you may be able to select to obtain a successful design. Give a rough sketch of the subsystem chosen by you and express the constraints, requirements, etc., mathematically, as far as possible.

2.23 Let us assume that your design group, working in an industrial concern, has completed the design of the following thermal systems, using several new ideas and materials. What are the important means of communicating these designs and to which groups within or outside the company do you need to make appropriate presentations?

a. A very efficient room air conditioning system
b. A new radiator design for an automobile
c. A substantially improved and efficient household refrigerator

2.24 For the thermal systems in the preceding problem, outline the main design steps employed by you and your design group to reach optimal solutions.

2.25 You have just joined the design and development group at Panasonic, Inc. The first task you are given is to work on the design of a thermal system to anneal television glass screens. Each screen is made of semi-transparent glass and weighs 10 kg. You need to heat it from a room temperature of 25°C to 1100°C, maintain it at this temperature for 15 minutes, and then cool slowly to 500°C, after which it may be cooled more rapidly to room temperature.

The allowable rate of temperature change with time, $\partial T/\partial t$, is given for heating, slow cooling, and fast cooling processes. Any energy source may be used and high production rates and uniform annealing are desired.

a. Give the sketch of a possible conceptual design for the system and of the expected temperature cycle. Briefly give reasons for your choice.
b. List the requirements and constraints in the problem.
c. Give the location and type of sensors you would use to control the system and ensure safe operation. Briefly justify your choices.
d. Outline a simple mathematical model to simulate the process.

2.26 You are asked to design the cleaning and filtration system for a round swimming pool of diameter D and depth H. The system must be designed to run the entire volume of water contained in the pool through the system in 5 hours, after which a given level of purity must be achieved.

a. Give the formulation of the design problem.
b. Provide a sketch of a possible conceptual design.
c. Suggest the location of two sensors for purity measurements.

2.27 As an engineer at General Motors Co., you are asked to design an engine cooling system. The system should be capable of removing 15 kW of energy from the engine of the car at a speed of 80 km/h and ambient temperature of 35°C. The system consists of the radiator, fan, and flow arrangement. The dimensions of the engine are given. The distance between the engine of the car and the radiator must not exceed 2.0 m and the dimensions of the radiator must not exceed 0.5 m × 0.5 m × 0.1 m.

a. Give the formulation of the design problem. No explanations are needed.
b. Give a possible conceptual design.
c. If you are allowed two sensors for safety and control, what sensors would you use and where would you locate these?

2.28 As an engineer employed by a company involved in designing and manufacturing food processing equipment, you are asked to design a baking oven for heating food items at the rate of 2 pieces per second. Each piece is rectangular, approximately 0.06 kg in weight, and less than 4 cm wide, 6 cm long, and 1 cm high. The length of the oven must not exceed 2.0 m and the height as well as the width must not exceed 0.5 m.

a. Sketch a possible conceptual design for the system. Very briefly give reasons for your selection.
b. List the design variables and constraints in the problem.
c. Which materials will you use for the outer casing, inner lining, and heating unit of the oven? Briefly justify your answers.

3 Modeling of Thermal Systems

3.1 INTRODUCTION

3.1.1 IMPORTANCE OF MODELING IN DESIGN

Modeling is one of the most crucial elements in the design and optimization of thermal systems. Practical processes and systems are generally very complicated and are thus difficult to analyze. Idealizations and approximations are used to simplify the problem to make it amenable to a solution. The process of simplifying a given problem so that it may be represented in terms of a system of equations, for analysis, or a physical arrangement, for experimentation, is termed *modeling*. By the use of models, relevant quantitative results are obtained for the design and optimization of processes, components, and systems. However, despite its importance, and even though analysis is taught in many engineering courses, very often little attention is given to modeling.

Modeling is needed for understanding and predicting the behavior and characteristics of thermal systems. Once a model is obtained, it is subjected to a variety of operating conditions and design variations. If the model is a good representation of the actual system under consideration, the outputs obtained from the model characterize the behavior of the actual system. This information is used in the design process as well as in the evaluation of a particular design to determine if it satisfies the given requirements and constraints. Modeling also helps in obtaining and comparing alternative designs by predicting the performance of each design, ultimately leading to an optimal design. Thus, the design and optimization processes are closely coupled with the modeling effort, and the success of the final design is very strongly influenced by the accuracy and validity of the model employed. Consequently, it is important to understand the various types of models that may be developed; the basic procedures that may be used to obtain a satisfactory model; validation of the model obtained; and its representation in terms of equations, important parameters, and relevant data on material properties.

3.1.2 BASIC FEATURES OF MODELING

The model may be *descriptive* or *predictive*. We are all very familiar with models that are used to describe and explain various physical phenomena. A working model of an engineering system, such as a robot, an internal combustion engine, a heat exchanger, or a water pump, is often used to explain how the device works. Frequently, the model may be made of clear plastic or may have a cutaway section to show the internal mechanisms. Such models are known as *descriptive* and are frequently used in classrooms and in marketing to explain the basic operation and underlying principles.

Predictive models are of particular interest to our present topic of engineering design because these can be used to predict the performance of a given system. The equation that describes the cooling of a hot metal sphere immersed in an extensive cold-water environment represents a predictive model because it allows us to obtain the temperature variation with time and determine the dependence of the cooling curve on physical variables such as initial temperature of the sphere, water temperature, and material properties. Similarly, a graph of the number of items sold versus the item price, such as that shown in Figure 1.6, represents a predictive model because it allows one to predict the volume of sales if the price is reduced or increased. Models such as the control mass and control volume formulations in thermodynamics, representation of a projectile as a point to study its trajectory, and enclosure models for radiation heat transfer are quite common in engineering analysis for understanding the basic principles and for deriving the characteristic equations. A few such models are sketched in Figure 3.1.

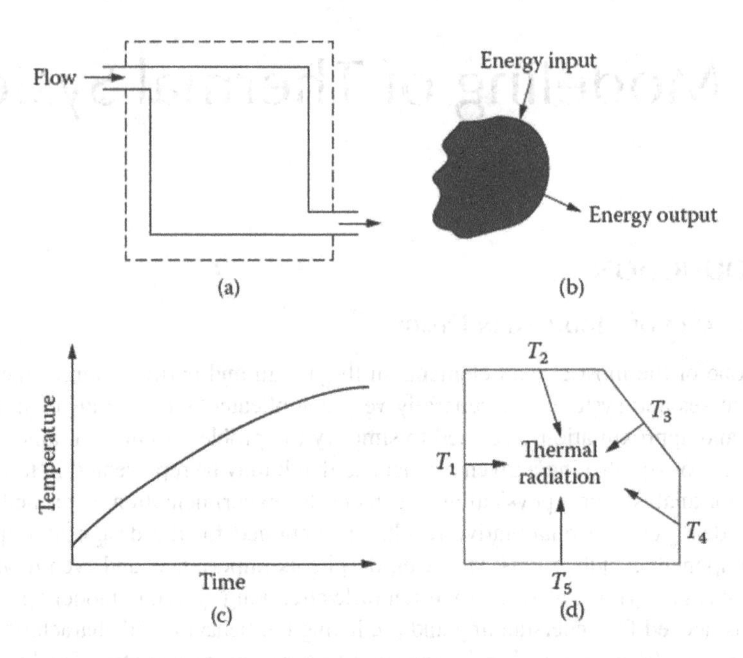

FIGURE 3.1 A few models used commonly in engineering: (a) Control volume, (b) control mass, (c) graphical representation, and (d) enclosure configuration for thermal radiation analysis.

Modeling is particularly important in thermal systems and processes because of the generally complex nature of the transport, resulting from variations with space and time, nonlinear mechanisms, complicated boundary conditions, coupled transport processes, complicated geometries, and variable material properties. As a result, thermal systems are often represented by sets of time-dependent, multidimensional, nonlinear partial differential equations with complicated domains and boundary conditions. Finding a solution to the full three-dimensional, time-dependent problem is usually an extremely involved process. In addition, the interpretation of the results obtained and their application to the design process are usually complicated by the large number of variables involved. Even if experiments are carried out to obtain the relevant input data for design, the expense incurred in each experiment makes it imperative to develop a model to guide the experimentation and to focus on the dominant parameters. Therefore, it is necessary to neglect relatively unimportant aspects, combine the effects of different variables in the problem, employ idealizations to simplify the analysis, and reduce the number of parameters that characterize the process or system. This effort also generalizes the problem so that the results obtained from one analytical or experimental study can be extended to other similar systems and circumstances.

Physical insight is the main basis for the simplification of a given system to obtain a satisfactory model. Such insight is largely a result of experience in dealing with a variety of thermal systems. Estimates of the underlying mechanisms and different effects that arise in a given system may also be used to simplify and idealize. Knowledge of other similar processes and of the appropriate approximations employed for these also helps in modeling. Overall, modeling is an innovative process based on experience, knowledge, and originality. Exact, quantitative rules cannot be easily laid down for developing a suitable model for an arbitrary system. However, various techniques such as scale analysis, dimensional analysis, and similitude can be employed to aid the modeling process. These methods are based on a consideration of the important variables in the problem and are presented in detail later in this chapter. However, modeling remains one of the most difficult and elusive, though extremely important, aspects in engineering design.

In many practical systems, it is not possible to simplify the problem enough to obtain a sufficiently accurate analytical or numerical solution. In such cases, experimental data are obtained, with help

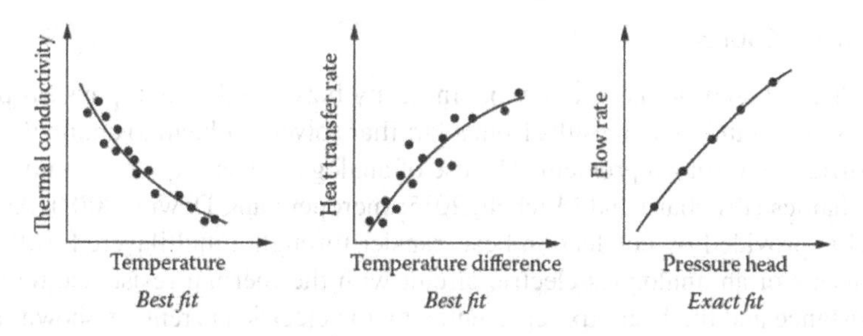

FIGURE 3.2 Examples of curve fitting in thermal processes.

from dimensional analysis to determine the important dimensionless parameters. Experiments are also crucial to the validation of the mathematical or numerical model and for establishing the accuracy of the results obtained. Material properties are usually available as discrete data at various values of the independent variable, e.g., density and thermal conductivity of a material measured at different temperatures. For all such cases, *curve fitting* is frequently employed to obtain appropriate correlating equations to represent the data. These equations can then serve as inputs to the model of the system, as well as to the design process. Curve fitting can also be used to represent numerical results in a compact and convenient form, thus facilitating their use. Figure 3.2 shows a few examples of curve fitting as applicable to thermal processes, indicating best and exact fits to the given data. In the former case, the curve does not pass through each data point but represents a close approximation to the data, whereas in the latter case the curve passes through each point. Curve fitting approaches the problem as a quantitative representation of available data. Though physical insight is useful in selecting the form of the curve, the focus in this case is clearly on data processing and not on the physical problem.

The *validation* of the model developed for a given system is another very important consideration because it determines whether the model is a faithful representation of the actual physical system and indicates the level of accuracy that may be expected in the predictions obtained from the model. Validation is based on the physical behavior of the model, application of the model to simpler and existing systems and processes, and comparisons with available experimental, analytical or numerical results. In addition, as mentioned in Chapter 2, modeling and design are linked so that the feedback from system simulation and design is used to improve the model. Models are initially developed for individual processes and components, followed by a coupling of these individual models to obtain the model for the entire system. This final model usually consists of the governing equations; correlating equations derived from experimental data; and curve-fit results from data on material properties, characteristics of relevant components, financial trends, environmental aspects, and other considerations relevant to the design.

3.2 TYPES OF MODELS

Several types of models can be developed to represent a thermal system. Each model has its own characteristics and is particularly appropriate for certain circumstances and applications. The classification of models as descriptive or predictive was mentioned in the preceding section. Our interest lies mainly in predictive models that can be used to predict the behavior of a given system for a variety of operating conditions and design parameters. Thus, we will consider only predictive models here, and modeling will refer to the process of developing such models. Four main types of predictive models are of interest in the design and optimization of thermal systems. These are:

1. Analog models
2. Mathematical models
3. Physical models
4. Numerical models

3.2.1 Analog Models

Analog models are based on the analogy or similarity between different physical phenomena and allow one to use the solution and results from a familiar solved problem to obtain the corresponding results for a different unsolved problem. The use of analog models is quite common in heat transfer and fluid mechanics (Pritchard and Mitchell, 2015; Incropera and Dewitt, 2001). An example of an analog model is provided by conduction heat transfer through a multilayered wall, which may be analyzed in terms of an analogous electric circuit with the thermal resistance represented by the electrical resistance and the heat flux represented by the electric current, as shown in Figure 3.3(a). The temperature across the region is the potential represented by the electric voltage. Then, Ohm's law and Kirchhoff's laws for electrical circuits may be employed to compute the total thermal resistance and the heat flux for a given temperature difference, as discussed in most heat transfer textbooks.

Similarly, the analogy between heat and mass transfer is often used to apply the experimental and analytical results from one transport process to another. The density differences that arise in room fires due to temperature differences are often simulated experimentally by the use of pure and saline water, the latter being more dense and, thus, representative of a colder region. The flows generated in a fire can then be studied in an analogous salt-water/pure-water arrangement, which is often easier to fabricate, maintain, and control. Figure 3.3(b) shows the analog modeling of a fire plume in an enclosure. The flow is closely approximated. However, the jet is inverted as compared to an actual fire plume, which is buoyant and rises; salt water is heavier than pure water and drops downward. A graph is itself an analog model because the coordinate distances represent the physical quantities plotted along the axes. Flow charts used to represent computer codes and process flow diagrams for industrial plants are all analog models of the physical processes they represent; see Figure 3.3(c).

Clearly, the analog model may not have the same physical appearance as the system under consideration, but it must obey the same physical principles. However, even though analog models are useful in the understanding of physical phenomena and in representing information, energy, or material flow, they have only a limited use in engineering design. This is mainly because the analog models themselves have to be solved and may involve the same complications as the original problem. For instance, an electrical analog model results in linear algebraic equations that are usually solved numerically. Therefore, it is generally better to develop the appropriate mathematical model for the thermal system rather than complicate the modeling by bringing in an analog model as well.

3.2.2 Mathematical Models

A mathematical model is one that represents the performance and characteristics of a given system in terms of mathematical equations. These models are the most important ones in the design

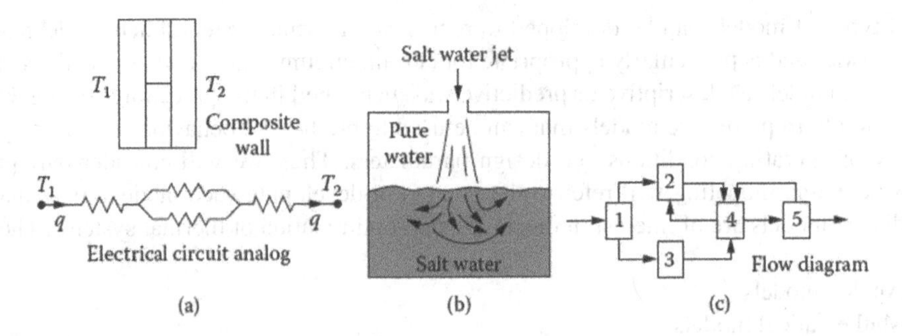

FIGURE 3.3 Analog models. (a) Conduction heat transfer in a composite wall; (b) analog model of plume flow in a room fire using salt water and pure water; and (c) flow diagram for material flow in an industry.

of thermal systems because they provide considerable versatility in obtaining quantitative results that are needed as inputs for design. Mathematical models form the basis for numerical modeling and simulation, so that the system may be investigated without actually fabricating a prototype. In addition, the simplifications and approximations that lead to a mathematical model also indicate the dominant variables in a problem. This helps in developing efficient experimental models, if needed. The formulation and procedure for optimization are also often based on the characteristics of the representative equations. For example, the sets of equations that describe the characteristics of a metal casting system or the performance of a heat exchanger, shown respectively in Figure 1.3 and Figure 1.5, would, therefore, constitute the mathematical models for these two systems. A solution to the equations for a heat exchanger would give, for instance, the dependence of the total heat transfer rate on the inlet temperatures of the two fluids and on the dimensions of the system. Similarly, the dependence of the solidification time in casting on the initial temperature and cooling conditions is obtained from a solution of the corresponding equations. Such results form the basis for design and optimization.

As mentioned earlier, the model may be based on physical insight or on curve fitting of experimental or numerical data. These two approaches lead to two types of models that are often termed as *theoretical* and *empirical*, respectively. Heat transfer correlations for convective transport from heated bodies of different shapes represent empirical models that are frequently employed in the design of thermal systems. The basic objective of mathematical modeling is to obtain mathematical equations that represent the behavior and characteristics of a given component, subsystem, process, or system. Mathematical models focus on the physical principles such as conservation laws to derive the representative equations. Curve fitting of data to obtain mathematical representations of experimental or numerical results, thus yielding empirical models, is discussed later.

3.2.3 Physical Models

A physical model is one that resembles the actual system and is generally used to obtain experimental results on the behavior of the system. An example of this is a scaled down model of a car or a heated body, which is positioned in a wind tunnel to study the drag force acting on the body or the heat transfer from it, as shown in Figure 3.4. Similarly, water channels are used to investigate the forces acting on ships and submarines. In heat transfer, a considerable amount of experimental data on heat transfer rates from heated bodies of different shapes and dimensions, in different fluids, and under various thermal conditions have been obtained by using such scale models. In fact, physical modeling is very commonly used in areas such as fluid mechanics and heat transfer and is thus of particular importance in thermal systems. The physical model may be a scaled down version of the actual system, as mentioned previously, a full-scale experimental model, or a prototype that is essentially the first complete system to be checked in detail before going into production. The development of a physical model is based on a consideration of the important parameters and mechanisms. Thus, the efforts directed at mathematical modeling are generally employed to facilitate physical modeling. This type of model and the basic aspects that arise are discussed in Section 3.4.

3.2.4 Numerical Models

Numerical models are based on mathematical models and allow one to obtain, using a computer, quantitative results on the system behavior for different operating conditions and design parameters. Only very simple cases can usually be solved by analytical procedures; numerical techniques are needed for most practical systems. Numerical modeling refers to the restructuring and discretization of the governing equations in order to solve them on a computer. The relevant equations may be algebraic equations, ordinary or partial differential equations, integral equations, or combinations of these, depending upon the nature of the process or system under consideration.

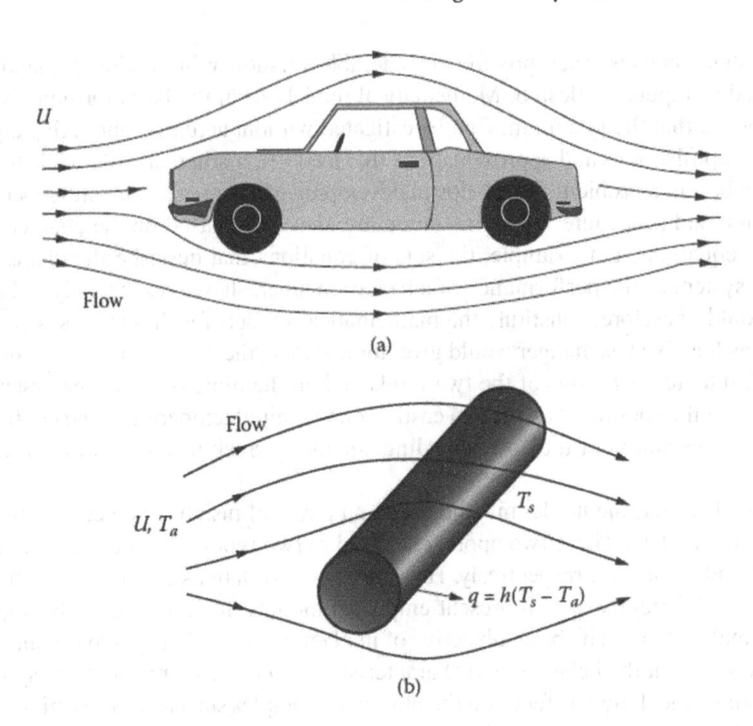

FIGURE 3.4 Physical modeling of (a) fluid flow over a car and (b) heat transfer from a heated body, these being located in a wind tunnel.

Numerical modeling involves selecting the appropriate method for the solution, for instance, the finite difference or the finite element method; discretizing the mathematical equations to put them in a form suitable for digital computation; choosing appropriate numerical parameters, such as grid size and time step; and developing the numerical code and obtaining the numerical solution; see, for instance, Gerald and Wheatley (2003), Recktenwald (2000), Mathews and Fink (2004), and Jaluria (2012). Additional inputs on material properties, heat transfer coefficients, component characteristics, etc., are entered as part of the numerical model. The validation of the numerical results is then carried out to ensure that the numerical scheme yields accurate results that closely approximate the behavior of the actual physical system (Roache, 2009). The numerical scheme for the solution of the equations that describe the flow and heat transfer in a solar energy storage system, for instance, represents a numerical model of this system. Because numerical modeling is closely linked with the simulation of the system, these two topics are presented together in the next chapter. Figure 3.5(a) shows a sketch of a typical numerical model for a hot water storage system in the form of a flowchart. Figure 3.5(b) shows the various components of the code, such as material properties, mathematical model, experimental data, and analytical methods that are linked together through the main numerical scheme to obtain the solution.

3.2.5 Interaction between Models

Even though the four main types of modeling of particular interest to design are presented as separate approaches, several of these frequently overlap in practical problems. For instance, the development of a physical scale model for a heat treatment furnace involves a consideration of the dominant transport mechanisms and important variables in the problem. This information is generally obtained from the mathematical model of the system. Similarly, experimental data from physical models may indicate some of the approximations or simplifications that may be used in developing a mathematical model. Although numerical modeling is based largely on the mathematical model,

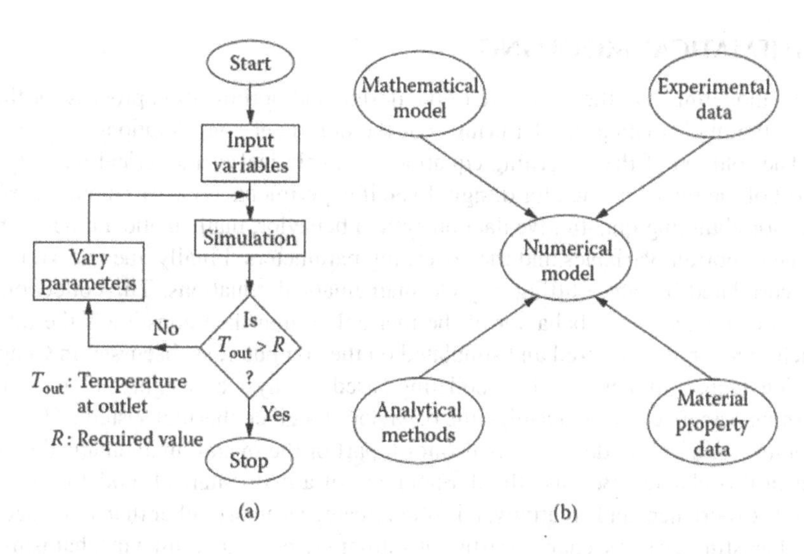

FIGURE 3.5 Numerical modeling. (a) A computer flowchart for a hot water storage system and (b) various inputs and components that constitute a typical numerical model for a thermal system.

outputs from the physical or analog models may also be useful in developing the numerical scheme. Mathematical modeling is generally the most significant consideration in the modeling of thermal systems and, therefore, most of the effort is directed at obtaining a satisfactory mathematical model. If an analytical solution of the equations obtained is not convenient or possible, numerical modeling is employed. Physical models are used if the numerical solution is not easy to obtain; they also provide important physical insight and validation data for the mathematical and numerical models.

3.2.6 Other Classifications

Several other classifications of modeling are frequently used to characterize the nature and type of the model. Thus, the model may be classified as steady state or dynamic, deterministic or probabilistic, lumped or distributed, and discrete or continuous.

A *steady-state* model is one whose properties and operating variables do not change with time. If time-dependent aspects are included, the model is *dynamic*. Thus, the initial, or startup, phase of a furnace would require a dynamic model, but this would often be replaced by a steady-state model after the furnace has been operating for a long time and the transient effects have died down. The development of control systems for thermal processes and devices generally requires dynamic models. *Deterministic* models generally predict the behavior of the system with certainty, whereas *probabilistic* models involve uncertainties in the system that may be considered as random or as represented by probability distributions. Models for supply and demand are often probabilistic, whereas typical thermal systems are analyzed with deterministic models. *Lumped* models use average values over a given volume, whereas *distributed* models provide information on spatial variation. *Discrete* models focus on individual items, whereas *continuous* models are concerned with the flow of material in a continuum. In a heat treatment system, for instance, a discrete model may be developed to study the transport and temperature variation associated with a given body, say a gear, undergoing heat treatment. The flow of hot gases and thermal energy, on the other hand, is studied as a continuum, using a continuous model. Both the discrete and the continuous models are commonly used in modeling thermal systems and processes (Rieder and Busby, 1986).

Once the model has been developed, its type may be indicated by using the classifications mentioned here. For instance, the model for a hot water storage system may be described as a dynamic, continuous, lumped, deterministic mathematical model. Similarly, the mathematical model for a furnace may be specified as steady state, continuous, distributed, and deterministic.

3.3 MATHEMATICAL MODELING

Mathematical modeling is at the very core of the design and optimization process for thermal systems because the mathematical model brings out the dominant considerations in a given process or system. The solution of the governing equations by analytical or numerical techniques usually provides most of the inputs needed for design. Even if experiments are carried out for validation of the model or for obtaining quantitative data on system behavior, mathematical modeling is used to determine the important variables and the governing parameters. Finally, the experimental results are usually correlated by curve fitting to yield mathematical equations. The collection of all the equations that characterize the behavior of the thermal system then constitutes the mathematical model, which is generally analyzed and simulated on the computer, as discussed in Chapter 4.

This section deals with mathematical modeling based on physical insight and on a consideration of the governing principles that determine the behavior of a given thermal system. The use of curve fitting to obtain empirical models, which also form part of the overall mathematical model, is discussed later in this chapter. Because the development of a mathematical model requires physical understanding, experience, and creativity, it is often treated as an art rather than a science. However, knowledge of existing systems, characteristics of similar systems, governing mechanisms, and commonly made approximations and idealizations provides substantial help in model development.

3.3.1 General Procedure

A general step-by-step procedure may be outlined for mathematical modeling of a thermal system. Such a procedure is given here, with simple illustrative examples, to indicate the application of various ideas. However, there is no substitute for experience and creativity, and, as one continues to develop models for a variety of thermal systems, the process becomes simpler and better defined. Generally, there is no unique model for a typical thermal system and the approach given here simply provides some guidelines that may be used for developing an appropriate model. Frequently, very simple models are initially developed and the model is gradually improved over time by including additional complexities.

3.3.1.1 Transient/Steady State

One of the most important considerations in modeling is whether the system can be assumed to be at steady state, involving no variations with time, or if the time-dependent changes must be taken into account. Because time brings in an additional independent variable, which increases the complexity of the problem, it is important to determine whether transient effects can be neglected. Most thermal processes are time-dependent, but for several practical circumstances, they may be approximated as steady. Thus, even though the hot rolling process, sketched in Figure 1.10(d), starts out as a transient problem, it generally approaches a steady-state condition as time elapses. Similarly, the solar heat flux incident on the wall of a house clearly varies with time. Nevertheless, over certain short periods, it may be approximated as steady. Several such processes may also be treated as periodic, with the conditions and variables repeating themselves in a cyclic manner.

Two main characteristic time scales need to be considered. The first, τ_r, refers to the response time of the material or body under consideration, and the second, τ_c, refers to the characteristic time of variation of the ambient or operating conditions. Therefore, τ_c indicates the time over which the conditions change. For instance, it would be zero for a step change and the time period τ_p for a periodic process, where $\tau_p = 1/f$, f being the frequency. As mentioned in Chapter 2 and discussed later in this chapter, the response time τ_r for a uniform-temperature (lumped) body subjected to a step change in ambient temperature for convective cooling or heating is given by the expression

$$\tau_r = \frac{\rho C V}{hA} \tag{3.1}$$

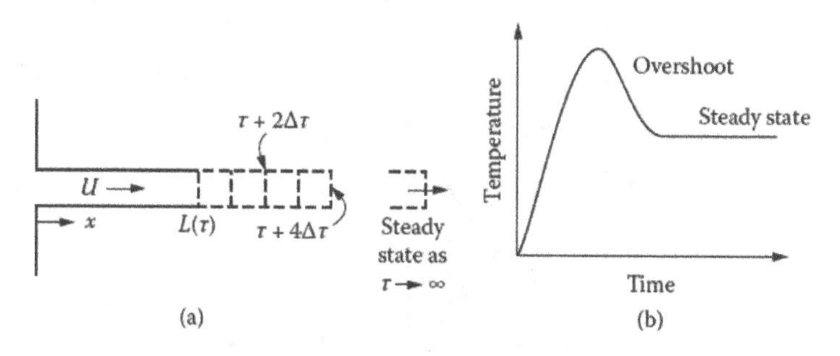

FIGURE 3.6 Attainment of steady-state conditions at large time. (a) Modeling of heated moving material, and (b) temperature variation of an electronic chip heated electrically.

where ρ is the density, C is the specific heat, V is the volume of the body, A is its surface area, and h is the convective heat transfer coefficient. Several important cases can be obtained in terms of these two time scales, as follows:

1. τ_c is very large, i.e., $\tau_c \rightarrow \infty$: In this case, the conditions may be assumed to remain unchanged with time and the system may be treated as steady state. At the start of the process, the variables change sharply over a short time and transient effects are important. However, as time increases, steady-state conditions are attained. Examples of this circumstance are the extrusion, wire drawing, and rolling processes, as sketched in Figure 3.6(a). Clearly, as the leading edge of the material moves away from the die or furnace, steady-state conditions are attained in most of the region away from the edge. Thus, except for the starting transient conditions and in a region close to the edge, the system may be approximated as steady. A similar situation arises in many practical systems where the initial transient is replaced by steady conditions at large time; for instance, in the case of an initially unheated electronic chip that is heated by an electric current and finally attains steady state due to the balance between heat loss to the environment and the heat input [see Figure 3.6(b)]. The transient terms, which are of the form $\partial\varphi/\partial\tau$, where φ is a dependent variable, are dropped and the steady-state characteristics of the system are determined.
2. $\tau_c \ll \tau_r$: In this case, the operating conditions change very rapidly, as compared to the response of the material. Then the material is unable to follow the variations in the operating variables. An example of this is a deep lake whose response time is very large compared to the fluctuations in the ambient medium. Even though the surface temperature may reflect the effect of such fluctuations, the bulk fluid is expected to show essentially no effect of temperature fluctuations. Then the system may be approximated as steady with the operating conditions taken at their mean values. Such a situation arises in many practical systems due to rapid fluctuations in the heat input or the flow rate, while the mean values remain unchanged. If the mean value itself varies with time, then the characteristic time of this variation is considered in the modeling. In addition, if the operating conditions change rapidly from one set of values to another, the system goes from one steady-state situation to another through a transient phase. Again, away from this rapid variation, the problem may be treated as steady.
3. $\tau_r \ll \tau_c$: This refers to the case where the material or body responds very quickly but the operating or boundary conditions change very slowly. An example of this is the slow variation of the solar flux with time on a sunny day and the rapid response of the collector. Similarly, an electronic component responds very rapidly to the turning on of the system, but the walls of the equipment and the board on which it is located respond much more slowly. Another example is a room that is being heated or cooled. The walls respond very slowly as compared to the items in the room and the air. It is then possible to take the

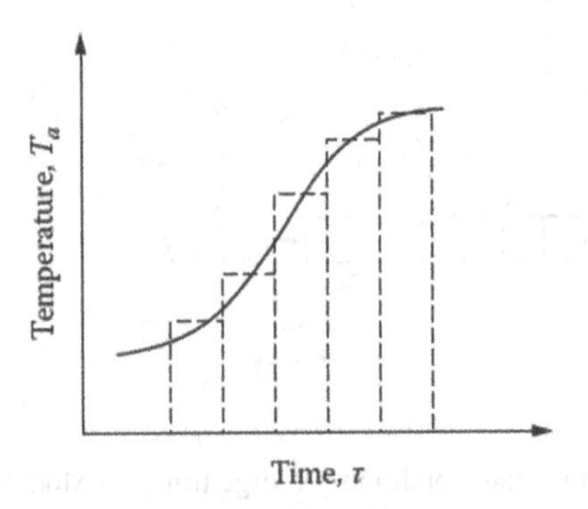

FIGURE 3.7 Replacement of the ambient temperature variation with time by a finite number of steps, with the temperature held constant over each step.

surroundings as unchanged over a portion of the corresponding response time. Therefore, in such cases, the transport may be modeled as *quasi-steady*, with the steady problem being solved at different times. This implies that the part or system goes through a sequence of steady states, each being characterized by constant operating or environmental conditions. Figure 3.7 shows a sketch of such quasi-steady modeling. This is a frequently employed approximation in time-dependent problems, because many practical systems involve such slowly varying operating, boundary, or forcing conditions.

4. Periodic processes: In many cases, the behavior of the thermal system may be represented as a periodic process, with the characteristics repeating over a given time period τ_p. Environmental processes are examples of this modeling because periodic behavior over a day or over a year is of interest in many of these systems. The modeling of solar energy collection systems, for instance, involves both the cyclic nature of the process over a day and night sequence, as well as over a year. Long-term energy storage, for instance, in salt-gradient solar ponds or in large water tanks, is considered as cyclic over a year. Similarly, many thermal systems undergo a periodic process because they are turned on and off over fixed periods. The main requirement of a periodic variation is that the temperature and other variables repeat themselves over the period of the cycle, as shown in Figure 3.8(a) for the temperature of a natural water body such as a lake. In addition, the net heat transfer over the cycle must be zero because, if it is not, there is a net gain or loss of energy.

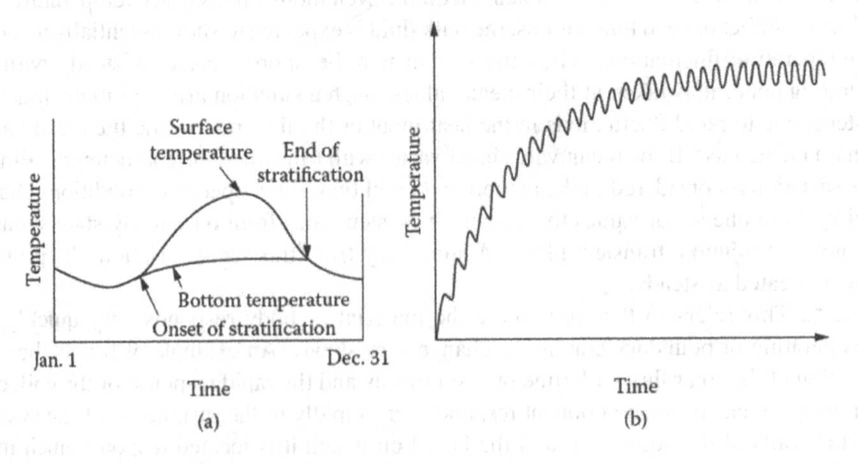

FIGURE 3.8 Periodic temperature variation in (a) a natural lake over the year, and (b) a body subjected suddenly to a periodic variation in the heat input.

This would result in a consequent increase or decrease of temperature with time and a cyclic behavior would not be obtained. These conditions may be represented as

$$\int_0^{\tau_p} Q(\tau)d\tau = 0 \tag{3.2}$$

$$(T)_\tau = (T)_{\tau+\tau_p} \tag{3.3}$$

where $Q(\tau)$ is the total heat transfer rate from a body as a function of time τ. For a deep lake with a large surface area, $Q(\tau)$ is essentially the surface heat transfer rate because very little energy is lost at the bottom or at the sides. Either one of the above conditions may be used in the modeling of a periodic process.

The main advantage of modeling a system as periodic is that results need to be obtained only over the time of a cycle. The conditions given by Equation (3.2) and Equation (3.3) can be used for validation as well as for the development of the numerical code. Frequently, the system undergoes a starting transient and finally attains a periodic behavior. This is typical of many industrial systems that are operated over fixed periods following a startup. Figure 3.8(b) shows the typical temperature variation in such a process. The time-dependent terms are retained in the equations and the problem is solved until the cyclic behavior of the system is obtained. Because of the periodic nature of the process, analytical solutions can often be obtained, particularly if the periodic process can be approximated by a sinusoidal variation (Gebhart, 1971; Eckert and Drake, 1972).

5. Transient: If none of the above approximations is applicable, the system has to be modeled as a general time-dependent problem with the transient terms included in the model. Because this is the most complicated circumstance with respect to time dependence, efforts should be made, as outlined above, to simplify the problem before resorting to the full transient, or dynamic, modeling. However, there are many practical systems, particularly in materials processing and manufacturing, that require such a dynamic model because transient effects are crucial in determining the quality of the product and in the control and operation of the system. Heat treatment and metal casting systems are examples in which a transient model is essential to study the characteristics of the system for design.

3.3.1.2 Spatial Dimensions

This consideration refers to the number of spatial dimensions needed to model a given system. Though all practical systems are three-dimensional, they can often be approximated as two- or one-dimensional to considerably simplify the modeling. Thus, this is an important simplification and is based largely on the geometry of the system under consideration and on the boundary conditions. As an example, let us consider the steady-state conduction in a solid bar of length L, height H, and width W, as shown in Figure 3.9. Let us also assume that the thermal boundary conditions are uniform, though different, on each of the six surfaces of the solid. Now the temperature distribution within the solid $T(x, y, z)$, where x, y, z are the three coordinate distances, as shown, is governed by energy balance given by the following partial differential equation, if the thermal conductivity is constant and no heat source exists in the material:

$$\frac{\partial^2 T}{\partial x^2} + \frac{\partial^2 T}{\partial y^2} + \frac{\partial^2 T}{\partial z^2} = 0 \tag{3.4}$$

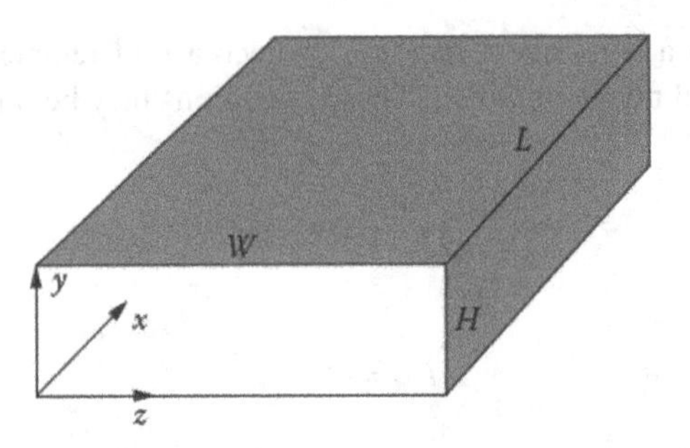

FIGURE 3.9 Three-dimensional conduction in a solid block.

This equation may be generalized by using the dimensionless variables

$$X = \frac{x}{L} \quad Y = \frac{y}{H} \quad Z = \frac{z}{W} \quad \theta = \frac{T}{T_{ref}} \tag{3.5}$$

to yield the dimensionless equation

$$\frac{\partial^2 \theta}{\partial X^2} + \frac{L^2}{H^2}\frac{\partial^2 \theta}{\partial Y^2} + \frac{L^2}{W^2}\frac{\partial^2 \theta}{\partial Z^2} = 0 \tag{3.6}$$

where T_{ref} is a reference temperature and may simply be the ambient temperature or the temperature at one of the surfaces. Other definitions of the nondimensional variables, particularly for θ, are used in the literature.

With this nondimensionalization, the second derivative terms in Equation (3.6) are all of the same order of magnitude because X, Y, and Z all vary from 0 to 1, and the variation in θ is also on the order of 1. Then, the magnitude of each term in this equation is determined by the magnitude of the coefficient. It can be seen that if $L^2/W^2 \ll 1$, the last term in Equation (3.6) becomes small and may be neglected, making the problem two-dimensional, with the temperature a function of only x and y, i.e., $T(x,y)$. If, in addition, L^2/H^2 is also much less than one, the second term may also be neglected, making the problem one-dimensional, with the temperature varying only with x, i.e., $T(x)$. Thus, the problem can be simplified considerably if the region of interest is much larger in one dimension as compared to the others with uniform boundary conditions at the surfaces.

Similarly, cylindrical configurations can be modeled as axisymmetric, i.e., symmetric about the axis, with the temperature and other dependent quantities varying only with the radial coordinate r and the axial coordinate z. If the cylinder is also very long, i.e., $L/R \gg 1$, where L is the length and R the radius, the problem becomes one-dimensional in r. Spherical regions can also be frequently approximated as one-dimensional radial problems. Similar results are obtained by using scale analysis, which is based on a consideration of the scales of the various quantities involved (Bejan, 1993).

The modeling of a given system as one-dimensional, two-dimensional, or axisymmetric, even though it is actually a three-dimensional problem, is an important simplification in modeling and is frequently used (Bergman et al., 2017). The approximation of a fin or an extended surface in heat transfer as one-dimensional, by assuming negligible temperature variation across its thickness, is commonly employed. Similarly, convective transport from a wide flat plate is modeled as two-dimensional and the developing flow in circular tubes as axisymmetric, leading to the results being independent of the angular position. Three-dimensional modeling is generally avoided unless absolutely essential because of the additional complexity in obtaining a solution to the basic equations. In addition, results from a three-dimensional model are not easy to interpret and special techniques are often needed just to visualize the flow and the temperature field. It is difficult to determine the exact values of the parameters, such as L^2/W^2 and L^2/H^2 (or correspondingly L/W and L/H) in

Equation (3.6), for which these approximations may be made for an arbitrary system. However, if these parameters are typically of order 0.1 or less, the approximations are expected to result in negligible loss of accuracy in the solution.

3.3.1.3 Lumped Mass Approximation

The preceding consideration may be continued to obtain a particularly simple model, termed as the *lumped mass* approximation. In this model, which is extensively used and is thus an important circumstance, the temperature, species concentration, or any other transport variable is assumed to be uniform within the domain of interest. Thus, the variable is lumped and no spatial variation within the region is considered. For steady-state conditions, algebraic equations are obtained instead of differential equations. Most thermodynamic systems, such as air conditioning and refrigeration equipment, internal combustion engines, and power plants, are analyzed assuming the conditions in the different components as uniform and, thus, as lumped (see Cengel and Boles, 2014).

For transient problems, the variables change only with time, resulting in ordinary differential equations instead of partial differential equations. Consider, for instance, a heated body at an initial temperature of T_o cooling in an ambient medium at temperature T_a by convection, with h as the convective heat transfer coefficient. Then, if the temperature T is assumed to be uniform in the body, the energy equation is the one that was given earlier in Example 2.5 and may be rewritten as

$$\rho CV\frac{dT}{d\tau} = -hA(T - T_a) \tag{3.7}$$

where the symbols were defined earlier for Example 2.5 and for Equation (3.1). If the temperature difference $(T - T_a)$ is substituted by θ, the energy equation and its solution are obtained as

$$\rho CV\frac{d\theta}{d\tau} = -hA\theta \tag{3.8}$$

$$\theta = \theta_o \exp\left(-\frac{hA\tau}{\rho CV}\right) \tag{3.9}$$

where $\theta_o = T_o - T_a$. The quantity $\rho CV/hA$ represents a characteristic time and is the time needed for the temperature difference from the ambient, $T - T_a$, to drop to $1/e$ of its initial value, where e (= 2.71828) is the base of the natural logarithm. This e-folding time is also known as the response time of the body, as given earlier in Equation (3.1). This model and the corresponding temperature variation are shown in Figure 3.10.

The applicability of the lumped body approximation is based on the ratio of the conductive resistance to the convective resistance for such a heat transfer process. If this ratio is much smaller

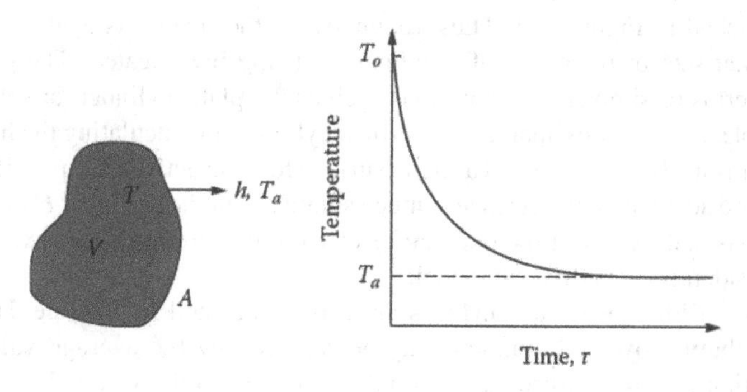

FIGURE 3.10 Lumped mass approximation of a heated body undergoing convective cooling.

than 1.0, the convective resistance dominates and the temperature variation in the material is negligible compared to that in the fluid. This ratio is expressed in terms of the Biot number Bi, where Bi = hL/k, L being the characteristic dimension given by V/A. Thus, if Bi $\ll$ 1, the lumped mass approximation may be used. Usually, a value of around 0.1 or less for Bi is adequate for this approximation. For conduction in layers of different materials, the corresponding thermal resistances may be calculated to determine if any of these could be approximated as lumped. For instance, a thin, highly conducting layer may be treated as lumped.

For radiative transport processes, an equivalent convective heat transfer coefficient h_r can often be derived to determine the Biot number and whether the lumped mass approximation is applicable. For instance, if the radiative heat transfer between two bodies at temperatures T_1 and T_2 varies as $S\left(T_1^4 - T_2^4\right)$, where S is a constant, this may be written as $S\left[\left(T_1^2 + T_2^2\right).\left(T_1 + T_2\right)\right].\left(T_1 - T_2\right)$, which may be approximated as $4ST_{avg}^3\left(T_1 - T_2\right)$, if T_1 and T_2 are close to each other. Then, the equivalent convective heat transfer coefficient is $4ST_{avg}^3$, where T_{avg} is the average of T_1 and T_2. Similarly, other heat transfer processes may be approximated.

The lumped mass approximation is used frequently in modeling because of the considerable simplification it generates and also because it accurately represents the process in many cases. A spherical metal ball being heat treated, a well-mixed water tank for hot water storage, the hot upper layer in a room fire that is often turbulent and well mixed, and a heated electronic component in an electrical circuit are all examples where the lumped mass approximation may be applicable.

The model may be used for other thermal boundary conditions as well, for instance, a constant heat flux input q or a combined convective-radiative heat loss, giving rise to the following equations:

$$\rho CV \frac{dT}{d\tau} = qA \tag{3.10}$$

$$\rho CV \frac{dT}{d\tau} = -hA\left(T - T_a\right) - \varepsilon\sigma A\left(T^4 - T_{surr}^4\right) \tag{3.11}$$

where ε is the surface emissivity, σ is the Stefan-Boltzmann constant, and T_{surr} represents the temperature of the surrounding environment. This simple radiative transport equation applies for a gray and diffuse body surrounded by a large or black enclosure. The first equation yields a linear variation of T with time for constant q and the second equation is a nonlinear equation that may be solved analytically or numerically.

3.3.1.4 Simplification of Boundary Conditions

Most practical systems and processes involve complicated, nonuniform, and time-varying boundary conditions. However, considerable simplification can be obtained, without significant loss of accuracy or generality, by approximating the boundaries as smooth, with simpler geometry and uniform conditions, as sketched in Figure 3.11. Thus, roughness of the surface is neglected unless interest lies in scales of that size or the effect of roughness is being investigated. The geometry may be approximated in terms of simpler configurations such as flat plate, cylinder, or sphere. The human body is, for example, often approximated as a vertical cylinder for calculating the heat transfer from it. A large cylinder is itself approximated as a flat surface for convective transport if the thickness of the boundary layer δ adjacent to the surface is much smaller than the diameter D of the cylinder, i.e., $\delta/D \ll 1$. Conditions that vary over the boundaries or with time are often approximated as uniform or constant to considerably simplify the model.

Isothermal and uniform heat flux surfaces are rarely obtained in practice. However, a given temperature distribution over a boundary may be replaced by the average value if the amplitude of the variation in temperature, ΔT, is small compared to the mean T_{avg}, i.e., $\Delta T/T_{avg} \ll 1$. Similar considerations may be applied to the surface heat flux and other boundary conditions.

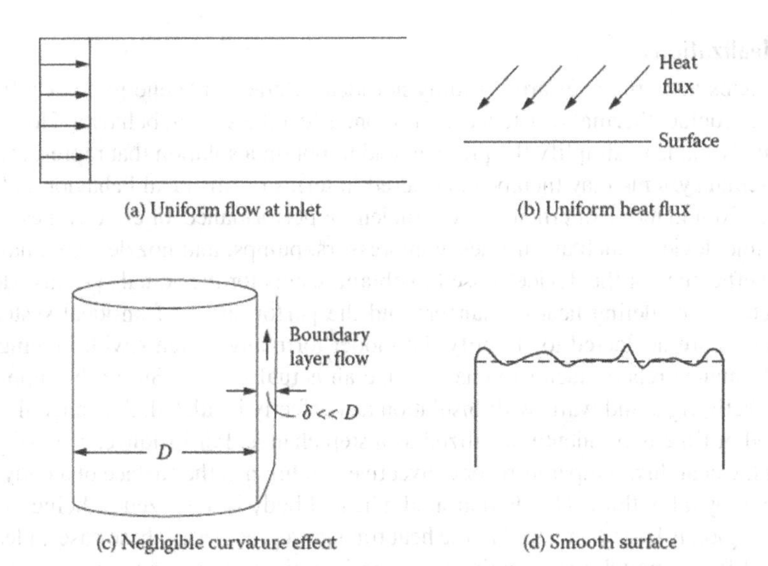

FIGURE 3.11 Several commonly used approximations. (a) Uniform flow at inlet to a channel; (b) uniform surface heat flux; (c) negligible curvature effects; and (d) negligible effect of surface roughness.

The assumption of uniform flow at the inlet to a circular tube or channel is commonly employed, while keeping the total flow rate at a specified value. The velocity distribution at the inlet is not very important for a long channel because the flow develops rapidly downstream. However, all such approximations must keep the total energy input, flow rate, etc., the same as those for the given profile. Such simplifications of the boundary conditions not only reduce the complexity of the model, but also make it easier to understand and generalize the results obtained from the model.

3.3.1.5 Negligible Effects

Major simplifications in the mathematical modeling of thermal systems are obtained by neglecting effects that are relatively small. Estimates of the relevant quantities are used to eliminate considerations that are of minor consequence. For instance, estimates of convective and radiative loss from a heated surface may be used to determine if radiation effects are important and need to be included in the model. If Q_c and Q_r are the convective and radiative heat transfer rates, respectively, these may be estimated for a surface of area A as

$$Q_c = hA(T - T_a) \quad \text{and} \quad Q_r = \varepsilon\sigma A\left(T^4 - T_{\text{surr}}^4\right) \tag{3.12}$$

where given or expected values of the surface temperature T may be employed to estimate the relative magnitudes of these transport rates. Clearly, at relatively low temperatures, the radiative heat transfer may be neglected and at high temperatures it may be the dominant mechanism. Such estimates are often based on available information from other similar processes and systems to quantify the range of variation of the relevant quantities, such as temperature in this case.

Similarly, the change in the volume of a material as it changes phase from, say, liquid to solid, may be neglected in several cases if this change is small. Changes in dimensions due to temperature variation are usually neglected, unless these changes are significant or lead to an important consideration in the problem. Potential energy effects are usually neglected, compared to the kinetic energy changes, in a gas turbine. Such approximations are well-known and extensively used in fluid mechanics, heat transfer, and thermodynamics.

3.3.1.6 Idealizations

Practical systems and processes are certainly not ideal. Undesirable energy losses, friction forces, fluid leakages, contact thermal resistance, and so on, affect the system behavior. However, idealizations are usually made to simplify the problem and to obtain a solution that represents the best performance. Actual systems may then be considered in terms of this ideal behavior and the resulting performance given in terms of efficiency, coefficient of performance, or effectiveness. For instance, thermodynamic devices, such as turbines, compressors, pumps, and nozzles, are analyzed as ideal and then the efficiency of the device is used to obtain results for the actual systems. Heat losses are often neglected in modeling heat exchangers and the performance of an ideal system is studied. Frictional losses are neglected to simplify the model for many systems with moving parts, again using a performance-related factor to characterize an actual system. Similarly, supports are often taken as perfectly rigid and walls with insulation as perfectly insulated. A change that occurs over a short period of time is frequently idealized as a step change. For instance, a step change is often assumed for the heat flux, temperature, or convective condition at the surface of a body being heated in a furnace or by a hot fluid. The fluid around a heated body is idealized as being extensive if the extent of the region is large compared to the heat transfer region. In all these cases, idealizations are made to simplify the model, focus on the main considerations, and avoid aspects that are often difficult to characterize such as frictional effects, leakages, and contact resistance. Figure 3.12 shows the schematics of some of the idealizations used in mathematical modeling.

3.3.1.7 Material Properties

For a satisfactory mathematical modeling of any thermal system or process, it is important to employ accurate material property data. The properties are usually dependent on physical variables such as temperature, pressure, and species concentration. In polymeric materials such as plastics, the viscosity of the fluid also depends on the shear rate and thus on the flow field. Even though the properties vary with temperature and other variables, they can be taken as constant if the change in the property, say thermal conductivity k, is small compared to the average value k_{avg}, i.e., $\Delta k/k_{avg} \ll 1$. Here, the change in the property is evaluated over the anticipated range of variables that affect the property.

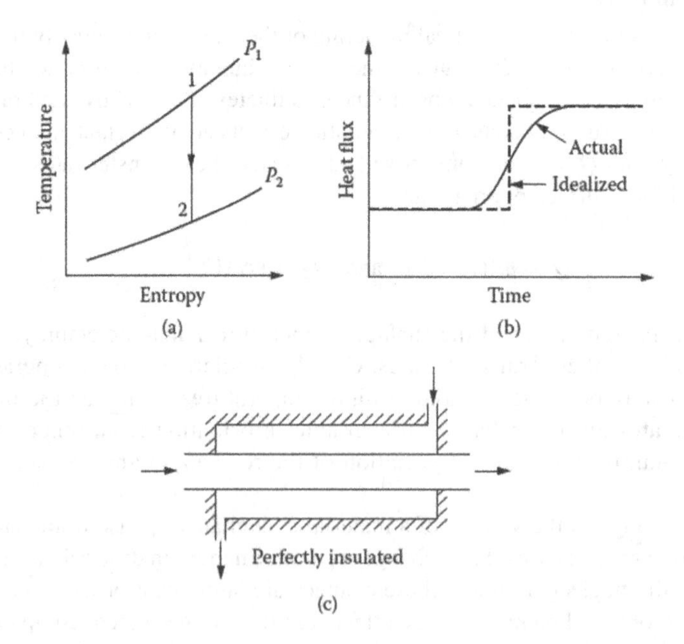

FIGURE 3.12 Idealizations used in mathematical modeling. (a) Ideal turbine behavior; (b) step change in heat flux; and (c) perfectly insulated outer surface of a heat exchanger.

However, in many practical circumstances the constant property approximation cannot be made because of large changes in these variables. In such cases, curve fitting is often used to represent the variation of the relevant properties. For instance, the variation of the thermal conductivity with temperature may be represented by a function $k(T)$, where

$$k(T) = k_o[1 + a(T - T_o) + b(T - T_o)^2] \tag{3.13}$$

Here, k_o is the thermal conductivity at a reference temperature T_o and a and b are constants obtained from the curve fitting of the experimental data on this property at different temperatures. Higher order polynomials and other algebraic functions may also be used to represent the property data. Similarly, curve fitting may be used for other properties such as density, specific heat, and viscosity. Such equations are very valuable in the mathematical modeling of thermal systems and in numerical modeling and simulation.

3.3.1.8 Conservation Laws

The conservation laws for mass, momentum, and energy form the basis for deriving the fundamental equations for thermal systems and processes. The equations are simplified by using the various considerations given in the preceding discussion. The resulting equations may be algebraic, differential, or integral.

Algebraic equations arise mainly from curve fitting, such as Equation (3.13), and also apply for steady-state, lumped systems. As mentioned earlier, thermodynamic systems are often approximated as steady and lumped (Howell and Buckius, 1992; Cengel and Boles, 2014), resulting in algebraic governing equations. In some cases, overall or global balances also lead to algebraic equations. For instance, the energy balance at a furnace wall, under steady-state conditions, yields the equation

$$\varepsilon\sigma\left(T_h^4 - T^4\right) = h(T - T_a) + \frac{k}{d}\left(T - T_s\right) \tag{3.14}$$

from a balance between the radiative heat transfer to the inner surface of the wall and the convective and conductive heat losses from the surface. Here, T_h is the temperature of the heater radiating to the inner surface at temperature T, T_a is the temperature of air adjacent to the inner surface, and T_s is the outer surface temperature of the wall. The temperature T at the wall may then be obtained by solving Equation (3.14), which is a nonlinear equation and will generally require iterative methods. For systems of algebraic equations as well as for a single nonlinear equation, numerical methods are generally needed to obtain the solution (Jaluria, 2012).

Differential approaches are the most frequently employed conservation formulation because they apply locally, allowing the determination of variations in time and space. Ordinary differential equations arise in a few idealized situations for which only one independent variable is considered. Therefore, if the lumped mass assumption can be applied and transient effects are important, Equation (3.7), Equation (3.10), and Equation (3.11) would be the relevant energy equations. If several lumped mass systems are considered as constituents of a thermal system, a set of simultaneous ordinary differential equations arises. For instance, the temperatures T_1, T_2, T_3, …, T_n of n components of a given system are represented by a system of equations of the form

$$\begin{aligned}
\frac{dT_1}{d\tau} &= F_1\left(T_1, T_2, T_3, \ldots, T_n, \tau\right) \\
\frac{dT_2}{d\tau} &= F_2\left(T_1, T_2, T_3, \ldots, T_n, \tau\right) \\
&\vdots \\
\frac{dT_n}{d\tau} &= F_n\left(T_1, T_2, T_3, \ldots, T_n, \tau\right)
\end{aligned} \tag{3.15}$$

where the F variables are functions of time and the temperatures and thus couple the equations. These equations can be solved numerically to yield the temperatures of the various components as functions of time τ (see Example 2.6).

Partial differential equations are obtained for distributed models. Thus, Equation (3.4) is the applicable energy equation for three-dimensional, steady conduction in a material with constant properties. Similarly, one-dimensional transient conduction in a wall, which is large in the other two dimensions, is given by the equation

$$\rho C\frac{\partial T}{\partial \tau} = \frac{\partial}{\partial x}\left(k\frac{\partial T}{\partial x}\right) \quad (3.16)$$

if the material properties are taken as variable. For constant properties, the equation becomes

$$\frac{\partial T}{\partial \tau} = \alpha\frac{\partial^2 T}{\partial x^2} \quad (3.17)$$

where $\alpha = k/\rho C$ is the thermal diffusivity. Similarly, equations for two- and three-dimensional cases may be written.

For convective transport, the flow field affects the thermal transport and the energy equation is obtained for a two-dimensional, constant-property, transient problem, with negligible viscous dissipation and pressure work, as

$$\rho C_p\left(\frac{\partial T}{\partial \tau} + u\frac{\partial T}{\partial x} + v\frac{\partial T}{\partial y}\right) = k\left(\frac{\partial^2 T}{\partial x^2} + \frac{\partial^2 T}{\partial y^2}\right) \quad (3.18)$$

where C_p is the specific heat of the fluid at constant pressure and u and v are the velocity components in the x and y directions, respectively. Convective transport equations are derived, presented, and discussed in most books on heat transfer, particularly those on convective heat transfer (Burmeister, 1993). Partial differential equations that describe most practical thermal systems are amenable to a solution by analytical methods in very few cases and numerical methods are generally necessary. Finite-difference and finite-element methods are the most commonly employed techniques for partial differential equations. Ordinary differential equations can often be solved analytically, particularly if the equation is linear.

The integral formulation is based on an integral statement of the conservation laws and may be applied to a small finite region, from which the finite-element and finite-volume methods are derived, or to the entire domain. For instance, conduction in a given region is governed by the integral equation

$$\rho C_p\frac{\partial}{\partial \tau}\int_V T dV = \int_S k\frac{\partial T}{\partial n} dS + \int_V q''' dV \quad (3.19)$$

where V is the volume of the region, A is its surface area, q''' is an energy source per unit volume in the region, and n is the outward drawn normal to the surface. This integral equation states that the rate of net energy generated in the region plus the rate of net heat conducted into the region at the surfaces equals the rate of increase in stored thermal energy in the region. Similar equations may be derived for convection in a given domain. Radiative transport often leads to integral equations because energy is absorbed over the volume of a participating fluid or material. In addition, the total radiative transport, in general, involves integrals over the area, wavelength, and solid angle. Figure 3.13 shows a few examples of integral and differential formulations for the mathematical modeling of thermal systems. More is said about the preceding equations as we consider different thermal systems in later chapters.

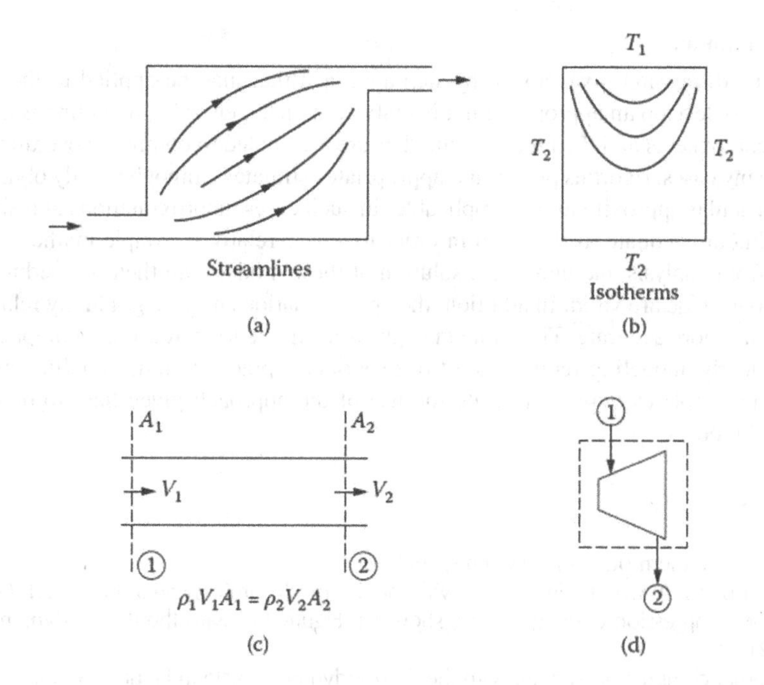

FIGURE 3.13 Differential formulations. (a) Flow in an enclosed region due to inflow and outflow of the fluid; and (b) temperature distribution due to conduction in a solid body. Also shown are a few integral formulations: (c) flow in a pipe and (d) flow through a turbine.

3.3.1.9 Further Simplification of the Basic Equations

After the basic equations from conservation principles are assembled, along with the relevant boundary conditions, employing the various approximations and idealizations outlined here, further simplification can sometimes be obtained by a consideration of the various terms in the equations to determine if any of them can be neglected. This is generally based on a nondimensionalization of the equations and evaluation of the important parameters, as given earlier in Equation (3.6).

For instance, the cooling of an infinite heated rod moving continuously at speed U along the axial direction x [Figure 1.10(d) and Figure 3.6(a)] is given by the dimensionless equation

$$\frac{\partial \theta}{\partial \tau'} + \text{Pe}\frac{\partial \theta}{\partial X} = \nabla^2 \theta \tag{3.20}$$

where ∇^2 is the Laplacian operator in cylindrical coordinates, similar to $\frac{\partial^2}{\partial x^2} + \frac{\partial^2}{\partial y^2}$ in a two-dimensional Cartesian system that would apply for a moving flat plate that is wide in the third direction z, see the right hand side of Equation (1.8). The energy equation is nondimensionalized by the rod diameter D, and the *Peclet number* Pe is given by $\text{Pe} = UD/\alpha$. The dimensionless temperature θ is defined as $\theta = T/T_{ref}$, where the reference temperature T_{ref} may be the temperature at $x = 0$. Also, dimensionless time τ' is defined here as $\tau' = \alpha\tau/D^2$, where τ is physical time. If Pe is very small, $\text{Pe} \ll 1$, the second term from the left, which represents convective transport due to rod movement, may be neglected, reducing the given circumstance to a simple conduction problem.

Similarly, at low Reynolds number Re in a flow, $\text{Re} \ll 1$, where $\text{Re} = UL/\nu$, ν being the kinematic viscosity and L a characteristic dimension, the inertia or convection terms can be neglected in the momentum equation. This is the creeping flow approximation, which is used in film lubrication modeling. Several such approximations are well-known and frequently employed in modeling, as discussed in standard textbooks on heat transfer, fluid mechanics, thermodynamics, and mechanics.

3.3.1.10 Summary

The preceding discussion gives a step-by-step approach that may be applied to the system or its parts in order to develop an appropriate mathematical model. Generally, modeling is first applied to the various components and then these submodels are assembled to obtain the overall model for the system. In many cases, rigorous proofs and appropriate estimates cannot be easily obtained to determine if a particular approximation is applicable. In such cases, approximations and simplifications are made without adequate justification in order to derive relatively simple mathematical models. The results from analysis and numerical solution of these models can then be used to verify if the approximations made are valid. In addition, the approximations may be gradually relaxed to obtain models that are more accurate. Thus, one may go from simple to increasingly complicated models, if needed. Clearly, modeling requires a lot of experience, practice, understanding, and creativity. The following simple examples illustrate the use of the approach given here to develop suitable mathematical models.

Example 3.1

Consider typical thermodynamic systems, such as

A power plant, shown in Figure 2.17, with the thermodynamic cycle in Figure 2.15(a).

A vapor compression cooling system, shown in Figure 1.8, with the thermodynamic cycle in Figure 2.21.

An internal combustion engine, with the thermodynamic cycle in Figure 2.15(b).

Discuss the development of simple mathematical models for these in order to calculate the energy transport rates and the overall performance.

SOLUTION

In all of these commonly used systems, as well as in many others like them, the major focus is on the heat input or removal rate and on the work done. Many of the details, such as the temperature and velocity distributions in the various components, are not critical. Similarly, though the transients are important in controlling the system as well as at startup and shutdown, the system performance under steady operation is of particular interest for system analysis and design.

Keeping the preceding considerations in mind, the two main assumptions that can be made for each component are:

Steady-state conditions
Lumped flow and temperature

This implies that time dependence is neglected and uniform conditions are assumed to exist within each component. Energy loss to or gain from the environment may be neglected for idealized conditions, which will yield the best possible performance and can thus be used for calculating the efficiency of actual systems.

Then, considering the vapor compression system of Figure 2.21, we obtain for a mass flow rate of $\dot{m}$

Heat rejected at the condenser = $\dot{m}(h_2 - h_3)$
Heat removed at the evaporator = $\dot{m}(h_1 - h_4)$
Work done on the compressor = $\dot{m}(h_2 - h_1)$

These expressions yield the coefficient of performance (COP), given in Equation (2.19), as $(h_1 - h_4)/(h_2 - h_1)$.

Similarly, for the power plant given by the cycle in Figure 2.15(a), the heat input in the boiler or condenser is $\dot{m}(h_{out} - h_{in})$ and work done by the turbine or the pump is $\dot{m}(h_{in} - h_{out})$, where *in* and *out* refer to conditions at the inlet and outlet of the component. Thus, boiler heat input is positive, condenser heat input is negative (heat rejected), work done by the turbine is positive, and work done by the pump is negative (work done on the pump).

The internal combustion engine and other thermodynamic systems may be similarly analyzed to yield the net heat input and work done, allowing subsequent design and optimization of the system. The design considerations are discussed in Chapter 5.

Example 3.2

For common heat exchangers, such as the parallel and counterflow heat exchangers shown in Figure 1.5, discuss the development of a simple mathematical model to analyze the system.

SOLUTION

In heat exchangers, the main physical aspect of interest is the overall heat transfer between the two fluids. Though the flow rates and inlet/outlet temperatures are of interest, the velocity and temperature distributions at various cross-sections of the heat exchanger are generally of little interest. Similarly, transient aspects, although important in some cases, are usually not critical. Thus, energy transfer under steady flow, as a function of the operating conditions and the heat exchanger design, is generally needed. With this in mind as the major consideration, we can assume the following:

The flow is lumped across the cross-sections of the channels or tubes.
The temperature is also uniform across these cross-sections.
Steady-state conditions exist.
The conduction along the axis is neglected.

With these assumptions, the temperature in, say, the inner tube or channel of the heat exchangers in Figures 1.5(a) and (b) varies only with distance in the axial direction. The overall energy balance is

$$\dot{m}_c C_{p,c}\left(T_{c,out} - T_{c,in}\right) = Q$$

where Q is the rate of heat input to the colder fluid, indicated by the subscript *c*, over the entire length. If energy loss to the ambient is neglected, we have for the hotter fluid, which is indicated by subscript *h*,

$$\dot{m}_h C_{p,h}\left(T_{h,in} - T_{h,out}\right) = Q$$

In addition, the total heat transfer *Q* may be written as

$$Q = q\,A = q(\pi DL) = h\,A(T_h - T_c)$$

where *q* is the heat flux per unit area due to the difference in the bulk temperatures, T_h and T_c, respectively, of the hot and cold fluids, *A* is the contact area, being πDL for a tube of diameter *D* and length *L*, and *h* is the convective heat transfer coefficient. Further details on the analysis and design of such heat exchangers are discussed in Chapter 5.

Example 3.3

In the design of a hot water storage system, it is given that a steady flow of hot water at 75°C and a mass flow rate $\dot{m}$ of 113.1 kg/h enters a long circular pipe of diameter 2 cm, with convective heat loss at the outer surface of the pipe to the ambient medium at 15°C with a heat transfer coefficient *h* of 100 W/m²K. The density ρ, specific heat at constant pressure C_p, and thermal conductivity *k* of water are given as 10^3 kg/m³, 4200 J/kgK, and 0.6 W/mK, respectively. Develop

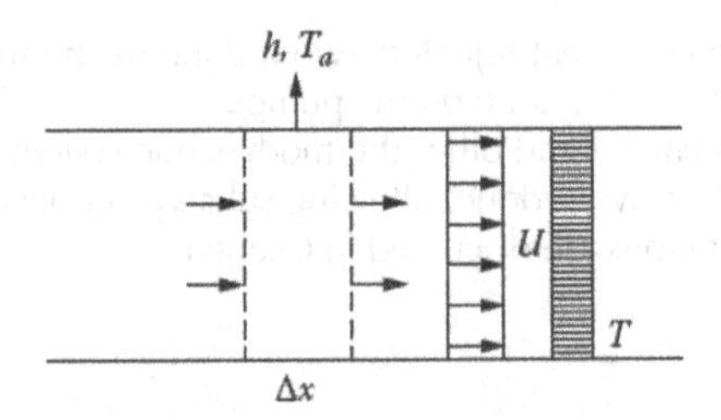

FIGURE 3.14 Assumption of uniform flow and temperature across the pipe cross-section in Example 3.3.

a simple mathematical model for this process and calculate the water temperature after the flow has traversed 10 m of pipe.

SOLUTION

The problem can be simplified considerably by assuming steady-state conditions and lumped velocity and temperature conditions across any cross-section of the pipe. This approximation applies for turbulent flow in a pipe of relatively small diameter. In addition, interest lies in the average temperature at any given x, where $x = 0$ is the inlet and x is the distance along the pipe, as shown in Figure 3.14. The average velocity U in the flow is

$$U = \frac{\dot{m}}{\rho\left(\pi D^2/4\right)3600} = 0.1\ m/s$$

where D is the pipe diameter. The Reynolds number $\mathrm{Re} = UD/\nu = (0.1)(0.02)/(5.5 \times 10^{-7}) = 3636$. Turbulent flow arises in the pipe at this high value of the Reynolds number. The Peclet number $\mathrm{Pe} = UD/\alpha = (0.1)(0.02)/(1.5 \times 10^{-7}) = 1.3 \times 10^4$. Therefore, convection dominates and axial diffusion effects may be neglected; see Equation (3.20).

With the above approximations, the governing equation for the temperature $T(x)$ is obtained from energy balance over a region of length Δx, as shown in Figure 3.14. The reduction in thermal energy transported in the pipe equals the convective loss to the ambient. This gives the decrease in temperature ΔT over an axial distance Δx as

$$\rho C_p U A \Delta T = -hP\,\Delta x (T - T_a)$$

Therefore, with $\Delta x \to 0$, we obtain the differential equation

$$\rho C_p U A \frac{dT}{dx} - hP(T - T_a)$$

where A is the cross-sectional area ($\pi D^2/4$) and P is the perimeter (πD). This gives the simple mathematical model for this problem. The inlet temperature is given as 75°C and the ambient temperature $T_a = 15$°C. This equation may be solved analytically to give

$$\theta = \theta_o \exp\left(-\frac{hP}{\rho C_p U A} x\right)$$

where $\theta = T - T_a$ and θ_o is the temperature difference at the inlet, i.e., 60°C. Therefore, at $x = 10$ m, we have

$$\theta = 60 \exp(-0.0476x) = 60 \exp(-0.476) = 37.276$$

Therefore, the temperature at 10 m is 15 + 37.276 = 52.276°C. Clearly, the temperature drops very slowly due to the high mass flow rate and relatively small heat loss rate. This is a simple model and is easy to solve. Models very similar to this one are frequently used for analysis of flow and heat transfer in pipes and channels, for example, in the design of residential heating and cooling systems and of car radiators.

The preceding three examples present relatively simple models of some commonly encountered thermal systems. These included thermodynamic systems like refrigeration and air conditioning systems and flows through channels as in heat exchangers. Steady-state conditions could be assumed in these cases, along with lumping to further simplify the models. The resulting models yielded algebraic equations and first-order ordinary differential equations, which could be easily solved analytically to yield the desired results. However, many practical thermal systems are more involved than these and spatial and temporal variations have to be considered. Then the resulting equations are partial differential equations, which generally require numerical methods for the solution. In a few cases, the partial differential equations can be simplified or idealized to obtain ordinary differential equations, which may again be solved analytically. The following two examples illustrate such problems that would generally need numerical methods for the solution and that may be idealized to obtain analytical results in some cases for validation of the numerical scheme.

Example 3.4

A large cylindrical gas furnace, 3 m in diameter and 5 m in height, is being simulated for design and optimization. Its outer wall is made of refractory material, covered on the outside with insulation, as shown in Figure 3.15. The wall is 20 cm thick and the insulation is 10 cm thick. The variations of

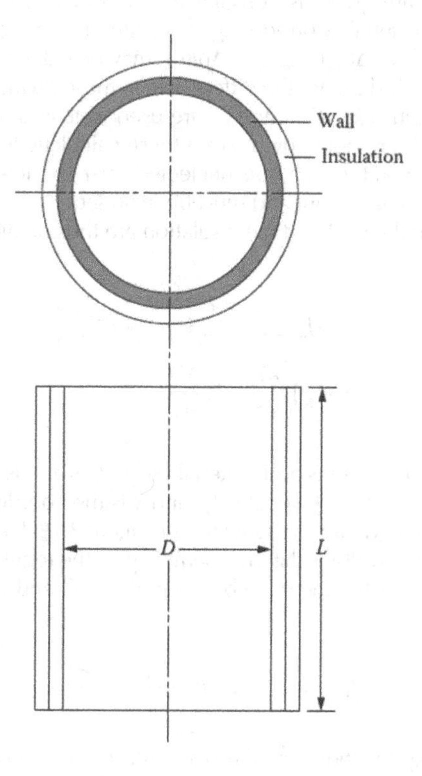

FIGURE 3.15 The cylindrical furnace, with the wall and insulation, considered in Example 3.4.

the thermal conductivity k, specific heat at constant pressure C_p, and density ρ of the wall material with temperature are represented by best fits to experimental data on properties as

$$k = 2.2\,(1+1.5\times10^{-3}\times\Delta T)$$
$$C_p = 900\,(1+10^{-4}\times\Delta T)$$
$$r = 2500\,(1+6\times10^{-5}\times\Delta T)$$

where ΔT is the temperature difference from a reference temperature of 300 K and all the values are in S.I. units. The temperature difference across the wall is not expected to exceed 200 K. The properties of the insulation may be taken as constant. Develop a mathematical model for the time-dependent temperature distribution in the wall and in the insulation. Solve the energy equations for the temperature distribution in the idealized steady-state circumstance, with the thermal conductivity of the insulation given as 1.0 W/mK, temperature $(T_w)_1$ at the inner surface of the wall as 500 K, and temperature $(T_i)_2$ at the outer surface of the insulation as 300 K.

SOLUTION

The ratio of the wall thickness to the furnace diameter is 0.2/3.0, which gives 0.067. Similarly, the ratio of the insulation thickness to the furnace diameter is 0.1/3, or 0.033. Because both of these ratios are much less than 1.0, the curvature effects can be neglected, i.e., the wall and insulation may be treated as flat surfaces.

The ratio of the furnace height to the wall thickness is 5.0/0.2, or 25, and that to the insulation thickness is 50. In addition, the circumference is much larger than these thicknesses. If there is good circulation of gases in the furnace, the thermal conditions on the inner surface of the wall can be assumed uniform. Then, the wall, as well as the insulation, may be modeled as one-dimensional, with transient diffusion occurring across the thickness and uniform conditions in the other two directions.

The material properties are given as constant for the insulation. However, these vary with temperature for the wall material. Considering a maximum temperature difference of 200 K across the wall, the ratios $\Delta k/k_o$, $\Delta C_p/(C_p)_o$, and $\Delta\rho/(\rho)_o$ may be calculated as 0.3, 0.02, and 0.012, respectively, where Δk, ΔC_p, and $\Delta\rho$ are the differences in these quantities due to the temperature difference. The reference values k_o, $(C_p)_o$, and ρ_o are used instead of the average values because the actual temperature levels are not known. From these calculations, it is evident that the variations of C_p and ρ with temperature may be neglected over the temperature range of interest. However, the variation of k is important and must be included.

The energy equations for the wall and the insulation are thus obtained as, respectively,

$$(\rho C_p)_w \frac{\partial T_w}{\partial \tau} = \frac{\partial}{\partial x}\left[k_w(T_w)\frac{\partial T_w}{\partial x}\right]$$
$$(\rho C_p)_i \frac{\partial T_i}{\partial \tau} = k_i \frac{\partial^2 T_i}{\partial x^2}$$

where the corresponding temperatures and material properties are used, denoted by subscripts w and i for the wall and the insulation, respectively, and x is the coordinate distance normal to the surface, i.e., in the radial direction for the furnace; see Figure 3.16. Heat transfer conditions at the inner and outer surfaces of the wall-insulation assembly give the required boundary conditions for these equations. In addition, at the interface between the wall and the insulation, the following conditions apply

$$T_w = T_i \quad \text{and} \quad k_w \frac{\partial T_w}{\partial x} = k_i \frac{\partial T_i}{\partial x}$$

Therefore, the governing equations for the wall and the insulation may be solved, with the appropriate boundary conditions, to yield the time-dependent temperature distributions in these

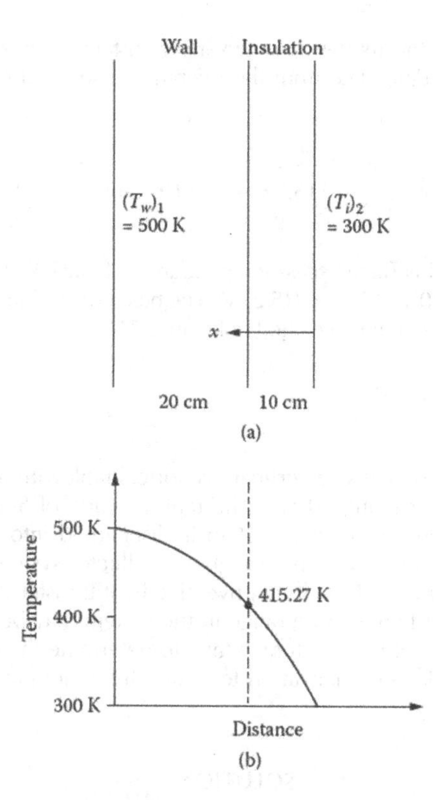

FIGURE 3.16 Boundary conditions and analytical solution obtained for the steady-state circumstance in Example 3.4.

two parts of the thermal system. Because of the variation of k_w with temperature, the two partial differential equations, which are coupled through the boundary conditions, are nonlinear. Therefore, numerical modeling will generally be needed to solve these equations.

The simpler steady-state problem, with temperatures specified at the inner and outer surfaces of the wall-insulation combination, is an idealized circumstance and may be solved analytically. The equations that apply in the wall and the insulation for this case are

$$\frac{d}{dx}\left(k_w \frac{dT_w}{dx}\right) = 0 \quad \text{and} \quad k_i \frac{d^2T_i}{dx^2} = 0$$

These equations may be solved analytically to yield

$$2.2(1+0.0015T_w)\frac{dT_w}{dx} = C_1 \quad \text{and} \quad T_i = C_2x + C_3$$

or

$$2.2T_w + 0.0033\frac{T_w^2}{2} = C_1x + C_4$$

All the temperatures may be taken as differences from the reference value of 300 K to simplify the analysis. The C's are constants to be determined from the boundary conditions shown in Figure 3.16. At the interface, the heat flux and the temperature in the two regions match, as given previously. The temperature distribution in the insulation is linear, with 0 K temperature difference from the reference value at the outer surface. The temperature distribution in the wall is nonlinear, with 200 K difference from the reference at the inner surface. The temperature distributions, in terms of the temperature difference from the reference value and denoted by overbars, are obtained as

$$2.2\bar{T}_w + 0.0033\frac{\bar{T}_w^2}{2} = 1152.7x + 160.19 \quad \text{and} \quad \bar{T}_i = 1152.7x$$

These equations give the interface excess temperature as 115.27 K. Therefore, the actual temperature at the interface is 300 + 115.27 = 415.27 K. The heat flux is obtained as 1152.7 W/m². The calculated temperature distribution is sketched in Figure 3.16.

Example 3.5

A hot water storage system consists of a vertical cylindrical tank with its height L to diameter D ratio given as 8, the diameter being 40 cm. The tank is made of 5-mm-thick stainless steel. Hot water from a solar energy collection system is discharged into the tank at the top and withdrawn at the bottom for recirculation through the collector system. The tank loses energy to the ambient air at temperature T_a with a convective heat transfer coefficient h at the outer surface of the tank wall. The temperature range in the system may be taken as 20°C to 90°C. Develop a mathematical model for the storage tank to determine the temperature distribution in the water. Also use nondimensionalization to obtain the main parameters. Then solve the steady-state problem.

SOLUTION

The temperature range being relatively small, the variation in material properties may be taken as negligible because parameters such as $\Delta\rho/\rho_{avg}$, $\Delta k/k_{avg}$, etc., where ρ is the density and k is the thermal conductivity, are much less than 1.0. Because of the thinness of the stainless steel wall and its high thermal conductivity compared to water, the ratio being 23.59, the energy storage and temperature drop in the wall may be neglected compared to those in water. This is justified from the ratio of the wall thickness, 5 mm, to the tank diameter, 40 cm.

A substantial simplification of the problem is obtained by assuming that the temperature distribution across any horizontal cross-section in the tank is uniform. This is based on axisymmetry, which reduces the original three-dimensional problem to a two-dimensional one and the effect of buoyancy forces that tend to make the temperature distribution horizontally uniform. Because hot water is discharged at the top, the water in the tank is stably stratified, with lighter warmer fluid lying above colder denser fluid. This curbs recirculating flow in the tank and promotes horizontal temperature uniformity. Therefore, the temperature T in the water is taken as a function only of the vertical location z, i.e., $T(z)$. The vertical velocity in the tank is also taken as uniform across each cross-section, by employing the average value. This is obviously an approximation because the velocity at the walls is zero due to the no-slip condition. However, the problem is substantially simplified because the flow field is taken as a uniform vertical downward velocity, which can easily be obtained from the flow rate. Without this simplification, the convective flow has to be determined, making the problem far more involved.

The energy equation for thermal transport in the water tank may be written with the above simplifications as

$$\rho C_p A\left[\frac{\partial T}{\partial \tau} + w\frac{\partial T}{\partial z}\right] = kA\frac{\partial^2 T}{\partial z^2} - hP(T - T_a)$$

where ρ is the density of the fluid, C_p is its specific heat at constant pressure, τ is the physical time, w is the average vertical velocity in the tank, k is the fluid thermal conductivity, A is the

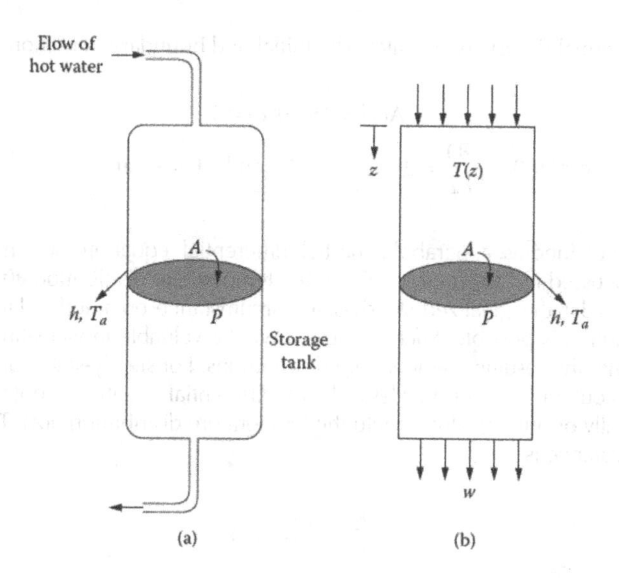

FIGURE 3.17 The hot water storage system considered in Example 3.5, along with the simplified model obtained.

cross-sectional area, and P is the perimeter of the tank; see Figure 3.17. The problem is treated as transient because the time-dependent behavior is generally important in such energy storage systems. The initial and boundary conditions may be taken as

$$\text{At } \tau = 0\text{: } T(z) = T_a$$

$$\text{For } \tau > 0\text{: } \quad \frac{\partial T}{\partial z} = 0 \text{ at } z = L \quad \text{and} \quad T = T_o \text{ at } z = 0$$

where T_o is the discharge temperature of hot water. Therefore, a one-dimensional, transient, mathematical model is obtained for the hot water storage system. The various assumptions made, particularly that of uniformity across each cross-section, may be relaxed later on for more accurate simulation. However, under the given conditions, this model is adequate for simulation and design of practical hot water storage systems.

The energy equation and the boundary conditions may be nondimensionalized by defining the dimensionless temperature θ, time τ', and vertical distance Z as

$$\theta = \frac{T - T_a}{T_o - T_a} \quad \tau' = \frac{\alpha\tau}{L^2} \quad Z = \frac{z}{L}$$

Then the dimensionless energy equation is obtained as

$$\frac{\partial \theta}{\partial \tau'} + W\frac{\partial \theta}{\partial Z} = \frac{\partial^2 \theta}{\partial Z^2} - H\theta$$

where the two dimensionless parameters W and H are

$$W = \frac{wL}{\alpha} \quad H = \frac{hPL^2}{Ak}$$

Here α is the thermal diffusivity of water. The initial and boundary conditions become

$$\text{At}\,\tau' = 0: \quad \theta(Z) = 0$$

$$\text{For}\,\tau' > 0: \quad \frac{\partial \theta}{\partial Z} = 0 \quad \text{at} \quad Z = 1 \quad \text{and} \quad \theta = 1 \quad \text{at} \quad Z = 0$$

The equation obtained is a parabolic partial differential equation, which may be solved numerically, as discussed in Chapter 4, to obtain the temperature distribution $\theta(\tau', Z)$.

Let us now consider the idealized steady-state circumstance obtained at large time and see if an analytical solution is possible. Such a solution will be valuable in validating the model and providing insight into the resulting temperature distributions. For steady-state conditions, the time dependence drops out and a second-order ordinary differential equation is obtained, which may be solved analytically or numerically to yield the temperature distribution $\theta(Z)$. The energy equation for this circumstance is

$$W\frac{d\theta}{dz} = \frac{d^2\theta}{dz^2} - H\theta$$

with the boundary conditions

$$\theta = 1 \quad \text{at} \quad Z = 0 \qquad \frac{d\theta}{dZ} = 0 \quad \text{at} \quad Z = 1$$

This second-order ordinary differential equation may be solved analytically to yield the solution

$$\theta = a_1 \exp(\alpha_1 Z) + a_2 \exp(\alpha_2 Z)$$

Here, a_1, a_2, α_1 and α_2 are constants, given by the expressions

$$a_1 = \frac{\alpha_2 \exp(\alpha_2)}{\alpha_1 \exp(\alpha_1) - \alpha_2 \exp(\alpha_2)} \qquad a_2 = \frac{\alpha_1 \exp(\alpha_1)}{\alpha_1 \exp(\alpha_1) - \alpha_2 \exp(\alpha_2)}$$

with

$$\alpha_1 = \frac{W + \sqrt{W^2 + 4H}}{2} \qquad \alpha_2 = \frac{W - \sqrt{W^2 + 4H}}{2}$$

If $W = 0$, $\alpha_1 = \sqrt{H}$ and $\alpha_2 = -\sqrt{H}$. If these values are substituted in the preceding expressions, the standard solution for conduction in a fin with an adiabatic end is obtained (Incropera and Dewitt, 2001). The solution for this case is

$$\theta = \frac{\cosh\left[\sqrt{H}(1 - Z)\right]}{\cosh\left[\sqrt{H}\right]}$$

Therefore, analytical solutions may be obtained in a few idealized circumstances. Numerical solutions are needed for most realistic and practical situations. However, these analytical results may be used for validating the numerical model. The preceding examples show some simple, as well as a few relatively complicated, thermal systems. The corresponding mathematical models are obtained and some characteristic results are given. Similar approaches are used for much more complex practical systems that may have many more components and subsystems. A few such practical systems are considered later in the book.

3.3.2 Final Model and Validation

The mathematical modeling of a thermal system generally involves modeling of the various components and subsystems that constitute the system, followed by a coupling of all these models to obtain the final, complete model for the system. The general procedure outlined in the preceding may be applied to a component and the applicable equations derived based on various simplifications, approximations, and idealizations that may be appropriate for the circumstance under consideration. These equations may be a combination of algebraic, differential, and integral equations. The differential equations may themselves be ordinary or partial differential equations. Though differential equations are the most common outcome of mathematical modeling, algebraic equations are obtained for lumped, steady-state systems and from curve fitting of experimental or material property data. In the mathematical model of the overall system, one component may be modeled as lumped mass, another as one-dimensional transient, and still another as three-dimensional. Thus, different levels of simplifications arise in different circumstances.

As an example, let us consider the furnace shown in Figure 3.18. The walls, insulation, heaters, inert gas environment, and material undergoing thermal processing may all be considered as components or constituents of the furnace. The general procedure for mathematical modeling may be applied to each of these components, using physical insight and estimates of the various transport mechanisms. Such an approach may indicate that the wall and the insulation can be modeled as one-dimensional, the gases as fully mixed, and the heaters and the solid body undergoing heat treatment as lumped, with all the temperatures being time-dependent. Similarly, the importance of variable properties and of radiation versus convection in the interior of the furnace may be evaluated.

Once the final model of the thermal system is obtained, we proceed to obtain the solution of the mathematical equations and to study the behavior and characteristics of the system. This is the process of simulation of the system under a variety of operating and design conditions. However, before proceeding to simulation, we must validate the mathematical model and, if needed, improve it in order to represent the physical system more closely. Several approaches may be applied for validating the mathematical model and for determining if it provides an accurate representation of the given thermal system. Three commonly employed strategies for validation are:

1. *Physical behavior of the system.* In this approach, the operating, ambient, and other conditions are varied and the effect on the system is investigated. It is ascertained that the behavior is physically reasonable. For instance, if the energy input to the heater in the furnace of

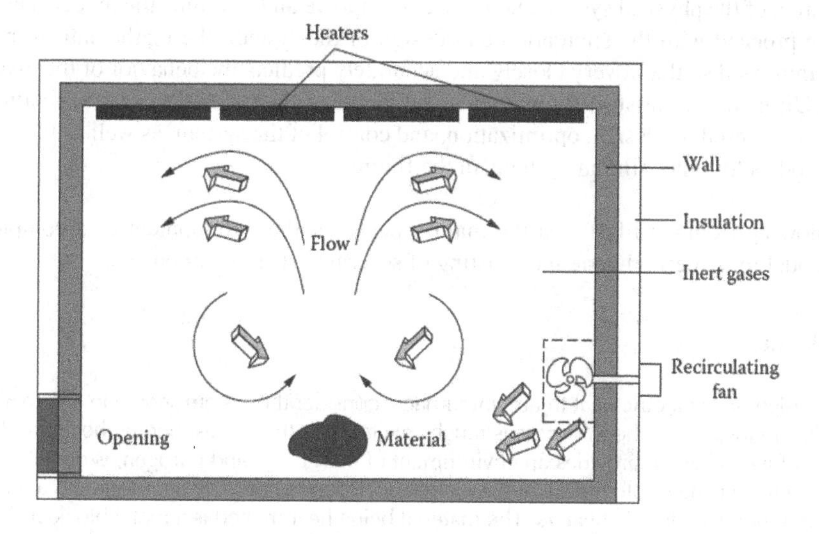

FIGURE 3.18 An electric furnace for heat treatment of materials.

Figure 3.18 is increased, the temperature levels are expected to rise. Similarly, if the wall or insulation thickness is increased, the temperatures within the furnace must increase due to reduced energy losses. An increase in the convective cooling at the outer surface of the furnace should lower the temperatures. Thus, the results from the solution of the equations that constitute the mathematical model must indicate these trends if the model is a satisfactory representation of the system.

2. *Comparison with results for simpler systems.* Usually experimental or numerical results are not available for the system under consideration. However, the mathematical model may be applied to simpler systems that may be studied analytically or experimentally to provide the relevant data for comparison. For instance, the model for a solar energy collection system may be applied to a simpler, scaled-down version, which could then be fabricated for experimentation. Usually the geometrical complexities of a given system are avoided to obtain a simpler version for validation. A fluid or material, whose characteristics are well-known, may be substituted for the actual one. For instance, a viscous Newtonian fluid such as corn syrup may be substituted for a more complicated non-Newtonian plastic material, whose viscosity varies with the shear rate in the flow, in order to simplify the model for validation. A fewer number of components of the system may also be considered for a simpler arrangement. The experimental study is then directed at validation and specific, well-controlled experiments are carried out to obtain the required data. However, it must be borne in mind that such a simpler system or fluid may lack some of the important characteristics of the actual system, thus limiting the value of such a validation.
3. *Comparison with data from full-scale systems.* Whenever possible, comparisons between the results from the model and experimental data from full-scale systems are made in order to determine the validity and accuracy of the model. The system available may be an older version that is being improved through design and optimization or it may be a system similar to the one under consideration. In addition, a prototype of the given system is often developed before going into production and this can be effectively used for validation of the model. Generally, older versions and similar systems are used first and the prototype is used at the final stage to ensure that the model is valid and accurate. In addition to the validation approaches given above, it must be remembered that the mathematical model is closely coupled with the numerical scheme, the system simulation, and the design evaluation and optimization. Therefore, the model provides important inputs for the subsequent processes and obtains feedback from them. This feedback indicates the accuracy of representation of the physical system and is used to improve and fine-tune the model. Therefore, as we proceed with the simulation and design of the system, the mathematical model is also improved so that it very closely and accurately predicts the behavior of the given system. Ultimately, a satisfactory mathematical model of the thermal system is obtained and this can be used for design, optimization, and control of the system, as well as for developing models for other similar systems in the future.

The following example illustrates the main aspects for the development of a complete mathematical model for a thermal system consisting of several parts or components.

Example 3.6

For the design of an electric heat treatment furnace, consider the system shown in Figure 3.18. For the walls and insulation, the thickness is much smaller than the corresponding height and width. The flow of gases, which provides an environment of inert gases and nitrogen, is driven by buoyancy and a fan, giving rise to turbulent flow in the enclosed region. The heat source is a thin metal strip with imbedded electric heaters. The material being heat-treated is a metal block and is small compared to the dimensions of the furnace. This thermal system is initially at room temperature T_r

and the material is raised to a desired temperature level, followed by gradual cooling obtained by controlling the energy input to the heaters. Discuss and develop a simple mathematical model for this system.

SOLUTION

The given thermal system consists of several parts or constituents that are linked to each other through energy transport. These parts, with the subscripts used to represent them, are:

1. Metal block, m
2. Heater, h
3. Gases, g
4. Walls, w
5. Insulation, i

Let us first consider each of these components separately and obtain the corresponding conservation equations and boundary conditions. Details are given to illustrate the basic considerations needed for model development.

Clearly, the time-dependent variation of the temperature in the material being heat-treated is of particular interest, making it necessary to retain the transient effects. Because this piece is made of metal and its size is given as small, the Biot number is expected to be small and it may be modeled as a lumped mass. If additional information is given, the Biot number may be estimated to check the validity of this assumption. Therefore, the energy equation for the metal block may be written, using simple radiation analysis, as

$$(\rho CV)_m \frac{dT_m}{d\tau} = h_m A_m (T_g - T_m) + \varepsilon_m \sigma A_m \left(F_{mh} T_h^4 + F_{mw} T_w^4 - T_m^4\right)$$

where ρ, C, V, A, and ε refer to the material density, specific heat, volume, surface area, and surface emissivity, respectively, and h_m is the convective heat transfer coefficient at its surface. F_{mh} and F_{mw} are geometrical view factors between the metal block and the heater and the metal block and the wall, respectively, and enclosure radiation analysis is used.

The wall and the heater are taken as black and the energy reflected at the block surface is assumed to be negligible, otherwise the absorption factor or radiosity method may be used (Jaluria and Torrance, 2003). The gases, being nitrogen and inert gases, are taken as nonparticipating. Only the initial condition is needed for this equation, this being written as $T_m = T_r$ at $\tau = 0$, where T_r is the room temperature. Depending on the temperatures, various terms in the preceding energy balance equation may dominate or may be negligible. For instance, the radiation from the material being heat treated may be small compared to that from the heater and may be neglected.

Similarly, the heater may be treated as a lumped mass because it is a thin metal strip. An equation similar to the one given previously for the metal block may be written, using the above assumptions, as

$$(\rho CV)_h \frac{dT_h}{d\tau} = Q(\tau) + h_h A_h \left(T_g - T_h\right) + \sigma A_h \left(F_{hw} T_w^4 - T_h^4\right)$$

where $Q(\tau)$ is the heat input into the heater, h_h is the convective heat transfer coefficient at the heater surface, and the F's are the view factors. Again, radiation from the heater may dominate over the other two radiation terms. The initial condition is $T_h = T_r$ at $\tau = 0$, when the heat input $Q(\tau)$ is turned on.

The gases are driven by a fan and by buoyancy in an enclosed region. As such, a well-mixed condition is expected to arise. Therefore, if a uniform temperature is assumed in the gases, an energy balance gives

$$(\rho CV)_g \frac{dT_g}{d\tau} = h_h A_h \left(T_h - T_g\right) - h_m A_m \left(T_g - T_m\right) - h_w A_w \left(T_g - T_w\right)$$

where convective heat transfer occurs at the heater, the walls, and the metal block. The gases gain energy at the heater and may gain or lose energy at the other two, depending on the temperatures. The initial temperature is again T_r and the heat input is due to $Q(\tau)$.

Because the thickness of the walls is much smaller than the other two dimensions, the thermal transport in the walls may be approximated as transient one-dimensional conduction, given by the equation

$$(\rho C)_w \frac{\partial T_w}{\partial \tau} = \frac{\partial}{\partial x}\left(k_w \frac{\partial T_w}{\partial x}\right)$$

where x is the coordinate distance normal to the wall surface, taken as positive toward the outside environment. The thermal conductivity k can often be taken as a constant for typical materials and temperature levels. If this is done, the term on the right-hand side becomes $k_w(\partial^2 T_w/\partial x^2)$, further simplifying the model. A similar energy equation applies for the temperature T_i in the insulation.

The boundary and initial conditions needed for these equations are

$$\text{at } \tau = 0: T_w = T_i = T_r$$

$$\text{For } \tau > 0: \; -k_w \frac{\partial T_w}{\partial x} = h_w\left(T_g - T_w\right) + \sigma A_h F_{hw}(T_h^4) \text{ at } x = 0$$

$$\text{and } \; T_w = T_i \; \text{ at } x = d_1$$

$$-k_i \frac{\partial T_i}{\partial x} = h_e\left(T_i - T_e\right) \text{ at } x = d_1 + d_2 \; \text{ and } \; -k_w \frac{\partial T_w}{\partial x} = -k_i \frac{\partial T_i}{\partial x} \; \text{ at } x = d_1$$

Here, $x = 0$ is the inner surface of the wall, d_1 is the wall thickness, d_2 is the insulation thickness, h_e is the heat transfer coefficient at the outer surface of the furnace, and T_e is the outside environmental temperature. Convection transport at the boundaries and continuity in temperature and heat flux at the interfaces is used to obtain these conditions. Similarly, the equations and boundary conditions may be derived for other similar thermal systems and processes.

The preceding system of equations, along with the corresponding boundary conditions, represents the mathematical model for the given thermal system. Many simplifications have been made, particularly with respect to the dimensions needed, radiative transport, and variable properties. These may be relaxed, if needed, for higher accuracy and more realistic representation of the system. In addition, the various convective heat transfer coefficients are assumed to be known. Heat transfer correlations available in the literature may be used for the purpose. In actual practice, these should be obtained by solving the convective flow in the gases for the given geometrical configuration. However, this is a far more involved problem and would require substantial effort with commercially available or personally developed computational software. The mathematical model derived is relatively simple, though several coupled differential equations are obtained. The model then provides the basis for a dynamic, or time-dependent, numerical simulation of the system. The equations can be solved to obtain the variation in the different components with time, as well as the temperature distribution in the wall and the insulation.

If all the components are taken as lumped, a system of ordinary differential equations is obtained instead of the partial differential equations derived here for a distributed model. This additional approximation considerably simplifies the model. Example 2.6 presented numerical results on a problem similar to this one when all the parts of the system are taken as lumped. The variation of the temperature with time was obtained in that example for the different components, indicating the fast response of the heater and the relatively slow response of the walls.

3.4 PHYSICAL MODELING AND DIMENSIONAL ANALYSIS

A physical model refers to a model that is similar to the actual system in shape, geometry, and other physical characteristics. Because experimentation on a full-size prototype is often impossible or very expensive, scale models that are smaller than the full-size system are of particular interest

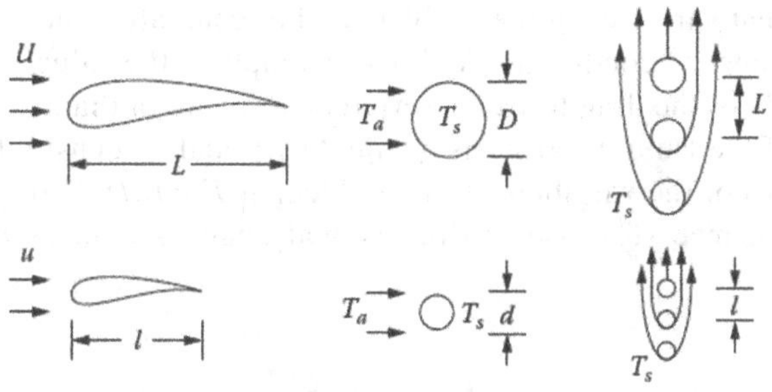

FIGURE 3.19 Scale models with geometric similarity for flow and heat transfer.

in design. The model may also be a simplified version of the actual system or may focus on particular aspects of the system. Experiments are carried out on these models and the results obtained are employed to represent the behavior and characteristics of the given component or system. Therefore, the information obtained from physical modeling provides better physical understanding of the processes involved and inputs for the design process, as well as data for the validation of the mathematical model. Figure 3.19 shows a few scale models used for investigating the drag force and heat transfer from heated bodies of different shapes.

Physical modeling is of particular importance in the design of thermal systems because of the complexity of the transport processes that arise in typical practical systems. In many cases, it is not possible or convenient to simplify the problem adequately through mathematical modeling and to obtain an accurate solution that closely represents the physical system. In addition, the validity of some of the approximations may be questionable. Experimental data are then needed for a check on accuracy and validity. In some cases, the basic mechanisms are not easy to model. For example, turbulent, separated, and unstable flows are often difficult to model mathematically. Experimental inputs are then needed for a satisfactory representation of the problem. However, experimental work is time-consuming and expensive. Therefore, it is necessary to minimize the number of experiments needed for obtaining the desired information. This is achieved through dimensional analysis by determining the dimensionless parameters that characterize the given system. This approach is widely used in fluid mechanics and heat and mass transfer (White, 2015; Incropera and Dewitt, 2001; Pritchard and Mitchell, 2015). A brief discussion of dimensional analysis is given here for completeness and to explain its relevance to the design process. The following section may be skipped if the reader is already well-versed in this material. In addition, the references cited and other textbooks in these areas may be consulted for further details.

3.4.1 Scale Up

An important consideration that arises in physical modeling is the relationship between the results obtained from the scale model and the characteristics of the actual system. Obviously, if the results from the model are to be useful with respect to the system, there must be known principles that link the two. These are usually known as scaling laws and are of considerable interest to industry because they allow the modeling of complicated systems in terms of simpler, scaled-down versions. Using these laws, the results from the models can be scaled up to larger systems. However, in many cases, different considerations lead to different scaling parameters and the appropriate model may not be uniquely defined. In such cases, similarity is achieved only for the few dominant aspects between the model and the system (Wellstead, 1979; Doebelin, 1980).

3.4.2 Dimensional Analysis

Dimensional analysis refers to the procedure to obtain the dimensionless governing parameters that determine the behavior of a given system. This analysis is carried out largely to reduce the number of independent variables in the problem and to generalize the results so that they may be used over wide ranges of conditions. A classic example of the value of dimensional analysis is provided by a study of the drag force F exerted on a sphere of diameter D in a uniform fluid flow at velocity V. If the fluid viscosity is denoted by μ and its density by ρ, the drag force F may be given in terms of the variables of the problem as $F = f_1(D, V, \mu, \rho)$, where f_1 represents the functional dependence. The use of dimensional analysis reduces this equation for F to the form

$$\frac{F}{\rho V^2 D^2} = f_2\left(\frac{\rho VD}{\mu}\right) \tag{3.21}$$

Here, $\rho VD/\mu$ is the well-known dimensionless parameter known as the Reynolds number Re. The functional dependence given by f_2 has to be obtained experimentally. But only a few experiments are needed to determine the functional relationship between the dimensionless drag force $F/(\rho V^2 D^2)$ and the Reynolds number Re. Equation (3.21) also indicates the variation of the drag force F with the different physical variables in the problem such as D, V, μ, and ρ.

Clearly, considerable simplification and generalization has been obtained in representing this problem so that once the function f_2 has been obtained through experiments, the results can be applied to spheres of different diameters, to different fluids, and to a range of fluid velocities V, as long as the flow characteristics remain unchanged, such as laminar or turbulent flow in this example. Similarly, heat transfer data in forced convection are correlated in terms of the Nusselt number $Nu = hD/k_f$, where k_f is the fluid thermal conductivity. The areas of fluid mechanics and heat transfer are replete with similar examples that demonstrate the importance and value of dimensional analysis.

There are two main approaches for deriving the dimensionless parameters in a given problem. These are:

Combinations of variables. This method considers all the variables in the problem and the appropriate basic dimensions, such as length, mass, temperature, and time, associated with them. Then the dimensionless parameters are obtained by forming combinations of these variables to yield dimensionless groups. The Buckingham Pi theorem states that for n variables, $(n - m)$ dimensionless ratios, or π parameters, can be derived, where m is usually, but not always, equal to the minimum number of independent dimensions that arise in all the variables. Thus, the number of important variables that characterize a given system must be determined on the basis of experience and physical interpretation of the problem. The primary dimensions are determined and combinations of the variables are formed to obtain dimensionless groups. If a particular group can be formed from others through multiplication, division, raising to a power, etc., it is not independent. Thus, independent dimensionless groups may be obtained. However, all the important variables must be included and the method does not yield the physical significance of the various dimensionless parameters.

Characterizing equations. This approach is based on the nondimensionalization of the equations and boundary conditions that characterize the given system and that are obtained by mathematical modeling. The equations are first written in terms of the physical variables, such as time, spatial coordinates, temperature, and velocity. Characteristic quantities are then chosen based on experience and the physical nature of the system. These are used to nondimensionalize the variables that arise in the equations, which are thus transformed into dimensionless equations, with all the dimensionless groups appearing as coefficients in the equation. Similarly, the boundary conditions are nondimensionalized and these may yield additional dimensionless parameters.

For steady-state, three-dimensional conduction in the solid bar shown in Figure 3.9, Equation (3.6) gave the nondimensional energy equation, with L^2/H^2 and L^2/W^2, or simply L/H and L/W, as the two dimensionless parameters that characterize the temperature distribution in the solid. For this problem, Equation (3.5) gave the transformation used for nondimensionalization, with L, H, and W as the characteristic length scales.

No additional parameters arise from the boundary conditions if the temperatures are specified as constant at the surfaces. But if a convective boundary condition of the form

$$-k\frac{\partial T}{\partial x} = h(T - T_a) \tag{3.22}$$

is given at, say, the surface $x = L$, this equation may be nondimensionalized, using Equation (3.5), to give

$$-\frac{\partial \theta}{\partial X} = \frac{hL}{k}\theta, \quad \text{where } \theta = \frac{T - T_a}{T_s - T_a} \tag{3.23}$$

Here, T_s is a reference temperature and could be taken as the specified temperature at $x = 0$. Therefore, hL/k arises as an additional dimensionless parameter from the boundary conditions. This parameter is referred to as the Biot number Bi, mentioned and defined earlier.

Similarly, the physical temperature T in an infinite rod moving at speed U along the x direction, which is taken along its axis, and losing energy by convection at the surface, as sketched in Figure 1.10(d) and Figure 3.6(a), is determined by the equation

$$\rho C\left(\frac{\partial T}{\partial \tau} + U\frac{\partial T}{\partial x}\right) = k\nabla^2 T \tag{3.24}$$

where ∇^2 is the Laplacian, given by $\nabla^2 = \partial^2/\partial x^2 + \partial^2/\partial y^2 + \partial^2/\partial z^2$.

It can be easily confirmed that, using the nondimensional variables given earlier, the dimensionless equation obtained from Equation (3.24) is Equation (3.20), where the Peclet number Pe was seen to arise as the only governing dimensionless parameter in the equation. Other parameters such as the Biot number Bi may arise from the boundary conditions. Similarly, other circumstances of interest in thermal systems may be considered to derive the governing parameters.

Both the approaches for dimensional analysis outlined here are useful for thermal processes and systems. However, the nondimensionalization of equations does not require the listing of all the important variables in the problem and also leads to a physical interpretation of the dimensionless groups, as discussed for the two examples given previously. Therefore, the nondimensionalization of the equations is the preferred approach. However, characteristic quantities are needed and may be difficult to obtain if no simple scales are evident in the problem or if several choices are possible. Therefore, both the approaches may be employed to derive the relevant parameters for a given thermal system.

Several different dimensionless groups have been used extensively in thermal sciences and engineering, with a set of these characterizing a given system or process. Each group has a specific physical significance, often given in terms of the ratio of the orders of magnitude of two separate mechanisms or effects. Some of the important dimensionless parameters that arise in fluid mechanics and in heat and mass transfer are listed in Table 3.1, along with the ratio of forces, transport mechanisms, etc., that they represent.

In this table, V, L, and ΔT are the characteristic velocity, length, and temperature difference; g is the magnitude of gravitational acceleration; β is the coefficient of thermal expansion; k_f is the fluid thermal conductivity, as distinguished from the solid thermal conductivity k; D_{AB} is the mass

TABLE 3.1
Commonly Used Dimensionless Groups in Fluid Mechanics and Heat and Mass Transfer

Parameter	Definition	Ratio of Effects
Reynolds number	$\mathrm{Re} = \frac{\rho VL}{\mu}$	Inertia forces / Viscous forces
Froude number	$\mathrm{Fr} = \frac{V^2}{gL}$	Inertia forces / Gravitational forces
Mach number	$\mathrm{M} = \frac{V}{a}$	Flow speed / Sound speed
Weber number	$\mathrm{We} = \frac{\rho V^2 L}{\sigma}$	Inertia forces / Surface tension forces
Euler number	$\mathrm{Eu} = \frac{\Delta p}{\rho V^2}$	Pressure forces / Inertia forces
Prandtl number	$\mathrm{Pr} = \frac{\nu}{\alpha}$	Momentum diffusion / Thermal diffusion
Peclet number	$\mathrm{Pe} = \frac{VL}{\alpha}$	Convective transport / Conductive transport
Eckert number	$\mathrm{Ec} = \frac{V^2}{C_p \Delta T}$	Kinetic energy / Enthalpy difference
Biot number	$\mathrm{Bi} = \frac{hL}{k}$	Conductive resistance / Convective resistance
Fourier number	$\mathrm{Fo} = \frac{\alpha \tau}{L^2}$	Thermal diffusion / Thermal energy storage
Grashof number	$\mathrm{Gr} = \frac{g\beta \Delta T L^3}{\nu^2}$	Buoyancy forces / Viscous forces
Nusselt number	$\mathrm{Nu} = \frac{hL}{k_f}$	Convection / Diffusion
Sherwood number	$\mathrm{Sh} = \frac{h_m L}{D_{AB}}$	Convective mass transfer / Mass diffusion
Schmidt number	$\mathrm{Sc} = \frac{\nu}{D_{AB}}$	Momentum diffusion / Mass diffusion
Lewis number	$\mathrm{Le} = \frac{\alpha}{D_{AB}}$	Thermal diffusion / Mass diffusion

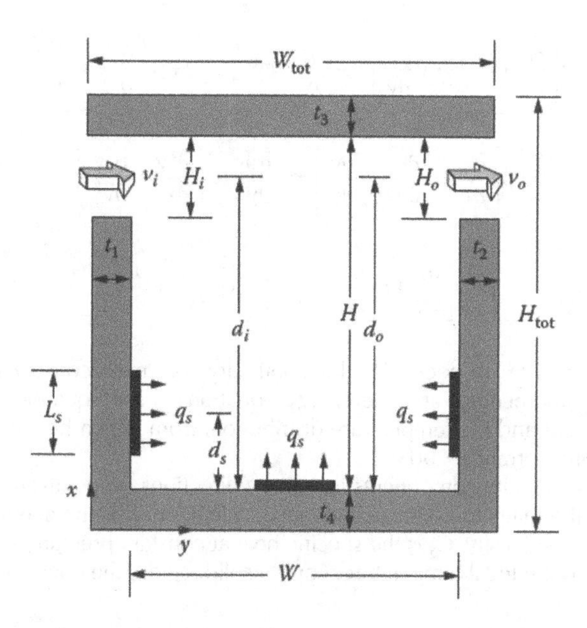

FIGURE 3.20 The electronic system, with three electrical components as heat sources and forced air cooling, considered in Example 3.7.

diffusivity; h_m is the mass transfer coefficient; a is the speed of sound in the given medium; and σ is the surface tension. The other symbols have been defined earlier.

There are many more dimensionless parameters that arise in the analysis and design of thermal processes and systems. The ratio of different effects in a particular dimensionless parameter is largely qualitative and may be used for an interpretation of the physical significance of the dimensionless group. For further details on the derivation of these dimensionless parameters and their use, textbooks in heat transfer and in fluid mechanics may be consulted.

Example 3.7

The electronic system shown in Figure 3.20 is cooled by the forced flow of ambient air driven by a fan through openings near the top of the enclosure. The dimension in the third direction is given as large and the problem may be treated as two-dimensional. The various dimensions in the system and the locations of three electronic components are indicated in the figure. The velocities are uniform over the inlet and outlet, with magnitudes v_i and v_o, respectively. The temperature at the inlet is T_i and developed conditions, i.e., $\partial T/\partial y = 0$, may be assumed at the exit. The outer wall of the system loses energy to ambient air at temperature T_i with a given convective heat transfer coefficient h. Write down the equations and boundary conditions that determine the temperature distributions in the system. Nondimensionalize these to obtain the various dimensionless parameters in the problem. Assume laminar flow and constant properties.

SOLUTION

This is clearly a fairly complicated problem and involves combined conduction and convection. Because of spatial and temporal variations, partial differential equations will be obtained. However, this example serves to indicate some of the major complexities that arise when dealing with practical thermal systems.

For the given two-dimensional laminar flow problem, the equations for convection in the enclosure may be written, in terms of the coordinate system shown, as (Burmeister, 1993)

$$\frac{\partial u}{\partial x} + \frac{\partial v}{\partial y} = 0$$

$$\rho\left[\frac{\partial u}{\partial \tau}+u\frac{\partial u}{\partial x}+v\frac{\partial u}{\partial y}\right]=-\frac{\partial p}{\partial x}+g\beta(T-T_i)+\mu\left(\frac{\partial^2 u}{\partial x^2}+\frac{\partial^2 u}{\partial y^2}\right)$$

$$\rho\left[\frac{\partial v}{\partial \tau}+u\frac{\partial v}{\partial x}+v\frac{\partial v}{\partial y}\right]=-\frac{\partial p}{\partial y}+\mu\left(\frac{\partial^2 v}{\partial x^2}+\frac{\partial^2 v}{\partial y^2}\right)$$

$$\rho C_p\left[\frac{\partial T}{\partial \tau}+u\frac{\partial T}{\partial x}+v\frac{\partial T}{\partial y}\right]=k\left(\frac{\partial^2 T}{\partial x^2}+\frac{\partial^2 T}{\partial y^2}\right)$$

The above equations are, respectively, the continuity, or mass conservation equation, the x-momentum, the y-momentum and the energy equation. These equations have to be solved to obtain the flow field and the temperature distribution, from which heat transfer rates may be calculated (Jaluria and Torrance, 2003).

Here, u and v are velocity components in x and y directions, respectively, p is the local pressure, τ is time, g is the magnitude of gravitational acceleration, β is the coefficient of volumetric thermal expansion, ρ is density, C_p is the specific heat at constant pressure, k is the fluid thermal conductivity, and T is the local temperature. For the solid region, the energy equation is

$$(\rho C)_s\frac{\partial T}{\partial \tau}=k_s\left(\frac{\partial^2 T}{\partial x^2}+\frac{\partial^2 T}{\partial y^2}\right)$$

where the subscript s denotes solid material properties. The *Boussinesq* approximations, which neglect density variation in the continuity equation and assume density variation with temperature to be linear, have been used for the buoyancy term in the x-direction momentum equation. The pressure work and viscous dissipation terms have been neglected. The boundary conditions on velocity are the no-slip conditions, i.e., zero velocity at the solid boundaries. At the inlet and outlet, the given velocities apply.

For the temperature field, at the inner surface of the enclosure, continuity of the temperature and the heat flux gives

$$T=T_s \quad \text{and} \quad \left(\frac{\partial T}{\partial n}\right)=k_s\left(\frac{\partial T}{\partial n}\right)_s$$

where n is the coordinate normal to the surface. Also, at the left source, an energy balance gives

$$Q_s=L_s\left(-k\frac{\partial T}{\partial y}+k_s\frac{\partial T}{\partial y}\right)$$

where Q_s is the energy dissipated by the source per unit width. Similar equations may be written for other sources. At the outer surface of the enclosure walls, the convective heat loss condition gives

$$-k_s\frac{\partial T}{\partial n}=h(T-T_i)$$

At the inlet, the temperature is uniform at T_i and at the outlet developed temperature conditions, $\partial T/\partial y=0$, may be used.

Therefore, the equations and boundary conditions that characterize this coupled conduction-convection problem are obtained. The main characteristic quantities in the problem are the conditions at the inlet and the energy input at the sources. The energy input governs the heat transfer processes and the inlet conditions determine the forced airflow in the enclosure. Therefore, v_i, H_i, T_i, and Q_s are taken as the characteristic physical quantities. The various dimensions in the problem are nondimensionalized by H_i and the velocity V by v_i. Time τ is nondimensionalized by H_i/v_i to give dimensionless time $\tau^* = \tau(v_i/H_i)$. The nondimensional temperature θ is defined as

$$\theta = \frac{T - T_i}{\Delta T}, \quad \text{where}\ \Delta T = \frac{Q_s}{k}$$

Here ΔT is taken as the temperature scale based on the energy input by a given source. The energy input by other sources may be nondimensionalized by Q_s. The pressure p is nondimensionalized by ρv_i^2, which comes from Bernoulli's equation and is commonly used in fluid mechanics.

The governing equations and the boundary conditions may now be nondimensionalized to obtain the important dimensionless parameters in the problem. The dimensionless equations for the convective transport in this problem are obtained as

$$\frac{\partial u^*}{\partial x^*} + \frac{\partial v^*}{\partial y^*} = 0$$

$$\frac{\partial u^*}{\partial \tau^*} + u^* \frac{\partial u^*}{\partial x^*} + v^* \frac{\partial u^*}{\partial y^*} = -\frac{\partial p^*}{\partial x^*} + \left(\frac{Gr}{\mathrm{Re}^2}\right)\theta + \frac{1}{\mathrm{Re}}\left(\frac{\partial^2 u^*}{\partial x^{*2}} + \frac{\partial^2 u^*}{\partial y^{*2}}\right)$$

$$\frac{\partial v^*}{\partial \tau^*} + u^* \frac{\partial v^*}{\partial x^*} + v^* \frac{\partial v^*}{\partial y^*} = -\frac{\partial p^*}{\partial y^*} + \frac{1}{\mathrm{Re}}\left(\frac{\partial^2 v^*}{\partial x^{*2}} + \frac{\partial^2 v^*}{\partial y^{*2}}\right)$$

$$\frac{\partial \theta}{\partial \tau^*} + u^* \frac{\partial \theta}{\partial x^*} + v^* \frac{\partial \theta}{\partial y^*} = \frac{1}{\mathrm{RePr}}\left(\frac{\partial^2 \theta}{\partial x^{*2}} + \frac{\partial^2 \theta}{\partial y^{*2}}\right)$$

The dimensionless energy equation for the solid is

$$\frac{\partial \theta}{\partial \tau^*} = \left(\frac{\alpha_s}{\alpha}\right)\left(\frac{1}{\mathrm{RePr}}\right)\left(\frac{\partial^2 \theta}{\partial x^{*2}} + \frac{\partial^2 \theta}{\partial y^{*2}}\right)$$

where the asterisk denotes dimensionless quantities. Therefore, the dimensionless parameters that arise are the Reynolds number Re, the Grashof number Gr, and the Prandtl number Pr, where these are defined as

$$\mathrm{Re} = \frac{v_i H_i}{\nu} \quad \mathrm{Gr} = \frac{g\beta H_i^3 \Delta T}{\nu^2} \quad \mathrm{Pr} = \frac{\nu}{\alpha}$$

The Peclet number Pe, mentioned earlier, is given by $Pe = Re\ Pr$. In addition, the ratio of the thermal diffusivities α_s/α arises as a parameter. Here, $\nu = \mu/\rho$ is the kinematic viscosity of the fluid. The Reynolds number determines the characteristics of the flow, particularly whether it is laminar or turbulent, the Grashof number determines the importance of buoyancy effects, and the Prandtl number gives the effect of momentum diffusion as compared to thermal diffusion and is fixed for a given fluid at a particular temperature.

Additional parameters arise from the boundary conditions. These are the ratio of the thermal conductivities k_s/k and the Biot number $\mathrm{Bi} = hH_i/k$. A perfectly insulated condition at the outer surface is achieved for $\mathrm{Bi} = 0$. In addition to these, several geometry parameters arise from the dimensions of the enclosure (see Figure 3.20), such as d_i/H_i, H_o/H_i, L_s/H_i, etc. Heat inputs at different sources lead to parameters such as $(Q_s)_2/Q_s$, $(Q_s)_3/Q_s$, etc., where $(Q_s)_2$ and $(Q_s)_3$ are the heat inputs by different electronic components.

The preceding considerations yield the dimensionless equations and boundary conditions, along with all the dimensionless parameters that govern the thermal transport process. Clearly, a large number of parameters are obtained. However, if the geometry, fluid, and heat inputs at the

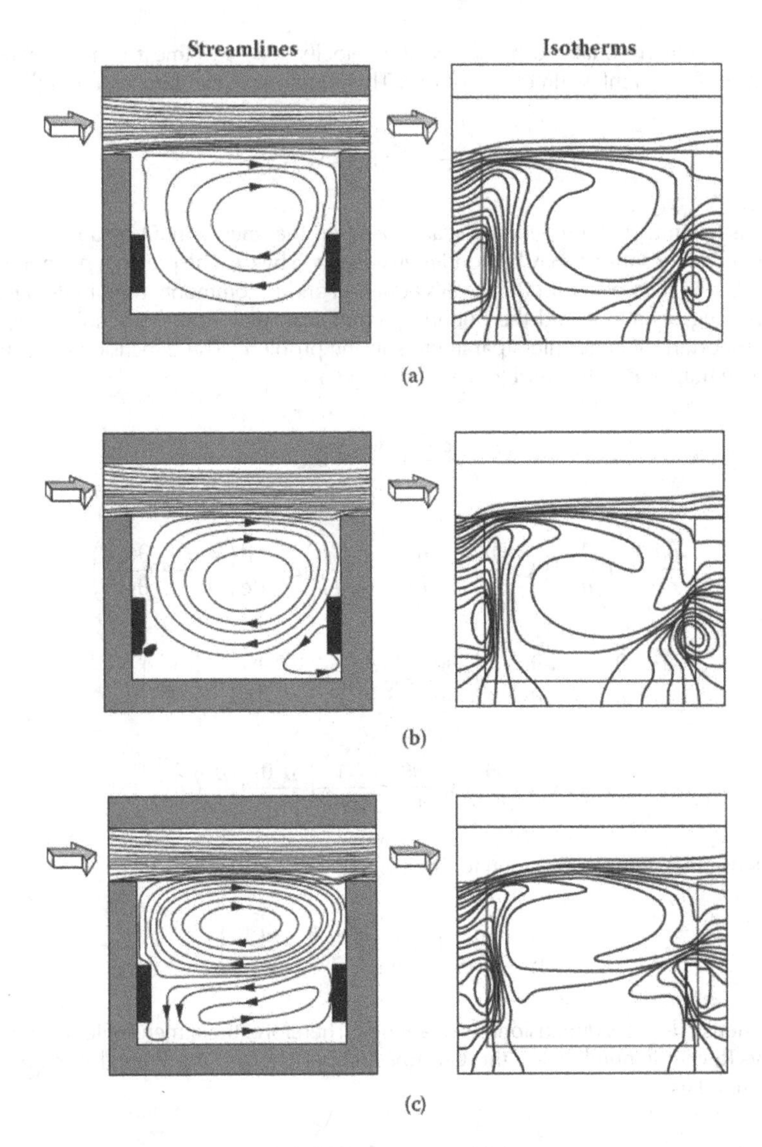

FIGURE 3.21 Calculated streamlines and isotherms for the steady solutions obtained in the LR configuration for the problem considered in Example 3.7 at Re = 100 and Gr/Re^2 values of (a) 0.1, (b) 1.0, and (c) 10.0. (Adapted from Papanicolaou and Jaluria, 1994.)

sources are fixed, the main governing parameters are Re, Gr, Bi, and the material property ratios α_s/α and k_s/k. These may be varied in the simulation of the given system to determine the effect of materials used and the operating conditions. Similarly, geometry parameters may be varied to determine the effect of these on the performance of the cooling system, particularly on the temperature of the electronic components, whose performance is very temperature-sensitive.

Some typical results, obtained by the use of a finite-volume-based numerical scheme for solving the dimensionless equations, are shown in Figures 3.21 and 3.22 from a detailed numerical simulation carried out by Papanicolaou and Jaluria (1994). Two electronic components are taken, placing these on the left wall (L), the right wall (R), or the bottom (B). The flow field, in terms of streamlines, and the temperature field, in terms of isotherms, i.e., constant temperature contours, are shown for one, LR, configuration. Such results are used to indicate if there are any stagnation regions or hot spots in the system. The configuration may be changed to improve the flow and temperature distributions to obtain greater uniformity and/or lower temperatures. Figure 3.22

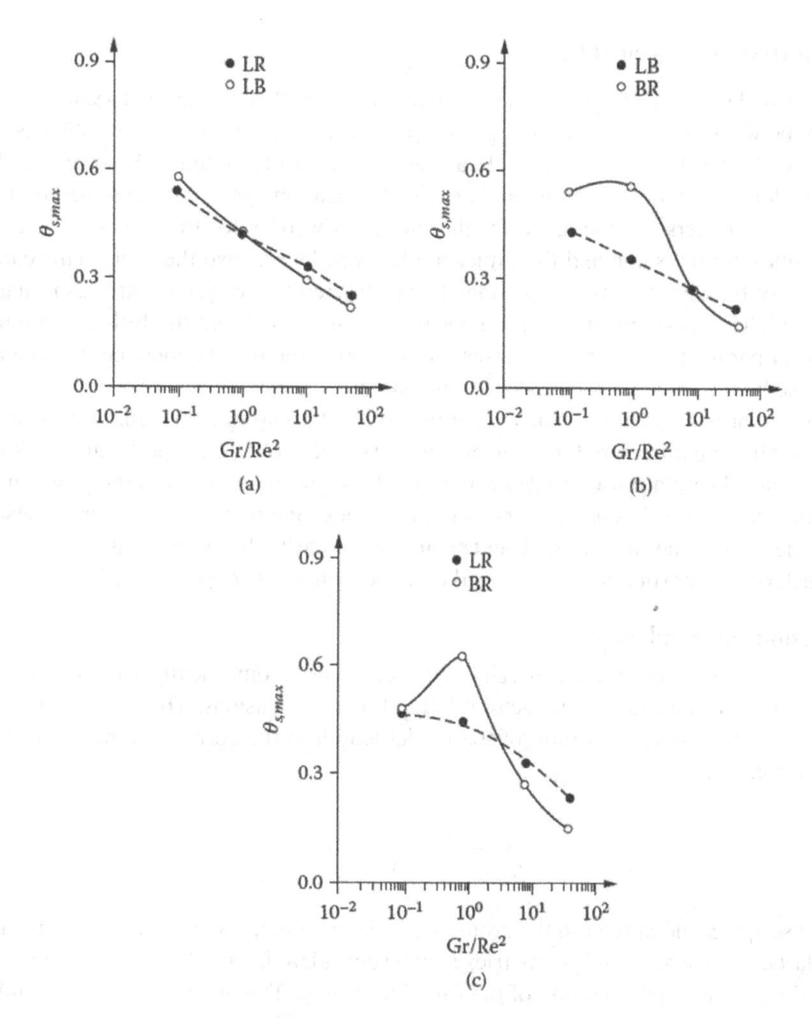

FIGURE 3.22 Calculated maximum temperature for different source locations in the configurations considered, at various values of Gr/Re^2 for (a) left wall location, (b) bottom wall location, and (c) right wall location. (Adapted from Papanicolaou and Jaluria, 1994.)

shows the maximum temperatures of the electronic components for different configurations as functions of the parameter Gr/Re^2. We can use these results to determine if the allowable temperatures are exceeded in a particular case and also to vary the configuration and flow rate to obtain an acceptable design. Thus, the simulation results may be used to change the design variables over given ranges in order to obtain an acceptable or optimal design of the system.

This is clearly a fairly complicated problem because transient effects and spatial variations are included. However, steady-state operation of such systems is generally of interest and the transient effects need not be considered. Many practical systems involve complicated coupled system of equations and complex geometry. Finite-element methods are particularly well-suited for generating the numerical results needed for the design and optimization of the system. In this problem, we may be interested in finding the optimal location of the heat sources, appropriate dimensions, airflow rate, wall thickness, and materials for the given electronic circuitry. This example is given mainly to illustrate some of the complexities of practical thermal systems and the derivation of governing dimensionless parameters. The results indicate typical outputs obtained and their relevance to system design. Cooling of electronic systems has been an important area for research and design over the past two to three decades. In many cases, commercially available software, such as Ansys, is used to simulate the system and obtain the results needed for design and optimization.

3.4.3 Modeling and Similitude

In order for a scale model to predict the behavior of the full-scale thermal system, there must be similarity between the model and the prototype. Scaling factors must be established between the two so that the results from the model can be applied to the system. These scaling laws and the conditions for similitude are obtained from dimensional analysis. As mentioned earlier, if the dimensionless parameters are the same for the model as well as for the prototype, the flow and transport regimes are the same and the dimensionless results are also the same. This can be seen easily in terms of the dimensionless equations that characterize the system, such as Equation (3.6) and Equation (3.20). These equations are the same for the model and the full-size system. If the nondimensional parameters for the two cases are the same, the results obtained, in dimensionless terms, will also be the same for the model and the system.

Several different mechanisms usually arise in typical thermal systems, and it may not be possible to satisfy all the parameters for complete similarity. However, each problem has its own specific requirements. These are used to determine the dominant parameters in the problem and thus establish similitude. Several common types of similarities may be mentioned here. These include geometric, kinematic, dynamic, thermal, and chemical similarity. It is important to select the appropriate parameters for a particular type of similarity (Schuring, 1977; Szucs, 1977).

3.4.3.1 Geometric Similarity

The model and the prototype are generally required to be geometrically similar. This requires identity of shape and a constant scale factor relating linear dimensions. Thus, if a model of a bar is used for heat transfer studies, the ratio of the model length to the corresponding prototype length must be the same, i.e.,

$$\frac{L_p}{L_m} = \frac{H_p}{H_m} = \frac{W_p}{W_m} = \lambda_1 \tag{3.25}$$

where the subscripts p and m refer to the prototype and the model, respectively, and λ_1 is the scaling factor. Similarly, other shapes and geometries may be considered, with the scale model representing a geometrically similar representation of the full-size system. This is the first type of similitude in physical modeling and is commonly required of the model. However, sometimes the model may represent only a portion of the full system. For example, a long drying oven may be studied with a relatively short model that is properly scaled in terms of the cross-section but is only a fraction of the oven length. Models of solar ponds often scale the height but not the large surface area of typical ponds. In these cases, the model is chosen to focus on the dominant aspects.

3.4.3.2 Kinematic Similarity

The model and the system are kinematically similar when the velocities at corresponding points are related by a constant scale factor. This implies that the velocities are in the same direction at corresponding points and the ratio of their magnitudes is a constant. The streamline patterns of two kinematically similar flows are related by a constant factor, and, therefore, they must also be geometrically similar. The flow regime, for instance, whether the flow is laminar or turbulent, must be the same for the model and the prototype. Thus, if u, v, and w represent the three components of velocity in a model of a system, kinematic similarity requires that

$$\frac{u_p}{u_m} = \frac{v_p}{v_m} = \frac{w_p}{w_m} = \lambda_2 \tag{3.26}$$

where λ_2 is the scale factor and the subscripts p and m again indicate the prototype and the model. For kinematic similarity, the model and the prototype must both have the same length-scale ratio and the same time-scale ratio. Consequently, derived quantities such as acceleration and volume flow rate also have a constant scale factor. For a given value of the magnitude of the gravitational acceleration g, the Froude number Fr (Table 3.1) represents the scaling for velocity and length. Therefore, this kinematic parameter is used for scaling wave motion in water bodies.

3.4.3.3 Dynamic Similarity

This requires that the forces acting on the model and on the prototype are in the same direction at corresponding locations and the magnitudes are related by a constant scale factor. This is a more restrictive condition than the previous two and, in fact, requires that these similarity conditions also be met. All the important forces must be considered, such as viscous, surface tension, gravitational, and buoyancy forces. If dynamic similarity is obtained between the model and the prototype, the results from the model may be applied quantitatively to determine the prototype behavior. The various dimensionless parameters that arise in the momentum equation or that are obtained through the Buckingham Pi theorem may be used to establish dynamic similarity. For instance, in the case of the drag on a sphere, given by Equation (3.21), if the Reynolds numbers for the model and the prototype are equal, the dimensionless drag forces, given by $F/(\rho V^2 D^2)$, are also equal. Then the results obtained from the model can be used to predict the drag force on the full-size component. Clearly, the tests could be carried out with different fluids, such as air and water, and over a convenient velocity range, as long as the Reynolds numbers are matched. In fact, the model can be used in a wind or water tunnel to determine the functional dependence given by f_2 in Equation (3.21). Then this equation can be used for predicting the drag for a wide range of diameters, velocities, and fluid properties. Figure 3.23 shows the sketches of a few examples of physical modeling of the flow to obtain similitude.

3.4.3.4 Thermal Similarity

This similarity is of particular relevance to thermal systems. Thermal similarity requires that the temperature profiles in the model and the prototype be geometrically similar at corresponding times. If convective motion arises, kinematic similarity is also a requirement. Thus, the temperatures are related by a constant scale factor and the results from a model may be applied to obtain quantitative predictions on the temperatures in the prototype. The Nusselt number Nu characterizes the heat transfer in a convective process. Thus, in forced convection, if two flows are geometrically and kinematically similar and the flow regime, as determined by the Reynolds number Re, is the same, the Nusselt number is the same if the fluid Prandtl number Pr is the same. The Grashof number Gr arises as an additional parameter if buoyancy effects are significant. This relationship can be expressed as

$$\mathrm{Nu} = f_3(\mathrm{Re}, \mathrm{Gr}, \mathrm{Pr}) \tag{3.27}$$

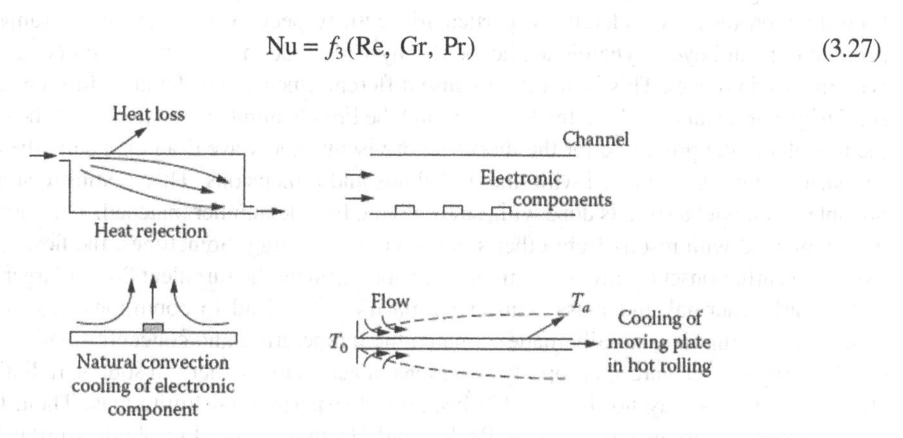

FIGURE 3.23 Experiments for physical modeling of thermal processes and systems.

where f_3 is obtained by analytical, numerical, or experimental methods. For conduction in a heated body with convective loss at the surface, the Biot number Bi arises as an additional dimensionless parameter from the boundary condition, as seen earlier in Equation (3.23).

Thus, thermal similarity is obtained if these parameters are the same between the model and the system. As mentioned earlier, the dimensionless applicable equations and corresponding boundary conditions indicate the dimensionless parameters that must be kept the same between the model and the system in order to apply the model-study results to the system. Experiments may be carried out to obtain the functional dependence, such as f_3 in Equation (3.27).

Radiative transport is often difficult to model because of the T^4 dependence of heat transfer rate on temperature. Similarly, temperature-dependent material properties and thermal volumetric sources are difficult to model because of the often arbitrary, nonlinear variations with temperature that arise. Consequently, physical modeling of thermal systems is often complicated and involves approximations similar to those discussed with respect to mathematical modeling. Relatively small effects are neglected to obtain similarity.

3.4.3.5 Mass Transfer Similarity

This similarity requires that the species concentration profiles for the model and the system be geometrically similar at corresponding times. At small concentration levels, the analogy between heat and mass transfer may be used, resulting in expressions such as Equation (3.27), which may be written for mass transfer systems as

$$\mathrm{Sh} = f_4(\mathrm{Re}, \mathrm{Gr}_c, \mathrm{Sc}) \tag{3.28}$$

where Sh is the Sherwood number, Sc is the Schmidt number (Table 3.1), and Gr_c is based on the concentration difference ΔC, instead of the temperature difference ΔT in Gr. Thus, the conditions for mass transfer similarity are close to those for thermal similarity in this case. If chemical reactions occur, the reaction rates at corresponding locations must have a constant scale factor for similitude between the model and the prototype. Because reaction rates are strongly dependent on temperature and concentration, the models are usually studied under the same temperature and concentration conditions as the full-size system.

3.4.4 Overall Physical Model

Based on dimensional analysis, which indicates the main dimensionless groups that characterize a given system, and the appropriate similarity conditions, a physical model may be developed to represent a component, subsystem, or system. However, even though a substantial amount of work has been done on these considerations, particularly with respect to wind and water tunnel testing for aerodynamic and hydrodynamic applications, physical modeling of practical processes and systems is an involved process. This is mainly because different aspects may demand different conditions for similarity. For instance, if both the Reynolds and the Froude numbers are to be kept the same between the model and the prototype for the modeling of viscous and wave drag on a ship, the conditions of similarity cannot be achieved with practical fluids and dimensions. Then complete similarity is not possible and model testing is done with, say, only the Froude number matched. The data obtained are then combined with results from other studies on viscous drag. Sometimes, the flow is disturbed to induce an earlier onset of turbulence in order to approximate the turbulent flow at larger Re.

Similarly, thermal and mass transfer similarities may lead to conditions that are difficult to match. An attempt is generally made to match the temperature and concentration levels in order to satisfactorily model material property variations, reaction rates, thermal source, radiative transport, etc. However, this may not be possible because of experimental limitations. Then, the matching of the dimensionless groups, such as Pr, Re, and Gr, may be used to obtain similarity and hence the desired information. Again, the dominant effects are isolated and physical modeling involves matching these between the system and the model. Because of the complexity of typical thermal systems, the physical model is rarely defined uniquely and approximate representations are generally used to provide the inputs needed for design. However, scale-up and scale-down are of particular interest in practical systems and processes. The preceding discussion provides the guidelines for achieving this accurately.

3.5 CURVE FITTING

An important and valuable technique that is used extensively to represent the characteristics and behavior of components, materials, and systems is *curve fitting*. Results are obtained at a finite number of discrete points by numerical computation and/or experimentation. If these data are represented by means of a smooth curve, which passes through or as close as possible to the points, the equation of the curve can be used to obtain values at intermediate points where data are not available and also to model the characteristics of the system. Physical reasoning may be used in the choice of the type of curve employed for curve fitting, but the effort is largely a data-processing operation, unlike mathematical modeling, which is based on physical insight and experience. The equation obtained as a result of curve fitting then represents the performance of a given equipment or system and may be used in system simulation and optimization. This equation may also be employed in the selection of equipment such as blowers, compressors, and pumps from items readily available from manufacturers. Curve fitting is particularly useful in representing calibration results and material property data, such as the thermodynamic properties of a substance, in terms of equations that form part of the mathematical model of the system.

There are two main approaches to curve fitting. The first one is known as an *exact fit* and determines a curve that passes through every given data point. This approach is particularly appropriate for data that are very accurate, such as computational results, calibration results, and material property data, and if only a small number of data points are available. If a large amount of data is to be represented, and if the accuracy of the data is not very high, as is usually the case for experimental results, the second approach, known as the *best fit*, which obtains a curve that does not pass through each data point but closely approximates the data, is more appropriate. The difference between the values given by the approximating curve and the given data is minimized to obtain the best fit. Sketches of curve fitting using these two methods were seen earlier in Figure 3.2. Both of these approaches are used extensively to represent results from numerical simulation and experimental studies. The availability of correlating equations from curve fitting considerably facilitates the design and optimization process.

3.5.1 Exact Fit

This approach for curve fitting is somewhat limited in scope because the number of parameters in the approximating curve must be equal to the number of data points for an exact fit. If extensive data are available, the determination of the large number of parameters that arise becomes very involved. The curve obtained is not very convenient to use and may be ill-conditioned. In addition, unless the data are very accurate, there is no reason to ensure that the curve passes through each data point. However, there are several practical circumstances where a small number of very accurate data are available and an exact fit is both desirable and appropriate.

Many methods are available in the literature for obtaining an exact fit to a given set of data points (Jaluria, 2012). Some of the important ones are:

1. General form of a polynomial
2. Lagrange interpolation
3. Newton's divided-difference polynomial
4. Splines

A polynomial of degree n can be employed to exactly fit $(n+1)$ data points. The general form of the polynomial may be taken as

$$f(x) = a_0 + a_1 x + a_2 x^2 + a_3 x^3 + \ldots + a_n x^n \tag{3.29}$$

where y is the dependent variable, x is the independent variable, and the a's are constants to be determined by curve fitting of the data. If (x_i, y_i), where $i = 0, 1, 2, \ldots, n$, represent the $(n + 1)$ data points, y_i being the value of the dependent variable at $x = x_i$, these values may be substituted in Equation (3.29) to obtain $(n + 1)$ equations for the a's. Thus,

$$y_i = a_0 + a_1 x_i + a_2 x_i^2 + a_3 x_i^3 + \ldots + a_n x_i^n \quad for\ i = 0, 1, 2, \ldots, n \tag{3.30}$$

Because x_i and y_i are known for the given data points, $(n + 1)$ equations are obtained from Equation (3.30). These linear equations can be solved for the unknown constants in Equation (3.29). Thus, two data points yield a straight line, $y = a_0 + a_1 x$, three points a second-order polynomial, $y = a_0 + a_1 x + a_2 x^2$, four points a third-order polynomial, and so on. The method is appropriate for small sets of very accurate data, with the number of data points typically less than ten. For larger data sets, higher-order polynomials are needed, which are often difficult to determine, inconvenient to use, and inaccurate because of the many small coefficients that arise for higher-order terms. A Matlab program for fitting an exact curve to given data, using the general form of the polynomial, is given in Appendix A.M.5.1.

Different forms of interpolating polynomials are used in other methods. In *Lagrange interpolation*, the polynomial used is known as the Lagrange polynomial and the nth-order polynomial is written as

$$\begin{aligned} y = f(x) = {} & a_0 (x - x_1)(x - x_2)\ldots(x - x_n) + a_1 (x - x_0)(x - x_2)\ldots \\ & (x - x_n) + \ldots + a_n (x - x_0)(x - x_1)\ldots(x - x_{n-1}) \end{aligned} \tag{3.31}$$

The coefficients a_i, where i varies from 0 to n, can be determined easily by substitution of the $(n + 1)$ data points into Equation (3.31). Then the resulting interpolating polynomial is

$$y = f(x) = \sum_{i=0}^{n} y_i \prod_{\substack{j=0 \\ j \neq i}}^{n} \left(\frac{x - x_j}{x_i - x_j} \right) \tag{3.32}$$

where the product sign Π denotes multiplication of the n factors obtained by varying j from 0 to n, excluding $j = i$, for the quantity within the parentheses. It is easy to see that this polynomial may be written in the general form of a polynomial, Equation (3.29), if needed. Lagrange interpolation is applicable to an arbitrary distribution of data points, and the determination of the coefficients of the polynomial does not require the solution of a system of equations, as was the case for the general polynomial. Because of the ease with which the method may be applied, Lagrange interpolation is extensively used for engineering applications.

In *Newton's divided-difference method*, the nth-order interpolating polynomial is taken as

$$\begin{aligned} y = f(x) = {} & a_0 + a_1 (x - x_0) + a_2 (x - x_0)(x - x_1) \\ & + \ldots + a_n (x - x_0)(x - x_1)\ldots(x - x_{n-1}) \end{aligned} \tag{3.33}$$

A recursive formula is written to determine the coefficients. The higher-order coefficients are determined from the lower-order ones. Therefore, the coefficients are evaluated starting with a_0 and successively calculating a_1, a_2, a_3, and so on, up to a_n. Once these coefficients are determined, the interpolating polynomial is obtained from Equation (3.33). Several simplified formulas can be derived if the data are given at equally spaced values of the independent variable x. These include the Newton-Gregory forward and backward interpolating polynomials. This method is particularly well-suited for numerical computation and is frequently used for an exact fit in engineering problems (Carnahan, et al., 1969; Hornbeck, 1975; Gerald and Wheatley, 2003; Jaluria, 2012).

Splines approach the problem as a piecewise fit and, therefore, can be used for large amounts of accurate data, such as those obtained for the calibration of equipment and material properties. Spline functions consider small subsets of the data and fit them with lower-order polynomials, as sketched in Figure 3.24. The cubic spline is the most commonly used function in this exact fit, though polynomials of other orders may also be used. Spline interpolation is an important technique used in a wide range of applications of engineering interest. Measurements of material properties such as density, thermal conductivity, mass diffusivity, reflectivity, and specific heat, as well as the results from calibrations of equipment and sensors such as thermocouples, often give rise to large sets of very accurate data. All these interpolation techniques are conveniently available in many commercially available software products, such as Matlab, and can be used to obtain results at

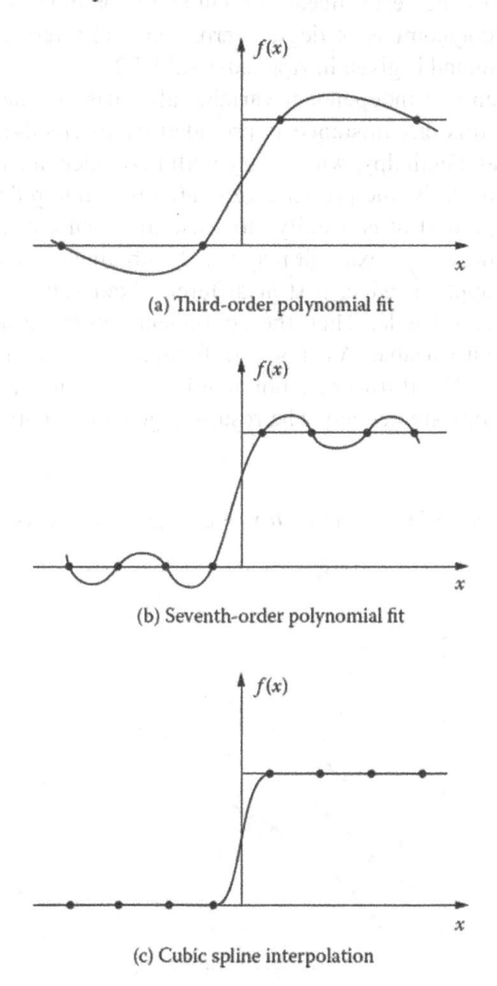

FIGURE 3.24 Interpolation with single polynomials over the entire range and with piecewise cubic splines for a step change in the dependent variable.

intermediate points by interpolation or predict the trends beyond the given domain by extrapolation (Recktenwald, 2000; Jaluria, 2012; Chapra and Canale, 2014). However, extrapolation comes with considerable uncertainty.

Use of Matlab commands considerably simplifies interpolation and curve fitting. The *interpl* command is useful in obtaining interpolated values from the given data set, using a specified interpolating polynomial, such as linear, cubic, and spline. The *interp2* command is used for two-dimensional curve fitting and *interp3* for three-dimensional. Therefore, the following commands would provide spline interpolation

```
interp1(v,t,vp,'spline')
```

or simply

```
yp = spline(v,t,vp)
```

where v and t are the two arrays of data for the independent and dependent variables, respectively and vp is the array of v values where interpolated results yp are desired, using spline interpolation. Also, 'spline' is replaced by 'nearest', 'linear', or 'cubic' to obtain the nearest-neighbor, linear, or cubic interpolation with polynomials of degree zero, one, and three respectively. A program to interpolate using this command is given in Appendix A.M.5.2.

Functions of more than one independent variable also arise in many problems of practical interest. An example of this circumstance is provided by thermodynamic properties such as density, internal energy and enthalpy, which vary with two independent variables, such as temperature and pressure. Similarly, the pressure generated by a pump depends on both the speed and the flow rate. Again, a best fit is usually more useful because of the inaccuracies involved in obtaining the data. However, an exact fit may also be obtained. Curve fitting with the chosen order of polynomials is applied twice, first at different fixed values of one variable to obtain the curve fit for the other variable. Then the coefficients obtained are curve fitted to reflect the dependence on the first variable. As shown in Figure 3.25, nine data points are needed for second-order polynomials. For third-order polynomials, 16 points are needed, and for fourth-order polynomials, 25 points are needed. The resulting general equation for the curve fit shown in Figure 3.25 is

$$y = \left(a_0 + a_1 x_2 + a_2 x_2^2\right) + \left(b_0 + b_1 x_2 + b_3 x_2^2\right) x_1 + \left(c_0 + c_1 x_2 + c_2 x_2^2\right) x_1^2 \tag{3.34}$$

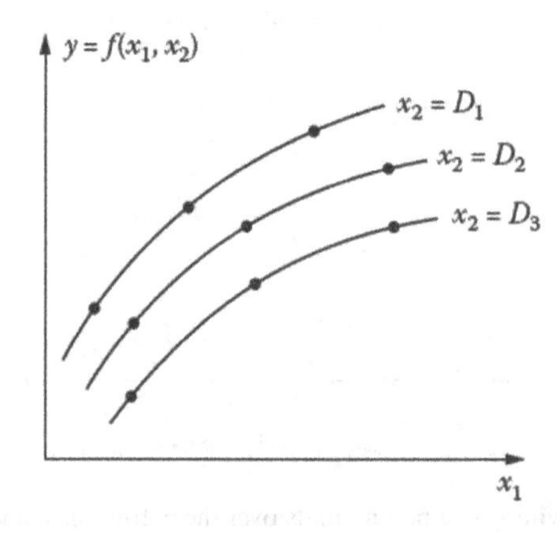

FIGURE 3.25 A function $f(x_1, x_2)$ of two independent variables x_1 and x_2, showing the nine data points needed for an exact fit with second-order polynomials.

3.5.2 Best Fit

The data obtained in many engineering applications have a significant amount of associated error. Experimental data, for instance, would generally have some scatter due to error whose magnitude depends on the instrumentation and the arrangement employed for the measurements. In such cases, requiring the interpolating curve to pass through each data point is not appropriate. In addition, large data sets are often generated and a single curve for an exact fit leads to high-order polynomials that are again not satisfactory. A better approach is to derive a curve that provides a best fit to the given data by somehow minimizing the difference between the given values of the dependent variable and those obtained from the approximating curve. Figure 3.26 shows a few circumstances where a best fit is much more satisfactory than an exact fit. The curve from a best fit represents the general trend of the data, without necessarily passing through every given point. It is particularly useful in deriving correlating equations to quantitatively describe the

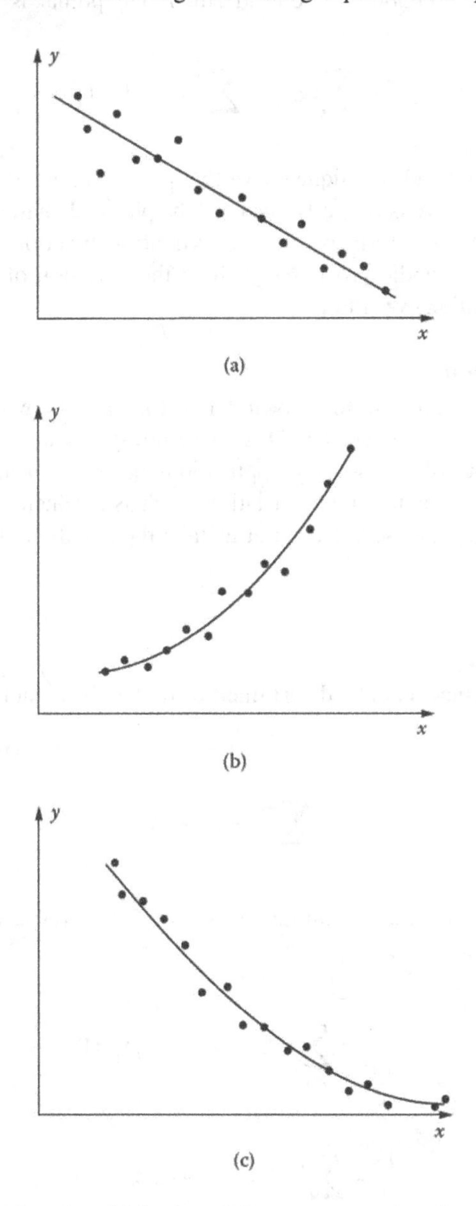

FIGURE 3.26 Data distributions for which a best fit is more appropriate than an exact fit.

thermal system or process under consideration. For instance, correlating equations derived from experimental data on heat and mass transfer from bodies of different shapes are frequently used instead of solving the relevant convection problem. Similarly, correlating equations representing the behavior of an internal combustion engine under various fuel-air mixtures are useful in the analysis and design of engines.

Several criteria can be used to derive the curve that best fits the data. If the approximating curve is denoted by $f(x)$ and the given data by (x_i, y_i), as before, the error e_i is given by $e_i = y_i - f(x_i)$. Then, one method for obtaining a best fit to the data is to minimize the sum of these individual errors; that is, minimize Σe_i. Because errors tend to cancel out in this case, being positive or negative, the sum of absolute values of the error, $\Sigma|e_i|$, may be minimized instead. Another approach is to minimize the largest error. However, all these approaches may not yield a unique curve. The most commonly used approach for a best fit is the method of least squares, in which the sum S of the squares of the errors is minimized. The expression for S, considering n data points, is

$$S = \sum_{i=1}^{n} (e_i)^2 = \sum_{i=1}^{n} [y_i - f(x_i)]^2 \tag{3.35}$$

This approach generally yields a unique curve that provides a good representation of the given data, if the approximating curve is properly chosen. The physical characteristics of the given problem may be used to choose the form of the approximating function. For instance, a sinusoidal function may be used for periodic processes such as the variation of the average daily ambient temperature at a given location over the year.

3.5.2.1 Linear Regression

The procedure of obtaining a best fit to a given data set is often known as *regression*. Let us first consider fitting a straight line to a data set. This curve fitting is known as *linear regression* and is important in a wide variety of engineering applications because linear approximations are often desirable and also because many nonlinear variations such as exponential and power-law forms can be reduced to a linear best fit, as seen later. Let us take the equation of the straight line for curve fitting as

$$f(x) = a + bx \tag{3.36}$$

where a and b are the coefficients to be determined from the given data. For a best fit, the sum S is to be minimized, where

$$S = \sum_{i=1}^{n} [y_i - (a + bx_i)]^2 \tag{3.37}$$

The minimum occurs when the partial derivatives of S with respect to a and b are both zero. This gives

$$\frac{\partial S}{\partial a} = \sum_{i=1}^{n} [-2(y_i - a - bx_i)] = 0 \tag{3.38a}$$

$$\frac{\partial S}{\partial b} = \sum_{i=1}^{n} [-2(y_i - a - bx_i)x_i] = 0 \tag{3.38b}$$

These equations may be simplified and expressed as

$$\sum y_i - \sum a - \sum bx_i = 0 \quad \text{and} \quad \sum y_i x_i - \sum ax_i - \sum bx_i^2 = 0$$

These equations may be written for the unknowns a and b as

$$na + b\sum x_i = \sum y_i \tag{3.39}$$

$$a\sum x_i + b\sum x_i^2 = \sum x_i y_i \tag{3.40}$$

Here, all the summations are over the n data points, from $i = 1$ to $i = n$. These two simultaneous linear equations may be solved to obtain the coefficients a and b. The resulting equation $f(x) = a + bx$ then provides a best fit to the given data by a straight line, as sketched in Figure 3.26(a).

The spread of the data before regression is applied is given by the sum S_m where

$$S_m = \sum_{i=1}^{n} (y_i - y_{avg})^2$$

y_{avg} being the average, or mean, of the dependent variable in the given data. Then the extent of improvement due to curve fitting by a straight line is indicated by the reduction in the spread of the data, given by the expression

$$r^2 = \frac{S_m - S}{S_m} \tag{3.41}$$

Here, r is known as the *correlation coefficient*. A good correlation for linear regression is indicated by a high value of r, the maximum of which is 1.0. The given data may also be plotted along with the regression curve in order to demonstrate how good a representation of the data is provided by the best fit, as seen in Figure 3.26.

3.5.2.2 Polynomial Best Fit

In general, an mth-order polynomial may also be used to fit the data. Then $m = 1$ refers to the linear regression presented in the preceding section. Let us consider a polynomial given as

$$f(x) = c_0 + c_1 x + c_2 x^2 + \ldots\ldots + c_m x^m \tag{3.42}$$

Then the sum S of the squares of the differences between the data points and the corresponding values from the approximating polynomial is given by

$$S = \sum_{i=1}^{n} \left[y_i - \left(c_0 + c_1 x_i + c_2 x_i^2 + \ldots\ldots + c_m x_i^m \right) \right]^2 \tag{3.43}$$

The coefficients c_0, c_1, …, c_m are determined by extending the procedure outlined earlier for linear regression. Therefore, S is differentiated with respect to each of the coefficients and the partial derivatives are set equal to zero in order to minimize S. The following system of $(m + 1)$ equations is then obtained for the unknown coefficients:

$$\begin{aligned} nc_0 + c_1\sum x_i + c_2\sum x_i^2 + \ldots + c_m\sum x_i^m &= \sum y_i \\ c_0\sum x_i + c_1\sum x_i^2 + c_2\sum x_i^3 + \ldots + c_m\sum x_i^{m+1} &= \sum x_i y_i \\ &\vdots \\ c_0\sum x_i^m + c_1\sum x_i^{m+1} + c_2\sum x_i^{m+2} + \ldots + c_m\sum x_i^{2m} &= \sum x_i^m y_i \end{aligned} \tag{3.44}$$

where all the summations in the preceding equations are carried out over all the n data points, $i = 1$ to $i = n$.

A solution to these equations yields the desired polynomial for a best fit, as given by Equation (3.42). For most practical problems, m is restricted to a small number, generally from 1 to 4, in order to simplify the calculations and to obtain simple correlating curves that approximate the data. The correlation coefficient r is again defined by Equation (3.41) and is calculated to determine how good a fit to the given data is obtained by the resulting polynomial. Appendix A.M.5.3 gives a Matlab program for best fit of given data, using different polynomials.

3.5.2.3 Nonpolynomial Forms and Linearization

The method of least squares is not restricted to polynomials for curve fitting and may easily be extended to various other forms in which the constants of the function appear as coefficients. This substantially expands the applicability and usefulness of a best fit. Important examples of such nonpolynomial forms are provided by periodic processes, which are of particular interest in environmental processes and systems. For instance, the following function may be used for curve fitting of data in a periodic process, with ω as the frequency in radians/s:

$$f(x) = A\sin(\omega x) + B\cos(\omega x) \tag{3.45}$$

The sum S is defined and then differentiated with respect to the coefficients A and B, setting these derivatives equal to zero. This gives rise to two linear equations that are solved for A and B. Similarly, other nonpolynomial forms may be employed, their choice being guided by the expected physical behavior of the process.

Several important nonpolynomial forms of the function for curve fitting can be linearized so that linear regression can be applied. Among these, the most common forms are exponential and power-law variations, which may be defined as

$$f(x) = Ae^{ax} \quad f(x) = Bx^b \tag{3.46}$$

The corresponding linearized forms are given by, respectively,

$$\ln[f(x)] = \ln(A) + ax \quad \ln[f(x)] = \ln(B) + b\ln(x) \tag{3.47}$$

Here, $\ln(x)$ represents the natural logarithm of x. In these two cases, if a dependent variable Y is defined as $Y = \ln[f(x)]$ and an independent variable X as $X = x$ in the first case and $X = \ln(x)$ in the second, then the two equations become linear in terms of X and Y. These equations may be written as $Y = C + DX$ and linear regression may be applied with the new variables X and Y to obtain the intercept C, which is $\ln(A)$ or $\ln(B)$, and the slope D, which is a or b for the two cases. Then, A or B is given by $\exp(C)$ and a or b by D. Therefore, from this linear fit, the constants A, a, B, and b can be determined.

Similarly, other nonpolynomial forms such as

$$f(x)=\frac{ax}{b+x} \qquad f(x)=a+\frac{b}{x} \qquad f(x)=\frac{a}{b+x} \tag{3.48}$$

can be written as

$$\frac{1}{f(x)}=\frac{1}{a}+\left(\frac{b}{a}\right)\frac{1}{x} \qquad f(x)=a+b\left(\frac{1}{x}\right) \qquad \frac{1}{f(x)}=\left(\frac{b}{a}\right)+\left(\frac{1}{a}\right)x$$

and linearized as

$$Y=\left(\frac{1}{a}\right)+\left(\frac{b}{a}\right)X \qquad Y=a+bX \qquad Y=\left(\frac{b}{a}\right)+\left(\frac{1}{a}\right)X \tag{3.49}$$

by substituting Y for $f(x)$ in the second case and for $1/f(x)$ in the first and third cases. X is substituted for $1/x$ in the first and second cases and for x in the last case. Linear regression may be applied to these equations to obtain the coefficients a and b.

Many problems of interest in the design of thermal systems are governed by exponential, power-law, and other forms (such as those just given), and linear regression may be employed to obtain the best fit to such data. For instance, many heat transfer correlations can be taken as power-law variations in terms of parameters such as Reynolds, Prandtl, and Grashof numbers. Such curve fitting is of considerable value because the resulting expressions can be easily employed in design as well as in optimization, as will be seen in later chapters.

3.5.2.4 More than One Independent Variable

Multiple linear regression may be developed in a very similar manner to that outlined earlier for a single independent variable. Consider, for instance, the dependent variable y as a linear function of independent variables x_1 and x_2, given by

$$y=f(x_1,x_2)=c_0+c_1x_1+c_2x_2 \tag{3.50}$$

where c_0, c_1, and c_2 are constants to be determined to obtain the best fit. We can define the sum S as before and differentiate it with respect to these coefficients, setting the derivatives equal to zero. This gives rise to the following equations for the coefficients:

$$nc_0+c_1\sum x_{1,i}+c_2\sum x_{2,i}=\sum y_i \tag{3.51}$$

$$c_0\sum x_{1,i}+c_1\sum (x_{1,i})^2+c_2\sum x_{1,i}x_{2,i}=\sum x_{1,i}y_i \tag{3.52}$$

$$c_0\sum x_{2,i}+c_1\sum x_{1,i}x_{2,i}+c_2\sum (x_{2,i})^2=\sum x_{2,i}y_i \tag{3.53}$$

These simultaneous linear equations may be solved for c_0, c_1, and c_2 to obtain the best fit. A regression plane is obtained instead of a line because y varies with two independent variables x_1 and x_2. The procedure can be extended to multiple linear regression with more than two independent variables. Similarly, multiple polynomial regression can also be derived for a best fit.

Linearization of nonlinear functions such as exponential and power-law variations can also be carried out for multiple independent variables in many cases, following the procedure outlined for a single independent variable. Thus, if y is assumed to be of the general form

$$y = c_0 x_1^{c1} x_2^{c2} x_3^{c3} \ldots\ldots x_m^{cm} \tag{3.54}$$

the equation may be transformed into a linear one by taking its natural logarithm to give

$$\ln(y) = \ln(c_0) + c_1 \ln(x_1) + c_2 \ln(x_2) + \ldots + c_m \ln(x_m) \tag{3.55}$$

Multiple linear regression may now be applied to obtain the coefficients for a best fit to the given data.

3.3.2.5 Concluding Remarks

Curve fitting is important in the design and optimization of thermal systems because it allows data obtained from experiments and from numerical simulations to be cast in useful forms from which the desired information can be extracted with ease. Equations representing material properties, heat transfer data, characteristics of equipment such as pumps and compressors, results from computational runs, cost and pricing information, etc., are all valuable in the design process as well as in formulating and solving the optimization problem. Though an exact fit of the data is used in some cases, particularly spline functions for material property representations, the best fit is much more frequently used because of the errors associated with the data and large sets of data that are often of interest. The following examples illustrate the use of the preceding analysis to obtain appropriate functions for best fit.

Example 3.8

The temperature T of a small copper sphere cooling in air is measured as a function of time τ to yield the following data:

τ (s)	0.2	0.6	1.0	1.8	2.0	3.0	5.0	6.0	8.0
T (°C)	146.0	129.5	114.8	90.3	85.1	63.0	34.6	25.6	14.1

An exponential decrease in temperature is expected from lumped mass modeling. Obtain a best fit to represent these data.

SOLUTION

The given temperature-time data are to be best fitted using an exponential variation, as obtained earlier for a lumped mass in convective cooling. Let us take the equation for the best fit to be

$$T = Ae^{-a\tau}$$

where A and a are constants to be determined.

Taking natural logarithms of this equation, we obtain

$$\ln(T) = \ln(A) - a\tau$$

This equation may be written as

$$Y = C_1 + C_2 X$$

where $Y = \ln(T)$, $C_1 = \ln(A)$, $C_2 = -a$, and $X = \tau$. Therefore, linear regression may be applied to the given data by employing the variables Y and X. The two equations for C_1 and C_2 are obtained as

$$nC_1 + C_2 \sum X_i = \sum Y_i$$

$$C_1 \sum X_i + C_2 \sum X_i^2 = \sum X_i Y_i$$

Here, n is the number of data points, being nine here, and the summation is over all the data points. Therefore, these summations are obtained, using $\ln(T)$ and τ as the variables, and C_1 and C_2 are calculated from these equations as

$$C_1 = \frac{\Sigma Y_i \Sigma X_i^2 - \Sigma X_i \Sigma X_i Y_i}{n\Sigma X_i^2 - (\Sigma X_i)^2} \quad \text{and} \quad C_2 = \frac{n\Sigma X_i Y_i - \Sigma X_i \Sigma Y_i}{n\Sigma X_i^2 - (\Sigma X_i)^2}$$

These two equations may be solved analytically or a simple computer program may be written to carry out these computations. A numerical scheme provides flexibility and versatility so that different data sets can easily be considered for best fit. The resulting values of C_1 and C_2 are

$$C_1 = 5.0431 \text{ and } C_2 = -0.2998$$

Therefore, $A = \exp(5.0431) = 154.948$ and $a = 0.2998$. This gives the equation for the best fit to the given data as

$$T = 154.948 \exp(-0.2998\tau)$$

This may be approximated as $T = 154.95 \exp(-0.3\tau)$ to simplify the calculations. The given data may be compared with the values obtained from this equation. The nine values of T from this equation are calculated as 145.93, 129.44, 114.81, 90.33, 85.07, 63.03, 34.61, 25.64, and 14.08. Therefore, the given data are closely represented by this equation.

As mentioned previously, a computer program may be developed to calculate the summations needed for generating the two algebraic equations for C_1 and C_2, using programming languages like Fortran90 and C++. However, Matlab is particularly well-suited for such problems because the command *polyfit* yields the best fit to a chosen order of the polynomial for curve fitting (see Appendix A.M.5.4). For instance, the following program may be used:

```
%Input Data
 tau=[0.2 0.6 1.0 1.8 2.0 3.0 5.0 6.0 8.0];
 t0=[146.0 129.5 114.8 90.3 85.1 63.0 34.6 25.6 14.1];
 t=log(t0);
% Curve Fit
 t1=polyfit(tau,t,1);
 a=t1(1)
 A=exp(t1(2))
```

Here the input data are entered and the chosen exponential function is linearized by the use of the natural logarithm. Then the *polyfit* command is used with the two variables and the order of the polynomial given as 1, or linear. Matlab specifies polynomials in descending order of the independent variable, so the *polyfit* command yields the two constants in the order C_2 and C_1, i.e., the slope first and then the intercept. These are given by t1(1) and t1(2) in the program. Then a is simply t1(1) and A is the exponential of t1(2). The program yields the same results for a and A as given above. Further details on such algorithms in Matlab may be obtained from Recktenwald (2000), Mathews and Fink (2004), and Jaluria (2012).

Example 3.9

The flow rate Q in circular pipes is measured as a function of the diameter D and the pressure difference Δp. The data obtained for the flow rate in m^3/s are

D(*m*)	0.3	0.5	1.0	1.4
Δp (atm)				
0.5	0.13	0.43	2.1	4.55
0.9	0.25	0.81	4.0	8.69
1.2	0.34	1.12	5.5	11.92
1.8	0.54	1.74	8.59	18.63

Obtain a best fit to these data, assuming a power-law dependence of Q on the two independent variables D and Δp.

SOLUTION

The variation of Q with D and Δp may be written for a power-law variation as

$$Q = BD^a(\Delta p)^b$$

Taking the natural logarithm of this equation, we obtain

$$\ln(Q) = \ln(B) + a\ \ln(D) + b\ \ln(\Delta p)$$

This equation may be written as

$$Y = C_1 + C_2X_1 + C_3X_2$$

where $Y = \ln(Q)$, $C_1 = \ln(B)$, $C_2 = a$, $C_3 = b$, $X_1 = \ln(D)$, and $X_2 = \ln(\Delta p)$. Therefore, multiple linear regression, as presented in the text, may be applied with $\ln(Q)$ taken as the dependent variable Y and $\ln(D)$ and $\ln(\Delta p)$ taken as the two independent variables X_1 and X_2. A computer program may be written to enter the data and calculate the summations over the given 16 data points.

The resulting equations for the constants C_1, C_2, and C_3 are obtained as

$$16C_1 - 6.243C_2 - 0.114C_3 = 9.555$$
$$-6.243C_1 + 8.173C_2 + 0.044C_3 = 9.490$$
$$-0.114C_1 + 0.044C_2 + 3.481C_3 = 3.762$$

These equations are solved to yield the three constants as

$$C_1 = 1.5039 C_2 = 2.3039 C_3 = 1.1005$$

Therefore, $B = \exp(C_1) = 4.4991$, $a = C_2 = 2.3039$, and $b = C_3 = 1.1005$. Rounding these off to the second place of decimal, the best fit to the given data is given by the equation

$$Q = 4.5D^{2.3}(\Delta p)^{1.1}$$

It can be easily shown that the best fit is a close representation of the data by comparing the values obtained from this equation with the given data.

3.6 SUMMARY

This chapter discusses the modeling of thermal systems, this being a crucial element in the design and optimization process. Because of the complexity of typical thermal systems, it is necessary to simplify the analysis so that the inputs needed for design can be obtained with the desired accuracy and without spending exorbitant time and effort on computations or experiments. The model also allows one to minimize the number of parameters that describe a given system or process and to generalize the results so that these may be used for a wide range of conditions.

Several types of models are considered, such as analog, mathematical, physical, and numerical. Analog models are of limited value because these models themselves have to be ultimately solved by mathematical and numerical modeling. This chapter considers mathematical and physical modeling in detail, leaving numerical modeling, in which the applicable equations are solved by digital computation, for the next chapter, which also presents numerical simulation. In mathematical modeling, both theoretical models, derived on the basis of physical insight, and empirical models, which simply curve fit available data, are considered because both of these lead to mathematical equations that characterize the behavior of a given system. Curve fitting is discussed in detail following physical modeling because it is used to develop equations from experimental data as well as from numerical results.

Mathematical modeling is at the very core of modeling of thermal systems because it brings out the basic considerations with respect to the given system, focusing on the dominant mechanisms and neglecting smaller aspects. It simplifies the problem by using approximations and idealizations. Conservation laws are used to derive the applicable equations, which may be algebraic equations, integral equations, ordinary differential equations, partial differential equations, or combinations of these. These equations can frequently be simplified further by neglecting terms that are relatively small, often employing nondimensionalization of the equations to determine which terms are negligible.

Physical modeling refers to the process of developing a model that is similar in shape and geometry to the given component or system. The given system is often represented by a scaled-down version on which experiments are performed to provide information that is not easily available through mathematical modeling. Dimensional analysis is employed to determine the important dimensionless groups that determine the behavior of the given system to reduce the experimental effort. These parameters are also used to establish similitude between the model and the actual system or prototype. Various kinds of similarity are outlined, including geometric, kinematic, dynamic, thermal, and mass transfer. The conditions needed for these types of similarity are presented.

The results from experiments and mathematical modeling are often obtained at discrete values of the variables. These data can be obtained in a more useful form by curve fitting, which yields mathematical equations that represent the data. In an exact fit, the curve passes through each data point, yielding the exact value at these points. It is particularly well-suited for relatively small but very accurate data sets. A best fit provides a close approximation to the given data without requiring the curve to pass through each data point. Thus, a best fit is appropriate for large data sets with significant error in the results. The method of least squares, which minimizes the sum of the squares of the differences between the data and the predicted values from the curve fit, is the most commonly used approach. A polynomial best fit, including linear regression, is extensively used for engineering systems. Nonpolynomial forms such as exponential and power-law variations are linearized and the curve fit is obtained by linear regression. Multiple linear regression is used for functions of more than one independent variable.

With the help of a suitable model, the behavior of the system may be studied under a variety of operating and design conditions, making it possible to consider and evaluate different designs. The model may be improved by employing the results from simulation and design. A relatively simple model may be used at the beginning; subsequently, the assumptions made can be relaxed and the model can be gradually transformed into a more sophisticated and accurate one.

REFERENCES

Bejan, A. (1993). *Heat transfer.* New York: Wiley.
Bergman, T.L., Lavine, A.S., Incropera, F.P., & Dewitt, D.P. (2017). *Fundamentals of heat and mass transfer* (8th ed.). New York: Wiley.
Burmeister, L.C. (1993). *Convective heat transfer* (2nd ed.). New York: Wiley.
Carnahan, B.H., Luther, H.A., & Wilkes, J.O. (1969). *Applied numerical methods.* New York: Wiley.
Cengel, Y.A., & Boles, M.A. (2014). *Thermodynamics: an engineering approach* (8th ed.). New York: McGraw-Hill.
Chapra, S.C., & Canale, R.P. (2014). *Numerical methods for engineers* (7th ed.). New York: McGraw-Hill.
Doebelin, E.O. (1980). *System modeling and response.* New York: Wiley.
Eckert, E.R.G., & Drake, R.M. (1972). *Analysis of heat and mass transfer.* New York: McGraw-Hill.
Gebhart, B. (1971). *Heat transfer* (2nd ed.). New York: McGraw-Hill.
Gerald, C.F., & Wheatley, P.O. (2003). *Applied numerical analysis* (7th ed.). New York: Pearson.
Hornbeck, R.W. (1975). *Numerical methods.* Englewood Cliffs, NJ: Prentice-Hall.
Howell, J.R., & Buckius, R.O. (1992). *Fundamentals of engineering thermodynamics* (2nd ed.). New York: McGraw-Hill.
Incropera, F.P., & Dewitt, D.P. (2001). *Fundamentals of heat and mass transfer* (5th ed.). New York: Wiley.
Jaluria, Y. (2012). *Computer methods for engineering with Matlab applications*, Boca Raton, FL: CRC Press.
Jaluria, Y. and Torrance, K.E. (2003). *Computational heat transfer* (2nd ed.). Washington, DC: Taylor & Francis.
Mathews, J.H., & Fink, K.D. (2004). *Numerical methods using MATLAB* (4th ed.). Upper Saddle River, NJ: Prentice-Hall.
Papanicolaou, E., & Jaluria, Y. (1994). Mixed convection from simulated electronic components at varying relative positions in a cavity, *Journal of Heat Transfer*, 116, 960-970.
Pritchard, P.J., & Mitchell, J.W. (2015). *Fox and McDonald's introduction to fluid mechanics* (9th ed.). New York: Wiley.
Recktenwald, G. (2000). *Numerical methods with MATLAB.* Upper Saddle River, NJ: Prentice-Hall.
Rieder, W.G., & Busby, H.R. (1986). *Introductory engineering modeling emphasizing differential models and computer simulation.* New York: Wiley.
Roache, P.J. (2009). *Fundamentals of verification and validation.* Albuquerque, NM: Hermosa Pub.
Schuring, D.J. (1977). *Scale models in engineering.* New York: Pergamon.
Szucs, E. (1977). *Similitude and modeling.* New York: Elsevier.
Wellstead, P.E. (1979). *Introduction to physical system modeling.* New York: Academic Press.
White, F.M. (2015). *Fluid mechanics* (8th ed.). New York: McGraw-Hill.

PROBLEMS

Note: In all problems dealing with model development, list the assumptions, approximations, and idealizations employed; give the resulting equations; and, whenever possible, give the analytical solution. Symbols may be used for the appropriate physical quantities. Matlab commands such as *polyfit* and *interp1* may be used to check on the results obtained from numerical calculations.

3.1 An energy storage system consists of concentric cylinders, the inner being of radius R_1, the outer of radius R_2 and both being of length L, as shown in Figure P3.1. The inner cylinder is heated electrically and supplies a constant heat flux q to the material in the outer cylinder, as shown. The annulus is packed with high-conductivity metal pieces. Assuming that the system is well-insulated from the environment and that the annular region containing the metal pieces may be taken as isothermal,

a. Obtain a mathematical model for the system.
b. If the maximum temperature is given as T_{max}, obtain the time for which heating may be allowed to occur, employing the usual symbols for properties

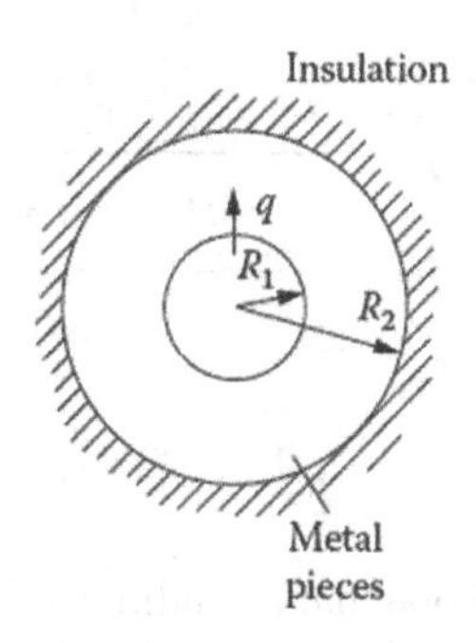

FIGURE P3.1

3.2 Solid plastic cylinders of diameter 1 cm and length 30 cm are heat treated by moving them at constant speed U through an electric oven of length L, as shown in Figure P3.2. The temperature at the oven walls is T_s and the air in the oven is at temperature T_a. The convective heat transfer coefficient at the plastic surface is given as h and the surface emissivity as ε. The cylinders are placed perpendicular to the direction of motion and are rotated as they move across the oven. Develop a simple mathematical model for obtaining the temperature in the plastic cylinders as a function of the temperatures T_s and T_a, h, L, and U, for design of the system. Clearly indicate the assumptions and approximations made.

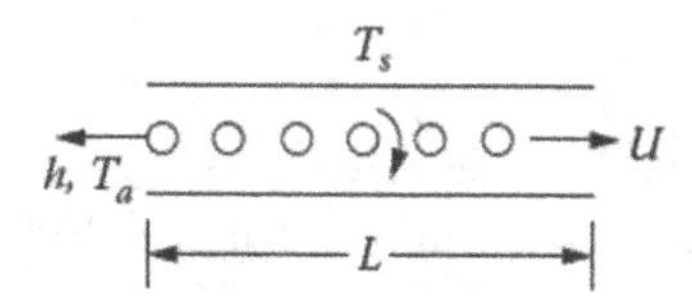

FIGURE P3.2

3.3 A chemical industry needs hot water at temperature $T_c \pm \Delta T_c$ for a chemical process. For this purpose, a storage tank of volume V and surface area A is employed. Whenever hot water is withdrawn from the tank, cold water at temperature T_a, where T_a is the ambient temperature, flows into the tank. A heater supplying energy at the rate of Q turns on whenever the temperature reaches $T_c - \Delta T_c$ and turns off when it reaches $T_c + \Delta T_c$. The heater is submerged in the water contained in the tank. Assuming uniform temperature in the tank and a convective loss to the environment at the surface, with a heat transfer coefficient h, obtain a mathematical model for this system. Sketch the expected temperature T of water in the tank as a function of time for a given flow rate $\dot{m}$ of hot water and also for the case when there is no outflow, $\dot{m} = 0$.

3.4 Consider a cylindrical rod of diameter D undergoing thermal processing and moving at a speed U as shown in Figure P3.4. The rod may be assumed to be infinite in the direction of motion. Energy transfer occurs at the outer surface, with a constant heat flux input q and convective loss to the ambient at temperature T_a and heat transfer coefficient h. Assuming one-dimensional (uniform temperature over any cross-section), steady transport, obtain the energy equation and the relevant boundary conditions. By nondimensionalization, determine the main dimensionless parameters. Finally, obtain $T(x)$ for (*a*) $h = 0$ and (*b*) $q = 0$.

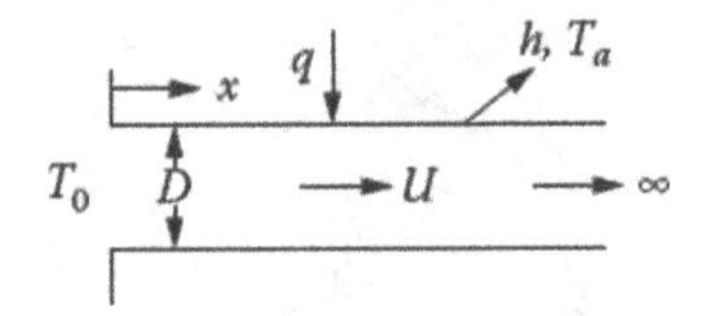

FIGURE P3.4

3.5 Give the energy equation and boundary conditions for the steady-state, two-dimensional (axisymmetric) case for the preceding physical problem. Derive the dimensionless parameters that arise using the nondimensionalization of the energy equation and the boundary conditions.

3.6 During the heat treatment of steel bolts, the bolts are placed on a conveyor belt that passes through a long furnace at speed U as shown in Figure P3.6. In the first section, the bolts are heated at a constant heat flux q. In the second and third sections, they lose energy by convection to the air at temperature T_a at convective heat transfer coefficients h_1 and h_2 in the two sections, respectively.

   a. Assuming lumped mass analysis is valid, obtain the governing equations for the three sections and outline the mathematical model thus obtained.
   b. Sketch the temperature variation qualitatively as a function of distance x from the entrance.

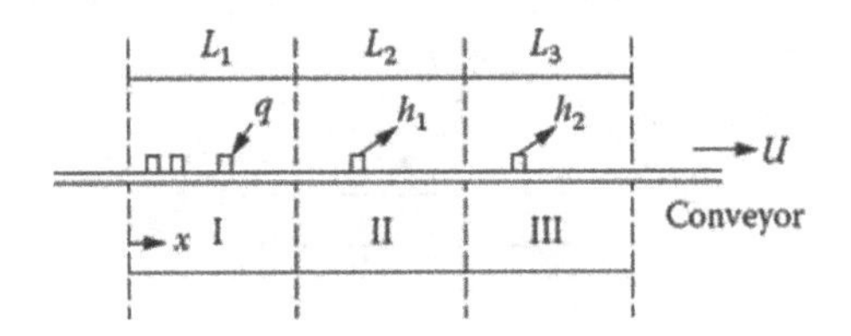

FIGURE P3.6

3.7 In a manufacturing process, a metal block of surface area A and volume V is melted in a furnace. The initial temperature of the block is T_i, the melting point is T_m, and the final temperature is T_f, where $T_f > T_m > T_i$. The block is exposed to a constant heat flux input q due to radiation and also loses energy by convection to the surrounding air at T_a with a convective heat transfer coefficient h. Employing the usual symbols for the properties and assuming no temperature variation in the block:

   a. Obtain a suitable mathematical model for the process.
   b. Qualitatively sketch the temperature variation with time.
   c. If the temperatures T_i, T_m, and T_f are given, what are the variables for operation and design?

3.8 A water cooler is to be designed to supply cold drinking water with a given time-dependent mass flow rate $\dot{m}$. Assume a cubical tank of cold water enclosed in an insulation of uniform thickness. Water at the ambient temperature flows into the tank to make up the cold water outflow. The refrigeration unit turns on if the water temperature reaches a value T_{max} and turns off when it drops to T_{min}, thus maintaining the temperature between these two values. Develop a simple mathematical model for this system.

3.9 It is necessary to model and simulate a hot water distribution system consisting of a tank, pump, and pipes. The heat input Q to the tank is given and the ambient temperature is T_a, with h as the heat transfer coefficient for heat loss. Develop a simple mathematical model for this system.

3.10 We wish to model a vapor compression cooling system, such as the one shown in Figure 1.8(a). For a simple model based on the thermodynamic cycle, list the main approximations and idealizations you would employ to obtain the model. Justify these in a few sentences.

3.11 A mathematical model is to be developed to simulate a power plant, such as the one shown in Figure 2.17. For a simple model based on the thermodynamic cycle, list the approximations and idealizations you would employ to obtain the model. Briefly justify these in a few sentences.

3.12 In the hot water storage system considered in Example 3.5, if the ambient temperature is 20°C and the heat transfer coefficient is 20 W/m²K, sketch the temperature distribution in the steady-state case. What are the important parameters in this problem? How does the solution vary with these parameters?

3.13 In a heat treatment furnace, a thin metallic sheet of thickness d, height L, and width W is employed as a shield. On one side of the sheet, hot flue gases at temperature $T_f(x)$ exchange energy with an overall heat transfer coefficient h_f. On the other side, inert gases at temperature $T_g(x)$ have a heat transfer coefficient h_g, as shown in Figure P3.13. The sheet also loses energy by radiation. If $L \gg d$ and $W \gg d$, obtain a mathematical model for calculating the temperature T in the sheet. Assume that T_f and T_g are known functions of height x. Also, take h_f and h_g as known constants. Give the resulting energy equation and its solution, if easily obtainable analytically.

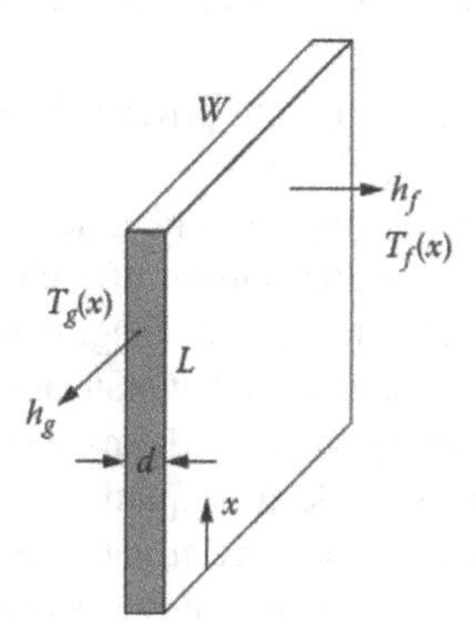

FIGURE P3.13

3.14 For the following systems, consider and briefly discuss the various approximations and idealizations that can be made to simplify the mathematical model. When are these approximations valid and how would you relax them? Outline the nature and type of the equations that you expect to obtain for the different systems.

a. Food-freezing plant to chill vegetables to –10°C by circulating chilled air past the vegetables.
b. A shell and tube heat exchanger, with hot and cold water as the two fluids.
c. A system consisting of pumps and pipe network to transport water from ground level to a tank 100 m high.
d. A vapor compression system for cooling a cold storage room.
e. Flow equipment such as compressors, fans, pumps, and turbines.

3.15 In the electronic system considered in Example 3.7, if the geometry and heat inputs are fixed, what are the design variables in terms of dimensionless parameters? If the maximum temperature in the electronic components is to be restricted for an acceptable design, what physical quantities may be adjusted to reach an acceptable design?

3.16 In a counterflow heat exchanger, the heat loss to the environment is to be included in the mathematical model. Considering the case of the hot fluid on the outside and the colder fluid on the inside, as shown in Figure P3.16, sketch qualitatively the change that the inclusion of this consideration will have on the temperature distribution in the heat exchanger. Also, give the energy equation taking this loss into account.

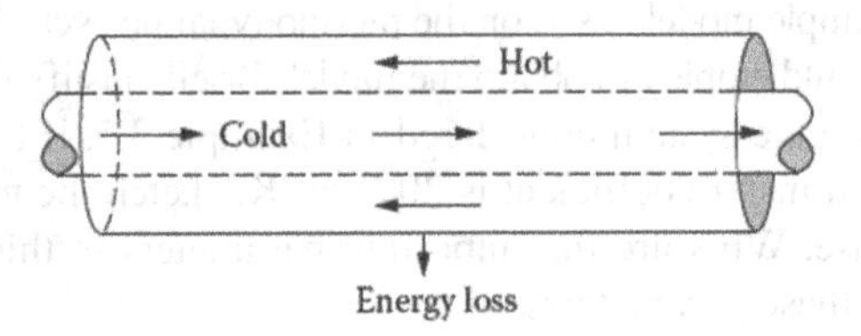

FIGURE P3.16

3.17 Scale-up from a laboratory system to a full-size version is an important consideration in industry. For the problems considered in Example 3.5 and Example 3.6, determine the important parameters that may be used for scale-up and whether it is possible to achieve the desired similarity.

3.18 A scaled-down version of a shell and tube heat exchanger is to be used to simulate the actual physical system to be used in a chemical plant. Determine the dimensionless parameters that must be kept the same in order to ensure similarity between the full-size and scaled-down systems.

3.19 Obtain the dimensionless parameters that govern the scale-down and scale-up of a vapor compression refrigeration system.

3.20 Consider the condensation soldering system discussed in Chapter 2 (Figure 2.4 and Figure 2.6), with boiling liquid at the bottom of a chamber and water-cooled condensing coils at the top, generating a condensing vapor region in the tank. A large electronic circuit board that may be approximated as a thick flat plate at room temperature is immersed in the chamber at time zero. Develop a simple mathematical model to compute the temperature distribution in the plate, giving the relevant equation(s) and boundary and initial conditions. Also, write down the global energy balance equation to determine the energy input into the liquid needed before and after immersion of the board. Make suitable simplifications and assumptions, indicating these in your answer.

3.21 A flat steel (ρ = 10,000 kg/m^3, C = 500 J/kg·K, k = 100 W/m·K) sheet emerges from a furnace at 10 cm/s and 800°C. At distances of 10 m each, there are three rolling dies; see Figure 1.10(d). The initial thickness of the sheet is 2 cm and at each die, a reduction of 20% in thickness occurs. In addition, a temperature rise of 50°C occurs due to friction at each rolling die. The sheet loses energy to the environment, at 20°C, at an overall heat transfer coefficient of 120 W/m^2·K. It is necessary to maintain the temperature of the material higher than 700°C. Using a simple mathematical model of the process, determine the level of heating, or cooling, needed between the rolling stations.

3.22 The average daily temperature in New Brunswick, New Jersey, is obtained by taking data over several years. The results are given as 365 data points, with each point corresponding to a day during the year. A curve fit to these data is to be obtained for the design of air conditioning systems. Will an exact or a best fit be more appropriate? Suggest a suitable form of the function for curve fitting.

3.23 The average daily air temperature at a location is available for each day of 2005. We wish to obtain a best fit to these data and use the equation obtained in a computer model for an environmental thermal system. Choose an appropriate form of the equation that may be employed to curve fit the data and outline the reasons for your choice. Outline the mathematical procedure to determine the constants of the equation chosen for the curve fit.

3.24 A steel sphere at initial temperature T_o is immersed in a cold fluid at temperature T_a and allowed to cool rapidly for hardening. At 20 time intervals τ_i, the corresponding temperature T_i in the sphere is measured, where $i = 1, 2, 3, \ldots, 20$. The temperature variation across the sphere may be taken as negligible. We wish to obtain the best fit to the data collected. What function $f(\tau)$, where $T = f(\tau)$, will you employ for the purpose? Justify your answer.

3.25 In a heat treatment process, a metal cube of side 2 cm, density 6000 kg/m³, and specific heat 300 J/kg·K is heated by convection from a hot fluid at temperature $T_f = 220°C$. The initial temperature of the cube is $T_i = 20°C$. If the temperature T within the cube may be taken as uniform, write down the equation that governs the temperature as a function of time τ. Obtain the general form of the solution. The measured temperature values at different times are given as

τ (min)	0	0.5	1.0	2.0	3.0	6.0
$\frac{T - T_f}{T_i - T_f}$	1.0	0.85	0.72	0.5	0.4	0.14

Obtain a best fit to these data using information from the analytical solution for $T(\tau)$. Sketch the resulting curve and plot the original data to indicate how good a representation of the data is obtained by this curve. From the results obtained, compute the heat transfer coefficient h.

3.26 Obtain a linear best fit to the data given below from a chemical reactor by using the method of least squares:

Concentration (g/m³)	0.1	0.2	0.5	1.0	1.2
Reaction Rate (g/s)	1.75	1.91	2.07	2.32	2.4

Is a linear fit satisfactory in this case?

3.27 The temperature variation with height in the large oil fires in Kuwait several years ago was an important consideration. Measurements of the temperature T versus the height H were taken and presented in dimensionless terms as

H:	1.0	2.0	3.0	4.0	5.0
T:	10.0	7.9	6.9	6.3	5.9

It is given that T varies as $T = A(H)^a$. Using linear regression methods, as applied to such equations, obtain the values of A and a from these data. How accurate is your correlation?

3.28 Experimental runs are performed on a compressor to determine the relationship between the volume flow rate Q and the pressure difference P. It is expected that Q will vary as P^b, where b is a constant. The measurements yield the mass flow rate Q for different pressure differences P as

P (atm)	5.0	10.0	15.0	20.0	25.0	30.0
Q (m³/h)	7.4	13.3	16.5	19.0	20.6	24.3

It is known that there is some error in the data. Will you use a best or an exact fit? Use the appropriate fit to these data and determine the coefficients. Is your equation a good fit?

3.29 Tests are performed on a nuclear power system to ensure safe shutdown in case of an accident. The measurements yield the power output P versus time τ in hours as

τ (hours)	1	3	5	9	10	12
P (MW)	13.0	7.0	5.4	4.7	4.5	4.2

From theoretical considerations, the power is expected to vary as $a + b/\tau$, where a and b are constants. It is also known that there is experimental error in the data. Will you use a best or an exact fit? Use an appropriate fit to these data points and determine the relevant constants. Is it a good curve fit? Briefly explain your answer.

3.30 Experiments are carried out on a plastic extrusion die to determine the relationship between the mass flow rate $\dot{m}$ and the pressure difference P. We expect the relationship to be of the form $\dot{m} = AP^n$, where A and n are constants. The measurements yield the mass flow rate $\dot{m}$ for different pressure differences P as

$\dot{m}$(kg/h)	12.8	15.5	17.5	19.8	22.0
P (atm)	10.0	15.0	20.0	25.0	30.0

Obtain a best fit to these data and determine the coefficients A and n. Is this a good best fit, or should we consider other functional relationships?

3.31 Use *polyfit* in MATLAB to get the best fit to the following data, considering first-, second-, and third-order polynomials. Then plot the data as well as the three best-fit curves obtained. Which is the best fit?

x:	0	0.1	0.2	0.3	0.4	0.5	0.6	0.8	1.0	1.2
y:	0	0.87	1.82	2.86	4.0	5.26	6.65	9.88	13.8	18.52

3.32 The flow rate F is given at various values of the pressure P as

P	0.025	0.05	0.1	0.2	0.3	0.4	0.5
F	1.41	2.54	4.2	5.9	6.9	7.6	7.8

Use the last five points to get an exact fit. Use extrapolation with this fit to obtain values at 0.025 and 0.05. Compare with the given data at these points. Comment on the results.

3.33 Obtain the first-, second-, and third-order best fits to the preceding data. Plot the three curves and the data to determine the best curve to use.

3.34 In a chemical reaction, the effect of the concentration C of a catalyst on the reaction rate is investigated and the experimental results are tabulated as

C (g/m³)	0.1	0.2	0.5	1.0	1.2	1.8	2.0	2.6	3.5	4.0
R (g/s)	1.75	1.91	2.07	2.32	2.40	2.54	2.56	2.53	2.03	1.24

Using the method of least squares and considering polynomials up to the fifth order, obtain a best fit to these data. Which curve provides the best approximation to the given data? Also, compare the results with those obtained in Problem 3.26.

3.35 A small heated metal block cools in air. Its temperature T is measured as a function of time τ and the results are given as

τ (s)	1	2	5	10	15	20	25	30
T (°C)	109.5	99.25	73.78	45.15	26.78	17.24	9.85	6.97

From the physical considerations of this problem, the temperature is expected to decay exponentially, as sketched in Figure P3.35. Obtain a best fit to the given data and determine the two constants A and a.

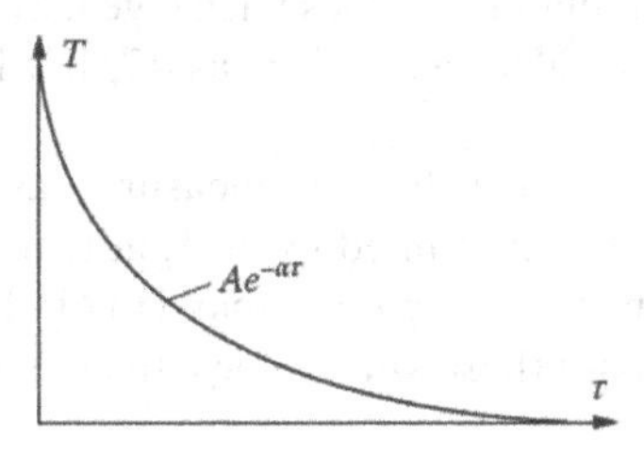

FIGURE P3.35

3.36 The displacement x of a particle in a flow is measured as a function of time τ. The data obtained are

τ (sec):	0.0	1.0	2.0	3.0	4.0	5.0
x (m):	0.0	2.0	8.0	20.0	40.0	62.0

Obtain a linear best fit to these data. From this fit calculate the values at $\tau = 2.0$ and 4.0. Compare these with the given data and comment on the difference. How would you improve the accuracy of the curve fit?

3.37 In an experiment, the signal from a sensor is measured over the velocity range of 0 – 3 m/s. If the signal E is measured as 2, 9, 24, and 47 volts at the velocity V of 0, 1, 2, and 3 m/s, respectively:

a. What is the highest-order polynomial $E(V)$ that exactly fits the given data?
b. Obtain the best linear fit, employing the method of least squares.
c. Determine the value of E at $V = 5$ as calculated from the two curves obtained above and comment on the difference between the two results.

3.38 A thermocouple is being calibrated for temperature measurements by measuring its voltage output V in millivolts and the corresponding fluid temperature T in °C, using a calibration device. For voltage values of 0, 0.1, 0.2, and 0.3 millivolts, the temperature is measured as 15°C, 18.5°C, 24°C, and 31.5°C. Determine the highest-order polynomial that exactly fits the data and give the result as $T = F(V)$. Also, obtain a linear best fit to these data using the method of least squares. Compare the two expressions obtained and comment on the difference.

3.39 In a heat transfer experiment, the heat flux q is measured at four values of the flow velocity, which is related to the fluid flow rate. The velocity V is measured as 0, 1, 2, 3, and 4 m/s and the corresponding heat flux as 1, 2, 9, 29, and 65 W/m^2. It is desired to fit a polynomial to these points so that q may be expressed as $q = f(V)$. What is the highest-order polynomial that may be obtained from these data? Also determine a linear best fit to the given data.

3.40 The volume flow rate Q in m^3/s of water in an open channel with a slight downward slope S and a hydraulic radius R is measured to yield the following data:

R (m)	**0.5**	**1.0**	**1.5**	**2.0**
S				
1.5×10^{-3}	1.91	3.10	4.11	5.03
5×10^{-3}	3.48	6.66	7.51	9.19
9×10^{-3}	4.67	7.59	10.08	12.33

It is expected from theoretical considerations that Q varies with R and S as AR^bS^c, where A, b, and c are constants. Obtain a best fit to the given data and determine these constants.

3.41 Repeat Example 3.9 if the values of the pipe diameter D were given as 0.5, 0.8, 1.4, and 1.9, instead, with all the remaining values unchanged. Similarly, solve the problem again if the pressure difference Δp values were given as 0.7, 1.2, 1.5, and 2.1, instead, with all other values unchanged.

3.42 The kinematic viscosity ν of a fluid is measured as 20.92, 32.39, 38.79, 45.57, and 62.21 m^2/s, with each value multiplied by 10^{-6}, at temperatures 350, 450, 500, 550, and 650 K, respectively. Using the *interp1* command in Matlab, obtain the interpolated values at 400 and 600 K. Compare these with values given in the literature as 26.41×10^{-6} and 52.69×10^{-6} m^2/s.

3.43 The calibration table given below for a copper-constantan thermocouple, which is employed for temperature measurements, gives the temperature T in °C for different values of the output voltage V in millivolts (mV). Using the *interp1* command in Matlab for splines, calculate the interpolated values at 0.9 and 1.75 mV.

T (°C)	10	20	30	40	50	60	70	80
V (mV)	0.391	0.789	1.196	1.611	2.035	2.467	2,908	3.357

4 Numerical Modeling and Simulation

In the preceding chapter, the modeling of thermal systems was presented, and different types of models were discussed. The main focus was on mathematical modeling, which employs approximations, simplifications, and idealizations to obtain a set of mathematical equations that describe a given component, subsystem, or the overall system. Mathematical modeling also brings out the dominant mechanisms and determines the important dimensionless parameters that may be varied in an experimental or analytical study to characterize the behavior of the given thermal system. Physical modeling, which involves experimentation on a scale model of the system, is used as a means to obtain results that are not easily extracted from mathematical modeling. Curve fitting is often used to derive algebraic equations to represent experimental or numerical results, as well as data on material properties, environmental conditions, financial trends, and equipment characteristics.

As a consequence of mathematical and physical modeling, along with curve fitting, mathematical equations that describe the behavior of the thermal system are obtained. These equations are generally linked to each other through material properties, boundary conditions, flow of material and energy, and interaction between the various components of the system. Interest lies in obtaining solutions to this coupled set of equations to determine the behavior and characteristics of the system for wide ranges of design variables and operating conditions. Because of the coupled nature of these equations and because nonlinear algebraic and differential equations, including both ordinary and partial differential equations, commonly arise in thermal systems, analytical solutions are rarely possible and numerical techniques are employed to obtain the desired results.

A *numerical model* of the thermal system refers to a computational or numerical representation of the system on a computer, which may be used to approximate the behavior and characteristics of the system. It consists of a numerical scheme or procedure that would yield a solution to the applicable mathematical equations, with numerically imposed boundary and initial conditions, relevant property data, component characteristics, and other inputs needed for representing the entire system. The numerical algorithm, as well as its implementation on a computer, constitutes the numerical model. Once the model is confirmed to be a valid and accurate representation of the system, the model is subjected to changes in the design variables and operating conditions. This process of studying the behavior of the system by means of a model, rather than by fabricating a prototype, is known as *simulation*. The results obtained allow us to consider many different design possibilities as well as a variety of operating conditions. Different designs may thus be evaluated to choose an acceptable design and safe levels may be established for the operating conditions. These results are also used for optimization of the system design as well as optimizing the operating conditions. Therefore, the success of the design and optimization process is strongly dependent on the numerical modeling and simulation of the system. The basic considerations in the development of a numerical model are first presented in this chapter, followed by a discussion on numerical simulation.

4.1 NUMERICAL MODELING

The numerical solution of mathematical equations that are commonly encountered in engineering applications is covered in a variety of courses dealing with numerical analysis. A large number of books available in this area discuss various methods for different types of equations, along with important aspects such as accuracy, convergence, and stability of these methods (Smith, 1965; Hornbeck, 1975; Atkinson, 1989; Gerald and Wheatley, 2003; Ferziger, 1998). A few others are concerned with problems of engineering interest and discuss the implementation of the algorithm on the computer (Carnahan et al., 1969; James et al., 1985; Jaluria, 1996). There has also been a substantial increase in interest in the solution of the relevant mathematical problems by the use of Matlab, as presented by Recktenwald (2000), Mathews and Fink (2004), Chapra (2017), Jaluria (2012), and others. This chapter presents a brief discussion of numerical modeling in order to indicate the main concerns with respect to thermal systems and the commonly used techniques. For further details, the aforementioned books and others available on this subject may be consulted.

4.1.1 General Features

The main purpose of numerical modeling is to develop a computational code, implemented on a digital computer, which provides a physically valid and accurate representation of the real system and allows the behavior of the system to be determined under different conditions. Thus, a one-to-one correspondence is established between the physical thermal system and the numerical model so that the desired information on system characteristics and behavior can be obtained by subjecting the numerical model to different conditions. As shown schematically in Figure 4.1, the inputs into the physical system, arising due to changes in the design variables or in the operating conditions, are given as corresponding mathematical inputs to the numerical model. The outputs from the model then indicate the expected outputs from the actual physical system, if such a system were to be fabricated and tested.

Numerical modeling starts with the solution of individual equations. The numerical schemes for different equations are then assembled to yield the solution procedure for the set of equations that represent a given part, subsystem, or the complete system. The main steps that may be followed in the numerical modeling of a system are:

1. Determination of the nature and characteristics of each equation to be solved
2. Selection of a numerical scheme for solving each equation
3. Development of numerical code for the solution of each equation
4. Assembly of solution procedures for different coupled equations to model a component or part of the system
5. Validation and estimation of accuracy of numerical model for each system component
6. Compilation of numerical models for system parts to model complete system
7. Check on accuracy of overall system model and its validation

Therefore, a systematic approach is used to obtain a satisfactory numerical model for the complete thermal system. A numerical model is built up for each part, component, or subsystem of the system. Individual models for the different parts or subsystems are finally assembled and the overall numerical model for the system is obtained and validated.

An example. Consider a shell and tube heat exchanger, as shown in Figure 1.5(e). Even if the shell and the tubes are taken as the only parts of the system, the equations that represent different portions or sections of these two may be different. The ends of the shell, the regions near the inlet and outlet, the flow within the shell, and the baffles may all be modeled separately to focus on the dominant transport mechanisms in these regions and to take advantage of possible approximations. Similarly,

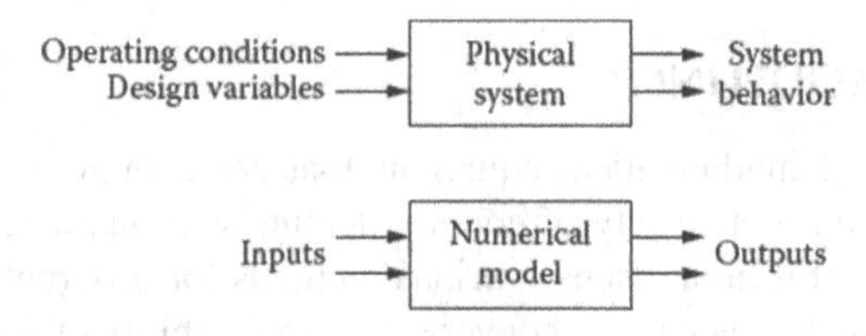

FIGURE 4.1 Representation of an actual physical system by a numerical model.

different portions on the tube side, with single or multiple passes, may be modeled. Depending on the level of sophistication and accuracy needed for predicting the temperature distributions and the heat transfer rates in the heat exchanger, different mathematical and numerical models may be developed for this problem. The coupled equations are solved to obtain the flow and the associated temperature distributions, from which the heat transfer rates may be determined.

The preceding example represents a complicated numerical modeling problem. However, it may be considerably simplified if inputs from experimental results are employed to circumvent analysis. For instance, the transport processes can be modeled easily if the overall heat transfer coefficient U is taken as known, even though it is a function of the flow, which, in turn, depends on the geometry, dimensions, fluids, and flow rates (Incropera and Dewitt, 2001). But, experimental results and simplified analysis are often used to avoid solving the full convective heat transfer equations because of the complexity of the resulting problem. In addition, there are additional phenomena that may be very hard to model. Fouling of heat exchangers, resulting from deposition of a film or scale on the heat transfer surfaces, is one such consideration that often needs inputs from experimental data.

4.1.1.1 Accuracy and Validity of the Model

The most important concern in numerical modeling is how closely the model represents the real system. This consideration refers to the physical behavior of the system, as predicted by the model, as well as to the accuracy of the results obtained from the model. It is critical to ensure that the model is a valid and accurate representation of the system. Therefore, the results obtained should be independent of the numerical scheme and its implementation. If arbitrary numerical parameters, such as grid size, time step, convergence criterion, and starting conditions, are introduced in order to obtain a solution, the results must not be significantly affected by the values chosen. This is sometimes referred to as *verification* of the numerical scheme. An important check on the validity of the model is also provided by the physical characteristics of the outputs, which must follow trends expected for the actual system. Comparisons of the results with experimental data, whenever possible, are employed to further validate the model and estimate the accuracy of the outputs derived from the model (Roache, 2009).

As a result of the various considerations and steps just outlined, a numerical model that accurately represents a given thermal system is obtained. Several useful software features that have been developed in the last three decades may be built into the computer code for added convenience. For instance, the outputs from the numerical model are commonly displayed in graphical form. Inputs, representing design variables and operating conditions, may also be entered graphically. Interactive use of the computer is frequently employed. Menu-driven selection of different solution methods, whose algorithms are stored in the system, and of other standard software may be employed. For example, Matlab, Comsol Multiphysics, Maple, Mathcad, and other software are commercially available to solve different types of mathematical equations. Information storage and retrieval may be used to access material property data, empirical heat transfer correlations, and data on the characteristics of selected components, such as pumps and compressors. These features make it easy to use the model for studying different conditions.

4.1.2 Development of a Numerical Model

Numerical models are first developed for each part, component, or subsystem of the given thermal system. It is assumed that the mathematical model has been obtained, resulting in the relevant equations that must be solved to predict the behavior. These equations may be algebraic, differential, or integral and may be linear or nonlinear. Different solution procedures are needed for different types of equations. All these procedures are linked to each other to solve the coupled system of equations.

The numerical methods available for a given problem depend on the nature and type of equation involved. The implementation on the computer requires selection of a computer system, along with an operating system, an appropriate programming language, and any available software or programs that are to be used. Unix and Linux are popular choices for the operating system because of their versatility. Though Fortran remains a popular programming language for engineers, particularly in its recent versions of Fortran 2003, 2008, and 2018, the use of C has increased rapidly because of its advantageous features in flow control and data structures (Kernighan and Ritchie, 1988). Variations in C, such as C++, have become very popular (Stroustrup, 2013). Python and Java are additional languages that have grown in popularity in recent years for different applications. Symbolic languages, such as Lisp and Prolog, are needed for knowledge-based design, as discussed in Chapter 11.

Though many of the current programs are written for a single central processing unit (CPU), parallel machines in which several processors are employed simultaneously have also grown in popularity because of the large computational speeds obtained at relatively small cost (Boyle et al., 1987; Kirk and Hwu, 2016). Due to substantial recent advancements in computer software, implementation on the computer has been considerably simplified through the use of efficient operating systems, editors, and standard programs, for example, those on graphics, interpolation, and matrix algebra.

Numerical errors include *round-off errors*, which arise due to the finite precision of the computer system, and *truncation errors*, which result from the approximations used in the numerical scheme, such as those that replace differential changes with finite ones. The convergence of an iterative scheme, which starts with an initial, guessed value and progresses toward a solution, depends on the equations, the starting point, the convergence criterion, and the numerical scheme. The instability of the scheme implies an unbounded growth of numerical errors as computation proceeds. The conditions under which the scheme becomes unstable or divergent must be determined and avoided. In order to obtain useful results from the numerical model, it is necessary to obtain a convergent, stable, and accurate numerical solution to a given mathematical equation.

4.1.3 Available Software

It is not necessary to develop the numerical scheme for each aspect of the given design problem because extensive libraries of computational software are usually available and may be used to simplify the model development process. The computer programming would obviously be considerably simplified if such available programs were used. In engineering practice, the use of such available software is quite prevalent because interest often lies in obtaining the desired results as quickly as possible. If a particular computer program has been successfully employed in the past, it is a good idea to use it for future applications. However, it is necessary to understand the algorithm adopted by the available software so that modifications, if needed, can be made and the inputs/outputs adjusted for the problem under consideration. It is also important to be aware of the limitations of the program and the expected accuracy of the numerical results.

Many computer programs are available in the public domain and may simply be adapted to the computer system and the overall numerical model. Such programs include methods for matrix inversion, curve fitting, numerical integration, and for solving ordinary differential equations and sets of linear algebraic equations. These programs are usually well-tested and may be used effectively in the development of the numerical model. Some free software is also available for certain fields. For instance, *OpenFoam* is a useful free open source software for computational fluid dynamics. Commercially available software comes with instructions on how to use it, but frequently without adequate details on the numerical procedures employed. Although such programs appear attractive because of the often exaggerated claims made in terms of their applicability, one must judge each program very carefully on the basis of accuracy, flexibility, efficiency, ease with which modifications may be made, algorithm used, and, of course, cost. General-purpose codes such as Fidap, Fluent, Ansys, and Phoenics are widely used in industrial applications. Special-purpose software directed at specific applications such as thermal management of electronics, air conditioning, injection molding, and combustion are also available. These are often expensive but, nevertheless, are used extensively by industry to model practical processes and systems.

A particular popular software is Matlab, which is discussed in some detail in Appendix A and which is used extensively to solve various mathematical equations. Matlab provides an interactive computational environment that can be used effectively to solve a wide range of mathematical equations. Several applications of Matlab are discussed in this book in order to obtain the inputs needed for design and optimization.

4.2 SOLUTION PROCEDURES

As seen in the preceding chapter, the mathematical modeling of thermal systems leads to different types of appropriate equations. Among the most common ones are

1. Set of linear algebraic equations
2. Single nonlinear algebraic equation
3. Set of nonlinear algebraic equations
4. Ordinary differential equations
5. Partial differential equations
6. Integral equations

In addition to the solution procedures for these types of equations, numerical methods are needed for differentiation and integration, and for curve fitting. The analytical and numerical methods for curve fitting of data were discussed in the preceding chapter. Some of the commonly used numerical procedures for solving the various equations just listed are presented here. This brief discussion will serve to present the nature and characteristics of the different types of equations as well as the relevant solution techniques for individual and sets of equations. Many of these methods and ideas will be referred to in later chapters. For further information on these methods, the references given earlier may be consulted.

The main purpose of discussing the solution procedures is to present the wide range of options available to an engineer or designer to solve the equations that describe a given thermal process or system and thus obtain the information and inputs needed for design and optimization. The options range from developing the code, modifying and/or using software available as open access, using computational environments such as Matlab or employing commercially available codes for particular topics and applications.

4.2.1 Linear Algebraic Systems

The solution of simultaneous linear algebraic equations is extremely important in the numerical modeling of thermal systems because the solution procedures for nonlinear algebraic equations and for differential equations often end up requiring the solution of sets of linear algebraic equations. In addition, many applications, such as those concerned with fluid flow circuits, chemical reactions, steady conduction heat transfer, and data analysis, are often governed by linear systems. A system of n linear equations may be written in the general form

$$\begin{aligned}
a_{11}x_1 + a_{12}x_2 + a_{13}x_3 + \ldots + a_{1n}x_n &= b_1 \\
a_{21}x_1 + a_{22}x_2 + a_{23}x_3 + \ldots + a_{2n}x_n &= b_2 \\
&\vdots \\
a_{n1}x_1 + a_{n2}x_2 + a_{n3}x_3 + \ldots + a_{nn}x_n &= b_n
\end{aligned} \tag{4.1}$$

Here, the a's represent the n^2 coefficients, the x's represent the n unknowns, and the b's represent the n constants on the right-hand side of the equations. The system may also be written more concisely in matrix notation as

$$(\mathbf{A})(\mathbf{X}) = (\mathbf{B}) \tag{4.2}$$

where (**A**) is an $n \times n$ square matrix of the coefficients, (**X**) is a column matrix of the n unknowns, and (**B**) is a column matrix of the constants that appear on the right-hand side of the n equations. Thus, a_{ij} represents an element of matrix (**A**), x_j an unknown element in (**X**), and b_i an element of vector (**B**). Two main approaches, direct and iterative, may be adopted for solving this set of linear algebraic equations.

4.2.1.1 Direct Methods

These methods solve the equations exactly, except for the round-off error, in a finite number of operations. Some of the methods are based on matrix inversion, because the solution of Equation (4.2) may be written as

$$(\mathbf{X}) = (\mathbf{A}^{-1})(\mathbf{B}) \tag{4.3}$$

where ($\mathbf{A}^{-1}$) is the inverse of matrix (**A**). The *determinant* of (**A**) must not be zero for this inverse to exist. If the determinant is zero, matrix (**A**) is said to be singular and nontrivial solutions can be obtained only if the column vector (**B**) is also zero, i.e., the equations are homogeneous. A common example of this circumstance is the eigenvalue problem, which is of particular interest in vibrations and in the stability of systems and flows. Many of the direct methods are based on elimination and reduction of variables, so that the given set of equations is reduced to a form that is amenable to a solution by simple algebraic analysis. The important direct methods available in the literature are:

1. Gaussian elimination
2. Gauss-Jordan elimination
3. Matrix decomposition methods
4. Matrix inversion methods

Gaussian elimination is used in a wide variety of engineering problems. The method reduces the matrix (**A**) to an upper triangular matrix so that the bottom row has only one element, as shown in Figure 4.2(a). Then the equation corresponding to this bottom row is a linear equation with

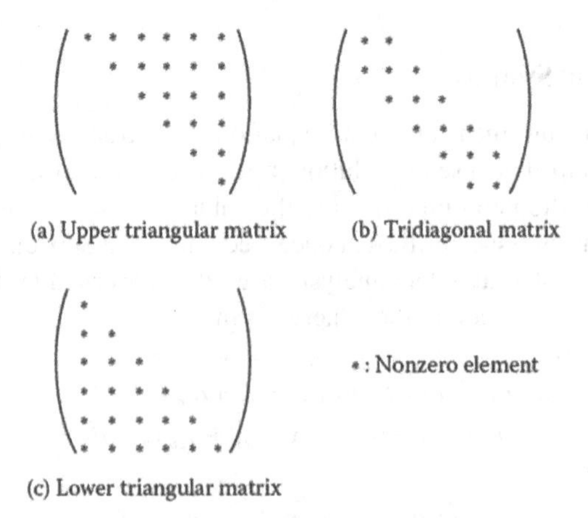

FIGURE 4.2 Special types of matrices that are of interest in the solution of sets of linear algebraic equations.

only one unknown, which is easily determined. The remaining unknowns are obtained by back-substitution, going from the bottom row to the top while considering each row in turn, so that each has only one unknown. For the numerical solution, an augmented matrix is formed by placing the column vector (**B**) at the end of matrix (**A**). The first step involves eliminating the element in the first column of all the rows below the first row, which is termed the *pivot row* and the element in the first column the *pivot element.* In the second step, the second row becomes the pivot row and the element in the second column the pivot element. Again, all the elements in the second column for rows below the second row are eliminated. The process is repeated until the matrix (**A**) is replaced by an upper triangular matrix.

The preceding elimination procedure may be written mathematically as

$$a_{ij}^{(r)} = a_{ij}^{(r-1)} - \frac{a_{ir}^{(r-1)}}{a_{rr}^{(r-1)}}\left[a_{rj}^{(r-1)}\right] \quad \text{for } r+1 \le i \le n \quad \text{and} \quad r \le j \le n+1 \tag{4.4}$$

where the superscripts within parentheses indicate the elimination step and r represents the pivot row, which thus varies from 1 to $n - 1$. Once the reduced matrix is obtained, the unknown x_n is obtained from

$$x_n = \frac{a_{n,n+1}}{a_{nn}}$$

where the subscript $(n + 1)$ refers to the last or augmented column in the matrix and thus to the modified constants on the right-hand side of the given set of equations. The other unknowns are then obtained by back-substitution, considering first the row above the bottom one to calculate x_{n-1}, then the row above it, and so on. Thus,

$$x_i = \frac{a_{i,n+1} - \sum_{j=i+1}^{n} a_{ij} x_j}{a_{ii}} \quad \text{where } i = n-1, n-2, \ldots, 2, 1 \tag{4.5}$$

Therefore, by applying the generalized procedure just given, the set of linear equations, Equation (4.1), is solved to yield the unknowns x_i, where i varies from 1 to n.

Problems arise if the pivot element is zero because this leads to division by zero in the preceding process. In addition, if it is a relatively small number, inaccurate results may be obtained. To avoid such problems, *partial pivoting*, in which the rows below and including the pivot row are interchanged at each step to employ the one with the largest pivot element as the pivot row, is commonly used. *Complete pivoting*, in which both rows and columns are interchanged to use the largest pivot element at each step, is also employed, though it is more involved than partial pivoting.

In several engineering problems, particularly in the numerical solution of partial differential equations (PDEs), the coefficient matrix (**A**) is tridiagonal, or banded, as shown in Figure 4.2(b). In this case, only the diagonal elements b_i and those on either side of it, a_i being on the left and c_i on the right side of the diagonal, are nonzero. Only two elements exist in the top and bottom rows of the matrix. Gaussian elimination becomes particularly simple for this case and requires very few mathematical operations. The corresponding scheme is known as the *Thomas algorithm* or tridiagonal matrix algorithm (TDMA) and is extensively used in engineering. The number of arithmetic operations needed to solve a tridiagonal system is on the order of n, i.e., $O(n)$, as compared to $O(n^3/3)$ for Gaussian elimination applied to an arbitrary system. Therefore, much smaller CPU times and much smaller round-off errors arise in the solution of tridiagonal systems. Effort is often made to cast the algebraic equations in tridiagonal form so that the solution may be obtained efficiently and accurately by this method.

Several other direct methods are available for solving sets of linear algebraic equations. In the *Gauss-Jordan* elimination method, which is a variation of the Gaussian elimination procedure, the coefficient matrix (**A**) is reduced to an identity matrix (**I**), with the only nonzero elements being unity along the diagonal. Then the unknown vector (**X**) is simply given by the modified constant vector (**B′**), because (**I**)(**X**) = (**B′**) yields (**X**) = (**B′**). No back-substitution is needed. At each step in the elimination process, the unknown is eliminated from both above and below the pivot equation and the pivot equation is normalized by dividing it by the pivot element, yielding an identity matrix at the end of the process. The number of arithmetic operations needed is $O(n^3/2)$. Therefore, this method is somewhat less efficient than Gaussian elimination. Nevertheless, it is particularly well-suited for matrix inversion methods based on Equation (4.3), since the inverse ($\mathbf{A}^{-1}$) is obtained directly without back-substitution if (B) is taken as an identity matrix because $(A)(A^{-1}) = (I)$.

Matrix inversion is often used because of available software to invert a matrix, because interest lies in the inverse itself in order to study the behavior of the system, or because solutions for different values of (**B**) with the same coefficient matrix (**A**) are to be obtained. However, this approach is much less efficient than Gaussian elimination, requiring $O(4n^3/3)$ arithmetic operations. Similarly, several efficient methods based on matrix decomposition into upper (U) and lower (L) triangular matrices (see Figure 4.2) are available. This is known as LU decomposition and various numerical procedures, such as Crout's and Cholesky's methods, may be employed for the purpose.

4.2.1.2 Iterative Methods

In engineering systems, particularly in the solution of differential equations by finite-difference and finite-element methods, we frequently encounter large sets of linear equations that are generally sparse, with only a few nonzero elements in each equation. Iterative methods use this sparseness advantageously because only the nonzero terms are considered. In addition, the round-off error after each iteration simply gives a less accurate input for the next iteration and the error in the solution is only what arises in the final iteration. There is no accumulation of round-off error as is the case in direct methods. The solution is not exact but is obtained to an arbitrary, specified, convergence criterion.

The iterative scheme may be obtained from Equation (4.1) by solving for the unknowns, starting with x_1 and successively obtaining $x_2, x_3, \ldots, x_i, \ldots, x_n$. A commonly used scheme is the Gauss-Seidel method, which is given as

$$x_i^{(l+1)} = \frac{b_i - \sum_{j=1}^{i-1} a_{ij} x_j^{(l+1)} - \sum_{j=i+1}^{n} a_{ij} x_j^{(l)}}{a_{ii}} \quad \text{for } i = 1,2,3,\ldots,n \tag{4.6}$$

The iteration starts with initially guessed values of the unknowns, denoted by $x_i^{(0)}$, and subsequent iterative values, denoted by the superscript in the preceding equation, are calculated. Only the latest values of the unknowns are stored and used in subsequent calculations. The iteration is terminated when a convergence criterion such as

$$\left| x_i^{(l+1)} - x_i^{(l)} \right| \le \varepsilon \quad \text{or} \quad \left| \frac{x_i^{(l+1)} - x_i^{(l)}}{x_i^{(l)}} \right| \le \varepsilon \quad \text{for } i = 1,2,3,\ldots,n \tag{4.7}$$

is satisfied. Here, ε is a convergence parameter chosen such that, if it is reduced further, the results are essentially unaffected. The second criterion considers the normalized change in x_i and is appropriate if none of the unknowns is expected to be close to zero. The rate of convergence depends on the initial guess and the scheme is guaranteed to converge if the magnitude of each diagonal element a_{ii} of the coefficient matrix is larger than the sum of the magnitudes of the other elements in the row. This implies that this iterative scheme would converge for linear systems if

$$|a_{ii}| > \sum_{j=1, j \neq i}^{n} |a_{ij}| \tag{4.8}$$

The system is then said to be *diagonally dominant*. However, convergence is generally obtained with much weaker diagonal dominance. Therefore, the equations must be arranged in order to have the unknown with the largest coefficient at the diagonal in order to achieve convergence. This implies solving each equation for the unknown with the largest coefficient in magnitude.

The convergence characteristics of the Gauss-Seidel method can often be significantly improved by the use of point relaxation, given by

$$x_i^{(l+1)} = \omega \left[x_i^{(l+1)} \right]_{GS} + (1-\omega) x_i^{(l)} \tag{4.9}$$

where ω is a constant in the range $0 < \omega < 2$ and $[x_i^{(l+1)}]_{GS}$ is the value of x_i obtained for the $(l + 1)$th iteration by using the Gauss-Seidel iteration. If $0 < \omega < 1$, the scheme is known as *successive under-relaxation* (SUR), and if $1 < \omega < 2$, it is known as *successive over-relaxation* (SOR). For linear equations, an optimum value ω_{opt} of the relaxation factor can be found at which the convergence is much faster than that for the Gauss-Seidel method, $\omega = 1$. Figure 4.3 shows the typical dependence of the number of iterations to convergence on the relaxation factor ω and an optimum value ω_{opt} at which convergence is fastest. Therefore, SOR is widely used for solving linear systems. SUR is generally used for nonlinear equations.

Both direct and iterative methods are used in the numerical modeling of thermal systems. Large sets of equations that arise in the finite-difference and finite-element solutions of PDEs are generally solved by iterative methods, unless they can be obtained in the form of a tridiagonal system. For smaller sets of linear equations, direct methods are more efficient and accurate. However, thermal systems usually lead to nonlinear equations that have to be solved by iteration.

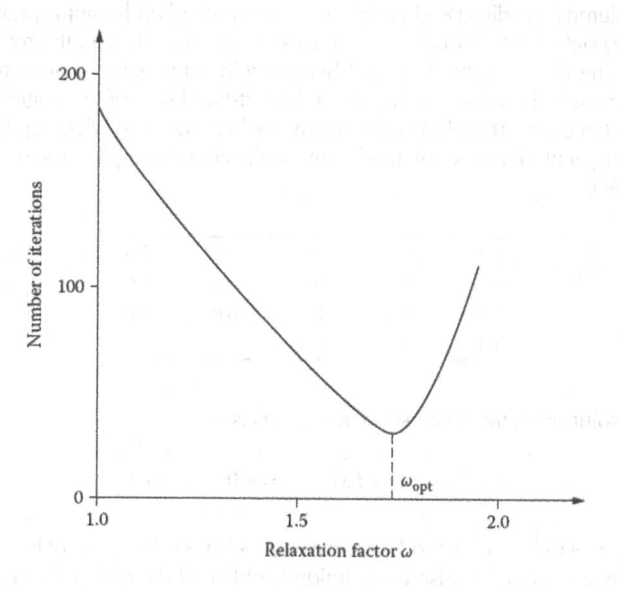

FIGURE 4.3 Typical variation of the number of iterations needed for convergence of a linear system with the relaxation factor ω.

Example 4.1

An industrial organization produces four items x_1, x_2, x_3, and x_4. A portion of the amount produced for each is used in the manufacture of other items, and the net product is sold. The balance between the output and the production rate, resulting from various inputs, gives rise to the following four linear equations:

$$2x_1 + x_2 + 6x_4 = 64$$
$$5x_1 + 2x_2 = 37$$
$$7x_2 + 2x_3 + 2x_4 = 66$$
$$8x_3 + 9x_4 = 104$$

Solve this set of equations by the Gauss-Jordan elimination method.

SOLUTION

The given equations are linear and the Gauss-Jordan method may be used to convert the coefficient matrix to an identity matrix so that the constants on the right-hand side of the equations are transformed into the desired solution, as discussed earlier. Partial pivoting should be used in the scheme because several coefficients are zero. This will also lead to higher accuracy in the solution.

The given set of equations may be written as (**A**)(**X**) = (**B**) and the Gauss-Jordan method applied to the augmented matrix, which is

2.0	1.0	0.0	6.0	64.0
5.0	2.0	0.0	0.0	37.0
0.0	7.0	2.0	2.0	66.0
0.0	0.0	8.0	9.0	104.0

The Gauss-Jordan elimination method uses normalization of the pivot row and elimination of elements above and below the pivot element, in a series of steps, to yield an identity matrix (**I**) in the first four columns. As discussed earlier, rows are multiplied by appropriate constants and added to eliminate coefficients. Partial pivoting is used, i.e., at each step, the row with the largest pivot element, among the rows that have not been employed thus far as pivot row, is chosen as the pivot row. Ultimately, in the 4 × 4 matrix on the left-hand side of the augmented matrix, the diagonal elements become 1.0 and the other elements become zero, yielding the solution in the last column. The augmented matrix obtained numerically after the application of the Gauss-Jordan elimination scheme is

1.0	0.0	0.0	0.0	5.0
0.0	1.0	0.0	0.0	6.0
0.0	0.0	1.0	0.0	4.0
0.0	0.0	0.0	1.0	8.0

Therefore, the solution to the given set of equations is

$$x_1 = 5.0; \quad x_2 = 6.0; \quad x_3 = 4.0; \quad x_4 = 8.0$$

The Gauss-Jordan elimination method is used extensively for solving sets of linear algebraic equations and for matrix inversion because the solution is obtained directly, without back-substitution, which is needed for Gaussian elimination.

This problem and other similar sets of linear equations with a larger number of unknowns can be solved very easily by using a computer program, as discussed in Appendix A. A few typical programs based on the methods discussed here are also given. Matlab is particularly well-suited to matrix algebra and available commands may be used in a Matlab environment to obtain the solution. In this problem, for instance, the matrices (**A**) and (**B**) are entered in Matlab as

```
a = [2 1 0 6; 5 2 0 0; 0 7 2 2; 0 0 8 9];
b = [64; 37; 66; 104];
```

Then the solution (**X**) is obtained simply by using Equation (4.3) as

```
x = inv(a)*b
```

or as

```
x = a\b
```

which uses the internal logic of the \ operator in Matlab to indicate multiplication on the left by the inverse of **A**. The solution can also be obtained by **LU** decomposition by the commands

```
[l,u,p] = lu(a);
y = l\(p*b);
x = u\y
```

where p is the permutation matrix that stores the information on row exchanges for partial pivoting. When any of these approaches is used, the solution vector is obtained as [5; 6; 4; 8]. The corresponding Matlab program for the preceding solution is given in Appendix A.M.4.1. The program for solving the same problem by the Gauss-Seidel method is given in Appendix A.M.4.2.

4.2.2 Nonlinear Algebraic Systems

The mathematical modeling of thermal systems frequently leads to nonlinear algebraic and differential equations. This is because of nonlinear transport mechanisms such as radiative heat transfer and variable material properties. The solution of nonlinear equations is much more involved than of linear equations. In addition, multiple solutions may be obtained, requiring additional inputs, particularly from the physical nature of the problem, to choose the right solution. Except for a few special cases such as the quadratic equation, direct solution of the equations is not possible and iterative methods are needed. In fact, the nonlinear problem is generally linearized to obtain a linear problem, which is used in the iteration process to yield the solution. Let us first consider the solution of a single nonlinear algebraic equation and then extend the solution strategy to a system of nonlinear equations.

4.2.2.1 Single Nonlinear Algebraic Equation

This is the problem of finding the solution or roots of a single nonlinear equation such as $f(x) = 0$. We need to determine the values of x that would satisfy the given equation, which may be a polynomial equation of the form

$$f(x) = x^n + a_1 x^{n-1} + a_2 x^{n-2} + \ldots + a_{n-1} x + a_n = 0 \tag{4.10}$$

where n is the degree of the polynomial and the a's are real coefficients. This equation has n roots, which may be real or complex. The equation may also be a transcendental one, such as $x \tan x - 1 = 0$ or $2e^x + 3x - 5 = 0$, involving exponential, logarithmic, trigonometric, and other such functions. Though real, single-valued roots are generally of interest in thermal systems, complex roots and those with multiple values are also sometimes sought. Generally, the nature of the root and the approximate range in which it lies is known from the physical background of the problem under consideration. For instance, if an algebraic equation governing the heat balance at the surface of a body is being solved to obtain the temperature, the maximum and minimum

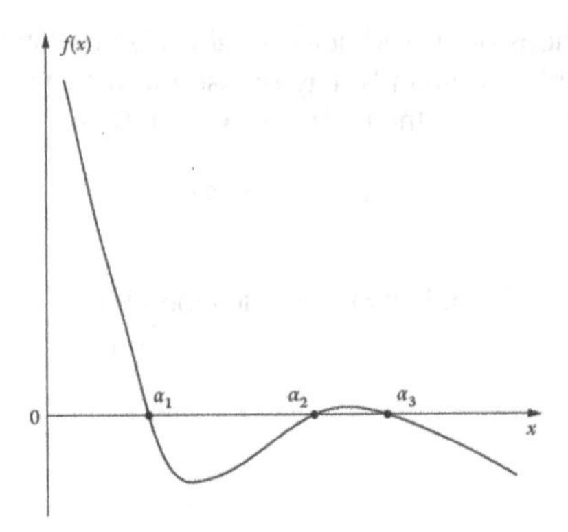

FIGURE 4.4 Plot of a function $f(x)$ versus x, indicating three roots a_1, a_2, and a_3 of the equation $f(x) = 0$, at which the function changes sign.

temperatures in the system may be known, yielding the range in which the surface temperature, which must be a real root, lies.

Several methods are available for finding the roots of an algebraic equation. Many of these, known as *bracketing* methods, are based on the change in the sign of $f(x)$ as it crosses the x-axis at the root, as shown in Figure 4.4. A change in the sign of $f(x)$ as x is incremented by Δx indicates the presence of a real root within the interval Δx. This interval is gradually reduced by subdividing it into smaller divisions and again finding the interval in which the root lies. The process is continued until the root is obtained to the desired level of accuracy, as given by the final interval containing the root. This method is known as the *search* method and is guaranteed to converge if a real root is present and if the function $f(x)$ changes sign. It does not apply to complex roots or to multiple roots resulting from the graph of $f(x)$ versus x being tangent to the x-axis. This method, along with a few others that bracket the root and are based on a sign change of $f(x)$, are listed as:

1. Search method
2. Bisection method
3. *Regula falsi*, or false position, method

In the *bisection* method, an approximation to the root x_3 is obtained by taking an average of the two end points, x_1 and x_2, of the interval in which the root lies, i.e., $x_3 = (x_1 + x_2)/2$. At each iteration, the new, reduced interval in which the root lies is determined and a new approximation to the root computed. In the regula falsi method, interpolation is used, employing the end points of the interval at a given iteration, to approximate the root as $x_3 = [x_1 f(x_2) - x_2 f(x_1)]/[f(x_2) - f(x_1)]$. Sign change between x_3 and x_1 or between x_3 and x_2 is used to choose the interval containing the root. The iterative process is continued, reducing the interval at each step, until the change in the approximation to the root from one iteration to the next is less than a chosen convergence criterion, as given by

$$\left| x^{(l+1)} - x^{(l)} \right| \leq \varepsilon \quad \text{or} \quad \left| \frac{x^{(l+1)} - x^{(l)}}{x^{(l)}} \right| \leq \varepsilon \tag{4.11}$$

where $x^{(l+1)}$ and $x^{(l)}$ represent approximations to the root after the $(l + 1)$th and (l)th iterations, respectively, and ε is the chosen convergence parameter.

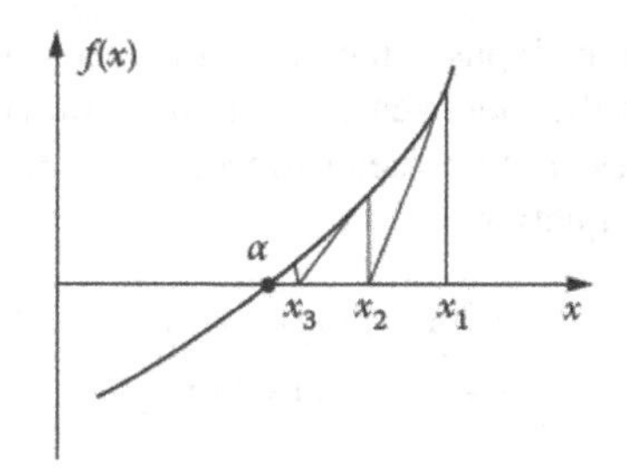

FIGURE 4.5 The Newton-Raphson iterative method for solving an algebraic equation $f(x) = 0$.

Probably the most important and widely used method for root solving is the *Newton-Raphson* method, in which the iterative approximation to the root x_i is used to calculate the next iterative approximation to the root x_{i+1} as

$$x_{i+1} = x_i - \frac{f(x_i)}{f'(x_i)} \tag{4.12}$$

where $f'(x_i)$ is the derivative of $f(x)$ at $x = x_i$. This equation gives an iterative process for finding the root, starting with an initial guess x_1. The process is terminated when the convergence criterion, given by Equation (4.11), is satisfied.

The Newton-Raphson method can be used for real as well as complex roots, employing complex algebra for the functions, for their derivatives, and for x. It can also be used for multiple roots where the graph of $f(x)$ versus x is tangent to the x-axis, with no sign change in $f(x)$. When the scheme converges, it converges very rapidly to the root. It can be shown that it has a *second-order convergence*, implying that the error in each iteration varies as the square of the error in the previous iteration and thus reduces very rapidly. However, the iteration process may diverge, depending on the initial guess and nature of the equation. Figure 4.5 shows graphically the iterative process in a convergent case. The tangent to the curve at a given approximation is used to obtain the next approximation to the root. Figure 4.6 shows a few cases in which the method diverges. If the scheme diverges, a new starting point is chosen and the process repeated.

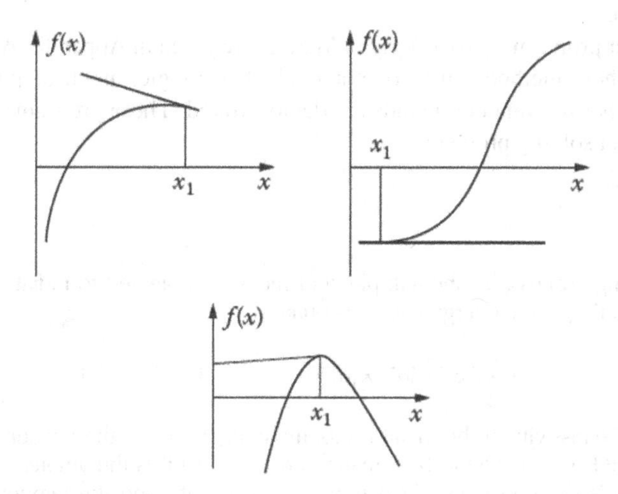

FIGURE 4.6 A few cases in which the Newton-Raphson method might not converge.

A method similar to the Newton-Raphson method is the *secant* method, which uses interpolation and extrapolation to approximate the root in each iteration, employing the last two iterative values in the approximation. No derivatives are to be determined, as was the case for Newton's method. The iterative scheme is given by the equation

$$x_{i+1} = \frac{x_{i-1} f(x_i) - x_i f(x_{i-1})}{f(x_i) - f(x_{i-1})} \tag{4.13}$$

where the subscripts indicate the order of the iteration, starting with x_1 and x_2 as the first two approximations to the root. The iterative process is continued until Equation (4.11) is satisfied. Again, the iterative scheme may diverge, depending on function $f(x)$ and the starting values. If the method diverges, new starting values are taken and the process is repeated.

A particularly simple method for root solving is the *successive substitution* method, in which the given equation $f(x) = 0$ is rewritten as $x = g(x)$. At the root, $\alpha = g(\alpha)$, where α is the root of the original equation and thus $f(\alpha) = 0$. The modified equation yields an iterative scheme given by

$$x_{i+1} = g(x_i) \tag{4.14}$$

Therefore, the iteration starts with an initial approximation to the root x_1, which is substituted on the right-hand side of this equation to yield the next approximation, x_2. Then x_2 is substituted in the equation to obtain x_3, and so on. The process is continued until Equation (4.11) is satisfied. The scheme is a very simple one and is based on the successive substitution of the approximations to the root to obtain more accurate values. However, convergence is not assured and depends on the initial guess as well as on the choice of the function $g(x)$, which can be formulated in many ways and is not unique. It can be shown that if $|g'(\alpha)| < 1$, the method converges to the root in a region close to the root. Here, $g'(\alpha)$ is the derivative of $g(x)$ at the root and is known as the *asymptotic convergence factor.* The convergence characteristics of the method may be improved by employing the recursion formula

$$x_{i+1} = (1 - \beta)x_i + \beta g(x_i) \tag{4.15}$$

where β is a constant. A value of β less than 1.0 reduces the change in each iteration and helps in the convergence of the scheme. This is similar to the SUR method. The choices for $g(x)$ and β depend on the function $f(x)$.

Several computer programs, particularly in Matlab, are given in Appendix A.M.3 to demonstrate the application of these methods on a computer. The basic logic, input, output, the mathematical calculations, and other relevant commands are demonstrated. These programs can easily be modified for different root solving problems.

Example 4.2

In a manufacturing process, a spherical piece of metal is subjected to radiative and convective heat transfer, resulting in the energy balance equation

$$0.6 \times 5.67 \times 10^{-8} \times [(850)^4 - T^4] = 40 \times (T - 350)$$

Here, the surface emissivity of the metal is 0.6, the temperature of the radiating source is 850 K, 5.67×10^{-8} W/(m^2·K^4) is the Stefan-Boltzmann constant, 350 K is the ambient fluid temperature, and 40 W/(m^2·K) is the convective heat transfer coefficient. Find the temperature T using the secant method.

SOLUTION

This problem involves determining the root of the given nonlinear algebraic equation, which may be rewritten as

$$f(T) = 0.6 \times 5.67 \times 10^{-8} \times [(850)^4 - T^4] - 40 \times (T - 350) = 0$$

in order to apply the root solving methods given earlier. Here, the highest temperature in the heat transfer problem considered is 850 K and the lowest is 350 K. Therefore, the desired root lies between these two values and should be positive and real. The recursion formula for the secant method may be written as

$$T_{i+1} = \frac{T_{i-1} f(T_i) - T_i f(T_{i-1})}{f(T_i) - f(T_{i-1})}$$

where the subscripts $i - 1$, i, and $i + 1$ represent the values for three consecutive iterations. The starting values are taken as $T_{i-1} = T_1 = 350$ and $T_i = T_2 = 850$. The equation just given is used to calculate $T_{i+1} = T_3$. Then T_2 and T_3 are used to calculate T_4, and so on. The iteration is terminated when

$$\left| \frac{T_{i+1} - T_i}{T_i} \right| \leq \varepsilon$$

where ε is a chosen small quantity. Generally, ε should be varied to ensure that the results do not depend on the value chosen. For the preceding equation, a value of around 10^{-3} or lower would typically be adequate. Thus, the relative change in T from one iteration to the next is used for the convergence criterion. The numerical results obtained from the secant method are shown below, indicating a few steps in the convergence to the desired root.

$T = 581.5302$	$f(T) = 4606.784180$
$T = 631.7920$	$f(T) = 1066.578125$
$T = 646.9347$	$f(T) = -77.774414$
$T = 645.9056$	$f(T) = 1.222656$
$T = 645.9215$	$f(T) = 0.005859$
$T = 645.9216$	$f(T) = -0.004883$

Therefore, the temperature T is obtained as 645.92 K, rounding off the numerical result to two decimal places. A fast convergence to the root is observed. The convergence parameter ε is taken as 10^{-5} here, and it was confirmed that the result was negligibly affected if a still smaller value of ε was employed. A significant change in the root was obtained if ε was increased to much larger values.

Though computer programs may be written in Fortran, C++, or other programming languages to solve this root-solving problem, as given in Appendix A, the Matlab environment provides a particularly simple solution scheme on the basis of the internal logic of the software. By rearranging $f(T)$, the corresponding polynomial p is given in terms of the coefficients a, b, c, d, and e, in descending powers of T, i.e., T^4, T^3, T^2 and T, as:

```
a = 0.6*5.67*10^-8;
b = 0;
c = 0;
d=40.0;
e=-40.0*350.0-0.6*5.67*(10^-8)*(850^4);
p=[a b c d e];
```

Then the roots are obtained by using the command

```
r=roots(p)
```

This yields four roots because a fourth-order polynomial is being considered. It turns out that, when the above scheme is used, one negative and two complex roots are obtained in addition to one real root at 645.92, which lies in the appropriate range and is the correct solution.

4.2.2.2 System of Nonlinear Algebraic Equations

The mathematical modeling of thermal systems frequently leads to sets of nonlinear equations. The solution of these equations generally involves iteration and combines the strategies for root solving and those for linear systems. Two important approaches for solving a system of nonlinear algebraic equations are based on Newton's method and on the successive substitution method. If $x_1, x_2, \ldots, x_n$ are the unknowns and $f_1(x_1, x_2, \ldots, x_n) = 0, f_2(x_1, x_2, \ldots, x_n) = 0, \ldots, f_n(x_1, x_2, \ldots, x_n) = 0$ are the nonlinear equations, Newton's method gives the solution as

$$\begin{aligned} x_1^{(l+1)} &= x_1^{(l)} + \Delta x_1^{(l)} \\ x_2^{(l+1)} &= x_2^{(l)} + \Delta x_2^{(l)} \\ &\vdots \\ x_n^{(l+1)} &= x_n^{(l)} + \Delta x_n^{(l)} \end{aligned} \tag{4.16}$$

where the superscripts (l) and $(l + 1)$ represent the values after l and $l + 1$ iterations. Extending Newton's method given by Equation (4.12) for a single equation to a system of nonlinear equations, the increments Δx_i are obtained from the following system of equations:

$$\begin{bmatrix} \frac{\partial f_1}{\partial x_1} & \frac{\partial f_1}{\partial x_2} & \frac{\partial f_1}{\partial x_3} & \cdots & \frac{\partial f_1}{\partial x_n} \\ \frac{\partial f_2}{\partial x_1} & \frac{\partial f_2}{\partial x_2} & \frac{\partial f_2}{\partial x_3} & \cdots & \frac{\partial f_2}{\partial x_n} \\ \vdots & \vdots & \vdots & \vdots & \vdots \\ \frac{\partial f_n}{\partial x_1} & \frac{\partial f_n}{\partial x_2} & \frac{\partial f_n}{\partial x_3} & \cdots & \frac{\partial f_n}{\partial x_n} \end{bmatrix} \begin{bmatrix} \Delta x_1 \\ \Delta x_2 \\ \vdots \\ \Delta x_n \end{bmatrix} = \begin{bmatrix} -f_1 \\ -f_2 \\ \vdots \\ -f_n \end{bmatrix} \tag{4.17}$$

These equations are linear because the functions and their derivatives are known for given x_i and the unknowns are the increments.

The iterative scheme starts with an initial guess of the values of the unknowns, $x_i^{(1)}$. From these values, the functions $f_i^{(1)}$ and their derivatives needed for Equation (4.17) are calculated. Then the linear system given by Equation (4.17) is solved for the increments $\Delta x_i^{(1)}$, which are employed in Equation (4.16) to obtain the next iteration, $x_i^{(2)}$. This process is continued until the unknowns do not change from one iteration to the next, within a specified convergence criterion, such as that given by Equation (4.7).

Clearly, this scheme is much more involved than that for a system of linear equations. In fact, a system of linear equations has to be solved for each iteration to update the values of the unknowns. In addition, the derivatives of the functions have to be determined at each step. Therefore, the method is appropriate for relatively small sets of nonlinear equations, typically less than ten, and for cases where the derivatives are continuous, well-behaved, and easy to compute. The scheme may diverge if the initial guess is too far from the exact solution. Usually, the physical nature of the problem and earlier solutions are employed to guide the selection of the initial guess.

The system of nonlinear equations may also be solved using the successive substitution approach given by Equation (4.14) or (4.15) for a single nonlinear equation. Each unknown is computed in turn and the value obtained is substituted into the corresponding equations to generate an iterative scheme. Therefore, if the system of equations is rewritten by solving for the unknowns, we obtain

$$x_i = G_i[x_1, x_2, x_3, \ldots, x_i, \ldots, x_n] \quad \text{for } i = 1, 2, \ldots, n \tag{4.18}$$

The unknown x_i is also retained on the right-hand side in this case, because these are nonlinear equations and x_i may appear as a product with other unknowns or as a nonlinear function. Again, the function G_i can be formulated from the given equation $f_i = 0$ in many different ways. An iterative scheme similar to the Gauss-Seidel method may be developed as

$$x_i^{(l+1)} = G_i\left[x_1^{(l+1)}, x_2^{(l+1)}, \ldots, x_{i-1}^{(l+1)}, x_i^{(l)}, \ldots, x_n^{(l)}\right] \quad \text{for } i = 1, 2, \ldots, n \tag{4.19}$$

Here, the unknowns are calculated for increasing i, starting with x_1. The most recently calculated values of the unknowns are used in calculating the function G_i.

This scheme is often also known as the *modified Gauss-Seidel method*. It is similar to the successive substitution method for linear equations and is much simpler to implement than Newton's method because no derivatives are needed. The approach is particularly suitable for large sets of equations. However, Newton's method generally has better convergence characteristics than the successive substitution, or modified Gauss-Seidel, method. SUR is often used to improve the convergence characteristics of this method. Convergence of the iterative scheme for nonlinear equations is often difficult to predict because a general theory for convergence is not available as in the case of linear equations. Several trials, with different starting values and different formulations, are frequently needed to solve these equations. Newton's method and the successive substitution method also represent two different approaches to simulation, namely *simultaneous* and *sequential*, and are discussed later, along with a few solved examples.

4.2.3 Ordinary Differential Equations

Ordinary differential equations (ODEs), which involve functions of a single independent variable and their derivatives, are encountered in the modeling of many thermal systems, particularly for transient lumped modeling. A general nth-order ODE may be written as

$$\frac{d^n y}{dx^n} = F\left(x, y, \frac{dy}{dx}, \frac{d^2 y}{dx^2}, \ldots, \frac{d^{n-1} y}{dx^{n-1}}\right) \tag{4.20}$$

Here, x is the independent variable and $y(x)$ is the dependent variable. This equation requires n independent boundary conditions for a solution. If all these conditions are specified at one value of x, the problem is referred to as an *initial-value* problem. If the conditions are given at two or more values of x, it is referred to as a *boundary-value* problem. We shall first consider initial-value problems, followed by boundary-value problems.

4.2.3.1 Initial-Value Problems

The preceding equation can be reduced to a system of n first-order equations by defining new independent variables Y_i, where i varies from 1 to $(n - 1)$, as

$$Y_1 = \frac{dy}{dx} \quad Y_2 = \frac{d^2 y}{dx^2} \quad Y_3 = \frac{d^3 y}{dx^3} \; \ldots \; Y_{n-1} = \frac{d^{n-1} y}{dx^{n-1}}$$

Therefore, the system of n first-order equations becomes

$$\frac{dy}{dx} = Y_1 \quad \frac{dY_1}{dx} = Y_2 \quad \frac{dY_2}{dx} = Y_3 \ldots \frac{dY_{n-1}}{dx} = F(x, y, Y_1, Y_2, Y_3, \ldots Y_{n-1})$$

The n boundary conditions are given in terms of y and its derivatives, all these being specified at one value of x for an initial-value problem. The given nth-order ordinary differential equation may be linear or nonlinear. Linear equations can frequently be solved by analytical methods available in the literature. However, numerical methods are usually needed for nonlinear equations.

It is clear from the foregoing discussion that if we can solve a first-order ODE, we can extend the solution to higher-order equations and to systems of ODEs. Therefore, the numerical solution procedures are directed at the simple first-order equation written as

$$\frac{dy}{dx} = F(x, y) \tag{4.21}$$

with the boundary condition

$$y(x_0) = y_0 \tag{4.22}$$

where y_0 is the value of $y(x)$ at a given value of the independent variable, $x = x_0$. A numerical solution of this differential equation involves obtaining the value of the function $y(x)$ at discrete values of x, given as

$$x_i = x_0 + i\Delta x \quad \text{where } i = 1, 2, 3, \ldots \tag{4.23}$$

Therefore, the numerical scheme must provide the means for determining the values $y_1, y_2, y_3, y_4, \ldots$ for the dependent variable y corresponding to these discrete values of x. If the solution is sought for $x < x_0$, then x_i is taken as $x_i = x_0 - i\Delta x$ and a similar procedure is employed as for increasing x.

Several methods are available for the solution of a first-order ODE. Two main classes of methods are

1. *Runge-Kutta* methods
2. *Predictor-corrector* methods

In the Runge-Kutta methods, the derivative of the function y, as given by $F(x,y)$, is evaluated at different points within the interval x_i to $x_{i+1} = x_i + \Delta x$. A weighted mean of these values is obtained and used to calculate y_{i+1}, the value of the dependent variable at x_{i+1}. The simplest formula in these classes of methods is that of Euler's method, which has a cumulative truncation error of $O(\Delta x)$ up to a given x_i. Because the error varies as the first power of Δx, this is known as a first-order method. It is seldom used due to the large error in the solution. The computational formula for Euler's method is

$$y_{i+1} = y_i + \Delta x F(x_i, y_i) \quad \text{with } i = 0, 1, 2, 3, \ldots \tag{4.24}$$

Therefore, the solution can be obtained for increasing x, starting with $x = x_0$. Figure 4.7 shows this method graphically, indicating the accumulation of error with increasing x.

The most widely used method is the fourth-order Runge-Kutta method given by the computational formula

$$y_{i+1} = y_i + \frac{K_1 + 2K_2 + 2K_3 + K_4}{6} \tag{4.25a}$$

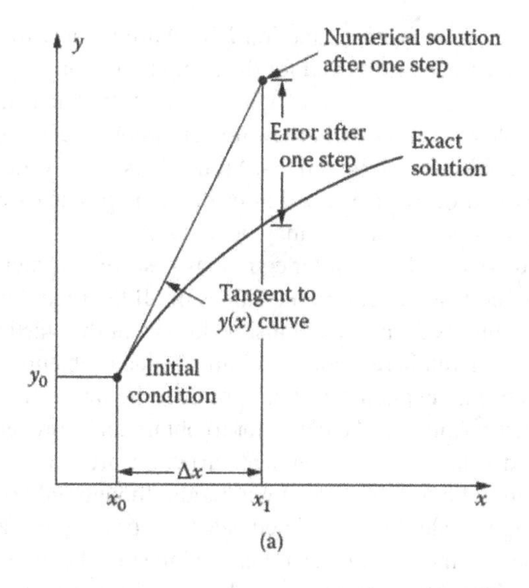

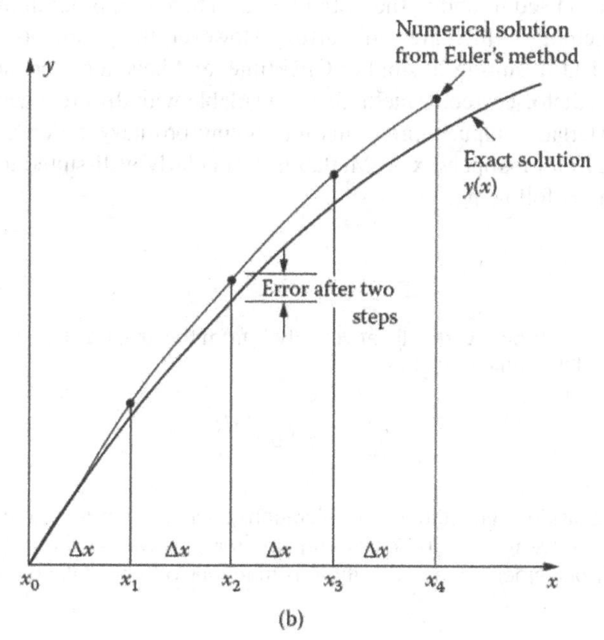

FIGURE 4.7 Graphical interpretation of Euler's method. (a) Numerical solution and error after the first step; (b) accumulation of error with increasing value of the independent variable x.

where

$$K_1 = \Delta x F(x_i, y_i) \tag{4.25b}$$

$$K_2 = \Delta x F\left(x_i + \frac{\Delta x}{2}, y_i + \frac{K_1}{2}\right) \tag{4.25c}$$

$$K_3 = \Delta x F\left(x_i + \frac{\Delta x}{2}, y_i + \frac{K_2}{2}\right) \tag{4.25d}$$

$$K_4 = \Delta x F(x_i + \Delta x, y_i + K_3) \tag{4.25e}$$

Therefore, four evaluations of the derivative function $F(x, y)$ are made within the interval $x_i \leq x \leq x_{i+1}$, and a suitable weighted average is employed for the computation of y_{i+1}. It is a fourth-order scheme because the total error up to a given value of x_i varies as $(\Delta x)^4$. The Runge-Kutta methods are self-starting, stable, and simple to use. As such, they are very popular and most computers have the corresponding software available for solving ODEs. Similarly, second-order schemes, such as Heun's method, which involves two derivative function evaluations in each step, and third-order schemes, which involve three function evaluations, may be derived.

For higher-order equations, a system of first-order equations is solved, as mentioned earlier. The computations are carried out in sequence to obtain the values of all the unknowns at the next step. All the conditions, in terms of y and its derivatives, must be known at the starting point to use this method. Therefore, the scheme, as given here, applies to initial-value problems.

Predictor-corrector methods use an explicit formula to predict the first estimate of the solution, followed by the use of an implicit formula as the corrector to obtain an improved approximation to the solution. Previously obtained values of the dependent variable y are extrapolated to obtain the predicted value, and the corrector equation is solved by iteration, though only one or two steps are generally needed for it to converge because the predicted value is close to the solution. These methods are not self-starting because the first few values are needed to start the predictor, and a method such as Runge-Kutta is used to obtain the initial points. Therefore, programming is more involved than Runge-Kutta methods, which are self-starting. However, the predictor-corrector methods are generally more efficient, resulting in smaller CPU time, and have a better estimate of the error at each step. Several predictor-corrector methods are available with different accuracy levels.

A few relevant Matlab computer programs for solving ordinary differential equations by different methods are given in Appendix A. Matlab is particularly well-suited to solving initial-value problems, as seen in the following.

Example 4.3

The motion of a stone thrown vertically at velocity V from the ground at $x = 0$ and at time $\tau = 0$ is determined by the differential equation

$$\frac{d^2x}{d\tau^2} = -g - 0.1\left(\frac{dx}{d\tau}\right)^2$$

where g is the magnitude of gravitational acceleration, given as 9.8 m/s², and the velocity is $dx/d\tau$, also denoted by V. Solve this second-order equation for x, as well as the first-order equation for V, to obtain the displacement x and velocity V as functions of time. Take the initial velocity V as 25 m/s.

SOLUTION

The second-order equation in terms of the displacement x is given above, with the initial conditions

$$\tau = 0: x = 0 \quad \text{and} \quad \frac{dx}{d\tau} = 25$$

The corresponding differential equation in terms of the velocity V is given by

$$\frac{dV}{d\tau} = -g - 0.1V^2$$

with the initial condition

$$\tau = 0: V = 25$$

Both these cases are initial-value problems because all the necessary conditions are given at the initial time, $\tau = 0$. Computer codes may be written for the two cases using high-level programming languages such as Fortran and C++ or a computational environment like Matlab or Comsol Multiphysics. However, Matlab can be used very easily for these problems by using the *ode23* and *ode45* built-in functions. Both are based on Runge-Kutta methods and use adaptive step sizes. Two solutions are obtained at each step, allowing the algorithm to monitor the accuracy and adjust the step size according to a given or default tolerance. The first method, ode23, uses second- and third-order Runge-Kutta formulas and the second one, ode45, uses fourth- and fifth-order formulas.

Considering the equation for the velocity, the following Matlab statements yield the solution in terms of V:

```
dvdt=inline('(-9.8-.1*v.^2)','t','v');
v0=25;
[t,v]=ode45(dvdt,1.4,v0)
```

The first command defines the first-order differential equation, the second defines the boundary condition, and the third allows time and velocity to be obtained. These can then be plotted, using Matlab plotting routines, as shown in Figure 4.8. The velocity decreases from 25 m/s to 0 with time. After the velocity becomes zero, the drag reverses direction and the differential equation changes, so the solution is valid only until $V = 0$.

Similarly, the equation for x may be solved. However, this is a second-order equation, which is first reduced to two first-order equations as

$$\frac{dx}{d\tau} = V$$

$$\frac{dV}{d\tau} = -g - 0.1V^2$$

First, the right-hand sides of these two equations are defined as

```
function dydt=rhs(t,y)
dydt=[y(2);-9.8-0.1*y(2)^2];
```

Thus, y is taken as a vector with the distance and velocity as the two components. Then the Matlab commands are given as

```
y0=[0;25];
[t,v]=ode45('rhs',1.4,y0)
```

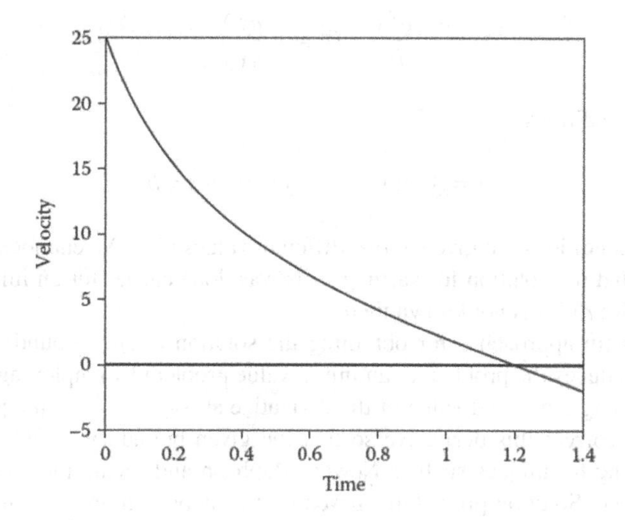

FIGURE 4.8 Velocity variation with time, as calculated by Matlab in Example 4.3.

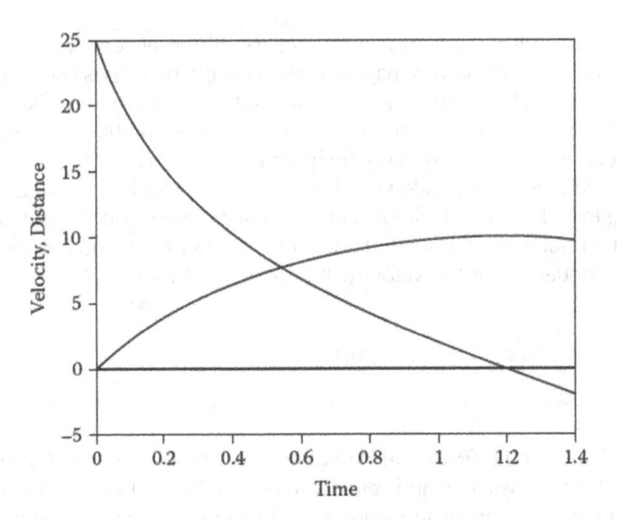

FIGURE 4.9 Variation of velocity v and distance x with time τ, as calculated by Matlab in Example 4.3.

Again, the initial conditions are given by the first line and the solution is given by the second. The results are obtained in terms of distance and velocity, which may be plotted, as shown in Figure 4.9. The calculated distance x and the velocity V are plotted against time. Clearly, the results in terms of the velocity V are the same by the two approaches. Thus, Matlab may be used effectively for solving initial-value problems, considering single equations as well as multiple and higher-order equations. The example considered here is from dynamics, but similar equations arise in thermal processes, as seen later. Further details on the use of Matlab for such mathematical problems are given in Appendix A.M.6.

4.2.3.2 Boundary-Value Problems

In the simulation of thermal systems, we are frequently concerned with problems in which the boundary conditions are given at two or more different values of the independent variable. Such problems are known as *boundary-value* problems. Because the number of boundary conditions needed equals the order of the ODE, the equation must at least be of second order to give rise to a boundary-value problem where the two conditions are specified at two different values of the independent variable. As an example, consider the following second-order equation:

$$\frac{d^2y}{dx^2} = F\left(x, y, \frac{dy}{dx}\right) \tag{4.26a}$$

with the boundary conditions

$$y = A, \text{ at } x = a \qquad y = B, \text{ at } x = b \tag{4.26b}$$

Therefore, the two conditions are given at two different values of x. We cannot start at either of the two locations and find the solution for varying x, as was done earlier for an initial-value problem, because the derivative dy/dx is not known there.

There are two main approaches for obtaining the solution to such boundary-value problems. The first approach reduces the problem to an initial-value problem by employing the first boundary condition and assuming a guessed value of the derivative at, say, $x = a$ for the preceding problem. Iteration is used to correct this derivative so that the given boundary condition at $x = b$ is also satisfied. Root solving techniques such as Newton-Raphson and secant methods may be used for the correction scheme. Solution procedures based on this approach are known as *shooting methods* because the adjustment of initial conditions to satisfy the conditions at the other location is

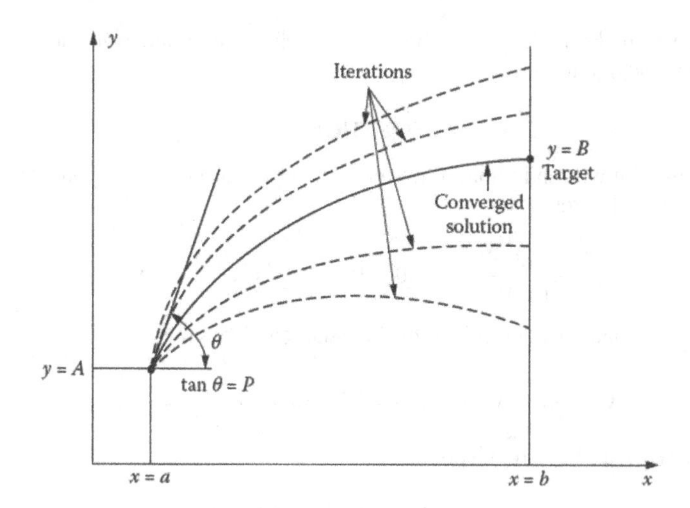

FIGURE 4.10 Iterations to the converged solution, employing a shooting method for solving a boundary-value ordinary differential equation.

similar to shooting at a target. Figure 4.10 shows a sketch of the shooting method. Thus, all of the methods discussed earlier for initial-value problems may be used, along with a correction scheme. The approach may easily be extended to higher-order equations and to different types of boundary conditions. The Matlab solution methods for initial-value problems, given earlier, can also be used along with an appropriate correction scheme.

The second approach is based on obtaining the finite-difference or finite-element approximation to the differential equation. In the former approach, the derivatives are replaced by their finite-difference approximations. This leads to a system of algebraic equations, which are solved to obtain the dependent variable at discrete values of the independent variable, as illustrated in the following example. These approaches are considered in greater detail for PDEs in the next section.

Example 4.4

The steady-state temperature $\theta(x)$ due to conduction in a bar, with convection at the surface and the assumption of uniform temperature across any cross-section, is given by the equation

$$\frac{d^2\theta}{dx^2} - G\theta = 0$$

where G is a constant and is given as 50.41 m^{-2}. Here, θ is the temperature difference from the ambient medium, which is at 20°C. The bar, which is 30 cm long, is discretized, as shown in Figure 4.11, using $\Delta x = 1$ cm and $x = i\Delta x$, where $i = 0, 1, 2, \ldots, 30$. It is given that the temperatures at the two ends, θ_0 and θ_{30}, are 100°C. Calculate the temperatures at the other grid points using

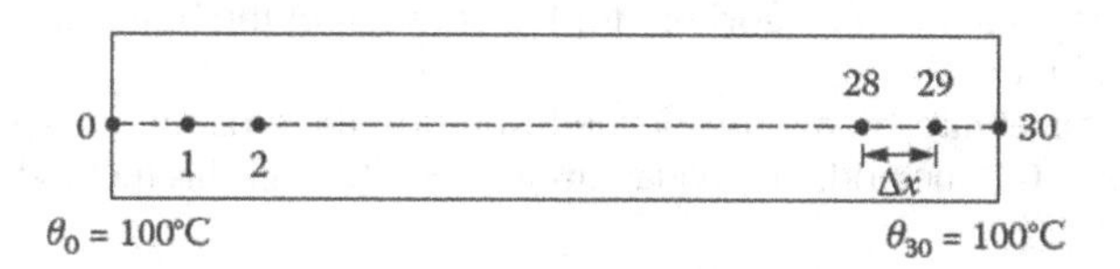

FIGURE 4.11 Physical problem considered in Example 4.4, along with the discretization.

the finite-difference method, along with the Gaussian elimination and SOR methods for solving the resulting algebraic equations.

SOLUTION

The given ODE may be written in finite-difference form by replacing the second-order derivative by the second central difference as

$$\frac{\theta_{i+1} - 2\theta_i + \theta_{i-1}}{(\Delta x)^2} - G\theta_i = 0 \quad \text{for } i = 1, 2, 3, \ldots, 29$$

Then the system of equations to be solved by Gaussian elimination is

$$\theta_{i+1} - [2 + G(\Delta x)^2]\theta_i + \theta_{i-1} = 0 \quad \text{for } i = 1, 2, 3, \ldots, 29$$

The equations for $i = 1$ and 29 are, respectively,

$$\theta_2 - S\theta_1 + \theta_0 = 0 \quad \text{and} \quad \theta_{30} - S\theta_{29} + \theta_{28} = 0$$

which give

$$S\theta_1 - \theta_2 = 100 \quad \text{and} \quad -\theta_{28} + S\theta_{29} = 100$$

where $S = 2 + G(\Delta x)^2$ and $\theta_0 = \theta_{30} = 100$. This system of equations may be written as

$$\begin{bmatrix} S & -1 & 0 & \cdots & & 0 \\ -1 & S & -1 & \cdots & & 0 \\ 0 & -1 & S & -1 & \cdots & 0 \\ \vdots & \vdots & \vdots & \vdots & & \vdots \\ 0 & 0 & 0 & 0 & -1 & S \end{bmatrix} \begin{bmatrix} \theta_1 \\ \theta_2 \\ \theta_3 \\ \vdots \\ \theta_{29} \end{bmatrix} = \begin{bmatrix} 100 \\ 0 \\ 0 \\ \vdots \\ 0 \\ 100 \end{bmatrix}$$

This is a tridiagonal system of linear equations and may be solved conveniently by Gaussian elimination, as outlined earlier. Computer programs in Fortran and Matlab are also given in Appendix A in order to present the algorithm. The same logic can be used to develop a program in other programming languages. Further details are given in Appendix A. The three nonzero elements in each row are denoted by *A*(I), *B*(I), and *C*(I). *B*(I) is the diagonal element and *A*(I), *C*(I) are elements on the left and right of the diagonal, respectively. Only two nonzero elements appear in the top and bottom rows. The constants on the right-hand side of the equations are denoted by *R*(I). Gaussian elimination is used to eliminate the left-most element in each row in one traverse from the top to the bottom row. Then the last row leads to an equation with only one unknown, which is calculated as *R*(29)/*B*(29), where both *R* and *B* are the new values after reduction. The other temperature differences are calculated by back-substitution, going up from the bottom to the top row. Figure 4.12 shows the computer output, in terms of the temperatures T_i, where $T_i = \theta + 20$, because the ambient temperature is given as 20°C. Clearly, the temperature distribution is symmetric about the midpoint. This numerical scheme, known as the *Thomas algorithm*, is extremely efficient, requiring $O(n)$ arithmetic operations for n equations. See Appendix A.M.6.5 for the Matlab program, which uses the tridiagonal matrix algorithm for the solution of the finite-difference equations obtained in this problem.

The set of linear algebraic equations obtained from the finite-difference approximation may also be solved by the SOR method. The equations are rewritten for this method as

$$\theta_i = \frac{\theta_{i+1} + \theta_{i-1}}{S} \quad \text{for } i = 1, 2, 3, \ldots, 29$$

The required temperatures are:

T(1) =	114.6583
T(2) =	109.7937
T(3) =	105.3816
T(4) =	101.3998
T(5) =	97.8282
T(6) =	94.6491
T(7) =	91.8461
T(8) =	89.4051
T(9) =	87.3139
T(10) =	85.5620
T(11) =	84.1405
T(12) =	83.0422
T(13) =	82.2616
T(14) =	81.7948
T(15) =	81.6395
T(16) =	81.7948
T(17) =	82.2616
T(18) =	83.0421
T(19) =	84.1404
T(20) =	85.5620
T(21) =	87.3140
T(22) =	89.4052
T(23) =	91.8462
T(24) =	94.6493
T(25) =	97.8286
T(26) =	101.4001
T(27) =	105.3819
T(28) =	109.7939
T(29) =	114.6584

FIGURE 4.12 Numerical results obtained in terms of the temperatures at the grid points by using the Thomas algorithm for the resulting tridiagonal set of equations in Example 4.4.

with $\theta_0 = \theta_{30} = 100$. Therefore, these equations may be solved for θ_i, varying i as $i = 1, 2, 3, \ldots, 29$. The SOR method may be written from Equation (4.9) as

$$\theta_i^{(l+1)} = \omega\left[\theta_i^{(l+1)}\right]_{GS} + (1-\omega)\theta_i^{(l)} \quad \text{for } i = 1, 2, 3, \ldots, 29$$

where

$$\left[\theta_i^{(l+1)}\right]_{GS} = \frac{\theta_{i+1}^{(l)} + \theta_{i-1}^{(l+1)}}{S} \quad \text{for } i = 1, 2, 3, \ldots, 29$$

The initial guess is taken as $\theta_i = 0$ and the temperature differences for the next iteration are calculated using the preceding equations. This iterative process is continued, comparing the values after each iteration with those from the previous iteration. Appendix A gives a sample program in Fortran for the Gauss-Seidel method, $\omega = 1$. Again, other programming languages or the Matlab environment may similarly be employed. The corresponding program in Matlab is given in Appendix A.M.4.3. The iteration is terminated if the following convergence criterion is satisfied:

$$\left|\theta_i^{(l+1)} - \theta_i^{(l)}\right| \le \varepsilon$$

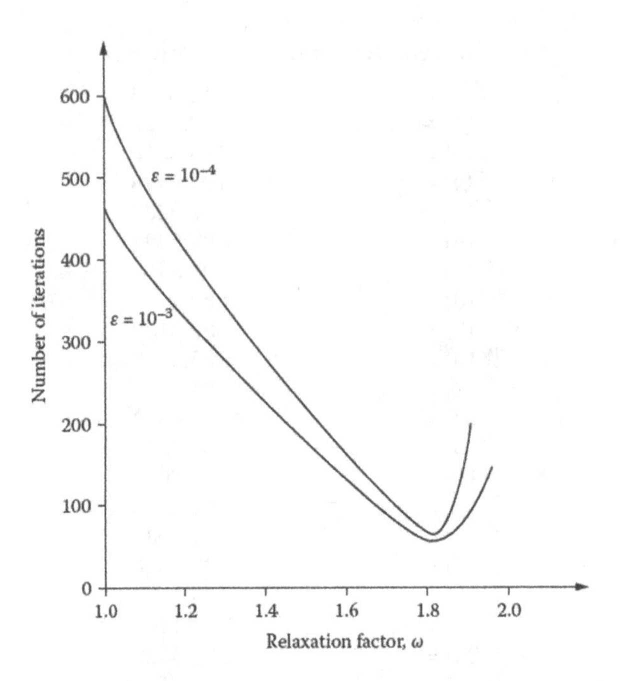

FIGURE 4.13 Variation of the number of iterations needed for convergence, in the solution of Example 4.4 by the SOR method, with the relaxation factor ω at two values of the convergence criterion ε.

where ε is a chosen small quantity. A value of 10^{-4} was found to be adequate. The relaxation factor ω was varied from 1.0 to 2.0 and the number of iterations needed for convergence determined. Figure 4.13 shows the results obtained for two values of ε and the optimum value of the relaxation factor ω_{opt}. The calculated numerical results for the temperature T_i are shown in Figure 4.14. Therefore, the results agree closely with the earlier ones from the tridiagonal matrix algorithm (TDMA). Both of these approaches are used extensively for solving differential equations, with the TDMA method being the preferred one for tridiagonal sets of equations.

4.2.4 Partial Differential Equations

A common circumstance in the numerical modeling of thermal systems is one in which the temperature, velocity, pressure, etc., are functions of the location and, possibly, of time as well. If the dependent variable is a function of two or more independent variables, the differential equations that represent such problems involve partial derivatives and are known as partial differential equations (PDEs). Two common PDEs that arise in thermal systems are given for a Cartesian coordinate system as

$$\frac{1}{a}\frac{\partial T}{\partial \tau} = \frac{\partial^2 T}{\partial x^2} \tag{4.27}$$

and

$$\frac{\partial^2 T}{\partial x^2} + \frac{\partial^2 T}{\partial y^2} = q'''(x, y) \tag{4.28}$$

Here, T is the temperature, x and y are the coordinate axes, τ is the time, q''' is a volumetric heat source, and α is the thermal diffusivity of the material. These equations, along with several others that are often encountered in thermal systems, have been given in earlier chapters.

Numerical results

EPS = 0.00010		EPS = 0.00001	
Number of Iterations = 600		Number of Iterations = 766	
T(1) =	114.6572	T(1) =	114.6578
T(2) =	109.7915	T(2) =	109.7928
T(3) =	105.3784	T(3) =	105.3803
T(4) =	101.3957	T(4) =	101.3981
T(5) =	97.8234	T(5) =	97.8263
T(6) =	94.6433	T(6) =	94.6467
T(7) =	91.8396	T(7) =	91.8434
T(8) =	89.3979	T(8) =	89.4022
T(9) =	87.3062	T(9) =	87.3109
T(10) =	85.5538	T(10) =	85.5588
T(11) =	84.1320	T(11) =	84.1371
T(12) =	83.0334	T(12) =	83.0388
T(13) =	82.2527	T(13) =	82.2581
T(14) =	81.7859	T(14) =	81.7914
T(15) =	81.6306	T(15) =	81.6360
T(16) =	81.7860	T(16) =	81.7914
T(17) =	82.2529	T(17) =	82.2581
T(18) =	83.0337	T(18) =	83.0388
T(19) =	84.1323	T(19) =	84.1371
T(20) =	85.5543	T(20) =	85.5588
T(21) =	87.3067	T(21) =	87.3109
T(22) =	89.3984	T(22) =	89.4022
T(23) =	91.8400	T(23) =	91.8434
T(24) =	94.6438	T(24) =	94.6467
T(25) =	97.8238	T(25) =	97.8263
T(26) =	101.3961	T(26) =	101.3981
T(27) =	105.3788	T(27) =	105.3803
T(28) =	109.7917	T(28) =	109.7928
T(29) =	114.6573	T(29) =	114.6578

FIGURE 4.14 Computer output for the solution of Example 4.4 by the SOR method for two values of ε (EPS).

The corresponding equations for other coordinate systems may similarly be written. We will consider only these two relatively simple equations to outline the numerical modeling of PDEs. The first equation is a *parabolic* equation, which can be solved by marching in time τ. Thus, the solution depends only on the results for previous time and not on those for the future time. It requires two boundary conditions in x and an initial condition in time. The second equation is an elliptic equation, which requires conditions on the entire boundary of the domain to be well-posed. The solution thus depends on all the conditions imposed at the boundaries. Several specialized books, such as those by Patankar (1980), Anderson et al. (2011), and Jaluria and Torrance (2003), are available on the numerical solution of PDEs that arise in fluid flow and heat transfer and may be consulted for details. Only a brief outline of the two main approaches, the finite-difference and the finite-element methods, is presented here.

4.2.4.1 Finite-Difference Method

In this approach, a grid is imposed on the computational domain so that a finite number of grid points is obtained, as seen in Figure 4.15. The partial derivatives in the given PDE are written in terms of the values at these grid points. Generally, Taylor series expansions are employed to derive the discretized forms of the various derivatives. These lead to finite-difference equations that are written for each grid point to yield a system of algebraic equations. Linear PDEs result in linear algebraic equations and nonlinear ones in nonlinear equations. The resulting system of algebraic equations is solved by the various methods mentioned earlier to obtain the dependent variables at the grid points. Iterative methods for solving algebraic equations are particularly useful because PDEs generally lead to large sets of equations with sparse coefficient matrices.

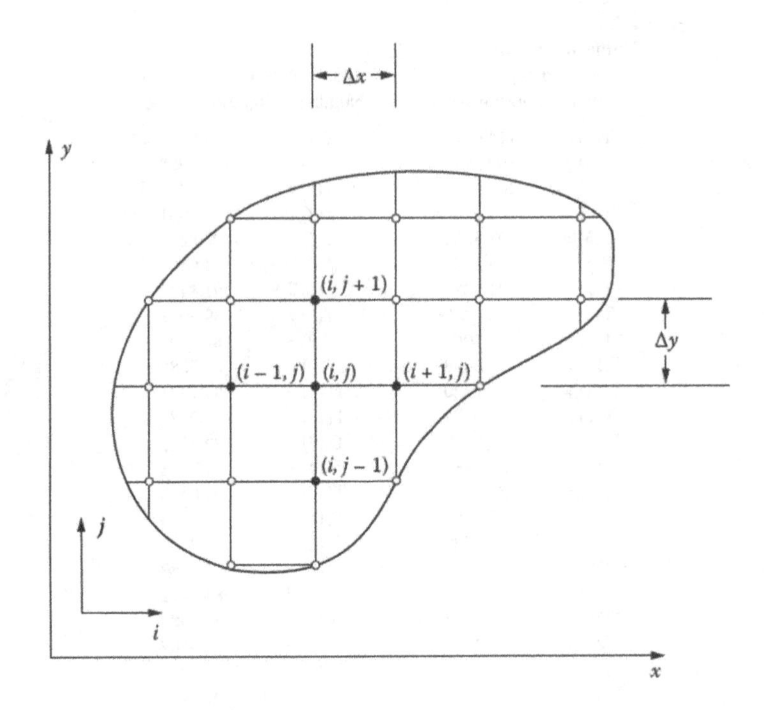

FIGURE 4.15 A two-dimensional computational region with a superimposed finite-difference grid.

Equation (4.27) may be written in finite-difference form as

$$\frac{T_{i+1,j} - T_{i,j}}{\Delta\tau} = \alpha\left[\frac{T_{i,j+1} - 2T_{i,j} + T_{i,j-1}}{(\Delta x)^2}\right] \tag{4.29}$$

where the subscript $(i + 1)$ denotes the values at time $(t + \Delta t)$ and i those at time τ. The spatial location is given by j. Here, $x = j\Delta x$ and $\tau = i\Delta\tau$. The truncation error, which represents the error due to terms that are neglected in the Taylor series for this approximation, is of order $\Delta\tau$ in time and $(\Delta x)^2$ in space. The second derivative is approximated as a central difference at time τ and a forward difference is taken for the first derivative in time. The resulting finite-difference equation may be derived from Equation (4.29) as

$$T_{i+1,j} = \left(1 - 2\frac{\alpha\Delta t}{(\Delta x)^2}\right)T_{i,j} + \frac{\alpha\Delta\tau}{(\Delta x)^2}\left(T_{i,j+1} + T_{i,j-1}\right) \tag{4.30}$$

This equation gives the temperature distribution at time $(\tau + \Delta\tau)$ at the grid point whose spatial coordinate is $x = j\Delta x$, in terms of temperatures at time τ at the grid points with coordinates $(x - \Delta x)$, x, and $(x + \Delta x)$. If the initial temperature distribution is given and the conditions at the boundaries, say, $x = 0$ and $x = a$, are given, the temperature distribution may be computed for increasing values of time τ. This is the *explicit* method, often known as the *forward time central space* (FTCS) method. However, the stability of the numerical scheme is assured only if $F = [\alpha\Delta\tau/\Delta x)^2] \leq 1/2$, where F is known as the *grid Fourier number*. This constraint on F ensures that the coefficients in Equation (4.30) are all positive, which has been found to result in stability of the scheme. Therefore, the method is conditionally stable.

In view of the constraint on $\Delta\tau$ due to stability in the explicit scheme, several *implicit* methods have been developed in which the spatial second derivative is evaluated at a different time, between τ and $\tau + \Delta\tau$. If it is evaluated midway between the two times, the scheme obtained is the popular

Crank-Nicolson method, which has a second-order truncation error, $O[(\Delta\tau)^2]$, in time as well and is more accurate than the FTCS method. If the derivative is evaluated at time $(\tau + \Delta\tau)$, the fully implicit method or *Laasonen* method is obtained. These methods do not have a restriction on $\Delta\tau$ due to stability considerations for linear equations, such as Equation (4.27), for a chosen value of Δx. The resulting finite-difference equation is

$$\frac{T_{i+1,j} - T_{i,j}}{\Delta\tau} = \alpha\left[\gamma\frac{T_{i+1,j+1} - 2T_{i+1,j} + T_{i+1,j-1}}{(\Delta x)^2} + (1-\gamma)\frac{T_{i,j+1} - 2T_{i,j} + T_{i,j-1}}{(\Delta x)^2}\right] \quad (4.31)$$

where γ is a constant, being 0 for the FTCS explicit, 1/2 for the Crank-Nicolson, and 1.0 for the fully implicit methods. Corresponding Matlab computer programs are given in Appendix A.M.7.

Multidimensional problems commonly arise in thermal systems. For instance, two-dimensional, unsteady conduction at constant properties is given by the following equation:

$$\frac{\partial T}{\partial \tau} = \alpha\left(\frac{\partial^2 T}{\partial x^2} + \frac{\partial^2 T}{\partial y^2}\right) \quad (4.32)$$

The methods for the one-dimensional problem may be extended to this problem. Stability considerations again pose a limitation of the form $[\alpha\Delta\tau/\Delta x)^2] \leq 1/4$, if $\Delta x = \Delta y$. A particularly popular method is the *alternating direction implicit* (ADI) method, which splits the time step into two halves, keeping one direction as implicit in each half-step and alternating the directions, giving rise to tridiagonal systems in the two cases.

For the elliptic problem, such as the one given by Equation (4.28), the computational domain is discretized with $x = i\Delta x$ and $y = j\Delta y$. Then the mathematical equation may be written in finite-difference form as

$$\frac{T_{i+1,j} - 2T_{i,j} + T_{i-1,j}}{(\Delta x)^2} + \frac{T_{i,j+1} - 2T_{i,j} + T_{i,j-1}}{(\Delta y)^2} = q'''_{i,j} \quad (4.33)$$

If this finite-difference equation is written out for all the grid points in the computational domain, where the temperatures are unknown, a system of linear algebraic equations is obtained. At the boundaries, the conditions are given, which may specify the temperature (*Dirichlet* conditions), the temperature derivative (*Neumann* conditions), or give a relationship between the temperature and its derivative (mixed conditions). Thus, special equations are obtained for temperature at the boundaries. The overall system of equations is generally a large set, particularly for three-dimensional problems, because of the usually large number of grid points employed. The coefficient matrix is also sparse, making iterative schemes such as SOR appropriate for the solution. Many specialized and efficient methods have been developed to solve specific elliptic equations such as the one considered here, which is a Poisson equation. For $q''' = 0$, it is known as the *Laplace* equation. If the given PDE is nonlinear, the resulting algebraic equations are also nonlinear. These are solved by the methods outlined earlier for sets of nonlinear algebraic equations. Obviously, the solution in this case is considerably more involved than that for linear equations.

4.2.4.2 Finite-Element Method

Finite-element methods are extensively used in engineering because of their versatility in the solution of a wide range of practical problems. Finite-difference methods are generally easier to understand and apply, as compared to finite-element methods; they also have smaller memory and computational time requirements. Thus, these are easier to develop and to program. However, practical problems generally involve complicated geometries, complex boundary conditions, material property variations, and coupling between different domains. Finite-element methods are particularly well-suited for such circumstances because they have the flexibility to handle arbitrary variations in boundaries and properties. Consequently, much of the software developed for engineering systems and processes in the last two decades has been based on the finite-element method (Huebner and Thornton, 2001; Reddy, 2018). Available software is used extensively in finite-element solutions of engineering problems because of the tremendous effort generally needed for the development of the computer program. Finite-difference methods continue to be popular for simpler geometries and boundary conditions. Also, a wide range of grids and discretization methods has been developed to apply the finite-difference method to more complex geometries and boundary conditions.

The finite-element method is based on the integral formulation of the conservation principles. The computational domain is divided into a number of finite-elements, several types and forms of which are available for different geometries and applicable equations. Linear elements for one-dimensional cases, triangular elements for two-dimensional problems, and tetrahedral elements for three-dimensional problems are commonly used (see Figure 4.16). The variation of the dependent variable is generally taken as a polynomial and frequently as linear within the elements. Integral equations that apply for each element are derived and the conservation principles are satisfied by minimization of the integrals or by reducing their residuals to zero. A method of weighted residuals that is very commonly used for thermal processes and systems is the *Galerkin method* (Jaluria and Torrance, 2003).

The ultimate result of applying the finite-element method to the computational domain and the given PDE is a system of algebraic equations. The overall set of equations, known as global equations, is formed by assembling the contributions from each element. Interior nodes are removed from the assembled system by a process called *condensation*. A solution of the set of equations then leads to the values at the nodes from which values in the entire domain are obtained by using the interpolation functions. The method is capable of handling complicated geometries by a proper choice and placement of finite elements. Arbitrary boundary conditions and material property variations can be easily incorporated. The same scheme can be used for different problems, making the method very versatile. Because of all these advantages, finite-element methods, largely in the form of available computer codes, are widely used in the simulation and analysis of engineering systems. In simpler cases, finite-difference methods may be used advantageously.

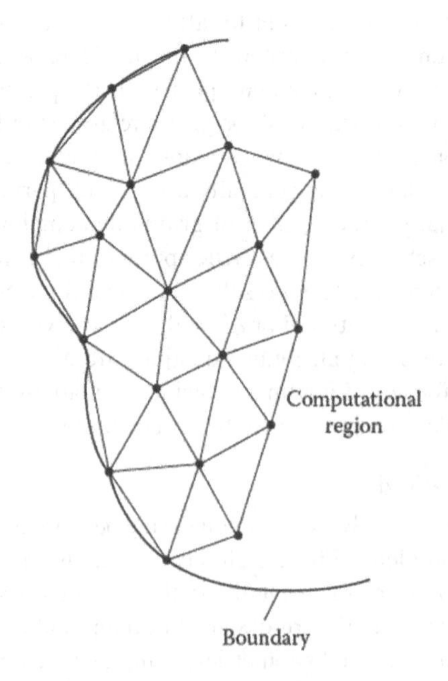

FIGURE 4.16 Finite-element discretization of a two-dimensional region, employing triangular elements.

4.2.4.3 Other Methods

Several other methods have been developed for solving PDEs. These include *finite volume*, *boundary element*, and *spectral* methods. In finite volume methods, the integral formulation is used with simple approximations for the values within the volume and at the boundaries. Therefore, this is a particular case of the finite-element method and is consequently not as versatile, though the programming is much simpler and is similar to that for finite-difference methods. It is quite a popular method, which combines the simplicity of the finite-difference method and the versatility of the finite-element method. Many common programs, such as the *Simpler* code of Patankar (1980) and commercial software, such as Ansys, are based on the finite-volume method.

In boundary element methods, the volume integral from the conservation postulate is converted into a surface integral using mathematical identities. This leads to discretization of the surface for obtaining the desired solution in the region. It is particularly useful for complicated geometries and complex boundary conditions (Brebbia et al., 2013). In spectral methods, the solution is approximated by a series of functions, such as sinusoidal functions. For particular equations such as the Poisson equation, geometries such as cylindrical and spherical cases, and certain boundary conditions, very efficient spectral schemes have been developed and are used advantageously. Very accurate results can often be obtained with a relatively small amount of effort for many heat transfer and fluid flow problems.

Example 4.5

The dimensionless temperature θ in a flat plate is determined by the PDE

$$\frac{\partial^2\theta}{\partial X^2} = \frac{\partial\theta}{\partial\tau}$$

The initial and boundary conditions are given as

$$\theta(X,0) = 0 \qquad \theta(0,\tau) = 1 \qquad \frac{\partial\theta}{\partial X}(1,\tau) = 0$$

where X and τ are the dimensionless coordinate distance and time, respectively. Solve this problem by the Crank-Nicolson method to obtain $\theta(X, \tau)$.

SOLUTION

The given PDE is a parabolic equation and can be solved by marching in time τ, starting with the initial conditions. The coordinate distance X varies from 0 to 1, with the temperature given as 1 at $X = 0$ and the adiabatic condition, $\partial\theta/\partial X = 0$, applied at $X = 1$. The finite-difference equation for the Crank-Nicolson method is

$$-F\theta_{i+1,j+1} + 2(1+F)\theta_{i+1,j} - F\theta_{i+1,j-1} = F\theta_{i,j+1} + 2(1-F)\theta_{i,j} + F\theta_{i,j-1} \qquad \text{(a)}$$

where $F = \Delta\tau/(\Delta X)^2$, i represents the time step, and j represents the spatial grid location. Therefore, $\tau = i\Delta\tau$ and $X = j\Delta X$, where i starts with 0 and increases to represent increasing time and j varies from 0 to n, with $n = 1/\Delta X$.

The finite-difference equation may be rewritten as

$$A\theta_{j-1} + B\theta_j + C\theta_{j+1} = D \qquad \text{(b)}$$

where the θ values are at the $(i + 1)$th time step and D is the expression on the right-hand side of Equation (a). Therefore, D is a function of the known θ values at the ith time step. The constants

A, *B*, and *C* are the coefficients on the left-hand side of Equation (a) and depend on the value of the grid Fourier number *F*. No constraints arise on $\Delta\tau$ due to stability considerations, though oscillations may arise in some cases at large *F*. It is evident from Equation (b) that the resulting set of algebraic equations is tridiagonal and can be solved conveniently by the Thomas algorithm discussed earlier and employed in Example 4.4.

The boundary condition at $X = 1$ is a gradient, or Neumann condition. One-sided second-order differences may be used to approximate it, giving an error of $O[(\Delta X)^2]$, as

$$\left(\frac{\partial \theta}{\partial X}\right)_{i,j} = \frac{\theta_{i,j-2} - 4\theta_{i,j-1} + 3\theta_{i,j}}{2\Delta X} \tag{c}$$

where *j* is replaced by *n* for the boundary at $X = 1$. Other approximations are also available (Jaluria and Torrance, 2003). A more accurate formulation is based on the energy balance for the finite volume represented by the grid point at the surface. The problem is solved by marching in time, with a time step $\Delta\tau$. At each time step, the tridiagonal set, represented by Equation (b), is solved to obtain the temperature distribution.

Because this problem has a steady state, the marching in time is carried out until a convergence criterion of the following form is satisfied for all *j*:

$$|\theta_{i+1,j} - \theta_{i,j}| \leq \varepsilon \tag{d}$$

where ε is a chosen small quantity. It must be ensured that the results are not significantly affected by changes in the grid size ΔX, time step $\Delta\tau$, and convergence parameter ε.

The numerical results obtained are shown in Figure 4.17 and Figure 4.18. The former shows the temperature distribution as a function of time, indicating the approach to steady-state conditions, which require the temperature distribution to become uniform at $\theta = 1$. Figure 4.18 shows the variation of the temperature at several locations in the plate with time τ. Again, the approach to steady state at large time is clearly seen. The Crank-Nicolson method is a very popular choice for such one-dimensional problems because of the second-order accuracy in time and space. Tridiagonal sets of equations are generated for one-dimensional problems, and these may be solved conveniently and accurately by the Thomas algorithm to yield the desired solution. See Appendix A for programs in Fortran and in Matlab for solving PDEs.

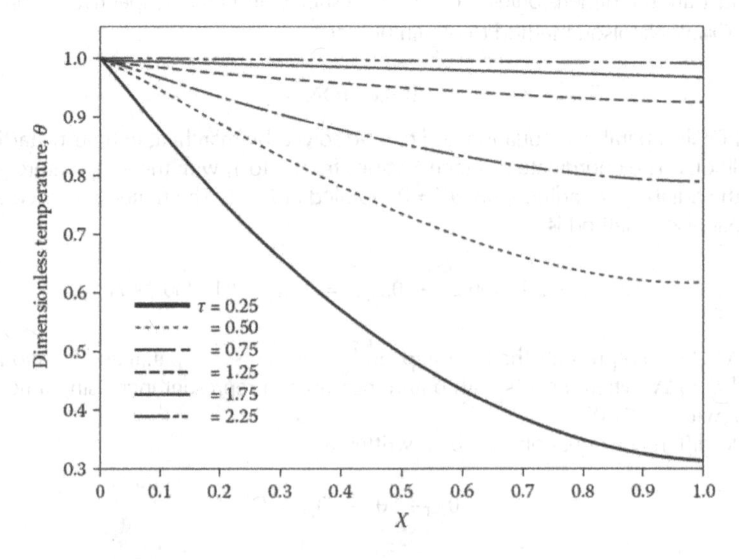

FIGURE 4.17 Computed temperature distribution at various time intervals for Example 4.5, using $\Delta\tau = 0.05$ and $\Delta X = 0.1$.

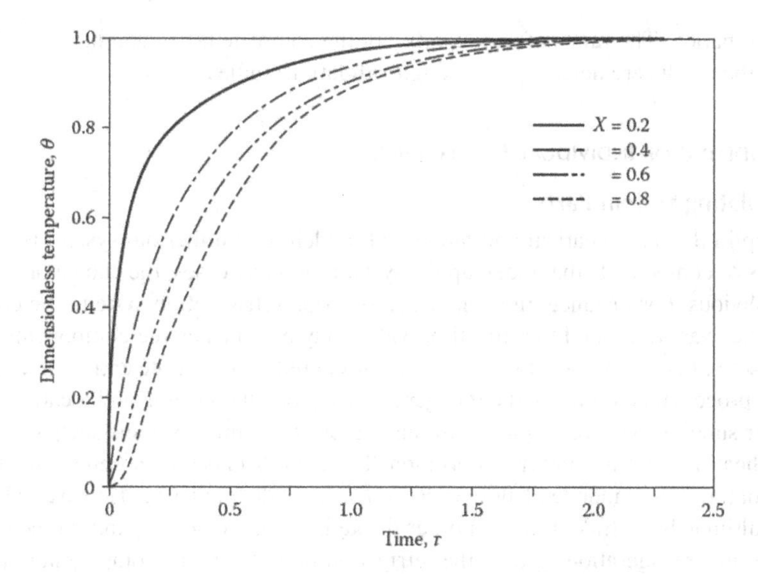

FIGURE 4.18 Variation of the temperature at several locations in the plate with dimensionless time τ for Example 4.5.

4.3 NUMERICAL MODEL FOR A SYSTEM

We now come to the numerical model for the overall system, which may comprise several parts, constituents, or subsystems. The model may be relatively simple, as is the case for systems with a small number of components such as a refrigerator, or may be very involved, as is the case for a major enterprise such as a power plant. The numerical model may be developed by the users themselves or it may be based on a commercially available general-purpose code such as Fidap, Ansys, Phoenics, Simpler, or Fluent. Specialized programs for specific applications are also available. Because the development of computer codes for large thermal systems is an elaborate and time-consuming process, it is often more convenient and efficient to use a commercially available program. Consequently, such codes are employed extensively in industry and form the basis for the numerical simulation and design of a variety of thermal systems, ranging from electronic packages to air conditioning and energy systems. However, it is important to be conversant with the algorithms used in the software and to be aware of their applicability, accuracy, limitations, and ease with which inputs may be given to simulate different circumstances.

Even if the numerical model is being developed indigenously, software available on the computer or in the public domain may be employed effectively. This is particularly the case for graphics programs and standard programs, such as matrix methods for solving sets of linear algebraic equations and the Runge-Kutta method for the solution of ODEs. Again, we must be familiar with the numerical approach used in the software and must have information on its accuracy and possible limitations. It is rarely necessary to develop the numerical code for graphics, because a wide variety of programs, such as Tecplot and Matlab graphics, are conveniently available and easy to use for different needs, ranging from line graphs to contour plotting. Similarly, programs for curve fitting are widely used for the analysis of experimental or numerical data and for the derivation of appropriate correlations.

In summary, the numerical model for the complete thermal system may contain programs that have been developed by the user, those in the public domain, standard programs available on the computer, and even commercially available general-purpose programs, with all of these linked to each other to simulate different aspects or components of the system. In addition to these programs, the numerical model may be linked with available information on material properties, characteristics of some of the devices or components in the system, heat transfer correlations, and other relevant information. The range of applicability of the complete numerical model and the expected accuracy of the results are determined through validation studies.

4.3.1 Modeling of Individual Components

4.3.1.1 Isolating System Parts

The first step in the mathematical and numerical modeling of a thermal system is to focus on the various parts or components that make up the system. In many cases, the choice of individual components is obvious. For instance, in a vapor compression refrigeration system, the compressor, the condenser, the evaporator, and the throttling value may be taken as the components of the system (see Figure 4.19). Each component here may be considered as a separate entity, in terms of the thermodynamic process undergone by the refrigerant and geometry, design, and location of the component. Similar subdivisions are employed in many thermodynamic systems such as those in energy generation, heating, cooling, and transportation. The components are chosen so that these are relatively self-contained and independent in order to facilitate the modeling. However, all such components must ultimately be linked to each other through energy, material, and momentum transport. For instance, in a refrigeration system, the refrigerant flows from one component to the other, conveying the energy stored in the fluid, as shown in Figure 1.8. In each component, energy exchanges occur, leading to the resulting thermodynamic state of the fluid at the exit of the component.

In many cases, the choice of the individual components is not so obvious. However, differences in geometry, material, function, thermodynamic state, location, and other such characteristics may be used to separate the components. For instance, the walls and the outside insulation in a furnace may be treated as different components because of the difference in material. The main thing to remember is that the component must be substantially separate or different from the others and must be amenable to modeling as an individual item.

The given system may also be broken down into subsystems, each with its own components. Then each subsystem is treated as a system for model development, with all the individual models being brought together at the end. For instance, the cooling system in an automobile, the boiler in a power plant, and the cooling arrangement in an electronic system may be considered as subsystems for modeling and design. Frequently, the subsystems are designed separately and the results obtained are employed in the design of the overall system, treating the subsystem simply as a component whose characteristics are known.

4.3.1.2 Mathematical Modeling

Once the individual components have been isolated, we can proceed to the development of the mathematical model for each. For this purpose, each component is treated separately, replacing its interaction with other components by known conditions that eliminate the coupling. For instance, for modeling a wall losing energy by convection to air in a room, as shown in Figure 4.20, the actual

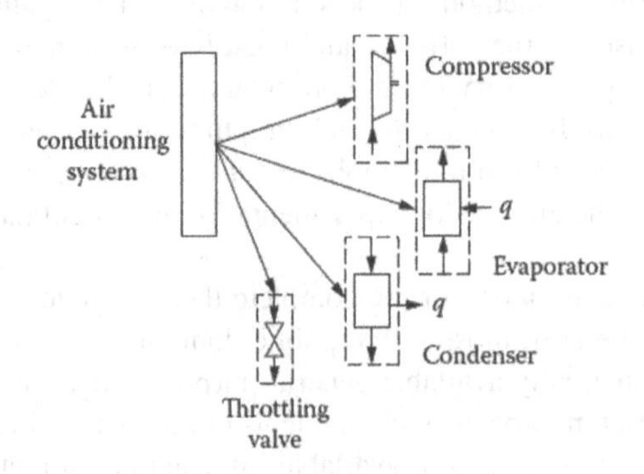

FIGURE 4.19 Isolating system parts or components for modeling.

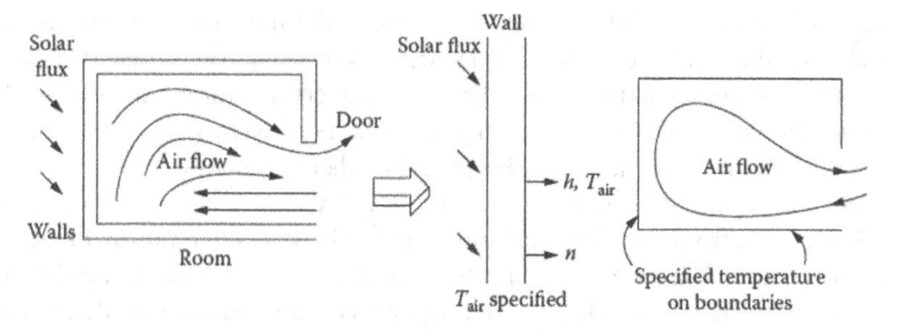

FIGURE 4.20 Decoupling a wall and enclosed air for modeling thermal transport in a room.

thermal coupling between the air flow and conduction in the wall may be replaced by a heat transfer coefficient h at the wall surface. This decouples the solutions for the two heat transfer regions because the conditions at the boundary

$$[T]_{\text{wall}} = [T]_{\text{air}} \qquad \left[-k\frac{\partial T}{\partial n}\right]_{\text{wall}} = \left[-k\frac{\partial T}{\partial n}\right]_{\text{air}} \tag{4.34a}$$

which require a solution for the flow and heat transfer in air, are replaced by

$$\left[-k\frac{\partial T}{\partial n}\right]_{\text{wall}} = h(T_{\text{wall}} - T_{\text{air}}) \tag{4.34b}$$

Here, n is in the direction normal to the surface and T_{air} is a specified temperature. Similarly, the air is modeled separately for heat transfer with a specified wall temperature. Then, the two regions are modeled as separate entities without linking the two. Similarly, the condenser in a home air conditioning system may be modeled using given, fixed inflow conditions of the refrigerant to decouple it from the compressor that provides the input to the condenser in an actual system. Then each component can be subjected to mathematical modeling procedures and the resulting mathematical equations derived.

Different simplifications and idealizations may apply for different components, resulting in different types of applicable equations. For example, one component may be modeled as lumped mass, giving rise to an ODE for the temperature as a function of time, while another component may be modeled as a one-dimensional transient problem, represented by Equation (4.27). A single nonlinear algebraic equation may arise from an energy balance for determining the temperature at the surface of a body. The continuity, momentum, and energy equations may be needed for modeling the flow.

The different mathematical models obtained for the various system parts are based on simplifications, approximations, and idealizations that are, in turn, based on the material, geometry, transport processes, boundary conditions, and estimates of the contributions of the various transport mechanisms, as discussed in Chapter 3. However, the mathematical models derived are not unique and further improvements may be needed, depending on the numerical results from simulation and on comparisons with experimental data. It is important to maintain the link between mathematical and numerical models and to be prepared to improve both as the need arises. It is generally best to start with the simplest possible mathematical and numerical models and to improve these gradually by including effects that may have been neglected earlier.

4.3.1.3 Numerical Modeling

The mathematical equations that represent each component must be solved to study its behavior. Numerical algorithms applicable to the different types of equations that arise are employed to solve these equations, as discussed earlier. Thus, a numerical model, which is decoupled from the others, is obtained for each component. The results from this model indicate the basic characteristics of the component under the idealized or approximated boundary conditions used. The behavior of the component, as some of these conditions or related parameters are varied, may be studied in order to ensure that the individual model is physically realistic. For instance, the flow rate of the colder fluid in a heat exchanger may be increased, keeping the other variables fixed. It is expected that the temperature rise of this fluid in the heat exchanger will decrease because a larger amount of fluid is to be heated. The results from the numerical model must show this trend if the model is physically valid. Grid refinement is also done to ensure the accuracy of the results. Some simple cases may be considered to obtain analytical solutions and thus provide a method of validating the individual numerical models.

4.3.2 Merging of Different Models

Once all the individual numerical models for the various components or parts of the given thermal system have been obtained and tested on the basis of physical reasoning and analytical results, these must be merged to obtain the model for the overall system. Such a merging of the models requires bringing back the coupling between the different parts that had been neglected in the development of individual models.

For instance, if two parts A and B of the system exchange energy by radiation, their temperatures T_A and T_B are coupled through a boundary condition of the form

$$Q = \frac{\sigma A_A\left(T_A^4 - T_B^4\right)}{\dfrac{1}{\varepsilon_A} + \left(\dfrac{1}{\varepsilon_B} - 1\right)\dfrac{A_A}{A_B}} \tag{4.35}$$

where Q is the radiative energy lost by A and gained by B, ε_A and ε_B are the emissivities, A_A and A_B are the two surface areas, and σ is the Stefan-Boltzmann constant. This equation applies for A completely surrounded by B, with no radiation from A falling directly on itself. For developing the numerical model for, say, part A, this radiative heat transfer may be idealized as a constant heat flux q at the surface of A or as a constant temperature environment in radiation exchange with the surface. Thus, the temperature of part B is eliminated from the model for part A, which may then be modeled separately. Similarly, such approximations with constant or known parameters in the boundary conditions may be used for modeling part B. In the process of merging the two models, these approximations must be replaced by the actual boundary condition, Equation (4.35), which couples the temperatures of the two parts.

Similarly, the approximation made in Equation (4.34b) is removed by replacing the convective condition given by a specified heat transfer coefficient h by the correct boundary conditions given by Equation (4.34a). This couples the transport in the wall with that in the air. For the air conditioning system considered earlier, the temperature and pressure at the inlet of the condenser are set equal to the corresponding values at the exit of the compressor to link these two parts of the system. Proceeding in this way, other parts are also coupled through the boundary and inflow/outflow conditions.

This approach of modeling individual parts and then coupling them may appear to be an unnecessarily complicated way of deriving the numerical model for the entire system. Indeed, for relatively simple systems consisting of a small number of parts, it is often more convenient and efficient to develop the numerical model for the system without considering individual parts separately. However, if the system has a large number of parts, it is preferable to develop individual numerical models and to test and validate them separately before merging them to obtain the model for the system. This allows a complicated problem to be broken down into simpler ones that may be individually treated and tested before the final assembly. This approach is used extensively in industry to

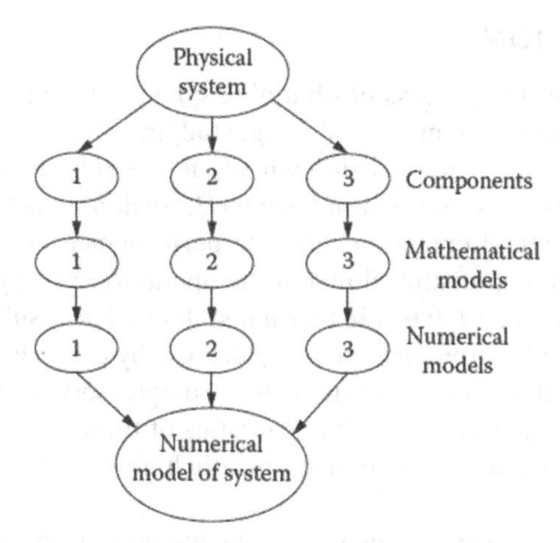

FIGURE 4.21 Schematic of the general approach of developing an overall model for a thermal system.

model complex systems. A direct modeling of the entire system has little chance of success because many coupled equations are involved. Figure 4.21 shows a schematic of the process described here for a general thermal system.

4.3.3 Accuracy and Validation

We discussed the validation of the mathematical model and the numerical scheme earlier. Numerical models for individual parts of the system are similarly tested and validated before merging them to yield the overall numerical model of the thermal system under consideration. The validation of this complete model, therefore, is based largely on the testing and validation employed at various steps along the process. The main considerations that form the basis for validation of the numerical model of the entire system are, as before,

1. Results should be independent of arbitrary numerical parameters
2. Demonstrates expected physical behavior
3. Comparison with analytical and experimental results
4. Comparisons with prototype results

The arbitrary numerical parameters refer to the grid, time step, and other quantities chosen to obtain a numerical solution. It is important to ensure that the results from the model are essentially independent of these parameters, as was done earlier for the numerical solution of individual equations and components. The physical behavior now refers to the thermal system, so that the results from the complete model are considered in terms of the expected physical trends to ascertain that the model does indeed yield physically realistic characteristics. The numerical model is subjected to a range of operating conditions and the results obtained examined for physical consistency.

Analytical and experimental results are rarely available for validation. However, as discussed earlier, analytical results may be obtained for a few highly idealized situations. Similarly, experimental data may be obtained or may be available for a few simple geometries and conditions. Such analytical results and experimental data are used for validating mathematical and numerical models for individual parts of the system. For the overall system, experimental data may be available from existing systems. For instance, existing cooling and heating systems may be numerically modeled in order to compare the results against data available on these systems. Before going into production, a prototype may be developed to test the model and the design. This provides the best information for the quantitative validation of the model and a check on the accuracy of the results obtained from the model.

4.4 SYSTEM SIMULATION

System simulation refers to the process of obtaining quantitative information on the behavior and characteristics of the real system by analyzing, studying, or examining a model of the system. The model may be a physical, scaled-down version of the given system, derived on the basis of the similarity principles outlined in Chapter 3. Such a model is subjected to a variety of operating and environmental conditions and the performance of the system is determined in terms of variables such as pressure, flow rate, temperature, energy input/output, and mass transfer rate that are of particular interest in thermal systems. The results from such a simulation may be expressed in terms of correlating equations derived by curve-fitting techniques. Physical modeling and testing of full-size components such as compressors, pumps, and heat exchangers are often used to derive the performance characteristics of these components. This approach is rarely used for the entire system because of the typically high cost and effort involved in fabrication and experimentation.

Sometimes the given physical system may be simulated by investigating another system that is represented by the same equations and that may be easier to fabricate or assemble. Such a model, called an analog model in the preceding chapter, also has a limited range of applicability, and, therefore, this simulation is not often used in the design of thermal systems. Electrical circuits used to simulate fluid flow systems, consisting of pipes, fittings, valves, and pumps, and conduction heat transfer through a multilayered wall are examples of analog simulation.

In the remaining portion of this chapter, we will consider only system simulation based on mathematical and numerical modeling. Therefore, the equations obtained from the mathematical model are solved by analytical or numerical methods to yield the system behavior under a variety of operating conditions as well as for different design variables, in order to provide the quantitative inputs needed for achieving acceptable designs and for optimization. Analytical solutions are obtained in only a few, often highly idealized, circumstances, and numerical modeling is generally needed to obtain the desired results for practical problems. Performance characteristics of components, as obtained from separate physical modeling and tests, as well as material properties, form part of the overall model and are assumed to be known. These may be available in the form of data or as correlating equations. The system may be represented by algebraic equations, ordinary or partial-differential equations, integral equations, or a combination of these. A numerical model is developed to solve the resulting simultaneous equations, many of which are typically nonlinear for thermal systems. Simulation of the system is carried out by means of this model.

4.4.1 Importance of Simulation

System simulation is one of the most important elements in the design and optimization of thermal systems. Because experimentation on a prototype of the actual thermal system is generally very expensive and time consuming, we usually have to depend on simulation based on a model of the given system to obtain the desired information on the system behavior under different conditions. A one-to-one correspondence is established between the model and the physical system by validation of the model, as discussed earlier. Then the results obtained from a simulation of the model are indicative of the behavior of the actual system.

There are several reasons for simulating the system. Simulation can be used to

1. Evaluate different designs for selection of an acceptable design
2. Study system behavior under *off-design* conditions
3. Determine safety limits for the system
4. Determine effects of different design variables for optimization
5. Improve or modify existing systems
6. Investigate sensitivity of the design to different variables

4.4.1.1 Evaluation of Design

Evaluation of different designs is an extremely important use of simulation because several designs are typically generated for a given application. If each of these were to be fabricated and tested for acceptability, the cost would be prohibitive. System simulation is employed effectively to investigate each design and to determine if the given requirements and constraints are satisfied, thus yielding an acceptable or workable design. For instance, several different designs, employing different geometry, materials, and dimensions, may be developed for a heat exchanger involving given fluids and given requirements on the temperatures or the heat transfer rates. Instead of fabricating each of these heat exchanger designs, mathematical and numerical modeling may be employed to obtain a satisfactory and accurate model. This model is then used for simulating the actual system in order to obtain the desired outputs in terms of heat transfer rates and temperatures. Operating conditions for which the system is designed are considered first to determine if the design meets the given requirements and constraints. These conditions are often termed design conditions because they form the basis for the design. Even if only one design has been developed for a given application, it must be evaluated to ensure that it is acceptable.

4.4.1.2 Off-Design Performance and Safety Limits

Predicting the behavior of the system under off-design conditions, i.e., values beyond those used for the design, is another important use of system simulation. Such a study provides valuable information on the operation of the system and how it would perform if the conditions under which it operates were to be altered, as under overload or fractional-load circumstances. Systems seldom operate at the design conditions and it is important to determine the range of operating conditions over which they would deliver acceptable performance. The deviation from design conditions may occur due to many reasons, such as variations in energy input, differences in raw materials fed into the system, changes in the characteristics of the components with time, changes in environmental conditions, and shifts in energy load on the system. The results obtained from simulation under off-design conditions would indicate the versatility and robustness of the system. It is obviously desirable to have a wide range of off-design conditions for which the system performance is satisfactory. A narrow range of acceptability is generally not suitable for consumer products because large variations in the operating conditions are often expected to arise. For instance, a residential hot water system designed for a particular demand and given inlet temperature must be able to perform satisfactorily if either of these were to vary substantially. In manufacturing processes, it is common to encounter variations in the shape, dimensions, and material properties of the items undergoing thermal processing.

These outputs also indicate the safety limits of the system. It is important to determine the maximum thermal load an air conditioner can take, the maximum power input to a furnace that can be given, and so on, without damage to the system or the user. Safety features can then be built into the system, such as mandatory shutdown of the system if the safety levels are exceeded or warning lights to indicate possible damage to the system. We are all familiar with such features in cars, lawn mowers, and other systems in daily use. Some of these aspects were also considered earlier in Section 2.3.6.

4.4.1.3 Optimization

System simulation plays an important role in the optimization of the system. As will be seen in later chapters, the outputs from the system must be obtained for a range of design variables in order to select the optimum design. The optimization of the system may involve minimization of objective functions such as cost per item, weight, and energy consumption per unit output or maximization of quantities such as output, return on investment, and rate of energy removal. Whatever the criterion for optimization, it is essential to change the variables over the design domain, determined by physical limitations and constraints, and to study the system behavior. Then, using the various techniques for optimization presented later, the optimal design is determined. The results obtained from simulation may sometimes be curve fitted to yield algebraic equations, which greatly facilitate the optimization process. For instance, if the cooling system for some electronic equipment using a fan has been simulated, different locations, flow rates, and dimensions of the fan may be considered to derive algebraic equations to represent the dependence of the heat removal rate on these variables. Then, an optimal configuration that delivers the most effective cooling per unit cost may be obtained easily.

4.4.1.4 Modifications in Existing Systems

The use of simulation for correcting a problem in an existing system or for modifying the system for improving its performance is also an important application. Rather than changing a particular component in order to correct the problem or improve the system, simulation is first used to determine the effect of such a change. Because the simulation closely represents the actual physical system, the usefulness of the proposed change can be determined without actually carrying out the change. For instance, if a flow system is unable to deliver the expected flow rates, the problem may lie with various sections of the piping, pipe fittings, valves, or pumps. Instead of proceeding to change a given valve or pump, simulation may be used to determine if indeed the problem is caused by a particular item and if an improved version of the item will be worthwhile. It may be shown by the simulation that the lack of flow at a given point is due to some other cause, such as blockage in a particular section of the piping. Clearly, considerable savings may be obtained by using simulation in this manner.

4.4.1.5 Sensitivity

A question that arises frequently in design is the effect of a given variable or component on the system performance. For instance, if the dimensions of the channels or of the collectors in a solar collection system were varied, what would be the overall effect on the system? Similarly, if the capacity of the fan or blower in a cooling system were varied, how would it affect the heat removal rate? Such questions relate to the sensitivity of the system performance to the design variables and are important from a practical viewpoint. A substantial reduction in the cost of the system may be obtained by slight changes in the design in order to use standard items available in the market. Pipes and tubings are usually available at fixed dimensions and if these could be employed in the system, rather than the exact custom-made dimensions, substantial savings may result. Similarly, fluid flow components such as blowers, pumps, and fans are often cheaply and easily available for given specifications. At different values, these may have to be fabricated individually, raising the price substantially. System simulation is used to determine the sensitivity of the system performance to such variables and to decide if slight alterations can be made in the interest of reducing the cost without significant sacrifice in system characteristics.

4.4.2 Different Classes

Several types of simulation are used for thermal systems. We have already mentioned analog and physical simulations, which are based on the corresponding form of modeling, as discussed in Chapter 3. In this chapter, we have focused on numerical modeling and numerical simulation, which are based on the mathematical modeling of the thermal system. Three main classes of this type of system simulation are discussed here.

4.4.2.1 Dynamic or Steady State

The simulation of a system may be classified as *dynamic* or *steady state*. The former refers to circumstances where changes in the operating conditions and relevant system variables occur with respect to time. Many thermal systems are time-dependent in nature and a dynamic simulation is essential. This is particularly true for the startup and shutdown of the system. Also, in most manufacturing processes the temperature and other attributes of the material undergoing thermal processing vary with time, as shown in Figure 2.1. The system itself may vary with time over

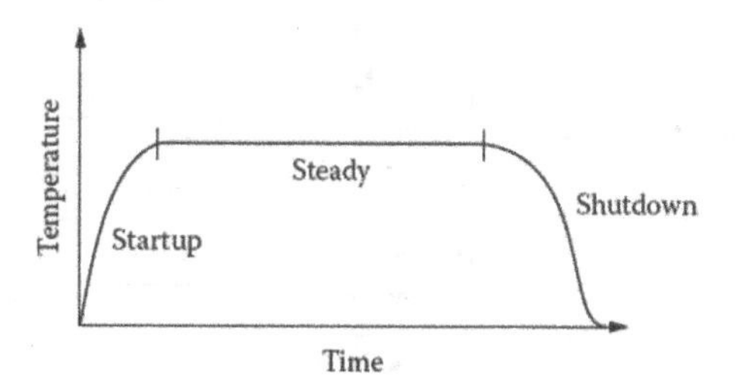

FIGURE 4.22 Temperature variation in a typical thermal system that is steady over most of the duration of operation and is time-dependent only near startup and shutdown.

the duration of interest due to energy input as in welding, gas cutting, heat treatment, and metal forming. In processes such as crystal growing, ingot casting, and annealing, the system varies with time along with the temperature of the material being processed. Dynamic simulation is also needed to study the response of the system to changes in the operating conditions such as a sharp increase in the heat load on a food freezing plant. The results obtained from a dynamic simulation are also useful in the design and study of the control scheme for satisfactory and safe operation of the system.

Steady-state simulation refers to situations where changes with respect to time are negligible or do not occur. Because the dependence of the variables on time is eliminated, a steady-state simulation is much simpler than the corresponding dynamic simulation. In addition, the steady-state approximation can be made in a large number of practical cases, making steady-state simulation of greater interest and importance in thermal systems. Except for times close to startup and shutdown, many systems behave as if they are under steady-state conditions. Thus, a blast furnace may be treated as essentially steady over much of its operation. A typical system that is transient at the beginning and end of its operation and steady over the rest is shown in Figure 4.22. In addition, the system itself may be approximated as steady even though the temperature of the material undergoing thermal processing varies with time. An example of this is a circuit board being baked. The baking oven may be approximated as being unchanged and operating under steady-state conditions while the board undergoes a relatively large temperature change as it moves through the oven.

4.4.2.2 Continuous or Discrete

In many systems such as refrigeration and air conditioning systems, power plants, internal combustion engines, and gas turbines, the flow of the fluid may be taken as continuous, with no finite gaps in the fluid stream. Thus, a continuum of the material or the fluid is assumed, with the conservation laws based on this assumption. This implies the use of continuity, momentum, and energy equations from fluid mechanics and heat transfer for transport processes in a continuum. Particles, if present in the flow, are not treated separately but as part of the average properties of the fluid. Most thermodynamic systems can be simulated as continuous because energy and fluid flow are generally continuous [see Figure 4.23(a)].

On the other hand, if discrete pieces, such as ball bearings, fasteners, and gears, undergoing thermal processing, are considered, the simulation focuses on a finite number of such items. In the manufacture of television sets, individual glass screens are heat treated as they pass through a furnace on a conveyor belt, as sketched in Figure 4.23(b). In such cases, the energy balance of each item is considered separately to determine, for instance, the temperature as a function of time. An aggregate or an average may be obtained at the end to quantify the process, if desired. An example of this class of simulation is provided by the modeling of the movement of a plastic piece as it goes from the hopper to the die in an extruder to determine the time taken (generally known as *residence time*). An average residence time may be defined by considering several such pieces.

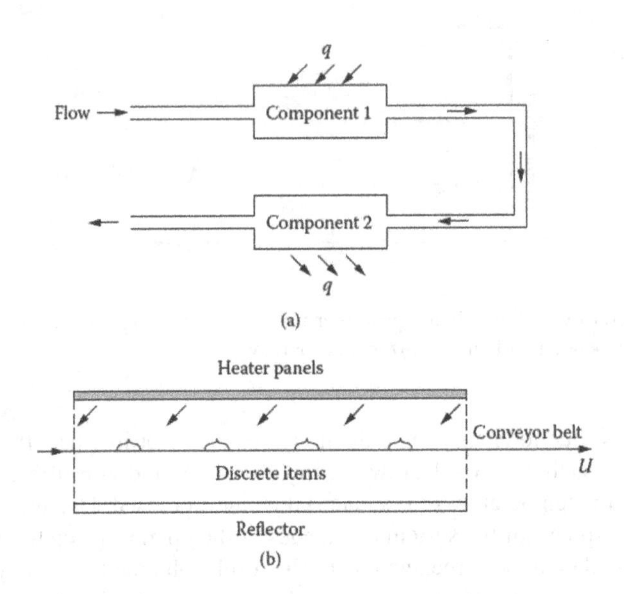

FIGURE 4.23 (a) Continuous and (b) discrete simulation.

4.4.2.3 Deterministic or Stochastic

In most of the examples of thermal processes and systems considered thus far, the variables in the problem are assumed to be specified with precision. Such processes are known as *deterministic*. However, there are several cases where the input conditions are not known precisely and a probability distribution may be given instead, with a dominant frequency, an average, and an amplitude of variation. The conditions may also be completely random with an equal probability of attaining any value over a particular range. When dealing with consumer demands for power, hot water, supplies, etc., probabilistic descriptions are often employed and the corresponding simulation, known as *stochastic,* is carried out to determine the appropriate design variables. A useful simulation method, known as the *Monte Carlo* method, uses the randomness of the process along with appropriate probability distributions to simulate the system to determine the resulting average output, transport rate, time taken, and other characteristics. Employing random numbers generated on the computer, events are selected from the probability distributions to simulate the randomness in selection. The various steps in the overall process are followed to obtain the result for a given starting point. An aggregate of several such simulations yields the expected average behavior of the system. This approach is often used in manufacturing systems to account for statistical variations at different stages of the process (Dieter, 2000; Ertas and Jones, 1996).

4.4.3 Flow of Information

A useful concept in the simulation of thermal systems is that of information flow between the different parts, components, or subsystems that make up the system. The flow of information from one part to another is considered in terms of quantities that are of particular interest to thermal systems, such as temperature, velocity, flow rate, and pressure. This flow of information also indicates the nature of coupling between two components. The input to a given item undergoing thermal processing or to a continuous flow may be provided by the system parts and, similarly, the output from this item or flow may be fed to these parts. Clearly, different types of information flow arrangements may arise, depending on the thermal system. Such arrangements are often shown as information-flow diagrams in which the various parts of the system are linked through inputs and outputs. The strategy for simulating the system is often guided by the nature and characteristics of information flow between the different parts of the system.

4.4.3.1 Block Representation

Each component may itself be represented as a block with its own characteristics, as well as inputs and outputs. For instance, a compressor may be represented by a block with the inlet pressure p_1 and the mass flow rate $\dot{m}$ as inputs and the outlet pressure p_2 as the output. The efficiency and discharge rate may also be taken as outputs. The equation that expresses the relationship between these variables may be written within the block to indicate the characteristics of the component. Again, this equation may be an algebraic, differential, or integral equation, or a combination of these. For steady-state problems, assuming uniformity within each component, the characteristic equation linking the inputs and outputs will be an algebraic equation. ODEs arise for transient cases, if uniformity or lumping is still assumed within each component or part. PDEs arise for general, distributed, or nonuniform cases.

Similarly, a heat exchanger may be represented by a block, with flow rates $\dot{m}_1$ *and* $\dot{m}_2$ of the two fluid streams and inlet temperatures $T_{1,i}$ and $T_{2,i}$ as the inputs. Then, the outlet temperature $T_{2,0}$ of one of the streams may be taken as the output, with $f\left(\dot{m}_1, \dot{m}_2, T_{1,i}, T_{2,i}, T_{2,0}\right) = 0$ representing the relationship between the inputs and the outputs. Figure 4.24 shows typical blocks that may be used to represent a few components of interest to thermal systems. Obviously, other combinations of inputs and outputs are possible and may be employed, depending on the specific application. The use of such blocks to represent the system components and subsystems facilitates the representation of the overall system in terms of the information flow between different parts.

4.4.3.2 Information Flow

A diagram showing the flow of information for a system is constructed using blocks for the various parts of the system. Let us consider the relatively complex problem of a plastic screw extrusion system, as shown in Figure 1.10(b). The plastic material is conveyed, heated, melted, and forced through the die due to the rise in pressure Δp and temperature ΔT in the extruder. The transport processes are governed by nonlinear PDEs, further complicated by material property variations, complex geometry, and the phase change. However, the extruder may be numerically simulated to study the dependence of temperature and pressure on the inputs such as mass flow rate $\dot{m}$, speed N in revolutions per minute, and barrel temperature T_b. The simulation results may be curve fitted to obtain algebraic equations. As an example, if the hopper, extruder, and die are taken as the three parts of the system, Figure 4.25 shows the information-flow diagram for this circumstance. The following equations express, respectively, the relationships between the inputs and outputs for these three parts:

$$
\begin{aligned}
f_1(D, \dot{m}) &= 0 \\
f_2(\dot{m}, N, T_b, \Delta p, \Delta T) &= 0 \\
f_3(\Delta p, \Delta T, \dot{m}) &= 0
\end{aligned}
\tag{4.36}
$$

Here, the extruder and the die are taken as fixed, so that only the operating conditions are varied. D is the diameter of the opening of the hopper and, thus, represents its geometry. Similarly, die diameter d and screw geometry may be included as variables.

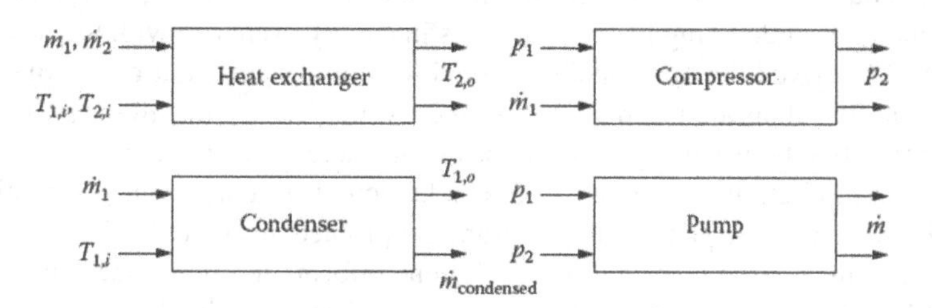

FIGURE 4.24 Block representations of a few common components used in thermal systems.

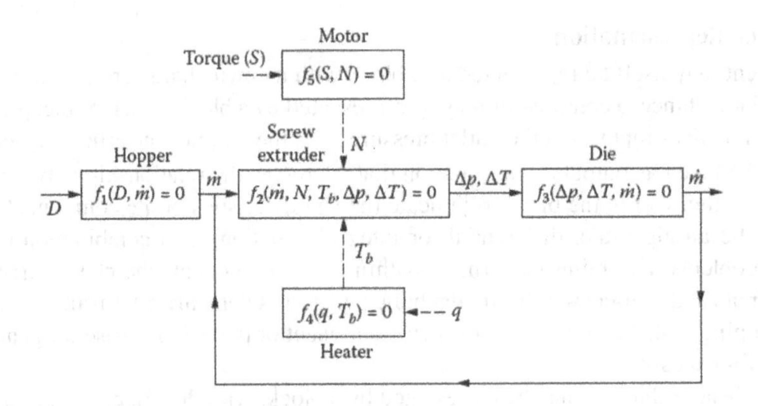

FIGURE 4.25 Information-flow diagram for a screw extrusion system.

The pressures at the entrance to the hopper and at the exit of the die are taken as atmospheric. Here, f_1 and f_2 represent the characteristics of the hopper and the die, respectively. The mass flow rate into the die must equal that emerging from the extruder from mass conservation considerations for steady-state operation. The motor that provides the torque and the heater that supplies the heat input to the barrel are additional parts that may be included, as shown by dotted lines in the figure. The extruder itself may be considered to consist of different zones that are modeled separately, such as the solid conveying, melting, and metering sections. Again, these are coupled through the flow rate $\dot{m}$ and the temperature and pressure continuity. For further details on this system, specialized books, such as those by Tadmor and Gogos (1979) and Rauwendaal (2014), may be consulted.

The information-flow diagram for a vapor-compression cooling system, shown in Figure 1.8(a), may be similarly drawn. The characteristics of the compressor, condenser, throttling valve, and evaporator can be expressed by the equations

$$\begin{aligned} f_1(p_1, h_1, p_2, h_2) &= 0 \\ f_2(p_2, h_2, h_3) &= 0 \\ f_3(p_1, p_2) &= 0 \\ f_4(p_1, h_1, h_3) &= 0 \end{aligned} \tag{4.37}$$

The different states are shown on a p-h plot of the thermodynamic cycle in Figure 4.26(a). The information-flow diagram is shown in Figure 4.26(b). Conservation of enthalpy h in the throttling process, $h_3 = h_4$, is employed and the characteristic equations are derived for a given refrigerant fluid such as a hydrofluorocarbon (HFC) such as R-410A or R-134a. The inlet into each part corresponds to the exit from the preceding one in this closed cycle.

The information-flow diagram for the electric furnace shown in Figure 4.27(a) is more involved than the preceding two cases because energy transfer occurs between different parts of the system simultaneously. Thus, the heater exchanges thermal energy with the walls, gases, and the material. Similarly, the material undergoing heat treatment is in energy exchange with the heater, walls, and gases. Example 3.6 derived the mathematical model for this system. Figure 4.27(b) shows a sketch of the information-flow diagram for this circumstance, strongly coupling all the parts of the system. The flow of information between any two parts is due to energy transfer that involves combinations of the three modes: radiation, convection, and conduction. The representative equations are ordinary and partial-differential equations for the transient problem, as modeled in Example 3.6.

In each of the three cases just outlined, different information-flow diagrams are obtainable by choosing different starting points and different inputs/outputs. The diagrams indicate the link between different parts of the system and thus suggest ways of approaching the simulation. In the

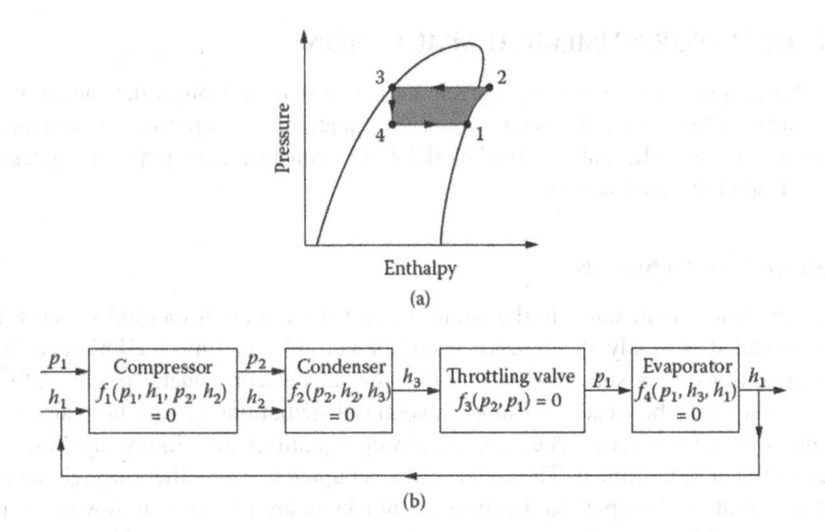

FIGURE 4.26 A vapor-compression cooling system. (a) Thermodynamic cycle; (b) information-flow diagram.

first two examples of the extruder and the air conditioning system, the output from one component feeds into the next as an input. The overall arrangement is sequential because one part depends only on the preceding one. However, in the furnace, a part is simultaneously coupled with several others through the energy exchange mechanisms. Therefore, in the former circumstances a sequential calculation procedure is appropriate for the simulation, whereas a simultaneous solution procedure is essential for the latter.

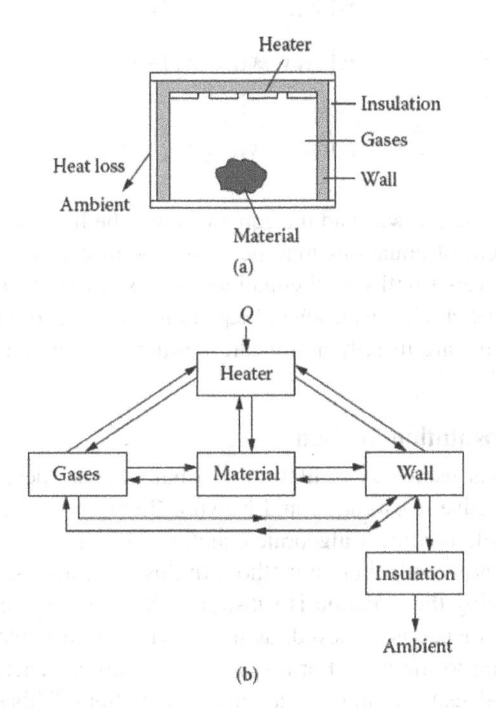

FIGURE 4.27 (a) An electric furnace; (b) information-flow diagram showing the coupling between different parts of the system.

4.5 METHODS FOR NUMERICAL SIMULATION

The method appropriate for simulating a given thermal system is strongly dependent on the nature of the system and, thus, on the characteristics of the equations that represent the system. These are best specified in terms of the mathematical model. Let us consider a few important types of systems and their corresponding simulations.

4.5.1 STEADY LUMPED SYSTEMS

This is the simplest circumstance in the modeling and simulation of thermal systems. If the system can be modeled as steady, the dependence of the variables on time is eliminated. In addition, if uniform conditions are assumed to exist within the various components or parts of the system, implying that they may be treated as lumped, spatial distributions of variables such as temperature and pressure do not arise. Then the governing equations are simply algebraic equations, which may be linear or nonlinear. The equations are coupled to each other through the unknowns, which are the inputs and outputs to the different blocks in an information-flow diagram, as seen earlier. The external inputs to the overall system are given or may be varied in the simulation and the overall outputs from the system represent the information desired from the simulation.

Most thermodynamic systems, such as power plants, air conditioners, internal combustion engines, gas turbines, compressors, pumps, etc., can usually be modeled as steady and lumped without much sacrifice in the accuracy of the simulation (Howell and Buckius, 1992; Moran and Shapiro, 2014). Similarly, fluid flow systems, such as a network of pipes and storage tanks, can often be treated as steady and lumped.

The governing algebraic equations in steady lumped systems may be written as

$$\begin{aligned}
f_1\left(x_1, x_2, x_3, \ldots, x_n\right) &= 0 \\
f_2\left(x_1, x_2, x_3, \ldots, x_n\right) &= 0 \\
f_3\left(x_1, x_2, x_3, \ldots, x_n\right) &= 0 \\
&\vdots \\
f_n\left(x_1, x_2, x_3, \ldots, x_n\right) &= 0
\end{aligned} \tag{4.38}$$

where the x's represent the unknowns and the equations may be linear or nonlinear. If all the equations are linear, the system of equations may be solved by direct or iterative methods, using the former approach for relatively small sets of equations and for tridiagonal systems, and the latter for large sets, as discussed earlier. However, several equations in the set of algebraic equations governing typical thermal systems are usually nonlinear, making the solution much more involved than that for linear equations.

4.5.1.1 Successive Substitution Method

The two main approaches used for simulating thermal systems described by nonlinear equations are based on successive substitution and Newton-Raphson methods, which were discussed in Section 4.2.2 for a single nonlinear algebraic equation and for a set of nonlinear equations. Let us first consider the successive substitution method. In this case, for a single equation, the iterative solution obtained by solving the equation is substituted back into the equation and the iterations are continued until convergence is achieved, as indicated by an acceptable small variation in the solution from one iteration to the next. For a system of equations, each equation is solved for an unknown using known values from previous iterative calculations. This solution is then substituted into the next equation, which is solved to obtain another unknown. This is again substituted into the next equation, in succession, and the process is continued until the solution obtained does not vary significantly from one iteration to the next. A scheme such as the modified Gauss-Seidel method given in Equation (4.19) may be used effectively to simulate the system.

The numerical algorithm is quite simple for this method. Relaxation may also be used to improve the convergence. SUR is particularly useful in obtaining convergence in nonlinear equations. Thus, if the relaxation factor ω is in the range $0 < \omega < 1$, SUR is applied as

$$x_i^{(l+1)} = \omega G_i\left[x_1^{(l+1)}, x_2^{(l+1)}, \ldots, x_{i-1}^{(l+1)}, x_i^{(l)}, \ldots, x_n^{(l)}\right] + (1-\omega)x_i^{(l)} \quad \text{for } i = 1, 2, \ldots, n \tag{4.39}$$

The successive substitution method is particularly well-suited for sequential information-flow diagrams, such as those shown in Figure 4.25 and Figure 4.26. The equations corresponding to the different parts of the system are solved in succession, using the inputs from the preceding part or equation, until convergence of the iteration is achieved to a chosen convergence criterion, as given by Equation (4.7). The method can be applied to large systems involving large sets of nonlinear algebraic equations. Computer storage requirements are small and computer programming is fairly simple. As a result, this approach is extensively used in industry for simulating steady-state thermal systems in mechanical and chemical engineering processes.

The main problem with the successive substitution method is difficulty in convergence of the iterative process. For sequential information-flow diagrams, convergence is much more easily obtained than for simultaneous information-flow cases, such as the one sketched in Figure 4.27, because the numerical simulation follows the physical characteristics of the system in the former circumstance. Convergence is strongly dependent on the starting point and on the arrangement of the equations for solution. As seen earlier for linear systems, diagonal dominance is needed to assure convergence. Therefore, in linear systems the equations are arranged in order to place the dominant coefficient at the diagonal of the coefficient matrix, implying that each equation is solved for the unknown with the largest coefficient. Though the corresponding convergence characteristics are not available for nonlinear equations, a change in the arrangement of the equations can affect the convergence substantially. Generally, information blocks should be positioned so that the effect on the output is small for large changes in the input. The equations may be rewritten to achieve this. Stoecker (1989) gives a few examples in which the sequence of the equations and thus of the unknowns being solved can be changed to obtain convergence. Similarly, the starting values should be picked based on available information on the given system so that these are realistic and as close as possible to the final solution. Again, SUR may be used for cases where convergence is a problem. A few examples are included later in this chapter to illustrate these strategies.

4.5.1.2 Newton-Raphson Method

The second approach for solving a set of nonlinear algebraic equations is the Newton-Raphson method discussed in Section 4.2.2 for a single nonlinear equation and then extended to a set of nonlinear algebraic equations. This method is appropriate for an information-flow diagram in which a strong interdependence arises between the different parts of the system, such as that shown in Figure 4.27. The convergence characteristics are generally better than those for the successive substitution, or modified Gauss-Seidel, method. However, the method is much more complicated because the matrix of derivatives, given in Equation (4.17) and generally known as the *Jacobian*, has to be calculated at each iterative step. Because the computed variation of each of the functions f_i with the unknowns x_i is employed in determining the changes in x_i for the next iteration, the iterative scheme is much better behaved than the successive substitution method, which does not use any such quantitative measure of the change in f_i with x_i. However, the derivatives may not be easily obtainable by analysis and numerical differentiation may be necessary, further complicating the procedure. Generally, the Newton-Raphson method is useful for relatively small sets of nonlinear equations and for cases where the derivatives can be obtained easily. However, different approaches have been developed in which the derivatives are computed numerically by efficient algorithms as the iterative scheme proceeds. For instance, the partial derivative $\partial f_i/\partial x_j$ may be computed as

$$\frac{\partial f_i}{\partial x_j} = \frac{f_i(x_1, x_2, \ldots, x_j + \Delta x_j, \ldots, x_n) - f_i(x_1, x_2, \ldots, x_j, \ldots, x_n)}{\Delta x_j} \tag{4.40}$$

where Δx_j is a chosen increment in x_j. Thus, all the partial derivatives needed for Equation (4.17) are computed at each iteration and the next approximation to the solution obtained, carrying out this procedure until convergence to the chosen convergence criterion is achieved.

The Newton-Raphson approach links all the parts of the system simultaneously through the derivatives. Thus, changes in one component will affect all others at the same time and the effect is dependent on the corresponding derivatives. Therefore, the approach follows the physical behavior of systems where all the parts are strongly linked with each other. Several thermal systems, such as those in manufacturing and cooling of electronic equipment, have strong coupling between the different parts, making it desirable to use the Newton-Raphson method. However, if the systems have a large number of parts, leading to a large number of algebraic equations, it may be more advantageous to use the successive substitution approach, employing under-relaxation for better convergence. The examples that follow illustrate these methods of simulation.

Example 4.6

In the ammonia production system sketched in Figure 4.28, a mixture of 90 moles/s of nitrogen, 270 moles/s of hydrogen, and 0.9 moles/s of argon enters the plant and is combined with the residual mixture crossing a bleed valve. Argon is an impurity and adversely affects the reaction (Stoecker, 1989; Parker, 1993). In the chemical reactor, a fraction of the entering mixture combines to form ammonia, which is removed by condensation. The bleed valve removes 23.5 moles/s of the mixture to avoid build-up of argon. The fraction of the mixture that reacts to give ammonia in the reactor is 0.57 exp(−0.0155 F_1), where F_1 is the amount of argon entering the reactor in moles per second. Solve the resulting set of algebraic equations by the successive substitution approach to obtain the flow rates and the amount of ammonia produced.

SOLUTION

The system of algebraic equations for the given process may be derived on the basis of mass conservation. If F_1 and F_2 are the flow rates of argon and nitrogen, respectively, in moles per second

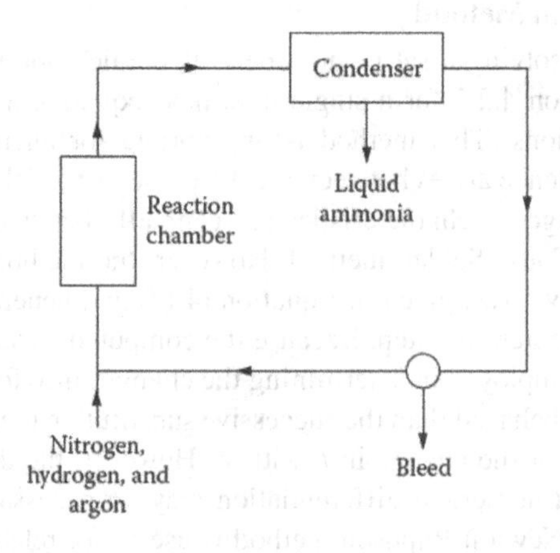

FIGURE 4.28 The ammonia production system considered in Example 4.6.

entering the reactor, the flow rate of hydrogen is $3F_2$ moles/s from the chemical reaction, which yields ammonia. Then, we obtain

$$F_1 = \frac{0.9}{(1-B)}$$

$$F_2 = \frac{90}{(1-BP)}$$

where

$$P = 1 - 0.57\exp(-0.0155\,F_1)$$

$$B = 1 - \frac{23.5}{4F_2P + F_1}$$

Here, P represents the fraction of unconverted mixture of nitrogen and hydrogen and B represents the fraction of mixture that goes past the bleed valve. The chemical reaction for producing ammonia is

$$N_2 + 3H_2 = 2NH_3$$

Therefore, the amount of ammonia produced D is given by

$$D = 2F_2[0.57\exp(-0.0155\,F_1)]$$

The preceding four nonlinear algebraic equations may be rewritten to solve for the four unknowns F_1, F_2, P, and B in sequence, terminating the iteration when values do not change significantly from one iteration to the next. Obviously, the equations and unknowns can be arranged in several ways to apply the successive substitution method. Many of these do not converge and solving the equations in a different order, as well as using different starting values, is tried. The computer programming is very simple, but the scheme diverges in many cases.

The method is found to converge if a starting value of B is taken and the four equations are solved in the following sequence:

$$F_1 = \frac{0.9}{(1-B)}$$

$$P = 1 - 0.57\exp(-0.0155F_1)$$

$$F_2 = \frac{90}{(1-BP)}$$

$$B = 1 - \frac{23.5}{4F_2P + F_1}$$

The convergence criterion may be applied to B or to the total flow rate F entering the reactor, where $F = F_1 + 4F_2$.

Applying the convergence criterion to F with the convergence parameter ε taken as 10^{-4}, the numerical results obtained are shown in Figure 4.29. The argon flow rate and the total flow rate entering the reactor are computed, along with the amount of ammonia produced, for each iteration. The convergence is found to be slow because of the first-order convergence of the method. It is ensured that the results are essentially independent of the value of ε chosen by varying ε. The numerical method is quite simple and is frequently applied to solve sets of nonlinear equations that arise in such thermal systems. The main problem is convergence, and different starting values

Results

ARGON:	1.00000	FLOW:	377.52020	NH3:	105.65780
ARGON:	6.36528	FLOW:	621.94090	NH3:	158.95630
ARGON:	11.64363	FLOW:	708.79150	NH3:	165.87840
ARGON:	14.43962	FLOW:	749.70330	NH3:	167.52770
ARGON:	15.88013	FLOW:	770.51630	NH3:	168.14510
ARGON:	16.62993	FLOW:	781.34420	NH3:	168.42190
ARGON:	17.02342	FLOW:	787.03170	NH3:	168.55670
ARGON:	17.23091	FLOW:	790.03320	NH3:	168.62510
ARGON:	17.34062	FLOW:	791.62100	NH3:	168.66060
ARGON:	17.39871	FLOW:	792.46190	NH3:	168.67910
ARGON:	17.42950	FLOW:	792.90770	NH3:	168.68890
ARGON:	17.44583	FLOW:	793.14420	NH3:	168.69410
ARGON:	17.45448	FLOW:	793.26920	NH3:	168.69680
ARGON:	17.45906	FLOW:	793.33560	NH3:	168.69830
ARGON:	17.46149	FLOW:	793.37080	NH3:	168.69900
ARGON:	17.46278	FLOW:	793.38960	NH3:	168.69950
ARGON:	17.46347	FLOW:	793.39960	NH3:	168.69970
ARGON:	17.46382	FLOW:	793.40450	NH3:	168.69980
ARGON:	17.46400	FLOW:	793.40720	NH3:	168.69990
ARGON:	17.46411	FLOW:	793.40890	NH3:	168.69990
ARGON:	17.46417	FLOW:	793.40990	NH3:	168.69990
ARGON:	17.46422	FLOW:	793.41040	NH3:	168.69990
ARGON:	17.46423	FLOW:	793.41060	NH3:	168.69990
ARGON:	17.46423	FLOW:	793.41060	NH3:	168.69990

FIGURE 4.29 Numerical results obtained for Example 4.6.

as well as different formulations and solution sequences of the algebraic equations may be tried to obtain convergence. SUR may also be used for improving the convergence characteristics. The method is popular because no derivatives are needed, as is the case for the Newton-Raphson method. Appendix A.M.4.5 gives a Matlab program for the solution of this problem.

Example 4.7

In a thermal system, the volume flow rate R of a fluid through a duct due to a fan is given in terms of the pressure difference P, which drives the flow as

$$R = 15 - 75 \times 10^{-6} \times P^2$$

with

$$P = 80 + 10.5R^{5/3}$$

Here, R is in m^3/s and P is in N/m^2. The first equation represents the characteristics of the fan and the second one that of the duct. Simulate this system by the successive substitution and Newton-Raphson methods to obtain the flow rate and pressure difference.

SOLUTION

From the physical nature of the problem, we know that both P and R must be real and positive. Figure 4.30 shows the characteristics of the fan-duct system in terms of the flow rate R versus pressure difference P graphs. As the pressure difference P needed for the flow increases, due to blockage or increased length of duct, the flow generated by the fan decreases, ultimately becoming zero at P of 447.2 N/m^2, as can be calculated from the equation for the flow rate by setting $R = 0$. The pressure difference P in the duct is smallest at zero flow and increases as the flow rate increases. In addition, the given equations indicate that P must be greater than 80 and R must be less than 15, giving ranges for these variables for selecting the starting values.

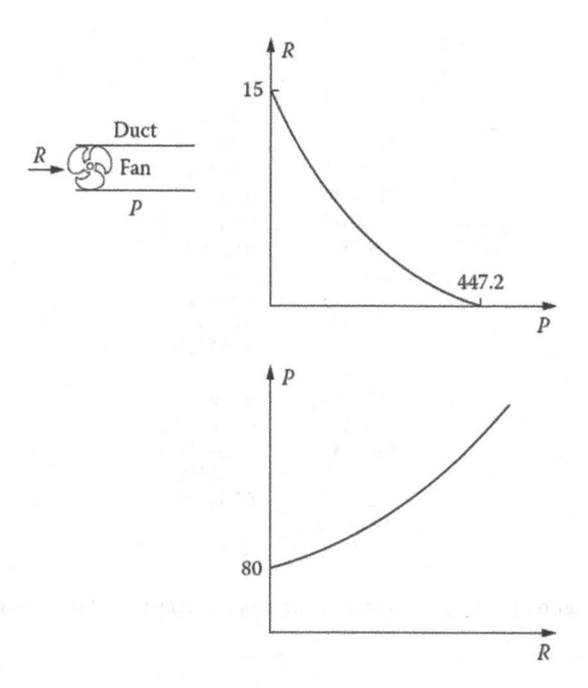

FIGURE 4.30 Characteristic curves, in terms of pressure difference versus flow rate graphs for the fan and the duct, respectively, for the system considered in Example 4.7.

The two equations are already given in the form $x_i = F_i(x_1, x_2, \ldots, x_i, \ldots, x_n)$, which is appropriate for the application of the successive substitution, or modified Gauss-Seidel, method. However, when the equations are employed as given, with starting values taken for P and R from their appropriate ranges, the scheme diverges rapidly. As mentioned earlier, if an equation $x = g(x)$ is being solved for the root α by the successive substitution method, the absolute value of the asymptotic convergence factor $g'(\alpha)$ must be less than 1.0 for convergence. An estimation of the corresponding values of g' in the given equations indicates that these values are much greater than 1.0.

Because both P and R are greater than 1.0, a reformulation of the equations may be carried out to yield

$$R = \left[\frac{P-80}{10.5}\right]^{3/5} \quad \text{and} \quad P = \left[\frac{15-R}{75\times10^{-6}}\right]^{1/2}$$

so that fractional powers are involved and g' becomes less than 1.0. The successive substitution scheme, when applied to these equations with starting values in the ranges $0 < R < 15$ and $447.2 > P > 80$, converges to yield the desired solution. The initial guesses for P and R are substituted on the right-hand sides of these equations to calculate the new values for R and P. These are resubstituted in the equations to obtain the values for the next iteration, and so on. The iterative process is terminated if

$$(R_{i+1} - R_i)^2 + (P_{i+1} - P_i)^2 \le \varepsilon$$

where the subscripts indicate the iteration number and ε is a chosen small quantity. The numerical results during the iteration are shown in Figure 4.31 for $\varepsilon = 10^{-6}$ and the starting values for P and R taken as 80 and 0, respectively. The scheme converges to $P = 332.0353$ and $R = 6.7314$, both variables being within their allowable ranges. The results were not significantly altered at still smaller values of ε.

This problem may also be solved by eliminating P or R from the two preceding equations for R and P to obtain a single equation. Then the problem may be solved as a root solving circumstance.

```
P = ? 80
R = ? 0
     447.2133      0
     447.2133      8.437959
     295.7935      8.437959
     295.7935      6.133516
     343.8311      6.133516
     343.8311      6.919616
     328.2355      6.919616
     328.2355      6.671212
     333.2424      6.671212
     333.2424      6.751625
     331.6299      6.751625
     331.6299      6.725796
     332.1485      6.725796
     332.1485      6.734112
     331.9817      6.734112
     331.9817      6.731439
     332.0353      6.731439
The solution is:
P = 332.0353          R = 6.731439
```

FIGURE 4.31 Computer output for the solution to the problem considered in Example 4.7 by the successive substitution method.

Using this approach, with R as the only unknown in the resulting nonlinear equation, successive substitution is applied to obtain the solution, as given by the Matlab program in Appendix A.M.3.4.

To apply the Newton-Raphson method, these equations are rewritten in terms of functions F and G as

$$F(P,R) = \left[\frac{P-80}{10.5}\right]^{3/5} - R = 0$$

$$G(P,R) = \left[\frac{15-R}{75\times10^{-6}}\right]^{1/2} - P = 0$$

Initial guesses are taken for P and R, as before, and the values for the next iteration, $i + 1$, are obtained from the values after the ith iteration as

$$P_{i+1} = P_i + (\Delta P)_i \quad \text{and} \quad R_{i+1} = R_i + (\Delta R)_i$$

where the increments $(\Delta R)_i$ and $(\Delta P)_i$ are calculated from the equations

$$\left(\frac{\partial F}{\partial R}\right)_i (\Delta R)_i + \left(\frac{\partial F}{\partial P}\right)_i (\Delta P)_i = -F(R_i, P_i)$$

$$\left(\frac{\partial G}{\partial R}\right)_i (\Delta R)_i + \left(\frac{\partial G}{\partial P}\right)_i (\Delta P)_i = -G(R_i, P_i)$$

The four partial derivatives in the above equations are calculated for the R and P values at the ith iteration, using analytical differentiation of the functions F and G. The iterative process is continued until a convergence criterion of the form $F^2 + G^2 \leq \varepsilon$ is satisfied. Figure 4.32 shows the computer output for $\varepsilon = 10^{-4}$ and starting values of 2 and 100 for R and P, respectively. The results obtained are very close to those obtained earlier by the successive substitution method. The program is simpler to write for the successive substitution method. However, the Newton-Raphson method converges at a faster rate, due to its second-order convergence. It usually converges if the

```
R = ? 2
P = ? 100
EPS = ? 0.0001
R = 9.873579        P = 290.2549
R = 6.86436         P = 338.1764
R = 6.732573        P = 332.0232
R = 6.732087        P = 332.0222
The required solution is:
The flow rate = 6.732087        Pressure = 332.0222
```

FIGURE 4.32 The results by the Newton-Raphson method for Example 4.7.

initial guessed values are not too far from the solution. Nevertheless, if divergence occurs, the initial guessed values may be varied and iteration repeated until convergence is achieved. Appendix A.M.4.6 gives a Matlab program for the solution of this problem in order to illustrate the use of Newton's method to solve a system of nonlinear equations.

4.5.2 Dynamic Simulation of Lumped Systems

Dynamic simulation of thermal systems is used for studying the system characteristics at startup and shutdown, for investigating the system response to changes in operating conditions, and for design and evaluation of a control scheme. We are interested in ensuring that the system does not go beyond acceptable limits under such transient conditions. For instance, at startup, the cooling system of a furnace may not be completely operational, resulting in temperature rise beyond safe levels. This consideration is particularly important for electronic systems because their performance is very sensitive to the operating temperature [see Figure 3.6(b)]. Similarly, at shutdown of a nuclear reactor, the heat removal subsystems must remain effective until the temperature levels are sufficiently low. In many cases, sudden fluctuations in the operating conditions occur due to, for instance, power surge, increase in thermal load, change in environmental conditions, change in material flow, etc. It is important to determine if the system exceeds safety limits and if acceptable performance is achieved under these conditions.

4.5.2.1 Analytical Solution

If the various parts of the system can be treated as lumped, the resulting equations are coupled ODEs. Modeling of a component as lumped was discussed in Chapter 3 and the resulting energy equations, such as Equation (3.7), Equation (3.10), and Equation (3.11), were given. For a lumped body represented by the equation

$$\rho CV \frac{dT}{d\tau} = qA - hA(T - T_a) \tag{4.41}$$

the temperature $T(\tau)$ is given by

$$\theta = T - T_a = \frac{q}{h} + \left(\theta_o - \frac{q}{h}\right) \exp\left(-\frac{hA}{\rho CV}\tau\right) \tag{4.42}$$

The symbols are the same as those employed for Equation (3.7) through Equation (3.10). In the analytical solution given by Equation (4.42), the steady-state temperature is q/h, obtained for time $\tau \rightarrow \infty$. The initial temperature at $\tau = 0$ is T_o, represented by $\theta_o = T_o - T_a$. This solution gives the basic characteristics of many dynamic simulation results in which the steady-state behavior is achieved at large time. If $q = 0$, the convective transport case of Equation (3.7) is obtained, with Equation (3.9) as the solution. The quantity $\rho CV/hA$ is the response time in that case, as given earlier by Equation (3.1). If convective heat loss is absent, only qA is left on the right-hand side of

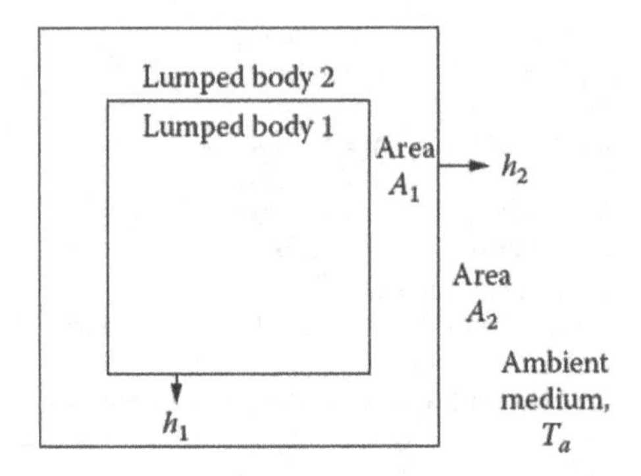

FIGURE 4.33 System consisting of two lumped bodies exchanging energy by convection with each other and with the ambient medium.

Equation (4.41) and the solution is $T - T_o = (qA/\rho CV)\tau$, indicating a linear increase with time if the heat input q is held constant.

The simulation of a system consisting of several parts, each of which is treated as lumped, involves a set of ODEs, rather than a single ODE. These equations may be linear or nonlinear. Most nonlinear equations, such as Equation (3.11), require a numerical solution. Even with linear equations, the presence of several coupled ODEs makes it difficult to obtain an analytical solution. As an example, let us consider two lumped bodies, denoted by subscripts 1 and 2, exchanging energy through convection. The energy equations for the two are obtained as

$$(\rho CV)_1 \frac{dT_1}{d\tau} = h_1 A_1 (T_2 - T_1) \tag{4.43a}$$

$$(\rho CV)_2 \frac{dT_2}{d\tau} = h_2 A_2 (T_a - T_2) - h_1 A_1 (T_2 - T_1) \tag{4.43b}$$

where T_a is the temperature of the ambient medium with which body 2 exchanges energy by convection. The convective heat transfer coefficients h_1 and h_2 refer to the inner and outer surfaces, as shown in Figure 4.33. The initial temperature is T_o at time t = 0.

Employing $\theta = T - T_a$, these equations may be written, with $\theta_1 = T_1 - T_a$, $\theta_2 = T_2 - T_a$, and $\theta_o = T_o - T_a$, as

$$\frac{d\theta_1}{d\tau} = H_1(\theta_2 - \theta_1) = F(\theta_1, \theta_2) \tag{4.44a}$$

$$\frac{d\theta_2}{d\tau} = -H_2\theta_2 - H_3(\theta_2 - \theta_1) = G(\theta_1, \theta_2) \tag{4.44b}$$

where

$$H_1 = \frac{h_1 A_1}{\rho_1 C_1 V_1}, \quad H_2 = \frac{h_2 A_2}{\rho_2 C_2 V_2} \quad H_3 = \frac{h_1 A_1}{\rho_2 C_2 V_2}$$

Then the analytical solution to these equations is obtained as

$$\frac{\theta_1}{\theta_o} = \frac{be^{a\tau}}{b-a} - \frac{ae^{b\tau}}{b-a} \tag{4.45a}$$

$$\frac{\theta_2}{\theta_o} = \frac{ab}{H_1(b-a)}\left[e^{a\tau} - e^{b\tau}\right] + \frac{\theta_1}{\theta_o} \tag{4.45b}$$

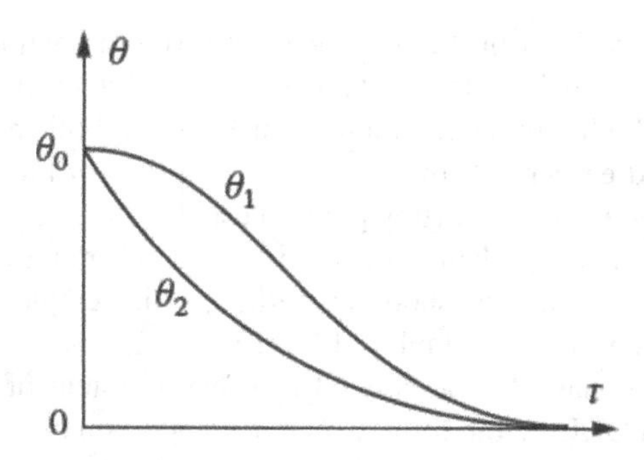

FIGURE 4.34 Variation of the temperatures of the two lumped bodies in Figure 4.33 with time.

where

$$a = \frac{-K_1 + K_2}{2} \quad \text{and} \quad b = \frac{-K_1 - K_2}{2}$$

$$\text{with } K_1 = H_1 + H_2 + H_3 \quad \text{and} \quad K_2 = \left[K_1^2 - 4H_1H_2\right]^{1/2}$$

Figure 4.34 shows the temperature variation with time for the two lumped bodies in this dynamic problem. The dimensionless temperatures start at θ_o, at time $\tau = 0$, and decay to zero with time because of heat loss to the environment. The gradient $d\theta_1/d\tau$ is zero at $\tau = 0$ because $\theta_2 = \theta_1 = \theta_o$ at the beginning of the process. Three first-order ODEs arise if three lumped bodies in energy exchange with each other are considered, four equations for four bodies, and so on. Analytical solutions may be obtained as given here or by using other analytical techniques such as the *Laplace* transform method, which is an integral transform used in many applications such as control of dynamic systems, for a few idealized cases, particularly if the equations are linear.

4.5.2.2 Numerical Solution

If the ordinary differential equations are nonlinear, analytical solutions are generally not possible and numerical methods must be employed for the simulation. The use of Runge-Kutta and predictor-corrector methods to solve a single ODE was discussed earlier. For solving a system of equations, such as that given by Equation (4.44), let us first consider the following two simultaneous first-order equations for dependent variables y and z:

$$\frac{dy}{d\tau} = F(\tau, y, z) \quad \text{and} \quad \frac{dz}{d\tau} = G(\tau, y, z) \tag{4.46}$$

Then the fourth-order Runge-Kutta method gives y_{i+1} and z_{i+1} as

$$y_{i+1} = y_i + \frac{K_1 + 2K_2 + 2K_3 + K_4}{6} \qquad z_{i+1} = z_i + \frac{K_1' + 2K_2' + 2K_3' + K_4'}{6} \tag{4.47}$$

where

$$\begin{aligned}
K_1 &= \Delta\tau F(\tau_i, y_i, z_i) & K_1' &= \Delta\tau G(\tau_i, y_i, z_i) \\
K_2 &= \Delta\tau F\left(\tau_i + \frac{\Delta\tau}{2}, y_i + \frac{K_1}{2}, z_i + \frac{K_1'}{2}\right) & K_2' &= \Delta\tau G\left(\tau_i + \frac{\Delta\tau}{2}, y_i + \frac{K_1}{2}, z_i + \frac{K_1'}{2}\right) \\
K_3 &= \Delta\tau F\left(\tau_i + \frac{\Delta\tau}{2}, y_i + \frac{K_2}{2}, z_i + \frac{K_2'}{2}\right) & K_3' &= \Delta\tau G\left(\tau_i + \frac{\Delta\tau}{2}, y_i + \frac{K_2}{2}, z_i + \frac{K_2'}{2}\right) \\
K_4 &= \Delta\tau F(\tau_i + \Delta\tau, y_i + K_3, z_i + K_3') & K_4' &= \Delta\tau G(\tau_i + \Delta\tau, y_i + K_3, z_i + K_3')
\end{aligned} \tag{4.48}$$

The computations are carried out in the sequence just given to obtain the values of y and z at the next step. This procedure may be extended to a system of three or more first-order differential equations, and thus also to higher-order equations, which can be broken down into coupled first-order equations as discussed earlier. All the conditions, in terms of the dependent variables and their derivatives, must be known at the starting point to use this method. Therefore, the scheme, as given here, applies to initial-value problems. If conditions at a different time must also be satisfied, a boundary-value problem arises and the shooting method, which employs a correction scheme to satisfy the boundary conditions, may be employed (Jaluria, 2012).

Finite-difference methods may also be applied to solve a system of ODEs. Algebraic equations are generated for each ODE by the finite-difference approximation and the combined set of equations is solved by the methods outlined earlier to obtain the desired simulation of the system. Considering, again, Equation (4.46), we may write the finite-difference equations as

$$\frac{y_{i+1} - y_i}{\Delta\tau} = F(\tau_i, y_i, z_i) \qquad \frac{z_{i+1} - z_i}{\Delta\tau} = G(\tau_i, y_i, z_i) \tag{4.49}$$

where time $\tau = i\Delta\tau$, subscripts i and $i + 1$ represent values at time τ and $\tau + \Delta\tau$, respectively, and the functions F and G may be linear or nonlinear. Therefore, values at $\tau + \Delta\tau$ may be determined explicitly from values at τ. This is the explicit formulation, which is particularly useful for nonlinear equations. However, F and G are often evaluated at $\tau + \Delta\tau$ or at some other time between τ and $\tau + \Delta\tau$, particularly at $\tau + \Delta\tau/2$, for greater accuracy and numerical stability, as mentioned earlier for PDEs. This is the implicit formulation that gives rise to a set of simultaneous algebraic equations, linear ODEs generating linear algebraic equations, and nonlinear ODEs generating nonlinear ones. Other, more accurate, finite-difference formulations are obviously possible for Equation (4.49). This set of equations is then solved to simulate the thermal system. Higher-order equations arise in many cases, particularly in the analysis of dynamic stability of systems. These may similarly be simulated using the finite-difference method.

Dynamic simulation is particularly valuable in areas such as materials processing, which inevitably involves variations with time. Lumping is commonly used in thermodynamic systems, such as energy conversion and refrigeration systems, and the simulation outlined here helps in ensuring that the system behavior and performance are satisfactory under time-varying conditions. The dynamic simulation of large systems such as power and steel plants is particularly important because of changes in demand and in the inputs to the systems. Some of these aspects are considered in detail, employing examples, in Chapter 5. A typical manufacturing system is considered in the following to illustrate these ideas.

Example 4.8

Numerically simulate the casting of a metal plate of thickness $L = 0.2$ m in a mold of wall thickness $W = 0.05$ m, assuming one-dimensional solidification, no energy storage in the solid formed, uniform temperature in the mold, and initial liquid temperature at the melting point $T_m = 1200$ K. A convective loss at heat transfer coefficient $h = 20$ W/(m²·K) occurs at the outer surface of the mold on both sides of the plate to an ambient medium at temperature $T_a = 20$°C. Find the total time needed for casting. Determine the effect of varying h, using values in the range of 10 to 40 W/(m²·K), and of varying W, using values in the range of 0.02 to 0.1 m. Take density, specific heat, and thermal conductivity of the cast material as 9000 kg/m³, 400 J/kg·K, and 50 W/m·K, respectively. The corresponding values for the mold are 8000, 500, and 200, respectively. The latent heat of fusion is 80 kJ/kg.

SOLUTION

The problem concerns solidification of a molten material in an enclosed region, as shown in Figure 1.3. However, a very simple, one-dimensional mathematical model is used, as sketched in

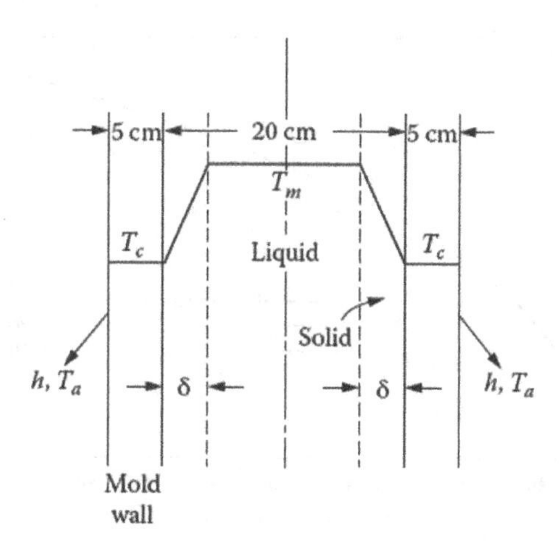

FIGURE 4.35 Simple mathematical model for the casting process considered in Example 4.8.

Figure 4.35. The liquid is at the melting temperature T_m, a linear temperature distribution exists in the solid because energy storage in it is neglected, and the mold is at uniform temperature $T_c(\tau)$, where τ is time. The applicable equations are obtained from energy balance as

$$(\rho C)_c \frac{dT_c}{d\tau} = k_s \frac{T_m - T_c}{\delta} - h(T_c - T_a)$$

$$\rho_s L_f \frac{d\delta}{d\tau} = k \frac{T_m - T_c}{\delta}$$

where the subscript c refers to the mold and s refers to the solid, δ is the thickness of the solid formed, and L_f is the latent heat of fusion. The first equation gives the energy balance for the mold, which gains energy from the solid and loses to the ambient medium. The second equation balances the energy removed by conduction in the solid to the latent heat for phase change.

Therefore, two coupled ODEs are obtained for this dynamic problem, one for T_c and the other for δ. The material property values are substituted in the equations, which are then rewritten in the form of Equation (4.46) as

$$\frac{dT_c}{d\tau} = F(\tau, T_c, \delta) \qquad \frac{d\delta}{d\tau} = G(\tau, T_c, \delta)$$

These can easily be solved by the Runge-Kutta method, as outlined in the preceding section. However, a small, finite, non-zero value of δ must be taken at time $\tau = 0$ to start the calculations. It must be ensured that the results are essentially independent of the value chosen. Some of the characteristic results obtained are shown in Figure 4.36 and Figure 4.37, in terms of the variation of T_c and δ with time. Casting is complete when $\delta = 0.1$ m because heat removal occurs on both sides of the plate. It is found that a variation in the heat transfer coefficient has no significant effect on the temperature or the solidification rate, over the range considered. However, the mold thickness W is an important design variable and substantially affects the solidification time and the temperature of the mold. From these results, the casting time at $h = 20$ W/(m²·K) and $W = 0.05$ m is 110 s. A thicker mold removes energy faster and thus reduces the casting time.

This example illustrates the use of dynamic simulation, which is particularly important for manufacturing processes and for modeling the time-dependent behavior of the system. Many more equations arise in large thermal systems; some may be ODEs and others may be PDEs. The solution to these equations may be obtained by extending the methods outlined here for simpler circumstances. However, for complex systems, the development of the computer codes is quite involved and commercially available general-purpose, as well as specialized, software is generally used to obtain the desired simulation results conveniently and rapidly.

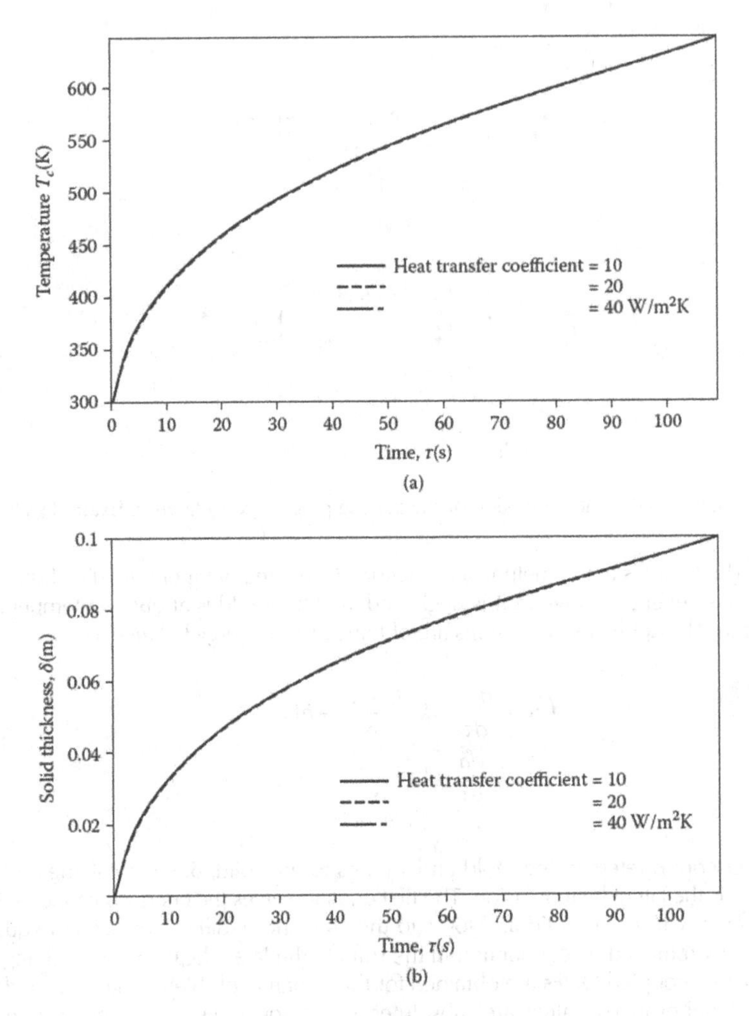

FIGURE 4.36 Calculated mold temperature T_c and solid region thickness δ as functions of time τ for different values of the heat transfer coefficient h, at $W = 0.05$ m, in Example 4.8.

4.5.3 Distributed Systems

In the preceding sections, we considered the relatively simple circumstances in which lumping may be employed for modeling the different parts of a given system. This approximation leads to algebraic equations in the steady-state case and to ODEs in the time-dependent or dynamic simulation case. Even though the assumption of lumping or uniform conditions in each system part has been and still is widely used because of the resulting simplification, the easy availability of powerful computers and versatile software has made it quite convenient to model and simulate the more general distributed circumstance in which the quantities vary with location and time. Of course, if the lumped model is appropriate for a given problem because of, say, the very low Biot number involved, there is no reason to complicate the analysis and the results by using a distributed model. However, there are many problems of practical interest in which large variations occur over the spatial domain and the lumped approximation cannot be used. Temperature variation in the wall and in the insulation of a furnace is an example of this circumstance. Similarly, the velocity and temperature fields in an electronic system, in the cylinder of an internal combustion engine, in the combustor of a gas turbine, and in the molten plastic in an injection molding process are strong functions of location and time, making it essential to simulate these as distributed, dynamic systems for accurate results.

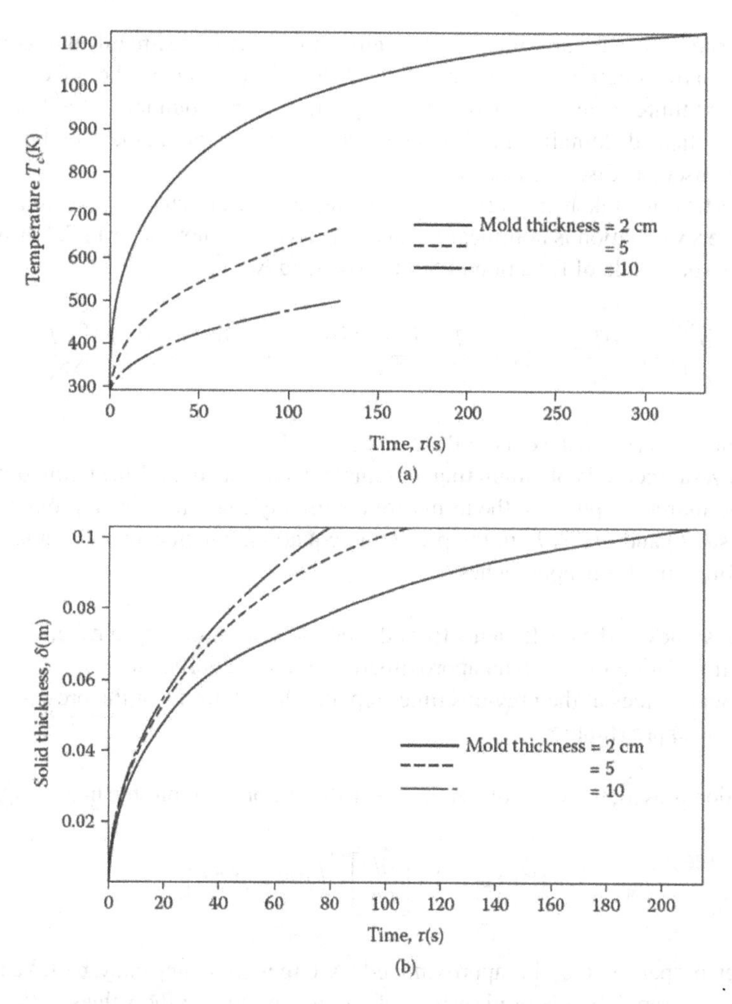

FIGURE 4.37 Calculated mold temperature T_c and solid region thickness δ as functions of time τ for different values of the mold wall thickness W, at $h = 20$ W/(m²·K), in Example 4.8.

The relevant equations for distributed systems are PDEs, which are frequently nonlinear due to material property changes, coupling with fluid flow, thermal buoyancy effects, and the presence of radiative transport. Several types of simple, linear PDEs, along with the corresponding solution procedures, were discussed in Section 4.2.4. Finite-difference, finite-element, and other approaches to obtain simultaneous algebraic equations from the PDEs and to solve these were outlined. Again, nonlinear PDEs lead to nonlinear algebraic equations and linear PDEs to linear algebraic equations. Once the set of algebraic equations is derived, the solution is obtained by the various methods for linear and nonlinear equations given earlier. Nonlinear equations are often linearized, as discussed in the following, so that new values may be calculated using the known values from previous time or iterations. In addition, commercially available software such as Ansys and CFD-Ace is generally employed in industry to simulate practical thermal systems.

4.5.3.1 Linearization

Consider the transient one-dimensional conduction problem represented by the equation

$$\rho(T)C(T)\frac{\partial T}{\partial \tau} = \frac{\partial}{\partial x}\left[k(T)\frac{\partial T}{\partial x}\right] \tag{4.50}$$

where the material properties are functions of temperature T. If these are taken as constant, the linear equation given by Equation (4.27) is obtained. Then this equation may be solved conveniently by explicit or implicit finite-difference methods if the geometry and boundary conditions are relatively simple. For complicated domains and boundary conditions, finite-element or boundary element methods may be used, as discussed earlier.

If the properties are taken as variable due to material characteristics or temperature range involved, the energy equation is nonlinear because the terms are nonlinear in T. For instance, if the term on the right-hand side of Equation (4.50) is expanded, we get

$$\frac{\partial}{\partial x}\left[k(T)\frac{\partial T}{\partial x}\right] = k(T)\frac{\partial^2 T}{\partial x^2} + \frac{\partial k(T)}{\partial x}\frac{\partial T}{\partial x} = k(T)\frac{\partial^2 T}{\partial x^2} + \frac{\partial k(T)}{\partial T}\left[\frac{\partial T}{\partial x}\right]^2 \tag{4.51}$$

indicating the nonlinearity that exists in the equation.

There are several methods of simulating systems in which such nonlinear equations arise. In an iterative or time-marching process, the terms are commonly linearized by approximating the coefficients, such as $k(T)$ and $\partial k(T)/\partial x$ in the preceding equation, which causes the nonlinearity in the terms, by the following three approaches:

1. Using the values of the coefficients from the previous iteration or time step
2. Using extrapolation to obtain an approximation of these coefficients
3. Starting with values at the previous time step and then iterating at the present time step to improve the approximation

If extrapolation is used, the value of k at the $(n+1)$th time or iteration step may be approximated as

$$k^{(n+1)} = k^{(n)} + \left[\frac{\partial k}{\partial T}\right]^{(n)} \left(T^{(n)} - T^{(n-1)}\right) \tag{4.52}$$

Similarly, other properties may be approximated. A larger time step may be taken for a desired accuracy level if extrapolation is used instead of simply employing the values at the previous time step. The third approach, which requires iteration, is more involved but allows still larger time steps for the same level of accuracy. Thus, nonlinear problems are linearized and then solved by the various methods discussed earlier for linear PDEs (Jaluria and Torrance, 2003).

4.5.4 Simulation of Large Systems

All the aspects considered in this chapter can easily be extended to large thermal systems that involve large sets of algebraic and differential equations. Such systems may range from a blast furnace for steel to an entire steel plant, from a cooling tower to a power plant, from the cooling system of a rocket to the entire rocket, and so on. Though many of the examples considered here involved relatively small sets of equations for simplicity and convenience, the basic ideas presented here are equally applicable to large and more complicated systems.

The two main features that distinguish large thermal systems from simpler ones are the presence of a large number of parts that lead to large sets of equations and relatively independent subsystems that make up the overall system. These aspects are treated by

1. Development of efficient approaches for solving large sets of equations
2. Better techniques for storing the relevant data
3. Subdivision of the system into subsystems that may be treated independently and then merged to obtain the simulation of the entire system

All these considerations have been discussed earlier in this chapter and need not be repeated. Methods such as Gauss-Seidel are particularly useful for handling large sets of algebraic equations while keeping the computer storage requirements small. Similarly, modularization of the simulation process has been stressed at several places because this allows building up of the system simulation package while ensuring that each subsystem is treated satisfactorily. Chapter 5 presents the overall design process for such large systems.

Several specialized computer languages have been developed for the simulation of engineering systems. These are usually designed for certain types of systems and, as such, are more convenient to use than a general-purpose language such as C or Fortran. Many of these simulation languages are particularly suited for manufacturing systems. The general-purpose simulation system (GPSS) is a simulation language suited for scheduling and inventory control applications dealing with different steps in a process. Other languages that may be mentioned are SIMAN, SIMSCIPT, MAST, and MAP. Each of these is particularly oriented to a specific application, making it easy to enter the relevant data for simulation and to obtain the desired outputs for design, operation, and control of the system. Other computational environments, such as those provided by specialized software for computer simulation and design (e.g., Matlab, MATHCAD, Maple, and other CAD programs) are also useful. Parallel computing, which involves a large number of processors, is particularly valuable in modeling and simulating large thermal systems.

4.5.5 Numerical Simulation versus Real System

It would be worthwhile to conclude the discussion on system simulation by stressing the most important aspect, namely, that the simulation must accurately and closely predict the behavior of the actual system. A satisfactory simulation of a system is achieved when the response of the simulated system to variations in operating conditions and to changes in the design hardware follows the expected physical trends and is a faithful representation of the given system. Unfortunately, the real system is rarely available to check the predictions of the simulation because one of the main uses of simulation is to study system behavior for a variety of designs without actually fabricating the system for these different designs. Therefore, other methods must generally be employed to validate the models and to ensure that accurate predictions of system behavior are obtained from the simulation.

As discussed in this and preceding chapters, the development of system simulation involves several steps. These include mathematical modeling, which generally also contains the correlating equations representing the results from physical modeling, material property data, and component characteristics; numerical solution to the representative equations; numerical modeling of different system parts; merging of separate models to yield an overall model for the system; variation of operating conditions to consider design and off-design conditions; and investigation of system behavior for different design parameters. Therefore, the validation of the simulation can be based on the validation of these ingredients that lead to overall system simulation and on any available results obtained from similar existing systems. Finally, when a prototype is developed and fabricated on the basis of the design obtained, the experimental results from the prototype can be employed to provide a valuable check on the accuracy of the predictions.

In conclusion, a validation of the numerical simulation of the system is carried out to confirm a close representation of the real system by considering the following:

1. Validation of mathematical model
2. Validation of numerical schemes
3. Validation of the numerical models for system parts
4. Physical behavior of the simulated system
5. Comparison of results from simulation of simpler systems with available analytical and experimental results

6. Comparison of results from simulation of existing systems with experimental data
7. Use of prototype testing results for final validation of simulation

It must also be reiterated that the results from future operation of the designed system are usually fed back into the model in order to continually make improvements for design and optimization efforts to be undertaken at later times. Once the system simulation is thoroughly validated and the accuracy of its predictions determined, it is used to obtain the numerical inputs needed for design and optimization, for studying off-design conditions, establishing safety limits, and investigating the sensitivity of the system to various design parameters.

4.6 SUMMARY

This chapter has considered the important topics of numerical modeling and system simulation. For most practical thermal systems, numerical methods are essential for obtaining a solution to the equations because of the inherent complexity in these systems arising from the nonlinear nature of transport mechanisms, complicated domains and boundary conditions, material property variations, coupling between flow and heat transfer mechanisms, transient and distributed nature of most processes, and a wide range of energy sources. Additional aspects such as those due to combined heat and mass transfer, phase change, chemical reactions as in combustion processes, strong coupling between material characteristics and the process, etc., further complicate the analysis of thermal processes and systems.

The basic considerations involved in numerical modeling, particularly those concerned with accuracy and validation, are discussed, first with respect to a part or component and then the entire system. Various aspects such as the use of computer programs available in the public domain and commercially available software are considered in the context of a thermal system. The decoupling of the parts of the system in order to develop the appropriate mathematical and numerical models, followed by a thorough validation, is presented as the first step in the development of the numerical model for the complete system. The numerical solution procedures for different types of mathematical equations such as algebraic and ordinary, as well as partial, differential equations are outlined, largely to indicate the applicability and limitations of the various commonly used approaches for solving these equations. Of particular interest here were the techniques for solving nonlinear equations that are commonly encountered in thermal systems.

The basic strategy for developing a numerical model for a thermal system is presented in detail, considering the treatment for individual parts and subsystems and the merging of these individual models to obtain the complete model. Again, the validation of the overall numerical model for the system is emphasized. The physical characteristics of the results obtained from the model, their independence of the numerical scheme and of arbitrarily chosen parameters, comparisons with available analytical and numerical results, and comparisons with experimental data from existing systems and finally from prototype testing are all discussed as possible approaches to ensure the accuracy and validity of the model. It is crucial to obtain a model that is a close and accurate representation of the actual, physical system under consideration.

The simulation of a thermal system is considered next. The importance and uses of simulation are presented. Of particular interest is the use of simulation to evaluate different designs and for optimization of the system. However, the application of simulation to investigate off-design conditions, to modify existing systems for improved performance, and to investigate the sensitivity to different variables is valuable in the design and implementation of the system as well. Various types of simulation are outlined, including physical and analog simulation. The focus of the chapter is on numerical simulation, which is discussed in detail. Different classes of thermal problems encountered in practice are considered. These include steady or transient cases and lumped or distributed ones, resulting in different types of governing equations and consequently different techniques for numerical modeling. Simulation of large systems is considered in terms of the basic strategy for modeling and simulation. The relationship between the simulation and the real system is a very important consideration and is confirmed at various stages of model development and simulation.

REFERENCES

Anderson, D.A., Tannehill, J.C., & Pletcher, R.B. (2011). *Computational fluid mechanics and heat transfer* (3rd ed.). Boca Raton, FL: CRC Press.
Atkinson, K. (1989). *An introduction to numerical analysis* (2nd ed.). New York: Wiley.
Boyle, J., Butler, R., Disz, T., Glickfeld, B., Lusk, E., Overbeek, R., . . . Stevens, R. (1987). *Portable programs for parallel processors*. New York: Holt, Rinehart and Winston.
Brebbia, C.A., Telles, J.C.F., & Wrobel, L.C. (2013). *Boundary element techniques: theory and applications in engineering, method for engineers*. Heidelberg, Germany: Springer.
Carnahan, B.H., Luther, H.A., & Wilkes, J.O. (1969). *Applied numerical methods*. New York: Wiley.
Chapra, S.C. (2017). *Applied numerical methods with Matlab for engineers and scientists* (4th ed.). New York: McGraw-Hill.
Chapra, S.C., & Canale, R.P. (2014). *Numerical methods for engineers* (7th ed.). New York: McGraw-Hill.
Dieter, G.E. (2000). *Engineering design: a materials and processing approach* (3rd ed.). New York: McGraw-Hill.
Ertas, A., & Jones, J.C. (1996). *The engineering design process* (2nd ed.). New York: Wiley.
Ferziger, J.H. (1998). *Numerical methods for engineering applications* (2nd ed.). New York: Wiley.
Gerald, C.F., & Wheatley, P.O. (2003). *Applied numerical analysis* (7th ed.). New York: Pearson.
Hornbeck, R.W. (1975). *Numerical methods*. Englewood Cliffs, NJ: Prentice-Hall.
Howell, J.R., & Buckius, R.O. (1992). *Fundamentals of engineering thermodynamics* (2nd ed.). New York: McGraw-Hill.
Huebner, K.H., & Thornton, E.A. (2001). *The finite element method for engineers* (4th ed.). New York: Wiley.
Incropera, F.P., & Dewitt, D.P. (2001). *Fundamentals of heat and mass transfer* (5th ed.). New York: Wiley.
Jaluria, Y. (1996). *Computer methods for engineering*. Washington, DC: Taylor & Francis.
Jaluria, Y. (2012). *Computer methods for engineering with Matlab applications* (2nd ed.). Boca Raton, FL: CRC Press.
Jaluria, Y., & Torrance, K.E. (2003). *Computational heat transfer* (2nd ed.). Washington, DC: Taylor & Francis.
James, M.L., Smith, G.M., & Wolford, J.C. (1985). *Applied numerical methods for digital computation* (3rd ed.). New York: Harper & Row.
Kernighan, B.W., & Ritchie, D.M. (1988). *The C programming language*. Englewood Cliffs, NJ: Prentice-Hall.
Kirk, D.B., & Hwu, W.W. (2016). *Programming massively parallel processors: a hands-on approach* (3rd ed.). Burlington, MA: Morgan Kaufmann.
Mathews, J.H., & Fink, K.D. (2004). *Numerical methods using MATLAB* (4th ed.). Upper Saddle River, NJ: Prentice-Hall.
Moran, M.J., & Shapiro, H.N. (2014). *Fundamentals of engineering thermodynamics* (8th ed.). New York: Wiley.
Parker, S.P. (Ed.). (1993). *Encyclopedia of chemistry* (2nd ed.). New York: McGraw-Hill.
Patankar, S.V. (1980). *Numerical heat transfer and fluid flow*. Washington, DC: Taylor & Francis.
Rauwendaal, C. (2014). *Polymer extrusion* (5th ed.). New York: Hanser.
Recktenwald, G. (2000). *Numerical methods with MATLAB*. Upper Saddle River, NJ: Prentice-Hall.
Reddy, J.N. (2018). *Introduction to the finite element method* (4th ed.). New York: McGraw-Hill.
Roache, P.J. (2009). *Fundamentals of verification and validation*. Albuquerque, NM: Hermosa Pub.
Smith, G.D. (1965). *Numerical solution of partial differential equations*. Oxford, UK: Oxford University Press.
Stoecker, W.F. (1989). *Design of thermal systems* (3rd ed.). New York: McGraw-Hill.
Stroustrup, B. (2013). *The C++ programming language* (4th ed.). Boston, MA: Addison-Wesley.
Tadmor, Z., & Gogos, C. (1979). *Principles of polymer processing*. New York: Wiley.

PROBLEMS

Note: In the following exercises, whenever possible, write the appropriate computer programs, using sample programs given in Appendix A. Matlab commands such as *roots, a\b, ode45,* and others that may be applied to find the roots, solve algebraic equations, solve ODEs, etc., can then be used to check the numerical results obtained. Analytical solutions, if available, may also be used to validate the numerical results.

4.1 The mass balance for three items x, y, and z in a chemical reactor is governed by the following linear equations:

$$2.2x+4.5y+1.1z=11.14$$
$$4.8x+y+2.5z=-1.62$$
$$-2.1x-3.1y+10.1z=15.57$$

Solve this system of equations by the Gauss-Seidel method to obtain the values of the three items. You may arrange the equations in any appropriate order. Do you expect convergence? Justify your answer. The initial guess may be taken as $x=y=z=0.0$ or 1.0.

4.2 An industrial system has three products whose outputs are represented by x, y, and z. These are described by the following three equations:

$$1.8x-3.1y+7.6z=12.2$$
$$4.8x+6y-1.1z=24.8$$
$$3.3x+1.7y+0.9z=13.0$$

a. Give the block representation for each of these subsystems.
b. Draw the information-flow diagram for the system.
c. Set up this system of equations for an iterative solution by any appropriate method, starting with an initial guess of $x=y=z=0$.
d. Show at least five iterative steps to obtain the solution to simulate the system.

4.3 The mass balance for three items a, b, and c in a reactor is given by the following linear equations:

$$4a+2b+2c=17$$
$$a-5b+c=-5$$
$$2a+3b-6c=-12$$

Solve this system of equations by the Gauss-Seidel iteration method. The initial guess may be taken as $a=b=c=0.0$ or 1.0.

4.4 Solve the following set of linear equations by the Gauss-Seidel iteration method. The initial guess may be taken as 0.0 or 1.0.

$$5x+y+2z=17$$
$$x+3y+z=8$$
$$2x+y+6z=23$$

Vary the convergence parameter to ensure that results are independent of the value chosen.

4.5 A firm produces four items, x_1, x_2, x_3, and x_4. A portion of the amount produced for each is used in the manufacture of the other items. The balance between the output and the production rate yields the equations

$$x_1+2x_2+5x_4=32$$
$$3x_1+2x_2+6x_3=36$$
$$3x_1+5x_2+2x_3+x_4=41$$
$$2x_2+10x_3+8x_4=58$$

Solve these equations by the SOR method and determine the optimum value of the relaxation factor ω. Obtain the production rates of the four items. Compare the number of iterations needed for convergence at the optimum w with that for the Gauss-Seidel method (ω = 1).

4.6 Using the successive substitution and the Newton-Raphson methods, solve the following equation for the value of x, which is known to be real and positive:

$$x^5 = [10(10-x)^{0.5} - 8]^3$$

The equation may be recast in any appropriate form for the application of the methods. Compare the solution and the convergence of the numerical scheme in the two cases.

4.7 The solidification equation for casting in a mold at temperature T_a, considering energy storage in the solid, is obtained as

$$C(T_m - T_a)/L\pi^{1/2} = \eta \exp\left(\eta^2\right)\text{erf}(\eta)$$

where C is the material specific heat, T_m is the melting point, L is the latent heat, and $\eta = \delta/[2(\alpha\tau)^{1/2}]$, δ being the thickness of the solidified layer, as shown in Figure P4.7, α is the thermal diffusivity, and τ is the time. Take $\alpha = 10^{-5}$ m²/s, L = 110 kJ/kg, T_m = 925°C, T_a = 25°C, and C = 700 J/kg·K for the material being cast. Approximate the error function as erf(η) = η, for 0 < η < 1, and erf(η) = 1.0 for η > 1. Solve this equation for η and calculate δ as a function of time τ. What is the solidification time for a 0.4-m-thick plate, with heat removal occurring on both sides of the plate?

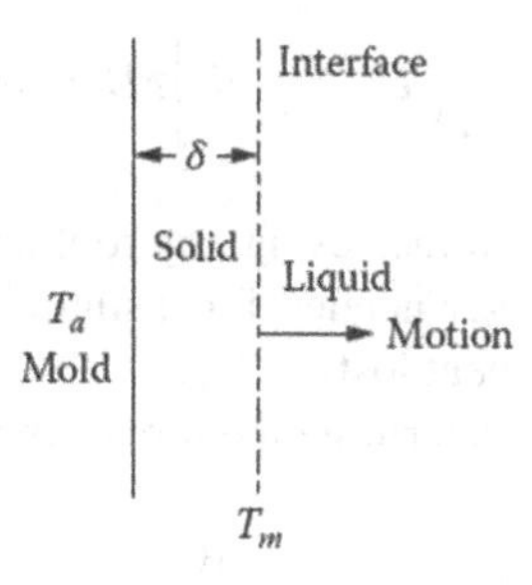

FIGURE P4.7

4.8 For the casting of a plate 10 cm thick, use the graphs presented in Example 4.8 to determine the total solidification times for the cases when the mold is 2 cm or 10 cm thick. Also, determine the time needed to solidify 75% of the plate. The heat transfer coefficient h is given as 40 W/m²K.

4.9 A spherical casting of diameter 10 cm has a total solidification time (TST) of 5 min. Assuming Chvorinov's model, TST = $C(V/A)^2$, where V is the volume, A is the surface area, and C is a constant, calculate the diameter of a long cylindrical runner with a TST of 12 min.

4.10 The speed V of a vehicle under the action of various forces is given by the equation

$$5.0 \exp(V/3) + 2.5V^2 + 2.0V = 20.5$$

Compute the value of V, using any appropriate method. Justify your choice of method. Suggest one other method that could also have been used for this problem.

4.11 The temperature T of an electrically heated wire is obtained from its energy balance. If the energy input into the wire, per unit surface area, due to the electric current is 1000 W/m², the heat transfer coefficient h is 10 W/(m²·K), and the ambient temperature is 300 K, as shown in Figure P4.11, the resulting equation is obtained as

$$1000 = 0.5 \times 5.67 \times 10^{-8} \times [T^4 - (300)^4] + 10 \times (T - 300)$$

Calculate the temperature of the wire by the secant method. Using this numerical simulation, determine the effect of the energy input on the temperature by varying the input by ±200 W/m². Also, vary the ambient temperature by ±50 K to determine its effect on the temperature. Do the results follow the expected physical trends?

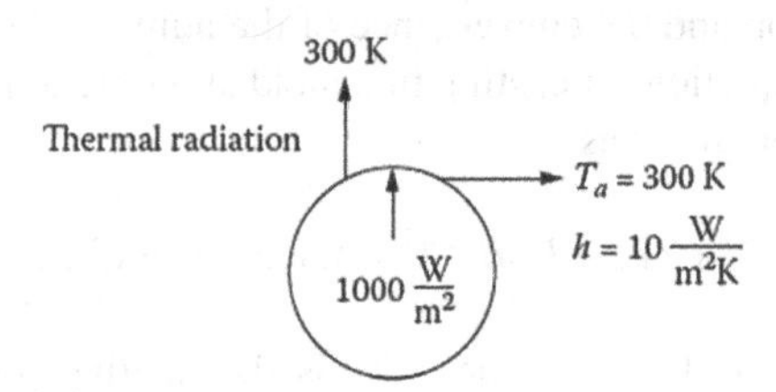

FIGURE P4.11

4.12 A cylindrical container of diameter D is placed in a stream of air and the energy transfer from its surface is measured as 100 W. The energy balance equation is obtained using correlations for the heat transfer coefficient as

$$\left[\frac{60}{D} D^{0.466} + 50\right] \pi D = 100$$

Find the diameter of the container using any root-solving method. Also, use this simulation to determine the diameter needed for losing a given amount of energy in the range 100 ± 20 W by varying the heat lost.

4.13 Use the bisection method to determine the root of the equation

$$x\left[1 - \exp\left(-\frac{10}{1+4x}\right)\right] - 1 = 0$$

4.14 Use the successive substitution method to determine the variable v from the equation

$$v = \left\{\left[\left(\frac{14 - v}{72 * 10^{-6}}\right)^{0.5} - 85\right] \Big/ 10.8\right\}^{0.65}$$

4.15 Use Newton's method or the secant method to solve the equation

$$\exp(x) - x^2 = 0$$

4.16 Use Newton's method to find the real roots of the equation

$$x^4 - 4x^3 + 7x^2 - 6x + 2 = 0$$

Discuss the nature and number of roots obtained. Also use the *roots* command in Matlab to find the roots and compare with the earlier results.

4.17 The root of the equation

$$[\exp(-0.5x)]x^{1.8} = 1.2$$

is to be obtained. It is given that a real root, which represents the location where the maximum heat flux occurs, lies between 0 and 6.0. Using any suitable method, find this root. Give reasons for your choice of method. What is the expected accuracy of the root you found?

4.18 The generation of two quantities, F and G, in a chemical reactor is governed by the equations

$$2.0F^2G^2 + 3.0G = 13.8$$
$$2.0G^3 + F^2 = 16.6$$

Solve this system of equations using the Newton-Raphson method and starting with $F = G = 1.0$ as the initial guess. What is the nature of these equations and do you expect the scheme to converge? Set the system up also as a root-solving problem. Suggest a method to solve it and obtain the solution.

4.19 In a chemical treatment process, the concentrations c_1, c_2, c_3, and c_4 in four interconnected regions are governed by the system of nonlinear equations

$$7c_1 + c_2^2 + c_3 + c_4 = 3.7$$
$$c + 8c_2 + 3c_3 - c_4 = 4.9$$
$$2c_1 - 2c_2 + 5c_3 + c_4^2 = 8.8$$
$$c_1 - c_2 + c_3^2 + 14c_4 = 18.2$$

Solve these equations by the modified Gauss-Seidel method to obtain the concentrations.

4.20 A manufacturing system consists of a hydraulic arrangement and an extrusion chamber. The two are governed by the following two equations:

$$1.3P - F^{2.1} = 0$$
$$P^{1.6} - (900 - 95F)^{0.5} + 10 = 0$$

Show the flow of information for this system. Set up the system of equations for an iterative solution, starting with an initial guess of $P = F = 1$. Show at least three iterative steps toward the solution.

4.21 Solve the following nonlinear system by Newton's method:

$$3X^3 + Y^2 = 11$$
$$2X^2 + Y - 4 = 0$$

Try solving these equations by the successive substitution method as well.

4.22 In a metal forming process, the force F and the displacement x are governed by

$$F = 70\left[1 - \exp\left\{\frac{1000}{21(5+20x)}\right\}\right]$$

$$250 = 4.2Fx$$

Solve for F and x, applying the Newton-Raphson method to this system of equations. Also, determine the sensitivity of the force F to a variation in the total input S, $\partial F/\partial S$, where S is the given value of 250, by slightly varying this input.

4.23 A copper sphere of diameter 5 cm is initially at temperature 200°C. It cools in air by convection and radiation. The temperature T of the sphere is given by the energy equation

$$\rho CV\frac{dT}{d\tau} = -\left[\varepsilon\sigma\left(T^4 - T_a^4\right) + h(T - T_a)\right]A$$

where ρ is the density of copper, C is its specific heat, V is the volume of the sphere, τ is the time, which is taken as zero at the start of the cooling process, ε is the surface emissivity, σ is the Stefan-Boltzmann constant, T_a is the ambient temperature, and h is the convective heat transfer coefficient. Compute the temperature variation with time using the Runge-Kutta method and determine the time needed for the temperature to drop below 100°C. The following values may be used for the physical variables: $\rho = 9000$ kg/m³, $C = 400$ J/(kg·K), $\varepsilon = 0.5$, $\sigma = 5.67 \times 10^{-8}$ W/(m²·K⁴), $T_a = 25$°C, and $h = 15$ W/(m²·K).

4.24 Consider the preceding problem for the negligible radiation case, $\varepsilon = 0$, with $h = 100$ W/(m²·K). Nondimensionalize this simpler problem and obtain the solution in dimensionless terms. Then, give the results for a 10-cm-diameter sphere.

4.25 The temperature variation in an extended surface, or fin, for the one-dimensional approximation, is given by the equation

$$\frac{d^2T}{dx^2} - \frac{hP}{kA}(T - T_a) = 0$$

where x is the distance from the base of the fin, as shown in Figure P4.25. Here P is the perimeter, being πD for a cylinder of diameter D, A is the cross-sectional area, being $\pi D^2/4$ for a cylindrical fin, k is the material thermal conductivity, T_a is the ambient temperature, and h is the heat transfer coefficient. The boundary conditions are shown in the figure and may be written as

$$\text{at } x = 0\text{: } T = T_o \quad \text{and} \quad \text{at } x = L\text{: } \frac{dT}{dx} = 0$$

Here, L is the length of the fin. Simulate this fin, which is commonly encountered in thermal systems, for $D = 2$ cm, $h = 20$ W/(m²·K), $k = 15$ W/m·K, $L = 25$ cm, $T_o = 80$°C, and $T_a = 20$°C. Nondimensionalize this problem to obtain the governing dimensionless parameters. Use the shooting method to obtain the temperature distribution, and discuss expected trends at different parametric values.

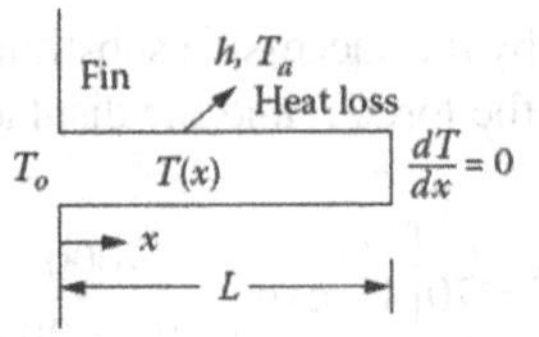

FIGURE P4.25

4.26 If radiative heat loss is included in the preceding problem, the energy equation becomes

$$\frac{d^2T}{dx^2} - \frac{hP}{kA}(T - T_a) - \frac{P}{kA}\varepsilon\sigma\left(T^4 - T_a^4\right) = 0$$

Solve this problem with $\varepsilon = 0.5$ and $\sigma = 5.67 \times 10^{-8}$ W/(m²·K⁴), using the finite-difference approach, and compare the results with the preceding problem. Increase the emissivity to 1.0 (black body) in this simulation and compare the results with those at 0.5. Are the observed trends physically reasonable?

4.27 The temperature distribution in a moving cylindrical rod, shown in Figure P4.27, is given by the energy equation

$$\frac{d^2T}{dx^2} - \frac{1}{\alpha}U\frac{dT}{dx} - \frac{2h}{kR}(T - T_a) = 0$$

where U is the velocity of the moving rod of radius R, α is the thermal diffusivity, and the other variables are the same as in the preceding problem. The boundary conditions are

$$\text{at } x = 0\text{: } T = T_0 \quad \text{and} \quad \text{as } x \to \infty\text{: } T \to T_\infty$$

Employing the finite-difference approach, compute $T(x)$. Take $U = 1$ mm/s, $h = 20$ W/(m²·K), $\alpha = 10^{-4}$ m²/s, $k = 100$ W/(m·K), $T_o = 600$ K, $T_a = 300$ K, and $R = 2$ cm. Numerically simulate $x \to \infty$ by taking a large value of x and ensuring that the results are independent of a further increase in this value. Also, nondimensionalize this problem and determine the governing dimensionless variables. If the material and dimensions are fixed, what are the main design variables? Discuss how these may be varied to control the temperature decay over a given distance.

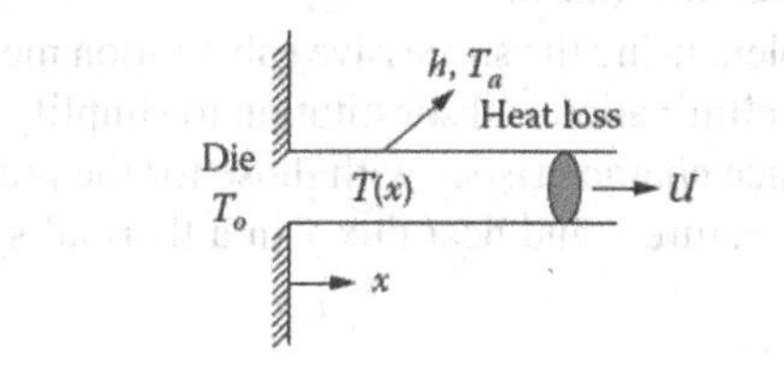

FIGURE P4.27

4.28 For the manufacturing process considered in Problem 3.6, set up the mathematical system for numerical simulation. Take each bolt to have surface area A, volume V, density ρ, and specific heat C. Outline a scheme for simulating the process. What outputs do you expect to obtain from such a simulation?

4.29 In the simulation of a thermal system, the temperatures in two subsystems are denoted by T_1 and T_2 and are given by

$$\frac{dT_1}{d\tau} = \left(\frac{hA}{\rho CV}\right)_1 (T_1 - T_a)$$

$$\frac{dT_2}{d\tau} = \left(\frac{hA}{\rho CV}\right)_2 (T_2 - T_a)$$

Under what conditions will the response of T_1 be much slower than that of T_2? Write down the finite-difference equations for solving this set of equations and outline the numerical procedure if the time step $\Delta\tau_1$, for T_1, is taken as much larger than the time step $\Delta\tau_2$, for T_2. Note that $(hA/\rho CV)$ is a function of temperature.

4.30 An experimental study is performed on a plastic screw extruder along with a die to determine the relationship between the mass flow rate $\dot{m}$ and the pressure difference P. The relationship for the die is found to be

$$\dot{m} = 0.5P^{0.5}$$

and the relationship for the screw extruder is

$$P = 2 + 3.5\dot{m}^{1.4} - 5\dot{m}^{2.2}$$

First, give the block representation for each of these subsystems. Then show the flow of information for the system. Set up this system of equations for an iterative solution by successive substitution, starting with an initial guess of $\dot{m} = 0$ and $P = 1$. Obtain the solution to simulate the extruder.

4.31 In an injection molding process, the flow of plastic in two parallel circuits is governed by the algebraic equations

$$\dot{m} = \dot{m}_1 + \dot{m}_2$$
$$\Delta p = 68 + 8\dot{m}^2 = 550 - 5\dot{m}_1 - 10\dot{m}_1^{2.5}$$
$$= 700 - 10\dot{m}_2 - 15\dot{m}_2^3$$

where $\dot{m}$ is the total mass flow rate, $\dot{m}_1$ and $\dot{m}_2$ are the flow rates in the two circuits, and Δp is the pressure difference. Simulate the system, employing the Newton-Raphson method. Study the effect of varying the zero-flow pressure levels (550 and 700 in the preceding equations) by ±10% on the total flow rate $\dot{m}$·

4.32 Solve the preceding problem using the successive substitution method. The number of equations may be reduced by elimination and substitution to simplify the problem. Compare the results and the convergence characteristics with those for the preceding problem.

4.33 The dimensionless temperature x and heat flux y in a thermal system are governed by the nonlinear equations

$$x^3 + 3y^2 = 21$$
$$x^2 + 2y = -2$$

Solve this system of equations by the Newton-Raphson and the successive substitution methods, comparing the results and the convergence in the two cases.

4.34 Simulate the ammonia production system discussed in Example 4.6 to determine the change in the ammonia production if the bleed (23.5 moles/s) and the entering argon flow (0.9 moles/s) are varied by ±25%. What happens if the bleed is turned off? Can this circumstance be numerically simulated?

4.35 The gross production of four substances by an engineering concern is denoted by a, b, c, and d. A balance between the net output and the production of each quantity leads to the following equations:

$$4a - 2b + 5d = 22$$
$$-2a + 8b - c = 16$$
$$3b + 4c - 3d = 30$$
$$-3c + 12d = 6$$

Solve these equations by the SOR method to simulate the system. The constants on the right-hand side of the equations represent the net production of the four items. If the net production of x_2 is to be increased from 16 to 24 (50% increase), calculate the gross production of all the items to achieve this.

4.36 A water pumping system consists of pipe connections and pumping stations, each of which has the following characteristics:

$$P = 1850 - 17.5\dot{m} - 0.7\dot{m}^2$$

where $\dot{m}$ is the mass flow rate of water and P is the pressure rise in each pumping station. The mass flow rate is measured as 32 kg/s with all eight pumping stations in the pipeline operating. The pressure drop in the pipe is given as proportional to the square of the mass flow rate. Obtain the governing equations to determine the flow rate if a few stations are inoperative and are bypassed. Then calculate the resulting flow rate if one or two stations fail.

4.37 The height H of water in a tank of cross-sectional area A is a function of time τ due to an inflow volume flow rate q_{in} and an outflow rate q_{out}. The governing differential equation is obtained from a mass balance as

$$A\frac{dH}{d\tau} = q_{\text{in}} - q_{\text{out}}$$

The initial height H at $\tau = 0$ is zero. Calculate the height as a function of time, with $A = 0.03$ m², $q_{\text{in}} = 6 \times 10^{-4}$ m³/s, and $q_{\text{out}} = 3 \times 10^{-4}\sqrt{H}$ m³/s. Use both Euler's and Heun's methods with a step size of 10 s. Plot H as a function of time τ. Give the times taken by the height to reach 2 m and 3.5 m in your answer. Why does the increase in H become very slow as time increases?

4.38 The temperature of a metal block being heated in an oven is governed by the equation

$$\frac{dT}{d\tau} = 10 - 0.05T$$

Solve this equation by Euler's and Heun's methods to get T as a function of time τ. Take the initial temperature as 100°C at $\tau = 0$.

4.39 A stone is dropped at zero velocity from the top of a building at time $\tau = 0$. The differential equation that yields the displacement x from the top of the building is (with $x = 0$ at t = 0)

$$\frac{d^2x}{d\tau^2} = g - 5\,\text{V}$$

where g is the magnitude of gravitational acceleration, given as 9.8 m/s², and V is the downward velocity $dx/d\tau$. Using Euler's method, calculate the displacement x and velocity V as functions of time, taking the time step as 0.5 s.

4.40 Simulate the hot water storage system considered in Example 3.5 for a flow rate of 0.01 m³/s, with the heat transfer coefficient h given as 20 W/(m²·K) and ambient temperature T_a as 25°C. The inlet temperature of the hot water T_o is 90°C. Obtain the time-dependent temperature distribution, reaching the steady-state conditions at large time. Study the effect of the flow rate on the temperature distribution by considering flow rates of 0.02 and 0.005 m³/s.

4.41 Consider one-dimensional conduction in a plate that is part of a thermal system. The plate is of thickness 3 cm and is initially at a uniform temperature of 1000°C. At time $\tau = 0$, the temperature at the two surfaces is dropped to 0°C and maintained at this value. The thermal diffusivity of the material is $\alpha = 5 \times 10^{-6}$ m²/s. Solve this problem by any finite-difference method to obtain the temperature distribution as a function of time.

4.42 A cylindrical rod of length 40 cm is initially at a uniform temperature of 15°C. Then, at time $\tau = 0$, its ends are raised to 100°C and held at this value. For one-dimensional conduction in the rod, the temperature distribution $T(x)$ is governed by the equation

$$\frac{1}{\alpha}\frac{\partial T}{\partial \tau} = \frac{\partial^2 T}{\partial x^2} - H(T - T_a)$$

where H is a heat loss parameter. Using any suitable method, solve this problem to obtain the time-dependent temperature distribution for $T_a = 15$°C, $\alpha = 10^{-6}$ m²/s, and $H = 100$ m^{-2}. Discuss how this simulation may be coupled with the modeling of fluid flow adjacent to the rod and other parts of the system to complete the model of a given system, without actually solving the flow.

4.43 Consider heat conduction in a two-dimensional, rectangular region of length 0.3 m and width 0.1 m. The dimension in the direction normal to this region may be taken as large. The dimensionless temperature is given as 1.0 at one of the longer sides and as 0.0 at the others. Solve the governing Laplace equation by the SOR method and determine the optimum relaxation factor. Discuss how, in actual practice, such a simulation may be linked with those for other parts of the system.

4.44 Consider the fan and duct system given in Example 4.7. Vary the zero-flow pressure, given as 80 in the problem, and the zero-pressure flow rate, given as 15 here, by ±20%. Discuss the results obtained. Are they consistent with the physical nature of the problem as represented by the equations?

4.45 Show the information-flow diagram for Problem 4.18. Also, draw the information-flow diagram for the simulation of Problem 4.35. Do not solve the equations; just explain what approach you will use.

5 Acceptable Design of a Thermal System: A Synthesis of Different Design Steps

5.1 INTRODUCTION

In the preceding chapters, we have discussed the main aspects involved in the design of a thermal system. An *acceptable* or *satisfactory* design must satisfy the given requirements for the system and must not violate the limitations or constraints imposed by the application, materials, safety, environmental effects, and other practical considerations. At this stage, we are not concerned with the optimization of the system and are largely interested in obtaining a *feasible* design. Though any design that meets the given requirements and constraints may be adequate for some applications, it is generally desirable to seek a domain of acceptable designs from which an appropriate design is selected on the basis of cost, ease of fabrication, availability of materials, convenience, marketability, etc.

The various considerations that are involved in the development of an acceptable design of a thermal system were discussed in Chapter 2. These led to the following main steps:

1. Formulation of the design problem
2. Conceptual design
3. Initial design
4. Modeling of the system
5. Simulation of the system
6. Evaluation of the design
7. Selection of an acceptable design

Optimization of the design follows the determination of a domain of acceptable designs and is not included here. Most of the other aspects, particularly problem formulation, conceptual design, modeling, and simulation, were discussed in detail in the preceding chapters. The first step in the foregoing list quantifies the design problem, and the second step provides the basic idea or concept to achieve the desired goals. The remaining steps constitute what might be termed as the detailed, quantitative design process, or simply the design process for convenience. These steps analyze the design and ensure that the problem statement is satisfied.

In this chapter, we will consider the synthesis of the different steps and stages that constitute the design effort in order to obtain an acceptable design. Individual aspects, such as modeling and simulation of thermal systems, which were discussed in detail earlier, will be considered as parts of the overall design strategy. The main purpose of this chapter is to link the different aspects that are involved in the design of a thermal system and to demonstrate the design procedure, starting with the problem formulation, proceeding through modeling and simulation, and ending with an acceptable design. Examples are employed to illustrate the coupling between the various steps and their combination to yield the desired design.

Several diverse thermal systems, ranging from those in materials processing and heating/cooling to those in energy and environmental systems, were considered in the previous chapters. It has been shown that the basic concerns, modeling, simulation, and system characteristics, vary significantly from one class of systems to another. For instance, lumped steady-state modeling is usually adequate for refrigeration and air conditioning systems, leading to algebraic equations, whereas distributed time-dependent modeling is generally needed for manufacturing processes and electronic equipment cooling, resulting in partial differential equations (PDEs). Consequently, the simulation procedures vary with the type of system under consideration. The design strategy itself may be affected by these considerations. Therefore, examples of thermal systems from different application areas are considered in this chapter and the corresponding design strategies presented. The systems considered range from relatively simple ones to fairly complicated ones in order to demonstrate the applicability of the basic ideas to the design of a wide variety of systems.

Before proceeding to the complete design process for typical thermal systems, an aspect that needs more detailed consideration is that of *initial design*. In many cases, the initial design is reached by considering the requirements and constraints of the problem and choosing the design variables, through approximate analysis and estimates, so that these satisfy the given problem statement. If different components are to be chosen and assembled for a thermal system, the choice of these components is guided by the requirements and constraints, so that the initial design is itself an acceptable design. Though redesign is obviously needed in case the initial design is not acceptable, it is important to employ the best possible initial design so that it is either acceptable by itself or the number of redesigns needed to converge to an acceptable design is small.

5.2 INITIAL DESIGN

The search for an initial design follows the formulation of the problem and the conceptual design. It is thus the first step in the quantitative design procedure. The analysis of the system, through modeling and simulation, and evaluation of the design for its acceptability are based on the initial design. The initial, starting, design affects the convergence of the iterative design process and often even influences the final acceptable or optimal design obtained. Therefore, the development of an initial design is a critical step in the design procedure, and considerable care and effort must be exerted to obtain a design that is acceptable or as close as possible to being acceptable.

Ideally, design variables should be selected so that the initial design satisfies the given requirements and constraints. Unfortunately, this is usually not possible for thermal systems because analysis only yields the outputs on system behavior for given inputs, rather than solve the *inverse problem* of yielding the inputs needed for a desired behavior. If the outputs and inputs were connected by simple relationships that could be inverted to obtain the inputs for required outputs, the problem would be considerably simplified. However, thermal systems usually involve complexities arising from nonlinear mechanisms, PDEs, coupled phenomena, and other complications, as discussed in Chapter 1. This makes it very difficult to solve the inverse problem in order to select the design variables, in an initial design, to satisfy all the requirements and constraints. Consequently, iteration is generally necessary to obtain a satisfactory design.

Several approaches may be adopted in the selection or development of the initial design. The approach that is appropriate for a given problem is a function of the nature of the thermal system under consideration, information available on previous design work, and the scope of the design effort. Some commonly used methods for obtaining an initial design are

1. Selection of components to meet given requirements and constraints
2. Use of existing systems
3. Selection from a library of previous designs
4. Use of current engineering practice and expert knowledge of the application

5.2.1 Selection of Components

In general, a combination of all the approaches given above is used to come up with the best initial design for practical thermal systems. However, each of these may also independently yield the desired starting point for iterative design. Selection of components is particularly valuable in thermodynamic systems, such as refrigeration, air conditioning, and heating systems, where the design of the overall system generally involves selecting the different components to meet the given requirements or specifications. An example of this is the air-cycle refrigeration system, based on the reverse *Brayton cycle* and shown in Figure 5.1 This system is commonly used aboard jet aircrafts to cool the cabin. The turbine, the compressor, and the heat exchanger may be selected based on the desired temperature and pressure in the cabin, along with the thermal load, to obtain an initial design.

An analysis of the thermodynamic cycle shown yields the appropriate specifications of the components for an ideal cycle or for a real one with given isentropic efficiencies (Reynolds and Perkins, 1977;

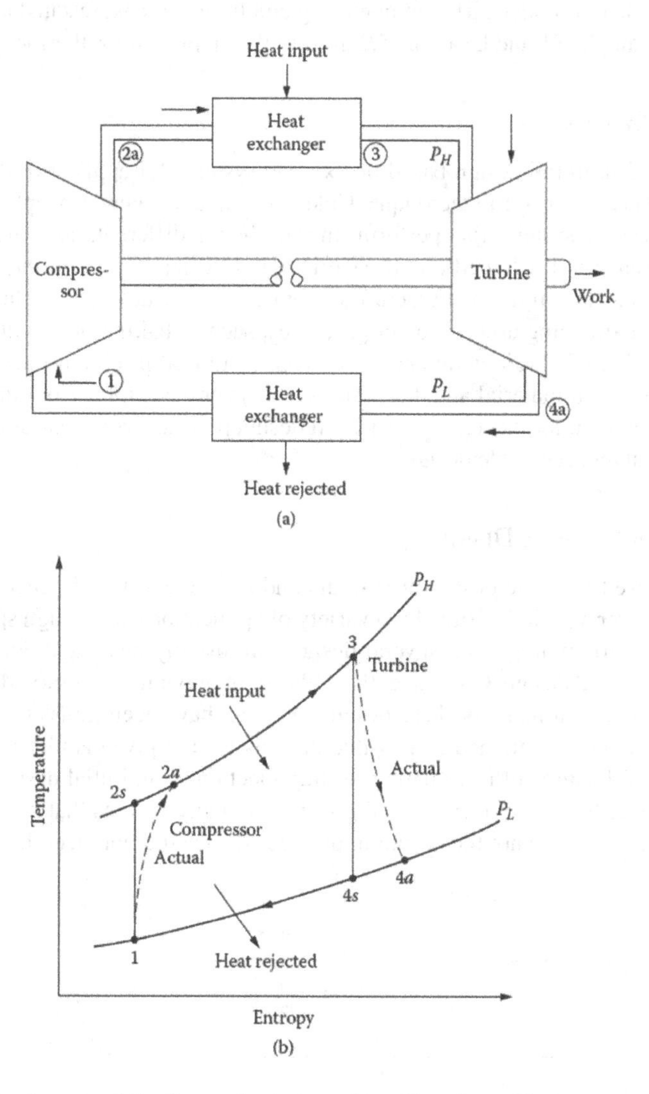

FIGURE 5.1 The hardware and the thermodynamic cycle, with real, nonideal compressor and turbine, for the Brayton cycle.

Howell and Buckius, 1992; Moran and Shapiro, 2014). For an ideal cycle, the efficiency η, which is the ratio of the work done to the energy input into the system, is given by the expression

$$\eta = 1 - \frac{1}{\left(\frac{P_H}{P_L}\right)^{\frac{\gamma-1}{\gamma}}} \tag{5.1}$$

where γ is the ratio of the specific heat at constant pressure C_p to that at constant volume C_v, and P_H and P_L are the high and low pressures in the system, respectively. The corresponding temperatures can be calculated for ideal constant entropy or isentropic processes and then for a real, nonideal system using the efficiencies. Any constraints on pressures or temperatures given in the problem can be taken care of by a proper choice of these components. A given range of desired efficiency for satisfactory systems may also be taken as a requirement. Therefore, the initial design itself satisfies the problem statement and is an acceptable design. This design may be modeled and simulated to study the system behavior under different operating conditions to ensure satisfactory performance in practical use. Example 5.1 and Example 5.2 discuss this approach for thermodynamic systems.

5.2.2 Existing Systems

The development of an initial design based on existing systems for applications similar to the one under consideration is a very useful technique. Unless a completely new concept is being considered for the given application, systems that perform similar, though different, tasks are usually available and in use. For instance, if a forced-air furnace is being designed for continuous heat treatment of silicon wafers as a step in the manufacture of semiconductor devices, as shown in Figure 5.2, similar systems that are being used for other processes, such as baking of circuit boards, drying of food items, and curing of plastic components, may be employed to obtain initial estimates of the heater specifications, wall material and dimensions, conveyor design, interior dimensions, etc. This provides the starting point for the iterative design-redesign process, which varies the relevant design variables to arrive at an acceptable design.

5.2.3 Library of Previous Designs

Any industry involved with the design of systems and equipment would generally develop many successful designs over a period of time for a variety of applications and design specifications. Even for the design of a particular system, several designs are usually generated during the process to obtain the best or optimal design. Consequently, a library of previous successful designs can be built up for future use. Note that many of these designs may not have been translated into actual physical systems and may have remained as possible designs for the given application. Such a library provides a very useful source of information for the selection of an initial design. For instance, an effort on the design of heat exchangers would give rise to many designs that may not be chosen for fabrication because they were not the optimum or because they did not meet the requirements for a

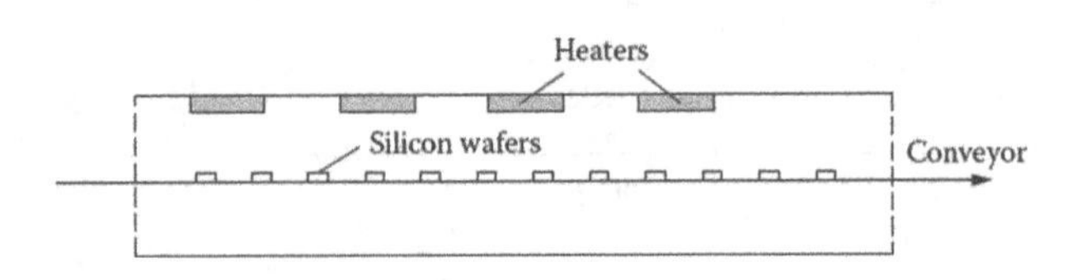

FIGURE 5.2 A thermal system for the heat treatment of silicon wafers in the manufacture of electronic components.

given application. However, for different design specifications, some of the earlier designs that were discarded might be satisfactory. Similarly, the design of an air compressor may yield many designs that are discarded because the pressure or the flow rate is too low. However, if this information is retained, it can be used for selecting an initial design for some other applications. Therefore, considerable effort is saved in the development of the initial design if such a library of earlier designs, along with the appropriate specifications, is available. As soon as a new design problem is initiated, the library may be employed to obtain a design with outputs as close as possible to the given requirements. For instance, if the total rate of heat transfer desired from the heat exchanger is given, a design that gives the closest heat transfer rate is chosen from the design library. This approach is particularly suitable for equipment, such as heat exchangers, heat pumps, boilers, and refrigerators.

5.2.4 Expert Knowledge

The last approach for developing an initial design is based on information available on the particular application and corresponding types of thermal systems employed, along with current engineering practice. Such an approach is very hard to quantify because the available information is often vague and may not have a solid analytical foundation. This is what is often termed as *expert knowledge*, i.e., the information obtained from an expert in the area. Several ideas developed over the years form the basis for such knowledge and play a major role in determining what is feasible. Information from earlier problems and attempts to resolve them is also part of this knowledge. Many aspects in thermal systems are difficult to analyze or measure, such as contact thermal resistance between surfaces, radiative properties of surfaces, surface roughness, fouling in heat exchangers, and losses due to friction. Similarly, random processes such as demand for power, changes in environmental conditions, and fluctuations in operating conditions are not easy to ascertain. In all such circumstances, current engineering practice and available information on the given application are used to come up with the initial design. These aspects are considered in greater detail in terms of *knowledge-based design* methodology in Chapter 11.

Example 5.1

A refrigeration system is to be designed to maintain the temperature in a storage facility in the range of –15°C to –5°C, while the outside temperature varies from 15°C to 22°C. The total thermal load on the storage unit is given as 20 kW. Obtain an initial design for a vapor compression cooling system.

SOLUTION

Because the lowest temperature in the storage facility is –15°C, the evaporator must operate at a temperature lower than this value. Let us select the evaporator temperature as –25°C. Similarly, the ambient temperature can be as high as 22°C. Therefore, the condenser temperature must be higher than this value to reject energy to the environment. Let us take the temperature at which the condenser operates as 30°C. The total thermal load is 20 kW, which is 20/3.517 = 5.69 tons. Therefore, the refrigeration system must be capable of providing this cooling rate. Because additional energy transfer may occur to the system and also for safe operation, let us design the system for 7.5 tons, which gives a safety factor of 7.5/5.69 = 1.32.

We must now choose the refrigerant. Because of environmental concerns with chlorofluorocarbons and because of the relatively large refrigeration system needed here, let us choose ammonia as the refrigerant. The various parts of the system are shown in Figure 1.8(a). All these parts, except for the compressor, usually have high efficiencies and may be assumed to be ideal. The compressor efficiency could range from 60% to 80%. Let us take this value as 65%. The thermodynamic cycle in terms of a temperature-entropy plot is shown in Figure 5.3. The fluid entering the throttling valve is assumed to be saturated liquid and that leaving the evaporator is assumed to be saturated vapor. These conditions are commonly employed in vapor compression

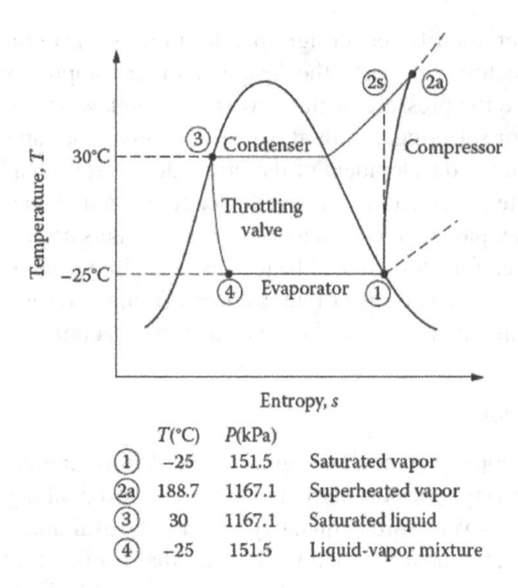

	T(°C)	P(kPa)	
①	−25	151.5	Saturated vapor
②a	188.7	1167.1	Superheated vapor
③	30	1167.1	Saturated liquid
④	−25	151.5	Liquid-vapor mixture

FIGURE 5.3 Thermodynamic cycle for the vapor compression refrigeration system considered in Example 5.1, along with the calculated conditions at various states.

refrigeration systems. The nonideal behavior of the compressor is seen in terms of an increase in entropy during compression.

For ammonia, the various pressures may be determined from available tables or charts (Van Wylen et al., 1994). Therefore, the pressure at the inlet to the compressor is obtained as 151.5 kPa. The pressure at the entrance to the throttling valve is determined as 1167.1 kPa, which is also the pressure at the exit of the compressor. The temperatures at the evaporator exit and valve entrance are −25°C and 30°C, respectively. The enthalpy at the compressor exit is obtained from

$$\eta = \frac{h_{2s} - h_1}{h_{2a} - h_1} = 0.65$$

where η is the compressor efficiency and the various states are shown in Figure 5.3. The entropy is constant between the states 2*s* and 1. Using this condition, the enthalpy h_{2s} is obtained as 1733 kJ/kg. Therefore, with h_1 = 1430.9 kJ/kg, h_{2a} is obtained from the preceding equation as h_{2a} = 1895.7 kJ/kg. This value of the enthalpy is used to determine the temperature at the compressor exit as 188.7°C. The coefficient of performance (COP) is obtained as

$$\text{COP} = \frac{\text{Heat removed}}{\text{Energy input}} = \frac{h_1 - h_4}{h_{2a} - h_1} = 2.34$$

Also, the heat removal rate, per unit mass flow rate of the refrigerant, $\dot{Q}/\dot{m}$, is

$$\frac{\dot{Q}}{\dot{m}} = h_1 - h_4 = 1430.9 - h_3 = 1430.9 - 342.5 = 1088.4 \text{ kJ/kg}$$

assuming enthalpy to remain unchanged in the throttling process, i.e., $h_3 = h_4$. Because the total required cooling rate is 7.5 tons = 26.38 kW, the mass flow rate $\dot{m}$ of the refrigerant is

$$\dot{m} = \frac{26.38}{1088.4} = 24.24 \times 10^{-3} \text{ kg/s} = 87.25 \text{ kg/hr}$$

Therefore, an initial design for the refrigeration system is obtained. It is seen that several design decisions had to be made during this process. Clearly, different values of the design variables could have been chosen, leading to a different initial design. This implies that the design obtained is not unique. In addition, because each part was chosen to satisfy the given problem statement, the initial design itself is an acceptable design. The fluid chosen is ammonia and the system capacity is 7.5 tons. The inlet and outlet conditions for each system part are obtained in terms of the inlet temperature and pressure, as given in Figure 5.3. The mass flow rate of the refrigerant is 87.25 kg/h. Thus, these items may be procured based on the given specifications. Because the items available in the market may have somewhat different specifications, the design may be adjusted to use available items, rather than have these custom made, in order to reduce costs. However, the system should be analyzed again if these items are changed to ensure that it meets the given requirements and constraints.

The preceding example illustrates the approach commonly used for a fairly wide range of thermal systems. The various components, dimensions, materials, and configuration are chosen on the basis of the conceptual design and the given problem statement for the design. The choices are also usually guided by the availability of various items so that the overall cost may be minimized.

Example 5.2

A remote town in Asia is interested in developing a 20 MW power plant, using the burning of waste material for heat input and a local river for heat rejection. It is found that temperatures as high as 350°C can be attained by this heat source, and typical temperatures in the river in the summer are around 30°C. Obtain a simple initial design for such a power plant.

SOLUTION

A Rankine cycle, such as the one shown in Figure 2.15(a), may be chosen without superheating the steam to simplify the system. This system has been analyzed extensively, as given in most textbooks on thermodynamics, and can be designed based on available information (see Moran and Shapiro, 2014). Water is chosen as the working fluid, again because of available property data, common use in typical power plants, and easy access to water at this location. Due to the temperature ranges given, the boiler temperature is taken as 300°C to ensure heating and boiling with energy input at 350°C. The condenser temperature is taken as 40°C to allow heat rejection to the river water, which is at 30°C. Then the initial temperature cycle of the proposed power plant may be drawn, as shown in Figure 5.4. The various states are given, with the idealized states indicated by subscript *s*, as in the previous example.

Now, we can proceed to first model the system and then analyze the thermodynamic cycle. All the components are taken as lumped, in order to simplify the model and because interest lies mainly in the energy transport and not in the detailed information for each component. The process is approximated as steady, which would apply for a steady operation of the power plant and not for the startup and shutdown stages or for power surges. The transient effects, which considerably complicate the analysis, may be considered later for designing the control system. Thus,

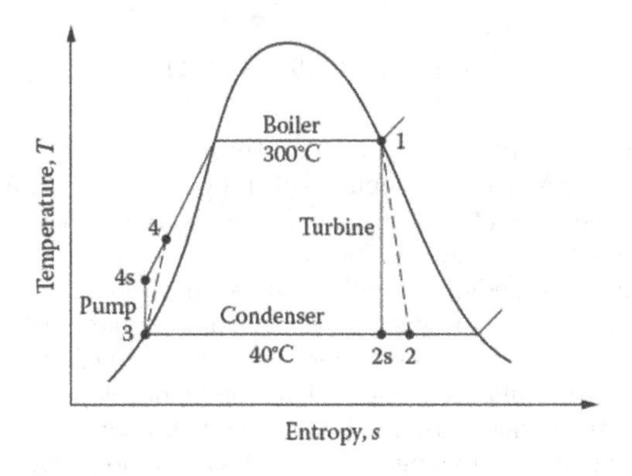

FIGURE 5.4 Thermodynamic cycle for the power plant design considered in Example 5.2.

the analysis with steady lumped components will lead to coupled algebraic equations, which can be solved to obtain the power delivered, water flow rate needed, heat input, and other relevant quantities.

Considering first the ideal cycle with isentropic turbine and pump, the steam tables are used to obtain properties at the relevant temperatures. We find that, for saturated steam, the enthalpy $h_1 = 2749$ kJ/kg and entropy $s_1 = 5.7045$ kJ/kg, which is equal to s_{2s} for an ideal turbine. Then the quality of the fluid x_{2s} is obtained as

$$x_{2s} = \frac{s_{2s} - s_f}{s_g - s_f} = \frac{5.7045 - 0.5725}{8.2570 - 0.5725} = 0.6678$$

where the subscripts f and g refer to saturated liquid and gas, respectively. This gives $h_{2s} = h_f + x_{2s}(h_g - h_f) = 167.57 + 0.6678 \times 2418.6 = 1782.71$ kJ/kg.

Similarly, for saturated liquid, $h_3 = 167.57$ kJ/kg, $s_3 = s_{4s} = 0.5725$ kJ/kg. The enthalpy h_{4s} is obtained by using the ideal pump work per unit mass, $v_3(p_4 - p_3)$, where v_3 is the specific volume at state 3 and the p's are the pressures. Thus,

$$\begin{aligned} h_{4s} &= h_3 + v_3(p_4 - p_3) = 167.57 + 1.0078 \times 10^{-3} \times (8.581 - 0.007384) \times 10^3 \\ &= 167.57 + 8.64 = 176.21 \text{ kJ/kg} \end{aligned}$$

where the pressures are in MPa and the multiplying factor 10^3 is used to obtain the work in kJ/kg. Then, the work done, or power output, for the ideal case is given by

$$W = \dot{m}\left(W_{Turbine,ideal} - W_{Pump,ideal}\right) = \dot{m}\left[(h_1 - h_{2s}) - (h_{4s} - h_3)\right]$$

where $\dot{m}$ is the mass flow rate of water/steam. It is calculated for the ideal cycle, using the preceding equation, as

$$\dot{m} = \frac{(20 \text{ MW})(1000 \text{ kW/MW})}{957.65 \text{ kJ/kg}} = 20.88 \text{ kg/s}$$

We can now include the effect of turbine and pump efficiencies. Taking them at typical values of 80%, we have

$$\frac{h_1 - h_2}{h_1 - h_{2s}} = 0.8$$

which gives h_2 as 1975.97 kJ/kg. Similarly, the pump work becomes

$$(W_{Pump,\ ideal})/0.8 = 10.8 \text{ kJ/kg}$$

This then gives $h_4 = 167.57 + 10.8 = 178.37$ kJ/kg. These values can now be used to obtain the water flow rate as 26.24 kg/s. The heat input is given by $\dot{m}(h_1 - h_4) = 26.24 \times (2749.0 - 178.37)$ kW = 67.45 MW. The heat rejected at the condensers is $\dot{m}(h_2 - h_3) = 47.45$ MW. The overall thermal efficiency is 20/67.45 = 0.2965, or 29.65%.

Thus, an acceptable initial design of the thermal system is obtained by choosing components and thermodynamic states based on given constraints and requirements. The efficiencies of the turbine and the pump can be adjusted if better information is available. As in Example 5.1, the design is not unique and several acceptable designs can be developed. The various components, such as the turbine, pump, condensers, and boiler, can be procured on the basis of the specified flow rate, pressure, and temperature ranges. The sensitivity of the design to variations in the components can be studied in order to choose available items instead of custom-made ones to reduce cost. These two examples demonstrate a frequently used approach for developing an initial design from the given problem statement so that an acceptable design is obtained. Once such an initial design is obtained, the operating conditions and component characteristics may be varied in the simulation to optimize the system, as discussed later.

5.3 DESIGN STRATEGIES

5.3.1 Commonly Used Design Approach

A strategy that is frequently employed for the design and optimization of thermal systems was discussed in earlier chapters. An initial design is developed based on the problem statement and the corresponding system is modeled, simulated, and evaluated. If the given requirements and constraints are satisfied, the initial design is acceptable; otherwise, a redesign process is undertaken until an acceptable design is obtained.

Clearly, this particular strategy is not unique, even though this is the most commonly used approach because of the systematic flow of information and the ease of implementation. In addition, as discussed earlier, the initial design may be based on existing systems and processes and thus result in a design that is very close to the final acceptable design. However, other strategies have been developed and are used for a variety of applications. Two strategies that are based on modeling and simulation are presented here.

5.3.2 Other Strategies

5.3.2.1 Adjusting Design Variables

Another approach is based on using the analysis, which incorporates modeling and simulation, to study a range of design variables and determine the resulting outputs from the system for a typical, fixed set of operating conditions. The basic concept is kept unchanged and the design variables, such as dimensions, components, geometrical configuration, and materials, are varied over their given ranges and the effect on the important quantities in the problem investigated. The resulting relationships between the outputs and the inputs may also be expressed in terms of correlating equations, using the curve-fitting techniques presented in Chapter 3. An acceptable design is then obtained by choosing the appropriate values for the various design variables based on the problem statement and quantitative simulation results.

5.3.2.2 Different Designs

Another strategy considers a collection of chosen designs and employs modeling and simulation to study the system behavior over the expected range of operating conditions. An initial design is not the starting point and simulation results are obtained for a variety of designs. An acceptable design is obtained from the various designs considered by comparing the simulation results with the problem statement, ensuring that all the requirements and constraints are satisfied.

Both of these strategies are shown schematically in Figure 5.5. The main difference between these and the approach discussed in detail earlier (Figure 2.13) is that an initial design is not the starting point for the design process. Extensive simulation results are obtained for a range of design variables for fixed operating conditions in one case and for a variety of designs under different operating conditions in the other. The desired acceptable designs are selected based on these results and the formulation of the design problem.

5.3.2.3 Examples

Let us consider the ingot casting system shown in Figure 1.3. Suppose the system is to be designed to obtain a solidification time τ_s smaller than a given value, without violating given constraints

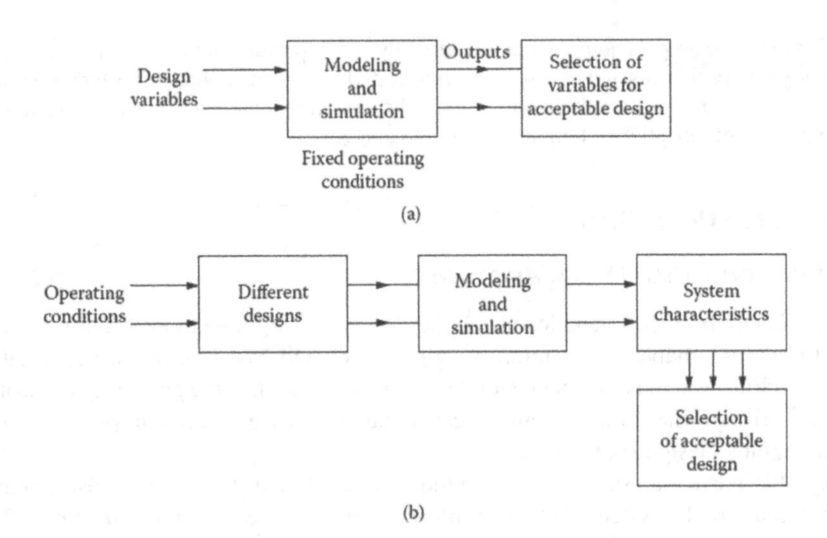

FIGURE 5.5 Two different strategies for design of a thermal system: (a) Using design variables as inputs for fixed operating conditions; (b) using operating conditions as inputs for different designs.

on temperature gradients in the materials. The solidification time is typically the time taken to solidify a given volume fraction of the melt, say 80%, because the ingot may be removed from the mold at this stage without waiting for the entire liquid region to solidify. A mathematical model and a simulation scheme may be developed for this process to compute the solidification time for different values of the design variables, keeping the molten material and dimensions of the enclosed region fixed. A simple one-dimensional model may be developed assuming negligible flow in the melt. Then, the numerical results on how solidification proceeds with time for a range of design variables, such as the wall material and thickness and convective heat transfer coefficient (representing a fan or circulating water for cooling the mold) at the outer surface of the wall, may be obtained. The governing equations for this simple model are (Viswanath and Jaluria, 1991)

$$\frac{\partial T_l}{\partial \tau} = \alpha_l \frac{\partial^2 T_l}{\partial y^2} \tag{5.2}$$

$$\frac{\partial T_s}{\partial \tau} = \alpha_s \frac{\partial^2 T_s}{\partial y^2} \tag{5.3}$$

$$\frac{\partial T_m}{\partial \tau} = \alpha_m \frac{\partial^2 T_m}{\partial y^2} \tag{5.4}$$

where the subscripts l, s, and m refer to the liquid, solidified region, and mold; α is the thermal diffusivity; and y is the coordinate distance, as shown in Figure 5.6.

Then numerical simulation may be employed to compute the location of the solid/liquid interface as a function of time, thus yielding the solidified region. Therefore, the time needed to solidify a given amount of material can be determined. For two- or three-dimensional problems, the progress of solidification from different sides may be determined to obtain the volume of the solidified material, if the solidification along different directions is assumed to be independent. Some of the typical results are shown in Figure 5.7, Figure 5.8, and Figure 5.9, indicating the effects of mold wall thickness $d = W_m - W_o$, thermal conductivity of mold material k_m

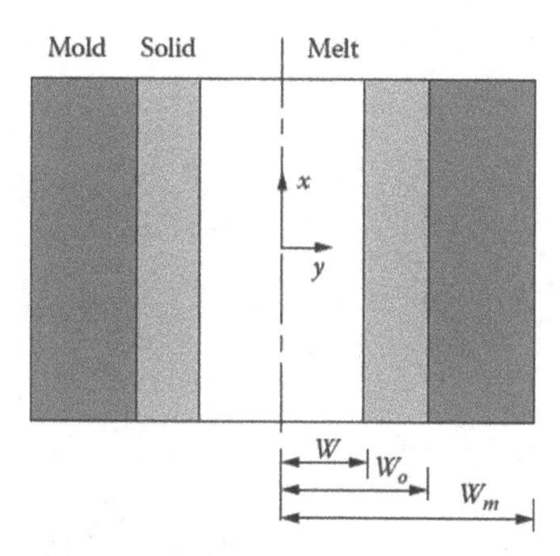

FIGURE 5.6 A one-dimensional model for solidification.

(normalized by k_s, the thermal conductivity of the solid), and convective heat transfer coefficient h at the outer wall of the mold. Thus, the effect of the different variables on solidification time τ_s is obtained. A greater mold wall thickness and thermal conductivity and a larger h all lead to faster solidification, as physically expected. Similarly, different operating conditions, such as ambient temperature, initial temperature of the mold, and initial temperature of the melt T_{pour} may be considered for a group of different designs, specified in terms of the design parameters. From these results, an acceptable design may be obtained to achieve the desired solidification time τ_s for solidifying 80% of the given volume. More sophisticated models have been developed and analysis of this system has been carried out by several investigators in recent years because of the importance of solidification in many manufacturing processes, such as crystal growing, alloy casting, three-dimensional printing, and thermal sprays.

Similarly, thermal systems arising from other application areas may be considered to illustrate the use of these two design strategies. For instance, the stratified water thermal energy storage

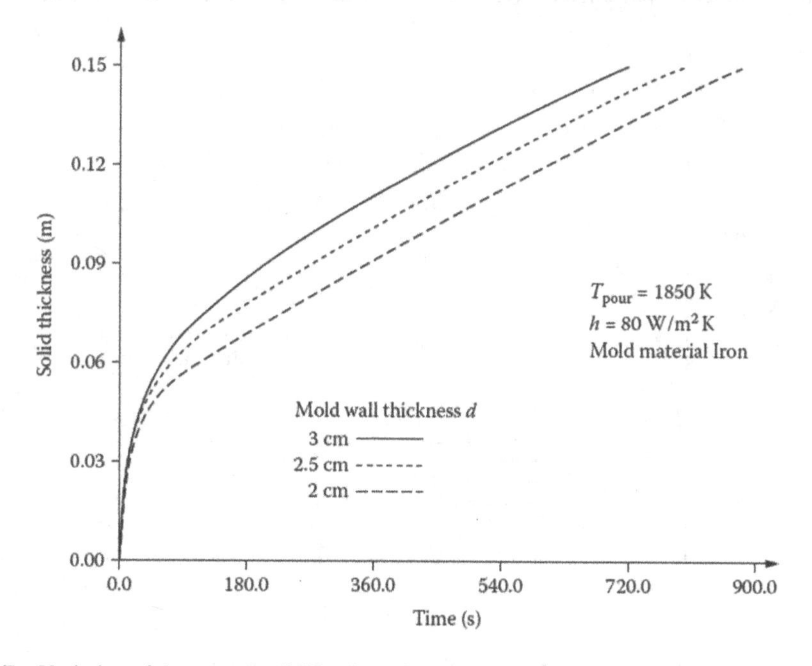

FIGURE 5.7 Variation of the rate of solidification with the mold wall thickness d.

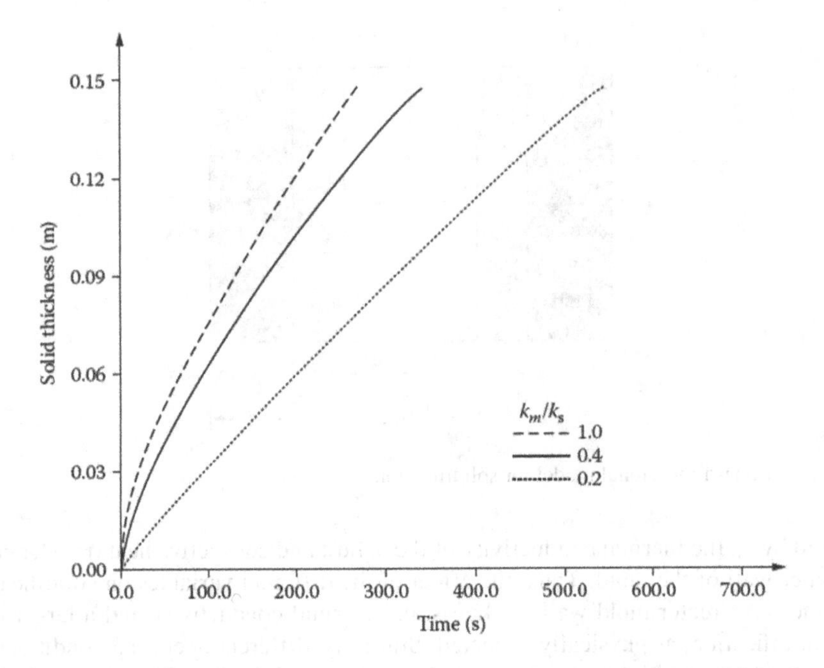

FIGURE 5.8 Effect of the thermal conductivity of the mold material on the rate of solidification.

system discussed in Example 3.5 may be taken. The simplified one-dimensional, vertical transport model yielded the governing equation

$$\frac{\partial \theta}{\partial \tau'} + W\frac{\partial \theta}{\partial Z} = \frac{\partial^2 \theta}{\partial Z^2} - H\theta \tag{5.5}$$

where all the terms in the equation were defined earlier and nondimensionalization was used to reduce the number of parameters. Here, W and H are the dimensionless vertical velocity and

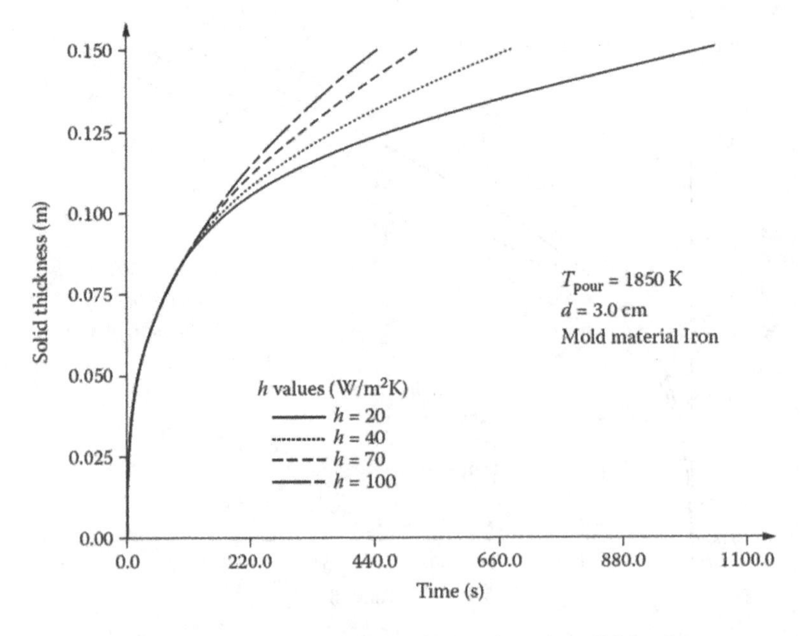

FIGURE 5.9 Effect of the convective heat transfer coefficient h on the solidification rate.

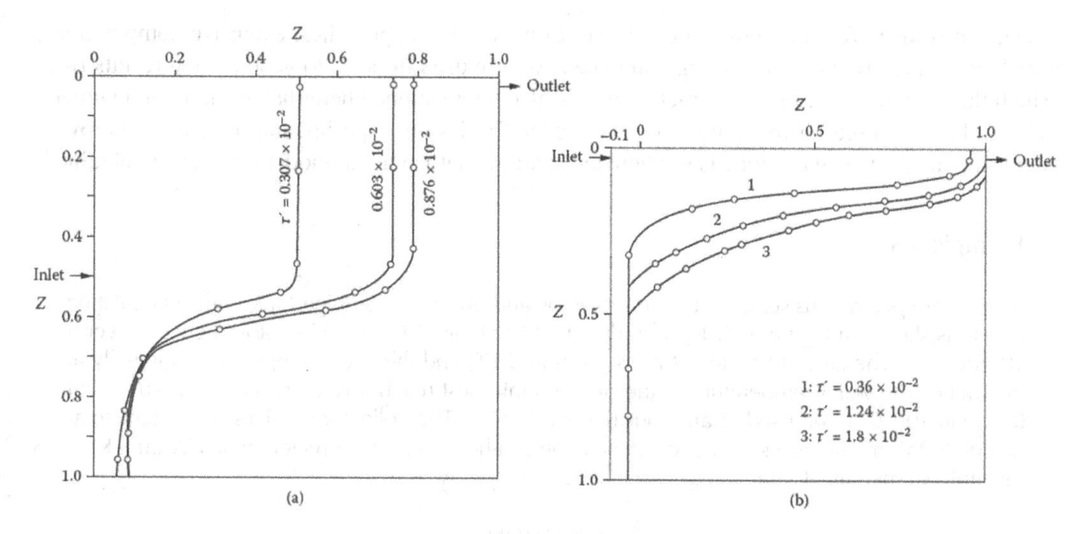

FIGURE 5.10 Calculated temperature profiles in an enclosed body of water for two inflow/outflow configurations at different values of dimensionless time τ'.

convective heat transfer coefficient, respectively. Therefore, the equation may be solved numerically for arbitrary values of the parameters W and H to yield the temperature distribution as a function of time. Figure 5.10 shows the results obtained from such a simulation for a typical energy storage system. Therefore, for given flow rate, inlet/outlet locations, and discharge temperature into the tank, the resulting temperature at the outlet may be calculated as a function of time. If hot water is to be supplied for a given duration at a specified minimum temperature level, the system may be designed by varying the dimensions, insulation, heat loss at the outer surface, etc., to meet this requirement. The simulation is used to generate results for a range of design variables and operating conditions. An acceptable design is then selected by comparing these results with the requirements and constraints.

5.3.2.4 Selection of Acceptable Design

Extensive work has been done on the analysis of a wide variety of thermal systems, and sophisticated models and simulation results are often available in the literature. However, the use of these results to obtain a satisfactory design is not a trivial exercise, even though most analyses claim that the results obtained will be valuable in design. As mentioned earlier, analysis is much simpler than design because the outputs resulting from given inputs are to be determined. In design, the inverse problem of finding the variables or conditions under which the desired outputs will be obtained is to be solved. By generating extensive simulation results, the attempt is to solve the inverse problem for design by correlating the outputs with the inputs.

Certainly, it is necessary to focus on some important parameters in order to obtain an acceptable design from simulation results. For instance, solidification time was taken as the main aspect in ingot casting. The duration for which water can be supplied without its temperature going below a minimum value may be the criterion for a water energy storage system. Then, such an output may be expressed in terms of the inputs by means of correlating equations, derived by the use of curve-fitting techniques. If such expressions are available, the design problem becomes relatively simple because the conditions needed for satisfying the requirements may be calculated easily from these expressions. Still, it must be noted that the inverse solution is generally not unique and optimization techniques are often used to narrow the domain from which an acceptable design may be selected.

In summary, different design strategies may be developed for different applications. The systematic approach represented by Figure 1.4 and Figure 2.13 is the most commonly employed strategy because it is also often the most efficient one. In most other approaches, extensive computations, which are generally time-consuming and expensive, are used in order to generate the results from which the appropriate design is extracted. It may also be mentioned here that, even though numerical simulation is used for most of the inputs needed for design, experimentation may also provide important data, particularly for cases where an accurate mathematical model is not easily obtained.

Example 5.3

A thermal system consisting of a solar collector and an energy storage tank with recirculating water, as shown in Figure 5.11, is to be designed to obtain 2.1×10^5 kJ of stored energy over a 10-hour day. The ambient temperature is given as 20°C and the water temperature is initially at this value. The water temperature in the storage tank must reach a value greater than 40°C, but less than 100°C, to be used in an industrial application. The collector receives a constant solar flux of 290 W/m² and loses energy by convection at a heat transfer coefficient h of 4 W/(m² · K) to the ambient medium. Obtain an acceptable design.

SOLUTION

A very simple mathematical model for this system is obtained by assuming that the convective heat loss q_c from the collector can be approximated as

$$q_c = hA\left(\frac{T_o + 20}{2} - 20\right)$$

where T_o is the maximum temperature attained over the day and A is the surface area of the collector, implying that an approximate average surface temperature is used to obtain the heat loss. Actually, the temperature varies nonlinearly with time and a differential equation needs to be solved to obtain the temperature variation. This approximation considerably simplifies the model. In addition, the storage tank is assumed to be well-mixed so that a uniform temperature exists across it. Heat loss from the tank is neglected.

With these assumptions, an energy balance for the collector yields

$$\left[290 - 4\left(\frac{T_o + 20}{2} - 20\right)\right]A(10 \times 3600) = 2.1 \times 10^5 \times 10^3$$

where a constant heat flux input of 290 W/m² into the collector arises over a 10-hour period. Both sides of the preceding equation are in Joules. An energy balance for the storage tank of volume V gives

$$1000 \times 4200 \times V \times (T_o - 20) = 2.1 \times 10^5 \times 10^3$$

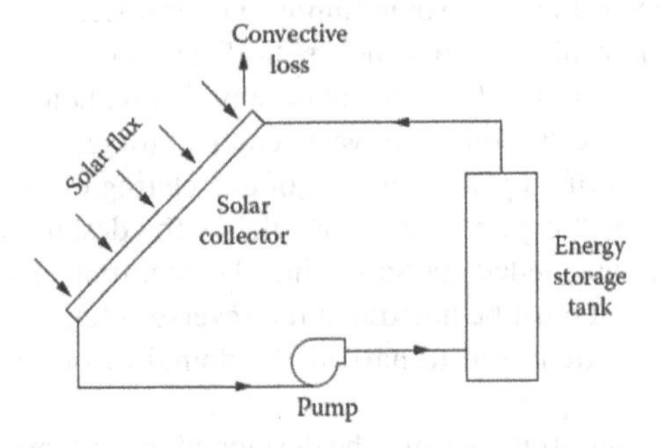

FIGURE 5.11 Solar collector and storage tank system considered in Example 5.3.

where the density of water is taken as 1000 kg/m^3 and the specific heat at constant pressure as 4200 J/(kg · K). The preceding two equations may be simplified to give

$$\left[290 - 2(T_o - 20)\right]A = 5833.3$$

$$T_o = \frac{50}{V} + 20$$

Therefore, these equations may be used to calculate the collector area A and the volume V of the storage tank for an acceptable design. The requirement of the total energy is already satisfied. The only other requirement is that $100 > T_o > 40°C$. Therefore, a domain of acceptable designs can be generated with these limitations. We may write these equations as

$$V = \frac{50}{T_o - 20} \quad \text{and} \quad A = \frac{5833.3}{290 - 100/V}$$

If T_o is chosen as 45°C, V is obtained as 2 m^3 and A as 24.3 m^2. This gives an acceptable design because it satisfies the given requirements and constraints. Similarly, if T_o is chosen as 95°C, V is 0.67 m^3 and A is 41.66 m^2. For T_o = 70°C, which is the average over the given range, V is 1 m^3 and A is 30.7 m^2.

Clearly, a unique solution is not obtained and an infinite number of designs can be generated in the domain given by the requirement $100 > T_o > 40°C$. If the system is optimized, with respect to cost or some other chosen criterion, this domain is substantially reduced, leading to an essentially unique solution in many cases. This is a small thermal system and approximations are used to develop a simple mathematical model. Models that are more complicated can also be developed for greater accuracy. However, this example illustrates a design strategy based on modeling and simulation, without using an initial design, to develop an acceptable design. It also indicates the crucial need to optimize the system.

5.3.3 Iterative Redesign Procedure

Iteration is an essential part of design in most design strategies and procedures because, as discussed earlier, an inverse problem is to be solved. In the analysis of thermal systems, the effort is directed at obtaining the output characteristics for given inputs such as operating conditions and design variables. However, in design, the requirements and constraints are given and the variables that result in a system that satisfies these are to be determined. As a result, the solution to the problem is not unique and several designs may have to be considered before obtaining one that satisfies the requirements and constraints.

5.3.3.1 Convergence

Any iterative procedure requires a criterion for convergence or termination of the iteration. In the design problem, because the given requirements and constraints may involve several variables and thus many criteria for convergence, it is useful to focus on a particular quantity or condition that is of particular significance to the problem at hand. This quantity may then be followed as iteration proceeds to ensure that the scheme is indeed converging and to stop the iteration when the desired results have been obtained or if a specified number of iterations have still not yielded a solution. For instance, in a cooling system, the rate of heat removal may be chosen as the main quantity of interest, even though the flow rates and temperatures are also important in the design. Similarly, the temperature of a material emerging from a heat treatment furnace may be selected as the criterion for following the iteration scheme.

Even though a particular parameter or quantity is considered with respect to the iteration scheme, the design obtained at convergence must be evaluated to ensure that all the design requirements and constraints have been satisfied. Because the quantity chosen for termination of the iteration is the most important aspect or a combination of dominant aspects in the design problem, there is a good possibility that the design obtained at convergence will be an acceptable design. However, if the design is not satisfactory, the design variables may be varied over a domain close to the converged design to seek an acceptable design. If, despite these efforts, a satisfactory design is not obtained, some of the requirements or constraints may have to be relaxed to obtain a solution.

If x_1, x_2, x_3, ..., x_n represent n quantities of interest in a thermal system to be designed, the requirements may be specified as

$$x_i = d_i, \quad x_i \le d_i, \quad \text{or} \quad x_i \ge d_i \tag{5.6}$$

which may be written as

$$x_i - d_i = 0, \quad x_i - d_i \le 0, \quad \text{or} \quad x_i - d_i \ge 0 \tag{5.7}$$

Here, any one of the preceding conditions may apply to a given quantity and d_i represents the given requirements, with $i = 1, 2, 3, \ldots, n$. The inequalities may be converted into equalities by assuming an acceptable tolerance level ε_i, as, for instance, $x_i - d_i = \varepsilon_i$, where ε_i may be positive or negative.

For example, in a heat exchanger, the given requirements relate to the flow rates, temperatures, and heat transfer rate. Thus, if the inlet flow rate and inlet temperatures of the hot and cold fluids are fixed, the outlet temperature of the cold fluid as well as the heat transfer rate may be taken as the requirements, with the configuration, dimensions, and materials used in the heat exchanger as design variables. If energy losses to the environment are included, the efficiency of the system may be defined as the ratio of the energy gained by the cold fluid to that lost by the hot fluid. An efficiency greater than a given value may then be a requirement. Several such requirements are generally associated with the design of a thermal system. However, the most important requirement, say the outlet temperature of the cold fluid in the present example, may be chosen as the criterion for convergence of the iterative redesign scheme.

If it is not possible to isolate a particular quantity for the iterative scheme, a combination of important variables or of their difference from the required values, such as

$$Y = x_1 + x_2 + x_3 \quad \text{or} \quad Y = (x_1 - d_1) + (x_2 - d_2) + (x_3 - d_3) \tag{5.8}$$

may be chosen and the function Y employed to keep track of the progress of the iteration. If both positive and negative values of the variables or of their differences from the requirements are considered, Y may be defined as

$$Y = |x_1| + |x_2| + |x_3| \quad \text{or} \quad Y = |(x_1 - d_1)| + |(x_2 - d_2)| + |(x_3 - d_3)| \tag{5.9}$$

or as

$$Y = x_1^2 + x_2^2 + x_3^2 \quad \text{or} \quad Y = (x_1 - d_1)^2 + (x_2 - d_2)^2 + (x_3 - d_3)^2 \tag{5.10}$$

A square root of the expressions on the right-hand sides of the two equations given in Equation (5.10) may also be employed. All the terms in the preceding equations for Y should generally be normalized by the required values, such as d_i, to make them of comparable magnitude.

Therefore, several different requirements may be included in a design parameter or quantity that is used to follow the iterative process and to determine its convergence. For instance, in the case of the heat exchanger discussed previously, the design parameter Y may be taken in terms of the cold fluid outlet temperature T_o and heat transfer rate Q as

$$Y = \left[\left(\frac{T_o - T_r}{T_r} \right)^2 + \left(\frac{Q - Q_r}{Q_r} \right)^2 \right]^{1/2} \tag{5.11}$$

where the subscript r refers to the required values. Then the desired value of Y for the given problem may be determined, being zero if differences from the requirements are employed as in Equation (5.11). Weighting factors may also be used to accentuate the importance of certain requirements over the others.

Therefore, the iterative redesign process becomes quite similar to the iterative procedures employed for solving nonlinear algebraic equations, as outlined in Chapter 4. The design parameter Y is defined in terms of the important requirements and the desired value obtained from the problem statement. As seen previously, neither the definition of Y nor its required value for a satisfactory design is unique. However, this approach does allow one to follow the iterative scheme and to terminate the iteration when Y attains the desired value Y_r to within a chosen tolerance level $\tilde{\varepsilon}$

$$|Y - Y_r| \le \tilde{\varepsilon} \tag{5.12}$$

Figure 5.12 shows the variation of Y as the iteration proceeds for a typical design problem. The value may go up or down locally. However, it is possible to determine if the scheme is approaching convergence in the long run, if divergence would occur, or if the results are simply oscillating without convergence.

A design parameter or criterion such as Y can also be used to determine the rate of convergence of the iterative scheme and to develop schemes that would accelerate convergence. Many of the ideas presented in Chapter 4 on the iterative convergence of nonlinear equations are applicable. Because each iteration is time-consuming for most practical thermal systems, it is important to reduce the number of iterations needed to obtain an acceptable design. Also, design variables that are particularly difficult to change, such as geometry, are often held constant while other variables are altered for reaching an acceptable design. A discussion of some of these aspects follows.

5.3.3.2 System Redesign

In the iterative redesign procedure, a given design is evaluated in terms of the problem statement, and, if it is found to be unacceptable, the system is redesigned by varying the design variables, keeping the conceptual design unchanged. This new design is again evaluated and the iterative process continued until a satisfactory design is obtained. As discussed previously, a single

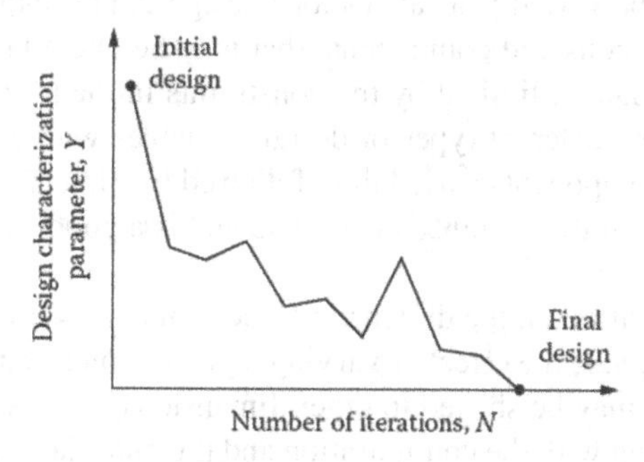

FIGURE 5.12 Variation of a parameter Y chosen to represent the acceptability or improvement of the design as a function of the number of iterations N.

important quantity or a parameter representing several important aspects in the problem may be employed to follow the iteration and to terminate it when a convergence criterion such as that given by Equation (5.12) is satisfied. We now wish to address how redesign is undertaken at each step of the iteration.

Redesign involves choosing different values of the design variables in the problem. The various types of design variables that are of interest in typical thermal systems are

1. Geometrical configuration
2. Materials employed
3. Dimensions of various parts
4. Characteristics or specifications of different components or devices used in the system

The performance of the system also depends on the operating conditions, which may be varied to obtain different product and system characteristics and for optimizing the operation of the system. However, in system design, we are largely interested in the hardware of the system and thus the listed design variables are considered for redesigning a system.

It is useful to follow a systematic approach in varying the design variables. Consider a simple household refrigerator. The configuration, materials, dimensions, and specifications of the components such as the compressor and condenser can be changed to obtain a new design. If all these are varied at each iterative step, it is hard to keep track of the progress made from one design to the next and to determine the effect of each variable on the system performance. One way of approaching redesign is to keep most design variables unchanged while one variable or a set of variables is altered. The geometrical configuration, materials, and dimensions may be kept constant while different compressors, condensers, etc., are considered. Similarly, the dimensions of the interior region, wall thickness, and other dimensions may be varied while the remaining design variables are held constant. The given constraints are invoked when any particular design variable is being changed or selected. Of course, the design variables may not be independent and may have to be varied together. For example, the condenser capacity and its surface area go together, linking the dimensions with the component specifications.

Similarly, we may consider the forced-air heat treatment oven discussed earlier and shown in Figure 2.28. Again, the geometry, materials, dimensions, and components, such as the heater and the fan, are the main design variables. The geometry and materials are often chosen on the basis of information available from existing or similar systems. The range of variation in these two parameters is generally limited by the application and by the availability and cost of materials. For instance, the configuration may be determined by the fact that a high side opening is needed to insert the material to be heated. Similarly, cost considerations may limit the material selection to steel and aluminum. In any case, the configuration and materials may initially be chosen to comply with such considerations related to the application. As the design process advances, even the geometry and the materials may be varied if a satisfactory design is not obtained. However, the initial efforts are directed at dimensions and components that may be altered somewhat more easily and that have wide ranges of variation, limited by the constraints in the problem. A schematic of such an approach, which considers different types of design variables with a predetermined priority, is shown in Figure 5.13, with components varied first, followed by dimensions, then by materials, and so on. This priority is based on the designer's expertise and is a good candidate for automation, as discussed in Chapter 11.

Even when attention is focused on the dimensions, these may be varied one at a time in order to determine the resulting effects. If the effect of varying a given dimension, say the wall thickness of the oven, is small, the effort may be shifted to other dimensions such as the height of the enclosed space. If the dimensions, along with the configuration and the materials, are held constant, different heaters and fans may be considered for the redesign. After each change, the design is evaluated in terms of a chosen quantity or parameter that characterizes the design to ascertain if the new design

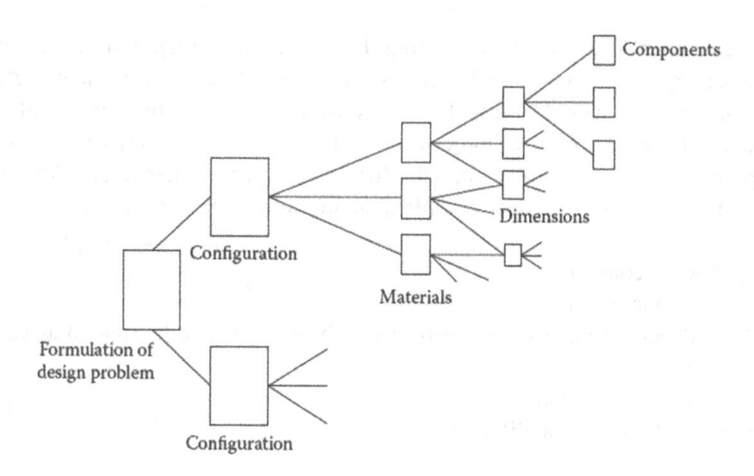

FIGURE 5.13 Priority for changing the design variables, considering the configuration, materials, dimensions, and components as variables.

is an improvement over the previous one. If the design appears to be becoming worse, the direction of the change is reversed.

The given constraints are taken care of in the selection of the design variables. As the iteration proceeds, the effect arising from each change is obtained and the *sensitivity* of the system performance to variation in different design variables is determined. This allows one to focus on the most important variables and thus converge to an acceptable solution more rapidly.

5.4 DESIGN OF SYSTEMS FROM DIFFERENT APPLICATION AREAS

We have considered the main aspects involved in the design of a thermal system, starting with conceptual design and proceeding through initial design, modeling, and simulation to design evaluation, redesign, and convergence to an acceptable design. It has also been shown that thermal systems arise in many diverse applications and vary substantially from one application to another. The examples considered thus far have similarly ranged from relatively simple systems, with a small number of parts, to complex ones that involve many parts and subsystems. In actual practice as well, the complexity of the design process is strongly dependent on the nature and type of thermal system under consideration.

Among the simplest design problems are those that involve selecting different components that make up the system and then simulating the system to ensure satisfactory performance for given ranges of operating conditions. The governing equations are generally nonlinear algebraic equations in such cases, and the various numerical techniques outlined in Chapter 4 may be used for the simulation. On the other hand, complex systems such as those in materials processing, aerospace applications, and electronic equipment cooling generally involve sets of PDEs that are coupled to each other and to other types of equations that govern different parts of the system. A few examples from some of the important areas of application are given here to illustrate the synthesis of the various ideas and design steps discussed earlier.

5.4.1 Manufacturing Processes

This is one of the most important areas in which thermal systems are of interest. Though manufacturing has always been of crucial significance in engineering, this area has become even more vital in recent years because of the development of new materials, applications, and processing techniques. Several important manufacturing processes were mentioned earlier, including processes such as plastic extrusion, heat treatment, casting, bonding, hot rolling, and optical fiber drawing. Many new and emerging processes and systems have been developed to meet the demands for advanced materials and devices. These include three-dimensional printing, chemical vapor deposition (CVD), thermal sprays, and laser processing. The thermal systems associated with different manufacturing processes are quite diverse, with different concerns, mathematical models, and governing mechanisms. They are generally complicated and involve features such as

1. Time-dependent behavior
2. Combined transport modes
3. Strong dependence on material properties, which often have to be treated as variable and not as constant
4. Sensitivity to operating conditions
5. Strong coupling between the different parts of the system.

Other characteristics may also be important in specific applications, as discussed in specialized books on manufacturing such as Ghosh and Mallik (1986), Kalpakjian and Schmid (2013), and Jaluria (2018).

The governing equations for manufacturing processes are typically PDEs that are coupled through the boundary conditions and material property variations. However, because the problem may vary substantially from one process to another, it is very difficult to develop a general approach to modeling, simulation, and design of these systems. A few examples of thermal systems in manufacturing were discussed in earlier chapters and a few others will be considered in the presentation on optimization. Here, in the following example, we shall discuss the design process for a typical system employed for thermal processing of materials. The problem is taken from an actual industrial process.

Example 5.4

Straight plastic (PVC) cords are to be made into a coil by thermoforming. The conceptual design involves winding the cord over a stainless steel mandrel and heating the plastic beyond its *glass transition temperature* of 250°F (121.1°C), without exceeding the maximum temperature of 320°F (160°C), followed by cooling to about 120°F (48.9°C) to make the shape permanent. The cords have a thickness of 0.1 in (2.54 mm) and the inner diameter of the coil must be 0.25 in (6.35 mm). The desired length of the final coil is 12 in (30.5 cm). Develop a mathematical model for the process and use the results from simulation to obtain an acceptable design of the thermal system. Suggest possible variations in the design that would improve the product and system performance.

SOLUTION

The design problem may be formulated easily from the preceding description of the process. The given quantities are some of the materials and dimensions that cannot be varied. Thus, the mandrel has a diameter of 6.35 mm and a length greater than 305 mm. The cord is wound tightly around it, giving an outer diameter of the composite cylinder assembly as 11.43 mm. The requirements are in terms of the desired temperature levels, with a constraint on the maximum allowable temperature, in the plastic. It is desirable to raise every point in the plastic cord to a temperature above 121.1°C, without exceeding the allowable value of 160°C.

Let us first consider a model for calculating the temperatures in the cord and the mandrel and then link it with the system. The typical properties of PVC and stainless steel are obtained from the literature (such as Appendix B) and are listed as, respectively,

density, ρ (kg/m^3)	958	8055
specific heat, C (J/kg.K)	2500	480
thermal conductivity, k (W/m.K)	0.3	15.1

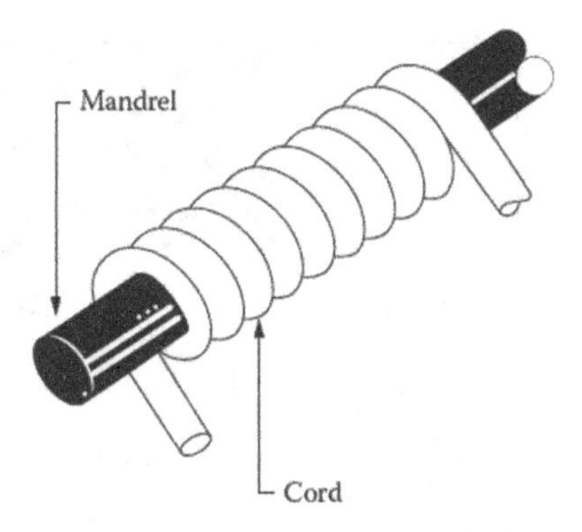

FIGURE 5.14 Plastic cord wound on a metal mandrel.

We can now apply the model development procedures discussed in Chapter 3. Because of the relatively small range of temperature variation, the ratio of the change in the various properties to their average values is small, e.g., $\Delta k/k_{avg} \ll 1$, allowing us to assume constant properties. It is also assumed that the cord is tightly wound on the mandrel so that no significant gaps are left along the cylinder axis. In addition, the length L (30.5 cm) is much greater than the outer diameter D (1.143 cm) of the cord-mandrel assembly, shown in Figure 5.14, i.e., $L/D \gg 1$. This implies that variation along the axial coordinate may be neglected. In addition, if the heat transfer at the surface is uniform, axisymmetry is assured. Therefore, the problem may be treated as a one-dimensional, radial, transport situation.

In addition to the aforementioned simplifications, the mandrel may be taken as lumped because of its small diameter and high thermal conductivity, compared to the plastic. If the *Biot number*, $Bi = h\,R/k$, is estimated, even for fairly high values of the convective heat transfer coefficient h, it is found to be smaller than 0.1, supporting the assumption of temperature uniformity in the mandrel. The problem reduces to that of heat transfer in a hollow plastic cylinder with given conditions at the inner and outer surfaces. Then the governing equation for the temperature $T(r, \tau)$, where r is the radial coordinate distance measured from the cylinder axis and τ is time, may be written as

$$\frac{\partial^2 T}{\partial r^2} + \frac{1}{r}\frac{\partial T}{\partial r} = \frac{1}{\alpha}\frac{\partial T}{\partial \tau} \quad \text{(a)}$$

with the boundary conditions (for $\tau > 0$)

$$k\frac{\partial T}{\partial r} = \varepsilon\sigma F\left(T_s^4 - T^4\right) - h(T - T_a), \quad \text{at } r = D/2 = R \quad \text{(b)}$$

$$\left(\rho CV\right)_i \frac{\partial T}{\partial \tau} = kA_i\frac{\partial T}{\partial r}, \quad \text{at } r = r_i \quad \text{(c)}$$

The initial temperature at $\tau = 0$ is simply taken at a uniform value of T_{init} for both the mandrel and the cord. Here, T_s is the temperature of a radiating source, F is a geometric factor, ε is the surface emissivity of the plastic, h is the convective heat transfer coefficient, T_a is the temperature of the fluid surrounding the cord, R is the radius of the core-mandrel assembly and subscript i refers to the mandrel and also the contact surface between the cord and the mandrel because of temperature uniformity in the mandrel. If $T_s \gg T$, the radiative transport may be replaced by a constant surface heat flux q_s. The convective heat transfer coefficient h is linked to the velocity of air U through correlations such as (Gebhart, 1971)

$$\text{Nu} = \frac{hD}{k_a} = 0.165\left[\frac{DU}{\nu}\right]^{0.466} \quad \text{(d)}$$

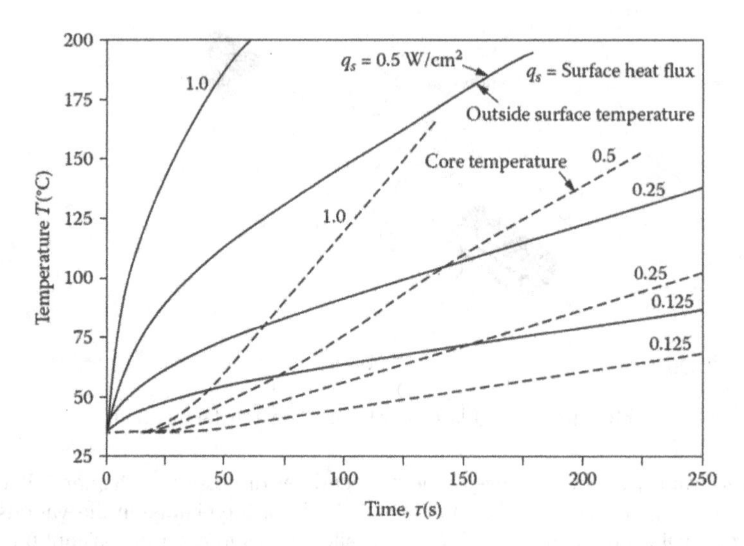

FIGURE 5.15 Transient temperature response to a constant heat flux input at the outer surface of the plastic cord.

where k_a is the thermal conductivity of the fluid and ν is its kinematic viscosity. Also, see Appendix D for various heat transfer correlations.

The preceding problem may be solved numerically to obtain the temperature variation in the plastic with time and location and that in the mandrel with time. Because the governing equation is *parabolic*, time marching may be used, starting with the initial conditions. The *Crank-Nicolson* method is appropriate because of the second-order accuracy in time and space and better stability characteristics as compared to the explicit *FTCS* method (Jaluria and Torrance, 2003). Sample computer programs, particularly those in Matlab, given in Appendix A may be used to develop the numerical model. Therefore, numerical results may be obtained for a variety of operating and design conditions. Based on these results, the desired acceptable design for the system may be obtained, as discussed earlier in terms of the design strategies, which are not based on an initial design.

Figure 5.15 shows the results in terms of the outer surface temperature and the inner surface or mandrel temperature for a constant heat flux input q_s at the surface, with no convection. Similarly, Figure 5.16 shows the corresponding results for convective heating, without radiative input. The

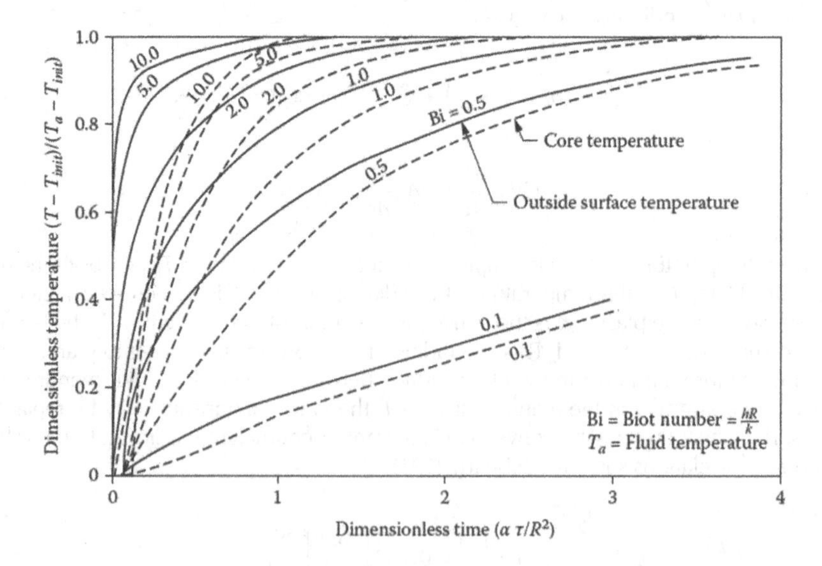

FIGURE 5.16 Temperature variation for convective heating.

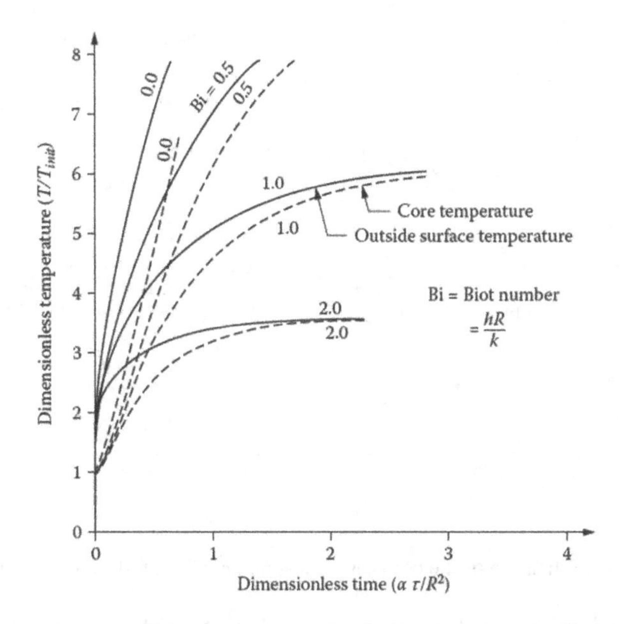

FIGURE 5.17 Temperature variation for combined convection and constant heat flux input q_s, with $q_sR/k = 5.0$.

results when both convection and radiation are present are shown in Figure 5.17. Different values of the heat flux and convection coefficient are taken in these calculations. The governing equation and the boundary conditions may also be nondimensionalized to generalize the problem and derive the governing dimensionless variables (Jaluria, 1976). The heat flux is nondimensionalized as q_sR/kT_{init} and the convective heat transfer coefficient as the Biot number *Bi*. In Figure 5.17, the dimensionless heat flux is kept at 5.0 and the Biot number is varied. Similarly, other numerical results may be obtained. It is clear from these results that a combination of radiation and convection gives the desired flexibility and control over the temperature levels.

Let us now consider the thermal system for this process. A continuous movement of the plastic cords, wound on the mandrel, in a wide channel with electric heaters and air flow driven by a fan may be designed, as shown in Figure 5.18. The mandrels are rotated to ensure uniform surface heating. The heaters are positioned over a chosen distance L_1, so that the plastic cords are heated up to this distance, and then cooled to room temperature over the remaining length. The time τ_1 in the heating region is L_1/V, where V is the speed at which the cords are traversed. The design variables are the fan, represented here in terms of U or h, the heater, represented in terms of the heat flux q_s, length of the heating region L_1, and length of the cooling region L_2.

The speed of the cords V, the ambient air temperature T_a, and the initial cord temperature T_{init} are operating conditions, with the design being obtained for chosen values of these. For off-design conditions, simulation can be used to determine the effect of these on the system performance. Both q_s and U may also be adjusted within the ranges available for the corresponding equipment. Clearly, the solution to this problem is not unique and many different design possibilities exist.

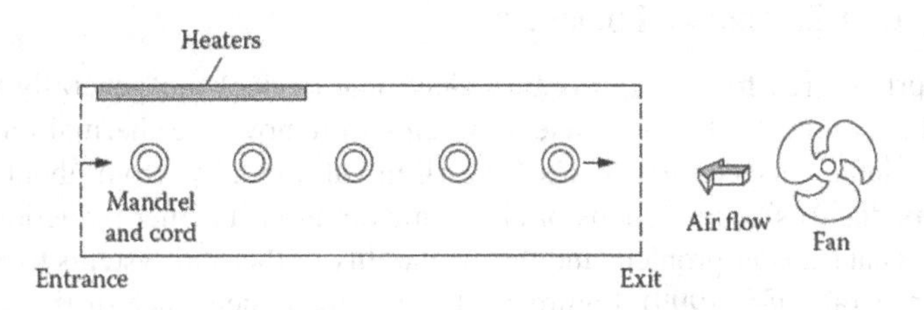

FIGURE 5.18 A possible conceptual design for the thermal system considered in Example 5.4.

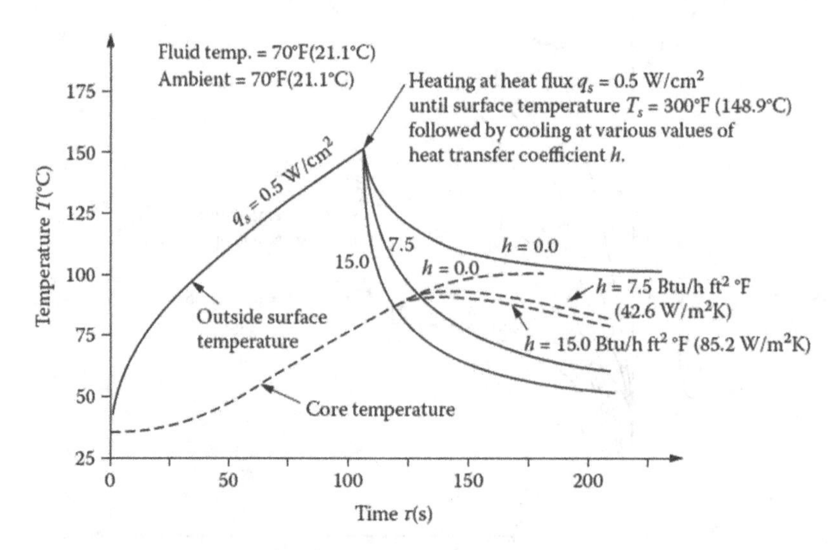

FIGURE 5.19 Results for heating of the cords at a constant heat flux, followed by convective cooling.

Figure 5.19 shows the circumstance when the cords are heated at constant heat flux in the heating zone and then cooled by convection. From these results different sets of design variables may be selected to satisfy the given requirements and constraints. For a chosen value of cord speed V, the lengths L_1 and L_2 may be determined for particular q_s and h. An acceptable design is thus obtained from this figure. For a mandrel traversing speed of 1 cm/s, a heating region length of 1.1 m and a cooling region length of 1.4 m are obtained from this figure if the maximum temperature is kept at 300°F (148.9°C) for safety. For higher maximum temperatures, the corresponding L_1 and L_2 may be determined. With these design variables, not every point in the cord reaches the required temperature for coiling, though most of the plastic does. For better and more uniform coiling, additional features may be needed.

Similarly, the heater and the fan may be varied for different designs. In this design problem, an initial design is not used. Instead, the modeling and simulation results are employed to guide us toward the appropriate acceptable design. In addition, this is obviously not the optimal design, for which a quantity such as cost or process time may be minimized.

The major problems encountered here are due to the low thermal conductivity of the plastic and the narrow temperature range in which the plastic must be maintained to avoid damage. This is typical of plastic thermoforming processes. The surface temperature easily reaches the maximum allowable value, while the inner surface is essentially unchanged. This suggests some changes in the system that may allow us to obtain greater temperature uniformity in the plastic. The mandrel may be made hollow to reduce the value of $(\rho CV)_i$ and thus diminish its effect on the inner surface temperature. It may also be made of a material whose thermal capacity, ρC, is less than that of stainless steel, such as molybdenum. Finally, a hollow mandrel may be used with flow of hot gases or with an electric heater located at the core in order to provide energy input at the inner surface of the plastic. Such changes would make the temperature distribution in the plastic cord more uniform than that for the earlier design.

5.4.2 Cooling of Electronic Equipment

This is an important area for design because electronic devices are generally very temperature-sensitive and it is crucial to design efficient systems to remove the thermal energy dissipated in electronic equipment. Surface heat fluxes have risen substantially, from about 10^2 to 10^6 W/m², over recent years due to size reductions of electronic circuitry. Further reduction in size is largely restricted by the heat transfer problem and the availability of thermal systems to effectively cool the equipment (Incropera, 1988, 1999). Figure 5.20 shows the dependence of the difference between the surface temperature of the electronic device and the ambient as a function of the input heat flux.

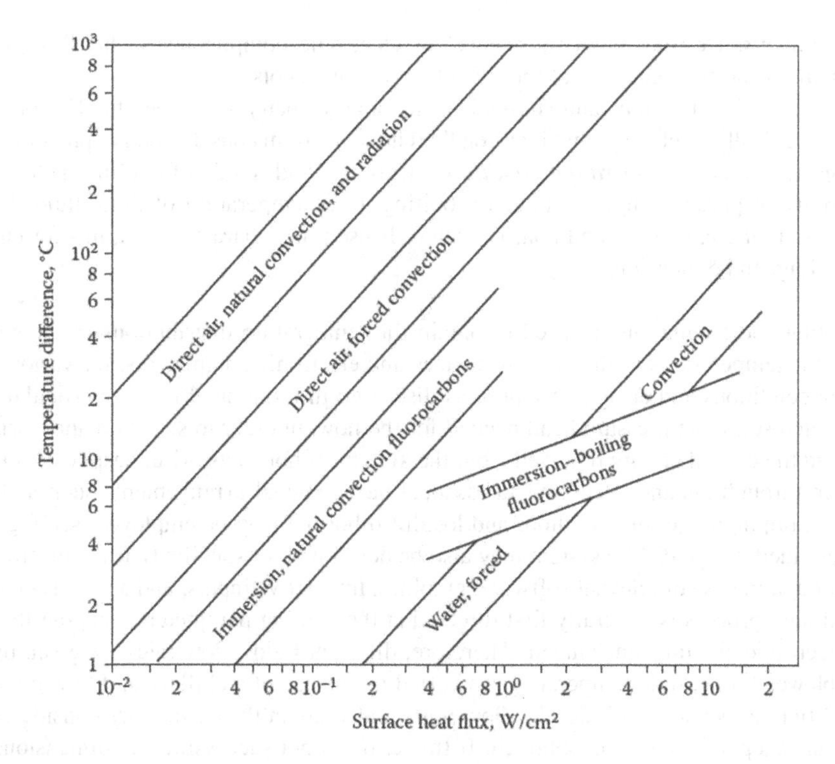

FIGURE 5.20 Temperature differences obtained in the cooling of electronic equipment for different modes of heat transfer. (Adapted from Kraus and Bar-Cohen, 1983.)

Various modes of heat transfer for removal of the dissipated energy are also indicated, with natural convection cooling in air applicable at very low heat flux levels and liquid cooling with boiling at very high levels. Cooling of electronic systems involves components and subsystems of various length scales, such as the chip, circuit board, server, rack, and data center. Different concerns arise for cooling at these different levels and a variety of cooling techniques are employed.

Though it is not possible to discuss all the different types of electronic systems and cooling methods employed in practice, the main characteristics of these systems are

1. Temperature-sensitive performance of circuitry, leading to tight temperature constraints
2. Strong dependence on geometry
3. Three-dimensional transport
4. Conjugate transport due to coupling between conduction in the solid and convection in the cooling fluid
5. Radiation heat transfer, which is often substantial in air cooling
6. Fluid must be electrically insulating if brought into direct contact with circuitry

Steady-state problems are usually of interest, though transient effects may be important at startup and shutdown. Other characteristics that arise for particular applications are discussed in specialized books in this area such as Steinberg (1991), Kraus and Bar-Cohen (1983), Seraphin et al. (1989), and Incropera (1999).

In the design of thermal systems for the cooling of electronic equipment, such as those sketched in Figure 1.12, typical inputs, requirements, constraints, and design variables are as follows:

Given quantities: Energy dissipated per component or thermal energy to be removed, number of components, basic geometry, and configuration of the circuitry

Requirements: Desired temperature level of electronic components such as chips. There should be no concentration of thermal effects, or hot spots.

Constraints: Material temperature limitations, size and geometry limitations, fluid in contact must be electrically insulating, limitations on fluid flow rate from consideration of pressure needed

Design variables and operating conditions: Cooling fluid, mode of cooling including possibility of phase change, particularly boiling, inlet temperature of fluid, fluid flow rate, location of components and boards, materials used, fan characteristics, fins for enhanced cooling, and dimensions

Modeling and simulation are used to obtain the temperature distributions in the system, particularly the temperatures in the various devices and electronic components, for various ranges of operating conditions and design variables, as discussed in Example 3.7 for a particular geometry. Also of interest are the pressure head needed for the flow, in order to select an appropriate blower or fan, and the overall energy removed from the system. If hot spots arise, despite efforts to eliminate them through enhanced cooling rates, local heat removal arrangements such as heat pipes, heat sinks, impinging jet of cold fluid, and localized boiling may be employed (see Figure 1.2). A computer-aided design (CAD) system may also be developed for specific types of electronic equipment, through the use of relevant software, graphics, interactive inputs, and appropriate databases.

The design process is generally first directed at the cooling parameters, keeping the geometry of the electronic circuitry unchanged. Therefore, different fluids, flow rates (as given by different fans or blowers), inlet fluid temperature (as varied by the use of a chiller), and flow configurations (due to different locations of inflow/outflow ports and vents in the casing) are considered to determine if an acceptable design is obtained. If this effort is not successful, the dimensions, number, and locations of the boards and the components may have to be varied within the given constraints. If even this does not lead to an acceptable design, the mode of cooling may be varied, for instance, going from natural to forced convection in air, to liquid immersion, or to boiling. Hydrofluorocarbon and fluorocarbon coolants are commonly used for immersion cooling and for boiling. The typical convective heat transfer coefficients for different modes and fluids are (in W/($m^2 \cdot K$))

Natural convection, air	5–10
Forced convection, air	10–50
Immersion cooling, liquids, natural convection	100–200
Immersion cooling, liquids, forced convection	200–500
Boiling, common liquids	1,000–2,000
Boiling, liquid nitrogen	5,000–10,000

Therefore, a considerable amount of control is obtained by varying the fluid and the mode of heat transfer. If adequate cooling is still not obtained, as indicated by the presence of hot spots and excessive temperatures in electronic components such as devices and chips, techniques to enhance local heat transfer rates and modifications in the design of the boards and the circuitry may be undertaken. Obviously, the latter approach involves a strong interaction with the designers of the electronic circuit for the given application. The design of the cooling system, therefore, may be directed at problems that arise at the component, board, server, rack or data center level. The thermal system to be designed is strongly influenced by the basic problem to be solved, the geometry, and the heat flux levels. A relatively simple design problem is shown in the following example to illustrate some of the basic considerations involved.

Example 5.5

Consider the forced convective cooling of the electronic system shown in Figure 5.21. Air is the cooling fluid and the vertical printed circuit boards contain electronic components, each of which dissipates 200 W. The height, width, and thickness of the board are given as 0.1, 0.1, and 0.02 m,

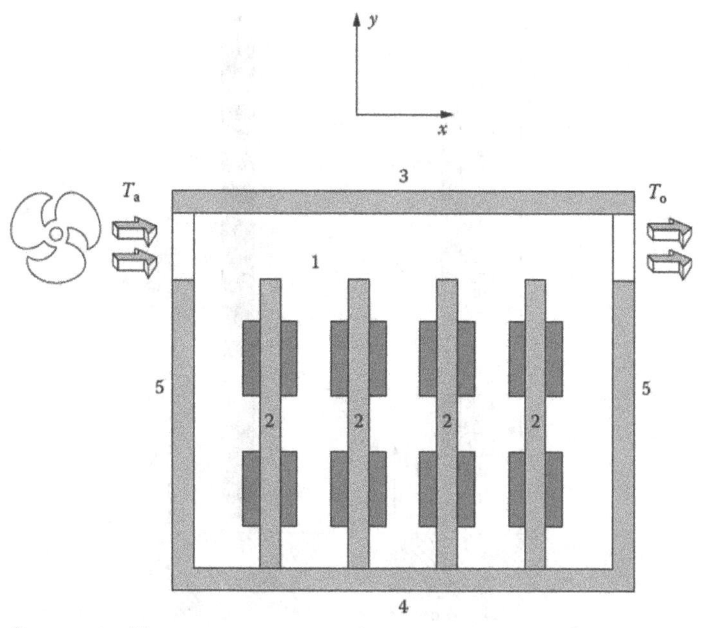

FIGURE 5.21 Physical arrangement of an electronic system being cooled by forced convection with air as the fluid.

respectively. Either copper or aluminum may be considered as representative of the material for the boards. The temperature of the components must not exceed 100°C, even if the air temperature in the enclosure containing the boards rises to a value as high as 55°C. Develop a mathematical model for this problem, assuming the convective heat transfer coefficient to be 20 W/(m²·K), and design the system to accommodate a given number of electronic components.

SOLUTION

By employing a given value of the convective heat transfer coefficient, the problem is considerably simplified because the fluid flow in the enclosure need not be determined. However, in general, the conjugate problem of conduction in the boards coupled with convective transport in the fluid has to be solved, requiring the solution of coupled nonlinear PDEs, as considered in Example 3.7. In the present case, a simple mathematical model is derived to determine the temperature distribution in the board.

The thermal conductivity k for copper is found from the literature as 391 W/(m·K) and that for aluminum as 226 W/(m·K). Because of the small thickness of the board, the Biot number Bi is found to be very small. For aluminum, $\mathrm{Bi} = 20 \times (0.02/2)/226 = 8.8 \times 10^{-4}$, where 0.02/2 m is the half-thickness of the board. Therefore, uniform temperature may be assumed across the board thickness. Because the width is much larger than the thickness, and because conditions do not vary in this direction, uniform temperature may also be assumed in the transverse direction, reducing the problem to that of a vertical extended surface, as shown in Figure 5.22. The bottom of this fin, or extended surface, is in contact with the base of the equipment, which is assumed to be at room temperature. One final approximation is made to complete the model. The total heat dissipated by the electronic components is assumed to be uniformly distributed over the total volume of the board, giving rise to a volumetric energy source q''' in W/m³ in the board, as shown.

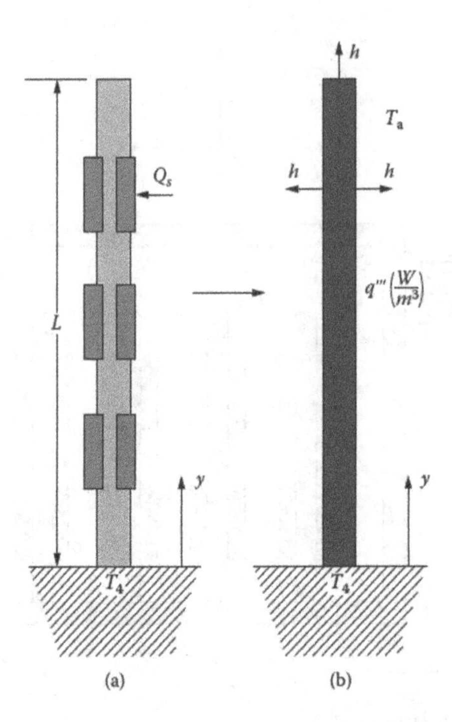

FIGURE 5.22 Model of an electric circuit board, as considered in Example 5.5.

The governing equation for this simplified conduction problem is

$$\frac{d^2T}{dy^2} - \left(\frac{hP}{kA}\right)(T - T_a) + \frac{q'''}{k} = 0 \quad \text{(a)}$$

where y is the vertical coordinate distance, as shown in Figure 5.22, P is the perimeter of the board, A is its cross-sectional area, and T_a is the average air temperature surrounding the board. Here, $A = wt$ and $P = 2(w + t)$, where w and t are the width and thickness of the printed circuit board, respectively. The volumetric heat generation q''' is given by

$$q''' = \frac{Q}{V} = \frac{Q}{Lwt} \quad \text{(b)}$$

where V is the volume of the board, L is its length, and Q is the total power dissipated in each board. If this dissipated power is due to a number of similar electronic components, n, each dissipating heat at the rate of Q_s, then $Q = nQ_s$, where Q_s is given as 200 W. The governing equation may also be written more conveniently as

$$\frac{d^2\theta}{dY^2} - \left(\frac{hP}{kA}\right)L^2\theta + \frac{q'''L^2}{k} = 0 \quad \text{(c)}$$

where $\theta = T - T_a$ and $Y = y/L$. The imposed boundary conditions are

$$\text{at } Y = 0\text{: } \theta = 0; \quad \text{and} \quad \text{at } Y = 1\text{: } \frac{d\theta}{dY} = -\frac{hL}{k}\theta \quad \text{(d)}$$

Therefore, the mathematical model yields a second-order ordinary differential equation, which may be solved analytically or numerically to obtain the temperature distribution in the board for a variety of operating and design conditions. Numerical modeling with

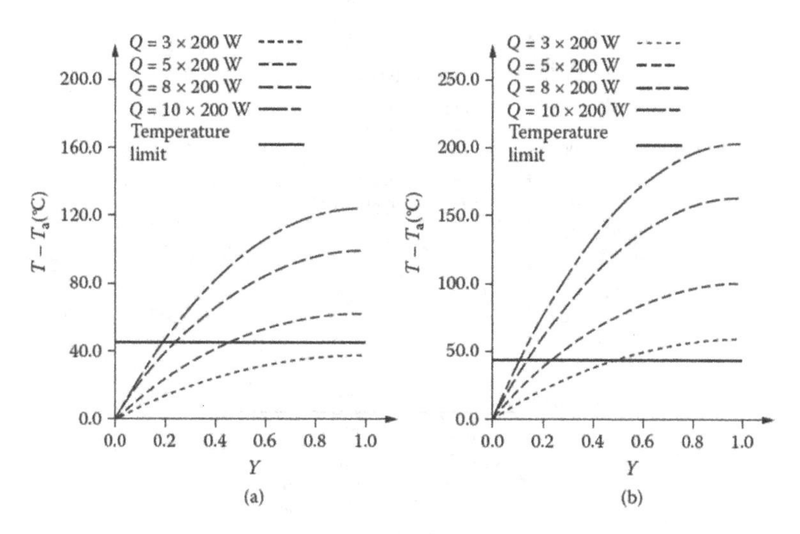

FIGURE 5.23 Temperature distribution for (a) copper and (b) aluminum board with different numbers of heat dissipating electronic components.

the Runge-Kutta method is particularly appropriate because of the high level of accuracy and versatility obtained by this method. A shooting scheme is needed for this boundary value problem to satisfy the conditions at $Y = 0$ and 1, as outlined in Chapter 4. Matlab may be employed, using an ODE solver such as *ode45*, as discussed earlier. A root solving method may be used for iterating the solution to satisfy the condition at $Y = 1$. The programs given in Appendix A may be easily modified to obtain the desired converged solution. The results obtained with a Newton-Raphson correction scheme applied to the fourth-order Runge-Kutta method are presented here.

Figure 5.23 shows the calculated temperature distributions in the boards for the two materials, considering different numbers of electronic components located on the board. The temperature limit, which is taken as $\theta = 45°C$ to reflect the worst condition with the largest air temperature, is also shown. It is seen that only three components located on a copper board yield a maximum temperature below the limiting value. This information leads to the selection of copper as the appropriate material for the board to obtain an acceptable design. The limitation on the number of components is also demonstrated by these results. For a larger number of components per board, other design variations are necessary. The effect of varying the width of the board for five components is shown in Figure 5.24. If the width can be increased to 0.3 m, five components can be accommodated without violating the temperature limit. Similarly, increase in the thickness and height of the board leads to reduction in temperature levels, allowing additional components to be located per board. In addition, higher convection heat transfer rates would be needed if higher component densities were desired.

Clearly, the mathematical model for this problem is a relatively simple one and the complexity due to coupling with convective flow has been avoided. Inclusion of the effect of geometry, turbulent flow, localized heat dissipation, and several other considerations that are important in practical systems requires a much more elaborate model (see Example 3.7). However, the results obtained from such an accurate model would again indicate the maximum temperatures encountered and the maximum number of components that may be located on a board without exceeding the temperature limit. These results are then employed to obtain an acceptable design that meets the given requirements and constraints.

5.4.3 Environmental Systems

Thermal systems involved in environmental problems have grown considerably in interest and importance in the recent years because of increasing concern with the environment and the need to design efficient systems for the disposal of rejected energy, chemical pollutants, and solid waste. Of

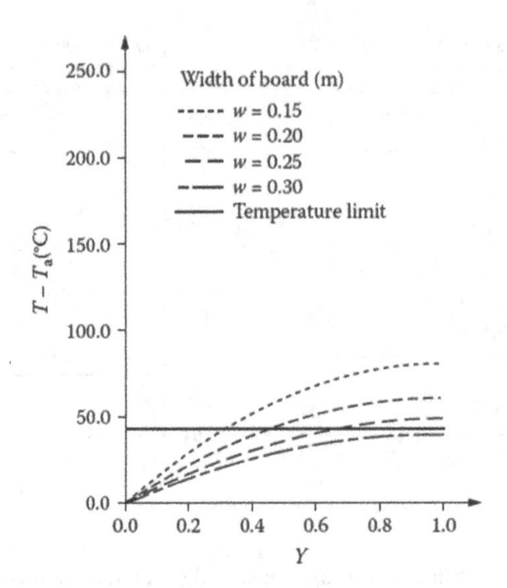

FIGURE 5.24 Effect of the variation of the board width w on the temperature distribution for a copper board with five electronic components.

particular concern is the discharge of thermal energy and chemicals into water bodies such as lakes and into the atmosphere. The decay and spread of the discharge determine the effect on the local environment as well as on a global scale. Sketches of the flows generated by such discharges into the environment are shown in Figure 5.25 (see also Figure 1.14).

Environmental processes are generally quite complex due to the strong dependence of the transport mechanisms on fluid flow. Some of the important characteristics of these processes are

1. Time-dependent, often periodic, phenomena
2. Generally turbulent flow
3. Combined modes of heat transfer, including phase change
4. Combined heat and mass transfer

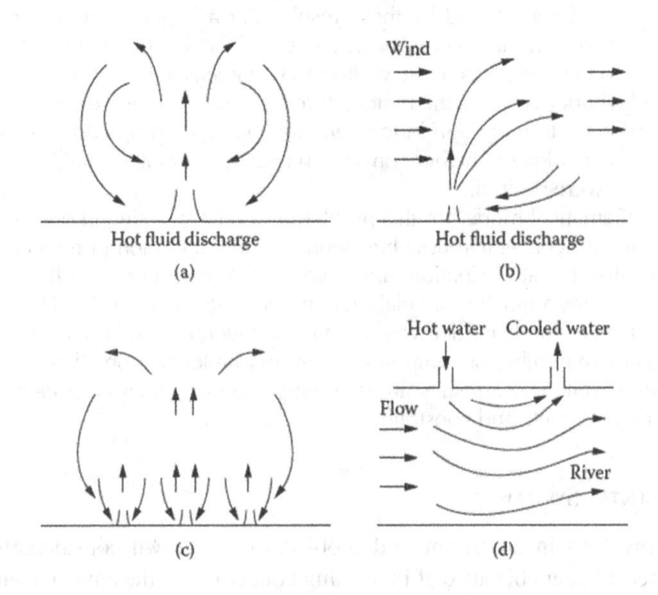

FIGURE 5.25 Flows generated in the ambient medium due to heat rejection.

5. Chemical reactions in some cases
6. Dependence on location, topology, and local ambient conditions
7. Stable thermal stratification of the air or water environment

The last one refers to colder or denser fluid lying below warmer or lighter fluid, impeding convective flow. All these aspects tend to complicate the modeling of environmental processes. Therefore, the design of the relevant thermal systems is quite involved.

The design problem varies from one application to the next. Let us first consider thermal or mass discharges to human-made ponds, lakes, cooling towers, etc., as shown in Figure 1.14. The design problem may then be formulated as

Given quantities: Total rate of energy or mass discharged, and geographical location, which fixes the average and time-dependent values of the local solar flux, wind speed, relative humidity, cloud cover, and ambient temperature.

Requirements: Temperature or concentration levels must not exceed specific values at the outfall or discharge into the water body or at a particular distance from it. Such requirements often arise from governmental regulations.

Constraints: Limits on maximum flow rate, maximum size of cooling pond, cooling tower, etc.

Design variables: Location of outfall or discharge, location of intake of water for a power plant or industrial unit, dimensions of inflow/outflow channels, hardware for varying the flow rate, temperature or concentration at outfall.

A similar formulation for the design problem in other environmental applications, such as solid waste disposal, may be obtained in terms of given quantities, requirements, constraints, and design variables. For instance, incineration as a means to dispose of solid waste involves a combustion furnace in which the waste material is burned at relatively high temperatures to avoid undesirable combustion products. The system design involves designing the furnace with the given requirements and constraints on temperatures, flow rates, and energy input/output.

The heat transfer from a cooling pond such as a lake involves

1. Solar flux absorbed in the pond
2. Heat loss due to evaporation
3. Heat transfer to the air due to convection
4. Radiative transport to the environment
5. Energy transfer at the bottom and sides

All these transport mechanisms are fairly involved and simplifications are generally used to estimate the resulting heat and mass transfer. The solar flux is assumed to be absorbed largely at the surface, heat losses at the bottom and sides are often neglected for deep lakes with large surface area, and so on. The resulting transport rates depend on the wind speed, relative humidity, cloud cover (varying from 0 for a clear sky to 1 for an overcast sky), location, time of day and year, and local topology. However, all the transport rates may be combined into a simple expression such as

$$q = h(T_s - T_e) \tag{5.13}$$

where h is an overall heat transfer coefficient; q is the total heat transfer rate at the surface, including evaporation; T_s is the surface temperature; and T_e is known as the *equilibrium temperature*, being the temperature that the surface must attain to make the heat transfer rate q become zero. This temperature T_e can often be represented as a sinusoidal variation, with appropriate values given for h over different seasons such as winter and summer (Moore and Jaluria, 1972). Because the basic process is periodic, the integral of the heat transfer over 365 days of the year is zero, i.e., for time τ in days,

$$\int_0^{365} q(\tau)\,d\tau = 0 \tag{5.14}$$

Therefore, a natural lake or pond may be modeled to compute the temperature distribution over the year. If thermal energy is discharged into the water body, its temperature must rise to get rid of the additional energy. In addition, the recirculating flow set up in the water body may result in a temperature increase at the intake for a power plant. Such a temperature rise increases the temperature for heat rejection and thus decreases the efficiency of the power plant. Let us consider an example of such a thermal system.

Example 5.6

A shallow pond of length 100 m, width 20 m, and depth 4 m is to be used to reject thermal energy from an industrial facility. The equilibrium temperature T_e of the pond is 25°C and the overall convective heat transfer coefficient h is given as 50 W/(m^2·K). This includes the effects of all the surface energy loss mechanisms. The difference in temperatures between the intake and the discharge is given as 10°C and the intake temperature must not rise beyond 2.5°C due to heat rejection, as limited by government environmental regulations. The turbulent transport may be modeled as an enhanced diffusive process with the eddy diffusivity and viscosity taken as 10^{-5} m^2/s over the flow region. Heat loss to the ground at the bottom may be neglected. Design a thermal system to reject 400 kW to the pond. How would this design change if higher energy levels were to be rejected to the pond?

SOLUTION

The given quantities are the pond dimensions, the total amount of heat rejected Q, the temperature difference ΔT between the intake and outfall, and the surface heat transfer parameters h and T_e that characterize the local ambient conditions. The requirement is that the temperature rise at the intake must not exceed 2.5°C. The main constraint is that the energy rejected to the pond must be rejected to the environment for a steady-state circumstance.

This is a fairly typical problem encountered in heat rejection to water bodies. The only design variables are the locations and dimensions of the intake and outfall channels. The dimensions of these channels will determine the flow velocities. In practice, limitations are generally imposed on the discharge velocity. Because the pond is given as shallow, with the depth H much less than the length L and width W, uniform conditions over the depth may be assumed. Then heat transfer from the pond occurs only at the surface and the total thermal energy rejected to the pond must be lost to the environment at the surface for steady-state conditions, which may be assumed to apply here.

Let us first consider a very simple one-dimensional model with uniformity assumed over the pond width as well, as shown in Figure 5.26. Then the total rate of heat rejected Q is given as

$$Q = \rho C_p u H W \Delta T = \int_0^L hW(T - T_e)\,dx \tag{a}$$

where u is the average discharge velocity in the x direction and $T(x)$ is the temperature distribution in the pond. The governing equation for $T(x)$ is

$$u\frac{dT}{dx} = \frac{d}{dx}\left(\varepsilon_h \frac{dT}{dx}\right) - \frac{h(T - T_e)}{\rho C_p H} \tag{b}$$

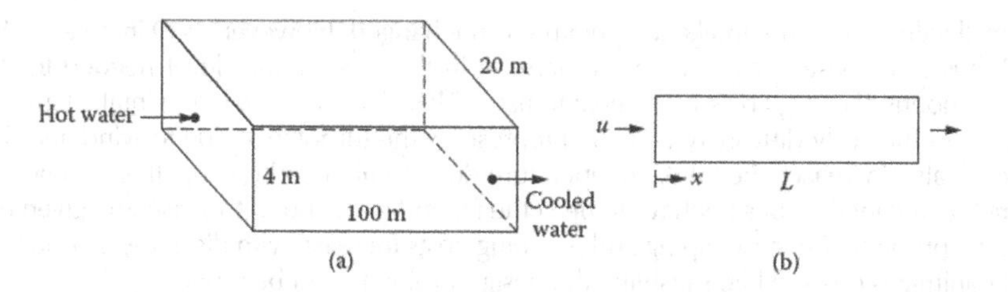

FIGURE 5.26 (a) Three-dimensional problem of heat rejection to a body of water, along with (b) a simplified one-dimensional model.

where ε_h is the eddy thermal diffusivity. The boundary conditions are

$$\text{at } x = 0: \ T = T_o = T_L + \Delta T \quad \text{and} \quad \text{at } x = L: \frac{dT}{dx} = 0 \tag{c}$$

where T_o is the temperature at the outfall, $x = 0$, and T_L is the temperature at the intake, $x = L$. It is assumed that there is no heat loss beyond $x = L$, giving the zero gradient condition.

This problem may be conveniently solved numerically by finite-difference methods, starting with T_L taken as T_e and T_o as $T_e + \Delta T$. The temperature distribution over the pond surface is calculated. If T_L increases above T_e, the new values of T_L and T_o are employed and the temperature distribution recalculated. This iterative process is carried out until the temperature distribution does not vary significantly from one iteration to the next. A typical convergence criterion would be $|T_L^{(n+1)} - T_L^{(n)}| \leq \tilde{\varepsilon}$ where the superscripts indicate the iteration number, and $\tilde{\varepsilon}$ is a chosen small quantity. Figure 5.27 shows the computed results for different values of the total rate of energy rejected to the pond. It is clearly seen that the temperature rise at the intake is less than the allowable value of 2.5°C for $Q = 400$ kW. Therefore, this is an acceptable design. In fact, the temperature rise at the intake is within the given limit even for $Q = 600$ kW. For still higher values of Q, the given requirements cannot be met and an additional heat rejection system, such as a cooling tower, will be needed.

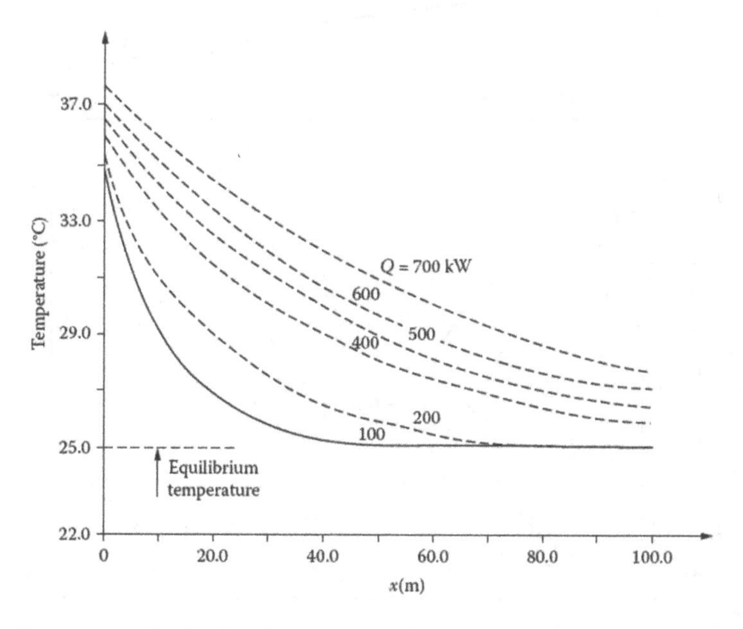

FIGURE 5.27 Temperature distribution from the one-dimensional slug flow model for different values of energy rejected Q, for Example 5.6.

For small values of Q, the intake temperature is unchanged. However, as Q increases, the flow rate and thus u increases, resulting in an increase in the temperature level needed to lose the increased amount of energy by surface heat transfer. This gives rise to a larger intake temperature. An increase in the eddy diffusivity ε_h, which represents the turbulence due to wind and the flow in the pond, also increases the intake temperature due to enhanced mixing. It is also noted from this figure that the intake does not have to be at the far end of the pond to satisfy the given requirements of the problem. Because piping and pumping costs increase with distance, an optimal solution that minimizes costs, while satisfying the design problem, can be found.

The model used is an extremely simple one, but it allows us to determine the location of the intake to restrict the temperature rise there. Eddy viscosity and diffusivity are dependent on the flow field and are, thus, not constants but vary with location. Information available in the literature may be used to represent the turbulent transport more accurately. Other turbulence closure models may also be used for higher accuracy (Shames, 2002). Two-dimensional models have the advantage of considering different locations of the intake and outfall over the surface of the pond and of varying the channel widths. For instance, two flow configurations are shown in Figure 5.28, along with typical flow results in terms of streamlines given for different values of the stream function Ψ. Uniformity is again assumed in the third direction and the governing convective transport equations are solved to obtain the temperature distribution over the pond surface. From such results, the temperature at the intake is determined for different flow configurations and intake/outfall channel dimensions. The intake temperature rises due to the flow recirculation, as well as due to the increased energy input into the pond. Again, acceptable designs may be obtained that satisfy the given problem statement.

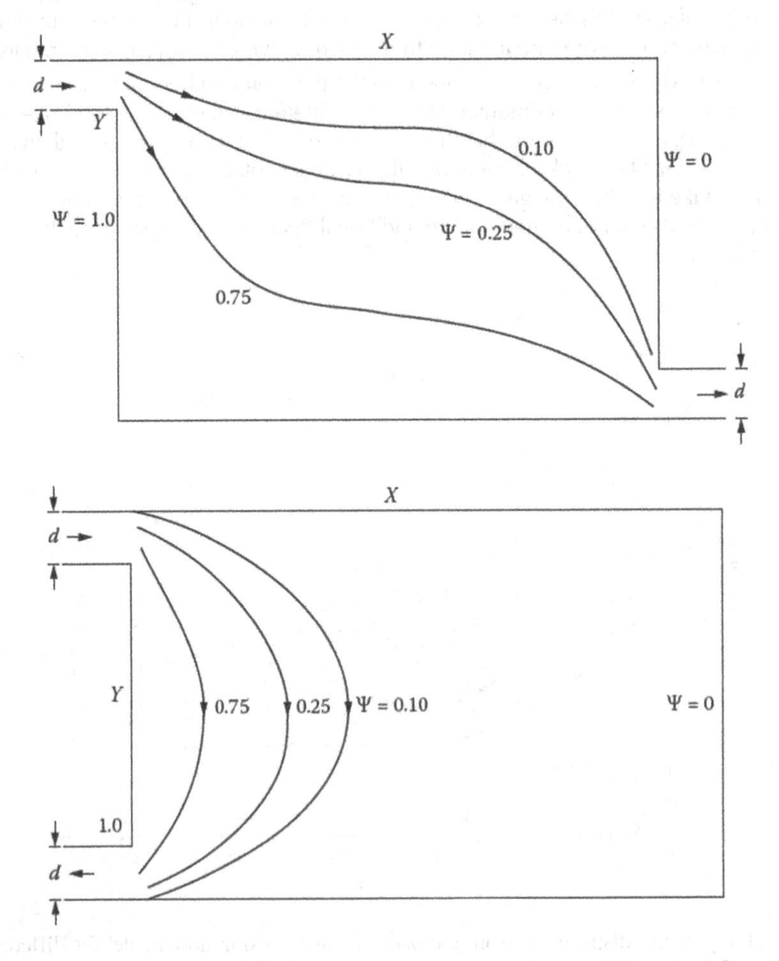

FIGURE 5.28 Two-dimensional surface flow due to heat rejection for two different configurations.

In actual practice, the problem of heat rejection to water bodies is quite complicated because of three-dimensional, turbulent, flows and combined heat transfer modes operating at the surface. Frequently, commercially available computer software is used to simulate the flow to obtain the temperature or concentration distributions. Some of the basic considerations involved in such problems are indicated by this simple example. This problem is particularly suitable for optimization in order to minimize the costs involved in pumping by keeping the intake and outfall as close as possible and with minimum piping lengths without violating the intake temperature requirements.

5.4.4 Heat Transfer Equipment

This is a particularly important topic in the design of thermal systems because heat transfer equipment, which includes heat exchangers, condensers, boilers, ovens, and evaporators, is used extensively in a wide variety of applications ranging from heating and cooling of buildings and automobiles to manufacturing processes. These items may be considered for design as subsystems or as systems that arise in specific applications. They may also be designed as general hardware that may be employed as components in a variety of systems. Once designed and manufactured, the components are available as ready-made items for given sets of specifications. This is particularly true of heat exchangers that are often designed and fabricated as separate items that may be incorporated into the overall design of the thermal system as components. Let us, therefore, consider the design of heat exchangers and outline some of the main concerns that arise.

5.4.4.1 Modeling and Simulation

The analysis of heat transfer processes in heat exchangers is given in most heat transfer textbooks, such as those by Lienhard and Lienhard (2011), Incropera and Dewitt (1990, 2001), and Bejan (1993). A few simple results are discussed here for the design of systems such as those shown in Figure 1.5. For a counter-flow heat exchanger, a simple mathematical model may be developed, as discussed in Chapter 3, by employing the following simplifications and assumptions:

1. Steady flow conditions
2. Uniform velocity distribution in pipes, i.e., slug flow
3. Temperature variation only in axial, x, direction; taken as lumped in other directions
4. Negligible conduction in the axial direction
5. Constant properties
6. Overall heat transfer coefficient U constant over heat transfer surface
7. Negligible energy loss to the environment

With these assumptions, the energy balance for a differential element in a counter-flow heat exchanger shown in Figure 2.19 may be written as

$$dQ = \dot{m}_1 C_{p1} dT_1 = \dot{m}_2 C_{p2} dT_2 = U\ \Delta T\ dA \tag{5.15}$$

where dQ is the rate of energy transfer across surface area dA between the two fluids, with dT_1 and dT_2 representing the corresponding temperature changes, with distance from one end, and ΔT is the local temperature difference $T_1 - T_2$. If these equations are integrated over the total heat transfer surface area A, we obtain the total heat transfer rate Q as

$$Q = \dot{m}_1 C_{p1}\left(T_{1,i} - T_{1,o}\right) = \dot{m}_2 C_{p2}\left(T_{2,o} - T_{2,i}\right) \tag{5.16a}$$

$$Q = UA\,\Delta T_m = UA\frac{\left(T_{1,i} - T_{2,o}\right) - \left(T_{1,o} - T_{2,i}\right)}{\ln\left[\left(T_{1,i} - T_{2,o}\right)/\left(T_{1,o} - T_{2,i}\right)\right]} \tag{5.16b}$$

Here, U is the overall heat transfer coefficient, and the remaining quantities are the same as those defined with respect to Figure 2.19. If Q is positive, fluid 1 is hotter than fluid 2; otherwise, fluid 2 is hotter. From these equations,

$$T_{1,o} = T_{1,i} - (T_{1,i} - T_{2,i}) \frac{1 - e^S}{(\dot{m}_1 C_{p1} / \dot{m}_2 C_{p2}) - e^S} \tag{5.17}$$

where

$$S = UA\left(\frac{1}{\dot{m}_1 C_{p1}} - \frac{1}{\dot{m}_2 C_{p2}}\right) \tag{5.18}$$

Therefore, the outlet temperature of a given fluid from the heat exchanger may be determined if the two entering temperatures are given. Similarly, the outlet temperature of the other fluid may be obtained by interchanging subscripts 1 and 2. For the special case of $\dot{m}_1 C_{p1} = \dot{m}_2 C_{p2} = \dot{m} C_p$, it can be shown that the temperature difference remains constant at $T_{1,o} - T_{2,i}$ and the outlet temperature of fluid 1 is

$$T_{1,o} = T_{1,i} - \frac{(T_{1,i} - T_{2,i})}{(\dot{m} C_p / UA) + 1} \tag{5.19}$$

The temperature difference ΔT_m in Equation (5.16b) is known as the *logarithmic mean temperature difference* (LMTD). For a parallel-flow heat exchanger, shown in Figure 1.5, this mean temperature difference is obtained as

$$\Delta T_m = \frac{(T_{1,i} - T_{2,i}) - (T_{1,o} - T_{2,o})}{\ln[(T_{1,i} - T_{2,i}) / (T_{1,o} - T_{2,o})]} \tag{5.20}$$

Therefore, the total heat transfer rate may be determined if all the temperatures are given. In addition, the outlet temperature of a particular fluid may be computed as just given for a counter-flow heat exchanger.

Another approach for the analysis of heat exchangers is based on the effectiveness ε, defined as

$$\varepsilon = \frac{Q}{Q_{max}} = \frac{Q}{(\dot{m} C_p)_{min} (T_{hot,in} - T_{cold,in})} \tag{5.21}$$

where Q_{max} is the maximum possible rate of heat transfer with the same inlet temperatures, fluids, and flow rates. It can be shown that Q_{max} is obtained when the fluid with smaller $\dot{m} C_p$ denoted here as $(\dot{m} C_p)_{min}$ goes through the maximum possible temperature difference. If fluid 1 is the fluid with smaller $\dot{m} C_p$, the effectiveness ε of a counter-flow heat exchanger can be derived from the energy balance equations to yield

$$\varepsilon = \frac{1 - e^S}{(M_{min} / M_2) - e^S} \quad \text{where } S = \frac{UA}{M_{min}}\left(1 - \frac{M_{min}}{M_2}\right) \tag{5.22}$$

Here, $M = \dot{m} C_p$ and UA/M_{min} is known as the *number of transfer units* (NTU). The dependence of the effectiveness ε on M_{min}/M_2, or M_{min}/M_{max}, and on the NTU has been studied for a variety

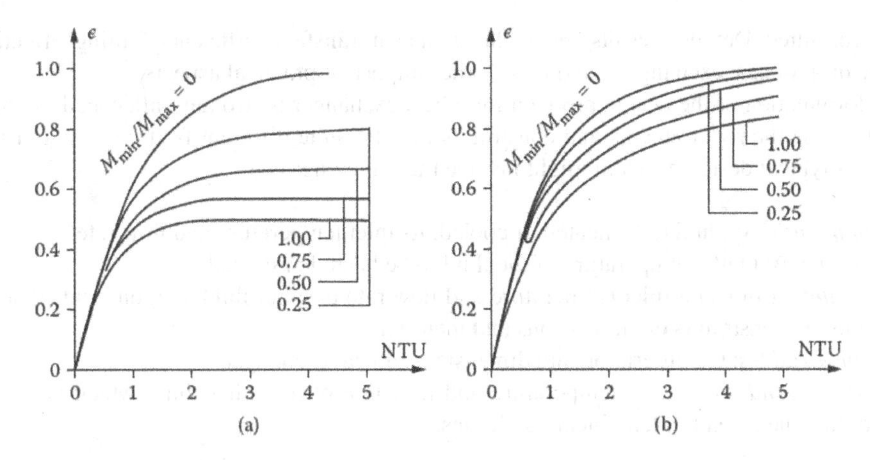

FIGURE 5.29 Effectiveness ε of (a) parallel-flow and (b) counter-flow heat exchangers in terms of the NTU and M_{min}/M_{max}. (Adapted from Incropera and Dewitt, 2001.)

of heat exchangers (Kakac et al., 1983; Kays and London, 1984). Figure 5.29 shows a few typical results presented in graphical form. Therefore, for given conditions, the NTU and M_{min}/M_{max} may be calculated and the appropriate charts or equations used to determine the effectiveness ε. The actual heat transfer is then obtained from Equation (5.21).

The overall heat transfer coefficient U is determined largely from heat transfer correlations for flow in channels and pipes. A commonly used correlation is the Dittus-Boelter equation, which gives

$$Nu = \frac{hD}{k_f} = 0.023(Re)^{0.8}(Pr)^n \quad \text{with } Re = \frac{VD}{\nu} \text{ and } Pr = \frac{\nu}{\alpha} \tag{5.23}$$

Here, Nu is the average Nusselt number, Re is the Reynolds number, Pr is the Prandtl number, D is the diameter, and n is 0.4 if the fluid is being heated and 0.3 if it is being cooled. The hydraulic diameter D_h, which is four times the flow cross-sectional area divided by the wetted perimeter, is used for annular regions, giving $D_h = D_o - D_i$, with D_o and D_i representing the outer and inner diameters, respectively. The various thermal resistances in the heat exchanger are added to obtain the overall thermal resistance and thus the heat transfer coefficient U. For a tubular heat exchanger, neglecting conductive resistances, this yields

$$U = \frac{1}{(1/hi)+(1/h_o)} \tag{5.24}$$

where h_i and h_o are convective heat transfer coefficients for the inner and outer fluids, respectively. Conductive resistances, if significant, may also be included. Fouling of heat exchangers leads to deposits on the surfaces and thus to a higher conductive resistance. This effect may be included as a fouling factor, which gives the additional resistance due to fouling for different fluids and operating conditions.

5.4.4.2 Design Problem

The foregoing discussion presented some of the salient points in the modeling and analysis of heat exchangers. Because of the importance of heat exchangers in engineering systems, extensive work has been done on different configurations, fluids, applications, and operating conditions. For further details on the results available in the literature, the references given earlier, along with several others concerned with thermal systems such as Boehm (1987), Stoecker (1989), and Janna (2014), may be consulted. Detailed results are available on heat transfer coefficients, fouling, effectiveness, pressure drop in heat exchangers, and many other important practical aspects.

The formulation of the design problem for a heat exchanger is strongly influenced by the application because the requirements and constraints may be quite different from one circumstance to another. A typical design problem might involve the following:

Given quantities: Fluid to be heated or cooled, its inlet temperature and flow rate.
Requirements: Outlet temperature of the fluid to be heated or cooled.
Constraints: Limits on inlet temperature and flow rate of other fluid, and on fluids that may be used. Constraints on dimensions and materials.
Design variables: Configuration and dimensions of heat exchanger.
Operating conditions: Inlet temperature and flow rate of the other fluid, which is heating/cooling the given fluid, ambient conditions.

Then an appropriate type of heat exchanger is selected and its size determined to obtain the surface area A that would lead to the required outlet temperature. The dimensions and materials are chosen with the given constraints in mind. Frequently, both fluids are given; otherwise, an appropriate fluid may be chosen for superior heat transfer characteristics, low fouling, low cost, lower viscosity that results in smaller pressure needed for the flow, and easy availability. Either of the two approaches given here, LMTD and NTU methods, may be used. The design problem may also require a particular heat transfer rate for specified fluids, flow rates, and inlet temperatures. Again, the equations given earlier and charts available in the literature may be used to design the system.

Generally, in the design of heat exchangers, the outer diameter is constrained due to size limitations. The inner tube diameter and the length are then the main design variables for parallel-flow and counter-flow heat exchangers. The flow rates are used to compute the Reynolds number, which is used to determine the convective heat transfer coefficient. The overall heat transfer coefficient U is then obtained by including the conductive resistances and fouling factors. From the calculated value of U, the heat transfer rate and the outlet temperatures are determined using the energy balance equations, given previously. Several problems are given at the end of this chapter to illustrate the modeling, simulation, and design of heat exchangers.

The model for a heat exchanger given here is fairly simple. Many of the approximations, such as negligible heat losses, radially lumped temperature distributions, negligible axial conduction, and uniform velocity, may be relaxed for more accurate results. Analytical or numerical solution of the governing equations may be employed to obtain the desired temperature variation in the system and the heat transfer rate. The convective problem is generally not solved and heat transfer correlations are used to determine the overall heat transfer coefficient. However, more accurate correlations as well as detailed simulations of the convective problem are available for use in designing these systems. The following example illustrates the use of the preceding analysis for the design of a heat exchanger.

Example 5.7

Design a counter-flow, concentric-tube heat exchanger to use water for cooling hot engine oil from an industrial power station, as shown in Figure 5.30. The mass flow rate of the oil is given as 0.2 kg/s and its inlet temperature as 90°C. The water is available at 20°C, but its temperature rise is restricted to 12.5°C because of environmental concerns. The outer tube diameter must be less than 5 cm and the inner tube diameter must be greater than 1.5 cm due to constraints arising from space and piping considerations. The engine oil must be cooled to a temperature below 50°C. Obtain a feasible design if the length of the heat exchanger must not exceed 200 m. Redesign the system if the length is restricted to 100 m. Even though the fluid properties vary with temperature, take these as constant for simplification, with the specific heat at constant pressure (C_p),

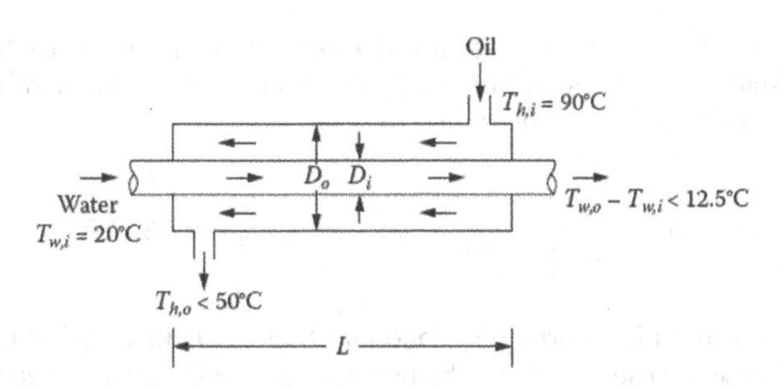

FIGURE 5.30 Counter-flow heat exchanger considered in Example 5.7.

viscosity (μ), and thermal conductivity (*k*) as 2100, 0.03, and 0.15 for the oil, and as 4179, 8.55×10^{-4}, and 0.613 for water, all in S.I. units.

SOLUTION

Several requirements and constraints are given as inequalities. Appropriate values may thus be chosen to satisfy these. Therefore, the outlet temperature of the oil may be taken as 45°C, the inner tube diameter as 2 cm, and the outer tube diameter as 4 cm. These values satisfy the inequalities, but may have to be adjusted if a feasible design is not obtained.

$$\text{Total energy lost by the oil} = Q = \dot{m}_h C_{p,h}\left(T_{h,i} - T_{h,o}\right)$$
$$= 0.2 \times 2100 \times (90 - 45) = 18.9 \text{ kW}$$

where the subscript *h* refers to the hot oil, *i* to the inlet, and *o* to the outlet. The mass flow rate is represented by $\dot{m}$ and the temperatures by *T*. Assuming zero heat loss to the ambient, the energy lost by the oil is gained by the water. Therefore,

$$\dot{m}_w C_{p,w}\left(T_{w,o} - T_{w,i}\right) = 18{,}900$$

Because the temperature rise is restricted to 12.5°C,

$$\dot{m}_w \geq \frac{18{,}900}{4179 \times 12.5} = 0.36 \text{ kg/s}$$

Let us choose the mass flow rate of water as 0.4 kg/s, which gives the outlet water temperature that satisfies the given constraint on temperature rise, as

$$T_{w,o} = 20 + \frac{18{,}900}{4179 \times 0.4} = 31.3\text{°C}$$

Therefore, the LMTD, ΔT_m, is

$$\Delta T_m = \frac{\left(T_{h,i} - T_{w,o}\right) - \left(T_{h,o} - T_{w,i}\right)}{\ln\left[\left(T_{h,i} - T_{w,o}\right)/\left(T_{h,o} - T_{w,i}\right)\right]} = \frac{58.7 - 25}{\ln(58.7/25)} = 39.5\text{°C}$$

Neglecting the conductive resistances due to tube walls and the layer formed by fouling, the overall heat transfer coefficient *U* is given by

$$U = \frac{1}{(1/h_i) + (1/h_o)}$$

where h_i and h_o are the convective heat transfer coefficients in the inner tube and the outer annulus. To determine if the flows are laminar or turbulent, the Reynolds numbers Re_D need to be determined. For water flow in the inner tube,

$$Re_D = \frac{4\dot{m}_w}{\pi D_i \mu} = \frac{4 \times 0.4}{\pi \times 0.02 \times 8.55 \times 10^{-4}} = 2.98 \times 10^4$$

Therefore, the flow is turbulent and a correlation such as the Dittus-Boelter equation (Incropera and Dewitt, 2001) may be used to obtain the heat transfer coefficient h_i. Therefore,

$$Nu_D = 0.023(Re_D)^{0.8}(Pr)^{0.4} = 0.023(29{,}800)^{0.8}(5.83)^{0.4} = 176.8$$

because the Prandtl number $Pr = \mu C_p/k = 5.83$ for water. This gives h_i as

$$h_i = Nu_D \frac{k}{D_i} = \frac{176.8 \times 0.613}{0.02} = 5418.9 \text{ W/}(\text{m}^2 \cdot \text{K})$$

For flow in the annulus, the hydraulic diameter $D_h = D_o - D_i = 0.02$ m. The Reynolds number Re_{D_h} is

$$Re_{D_h} = \frac{\rho u_m D_h}{\mu}$$

where u_m is the mean velocity, given by

$$u_m = \frac{4\dot{m}_h}{\rho\pi\left(D_o^2 - D_i^2\right)}$$

Therefore,

$$Re_{D_h} = \frac{4\dot{m}}{\pi(D_o + D_i)\mu} = \frac{4 \times 0.2}{\pi(0.04 + 0.02) \times 0.03} = 141.5$$

This implies that the flow in the annulus is laminar. For $D_i/D_o = 0.5$, the Nusselt number for developed annular flow with one surface isothermal and the other insulated may be employed to calculate the heat transfer coefficient at the inner surface of the annulus (Incropera and Dewitt, 1990, Table 8.2). The Nusselt number thus obtained is

$$Nu_{D_h} = \frac{h_o D_h}{k} = 5.74$$

Therefore,

$$h_o = \frac{5.74 \times 0.015}{0.02} = 43.1 \text{ W/}(\text{m}^2 \cdot \text{K})$$

and

$$U = \frac{1}{(1/5418.9) + (1/43.1)} = 42.8 \text{ W/}(\text{m}^2 \cdot \text{K})$$

Now, the total heat transfer is given by

$$Q = UA\,\Delta T_m$$

where A is the heat transfer area, being the inner surface of the annulus. Therefore, if L is the length of the heat exchanger,

$$Q = U\pi D_i L \Delta T_m = 42.8 \times \pi \times 0.02 \times L \times 39.5$$

This gives L as

$$L = \frac{18{,}900}{42.8 \times \pi \times 0.02 \times 39.5} = 177.9 \text{ m}$$

This satisfies the given requirement that the length be less than 200 m. Therefore, a feasible or acceptable design is obtained. Clearly, there are many other acceptable designs because several variables were chosen arbitrarily to satisfy the given constraints and requirements. This is typical of most design problems where there is considerable freedom in the choice of design variables, leading to a domain of acceptable designs from which an optimal design may be determined for a given objective function.

Let us now consider variations in this design to obtain a length less than 100 m. It is obvious from the preceding calculations that the heat transfer coefficient in the tube h_i is very high and has a small effect on the overall heat transfer coefficient U. Therefore, a reduction in D_i does not significantly affect U, but it reduces the area A, which leads to an increase in L. The outer diameter may be increased up to 5 cm, but this results in a reduction in h_o. The effect of changing other variables may similarly be considered. The best course of action is to reduce D_o while keeping D_i unchanged. If D_o is taken as 3 cm, $D_i/D_o = 0.667$, $Re_{D_h} = 169.8$, and Nu_{D_h} is obtained from Incropera and Dewitt (2001) as 5.45. This gives h_o as

$$h_o = \frac{5.45 \times 0.15}{0.01} = 81.8 \text{ W/(m}^2 \cdot \text{K)}$$

This yields the value of U as 80.6 W/(m$^2 \cdot$ K), which leads to $L = 94.5$ m. Therefore, this is an acceptable design because the length constraint of 100 m is satisfied. It is seen here that once an acceptable design is obtained, other designs can easily be generated by varying the design variables. In addition, the sensitivity of the results to changes in the variables, as obtained from the simulation or a sensitivity analysis, can be used effectively to obtain acceptable designs for other constraints and requirements.

Similar design procedures can be used for other heat exchanger designs. More accurate mathematical models may also be employed in cases where higher accuracy is desired and for more complex problems. Also, additional effects such as fouling may be included for practical systems. It must also be noted that numerical software may easily be developed for such problems and used effectively to consider wide ranges of design parameters and operating conditions.

5.4.5 Fluid Flow Systems

Fluid flow is an important part of thermal systems because the transport of mass and energy occurs due to the flow of fluids such as refrigerants, combustion products, water, and air. Although convective transport is relevant to many different types of thermal systems and has been considered for several applications in the preceding sections, fluid flow systems such as the pipe networks shown in Figure 1.20 are also important in many practical circumstances. Figure 5.31 shows a couple of fluid systems that involve a piping network as well as pumps to move the fluid. The design of such fluid flow systems generally involves the following two considerations:

1. Selection of fluid flow equipment
2. Design of the piping system for the flow

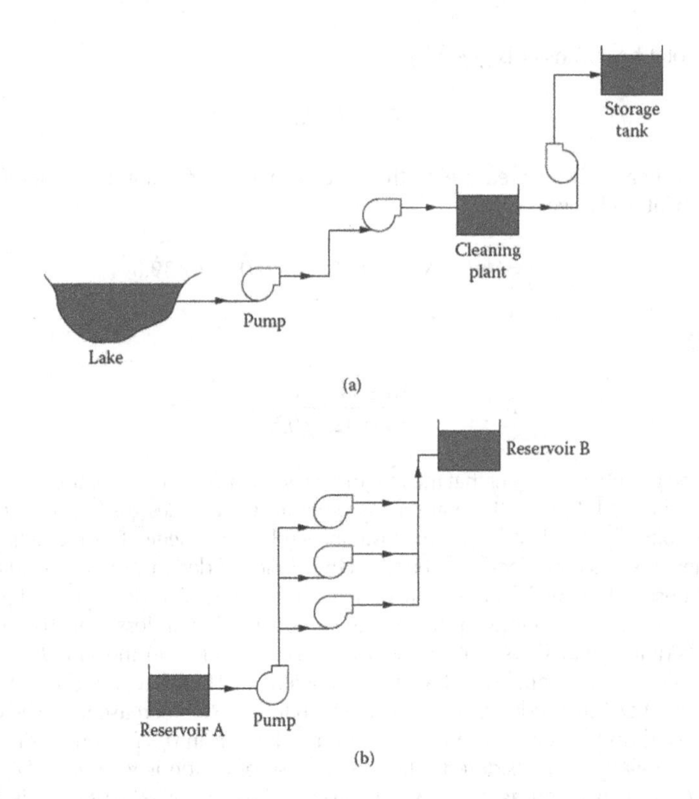

FIGURE 5.31 Two fluid flow systems, involving flow in pipes and pumps.

The first consideration is directed at the selection of equipment such as pumps, fans, blowers, compressors, valves, and storage vessels. Obviously, the design of these components may also be undertaken, depending on the application and the scope of the overall effort, as discussed previously for heat exchangers. The second aspect relates to pressure or head losses in pipes due to friction, bends, pipe fittings, joints, valves, etc., and the appropriate overall pressure difference that must be provided to maintain a given flow rate.

5.4.5.1 Selection of Equipment

In selecting appropriate equipment for a given application, it is necessary to know the desired flow rate, pressure head needed, and the fluid involved. Constraints on dimensions, flow velocity, system weight, etc., are also important considerations. The requirements and constraints are then matched with the specifications of the available hardware and a selection is made based on cost and characteristics of the equipment. A brief description of some of this equipment is given in the following for the sake of completeness.

A pump is a device used to move fluid by drawing the fluid into itself and then forcing it out through an exhaust port. It may be used to move liquids in pipelines, to lift water from a water processing plant to a storage tank high above the city, to empty a container, or to put an oil under pressure as in a hydraulic brake system. Many different types of pumps are available, often being classified as reciprocating, rotary, or centrifugal. Sketches of these three types of pumps are shown in Figure 5.32. In a reciprocating pump, an inlet valve, which opens at appropriate points during the motion of a piston, allows the lower pressure fluid to flow into a chamber. Then the back-and-forth movement of the piston is employed to push the fluid through an outlet valve. In a rotary pump, the rotating elements contain fluid that is physically pushed out. Both of these types of pumps employ fixed movements of the fluid and are thus positive displacement devices. The centrifugal pump raises the pressure by imparting kinetic energy to the fluid. The fluid picks up velocity as it flows in

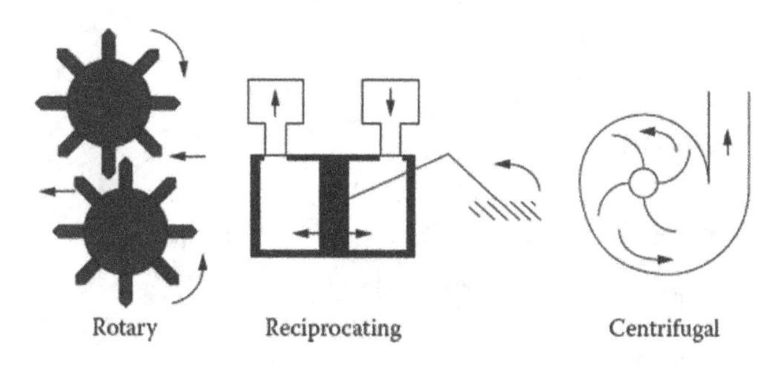

FIGURE 5.32 Different types of pumps. (Adapted from Boehm, 1987.)

the pump and, as it exits, a pressure rise is generated due to the centrifugal force. Further subdivision of centrifugal pumps as axial, radial, and mixed flow pumps is made according to the direction of fluid flow with respect to the axis of rotation. Several other types of pumps are available for fluid flow systems.

The characteristics of a pump may be written in terms of the pressure difference Δp generated by it and the mass flow rate $\dot{m}$ of a given fluid. Figure 5.33 shows a typical characteristic curve, indicating the decrease in the pressure due to friction and other losses as the flow rate increases. Curve fitting may be used to obtain a correlating equation, which represents such data on pump characteristics. This equation may be of the form

$$f(\Delta p, \dot{m}) = 0 \quad \textit{or, for example,} \quad \Delta p = A_0 - A_1\dot{m} - A_2\dot{m}^2 \tag{5.25}$$

where A_0, A_1, and A_2 are constants. A_0 gives the highest pressure generated, which arises under no-flow conditions. Similar correlations may be derived for different types of pumps if experimental data on their characteristics are given. Therefore, the constants, such as the A's in Equation (5.25), characterize a particular pump for a given fluid. The specifications of a pump may then be given in terms of the pressure generated for a specified flow rate of a particular fluid, the maximum flow rate that can be delivered, or the maximum pressure that can be generated. The requirements for an application can then be employed to select a pump for the purpose. See Pollak (1989), Warring (1984), and Boehm (1987) for further details.

A fan is also a device used to move fluids, though the pressure head is generally quite small. The flow is generated by producing a low compression ratio, as in ventilation. Blowers are fans that operate with most of the resistance to flow downstream of the fan. Exhausters are also fans that operate with most of the resistance upstream of the fan. Three main types of fans are usually defined. These are axial, propeller, and centrifugal, with the first two employing the angle of attack of the rotating blade to move the fluid. The housing plays an important role in controlling the flow rate in axial fans, whereas propeller fans are not good for controlling the flow. In centrifugal fans, the centrifugal force acting at the perimeter of the fan results in a pressure rise as the gas leaves the fan. The blade

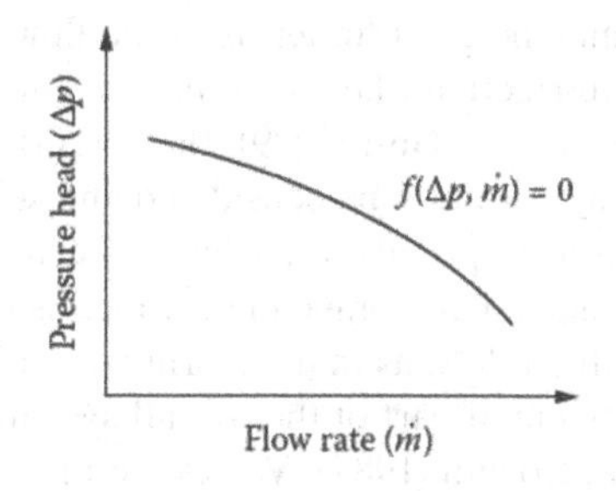

FIGURE 5.33 Typical graph representing the characteristics of a pump.

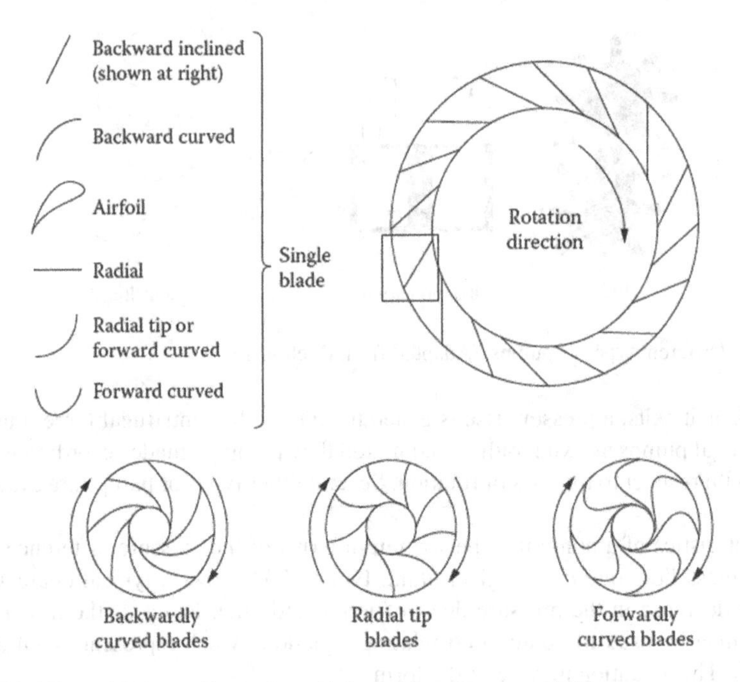

FIGURE 5.34 Different blade profiles in fans. (Adapted from Boehm, 1987.)

profiles, some of which are shown in Figure 5.34 from Boehm (1987), affect the performance of the fan significantly. The characteristics may again be given in terms of the pressure head Δp and mass flow rate $\dot{m}$. The diameter D and the revolutions per minute (RPM) N may also be included to obtain an equation of the general form

$$f\left(\Delta p, \dot{m}, D, N\right) = 0 \quad or, for\ example, \quad \Delta p = A\dot{m}^a D^b N^c \tag{5.26}$$

Here, A, a, b, and c are constants obtained from a curve fit of the data on the equipment. For further details on fans and their characteristics, see Thompson and Trickler (1983) and Avallone et al. (2007).

A compressor is a machine that increases the pressure of a gas or vapor by reducing the fluid specific volume as it passes through the equipment. Compressors are used in a wide variety of applications such as cleaning, pneumatic tools, paint spraying, refrigeration, and tire inflating. Again, there are several types of compressors such as reciprocating, rotary, centrifugal, jet, or axial flow, depending on the mechanical means used to compress the fluid. The thermodynamics of compressors are given in most textbooks on thermodynamics, such as Van Wylen et al. (1994), Howell and Buckius (1992), and Cengel and Boles (2014). The energy needed for an actual or real compressor is compared against ideal isothermal or adiabatic processes to yield the compression efficiency. The characteristics of a compressor may be given in terms of the flow rate and the pressure generated for a given fluid such as air or a refrigerant like ammonia. For further details on different types of compressors and their characteristics, see Gulf (1979), Bloch (2006), and Brown (2005).

Other fluid flow equipment may similarly be considered and selected for different applications. Storage vessels are commonly used to provide a buffer between the supply and the demand. An example of this is the household hot water storage tank, which is used to meet the demand for hot water when the outflow exceeds the inflow, as in the morning. Similarly, storage of thermal energy in solar energy utilization is an essential part of the overall system. Figure 5.35 shows a few common types of storage vessels from Boehm (1987). Valves are important ingredients in the successful design and operation of several thermal systems. Many types of valves, such as globe, gate,

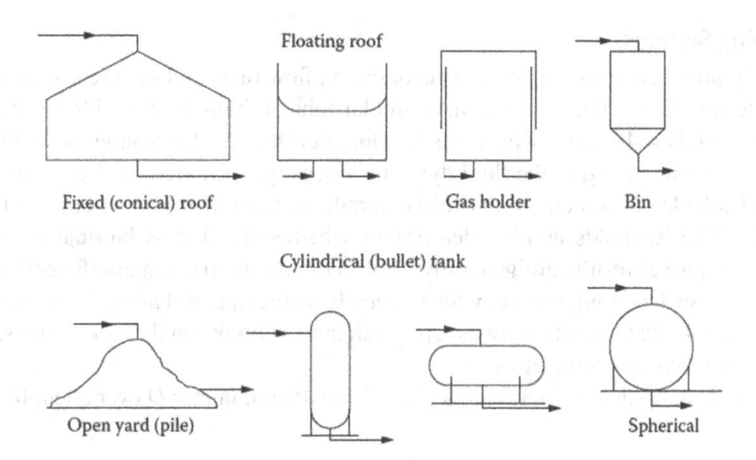

FIGURE 5.35 Common types of storage vessels. (Adapted from Boehm, 1987.)

butterfly, and ball, shown in Figure 5.36, are available to provide shut-off, open-flow, or throttling. The main consideration in throttling valves is the pressure drop as a function of the flow rate for a given fluid. The characteristics may therefore be given as $f(\Delta p, \dot{m}) = 0$. Check valves are used to obtain flow in one direction only. The flow rate is then given as a function of the opening size.

Extensive information is available in the literature on fluid flow equipment, particularly on the different types of devices and their basic characteristics. The manufacturers of these devices generally give the specifications in terms of the flow rates and the pressures, along with limitations on their use with respect to the fluid, temperatures, and environmental conditions. The characteristic curves obtained from prototype testing are also available in many cases. From this information and the needs of the given application, the appropriate equipment may be selected. However, it must be reiterated that these equipment designs may also be undertaken separately if the needs of the project require it.

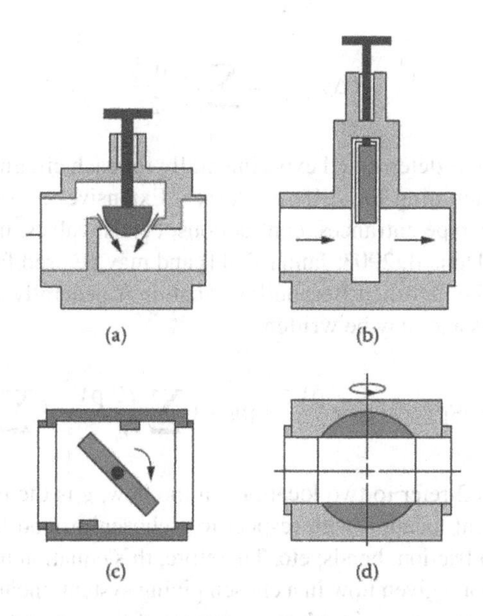

FIGURE 5.36 Schematic diagrams for (a) globe, (b) gate, (c) butterfly, and (d) ball valves. (Adapted from Boehm, 1987.)

5.4.5.2 Piping Systems

The design of piping networks and systems is based on flow rates and the pressure head needed to generate the desired flows. The flow rate in a circular tube of diameter D yields the Reynolds number Re as $Re = \rho VD/\mu = 4\dot{m}/\pi D\mu$, where ρ is the fluid density, V is the average velocity in the tube, $\dot{m}$ is the mass flow rate, and μ is the fluid dynamic viscosity. Similarly, the Reynolds number may be defined and calculated for noncircular tubes, annuli, and channels (Incropera and Dewitt, 2001; Shames, 2002). The Reynolds number determines whether the flow is laminar or turbulent (see Example 5.7). Empirical results are generally used to obtain the friction coefficient f as a function of Re and the pipe surface roughness e, which depends on the material and affects the pressure loss due to friction. The basic concepts involved are given in most books on fluid mechanics. Only a brief discussion is given here for completeness.

The pressure drop Δp due to friction in a pipe of constant diameter D over a length L is given by the expression

$$\Delta p = f \frac{L}{D} \frac{\rho V^2}{2} \tag{5.27}$$

where the friction factor f may be obtained for laminar and turbulent flows from, respectively,

$$f = \frac{64}{\text{Re}} \quad \text{and} \quad \frac{1}{f^{0.5}} = -2.0 \log\left(\frac{e/D}{3.7} + \frac{2.51}{\text{Re} f^{0.5}}\right) \tag{5.28}$$

Here, the second equation is the widely used *Colebrook formula* for f, and log represents the logarithm to base 10. Iteration is used to solve for the root f in this equation, using the techniques discussed in Chapter 4. Other equations and experimental results presented in graphical form, known as *Moody's chart*, are available for determining the friction factor f (Fox and McDonald, 2003). For noncircular channels, the hydraulic diameter D_h, defined earlier, is used instead of D.

The flow through a variety of fittings, bends, abrupt changes in area, joints, etc., also gives rise to pressure head losses, mainly due to flow separation. These losses are usually known as minor losses and are expressed as

$$(\Delta p)_{\text{minor}} = \sum K \frac{\rho V^2}{2} \tag{5.29}$$

where the loss coefficient K is determined experimentally for each circumstance and the total pressure loss is obtained by summing the different losses. Extensive information on experimentally determined coefficients for pipe entrances, contractions, bends, valves, fittings, etc., is available in the literature (Fox and McDonald, 2003; Janna, 2014) and may be used for calculating the pressure drop in a piping system. The modified Bernoulli's equation is generally used to include the effects of friction and minor losses and may be written as

$$p_1 + \frac{\rho V_1^2}{2} + \rho g z_1 = p_2 + \frac{\rho V_2^2}{2} + \rho g z_2 + \sum \frac{fL}{D_h} \frac{\rho V^2}{2} + \sum K \frac{\rho V^2}{2} \tag{5.30}$$

where the subscripts 1 and 2 refer to two locations in the flow, g is the magnitude of gravitational acceleration, z is the vertical location with respect to a chosen ground level, and the summations indicate head losses due to friction, bends, etc. Therefore, this equation may be applied to compute the pressure head needed for a given flow in a chosen piping system, including the effects of gravity involved in raising or lowering the fluid. Many examples of the application of this equation and of the calculations for head losses are given in most textbooks on fluid mechanics.

Example 5.8

A water distribution system consisting of two centrifugal pumps in two parallel-flow channels, as shown in Figure 5.37, is to be designed. The total mass flow rate $\dot{m}$ is the sum of the flow rates $\dot{m}_1$ and $\dot{m}_2$ in the two paths. Therefore,

$$\dot{m} = \dot{m}_1 + \dot{m}_2 \quad \text{(a)}$$

Also, the characteristics of the two pumps are given in terms of the pressure difference P and flow rates as

$$P = P_1 - A(\dot{m}_1)^{2.75} \text{ and } P = P_2 - B(\dot{m}_2)^{2.75} \quad \text{(b)}$$

where the pressures are in kPa and flow rates in kg/s. Here, P_1 and P_2 are the maximum pressure heads generated, for no-flow conditions, and A, B are constants. Curve fitting has been used to derive these equations from experimental data, as outlined in the preceding section and in Chapter 3. The energy balance, considering elevation change H and friction losses, is obtained from Bernoulli's equation as

$$P = H + C(\dot{m})^2 \quad \text{(c)}$$

The initial design values of P_1, P_2, A, B, and C are given as 500, 700, 7, 22, and 4.75, respectively. In addition, the pressure head H due to the elevation is fixed and given as 140 kPa. The required total flow rate is 6.5 kg/s. Determine if this initial design is satisfactory. If not, vary the design variables from their initial values up to ±35% of the initial values to obtain an acceptable design. From these results, determine conditions under which maximum flow rate is obtained.

SOLUTION

The mathematical model has been derived using the basic ideas presented in the preceding section. The fixed quantity is the elevation pressure head H, and the requirement is that the flow must be greater than 6.5 kg/s. The design variables that refer to the pump are P_1, P_2, A, and B, while C refers to the friction and other losses in the pipes. The constraints are placed on all the design variables as ±35% of the initial values.

We need to simulate this system by solving the given set of nonlinear algebraic equations to determine the total flow rate. Although all the equations may be used directly in the simulation, the problem may be simplified by eliminating $\dot{m}$ and P to obtain the following two equations for $\dot{m}_1$ *and* $\dot{m}_2$:

$$F(\dot{m}_1, \dot{m}_2) = H + C(\dot{m}_1 + \dot{m}_2)^2 - P_1 + A(\dot{m}_1)^{2.75} \quad \text{(d)}$$

$$G(\dot{m}_1, \dot{m}_2) = H + C(\dot{m}_1 + \dot{m}_2)^2 - P_2 + B(\dot{m}_2)^{2.75} \quad \text{(e)}$$

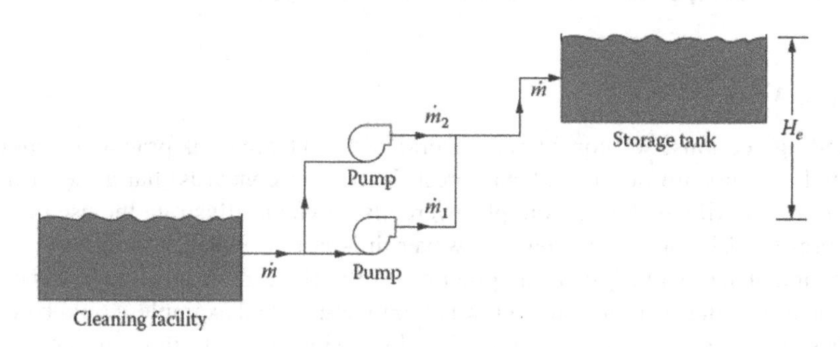

FIGURE 5.37 Fluid flow system considered in Example 5.8.

The Newton-Raphson method may be used conveniently to solve this system of two nonlinear equations, as discussed in Chapter 4. The resulting values of $\dot{m}_1$ and $\dot{m}_2$ would then yield the total flow rate from Equation (a). The pressure head P, if needed, may be calculated from the other equations.

If the values of $\dot{m}_1$ and $\dot{m}_2$ after the ith iteration are $\dot{m}_{1,i}$ and $\dot{m}_{2,i}$, the values for the next iteration are given by

$$\dot{m}_{1,i+1} = \dot{m}_{1,i} + \Delta\dot{m}_{1,i} \qquad \dot{m}_{2,i+1} = \dot{m}_{2,i} + \Delta\dot{m}_{2,i} \tag{f}$$

where

$$\left(\frac{\partial F}{\partial \dot{m}_1}\right)_i \Delta\dot{m}_{1,i} + \left(\frac{\partial F}{\partial \dot{m}_2}\right)_i \Delta\dot{m}_{2,i} = -F_i \tag{g1}$$

$$\left(\frac{\partial G}{\partial \dot{m}_1}\right)_i \Delta\dot{m}_{1,i} + \left(\frac{\partial G}{\partial \dot{m}_2}\right)_i \Delta\dot{m}_{2,i} = -G_i \tag{g2}$$

Here, the derivatives and functions are evaluated for the ith iteration and the equations are solved for $\Delta\dot{m}_{1,i}$ and $\Delta\dot{m}_{2,i}$. The derivatives are obtained mathematically from the expressions for F and G. Equation (f) then yields the new values for $\dot{m}_1$ and $\dot{m}_2$.

Using the approach just outlined, the individual as well as the total flow rates are computed. A convergence criterion of 10^{-4} is applied to the sum of the squares of the functions F and G, i.e., the iteration is terminated if $|F^2 + G^2| \le 10^{-4}$. It is ensured that the results are negligibly affected by a further reduction in the convergence parameter. Starting with initial, guessed values of 1.0 for both $\dot{m}_1$ and $\dot{m}_2$, convergence is achieved in seven iterations, yielding $\dot{m}_1 = 3.277$ and $\dot{m}_2 = 2.826$. The pressure head P is obtained as 316.932 and the total flow rate $\dot{m}$ as 6.103, which is less than the required value of 6.5. Therefore, the given initial design is not acceptable and the design variables are changed over the given ranges to obtain higher flow rates.

Figure 5.38 and Figure 5.39 show the computed results in terms of the total flow rate $\dot{m}$ over the range of variation of the design variables. In each case, one design variable is changed while the others are held constant at the base or initial values. It is easy to see that the required total flow rate $\dot{m}$ is 6.5 or larger if C is at its lowest value of 3.09. It exceeds 6.5 also if P_1 is greater than 605. In both cases, the other variables are at the base values. For changes in other design variables, the flow rate $\dot{m}$ is less than 6.5. Therefore, a domain of acceptable designs is obtained. Here, the design variables are changed one at a time in order to follow the basic trends and minimize the changes needed in the initial design.

Clearly, these results indicate that the highest total flow rate is obtained with the largest allowable values of P_1 and P_2, which represent the largest zero-flow pressures generated by the pumps, and the smallest values of A, B, and C, which indicate the smallest head losses. This result is physically expected. If the costs of making these changes in the design variables are also considered, an optimal design may be sought that minimizes the cost while meeting the given requirements and constraints. This aspect is considered in detail in later chapters.

5.4.6 Other Areas

In the preceding sections, we considered several different areas of practical application in which thermal systems are of particular interest. The main concerns that arise in the design of the system were outlined. A few examples were also given to illustrate the use of the design procedures presented in earlier chapters. However, it is not possible to consider every type of thermal system that arises in engineering practice. Similarly, even for the few areas considered in detail in the preceding sections, only a few salient features and examples could be discussed. However, these examples and the accompanying discussions serve to indicate the basic nature of the design process to obtain an acceptable thermal system for the particular application.

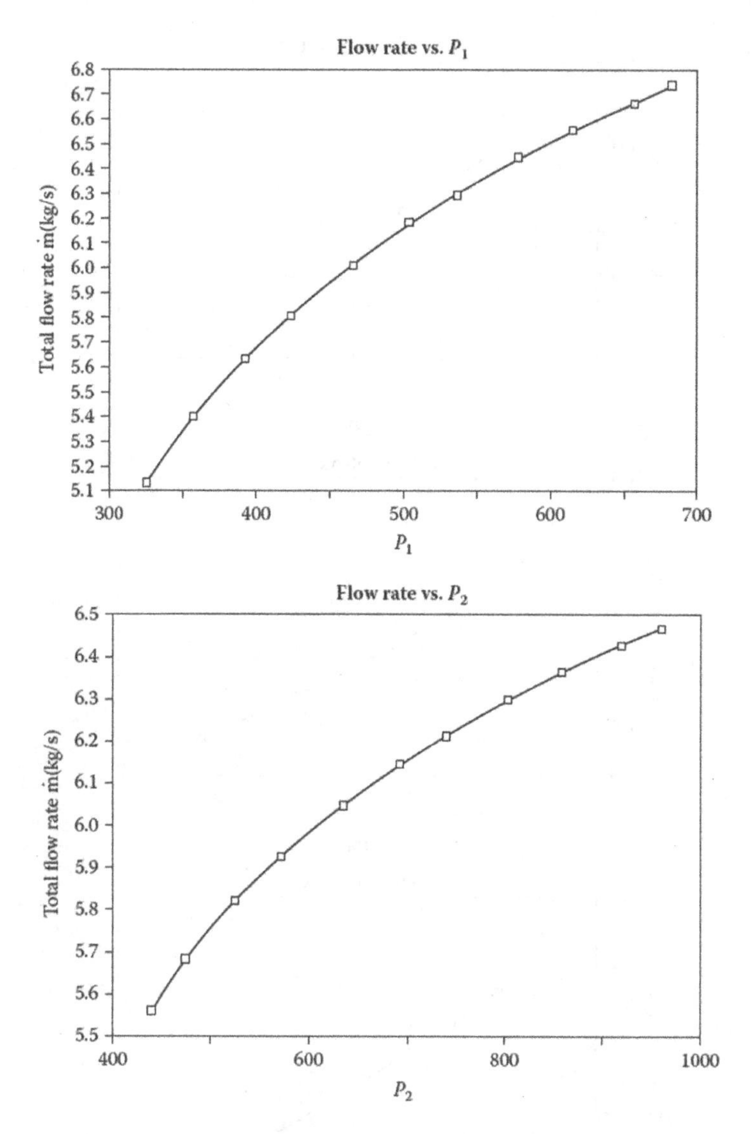

FIGURE 5.38 Effect of P_1 and P_2 on the total flow rate $\dot{m}$ in Example 5.8.

Starting with the formulation of the design problem and the conceptual design, the detailed, quantitative design process is illustrated, employing different strategies for converging to an acceptable design.

The different steps involved in the design process were discussed in earlier chapters and the coupling of all these aspects is illustrated here. The main considerations presented here are expected to apply to other types of thermal systems and to different problems. Some of these areas and problems are considered again in later chapters with respect to optimization.

5.4.7 Design of Components versus Design of Systems

Throughout this chapter, we have focused on thermal systems, ranging from small systems consisting of only a few parts to large systems consisting of many parts that interact with each other. Similarly, in earlier chapters, the treatment and discussions have been largely directed at systems. Not much has been said about the design of components, even though each system obviously consists

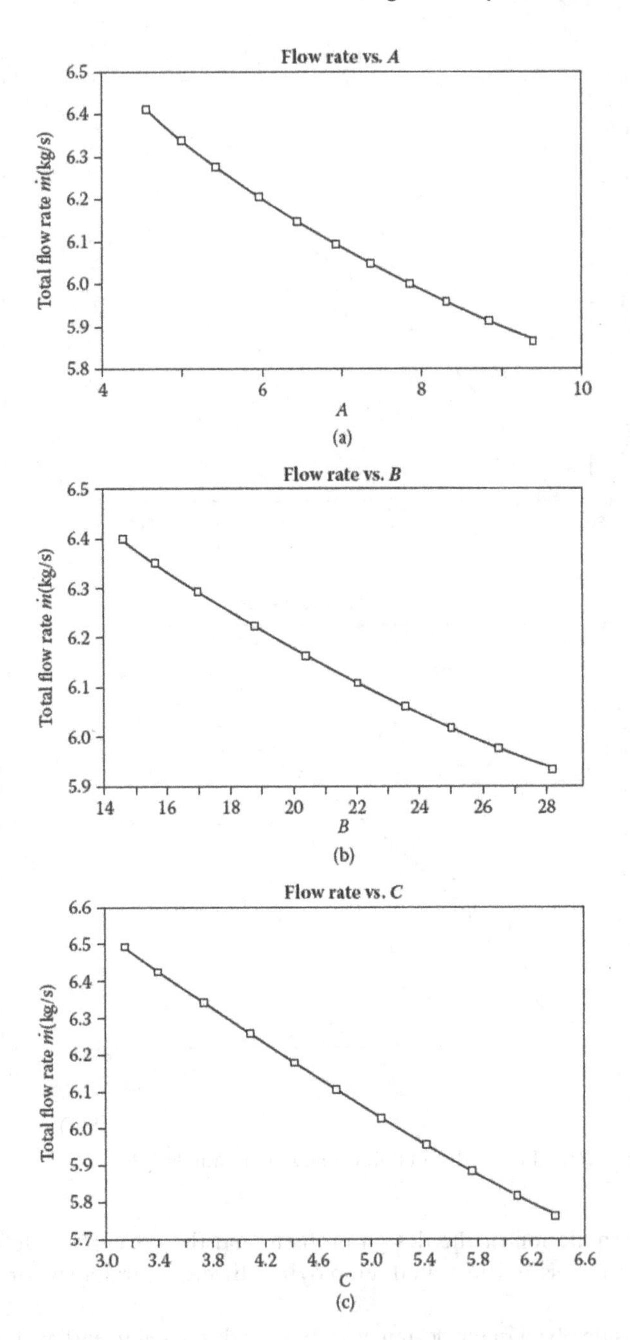

FIGURE 5.39 Effect of the parameters A, B, and C on the total flow rate $\dot{m}$ in Example 5.8.

of a number of components and the design of the system is often closely coupled with that of the components. In addition, in many cases, the components themselves consist of separate parts and may be considered as subsystems or systems for design purposes. Heat transfer and flow equipment such as heat exchangers, pumps, blowers, fans, and compressors are examples of items that are generally treated as components even though these involve interacting parts. Many of these components have been considered in preceding sections as parts of a larger system such as an air conditioning system or a water flow system. Therefore, it is worthwhile to clarify the design of a system as compared to that for a component.

A component is basically an independent item that is often available, ready-made, over wide ranges of specifications. The characteristics of each component, such as a pump or a blower, are also available from the manufacturer, who is obviously involved in the design and production of such components. Therefore, the design of the components precedes that of the system. However, we have employed the availability of standard items and their characteristics to design systems that are obtained by combining different components to obtain a desired thermal process. It is largely a question of focus and interest, these being related to the overall system in this book rather than to individual components. Some of the discussion in Section 5.4.4 and Section 5.4.5 has been directed toward components and their selection, with a few relevant references being given for additional information. The design process outlined for systems can also be employed for components that involve a number of interacting parts, such as heat transfer and flow equipment. For others, such as pipes, sensors, heaters, and valves, that do not contain interacting parts, modeling and simulation can again form the basis for design, but this problem has not been considered here. The components are largely treated as available items whose characteristics are employed in the modeling and simulation of the complete system and that can be selected for the overall design process.

5.5 ADDITIONAL CONSIDERATIONS FOR LARGE PRACTICAL SYSTEMS

In all our discussions on the design of thermal systems, we have focused on the thermal aspects arising from heat and mass transfer, fluid flow, and thermodynamics. This is obviously because of the types of systems that are of particular interest to us in this book. Thermal aspects are the dominant mechanisms in the processes and systems under consideration and, therefore, the design is largely based on these. However, the successful design and implementation of practical systems generally involve several additional considerations that must be included if the system is to perform satisfactorily.

Some of these additional considerations have been mentioned earlier and include

1. Safety
2. Control of the system
3. Environmental impact
4. Structural integrity and mechanical strength
5. Selection and availability of materials
6. Costs involved
7. Availability of facilities and utilities
8. Regulation, legal issues

Safety is a very important consideration and is generally addressed by providing sensors that monitor the levels of temperature, pressure, concentration, and other physical quantities that may affect the safety of material and personnel. In general, unsafe levels of such variables are given, and the system is turned down or turned off if these are exceeded. An alarm may also be used to alert the operator. In many cases, certain components of the system cannot be turned on if the specified conditions are not met. For instance, the heaters in a boiler may be set so that they can be turned on only if the water level is adequate. Many such safety features are usually built into the design to avoid damage to the system as well as to the user.

Control of the system is one of the most important technical aspects that must be included in the design for successful operation of the system. If a system is designed for a particular temperature at the surface or in the fluid, a control scheme is needed to ensure that these are maintained at these values, within an allowable tolerance. Sensors are used to monitor the appropriate physical quantity, and the control scheme makes the appropriate correction, as needed. Similarly, the flow rate of a fluid or a given material must be maintained at the design value in order to obtain the required product quality or production rate. A control strategy may be employed to preserve these within acceptable levels. Automation also depends strongly on the control scheme used. Many different control strategies are available and may be used effectively to ensure satisfactory performance of the system. These include on/off arrangements, which are very commonly used in thermal systems, proportional control, and integral control, among others. Control systems constitute an important area and, though beyond the scope of this book, generally form an important ingredient in the final implementation of the thermal system, as outlined in Chapter 2.

The environmental impact is generally an important consideration in system design today because of increasing concerns with pollution, depletion of the ozone layer, greenhouse effect, global warming, climate change, and solid waste disposal. Therefore, the impact of the designed system on the environment has to be evaluated for successful implementation. In particular, the types and amounts of pollutants discharged into the atmosphere or into bodies of water need to be estimated. The amount of solid waste generated and the procedures to dispose these must be determined. Similarly, if gases such as carbon dioxide or sulfur dioxide are being generated, it may be necessary to develop means to convert these into harmless byproducts.

One of the most important technical, though not necessarily thermal, considerations involved in the design of a system is that pertaining to the mechanical strength of the system. It is crucial that the structural integrity of the system be maintained and that the various elements that constitute the system do not fail under the temperatures, pressures, loads, and other forces acting on these. Therefore, the mechanical strength of the various materials must be considered in terms of stresses in different parts of the system, ensuring that these do not fail by fracture or excessive deformation. Even if a particular item satisfies the requirements and constraints imposed by the thermal aspects, it must still meet the strength requirements. However, in most cases, the constraints due to strength considerations may be translated into the appropriate limitations on temperature, pressure, speed, weight, etc. This is particularly true of thermal stresses that arise due to temperature gradients in the material. Thus, excessive thermal stresses can be expressed in terms of a constraint on the temperature gradient or difference across a given system part. Similarly, considerations such as wear, fatigue, and buckling are generally taken care of by limiting the speed, duration of daily usage, total time for which the system is employed, temperatures, pressures, and so on. Consequently, many of the constraints considered in this book may be the outcome of mechanical strength and structural integrity considerations.

The selection of materials is another important consideration, as discussed in Chapter 2. With the development of new materials such as composites, ceramics, alloys, and different types of polymers, the choice of materials has expanded substantially in recent years. The design would therefore be influenced by the availability of appropriate materials. In most cases, effort is made to choose the most suitable material with respect to the cost and the desired thermal properties. Economic considerations are always critical to the success of the design effort because the viability of a project depends strongly on the overall financial return. Therefore, it is important to evaluate the system with respect to the costs incurred for the hardware as well as for the operation of the system. The productivity in terms of the output can then be considered along with the price to determine the rate of return on the investment. Some of these aspects are presented in the next chapter. The availability of appropriate facilities and utilities such as power and water is an important consideration in the overall design process. Additional issues such as governmental regulations and legal matters also have to be satisfied for a successful system. The following example presents the design of a relatively large practical thermal system, considering mainly the thermal aspects that determine the product quality and consistency.

Example 5.9

Discuss the modeling, simulation, and design of the batch annealing furnace, shown in Figure 5.40, which is used for the annealing of steel sheets rolled up in the form of annular cylindrical coils.

SOLUTION

The problem considered is an actual industrial system that is used in the steel industry. Annealing is employed for relieving the stresses in the material, which has undergone a rolling process during

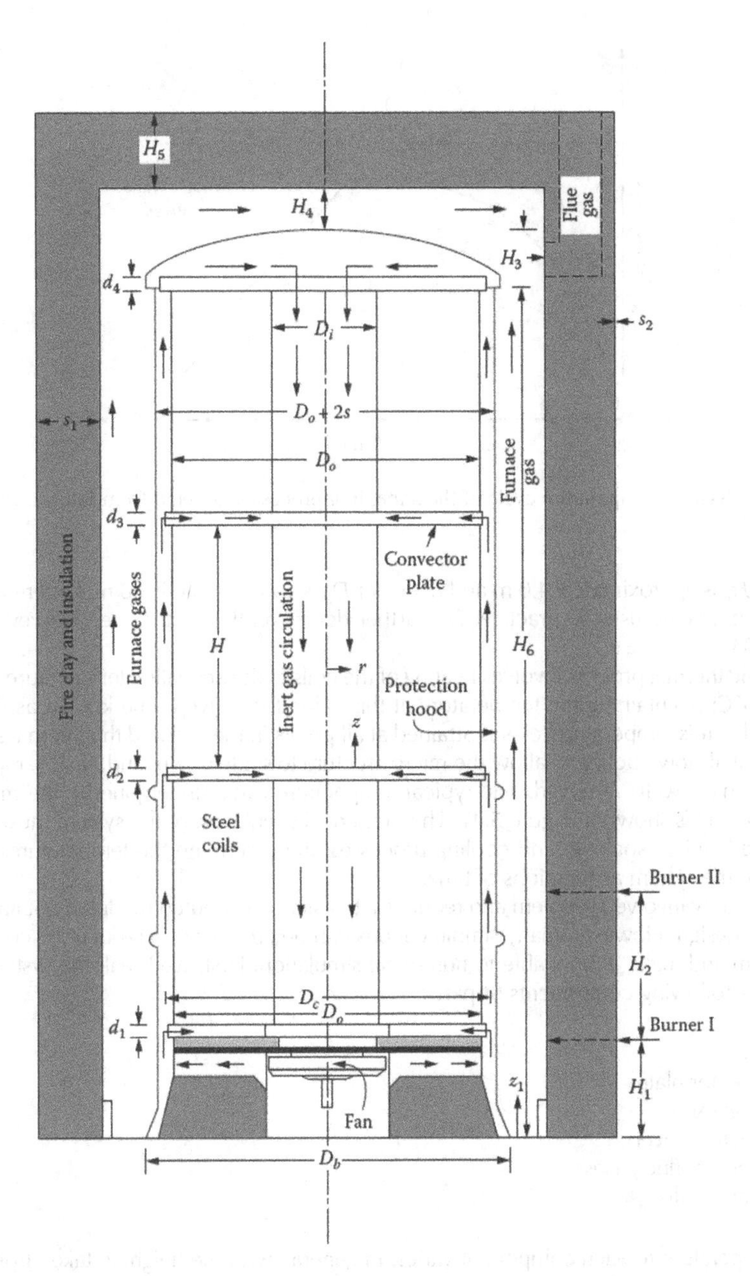

FIGURE 5.40 A batch annealing furnace for cylindrical coils of steel sheets. (Adapted from Jaluria, 1984.)

manufacture, such as that shown in Figure 1.10(d). The annealing process restores the ductility in the material for further machining and forming operations. As shown in Figure 5.40, the steel sheets are rolled into the form of three cylindrical coils and stacked vertically with convector plates, which aid the protective inert gas flow, at both ends of each coil. A stainless steel cover encloses the coils and an inert environment is maintained between the coils and the cover by the flow of gases such as nitrogen and helium. These gases are driven by a fan at the bottom, as shown. The region between the cover and the furnace walls contains flue gases, which are usually obtained from the blast furnace of the steel plant. These gases contain various combustion products, such as carbon dioxide, sulfur dioxide, and moisture, due to combustion occurring at the two burners. These burners are located circumferentially and the flow enters tangentially causing swirl in the flow. The dimensions in Figure 5.40 are shown in terms of symbols for generality. Typically,

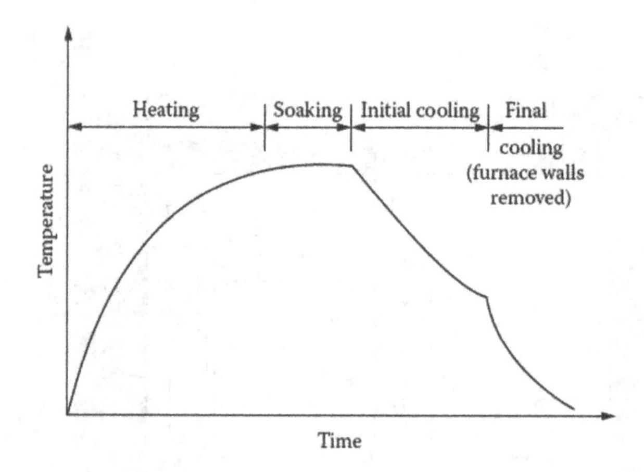

FIGURE 5.41 Typical temperature cycle of the annealing process. (Adapted from Jaluria, 1984.)

the height H_6 is approximately 4.0 m and diameter D_b is approximately 1.8 m. Different types and sizes of furnaces are used in practice. For further details on this system, see Harvey (1977) and Jaluria (1984).

The basic thermal process involves heating of the coils to the annealing temperature of approximately 723°C; maintaining the temperature at this value for a given time known as the *soaking period* so that this temperature level is attained at all points in the coil and the internal stresses are relieved; initial slow cooling to allow the microstructure to settle down; and, finally, rapid cooling with the furnace walls removed. The typical temperature cycle undergone by the material at a point in the coil is shown in Figure 5.41. The numerical simulation of the system must, therefore, include the heating, soaking, and cooling processes and determine the temperatures at various locations in the system as functions of time.

This is a fairly involved problem and requires a transient, distributed model to obtain the inputs needed for design. However, many simplifications can be employed to reduce the complexity of the problem and make it amenable to numerical simulation. First, we break the system down in terms of the following components or parts:

1. Coils
2. Convector plates
3. Inert gases
4. Protective cover
5. Furnace or flue gases
6. Furnace walls

The temperature in each component varies, in general, with the height z, taken from the base of the furnace, the radial distance r, taken from the axis of the coils, the circumferential location φ, and time τ. Mass, momentum, and energy balances lead to the governing equations for the different parts. All these equations are coupled to each other through the boundary conditions.

With respect to model development, it is first noted that axisymmetry may be assumed due to the cylindrical configuration of the system and the anticipated circumferential symmetry in the energy exchange mechanisms. This simplifies the problem to an axisymmetric transient circumstance. The coil is essentially a hollow cylinder with inner radius R_i and outer radius R_o. There are usually gaps, filled with inert gases, that exist between the different layers of the coil. As a result, the thermal conductivity in the radial direction k_r is generally much smaller than that in the axial direction k_z. Then the governing energy equation for the steel coils may be written as

$$\rho_m C_m \frac{\partial T_m}{\partial \tau} = \frac{1}{r}\frac{\partial}{\partial r}\left(rk_r \frac{\partial T_m}{\partial r}\right) + \frac{\partial}{\partial z}\left(k_z \frac{\partial T_m}{\partial z}\right)$$

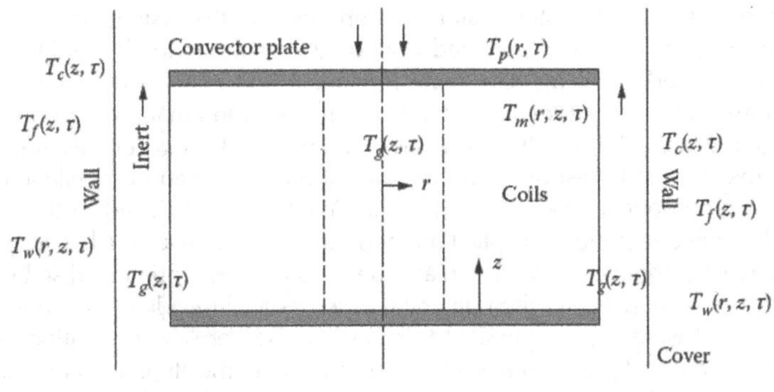

FIGURE 5.42 Components of the system for developing a mathematical model.

where the subscript m indicates the coil material and the other symbols have their usual connotations. The conductivities and other properties may be obtained from the available literature on this problem. All the properties depend on temperature. The initial and boundary conditions for the preceding equation are obtained from the initial temperature T_r, heat transfer with the cover and inert gases at the outer surface of the coils, and convective heat transfer with the inert gases at the inner surface of the coils.

Similarly, simplifying approximations are made for other components, shown schematically in Figure 5.42. Using the techniques discussed in Chapter 3, the temperature in the convector plate T_p is assumed to vary only with radial position and time, because the Biot number based on its thickness is small. For the cover, temperature variation across its thickness is neglected because of its small thickness and high conductivity. Thus, the temperature T_c of the cover varies with z and τ. The furnace wall is treated as an axisymmetric conduction problem, yielding the energy equation as

$$\rho_w C_w \frac{\partial T_w}{\partial \tau} = k_w \left[\frac{1}{r} \frac{\partial}{\partial r} \left(r \frac{\partial T_w}{\partial r} \right) + \frac{\partial}{\partial z} \left(\frac{\partial T_w}{\partial z} \right) \right]$$

where the subscript w denotes the wall.

For the gases, radial temperature uniformity is assumed because of turbulent mixing and only the variations with height and time are considered. The resulting energy equations for the inert gases in the annular and outer region of the coils are, respectively,

$$\rho_g \left(C_p \right)_g UA \frac{dT_g}{dz} = -Ph_i \left(T_m - T_g \right) \quad \text{at } r = D_i/2$$

$$\rho_g \left(C_p \right)_g UA \frac{dT_g}{dz} = Ph_o \left[\left(T_m - T_g \right) + \left(T_c - T_g \right) \right] \quad \text{at } r = D_o/2$$

Here, the subscripts g and c refer to the inert gas and the cover, respectively, U is the average velocity, A is the cross-sectional area, P is the perimeter for heat transfer, subscript i refers to the core region in the center, and subscript o to the region outside the coils between the coils and the cover. Similarly, the appropriate equations are written for the cover, flue gases, and convector plates.

As discussed in detail in Chapter 4, all the components of the system are modeled and simulated individually, with constant, specified boundary conditions to decouple them from each other. These uncoupled problems allow one to validate the mathematical model and the corresponding numerical scheme for each component. It was found that the various approximations made for the model are valid and that the gases and the cover have a very fast transient response. The coils are the slowest in response and the largest time step can be employed for simulating these. The various numerical schemes discussed in Chapter 4 can be used for the numerical simulation of the different components. Explicit methods are particularly useful because of the variable properties. However, implicit methods can also be used for better numerical stability.

The mathematical models and the numerical schemes for individual components are verified and validated by considering the physical trends obtained from the simulation, eliminating the dependence on numerical parameters such as grid size and time step, and comparing the results for a few idealized cases with analytical results. These individual numerical models are then coupled with each other by using the actual boundary conditions, arising from heat transfer between different components. The overall system is then simulated.

In practical systems, the overall annealing process is controlled by monitoring a thermocouple in contact with the base of the bottom coil. This is known as the *control thermocouple* and it is important to use numerical simulation to obtain the temperature cycle measured by this thermocouple.

Using typical values for the operating conditions, such as initial temperature, flow rates of the gases, and composition of the flue gases, the temperature variation with time was computed at various locations in existing furnaces. These results were compared with measurements. Figure 5.43 and Figure 5.44 show the comparison between the numerical results and experimental data, indicating fairly good agreement. The operating conditions were varied and it was confirmed that the behavior of the system follows expected trends. Large variations in the governing parameters and operating conditions were also tried to determine safe levels of operation and to generate system characteristics. Therefore, the annealing furnace is satisfactorily simulated numerically. For further details, consult the references cited here.

The results generated by the simulation may be used for the design and optimization of the system. Different materials, coil sizes, and heat treatment applications require different designs, in terms of configuration, dimensions, and heating/cooling arrangement. Different design strategies may be used to obtain an acceptable or optimal design. The results obtained may also be employed for the modification of existing systems to improve performance.

In the example considered here, considerable improvement in the process and the product was achieved simply by controlling the flow rate $\dot{m}$ of the flue gases entering the furnace. This controls the heat input because the heat released by combustion at the burners depends on the

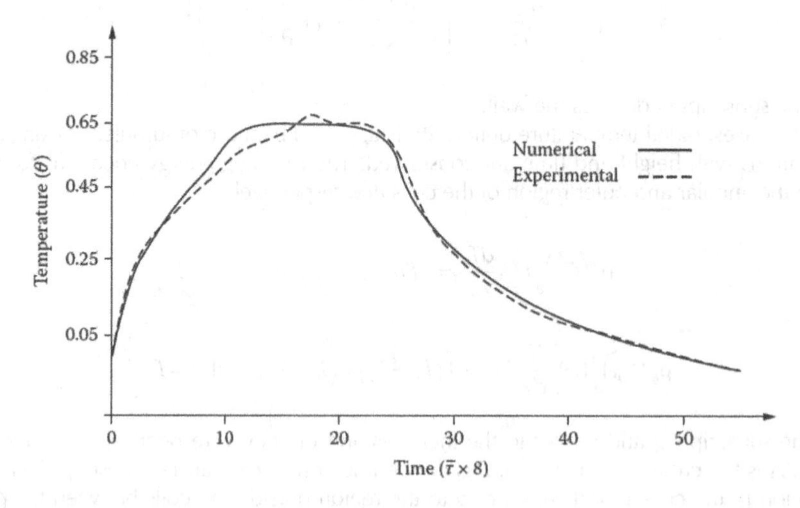

FIGURE 5.43 Comparison between the numerical simulation results and the temperature measurements of the control thermocouple. (Adapted from Jaluria, 1984.)

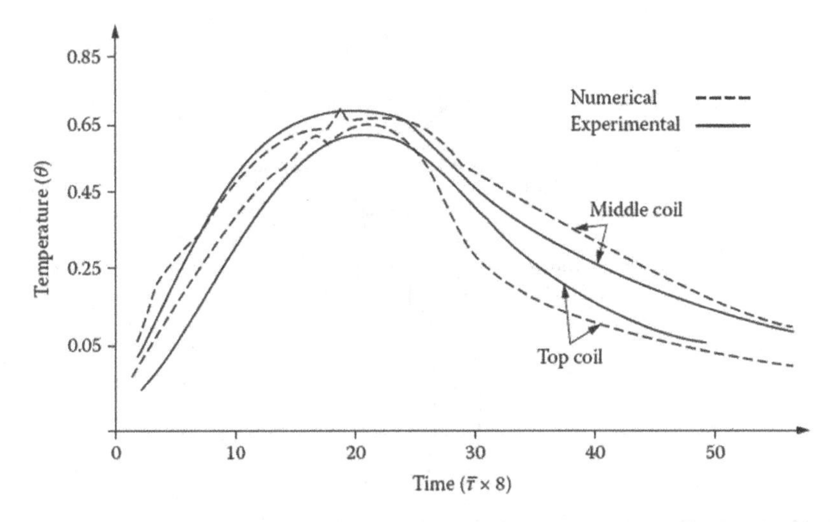

FIGURE 5.44 Comparison between the numerical simulation results and the temperature measurements in the steel coils. (Adapted from Jaluria, 1984.)

flow rate. The control thermocouple is set to follow a given temperature cycle and this, in turn, controls the gas flow rate at the burners. The temperature variation in the system is a strong function of the heat input and, therefore, the desired temperature cycle for heat treatment is obtained in the coils. The overall result of this effort was a more uniform annealing than that obtained earlier, and a consequent reduction in material wastage.

5.5.1 Other Large Systems

The preceding example presents a typical practical thermal system: a large one with many interacting parts and components and different flows and thermal transport processes. The sketch shown in Figure 5.40 is a schematic of the much more complicated industrial system, but it presents the main features of the thermal system under consideration. Additional subsystems that are involved due to safety, control, material loading, furnace top removal, etc., are important and must be included in the design and operation of the system. However, these are usually brought in after the essential thermal design of the system has been concluded.

Similarly many other thermal systems mentioned in the earlier chapters are large systems, even though the schematic diagrams shown indicate the main features of the system in a relatively simplified manner. An example of such a large system is the Czochralski crystal-growing process, shown schematically in Figure 5.45 (see also P2.1). A photograph of an industrial facility for the same process is shown in Figure 5.46, indicating the complexity of the system and the inclusion of many auxiliary arrangements for heating, feeding, control, safety, and other practical issues. However, the sketch in Figure 5.45 shows the main parts of the system and can be used to develop the mathematical model, using simplifications, approximations, and idealizations, and to simulate the system for wide ranges of the design variables and operating conditions. From these results, an acceptable design that meets the given requirements and constraints may be obtained. Optimization of the system, as well as of the operating conditions, may also be undertaken, following the determination of the domain of acceptable designs.

Another large system is the heat rejection system that was outlined in Example 5.6. The design problem involves a water body, which may be a natural one such as a lake or a constructed cooling pond, the condensers of the power plant, and the water pumping system consisting of pumps and piping network, as considered in Example 5.8. The requirements for a successful design involve both the recirculation, which raises the temperature at the intake, and the thermal effects on the

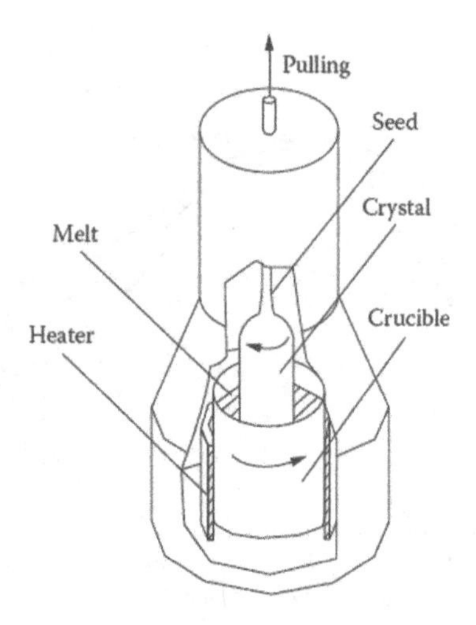

FIGURE 5.45 Schematic of the Czochralski crystal-growing process.

FIGURE 5.46 An industrial facility for the Czochralski crystal-growing process. (From Ferrofluids Corp.)

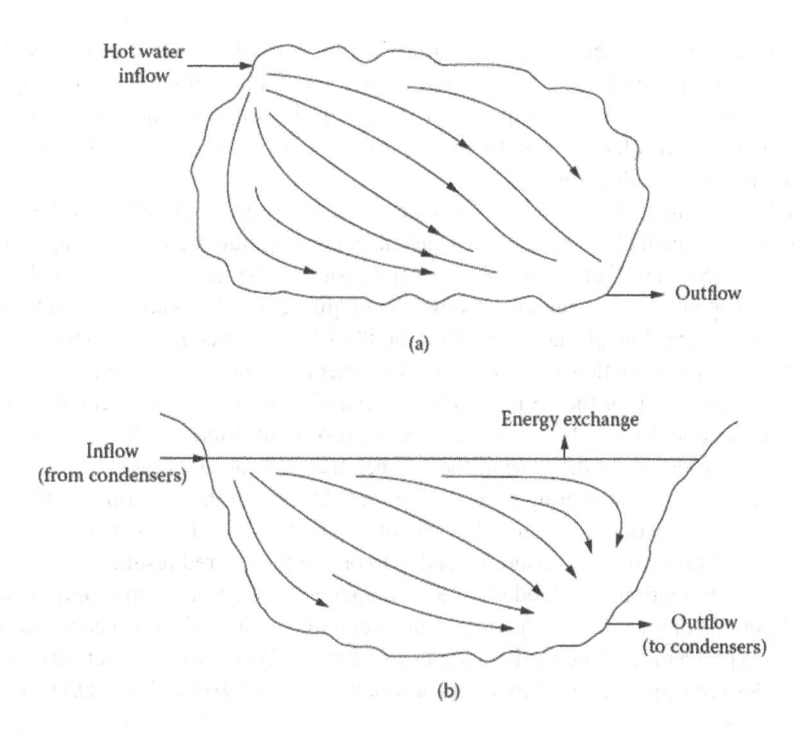

FIGURE 5.47 The flow system for power plant heat rejection to a body of water: (a) Top view, (b) front view.

body of water. Thus, the heat transfer and flow in the water body have to be modeled and coupled with the condensers and the pumping system. Design variables include the locations and dimensions of the inlet/outlet channels, as well as the dimensions and geometry of the pond itself. Operating conditions such as discharge temperature and flow rate must also be considered. Various flow configurations may be considered to curb the effects of recirculation (see Figure 5.47). For instance, the outlet may be moved as far away as possible from the inlet, or a wall may be placed as an impediment between the two. Computer simulations of the cooling pond, including the heat transfer to the surroundings and the flow due to intake and discharge, and of the pumping system are used to provide the necessary inputs for design and optimization. Because recirculation effects are reduced, but pumping costs increased, as the separation between the intake and discharge is increased, a minimization of the cost for acceptable temperature rise at the intake may be chosen as the objective function. Clearly, the design of the system is a major undertaking and involves many subsystems that make up the overall system. Over the years, the power industry has developed strategies to design and optimize such systems. Natural lakes, cooling ponds, rivers, and even the sea have been considered to minimize the recirculation, while maintaining low costs.

The design process is fundamentally the same for large and small systems. The basic approach presented in this chapter, as well as in the earlier chapters, can easily be applied to a wide variety of systems using the principles of modeling, simulation, design, and optimization presented in this book. However, additional aspects pertaining to safety, control, input/output, etc., need to be included in industrial systems before the prototype is developed and tested.

5.5.2 Inverse Problems

As discussed in detail in this chapter, one of the most important aspects in design of a thermal system is the choice of operating conditions and design parameters in order to achieve the desired thermal process. Thus, the result is known, whereas the conditions and parameters are not. As mentioned earlier, this is an inverse problem, which involves finding the conditions that would lead to a given result. As may be expected, several different conditions could lead to the same outcome. Thus, the solution of an inverse problem is not unique, though the domain of uncertainty may be reduced through optimization and other techniques. In most cases, the direct solution, with specified conditions, is solved and the results are used to develop a scheme to converge to a solution of the inverse problem in a fairly narrow domain.

An example of an inverse problem is the manufacturing process discussed in Example 5.9 for annealing of steel. A sketch of the desired time-dependent temperature variation, along with an acceptable envelope, was shown in Figure 5.41. The thermal system is the furnace for the batch annealing of steel sheets rolled up into cylindrical coils, as shown in Figure 5.40. The boundary conditions, particularly the time-dependent flow of flue gases from the blast furnace, need to be determined to achieve the desired temperature variation with time. This problem was numerically modeled, as discussed earlier, and the dependence of the temperature distributions on the boundary conditions was established. This information was used to solve the inverse problem to determine the temporal variation of a temperature sensor located at the base of the furnace that controls the flow rate of the hot gases in order to achieve the desired annealing process. Figure 5.43 showed the variation of this control thermocouple with time that would result in achieving the desired result. The calculated results were also confirmed by actual experimental variation needed to obtain the desired result.

Similarly, inverse solutions are needed for a wide range of design problems. Optimization is used to narrow the domain of uncertainty and obtain an essentially unique solution. The literature is replete with such inverse problems and the methodologies developed may be used to select various parameters to achieve the desired result and thus obtain an acceptable design (Ozisik, 2000; Orlande, 2012).

5.5.3 Uncertainties

An important consideration that arises in actual practice is that the design parameters and operating conditions are subject to uncertainties due to unforeseen variations. These must be considered for a practical solution because the failure of a given thermal system can be dangerous and expensive. The review paper by Lin et al. (2010) presents a systematic strategy for the modeling and optimization of a thermal system including the effects of uncertainties. An impingement type chemical vapor deposition reactor, which involves impingement of the reacting gases on a heated surface and deposition of a film resulting from chemical reactions, is taken as an example. Some of the major uncertainties that arise in this process are:

Uncertainties in Operating Conditions

- Inlet flow rate or velocity
- Temperature of deposition surface or heat flux input
- Mass fractions of reactive gases at inlet
- Initial conditions
- Environmental conditions

Uncertainties in System Design

- Dimensions of the system and components
- Location of inlet and heated surface
- Geometry, configuration, and symmetry
- Material and gas properties

The design and optimization with uncertainties is based on reliability and, thus, on the percentage failure rate of the design. Reliability-based design optimization (RBDO) algorithms are generally employed, with chosen distribution of uncertainty in the parameters. Requirements on deposition rate and quality may be used for design and optimization for this system. Probabilistic constraints are established, with respect to either normally or non-normally distributed random variables. Then the acceptable and optimal solutions are obtained, subject to the allowable level of failure probability. The failure is usually brought down to less than 0.13%, which is the commonly accepted level in RBDO. Due to the uncertainties, the acceptable or optimal design moves away from the one obtained for deterministic conditions, to satisfy this condition. Uncertainties are important in the design of most thermal systems because variations in operating conditions and in the fabrication of the system are common. These are particularly critical in the manufacture of microscale and nanoscale devices. Variations in the operating conditions or in the system hardware can have a significant effect on the product. However, this aspect has only recently been brought into detailed consideration for thermal systems.

5.6 SUMMARY

This chapter presents the synthesis of various design steps needed to obtain an acceptable design for a thermal system. Employing the basic considerations involved in design, as outlined in the earlier chapters, an overview of the design procedure is presented. Starting with the problem statement and the basic concept for the system, the various steps involved in design were given as initial design, modeling, simulation, evaluation, iterative redesign, and convergence to an acceptable design. Several of these aspects, particularly modeling and simulation, were presented in detail earlier and are applied in this chapter. Two ingredients in the design process that had not been discussed adequately earlier are the development of an initial design and different design strategies. These are presented in some detail in this chapter.

Initial design is an important element in the design process and is considered in terms of different methods that may be adopted to obtain a design that is as close as possible to an acceptable design. A range of acceptable designs may be obtained by changing the design variables, starting with the initial design values, in a domain specified by the constraints. The development of an initial design may be based on existing systems, selection of components to satisfy the given requirements and constraints, use of a library of designs from previous efforts, and current engineering practice for the specific application. In this way, the effort exerted to obtain an appropriate initial design is considerably reduced by building on available information and earlier efforts.

The main design strategy presented earlier was based on starting with an initial design and proceeding with an iterative redesign process until a converged acceptable design is obtained. This systematic approach is used quite extensively in the design of thermal systems. However, several other strategies are possible and are employed. In particular, extensive results on the system behavior and response to a variation in the design variables (for given operating conditions) as well as to different operating conditions (for selected designs) may form the basis for obtaining an acceptable design. Such strategies, though not as systematic as the previous one, are nevertheless popular because extensive results can often be obtained easily from numerical simulation. These strategies are also well-suited to systems with a small number of parts and those with only a few design variables. The methods to track the iterative redesign process and to study the convergence characteristics are also discussed.

In order to illustrate the coupling of the different aspects and steps involved in the design process, several important areas of application are considered and a few typical thermal systems that arise in these areas are considered as examples. This discussion is important for understanding the design process because the various steps involved in design had been discussed earlier as separate items. It is important to understand how these are brought together for an actual thermal system and how the overall process works.

Finally, this chapter presents additional considerations that are often important in the design and successful implementation of a practical thermal system. Included in this list are safety issues, control of the system, environmental effects, structural integrity of the system, material selection, costs involved, availability of facilities, governmental regulations, and legal issues. These considerations are important and must usually be included in the final design. A detailed discussion of these aspects is beyond the scope of this book. However, several of these aspects are included in the design process by a suitable choice of constraints for an acceptable design. The application of this process to large practical systems is outlined.

REFERENCES

Avallone, E.A., Baumeister, T., & Sadegh, A.M. (Eds.). (2007). *Marks' standard handbook for mechanical engineers* (11th ed.). New York: McGraw-Hill.

Bejan, A. (1993). *Heat transfer.* New York: Wiley.

Bloch, H. (2006). *A practical guide to compressor technology.* New York: Wiley.

Boehm, R.F. (1987). *Design analysis of thermal systems.* New York: Wiley.

Brown, R. (2005). *Compressors—selection and sizing* (3rd ed.). Houston, TX: Gulf Publishing Company.

Cengel, Y.A., & Boles, M.A. (2014). *Thermodynamics: an engineering approach* (8th ed.). New York: McGraw-Hill.

Fox, R.W., & McDonald, A.T. (2003). *Introduction to fluid mechanics* (6th ed.). New York: Wiley.

Gebhart, B. (1971). *Heat transfer* (2nd ed.). New York: McGraw-Hill.

Ghosh, A., & Mallik, A.K. (1986). *Manufacturing science.* Chichester, UK: Ellis Horwood.

Gulf Publishing Company. (1979). *Compressors handbook for the hydrocarbon processing industries.* Houston, TX: Gulf Publishing Company.

Harvey, G.F. (1977). Mathematical simulation of tight coil annealing. *Journal of the Australasian Institute of Metals, 22,* 28–37.

Howell, J.R., & Buckius, R.O. (1992). *Fundamentals of engineering thermodynamics* (2nd ed.). New York: McGraw-Hill.

Incropera, F.P. (1988). Convection heat transfer in electronic equipment cooling. *ASME J. Heat Transfer, 110,* 1097–1111.

Incropera, F.P. (1999). *Liquid cooling of electronic devices by single-phase convection.* New York: Wiley.

Incropera, F.P., & Dewitt, D.P. (1990). *Fundamentals of heat and mass transfer* (3rd ed.). New York: Wiley.

Incropera, F.P., & Dewitt, D.P. (2001). *Fundamentals of heat and mass transfer* (5th ed.). New York: Wiley.

Jaluria, Y. (1976). A study of transient heat transfer in long insulated wires. *Journal of Heat Transfer, 98,* 127–132, 678–680.

Jaluria, Y. (1984) Numerical study of the thermal processes in a furnace. *Numerical Heat Transfer, 7,* 211–224.

Jaluria, Y. (2018) *Advanced materials processing and manufacturing.* Cham, Switzerland: Springer.

Jaluria, Y., & Torrance, K.E. (2003). *Computational heat transfer* (2nd ed.). Washington, DC: Taylor & Francis.

Janna, W.S. (2014). *Design of fluid thermal systems* (4th ed.). Independence, KY: Cengage Learning.

Kakac, S., Shah, R.K., & Bergles, A.E. (Eds.). (1983). *Low Reynolds number flow heat exchangers.* Washington, DC: Taylor & Francis.

Kalpakjian, S., & Schmid, S.R. (2013). *Manufacturing engineering and technology* (7th ed.). New York: Pearson.

Kays. W.M., & London, A.L. (1984). *Compact heat exchangers* (3rd ed.). New York: McGraw-Hill.

Kraus. A.D., & Bar-Cohen, A. (1983). *Thermal analysis and control of electronic equipment.* Washington, DC: Hemisphere.

Lienhard, J.H. V, & Lienhard, J.H. IV. (2011). *A heat transfer textbook* (4th ed.). Mineola, NY: Dover Publications.

Lin, P.T., Gea, H.C., & Jaluria, Y. (2010). Systematic strategy for modeling and optimization of thermal systems with design uncertainties. *Frontiers Heat Mass Transfer, 1,* 013003-1-20.

Moore, F.K., & Jaluria, Y. (1972). Thermal effects of power plants on lakes. *ASME J. Heat Transfer, 94,* 163–168.

Moran, M.J., & Shapiro, H.N. (2014). *Fundamentals of engineering thermodynamics* (8th ed.). New York: Wiley.

Orlande, H.R.B. (2012). Inverse problems in heat transfer: new trends on solution methodologies and applications. *Journal of Heat Transfer, 134,* 031011-1-13.

Ozisik, M.N. (2000). *Inverse heat transfer: fundamentals and applications.* Philadelphia, PA: Taylor & Francis.

Pollak, F. (Ed.). (1989). *Pump users' handbook* (3rd ed.). Tulsa, OK: PennWell Books.
Reynolds, W.C., & Perkins, H.C. (1977). *Engineering thermodynamics* (2nd ed.). New York: McGraw-Hill.
Seraphin, D.P., Lasky, R.C., & Li, C.Y. (1989). *Principles of electronic packaging*. New York: McGraw-Hill.
Shames, I.H. (2002). *Mechanics of fluids* (4th ed.). New York: McGraw-Hill.
Steinberg, D.S. (1991). *Cooling techniques for electronic equipment* (2nd ed.). New York: Wiley-Interscience.
Stoecker, W.F. (1989). *Design of thermal systems* (3rd ed.). New York: McGraw-Hill.
Thompson, J.E., & Trickler, C.J. (1983). Fans and fan systems, *Chemical Engineering, 90* (March), 46-63.
Van Wylen, G.J., Sonntag, R.E., & Borgnakke, C. (1994). *Fundamentals of classical thermodynamics* (4th ed.), New York: Wiley.
Viswanath, R., & Jaluria, Y. (1991). Knowledge-based system for the computer aided design of ingot casting processes. *Engineering with Computers, 7*(2), 109–120.
Warring, R. (1984). *Pumps: selection, systems and applications* (2nd ed.). Houston, TX: Gulf Publishing Company.

PROBLEMS

Note: Appropriate assumptions, approximations, and inputs may be employed to solve the design problems in the following set. As seen in the examples given in this chapter, a unique solution is not obtained for an acceptable design in many of these problems, and the range in which the solution lies may be given wherever possible.

5.1 A refrigeration system is needed to provide 10 kW of cooling at 0°C, with the ambient at 25°C. Obtain a workable or acceptable design to achieve these requirements, assuming that a variation of ±5°C in both temperature levels is permissible. You may choose any appropriate fluid, component efficiencies in the range 75% to 90%, and a suitable thermodynamic cycle for the purpose.

5.2 Develop an acceptable design for a cooling system, using vapor compression, to achieve 0.5 ton of cooling at –10°C, with the ambient temperature as high as 40°C. The use of CFCs is not permitted because of their environmental effect. The efficiency of the compressor may be assumed to lie between 75% and 85%. Discuss any sensors that you might need for temperature control.

5.3 A heat pump is to be designed to obtain a heat input of 2 kW into a region that is at 25°C, as shown in Figure P5.3. The ambient temperature may be as low as 0°C. Obtain an acceptable design to satisfy these requirements, using efficiencies in the range 80% to 90% for the components. The only constraint is that the working fluid should not undergo freezing.

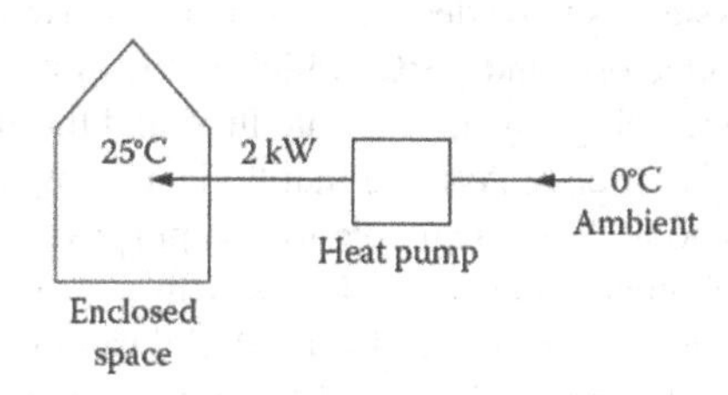

FIGURE P5.3

5.4 For the casting process considered in Problem 3.7, briefly discuss the simulation of the process and the anticipated results from the simulation. Develop a workable or acceptable design for a thermal system to achieve the desired heating.

5.5 In an oven, the support for the walls is provided by long horizontal bars, of length L and square in cross-section, attached to two vertical walls, as shown in Figure P5.5. A crossflow of ambient air, at velocity V and temperature T_a, cools the bars. The walls may be assumed to be at uniform temperature T_w. We can vary T_a, the material of the supporting bars, and the width H of the bars. The temperature at the midpoint A, T_A, must be less than a given value T_{max} due to strength considerations.

a. Develop a suitable mathematical model for this system, giving the governing equations and the relevant boundary conditions.
b. Sketch the expected temperature distribution in the bar.
c. What are the fixed quantities, requirements, and design variables in the problem?
d. Discuss the simulation of the system and obtain an acceptable design for this application.

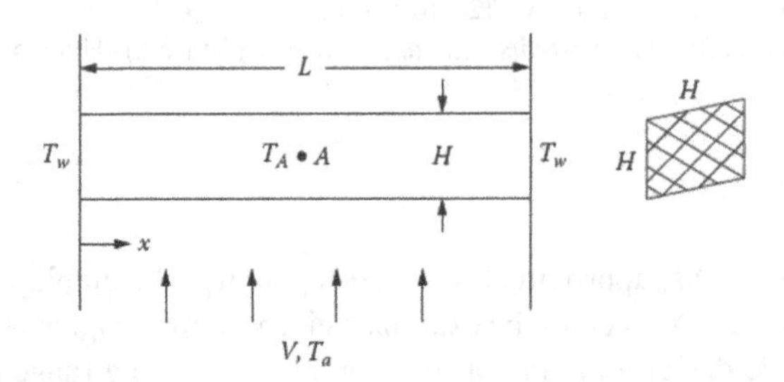

FIGURE P5.5

5.6 In the energy storage system consisting of concentric cylinders, considered in Problem 3.1, L and R_2 are given as fixed, while R_1 can be varied over a given range, $(R_1)_{min} < R_1 < (R_1)_{max}$. The approximations are the same as those given before. The metal pieces are to be heated without exceeding a maximum temperature T_{max} and interest lies in storing the maximum amount of energy.

a. Formulate the corresponding design problem, focusing on quantities that can be varied.
b. Simulate the system to determine the dependence of energy stored on the design variables.
c. Obtain an acceptable design.

5.7 If in the problem considered in Example 5.3, the hot water requirements are changed to 50°C to 75°C, determine the effect on the final results. Also, vary the ambient temperature to 30°C and determine the range of acceptable designs.

5.8 A solar energy power system is to be designed to operate between 90°C, at which hot water is available from the collectors, and 25°C, which is the ambient temperature, in order to deliver 200 kW of power. Using any appropriate fluid and thermodynamic cycle, obtain an acceptable design for this process. Assume that boilers, compressors, and turbines of efficiency in the range 70% to 80% are available for the purpose.

5.9 A cold storage room of inner dimensions 4 m × 4 m × 3 m and containing air is to be designed. The outside temperature varies from 40°C during the day to 20°C at night. The outside heat transfer coefficient is given as 10 W/(m^2·K) and that at the inner surface of the wall as 20 W/(m^2·K). A constant energy input of 4 kW may be assumed to enter the air through the door, as shown in Figure P5.9. A refrigerator system is used to extract energy from the enclosure floor. What are the important design variables in this problem? Develop a simple model for simulating the system and obtain the refrigeration capacity needed. The energy extracted by the refrigerator need not be constant with time. Also, determine the values of the other design variables to maintain a temperature of 5°C ± 2°C in the storage room. The wall thickness must not exceed 15 cm, and it is desirable to have the smallest possible refrigeration unit. Also, suggest any improvements that may be incorporated in your mathematical model for greater accuracy.

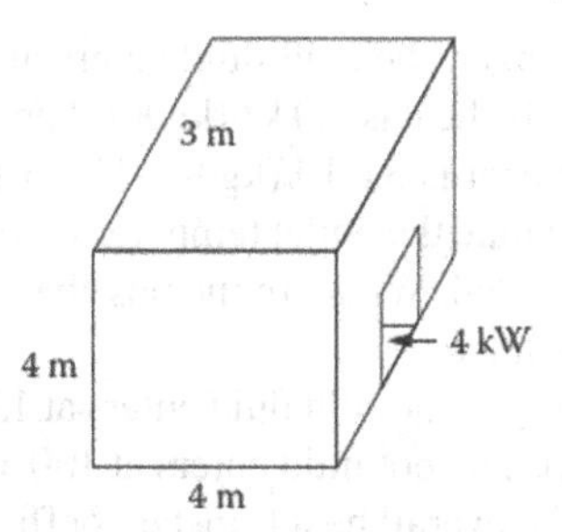

FIGURE P5.9

5.10 A piece of electronic equipment is to be designed to obtain satisfactory cooling of the components. The available air space is 0.45 m × 0.35 m × 0.25 m. The distance between any two boards must be at least 5 cm. The total number of components is 100, with each dissipating 20 W. The dimensions of a board must not exceed 0.3 m × 0.2 m. The heat transfer coefficient may be taken as 20 W/(m^2·K) if there is only one board. With each additional board, it decreases by 1 W/(m^2·K). Develop a suitable model for design of the system and obtain the minimum number of boards needed to satisfy the temperature constraint of 100°C in an ambient at 20°C. How can your model be improved for greater accuracy?

5.11 In Example 5.4, the use of hollow mandrels is suggested as an improvement in the design. Consider this change and determine the effect on the simulation and the design. However, the thickness of the wall of the mandrel should not be less than 0.5 mm from strength considerations. Also, consider the circulation of hot fluid through the mandrel to impose a higher temperature at the inner boundary of the plastic. Determine the effect of this change on the design.

5.12 In the cooling system for electronic equipment considered in Example 5.5, determine the effect on the design of allowing the board height to reach 0.2 m and of increasing the convective heat transfer coefficient to 40 W/(m^2·K) by improving the cooling process. Consider the two changes separately, taking the remaining variables as fixed. Discuss the implications of these results with respect to the design of the system.

5.13 In a condenser, water enters at 20°C and leaves at temperature T_o. Steam enters as saturated vapor at 90°C and leaves as condensate at the same temperature, as shown in Figure P5.13. The surface area of the heat exchanger is 2 m^2 and a total of 250 kW of energy is to be transferred in the heat exchanger. The overall heat transfer coefficient U is given by

$$U = \frac{\dot{m}}{0.05 + 0.2\dot{m}}$$

where $\dot{m}$ is the water mass flow rate in kg/s and U is in kW/(m^2·K). Obtain the algebraic equation that gives the water flow rate $\dot{m}$. Solve this equation by the Newton-Raphson method, starting with an initial guess between 0.5 and 0.9 kg/s. Also, calculate the outlet temperature T_o for water. Take the specific heat and density of water as 4.2 kJ/(kg·K) and 1000 kg/m^3, respectively. What are the main assumptions made in this model?

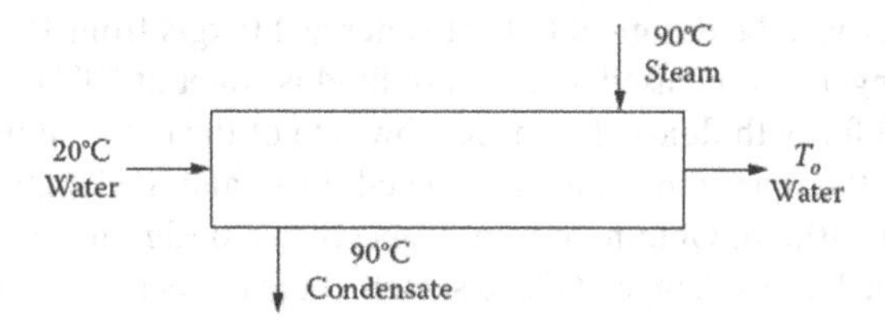

FIGURE P5.13

5.14 In a counter-flow heat exchanger, the cold fluid enters at 20°C and leaves at 60°C. Its flow rate is 0.75 kg/s and the specific heat is 4.0 kJ/(kg·K). The hot fluid enters at 80°C with a flow rate of 1.0 kg/s. Its specific heat is 3.0 kJ/(kg·K). The overall heat transfer coefficient U is given as 200 W/(m²·K). Calculate the outlet temperature of the hot fluid, the total rate of heat transfer Q, and the area A needed. What are the possible design variables in this problem, if the cold fluid conditions are fixed?

5.15 In a counter-flow heat exchanger, the cold fluid enters at 15°C. Its flow rate is 1.0 kg/s and the specific heat is 3.5 kJ/(kg·K). The hot fluid enters at 100°C at a flow rate of 1.5 kg/s. Its specific heat is 3.0 kJ/(kg·K). The overall heat transfer coefficient U is given as 200 W/(m²·K). It is desired to heat the cold fluid to 60°C ± 5°C. Outline a simple mathematical model for this system, giving the main assumptions and approximations. What are the design variables in the problem? Calculate the outlet temperature of the hot fluid, the total heat transfer rate q, and the area A needed.

5.16 In a counter-flow heat exchanger, cold water enters at 20°C and hot water at 80°C, as shown in Figure P5.16. The two flow rates are equal and denoted by $\dot{m}$ in kg/s. The specific heat is also given as the same for the cold and hot water streams and equal to 3.0 kJ/(kg·K). The value of the overall heat transfer coefficient U in kW/(m²·K) is given as

$$\frac{1}{UA} = 0.1 + \frac{0.2}{\sqrt{\dot{m}}}$$

where A is the surface area in square meters. Write down the relevant mathematical model and, employing the Newton-Raphson method for one equation, determine the value of $\dot{m}$ that results in a heat transfer rate of 300 kW. Start with an initial guess of $\dot{m}$ between 3 and 3.5 kg/s. Determine the sensitivity of the mass flow rate to the overall heat transfer rate by varying the latter from its given value of 300 kW.

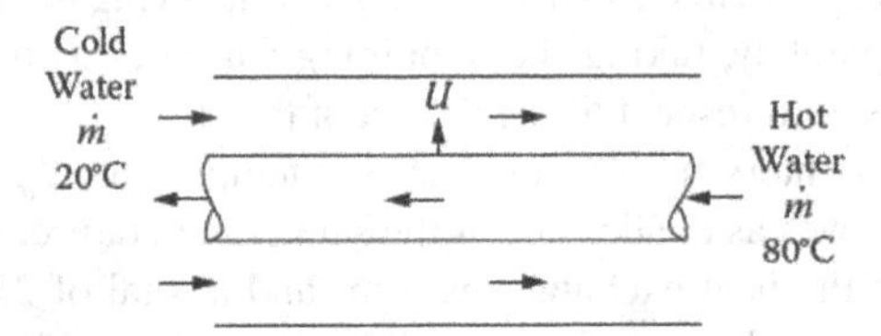

FIGURE P5.16

5.17 Water at 40°C flows at $\dot{m}$ kg/s into a condenser that has steam condensing at a constant temperature of 110°C. The UA value of the heat exchanger is given as 2.5 kW/K and the desired total heat transfer rate is 120 kW. The specific heat at constant pressure C_p for water may be taken as 4.2 kJ/(kg·K). Write the equation(s) to calculate $\dot{m}$ and, using any simulation approach, determine the appropriate value of $\dot{m}$ for the given heat transfer rate. If the total heat transfer rate varies as 120 ± 20 kW, determine the corresponding variation in $\dot{m}$.

5.18 A heat exchanger is to be designed to heat water at 1.0 kg/s from 15°C to 75°C. A parallel-flow heat exchanger is to be used and the hot fluid is water at 100°C. Take the specific heat as 4200 J/(kg·K) for both fluids. The mass flow rate of the hot fluid must not exceed 4 kg/s. The diameter of the inner pipe must not exceed 0.1 m and the length of the heat exchanger must be less than 100 m. Obtain an initial, acceptable design for this process and give the dimensions of the heat exchanger. Give a sketch of the temperature variation in the two fluid streams.

5.19 A condenser is to be designed to condense steam at 100°C to water at the same temperature, while removing 300 kW of thermal energy. A counter-flow heat exchanger is to be employed. Water at 15°C is available for flow in the inner tube and the overall heat transfer coefficient U is 2 kW/m^{2}K. The temperature rise of the cooling water must not be greater than 50°C, the inner tube diameter must not exceed 8 cm, and the length of the heat exchanger must not exceed 20 m. Obtain an acceptable design and give the corresponding mass flow rates, water temperature at the exit, and heat exchanger dimensions.

5.20 Choose a design parameter Y to follow the convergence of iterative redesign of a refrigeration system. Give reasons for your choice and sketch its expected variation as the compressor is varied to change the exit pressure.

5.21 Decide on a design parameter Y to study the convergence of an iterative design procedure for a shell and tube heat exchanger. If the design variables, such as tube and shell diameters, are varied to reach an acceptable design, how would you expect the chosen criterion Y to vary?

5.22 Take the refrigeration system considered in Example 5.1. If the storage facility is to be maintained in the temperature range of 0°C to 5°C, while the outside temperature range and the total thermal load remain unchanged, redesign the system to achieve these requirements.

5.23 Develop the initial, acceptable design for the problem considered in Example 5.2 if the maximum temperature obtainable from the heat source is only 290°C.

5.24 Redesign the solar energy storage system considered in Example 5.3 if the total amount of energy to be stored is halved, while the remaining requirements remain the same. Also, choose a design parameter Y that may be used to examine the convergence of the redesign process, giving reasons for your choice.

5.25 Redesign the heat exchanger considered in Example 5.7 for the requirements that the outer tube diameter be less than 6.0 cm and the inner tube diameter be greater than 2.0 cm, keeping the remaining conditions unchanged.

5.26 Redesign the heat exchanger in Example 5.7 to obtain a total length of less than 75.0 m, while keeping the outer tube diameter greater than 3.0 cm. No constraints are specified on the inner tube.

5.27 For a fluid flow system similar to the one considered in Example 5.8, take the design values of P_1, P_2, H, A, B, and C as 470, 700, 135, 10, 20, and 5, respectively, in the units given earlier. Simulate this system, employing the Newton-Raphson method. Study the effect on the total flow rate of varying the zero-flow pressure values (470 and 700 in the preceding) and the height (135) by ±20%. Find the maximum and minimum flow rates.

5.28 Determine the effect of varying the heat transfer coefficient to 100 W/(m$^2 \cdot$ K) and the equilibrium temperature T_e to 15°C in Example 5.6. Compare the results obtained with those presented earlier and discuss the implications for the design of a heat rejection system. What do such changes mean in actual practice?

5.29 A plastic (PVC) plate of thickness 2 cm is to be formed in the shape of an "N". For this purpose, it must be raised to a uniform temperature of 200°C and held at this temperature for 15 sec to complete the process. The temperature must not exceed the melting temperature, which is 300°C for this material. Develop a conceptual design and a mathematical model for this process. Obtain an acceptable design to achieve the desired temperature variation.

5.30 The surface of a thick steel plate is to be heat treated to a depth of 2.5 mm. A constant heat flux input of 10^6 W/m^2 is applied at the surface. The required temperature for heat treatment is 560°C, and the maximum allowable temperature in the material is 900°C. Can this arrangement be used to achieve an acceptable design? If so, determine the time at which the heat input must be turned off. Can you suggest a different or better design?

5.31 For the preceding problem, suggest a few conceptual designs and choose one as the most appropriate. Justify your choice.

6 Economic Considerations

6.1 INTRODUCTION

Among the most important indicators of the success of an engineering enterprise are the profit achieved and the return on investment. Therefore, economic considerations play a very important role in the decision-making processes that govern the design of a system. It is generally not enough to make a system technically feasible and to obtain the desired quality of the product. The costs incurred must be considered to make the effort economically viable. It is necessary to find a balance between the product quality and the cost, because the product would not sell at an excessive price even if the quality were exceptional. For a given item, there is obviously a limit on the price that the market will bear. As discussed in Chapter 1, the sales volume decreases with an increase in the price. Therefore, it is important to restrain the costs even if this means some sacrifice in the quality of the product. However, in some applications, the quality is extremely important and much higher costs are acceptable, as is the case, for instance, in racing cars, rocket engines, satellites, and defense equipment. Similarly, a poor-quality product even at a low price is not acceptable. The key aspect here is finding the proper balance between quality and cost for a given application.

Even if it can be demonstrated that a project is technically sound and would achieve the desired engineering goals, it may not be undertaken if the anticipated profit is not satisfactory. Because most industrial efforts are directed at financial profit, it is necessary to concentrate on projects that promise satisfactory return; otherwise, investment in a given company would not be attractive. Similarly, a large initial investment may make it difficult to raise the funds needed, and the project may have to be abandoned. Decisions at various stages of the design are also affected by economic considerations. The choice of materials and components, for instance, is often guided by the costs involved. The use of copper, instead of gold and silver, in electrical connections, despite the advantages of the latter in terms of corrosion resistance, is an example of such a consideration. The characteristics and production rate of the manufactured item are also affected by the market demand and the associated financial return.

Economic factors, though crucial in design and optimization, are not the only nonengineering ingredients in decision making. As mentioned earlier, several additional nontechnical aspects such as environmental, safety, legal, and political issues arise and may influence the decisions made by industrial organizations. However, several of these can also be considered as additional expenses and may again be cast in economic terms. For instance, pollution control may involve additional facilities to clean up the discharge from an industrial unit. The choice of forced draft cooling towers over natural draft ones may be made because of local opposition to the latter due to undesirable appearance, resulting in greater expense. Even political and legal concerns are often translated in terms of money and are included in the overall costs. Indeed, litigation has been one of the major hurdles in the expansion of the nuclear power industry. Providing transportation, housing, education, day care, and other facilities to workers satisfies important social needs, but these can again be treated as economic issues because of the additional expenses incurred.

Because of the crucial importance of economic considerations in most engineering decisions, it is necessary to understand the basic principles of economics and to apply these to the evaluation of investments, in terms of costs, returns, and profits. An important concept that is fundamental to economic analysis is the effect of *time* on the *worth* of money. The value of money increases as time elapses due to interest added on to the principal amount. An amount paid today is of greater value than the same amount paid 10 years later due to interest and this dependence on time must be taken into account. Similarly, inflation reduces the value of money because prices go up with time, decreasing the purchasing power of money. As we have often heard from our parents, what a dollar could buy 50 years ago is many times more than what it can buy today. Consequently, we generally consider economic aspects in terms of constant dollars at a given time, say 2000, in order to compare costs and returns. This involves bringing all the payments, expenditures, and returns to a common point in time so that the overall financial viability of an engineering enterprise can be evaluated.

This chapter first presents the basic principles involved in economic analysis, particularly the calculation of interest, the consequent variation of the worth of money with time, and the methods to shift different financial transactions to a common time frame. Different forms of payment, such as lumped sum and series of equal payments, and different methods of calculating interest that are used in practice are discussed. Taxes, depreciation, inflation, and other important factors that must generally be included in economic analysis are discussed. Thus, a brief discussion of economic analysis is presented here in order to facilitate consideration of economic factors in design and optimization. For further details on economic considerations, textbooks on the subject may be consulted. Some of the relevant books are those by Riggs and West (1986), Collier and Ledbetter (1988), Blank and Tarquin (2017), Thuesen and Fabrycky (2000), White et al. (2012), Newnan et al. (2017), Park (2012), and Sullivan et al. (2005).

An important task in the design of systems is the evaluation of different alternatives from a financial viewpoint. These alternatives may involve different designs, locations, procurement of raw materials, strategies for processing, and so on. Many of the important economic issues outlined in this chapter play a significant role in such evaluations. A few typical cases are included for illustration. The chapter also discusses the important issue of cost evaluation, considering different types of costs incurred in typical thermal systems.

6.2 CALCULATION OF INTEREST

A concept that is of crucial importance in any economic analysis is that of the worth of money as a function of time. The value increases with time due to interest accumulated, making the same payment or loan at different times leads to different amounts at a common point in time. Similarly, inflation erodes the value of money by reducing its buying capacity as time elapses. Both interest and inflation are important in analyzing and estimating costs, returns, and other financial transactions. Let us first consider the effect of interest on the value of a lumped sum, or given amount of money, as a function of time.

6.2.1 Simple Interest

The rate of interest i is the amount added or charged per year to a unit in the local currency, such as $1, of deposit or loan, respectively. This is known as the *nominal rate of interest*, and it is usually a function of time, varying with the economic climate and trends in the financial market. Frequently, the interest rate is given as a percentage, indicating the amount added per 100 of the local monetary unit. The total amount of the loan or deposit is known as the *principal*. If the interest is calculated only on the principal over a given duration, without considering the change in investment due to accumulation of interest with time and without including the interest with the principal for subsequent calculations, the resulting interest is known as *simple interest*. Then, the simple interest on the principal sum P invested over n years is simply Pni, and the final amount F consisting of the principal and interest after n years is given by

$$F = P(1 + ni) \tag{6.1}$$

Therefore, an investment of $1000 at 10% simple interest would yield $100 at the end of each year. At the end of 5 years, the total amount becomes $1500. The simple interest is very easy to calculate, but is seldom used because the interest on the accumulated interest can be substantial. In addition, one could invest the accumulated interest separately to draw additional interest. Therefore, interest on the interest generated is usually included in the calculations, and this is known as *compound interest.*

6.2.2 Compound Interest

The interest may be calculated several times a year and then added to the amount on which interest is computed in order to determine the interest over the next time period. This procedure is known as *compounding* and is frequently carried out monthly when the resulting amount, which includes the principal and the accumulated interest, is determined for calculating the interest over the next month. Compounding may also be done yearly, quarterly, daily, or at any other chosen frequency. For yearly compounding, the sum F after 1 year is $P(1 + i)$, which becomes the sum for calculating the interest over the second year. Therefore, the sum after 2 years is $P(1 + i)^2$, after 3 years $P(1 + i)^3$, and so on. This implies that for yearly compounding, the final sum F after n years is given by the expression

$$F = P(1+i)^n \tag{6.2}$$

Clearly, a considerable difference can arise between simple and compound interest as the duration of the investment or loan increases and as the interest rate increases. Figure 6.1 shows the resulting sum F for an investment of \$100 as a function of time at different interest rates, for both simple interest and annual compounding of the interest. While simple interest yields a linear increase in F with time, compound interest gives rise to a nonlinear variation, with the deviation from linear increasing as the interest rate or time increases. It is because of the considerable difference that can arise between simple and compound interest that the former is rarely used. In addition, different frequencies of compounding are often employed to yield wide variations in total interest.

If the interest is compounded m times a year, the interest on a unit amount in the time between two compoundings is i/m. Then the final sum F, which includes the principal and interest, is obtained after n years as

$$F = P\left(1+\frac{i}{m}\right)^{mn} \tag{6.3}$$

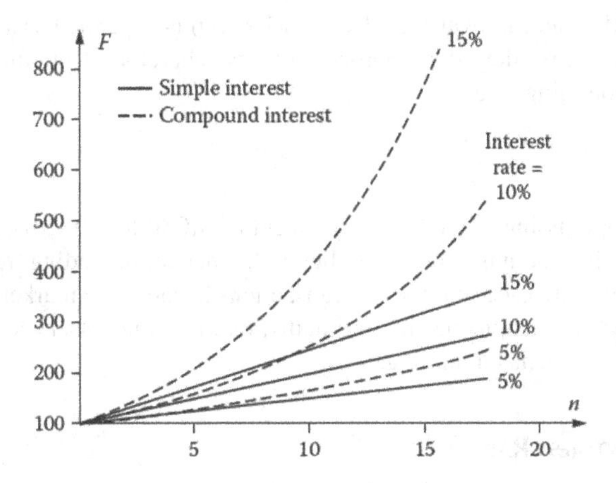

FIGURE 6.1 Variation of the sum F, consisting of the principal and accumulated interest, as a function of the number of years n, for simple interest and for annual compounding at different rates of interest.

Therefore, for a given lumped sum P, the final sum F after n years may be calculated from Equation (6.3) for different frequencies of compounding over the year. Monthly compounding, for which $m = 12$, and daily compounding, for which $m = 365$, are very commonly used by financial institutions.

It can be easily seen that a substantial difference in the accumulated interest arises for different compounding frequencies, particularly at large interest rates. For instance, an investment of $1000 becomes $2000 after 10 years at a simple interest of 10%, due to the accumulation of interest. The same investment after the same duration becomes $2593.74 if yearly compounding is employed, $2707.40 if monthly compounding is used, and $2717.91 if daily compounding is used. Therefore, a higher compounding frequency leads to a faster growth of the investment and is preferred when large financial transactions are involved.

6.2.3 Continuous Compounding

The number of times per year that the interest is compounded may be increased beyond monthly or even daily compounding to reflect the financial status of a company or an investment at a given instant. The upper limit on the frequency of compounding is *continuous compounding,* which employs an infinite number of compounding periods over the year. Thus, the interest is determined continuously as a function of time and the resulting sum at any given instant is employed in calculating the interest for the next instant. Then the total amount at a given instant is known, and investments and other financial transactions can be undertaken instantly based on the current financial situation.

As shown in the preceding section, the sum F after a period of n years with a nominal interest rate of i compounded m times per year is given by Equation (6.3). For continuous compounding, the frequency of compounding approaches infinity, i.e., $m \to \infty$, which gives

$$F = \left[P\left(1 + \frac{i}{m}\right)^{mn} \right]_{m \to \infty}$$

Therefore, taking the natural logarithm of both sides

$$ln\left(\frac{F}{P}\right) = mn\left[ln\left(1 + \frac{i}{m}\right)\right]_{m \to \infty} = mn\left[\frac{i}{m} - \frac{1}{2}\left(\frac{i}{m}\right)^2 + \frac{1}{3}\left(\frac{i}{m}\right)^3 + \cdots\right]_{m \to \infty} = ni$$

where ln represents the natural logarithm. Here, $\ln(1 + i/m)$ is expanded as a Taylor series in terms of the variable i/m and m is allowed to approach infinity. Therefore, from this equation, the sum F, for continuous compounding, is given by

$$F = Pe^{ni} \tag{6.4}$$

If continuous compounding is used, an investment of $1000 for 10 years would yield $2718.28, which is greater than the amounts obtained earlier with other compounding frequencies. Continuous compounding is commonly used in business transactions because the market varies from instant to instant and monetary transactions occur continually, making it necessary to consider the instantaneous value of money in decision making.

6.2.4 Effective Interest Rate

It is often convenient and useful to express the compounded interest in terms of an effective, or equivalent, simple interest rate. This allows one to calculate the resulting interest and to compare different investments more easily than by using the compound interest formula, such as Equation (6.3). The effective interest rate is also useful in analyzing economic transactions with different compounding frequencies, as seen later. If i_{eff} represents the effective simple interest for a given compounding scheme, the sum F, which includes the principal and interest at the end of the year, is simply

$$F = P(1 + i_{eff}) \tag{6.5}$$

Then i_{eff} is obtained from Equation (6.3), for interest being compounded m times per year, as

$$i_{eff} = \frac{F}{P} - 1 = \left(1 + \frac{i}{m}\right)^m - 1 \tag{6.6}$$

Similarly, for continuous compounding $i_{eff} = e^i - 1$.

It is also possible to obtain an equivalent interest rate over a number of years n. Then, from Equation (6.1),

$$F = P(1 + ni_{eff}) \tag{6.7}$$

which gives

$$i_{eff} = \frac{1}{n}\left(\frac{F}{P} - 1\right) = \frac{\left[\left(1 + \frac{i}{m}\right)^{mn} - 1\right]}{n} \tag{6.8}$$

The effective interest rate, therefore, allows an easy calculation of the total interest obtained on a given investment, as well as that charged on a loan, making it simple to compare different financial alternatives. It is common for financial institutions to advertise the effective interest rate, or yield, paid over the duration of an investment.

Example 6.1

Calculate the resulting sum F for an investment of \$100 after 1, 2, 5, 10, 20, and 30 years at a nominal interest rate of 10%, using simple interest as well as yearly, monthly, daily, and continuous compounding. From these results, calculate the effective interest rates over a year and also over 10 years.

SOLUTION

The resulting sum F for a given investment P is obtained for simple interest, compounding m times yearly and for continuous compounding from the following three equations, respectively, given in the preceding sections:

$$F = P(1 + ni)$$

$$F = P\left(1 + \frac{i}{m}\right)^{mn}$$

$$F = Pe^{ni}$$

Therefore, the resulting sum F for an investment of \$100 at a nominal interest rate of 10%, i.e., $i = 0.1$, after n years with different compounding frequencies, may be calculated. The results obtained are shown in Table 6.1. It is obvious that large differences in F arise over long periods of

TABLE 6.1
Effect of Compounding Frequency on the Resulting Sum for an Investment of $100 after Different Time Periods at 10% Nominal Interest Rate

Number of years	Simple Interest	Yearly Comp.	Monthly Comp.	Daily Comp.	Continuous Comp.
1	110.0	110.0	110.47	110.52	110.52
2	120.0	121.0	122.04	122.14	122.14
5	150.0	161.05	164.53	164.86	164.87
10	200.0	259.37	270.70	271.79	271.83
20	300.0	672.75	732.81	738.70	738.91
30	400.0	1744.94	1983.74	2007.73	2008.55

time, with continuous compounding yielding the largest amount. Simple interest, which considers the interest only on the initial principal amount, yields a considerably smaller amount because interest on the accumulated interest is not taken into account. Consequently, simple interest is not appropriate for most financial transactions and is generally not used. In addition, the difference between continuous and daily compounding is small. However, even this effect can be quite significant if large investments, expenditures, and payments are involved.

The effective interest rate i_{eff} is given by the equation

$$i_{eff} = \frac{F}{P} - 1$$

Therefore, i_{eff} may easily be obtained from Table 6.1 by using the calculated values of F after 1 year for the given investment of $100. It is seen that i_{eff} is equal to the nominal interest rate $i = 0.1$ for simple interest and for yearly compounding, as expected. For monthly, daily, and continuous compounding, i_{eff} is 0.1047, 0.1052, and 0.1052 (10.47%, 10.52%, and 10.52%), respectively.

The effective interest rates for yearly, monthly, daily, and continuous compounding over a period of 10 years may similarly be calculated using the equation

$$i_{eff} = \frac{1}{n}\left(\frac{F}{P} - 1\right)$$

Using the values given in Table 6.1, the effective interest rates for yearly, monthly, daily, and continuous compounding are obtained as 15.937%, 17.070%, 17.179%, and 17.183%, respectively, which are much higher than the nominal interest rate of 10%. These effective rates may be used to calculate the interest or sum after 10 years from Equation (6.7) with $n = 10$.

6.3 WORTH OF MONEY AS A FUNCTION OF TIME

It is seen from the preceding discussion that the value of money is a function of time. In order to compare or combine amounts at different times, it is necessary to bring these all to a common point in time. Once various financial transactions are obtained at a chosen time, it is possible to compare different financial alternatives and opportunities in order to make decisions on the best course of action. Different costs, over the expected duration of a project, and the anticipated returns can then be considered to determine the rate of return on the investment and the economic viability of the enterprise. Two approaches that are commonly used for bringing all financial transactions to a common time frame are the present and future worth of an investment, expenditure, or payment.

6.3.1 Present Worth

As the name suggests, the present worth (PW) of a lumped amount given at a particular time in the future is its value today. Thus, it is the amount that, if invested at the prevailing interest rate, would yield the given sum at the future date. Let us consider Equation (6.2), which gives the resulting sum F after n years at a nominal interest rate i. Then P is the present worth of sum F at the end of the given duration of n years. Therefore, the present worth of a given sum F may be written, for yearly compounding, as

$$\mathrm{PW} = P = F(1+i)^{-n} = (F)(P/F,i,n) \tag{6.9a}$$

where P/F is known as the *present worth factor* and is given by

$$P/F = (1+i)^{-n} \tag{6.9b}$$

This notation follows the scheme used in many textbooks on engineering economics; see, for example, Collier and Ledbetter (1988) and Stoecker (1989). In Equation (6.9a), the applicable interest rate i and the number of years n are included in the parentheses along with the present worth factor.

If the interest is compounded m times yearly, Equation (6.3) may be used to obtain the PW as

$$\mathrm{PW} = P = (F)\left(P/F, \frac{i}{m}, mn\right) \tag{6.10a}$$

where the present worth factor P/F is given by

$$P/F = \frac{1}{\left(1+\dfrac{i}{m}\right)^{mn}} = \left(1+\frac{i}{m}\right)^{-mn} \tag{6.10b}$$

Similarly, for continuous compounding

$$P/F = e^{-ni} \tag{6.11}$$

Therefore, the present worth factor P/F may be defined and calculated for different frequencies of compounding. The PW of a given lumped amount F representing a financial transaction, such as a payment, income, or cost, at a specified time in the future may then be obtained from the preceding equations.

The PW of the resulting sums shown in Table 6.1 for different duration and frequency of compounding is \$100. Therefore, the present worth of \$1983.74 after 30 years with monthly compounding at 10% interest is \$100, because the latter amount will yield the former if it is appropriately invested. In design, a specific amount in the future is commonly given and its PW is determined using the applicable interest rate and compounding frequency. An example of such a calculation is the expected expenditure on maintenance and repair of an industrial facility at a given time in the future. This financial transaction is then put in terms of the present in order to consider it along with other expenses.

The concept of PW is useful in evaluating different financial alternatives because it allows all transactions to be considered at a common time frame. It also makes it possible to estimate the expenses associated with a given system, financial outlay needed, and return on the investment, all these being usually based on the expected duration of the project. At the end of useful life of the system, it will be disposed or sold. This expense or financial gain is in the future and, therefore, it is usually brought to the present time frame, using the concept of PW, to include its effect in the overall financial considerations of the system. It is also possible to base such financial considerations on a point of time in the future, for which the concept of future worth, outlined in the following section, is used.

6.3.2 Future Worth

The future worth (FW) of a lumped amount P, given at the present time, may similarly be determined after a specified period of time. Therefore, the FWs of P after n years with an interest rate of i, compounded yearly or m times yearly, are given, respectively, by the following equations:

$$\mathrm{FW} = F = P(1+i)^n = (P)(F/P,i,n) \tag{6.12a}$$

$$\mathrm{FW} = F = P\left(1+\frac{i}{m}\right)^{mn} = (P)\left(F/P,\frac{i}{m},mn\right) \tag{6.12b}$$

where F/P is known as the *future worth factor* or *compound amount factor.* For continuous compounding, $F/P = e^{ni}$. Therefore, the FW of a given lumped sum today may be calculated at a specified time in the future if the compounding conditions and the interest rate are given.

Again, the FW of \$100 after different time periods and with different compounding frequencies, at a 10% nominal interest rate, may be obtained from Table 6.1. Using these results, the FW of \$1200 after 10 years of daily compounding at 10% interest is $12 \times 271.79 = \$3261.48$. Similarly, the FW for other lumped sums may be calculated for a specified future date, interest rate, and compounding. Table 6.2 shows the effect of interest rate on the FW with monthly compounding. As expected, the effect increases as the duration increases, resulting in almost a 20-fold difference between the FWs for 5% and 15% interest rates after 30 years.

As with PW, the concept of FW may be employed to bring all the relevant financial transactions to a common point in time. Frequently, the chosen time is the end of the design life of the given system. Therefore, if a telephone switching system is designed to last for 15 years, the end of this duration may be chosen as the point at which all financial dealings are considered. Once the net profit or expenditure is determined at this point, it can be easily moved to the present, if desired. The financial evaluation of a given design or system is independent of the time frame chosen for the calculations. Whether the present or the future time is employed is governed largely by convenience and by the time at which data are available. Clearly, if most of the data are available at the early stages of the project, it is better to use PW because the interest rates are better known close to the present. In addition, the duration of a given enterprise may not be specified or a definite time in the future may not be clearly indicated, making it necessary to use the PW as the basis for financial analysis and evaluation.

TABLE 6.2
Effect of Interest Rate on the Future Worth of \$100 with Monthly Compounding

	Interest Rate				
Number of Years	5 %	8 %	10 %	12 %	15 %
5	128.36	148.98	164.53	181.67	210.72
10	164.71	221.96	270.70	330.04	444.02
15	211.37	330.69	445.39	599.58	935.63
20	271.26	492.68	732.81	1089.26	1971.55
25	348.13	734.02	1205.69	1978.85	4154.41
30	446.77	1093.57	1983.74	3594.96	8754.10

Example 6.2

The design of the cooling system for a personal computer requires a fan. Three different manufacturers are willing to provide a fan with the given specifications. The first one, Fan A, is at $54, payable immediately on delivery. The second one, Fan B, requires two payments of $30 each at the end of the first and second years after delivery. The last one, Fan C, requires a payment of $65 at the end of 2 years after delivery. Because a large number of fans are to be purchased, the price is an important consideration. Consider three different interest rates, 6%, 8%, and 10%. Which fan is the best buy?

SOLUTION

In order to compare the costs of the three fans, the expenditure must be brought to a common time frame. Choosing the time of delivery for this purpose, the PW of the expenditures on the three fans must be calculated. The cost of Fan A is given at delivery and, therefore, its PW is $54. For the other two fans, the PWs at an interest rate of 6% are

$$\text{Fan B:} \quad PW = (30)(P/F, 6\%, 1) + (30)(P/F, 6\%, 2)$$

$$= \frac{30}{(1+0.06)} + \frac{30}{(1+0.06)^2} = 28.30 + 26.70 = \$55.00$$

$$\text{Fan C:} \quad PW = (65)(P/F, 6\%, 2) = \frac{65}{(1+0.06)^2} = \$57.85$$

Therefore, Fan A is the cheapest one at this interest rate.

At 8% interest rate, a similar calculation yields the PW of the cost for Fan B as $53.50 and that for Fan C as $55.73. Therefore, Fan B is the cheapest one at this rate. At 10% interest rate, the corresponding values are $52.07 for Fan B and $53.72 for Fan C. Again, Fan B is the cheapest, but even Fan C becomes cheaper than Fan A. This example illustrates the use of PW to choose between different alternatives for system design.

6.3.3 INFLATION

Inflation refers to the decline in the purchasing power of money with time due to increase in the price of goods and services. This implies that the return on an investment must be considered along with the inflation rate in order to determine the *real* return in terms of buying power. Similarly, labor, maintenance, energy, and other costs increase with time and this increase must be considered in the economic analysis of an engineering enterprise. For example, if the wages of a given worker increase from $10 per hour to $11 per hour, while the price of a loaf of bread goes from $1 to $1.10, the worker can still buy the same amount of bread and does not see a real increase in income. In order to obtain a real increase in income, the pay increase must be greater than the inflation rate. Thus, the buying power for a person may increase or decrease with time, depending on the rate of increase in income and the inflation rate. For industrial investments, it is not enough to have a rate of return that keeps pace with the inflation. The return must be higher to make a project financially attractive.

Inflation is often obtained from the price trends for groups of items that are of particular interest to a given industry or section of society. The most common measure of prices is the Consumer Price Index (CPI), obtained by the U.S. Department of Labor by tracking the prices of about 400 different goods and services. The current base year is 1983, at which point the CPI is assigned a value of 100. Table 6.3 gives the CPI from 1983 to 2018, along with the percent change from the previous year. The CPI is frequently used as a measure of the inflation rate. Note that the increase rate fluctuates from year to year. It represents the general trends in inflation, not the specific change for a particular

TABLE 6.3
Consumer Price Index (CPI)

Year	CPI	Percent Change	Year	CPI	Percent Change	Year	CPI	Percent Change
1983	100	3.8	1994	148.2	2.7	2006	201.6	3.2
1984	103.9	3.9	1995	152.4	2.5	2007	207.3	2.8
1985	107.6	3.8	1996	156.9	3.3	2008	215.3	3.8
1986	109.6	1.1	1997	160.5	1.7	2009	214.5	−0.4
1987	113.6	4.4	1998	163.0	1.6	2010	218.1	1.6
1988	118.3	4.4	1999	166.6	2.7	2011	224.9	3.2
1989	124.0	4.6	2000	172.2	3.4	2012	229.6	2.1
1990	130.7	6.1	2001	177.1	1.6	2013	233.0	1.5
1991	136.2	3.1	2002	179.9	2.4	2014	236.7	1.6
1992	140.3	2.9	2003	184.0	1.9	2015	237.0	0.1
1993	144.5	2.7	2004	188.9	3.3	2016	240.0	1.3
			2005	195.3	3.4	2017	245.1	2.1
						2018	251.1	2.4

Source: Monthly Labor Review, U.S. Dept. of Labor

item, situation, or application. Similarly, other cost measures, such as the Construction Cost Index and the Building Cost Index, representing cost of construction in terms of materials and labor, respectively, are employed to determine the inflation rate for the construction industry. The large inflation rates in the late 1970s and early 1980s significantly hampered the growth of the economy and have decreased to about 3% in recent years.

If the inflation rate is denoted by j, then the interest rate i must be equal to j for the buying power to remain unchanged, i.e., the future worth FW of a principal amount P must equal $P(1+j)^n$. If $i > j$, there is an increase in the buying power. Denoting this real increase in buying power by i_r, we may write by equating FW amounts,

$$FW = P(1+i)^n = P(1+j)^n\,(1+i_r)^n \tag{6.13}$$

Therefore,

$$i_r = \frac{1+i}{1+j} - 1 \tag{6.14}$$

This implies that, as expected, $i_r = i$ if $j = 0$, $i_r = 0$ if $i = j$, and i_r is positive for $i > j$. From this equation, we can also calculate the interest rate i needed to yield a desired real rate of increase in purchasing power, for a given inflation rate, as $i = (1+j) \times (1+i_r) - 1$. For example, if the inflation rate is 5% and the interest rate is 10%, the real interest rate, which gives the increase in the buying power, is $(1.1/1.05) - 1 = 0.0476$, or 4.76%. Similarly, if an 8% real return is desired with the same inflation rate, the interest rate needed is $(1.05)(1.08) - 1 = 0.134$ or 13.4%. Different compounding frequencies may also be considered by replacing $(1+i)^n$ in Equation (6.13) by $(1+i/m)^{mn}$ or by employing the effective interest rate, as illustrated in the following example.

Example 6.3

An engineering firm has to decide whether it should withdraw an investment that pays 8% interest, compounded monthly, and use it on a new product. It would undertake the new product if the real rate of increase in buying power from the current investment is less than 4%. The rate of inflation is given as 3.5%. Calculate the real rate of increase in buying power. Will the company decide to go for the new product? What should the yield from the current investment be if the company wants a 5% rate of increase in buying power?

SOLUTION

The real rate of increase in buying power i_r is given by the equation

$$i_r = \frac{1+i_{\text{eff}}}{1+j} - 1$$

where j is the inflation rate and the effective interest rate i_{eff} is given by

$$i_{\text{eff}} = \left(1+\frac{i}{m}\right)^m - 1$$

Here, the nominal interest rate is given as 8%. Therefore, for monthly compounding,

$$i_{\text{eff}} = \left(1+\frac{0.08}{12}\right)^{12} - 1 = 0.083$$

This gives the value of i_r as

$$i_r = \frac{1.083}{1.035} - 1 = 0.0464$$

Therefore, the real increase in purchasing power from the present investment is 4.64%. Because this is not less than 4%, the firm will continue this investment and not undertake development of the new product. However, if the inflation rate were to increase, the real rate will decrease and the company may decide to go for the new product.

To obtain a 5% real rate of increase in buying power from the current investment, the effective interest rate i_{eff} is governed by the equation

$$i_{\text{eff}} = (1+i_r)(1+j) - 1$$

which gives

$$i_{\text{eff}} = (1.05)(1.035) - 1 = 0.087$$

The nominal interest rate i may be obtained from the relationship between i and i_{eff}, given in the preceding, as

$$i = 12[(1+i_{\text{eff}})^{1/12} - 1] = 12(1.087^{1/12} - 1) = 0.0835$$

Therefore, a nominal interest rate of 8.35%, compounded monthly, is needed from the current investment to yield a real rate of increase in buying power of 5%.

6.4 SERIES OF PAYMENTS

A common circumstance encountered in engineering enterprises is that of a series of payments. Frequently, a loan is taken out to acquire a given facility and then this loan is paid off in fixed payments over the duration of the loan. Recurring expenses for maintenance and labor may be treated similarly as a series of payments over the life of the project. Both fixed and varying amounts of payments are important, the latter frequently being the result of inflation, which gives rise to increasing costs. The series of payments is also brought to a given point in time for consideration with other financial aspects. As before, the time chosen may be the present or a time in the future.

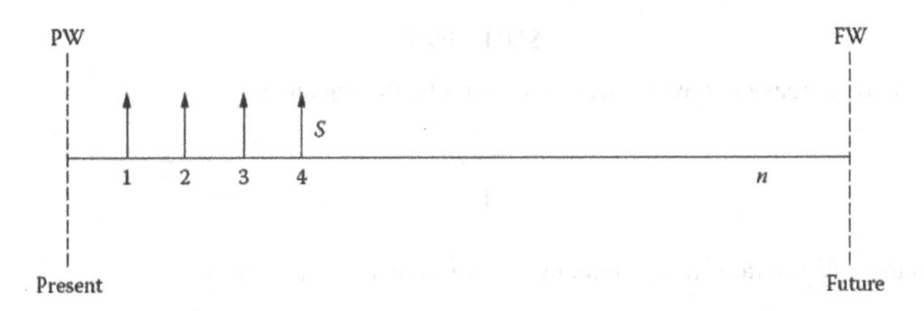

FIGURE 6.2 A uniform series of annual payments and locations of the present and future time frames, shown on the time coordinate axis in terms of number of years n.

6.4.1 Future Worth of Uniform Series of Amounts

Let us consider a series of payments, each of amount S, paid at the end of each year starting with the end of the first year, as shown in Figure 6.2. The future worth of this series at the end of n years is to be determined. This can be done easily by summing up the FWs of all these individual payments. The first payment accumulates interest for $n - 1$ years, the second for $n - 2$ years, and so on, with the second-to-last payment accumulating interest for 1 year and the last payment accumulating no interest. Therefore, if i is the nominal interest rate and yearly compounding is used, the future worth F of these series of payments is given by the expression

$$F = S[(1+i)^{n-1} + (1+i)^{n-2} + (1+i)^{n-3} + \cdots + (1+i) + 1] \tag{6.15}$$

Therefore, summing this geometric series, which has n terms and a factor of $(1 + i)$, we have

$$F = S\left[\frac{(1+i)^n - 1}{(1+i) - 1}\right] = S\left[\frac{(1+i)^n - 1}{i}\right] = (S)\left(\frac{F}{S}, i, n\right) \tag{6.16}$$

where F/S is often known as the *series future worth factor* or *the series compound amount factor*. It yields the FW of a series of payments of equal amount S when S is multiplied by this factor. The amount S of a series of payments to pay off an amount F due at a future date may also be calculated from Equation (6.16).

Different payment frequencies may similarly be considered. If m payments are made yearly, with compounding also done at this frequency, the FW is given by the expression

$$F = (S)\left(F/S, \frac{i}{m}, mn\right) = S\left[\frac{(1 + i/m)^{mn} - 1}{i/m}\right] \tag{6.17}$$

Therefore, the cumulative value of a series of payments on a future date, or the amount of each payment needed for achieving a given FW, may be calculated for different compounding frequencies. Other cases, where the payment and compounding schedules are different, are also possible and are discussed later.

6.4.2 Present Worth of Uniform Series of Amounts

The present worth of a series of equal amounts, paid at the end of the year for a number of years n starting at the end of the first year, as shown in Figure 6.2, is also obtained easily from the corresponding expression for the future worth by using the present worth factor P/F.

Therefore, for payments made at the end of each year and with annual compounding, the present worth P is given by

$$P = (F)(P/F,i,n) = S\left[\frac{(1+i)^n - 1}{i}\right](P/F,i,n) = S\left[\frac{(1+i)^n - 1}{i}\right]\frac{1}{(1+i)^n}$$

Therefore,

$$P = S\left[\frac{(1+i)^n - 1}{i(1+i)^n}\right] = (S)(P/S,i,n) \tag{6.18}$$

where P/S is the *series present worth factor.*

Similarly, if m payments are made each year with the same compounding frequency, the PW is obtained as

$$P = S\left[\frac{(1+i/m)^{mn} - 1}{(i/m)(1+i/m)^{mn}}\right] = (S)\left(P/S,\frac{i}{m},mn\right) \tag{6.19}$$

This is an important relationship because it yields the payment needed at the end of each month, year, or some other chosen time period, provided the interest compounding follows the same frequency, in order to pay off a loan taken at the present time. Therefore, if a company takes a loan to acquire a facility today, the payments at a chosen frequency over the duration of the loan can be calculated. These payments are then part of the expenses that are considered along with the income to obtain the profit.

The amount S of the uniform series of payments needed to pay off a loan taken now depends on the amount and duration of the loan, and on the rate of interest. This is given by the following expressions for yearly compounding and for compounding done m times a year, respectively:

$$S = P\left[\frac{i(1+i)^n}{(1+i)^n - 1}\right] = (P)(S/P,i,n) \tag{6.20a}$$

$$S = P\left[\frac{(i/m)(1+i/m)^{mn}}{(1+i/m)^{mn} - 1}\right] = \frac{P}{P/S} = (P)\left(S/P,\frac{i}{m},mn\right) \tag{6.20b}$$

where *S*/*P* is known as the *capital recovery factor* because it involves paying off the capital invested in the facility. It can be easily shown that the amount S of the series of payments decreases as the duration of the loan is increased and also as the interest rate is decreased. As expected, the total interest on the loan increases if the frequency of compounding is increased.

The uniform series of payments covers both the principal, or capital, and the accumulated interest. At the early stages of the loan, much of the payment goes toward the interest because the bulk of the capital accumulates interest. Near the end of the duration of the loan, very little capital is left and thus the interest is small, with most of the payment going toward paying off the capital. Therefore, the amount of unpaid capital decreases with time. It is often important to obtain the exact amount of outstanding loan at a given time so that a full payment may be made in case the financial situation of the company improves or if the current financial status of the company is to be determined for acquisitions, mergers, or other financial dealings. Note that this is the loan left in terms of the worth of money at a given time, not in terms of its PW. The calculation of the unpaid balance of the capital is demonstrated in a later example.

Example 6.4

In a food-processing system, the refrigeration and storage unit is to be purchased. A new unit can be obtained by paying \$100,000 on delivery and five annual payments of \$25,000 at the end of each year, starting at the end of the first year. A used and refurbished unit can be obtained by paying \$60,000 at delivery and 10 annual payments of \$20,000 at the end of each year. The salvage value of the new unit is \$75,000 and that of the used one is \$50,000, both being disposed of at the end of 10 years. The interest rate is 9%, compounded annually. Which alternative is financially more attractive?

SOLUTION

This problem requires bringing all the expenses and income to a common point in time. Choosing the delivery date as the present, we can move all the financial transactions to this time frame. Therefore, the present worth of the expenditure on a new unit is

$$(PW)new = 100{,}000 + (25{,}000)(P/S,\ 9\%,\ 5) - (75{,}000)(P/F,\ 9\%,\ 10)$$

$$= 100{,}000 + 25{,}000\left[\frac{(1.09)^5 - 1}{0.09(1.09)^5}\right] - 75{,}000\left[\frac{1}{(1.09)^{10}}\right]$$

$$= 100{,}000 + 97{,}241.28 - 31{,}680.81 = \$165{,}560.47$$

The PW of the expenditure on the used unit is

$$(PW)old = 60{,}000 + (20{,}000)(P/S,\ 9\%,\ 10) - (50{,}000)(P/F,\ 9\%,\ 10)$$

$$= 60{,}000 + 20{,}000\left[\frac{(1.09)^{10} - 1}{0.09(1.09)^{10}}\right] - 50{,}000\left[\frac{1}{(1.09)^{10}}\right]$$

$$= 60{,}000 + 128{,}353.15 - 21{,}120.54 = \$167{,}232.61$$

Therefore, the new unit has a smaller total expense and is preferred. If salvage values were not considered, the used unit would be cheaper. This example illustrates the use of time value of money and various economic factors to evaluate financial transactions in order to choose between alternatives and make other economic decisions. It must be noted that computer programs can easily be developed for such calculations so that variations in duration, interest rate, payments, salvage, etc., can efficiently be carried out to help in the decision-making process.

6.4.3 Continuous Compounding in a Series of Amounts

The concept of continuous compounding, presented earlier in Section 6.2.3, may also be applied to a series of lumped payments. Then Equation (6.15) may be replaced by

$$F = S\left[\left(e^i\right)^{n-1} + \left(e^i\right)^{n-2} + \left(e^i\right)^{n-3} + \cdots + \left(e^i\right) + 1\right] \tag{6.21}$$

which yields

$$F = S(F/S)_{\text{cont}} \quad \text{where}\ (F/S)_{\text{cont}} = \frac{e^{in} - 1}{e^i - 1} \tag{6.22}$$

This yields a higher future worth than that given by Equation (6.17), because continuous compounding is applied to the series of lumped amounts, rather than a finite compounding frequency. Similarly, if the annual payment amount S is divided into m equal amounts and applied uniformly over the year, with each amount drawing interest as soon as it is invested, the future worth of the series of payments becomes

$$F = \frac{S}{m}\left[\left(1+\frac{i}{m}\right)^{mn-1} + \left(1+\frac{i}{m}\right)^{mn-2} + \cdots + \left(1+\frac{i}{m}\right) + 1\right] \tag{6.23}$$

which gives

$$F = \frac{S}{m}\frac{\left(1+\frac{i}{m}\right)^{mn} - 1}{\frac{i}{m}} = S\frac{\left(1+\frac{i}{m}\right)^{mn} - 1}{i} \tag{6.24}$$

If now m is allowed to approach infinity, $(1 + i/m)^{mn}$ will approach e^{in}, as shown in Section 6.2.3. This yields

$$F = S(F/S)_{\text{cont,flow}} \quad \text{where } (F/S)_{\text{cont,flow}} = \frac{e^{\text{in}} - 1}{i} \tag{6.25}$$

Therefore, continuous compounding may be applied to a series of lumped payments or the payments themselves may be taken as a continuous flow, yielding additional factors that may be used for calculating the future worth or the present worth. This approach considers the payment and the accumulation of interest as a continuous flow, the worth of a given investment or financial transaction being obtained as a continuous function of time and thus providing the flexibility needed for making instantaneous economic decisions in a changing marketplace.

6.4.4 Changing Amount in Series of Payments

The amount in a series of payments may not be a constant, as considered previously, but may change with time. Such a variation may be the result of rising cost of labor, inflation, increasing rental charges, transportation costs, and so on. Because future changes in costs and expenditures are not easy to predict, a fixed amount of change C is often employed to consider such changes. Then the present or future worth of a series of amounts with a given annual increase C may be determined. A typical series of payments with a fixed increase in the amount is shown in Figure 6.3(a). This series may be considered as a combination of a series of uniform amounts, shown in Figure 6.3(b), and a gradient series, in which the amount is zero at the end of the first year and then increases by C each year, as shown in Figure 6.3(c). Because we have already considered the case of uniform amounts, let us consider the gradient case of Figure 6.3(c).

The PW of the gradient series shown in Figure 6.3(c) is given by the equation

$$\text{PW} = P = \frac{C}{(1+i)^2} + \frac{2C}{(1+i)^3} + \frac{3C}{(1+i)^4} + \cdots + \frac{(n-1)C}{(1+i)^n} = C\sum_{2}^{n}\frac{(n-1)}{(1+i)^n} \tag{6.26}$$

This series may be summed to yield

$$P = C\left\{\frac{1}{i}\left[\frac{(1+i)^n - 1}{i(1+i)^n} - \frac{n}{(1+i)^n}\right]\right\} = (C)(P/C,i,n) \tag{6.27}$$

where P/C is the *increment present worth factor,* which gives the PW of a series of amounts increasing by a fixed quantity each year. Then this expression, along with Equation (6.18) for a series of

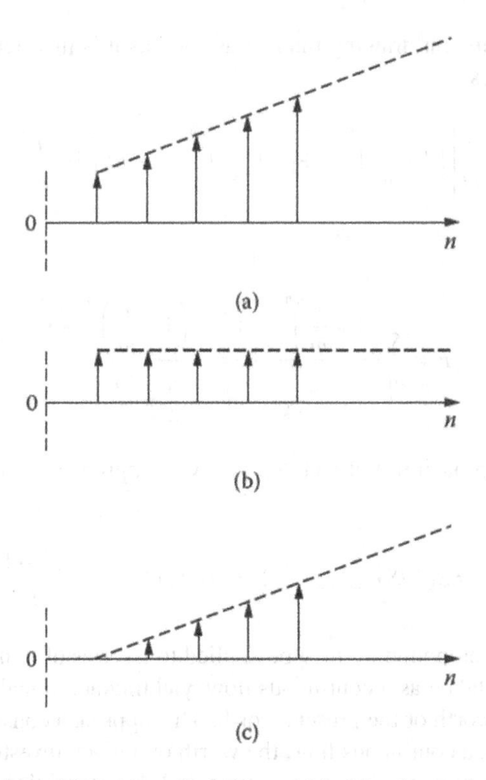

FIGURE 6.3 Sketches showing (a) a series of payments with a fixed amount of increase each year; (b) a series of uniform amounts; and (c) a gradient series of amounts.

uniform amounts, may be used to obtain the PW of a series of increasing amounts, shown in Figure 6.3(a). If the frequency of the payments is the same as the compounding frequency per year, but not annual compounding, Equation (6.27) may easily be modified by replacing i with i/m and n with mn, where m is the number of times compounding is done over the year. Table 6.4 summarizes many of the frequently used factors for economic analysis.

6.4.5 Shift in Time

If the first payment is made at the very onset of a loan, it effectively reduces the loan by the first payment amount. Therefore, payment usually starts at the end of the first time period. However, in some cases, such as payment for labor and utilities, payment is started immediately so that the payments are made at the beginning of each time period, which may be a day, month, year, etc. Then, the FW is obtained by simply adding an additional time period for the accumulation of interest for each payment. This implies multiplying the series in Equation (6.15) by $(1 + i)$. Therefore, for annual payments and compounding, with payments made at the beginning of each year, the FW is

$$F = S(1+i)\left[\frac{(1+i)^n - 1}{i}\right] \tag{6.28}$$

Similarly, for m payments each year, with each payment made at the beginning of each period and compounding done m times per year, the FW becomes

$$F = S(1+i/m)\left[\frac{(1+i/m)^{mn} - 1}{i/m}\right] \tag{6.29}$$

TABLE 6.4
Interest Factors

Factor	Purpose	Formula
F/P	Future worth of lumped sum at present	$F/P = (1+i)^n$
P/F	Present worth of lumped sum at future date	$P/F = \frac{1}{(1+i)^n}$
F/S	Future worth of series of uniform amounts	$F/S = \frac{(1+i)^n - 1}{i}$
P/S	Present worth of series of uniform amounts	$P/S = \frac{(1+i)^n - 1}{i(1+i)^n}$
P/C	Present worth of series of increasing amounts	$P/C = \frac{1}{i}\left[\frac{(1+i)^n - 1}{i(1+i)^n} - \frac{n}{(1+i)^n}\right]$
F/C	Future worth of series of increasing amounts	$F/C = \frac{(1+i)^n}{i}\left[\frac{(1+i)^n - 1}{i(1+i)^n} - \frac{n}{(1+i)^n}\right]$

6.4.6 Different Frequencies

We have considered several different compounding frequencies and payment schedules in the foregoing discussion. However, we assumed that the time period between the payments and that between compounding were the same, i.e., both were annual, quarterly, monthly, and so on. In actual practice, the two may be different, with the payment schedule based on convenience as monthly, quarterly, etc., whereas the interest is compounded more frequently or even continuously. In all such cases, the common approach is to determine the equivalent interest rate, as discussed in Section 6.2.4, and to use this rate for the subsequent calculations.

Let us consider a simple example to illustrate this procedure. If the interest is compounded monthly, whereas the uniform amount S is paid quarterly over n years, the equivalent or effective interest rate i_{eff} is obtained by equating the FWs after a year as

$$\left(1+\frac{i}{12}\right)^{12} = \left(1+\frac{i_{\text{eff}}}{4}\right)^{4}$$

which gives

$$i_{\text{eff}} = 4\left[\left(1+\frac{i}{12}\right)^{3} - 1\right] \tag{6.30}$$

Then the FW or PW of the series of amounts is determined using the effective interest rate i_{eff}. Therefore, the PW of this series of payments becomes

$$\text{PW} = P = S\left[\frac{(1+i_{\text{eff}}/4)^{4n} - 1}{(i_{\text{eff}}/4)(1+i_{\text{eff}}/4)^{4n}}\right] \tag{6.31}$$

Similarly, other frequencies of compounding and of the series may be considered, employing the preceding procedure to obtain the effective interest rate, which is then employed to calculate the relevant interest factors.

6.4.7 Changes in Schedule

The payment or withdrawal schedule for a given financial transaction is generally decided at the onset on the basis of the duration and the prevailing interest rate. However, changes in the needs or financial situation of a company may require adjustments in this schedule. For instance, the company may have problems meeting the payment and may want to reduce the amount by extending the duration of the loan. An improvement in the financial status of the company may make it possible to increase the payment amount and thus pay off the loan earlier. Significant changes in the interest rate may also require adjustments in the series of payments. Unexpected changes in inflation may make it necessary to increase the withdrawals to meet expenses. Acquisitions and other financial decisions could also affect the conditions within the company and, in turn, the strategy for payment of a loan or expenditure on a facility.

In all such cases that require a change in the schedule while the financial transaction is in progress, the best approach is to determine the worth of the loan or investment at the time of the change and then consider the new or changed conditions. For instance, if at the end of 5 years in a loan of 15 years, it is decided to accelerate payments so that the loan is paid off in 5 more years, the financial worth of the remaining loan at this point may be calculated and the new payment amount determined using the new duration and remaining loan. Because a lump sum may be moved easily from one time to another, using Equation (6.9) through Equation (6.12), all the pertinent amounts are obtained at the time when the change occurs and the calculations for the new payment or withdrawal schedule carried out. The following example illustrates the basic approach in such cases.

Example 6.5

A company acquires a packaging facility for \$250,000. It pays \$30,000 as down payment on delivery of the facility and takes a loan for the remaining amount. This loan is to be paid in 10 years, with monthly payments starting at the end of the first month. The rate of interest is 10%, compounded monthly. Calculate the monthly payment. After 5 years, the financial situation of the company is much better and the company wants to pay off the loan. Calculate the amount it has to pay at the end of 5 years to take care of the remaining loan. Also, calculate the monthly payment if the company wants to pay off the loan in the next 2 years instead.

SOLUTION

The monthly payment S that the company must pay toward the loan is obtained from Equation (6.20b), which gives

$$S = P\left[\frac{(i/m)(1+i/m)^{mn}}{(1+i/m)^{mn}-1}\right]$$

where P is the present worth of the loan, being \$250,000 – \$30,000 = \$220,000. In addition, the interest rate $i = 0.1$, number of years $n = 10$, and compounding frequency $m = 12$ for monthly compounding. Therefore,

$$S = (220,000)\left[\frac{(0.1/12)(1+0.1/12)^{120}}{(1+0.1/12)^{120}-1}\right] = \$2,907.32$$

This is the monthly payment needed to pay the loan in 10 years.

The future worth F_P of these monthly payments at the end of 5 years is calculated from Equation (6.17) with $n = 5$ as

$$F_P = S\left[\frac{(1+i/m)^{mn}-1}{i/m}\right] = (2,907.32)\left[\frac{(1+0.1/12)^{60}-1}{0.1/12}\right] = \$225,134.35$$

The future worth F_L of the loan after 5 years is given by Equation (6.12b) with $n = 5$ as

$$F_L = P\left(1+\frac{i}{m}\right)^{mn} = (220,000)(1+0.1/12)^{60} = \$361,967.97$$

Therefore, to pay off the loan at the end of 5 years, the company must pay

$$F_L - F_P = \$361,967.97 - \$225,134.35 = \$136,833.62$$

This implies that at the end of 5 years, which is half the duration of the loan, the amount needed to pay off the loan is almost 62% of the original loan. As mentioned earlier, the early payments go largely toward the interest and the outstanding loan decreases very slowly.

If the company wants to pay off the remaining loan in 2 more years, rather than the full amount now, the current value of the unpaid loan, \$136,833.62, is taken as the PW at this point in time from the preceding calculation. Monthly payments beyond this point in order to pay off this loan can be calculated from the formula given in Equation (6.20b). Then, $i = 0.1$, $n = 2$, and $m = 12$, and we obtain

$$S = (136,833.62)\left[\frac{(0.1/12)(1+0.1/12)^{24}}{(1+0.1/12)^{24}-1}\right] = \$6,314.18$$

Therefore, a monthly payment of \$6,314.18 will pay off the remaining loan in 2 more years and a payment of \$136,833.62 will pay off the loan in full at this stage. Other situations can similarly be considered and payments needed to pay off the loan can be calculated at various points in time.

6.5 RAISING CAPITAL

An important activity in the operation and growth of an industrial enterprise is that of raising capital. The money may be needed for replacing or improving existing facilities, establishing a new line of products, acquiring a new industrial unit, and so on. For example, the establishment of Saturn cars as a new division in General Motors represented a major investment for which raising capital was a critical consideration. Similarly, replacing existing injection molding machines with new and improved ones requires additional capital that may see a return in terms of higher productivity and thus greater profit. Though companies generally plan for routine replacement and upgrading of facilities, using internal funds for the purpose, new ventures and major expansions usually involve raising capital from external sources. A company may raise capital by many methods. For relatively small amounts, money may be borrowed from banks, the loan often being paid off as a series of payments over a chosen duration as discussed earlier. Among the most common means for raising large sums of money are bonds and stocks issued by the company.

6.5.1 BONDS

A bond is issued with a specific face value, which is the amount that will be paid by the company at the maturity of the bond, and a fixed interest rate to be paid while the bond is in effect. For instance, if a bond with a face value of \$1000 is issued for a duration of 10 years with an interest rate of 8% paid quarterly, an interest of $\$1000 \times 0.08/4 = \$20$ is paid after every 3 months for the duration of the bond and \$1000 is paid at maturity after 10 years. The company raises capital by selling a number of these bonds. The initial price of the bond, as well as the price at any time while the bond is in effect, may be greater or smaller than the face value, depending on the prevailing interest rate.

If the interest rate available in the market is higher than that yielded by the bond, the selling price of the bond drops below its face value because the same interest is obtained by investing a smaller amount elsewhere. Similarly, if the prevailing interest rate is lower than that paid by the bond, the seller of the bond can demand a price higher than the face value because the yield is larger than that available from other investments. If the selling price equals the face value, the bond is said to be sold *at par*. The stability of the company, the general economic climate in the country, the financial needs of the seller, etc., can play a part in the final sale price of a bond. The company that issued the bond to raise capital is generally not involved and continues to pay the dividend on the bond as promised.

In order to determine the appropriate current price of a bond, the basic principle employed is that the total yield from the bond equals that available from investment of the amount paid for the bond at the prevailing interest rate. If P_c is the current price to be paid for the bond, P_f is the face value of the bond, i_b is the interest rate on the bond paid m times a year, i_c is the current interest rate also compounded m times per year, and n is the number of years to the maturity of the bond, we may write

$$P_f + P_f\left(\frac{i_b}{m}\right)\left(F/S, \frac{i_c}{m}, mn\right) = (P_c)\left(F/P, \frac{i_c}{m}, mn\right) \tag{6.32}$$

where the FW of the investments is used as a basis for equating the two. The first term on the left-hand side is the face value paid at maturity. The second term gives the FW of the series of dividend payments from the bond, invested at the prevailing interest rate. This implies that the dividend yield from the bond is assumed to be invested immediately to obtain the current interest rate. The right-hand side simply gives the FW of the current price of the bond invested at the prevailing interest rate, which is assumed to remain unchanged over the remaining duration of the bond. Therefore, this equation may be written as

$$P_f\left[1+\left(\frac{i_b}{m}\right)\frac{(1+i_c/m)^{mn}-1}{i_c/m}\right] = P_c\left(1+\frac{i_c}{m}\right)^{mn} \tag{6.33}$$

It is easy to see that if $i_b = i_c$, $P_f = P_c$. Similarly, for $i_b > i_c$, it can be shown that $P_c > P_f$, and for $i_b < i_c$, $P_c < P_f$. Therefore, as the prevailing interest rate goes up or down, the selling price of the bond correspondingly goes down or up. This variation occurs because the yield of a bond is fixed, whereas the interest rate for an investment fluctuates due to the economic climate.

Frequently, the dividend is paid semiannually or quarterly, making $m = 2$ or 4, respectively. The frequencies of dividend payments and compounding may also be different. Such cases can easily be handled by the use of the effective interest rate i_{eff}, as illustrated in the following example.

Example 6.6

An industrial bond has a face value of \$1000 and has 6 years to maturity. It pays dividends at the rate of 7.5% twice a year. The current interest rate is 5%, compounded monthly. Calculate the sale price of the bond.

SOLUTION

The current sale price P_c of the bond is governed by Equation (6.33), which is written as

$$P_f\left[1+\left(\frac{i_b}{m}\right)\frac{(1+i_c/m)^{mn}-1}{i_c/m}\right] = P_c\left(1+\frac{i_c}{m}\right)^{mn}$$

if the frequencies of interest payment by the bond and compounding are the same. Here the face value $P_f = \$1000$, the current interest rate $i_c = 0.05$, compounding frequency is 12, and the interest rate of the bond $i_b = 0.075$. However, the number of times per year the bond pays interest is two.

Because the frequency of compounding is different from the frequency at which the interest from the bond is paid, we need to determine the effective interest rate over a 6-month period so that a common frequency of two per year may be used. Therefore,

$$P\left(1+\frac{i_{\text{eff}}}{2}\right)^2 = P\left(1+\frac{0.05}{12}\right)^{12}$$

which gives the effective interest rate over half a year as 0.0505. This effective interest rate is used in the equation given earlier for the sale price of the bond. Thus,

$$1{,}000+(1{,}000)\left(\frac{0.075}{2}\right)\left[\frac{(1+i_{\text{eff}}/2)^{12}-1}{i_{\text{eff}}/2}\right] = P_c\left(1+\frac{i_{\text{eff}}}{2}\right)^{12}$$

Here, the effective interest rate is used to obtain the same frequency as that of the dividends that are paid every 6 months, i.e., $m = 2$. This equation may be solved to obtain the sale price P_c. The resulting value of P_c is $1125.34. Because the current interest rate is lower than that paid by the bond, a sale price higher than the face value of the bond is obtained, as expected.

6.5.2 Stocks

Another important means used by industry to raise capital is by selling stock in the company. Stocks may be sold at the start of a company, when it goes public with its offering, or additional amounts may be offered at later stages to raise capital for new enterprises. The company obtains money only from such initial or additional stock offerings and not from later trading of the stocks on the various stock exchanges. Each stockholder thus shares the ownership of the company with other stockholders and the governing board is generally comprised of prominent stockholders and their nominees. Even though the company does not receive money from future trading of its stock, the stockholders are obviously interested in the worth of their stock. The progress and well-being of the company is judged by the value of its stock. In addition, if further stocks are offered, the demand, value, and number will depend on the current stock price. If the company wants to borrow money from other sources, or if it wants to acquire or merge with another company, the value of its stock is an important measure of its worth.

Because of all these considerations, considerable efforts are directed at avoiding a decrease in stock prices and at increasing their worth. Dividends are also paid depending on the profit made by the company. At the end of the year, the board of directors may decide that a dividend will be paid, as well as the rate of payment. However, very often companies simply invest the profits in the business or give additional stocks to the stockholders. Therefore, the long-term yield of a stock is much harder to determine than for a bond because the prices fluctuate, depending on the market, and the dividends are usually not fixed. However, stocks are very important for the company as well as for investors.

In order to determine the return on a stock, the initial price P_s, the final sale price, and the dividend, if any, must be considered. The dividends are assumed to be invested immediately at the prevailing interest rate, as done for the dividends from bonds, and the resulting total amount at the time when the final sale is made is calculated. Then the FW of the stock F_s consists of the sale price and the resulting amount from the dividends. The FW of any commission paid to the broker and other expenses F_c is subtracted to yield the final return from the stock. The rate of return r_s is then computed over the number of years n for which the stock is held as

$$r_s = \left[(F_s - F_c - P_s)/nP_s\right],$$

which may also be expressed as a percentage rate of return.

6.6 TAXES

The government depends heavily on taxes to finance its operations and to provide services. Most of this revenue comes from income taxes, which are levied on individuals as well as on companies. Because the income tax may vary from one location to another, states and cities with lower income taxes are popular locations for companies. In recent years, several organizations have moved their head offices from the Northeast (United States) to the Midwest and South in order to reduce the tax burden. Taxes on the facilities, through real estate taxes and other local taxes, are also important in deciding on the location of an establishment. An example of this is a company moving from New York to New Jersey to take advantage of the lower state and city taxes. Another interesting aspect is that various states, and even the federal government, may provide incentives to expand certain industries by giving tax breaks. The growth and use of solar energy systems were initially spurred, to a large extent, by tax incentives given by the government.

6.6.1 Inclusion of Taxes

It is necessary to include taxes in the evaluation of the overall return on the investment in an engineering enterprise and also for comparing different financial alternatives for a venture. As mentioned earlier, two main forms of taxation are of concern to an engineering company: income tax and real estate, or property, tax.

6.6.1.1 Income Tax

The overall profit made by a given company is the income that is taxed by the federal, state, and local governments. Though the federal taxation rate remains unchanged with location, the state and local taxes are strongly dependent on the location, varying from close to zero to as high as 20% across the United States. However, the federal tax may vary with the size of the company and the nature of the industry. Therefore, the tax bite on the profit of a company is quite substantial, generally being on the order of 40%–50% for typical industrial establishments.

Because the amount paid in taxes is lost by the company, diligent efforts are made to reduce this payment by employing different legal means. Certainly, locating and registering the company at a place where the local taxes are low is a common approach. Similarly, providing bonuses and additional benefits to the employees, expanding and upgrading facilities, and acquisition of new facilities or enterprises increase the expenses incurred and reduce the taxes owed by the company. Therefore, if a company finds itself with a possible profit of $6 million at the end of the year, it may decide to give away $1 million in bonuses to the employees, spend $1 million on providing additional health or residential amenities, $2 million on upgrading existing manufacturing facilities, and $2 million on acquiring a small manufacturing establishment that makes items of interest to the company. Thus, the net profit is zero and the company pays no taxes, while it improves its manufacturing capability and gains the goodwill of its employees, not to mention their well-being and efficiency. Such a move would also make the company more competitive and could result in an increase in the price of its stock.

6.6.1.2 Real Estate and Local Taxes

Taxes are also levied on the property owned by the company. These may simply be real estate taxes on the value of the buildings and land occupied by the company or may include charges by the local authorities to provide services, such as access roads, security, and solid waste removal. All these are generally included as expenses in the operation of the company. Different alternatives involve different types of expenses and, therefore, the design of the system may be affected by these taxes. For instance, a system that involves a smaller floor area and, therefore, a smaller building and lower real estate taxes is more desirable than one that requires a larger floor area. Similarly, the raw materials needed and the resulting waste are important in determining expenses for transportation and disposal, possibly making one system more cost effective than another.

6.6.2 Depreciation

An important concept with respect to the calculation of taxes is that of *depreciation*. Because a given facility has a finite useful life, after which it must be replaced, it is assumed to depreciate in value as time elapses until it is sold or discarded at its salvage value. In essence, an amount is allowed to be put aside each year for its replacement at the end of its useful life. This amount is the depreciation and is taken as an expense each year, thus reducing the taxes to be paid by the company.

The federal government allows several approaches to calculating depreciation. The simplest is *straight-line depreciation,* in which the facility is assumed to depreciate from its initial cost P to its salvage value Q at a constant rate. Therefore, the depreciation D in each year is given by

$$D = \frac{P - Q}{n} \tag{6.34}$$

where n is the number of years of tax life, which is the typical life of the facility in question based on guidelines available from the Internal Revenue Service. This approach allows a constant deduction from the income each year for the facility. The book value of the item is the initial cost minus the total depreciation charged up to a given point in time. Therefore, the book value B at the end of the jth year is given by

$$B = P - \frac{j}{n}(P - Q) \tag{6.35}$$

In actual practice, most facilities depreciate faster in the initial years than in later years, as anyone who has ever bought a new car knows very well. This is largely because of the lower desirability and unknown maintenance of the used item. As time elapses and the wear and tear are well-established, the depreciation usually becomes much smaller. Different distributions are used to represent this trend of greater depreciation rate in the early years. These include the *sum-of-years digits* (SYD), the *declining balance,* and the *modified accelerated cost recovery* methods.

In the SYD method, the depreciation D for a year n_1 under consideration is given by

$$D = \left[\frac{n - n_1 + 1}{n(n+1)/2}\right](P - Q) \tag{6.36}$$

where the denominator is the sum $n(n + 1)/2$ of the digits representing the years, i.e., 1, 2, 3, …, n. The numerator is the digit corresponding to the given year when the digits are arranged in reverse order, as n, n–1, n–2, and so on. By using this calculation procedure, the depreciation is larger than that obtained by the linear method in the early years and smaller in the later years. If the fractional depreciation D_f is defined as $D_f = D/(P - Q)$, the straight-line depreciation and the SYD methods give its value, respectively, as

$$D_f = \frac{1}{n} \quad \text{and} \quad D_f = 2\left[\frac{1}{n} - \frac{n_1}{n(n+1)}\right] \tag{6.37}$$

Figure 6.4 shows the fractional depreciation as a function of time for an item with a 15-year tax life, using these two approaches. Therefore, the deduction for depreciation is larger for the SYD method in early years, resulting in lower taxes, while the depreciation is smaller near the end of the tax life. However, because the value of money increases with time due to interest, it is advantageous to have a greater tax burden later in the life of the facility rather than at the early stages.

In the declining balance method, the depreciation D_j in the jth year is taken as a fixed fraction f of the book value of the item at the beginning of the jth year. Therefore, the depreciation in the first

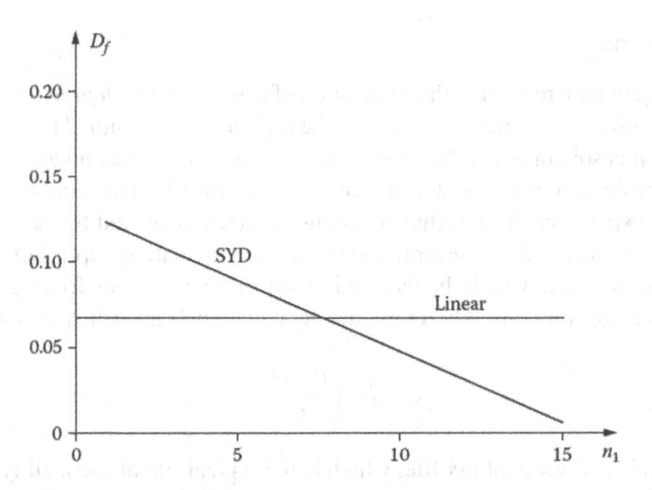

FIGURE 6.4 Variation of the fractional depreciation D_f with the number of years under consideration n_l for the linear and SYD depreciation calculation methods.

year is fP, giving a *book value*, or worth of the item, as $(1-f)P$ at the beginning of the second year. Similarly, the book value at the beginning of the third year is $(1-f)^2P$, and so on. This implies that the book value after n years is $(1-f)^nP$. Therefore, if the salvage value Q after n years is set equal to the book value, we can obtain f from the resulting equation as

$$f = 1 - \left(\frac{Q}{P}\right)^{1/n} \tag{6.38}$$

Also, the book value at the beginning of the jth year is $P(1-f)^{j-1}$. This gives the depreciation D_j in the jth year as

$$D_j = fP(1-f)^{j-1} \tag{6.39}$$

Therefore, an accelerated write-off is obtained in the early years.

In the modified accelerated cost recovery method, the depreciation D is calculated from the equation

$$D = rP \tag{6.40}$$

where r is termed the recovery rate and is obtained from tabulated values, as given in terms of percent in Table 6.5. The item is assumed to be put in service at the middle point of the first year. Therefore, only 50% of the first year depreciation is used for the first year and a half-year depreciation is used for the year $n+1$, where n is the total life of the item. The value of the item is assumed to be completely depreciated by the end of its useful life. The method starts out with the declining balance method and switches to the straight line method in later years.

Taxes must be included as an unavoidable part of any economic analysis. Property and other local taxes are included as expenditures and the income taxes are applied on the profit. Besides affecting the overall return on the investment, taxes may also influence the strategy for expenditures in the company. Increasing the spending on upgrading the facilities and on employee benefits were mentioned earlier as two possibilities. In addition, if two alternative facilities are available for a specific purpose, the selection may be influenced by the depreciation allowed and the corresponding effect on taxes.

TABLE 6.5
Recovery Rates r (%) Used in Modified Accelerated Cost Recovery Method for Calculating Depreciation

Year	$n = 3$	$n = 5$	$n = 7$	$n = 10$	$n = 15$
1	33.3	20.0	14.3	10.0	5.0
2	44.5	32.0	24.5	18.0	9.5
3	14.8	19.2	17.5	14.4	8.6
4	7.4	11.5	12.5	11.5	7.7
5		11.5	8.9	9.2	6.9
6		5.8	8.9	7.4	6.2
7			8.9	6.6	5.9
8			4.5	6.6	5.9
9				6.5	5.9
10				6.5	5.9
11				3.3	5.9
12–15					5.9
16					3.0

Source: G.E. Dieter (2000) *Engineering Design*, 3rd ed., McGraw-Hill, New York.

6.7 ECONOMIC FACTOR IN DESIGN

It is evident from the preceding discussions and examples that economic considerations play an important role in the planning, execution, and success of an engineering enterprise. Decisions on the upgrading of existing facilities, new ventures and investments, and the completion of ongoing projects are all strongly influenced by the underlying financial implications. Similarly, economic issues are addressed at various stages during the design of a thermal system and may affect the decisions concerning the selection of components, materials, dimensions, etc., of the system. Though economic considerations can influence the system design in many ways, the most important one relates to the evaluation of a potential investment or cost. Therefore, if two alternative methods are available to achieve a desired function or goal, an evaluation of the investment may be undertaken to determine which alternative is the preferred one. This determination may be based on the lowest cost or the highest return, depending on the particular application under consideration.

6.7.1 Cost Comparison

As discussed earlier, in the design process, it is common to make a decision between different alternatives, each of which satisfies the given requirements. For instance, different types and makes of heat exchangers are often available to transfer the desired amount of energy from one particular fluid to another. Similarly, different types of pumps may be employed for transporting water from one location to another. Different materials may be used for a given item in the system. In such cases, the choice is often made by comparing all the relevant costs. However, because of the time value of money, the costs must be compared on a similar basis with respect to time. Several approaches may be adopted for such cost comparisons, the most common being present worth, annual costs, and life-cycle savings. A detailed discussion follows.

6.7.1.1 Present Worth Analysis

If two alternatives for achieving a given function have the same time period of operation, a comparison may conveniently be made on the basis of the PW of all costs. Then, initial cost, salvage value, and maintenance costs are all brought to the common time frame of the present. If the items lead to savings or benefits, these may also be included in the comparison. Similarly, expenses on upgrading or refurbishing the item, if incurred during the period of operation, may be included. The following example illustrates the use of PW analysis to choose between two alternatives.

Example 6.7

A manufacturing system that is being designed needs a laser welding machine. Two machines, *A* and *B*, both of which are suitable for the manufacturing process, are being considered. The applicable costs in U.S. dollars are given as

	A	B
Initial cost	$20,000	$30,000
Annual maintenance cost	4,000	2,000
Refurbishing cost at end of 3 years	3,000	0
Annual savings	500	1,000
Salvage value	500	3,000

The useful life is 6 years for both machines, and the rate of interest is 8%, compounded annually. Determine which machine is a better acquisition.

SOLUTION

This illustration is typical of alternatives that frequently arise in the design of thermal systems. The machine with the lower initial cost has a larger maintenance cost and a smaller salvage value. It also needs to be refurbished at the end of 3 years. The savings provided by improvement in quality and in productivity are higher for the machine with the larger initial cost. Therefore, if only initial cost is considered, machine *A* is cheaper. However, the added expenses for maintenance and refurbishing, as well as lower salvage value and savings, may make machine *B* a better investment.

The PW of the expenses, minus the benefits or savings, for the two machines are calculated as

$$
\begin{aligned}
(\mathrm{PW})_A &= 20000 + [4000 - 500](P/S,\ 8\%,\ 6) + (3000)(P/F,\ 8\%,\ 3) \\
&\quad -(500)(\mathrm{P/F},\ 8\%\ 6) \\
&= 20000 + [3500](4.623) + 3000(0.794) - 500(0.63) = \$38,246.50 \\
(\mathrm{PW})_B &= 30000 + [2000 - 1000](P/S,\ 8\%,\ 6) - (3000)(P/F,\ 8\%,\ 6) \\
&= 30000 + [1000](4.623) - 3000(0.63) = \$32,732.37
\end{aligned}
$$

where the various factors, along with the interest rate and time period, are indicated within parentheses. Therefore, machine *B* is a better investment because the total costs are less by $5,514.13 on a PW basis. An economic decision based on this cost comparison leads to the selection of machine *B* over machine *A*. Unless other considerations, such as the availability of funds, are brought in, machine *B* is chosen for the desired application.

As mentioned earlier, even though such calculations can easily be carried out for economic analysis, in most practical situations, computer programs are developed so that all the given inputs may be entered to obtain the desired results conveniently and efficiently. Computational environments such as Matlab are particularly well-suited for developing interactive programs that may be used in the design process to obtain results on total expenditures, profits, payments, sales, etc.

6.7.1.2 Annual Costs

All the costs may also be considered on an annual basis for comparison. Thus, the initial cost, salvage value, savings, and additional expenses are put in terms of an annual payment or benefit.

This approach is particularly appropriate if the time periods of the two alternatives are different. Considering the preceding example of laser welding machines, the annual costs for the two machines are calculated as

$$\begin{aligned}C_A &= [4000-500]+20000/(P/S,\ 8\%,\ 6)-500/(F/S,\ 8\%,\ 6)\\&\quad+(3000)(\mathrm{P/F},\ 8\%,\ 3)/(P/S,\ 8\%,\ 6)\\&=[3500]+20000/4.623-500/7.336+2381.5/4.623=\$8273.31\\C_B &= [2000-1000]+30000/(P/S,\ 8\%,\ 6)-3000/(F/S,\ 8\%,\ 6)\\&=[1000]+30000/4.623-3000/7.336=\$7080.51\end{aligned}$$

Here, the refurbishing cost for machine A is first converted to its PW and then to the annual cost. The PW of the total costs, calculated earlier, could also be employed to calculate annual costs using the factor P/S, i.e., $C_A = (\mathrm{PW})_A/(P/S)$. As before, machine B is a better investment because the annual costs are lower. Similarly, the FW of the total costs at the end of the useful life of the facility may be used for selecting the better option.

6.7.1.3 Life-Cycle Savings

It is obvious that the comparison between any two alternatives is a function of the prevailing interest rate and the time period considered. Depending on the values of these two quantities, one or the other option may be preferred. The life-cycle savings considers the difference between the PW of the costs for the two alternatives and determines the conditions under which a particular alternative is advantageous. *Life-cycle savings*, or LCS, is given by the expression

$$\begin{aligned}\mathrm{LCS} &= (\text{Initial cost of } A - \text{Initial cost of } B)\\&\quad+[\text{Annual costs for } A - \text{Annual costs for } B](P/S,\ i,\ n)\\&\quad+[\text{Refurbishing cost of } A - \text{Refurbishing cost of } B](P/F,\ i,\ n_1)\\&\quad-[\text{Annual savings for } A - \text{Annual savings for } B](P/S,\ i,\ n)\\&\quad-[\text{Salvage value of } A - \text{Salvage value of } B](P/F,\ i,\ n)\end{aligned}$$

where n is the time period, i is the interest rate, and n_1 is the time when refurbishing is done.

Using the values given earlier for the laser cutting machines, the LCS is obtained as \$5514.13. Now, if either n is varied, keeping i fixed, or i is varied, keeping n fixed, the LCS changes, indicating the effect of these parameters on the additional cost of using machine A. If the LCS is positive, the costs are higher for machine A and savings are obtained if machine B is used. Therefore, LCS represents the savings obtained by using machine B.

Figure 6.5(a) shows that the LCS decreases with the interest rate, becoming zero at an interest rate of about 23.25% for the given time period of 6 years. This interest rate is sometimes referred to as the *return on investment.* If the prevailing interest rate is less than the return on investment, the LCS is positive and a greater return is obtained with machine B, because the costs for machine A are larger. If the actual interest rate is larger than the return on investment, machine A is a better choice. This implies that as long as the prevailing interest rate is less than the return on investment, the additional initial cost of machine B is recovered.

From Figure 6.5(b), the LCS is seen to increase with the number of years n, at the given interest rate of 8%, becoming zero at around 2.55 years. The time at which the LCS becomes zero is often termed as the *payback* time. If the actual time period is less than this payback time, the LCS is negative and machine A is recommended. For time periods larger than the payback time, machine B is preferred because positive savings are obtained due to larger costs for machine A. This implies that if the time period is greater than the payback time, there is enough time to recover the additional initial expense on machine B.

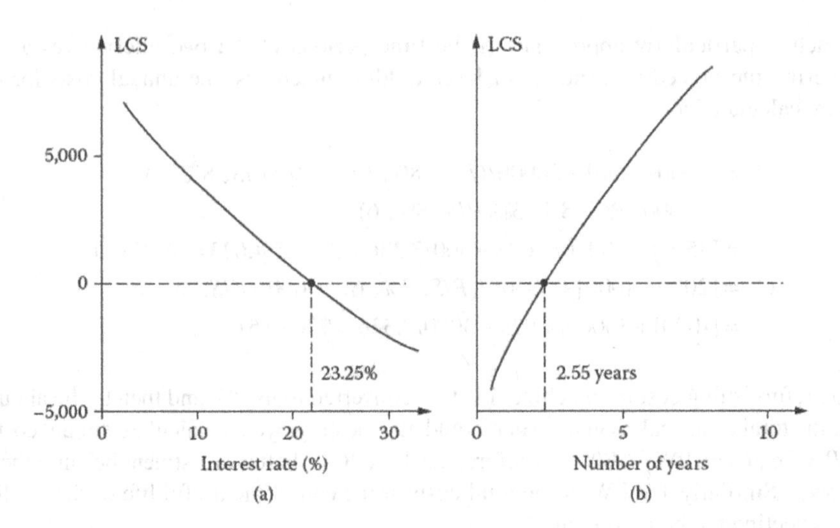

FIGURE 6.5 Variation of life-cycle savings (LCS) with interest rate and number of years for the problem considered in Example 6.7. The time period is held constant at 6 years for the first case and the interest rate is held constant at 8% for the second case.

6.7.2 Rate of Return

In the preceding section, we discussed cost comparisons for different courses of action in order to choose the least expensive one. These ideas can easily be extended to evaluate potential investments and to determine the most profitable investment. Thus, net present worth, payback period, and rate of return are commonly used methods for evaluating investments.

The *net present worth* approach calculates the benefits and the costs at time zero using the prevailing interest rate i or a minimum acceptable return on capital. Therefore, the following expression may be used for the net present worth (NPW):

$$\begin{aligned} \text{NPW} &= \text{Present worth of benefits} - \text{Present worth of costs} \\ &= [\text{Annual income} - \text{Annual costs}](P/S,\ i,\ n) \\ &\quad +[\text{Salvage value}](P/F,\ i,\ n) - \text{Initial cost} \end{aligned}$$

Preference is given to the project with the largest positive net present worth.

The *payback period* is the time needed to fully recover the initial investment in the enterprise. The prevailing interest rate may be used to obtain a realistic time period for recovery, as outlined in the preceding section. Therefore, in the above expression for the NPW, the value of n at which the NPW becomes zero is the payback time. If the NPW is set equal to zero, the resulting nonlinear equation may be solved by iteration to determine n. Root solving techniques given in Chapter 4 may be used to simplify the calculations. The investment with a shorter payback period is preferred.

The *rate of return* is an important concept in choosing different alternatives in the design process and in the consideration of the economic viability of an investment. Sometimes, the time value of money is not considered and the annual profit and expenses are employed, taking depreciation as an expense, to calculate the return. However, a more useful and widely used approach for calculating the rate of return is one that is similar to the concept of return on investment presented earlier. The rate of return is treated as an interest rate and is the rate at which the net present worth is zero. Thus, this rate of return, which is also known as *discounted cash flow* or *internal rate of return*, indicates the return on the investment as well as repayment of the original investment. All the costs and incomes are considered to calculate the rate of return, which is the interest rate at which the income and the costs balance out. The following example illustrates these calculations.

Example 6.8

Two plastic-forming facilities, A and B, are suitable for a plastic recycling system. The following financial data are given for the two facilities:

	A	B
Initial cost	\$50,000	\$80,000
Annual income	26,000	36,000
Annual maintenance and other costs	11,000	15,000
Salvage value	10,000	20,000

The life of the facilities is given as 5 years. Calculate the rates of return for the two cases. Also, include the effect of taxes, assuming a tax rate of 50% and using the straight-line method for depreciation.

SOLUTION

The equation to calculate the rate of return i_r is obtained by setting the NPW equal to zero to yield

$$[\text{Annual income} - \text{Annual costs}](P/S,\ i_r,\ n) + [\text{Salvage value}](P/F,\ i_r,\ n) - \text{Initial cost} = 0$$

The unknown in this equation is the rate of return i_r which is distinguished from the prevailing interest rate i. For machine A, we have

$$(26{,}000 - 11{,}000)\left[\frac{(1+i_r)^n - 1}{i_r(1+i_r)^n}\right] + \frac{10{,}000}{(1+i_r)^n} - 50{,}000 = 0$$

Similarly, the corresponding equation for machine B may be written. Therefore, two algebraic, nonlinear equations are obtained for the two cases and may be solved to obtain i_r. Because of the nonlinearity of the equation, iteration is needed, using root solving methods as discussed in Chapter 4, to determine i_r. The rates of return for the two cases are obtained as 19.05% and 15.16%, respectively, indicating that machine A is a better investment.

Taxes may also be included in the calculation for the rate of return by using the depreciation of the facility. For machine A, the annual profit is \$26,000 – \$11,000 = \$15,000. If the income tax rate is 50%, a tax of \$7500 has to be paid. However, if depreciation is included, the taxes are reduced. Using the straight-line method, the annual depreciation is (50,000 – 10,000)/5 = \$8000. With this depreciation, the annual income becomes \$7000, and the income tax is \$3500. Similarly, for machine B, the income tax is calculated with depreciation as \$4500. The income tax is an additional expense that reduces the rate of return. Adding the income tax to the annual costs, the rates of return for the two machines, A and B, are obtained as 9.86% and 7.79%, respectively. Therefore, the return after taxes is much lower and may even change the preferred alternative, depending on the depreciation.

Therefore, if the design of a thermal system involves selection of a component, such as a heat exchanger, pump, or storage tank, or of the materials to be employed, the rate of return on the investment may be used to choose the best alternative. In addition, the calculated rate of return may be used to decide whether the given investment should be undertaken at all or if another course of action should be pursued.

It is seen from the preceding example that the rate of return on an investment depends on the various costs, income, salvage value, time period, and depreciation. The return must be greater than the prevailing interest rate to make an investment worthwhile. If expenditures have to be undertaken as part of the project, cost comparisons may be used to select the least expensive course of action. The time value of money must be considered in order to obtain realistic costs or returns. The NPW and the payback period may also be employed, depending on circumstances. For machines A and B, the NPW is calculated as \$13,071.01 and \$12,024.95, respectively, using the expression given previously. Similarly, the payback period is obtained for the two cases as 3.53 and 3.98 years, respectively. Thus, all criteria point to machine *A* as the better investment.

6.8 APPLICATION TO THERMAL SYSTEMS

The preceding sections have presented economic analyses to take the time value of money into account for a variety of financial transactions. Comparisons of costs and income were also discussed, including the effects of inflation, taxes, depreciation, and salvage. As seen in some of the examples, these considerations are important in deciding whether a given project is financially viable and in choosing between different alternatives that are otherwise similar.

Economic aspects are also important in the design of thermal systems and are employed at various stages. Some of the important decisions based on economic considerations are

1. Whether to proceed with the project
2. Whether to modify existing systems or to develop new ones
3. Whether to design system parts and subsystems, such as heat exchangers and solar collectors, or to buy them from manufacturers
4. Which conceptual designs, materials, components, and configurations to use
5. Which heating and cooling methods, energy source, etc., to use
6. Effect of adjusting design variables to use standard items available in the market

Clearly, many decisions based on financial considerations pertain to the direction of the project and are made at high levels of management. However, many decisions are also made during the design process, particularly those covered by items 4 through 6 in the preceding list. The choice between various acceptable components and materials is made largely because of costs involved. Long-term energy costs would affect the decision on the energy source, such as electricity or natural gas. Costs are also invoked in the adjustment of design variables for the final design. In most cases, the output or performance is balanced against the cost so that an optimal design that maximizes the output/cost ratio is obtained. We have seen in Chapter 5 that a domain of acceptable designs that satisfy the given requirements and constraints is generally obtained. Costs then become an important factor in choosing the best or optimal design from this domain.

Cost evaluation involves determining the different types of costs incurred in the manufacture of a given product. It also concerns maintenance and operating costs of the system. This information is used in establishing the sale price of the product, in reducing manufacturing costs, and in advertising the product. The two main types of costs in manufacturing are fixed and variable. The former are essentially independent of the amount of goods produced, whereas the latter vary with production rate. Examples of these costs are

1. Fixed costs: Investment costs; equipment procurement; establishment of facilities; expenses on technical, management, and sales personnel; etc.
2. Variable costs: Labor, maintenance, utilities, storage, packaging, supplies and parts, raw materials, etc.

Estimating costs is a fairly complicated process and is generally based on information available on costs pertaining to labor, maintenance, materials, transportation, manufacturing, etc., as applicable to a given industry. Estimates have been developed for the time taken for different manufacturing processes and may be used to obtain the costs incurred in producing a given item (Dieter, 2000). Similarly, overhead charges may be applied to direct labor costs to take care of various fixed costs. Again, these charges depend on the industry and the company involved. Costs of different materials and components, such as blowers, pumps, fans, air conditioners, and heat exchangers, are also available from the manufacturers as well as from retailers. The costs obviously vary with the size, quality, and capacity of the equipment. In many cases, the cost versus size data may be curve fitted, using techniques given in Chapter 3, to simplify the calculations and facilitate the choice of a suitable item. Several such expressions are considered in the chapters on optimization.

Maintenance and operating costs for a system that has been designed are also not easy to estimate. Tests on prototype and actual systems are generally used to estimate the rate of consumption of energy. Accelerated tests are often carried out to determine the expected maintenance and service costs. Companies that manufacture thermal systems, such as refrigerators, air conditioners, automobiles, and plastic extruders, usually provide cost estimates regarding energy consumption and servicing. Such information is also provided by independent organizations and publications such as *Consumer Reports,* which evaluate different products and rate these in terms of the best performance-to-cost ratio. The sale price of a given system, as well as its advertisement, are strongly affected by such estimates of costs. Many of these aspects play an important role in system optimization.

6.9 SUMMARY

This chapter discusses financial aspects that are of critical importance in most engineering endeavors. Two main aspects are stressed. The first relates to the basic procedures employed in economic analysis, considering the time value of money; the second involves the relevance of economic considerations in the design of thermal systems. Therefore, calculations of PW and FW of lumped amounts as well as of a series of uniform or increasing payments, for different frequencies of compounding the interest, are discussed. The effects of inflation, taxes, depreciation, and different schedules of payment on economic analysis are considered. Methods of raising capital, such as stocks and bonds, are discussed. The calculation procedures outlined here will be useful in analyzing an enterprise or project in order to determine the overall costs, profits, and rate of return. This would allow one to determine if a particular effort is financially acceptable.

An important consideration, with respect to the design of thermal systems, is choosing between different alternatives based on expense or return on investment. Such a decision could arise at different stages in the design process and could affect the choice of conceptual design, components, materials, geometry, dimensions, and other design variables. Costs are very important in design and often form the basis for choosing between different options that are otherwise acceptable. Different methods for comparing costs are given and may be used to judge the superiority of one approach over another. Obviously, cost comparisons may indicate that a design that is technically superior is too expensive and may lead to a solution that is somewhat inferior but less expensive. Therefore, trade-offs have to be made to balance the technical needs of the project against the financial ones.

The economic analysis of the design could also indicate whether it is financially better to design a component of the system or to purchase it from a manufacturer. It could guide the modifications in existing systems by determining if the suggested changes are financially appropriate. The implementation of the final design is also very much dependent on the expenditures involved, funds available, and financial outlook of the market. All these considerations are time-dependent because the economic climate may vary within the company, the relevant industry, and the global arena. Therefore, economic decisions are made based on existing conditions as well as projections for the future. The analyses needed for such decisions are presented in this chapter along with several examples to illustrate the basic ideas involved. Such financial considerations are particularly important in the optimization of the system because we are often interested in maximizing the output per unit cost.

REFERENCES

Blank, L.T., & Tarquin, A.J. (2017). *Engineering economy* (8th ed.). New York: McGraw-Hill.

Collier, C.A., & Ledbetter, W.B. (1988). *Engineering economic and cost analysis* (2nd ed.). New York: Harper and Row.

Dieter, G.E. (2000). *Engineering design* (3rd ed.). New York: McGraw-Hill.

Newnan, D.G., Eschenbach, T.G., & Lavelle, J.P. (2017). *Engineering economic analysis* (13th ed.). Oxford, UK: Oxford University Press.
Park, C.S. (2012). *Fundamentals of engineering economics* (3rd ed.). New York: Pearson.
Riggs, J.L., & West, T. (1986). *Engineering economics* (3rd ed.). New York: McGraw-Hill.
Stoecker, W.F. (1989). *Design of thermal systems* (3rd ed.). New York: McGraw-Hill.
Sullivan, W.G., Wicks, E.M., & Luxhoj, J. (2005). *Engineering economy* (13th ed.). Upper Saddle River, NJ: Prentice-Hall.
Thuesen, G.J., & Fabrycky, W.J. (2000). *Engineering economy* (9th ed.). New York: Pearson.
White, J.A., Case, K.E., & Pratt, D.B. (2012). *Principles of engineering economic analysis* (6th ed.). New York: Wiley.

PROBLEMS

6.1 A steel plant has a hot rolling facility for steel sheets that is to be sold to a smaller company at $15,000 after 10 years. What is the PW of this salvage price if the interest is 8%, compounded annually? Also, calculate the PW for an interest rate of 12% with annual compounding. Will the PW be larger or smaller if the compounding frequency were increased to monthly? Explain the observed behavior.

6.2 A chemical company wants to replace its hot water heating and storage system. One buyer offers $10,000 for the old system, payable immediately on delivery. Another buyer offers $15,000, which is to be paid 5 years after delivery of the old system. If the current interest rate is 10%, compounded monthly, which offer is better financially?

6.3 A company wants to put aside $150,000 to meet its expenditure on repair and maintenance of equipment. Considering yearly, quarterly, monthly, and daily compounding, determine the total annual interest the company will get in these cases if the nominal interest rate is 7.5%.

6.4 For nominal interest rates of 8% and 12%, calculate the effective interest rates for yearly, quarterly, monthly, daily, and continuous compounding.

6.5 A company acquires a manufacturing facility by borrowing $750,000 at 8% nominal interest, compounded daily. The loan has to be paid off in 10 years with payments starting at the end of the first year. Calculate the effective annual rate of interest and the amount of the annual payment.

6.6 In the preceding problem, calculate the amount of the loan left after four and also after eight payments. Also, calculate the total amount of interest paid by the company over the duration of the loan.

6.7 A food processing company wants to buy a facility that costs $500,000. It can obtain a loan for 10 years at 10% interest or for 15 years at 15% interest. In both cases, yearly payments are to be made starting at the end of the first year.

   a. Which alternative has a lower yearly payment?
   b. What is the loan amount paid off after 5 years for the two cases? Calculate the amounts needed to pay off the entire loan at this time.

6.8 A company makes a profit of 10%. Calculate the real profit in terms of buying power for inflation rates of 4%, 6%, and 8%.

6.9 A firm wants to have an actual profit of 8% in terms of buying power. If the inflation rate is 11%, calculate the profit that must be achieved by the firm in order to achieve its goal.

6.10 A small chemical company wants to obtain a loan of $120,000 to buy a plastic recycling machine. It has the option of a loan at 6% interest for 10 years or a loan at 8% for 8 years, with monthly compounding and payment in both cases. Calculate the monthly payments in the two cases, assuming that the first payment is made at the end of the first month. Also, calculate the total interest paid in the two options.

6.11 A $1000 bond has 4 years to maturity and pays 8% interest twice a year. If the current interest is 6% compounded annually, calculate the sale price of the bond. Repeat the problem if the current interest is compounded daily.

6.12 A $5000 bond has 5 years to maturity and it pays 7% interest at the end of each year. If it is sold at $4500, calculate the current nominal interest.

6.13 A pharmaceutical company wants to acquire a packaging machine. It can buy it at the current price of $100,000 or rent it at $18,000 per year. The rental payments are to be made at the beginning of each year, starting on the date the machine is delivered. If the interest rate is 10%, compounded annually, and if the machine becomes the property of the company after 10 yearly payments, which option is better economically?

6.14 In the preceding problem, if the machine has a salvage value of $15,000 at the end of 10 years for the option of buying the facility, will the conclusions change? If the interest rate is 20%, with salvage, how will the results change?

6.15 An industrial concern wants to procure a manufacturing facility. It can buy an old machine by paying $50,000 now and ten yearly payments of $2000 each, starting at the end of the first year. It can also buy a new machine by paying $100,000 now and five yearly payments of $1000 each, starting at the end of the sixth year. The salvage value is $10,000 and $20,000 in the two cases, respectively. The nominal interest rate is 10%. Which is the better option, assuming that the performance of the two machines is the same?

6.16 As a project engineer involved in the design of a manufacturing facility, you need to acquire a polymer injection-molding machine. Two options are available from two different companies. The first one, option A, requires 15 payments of $8000 per year, paid at the beginning of each year and starting immediately. The second one, option B, requires eight payments of $15,000 per year, paid at the end of each year and starting at the end of the first year. Determine which option is better economically if the interest rate is 8%. Also, calculate the amounts needed to pay off the loan after half the number of payments have been made in the two options.

6.17 A company needs 1000 thermostats a year for a factory that manufactures heating equipment. It can buy these at $10 each from a subcontractor, with payment made at the beginning of each year for the annual demand. It can also procure a facility at $75,000, with $2000 needed for maintenance at the end of each year, to manufacture these. If the facility has a life of 10 years and a salvage value of $10,000 at the end of its life, which option is more economical? Take the interest rate as 8% compounded annually.

6.18 In the preceding problem, calculate the annual demand for thermostats at which the two options will incur the same expense.

6.19 You have designed a thermal system that needs a plastic part in the assembly. You can either buy the required number of parts from a manufacturer or buy an injection-molding machine to produce these items yourself. The number of items needed is 2000 every year. In the first option, you have to pay $12 per item for the yearly consumption at the beginning of each year. The chosen life of the project is 10 years. For the other option, you can lease a machine for $20,000 each year, paid at the end of each year for 10 years. The maintenance of the machine and raw materials cost $1000 at the end of the first year, $2000 at the end of the second year, and increasing by $1000 each year, until the last payment of $9000 is made at the end of the ninth year. Provide the payment schedule for the second option and determine which option is better financially. Take the interest rate as 10%, compounded annually.

6.20 A manufacturer of electronic equipment needs 10,000 cooling fans over a year. The company can buy these for $20 each, payable on delivery at the beginning of each year, or at $24, payable 2 years after delivery. Which is the better financial alternative if the interest rate is 9% compounded daily? Also, calculate the results if the interest rate drops to 8%.

6.21 A gas burner needed for a furnace can be purchased from three different suppliers. The first supplier wants \$100 for each burner, payable on delivery. The second supplier is willing to take payments of \$55 each at the end of 6 months and the year. The third supplier claims that its deal is the best and asks for \$110 at the end of the year. The current interest rate is 8.5%, compounded continuously. Because a large number of burners are to be bought, it is important to get the best financial deal. Whom would you recommend? Would your recommendation change if the interest rate were to go up, say to 12%?

6.22 A company acquires a manufacturing facility for \$300,000, to be paid in 15 equal annual payments starting at the end of the first year. The rate of interest is 8%, compounded annually. After six payments, the company is in good financial condition and wants to pay off the loan in four more equal annual payments, starting with the end of the seventh year, as shown in Figure P6.22. Calculate the first and the last payment (at the end of the tenth year) made by the company.

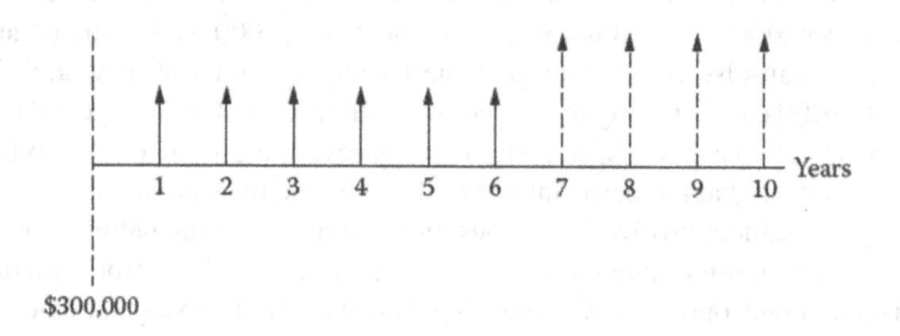

FIGURE P6.22

6.23 An industry takes a loan of \$200,000 for a machine, to be paid off in 10 years by annual payments beginning at the end of the first year. The rate of interest is 10%, compounded monthly. At the end of five payments, the company finds itself in a good financial situation and management decides to pay off the loan in the following year, as shown in Figure P6.23. How much does it have to pay at the end of the sixth year to end the debt? Also, calculate the amount of the annual payment in the first 5 years.

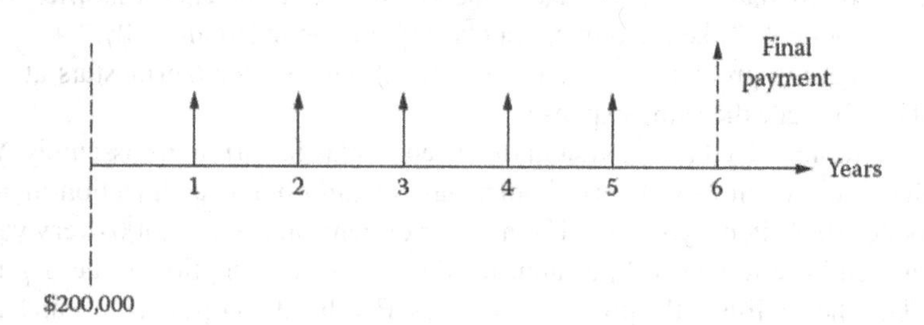

FIGURE P6.23

6.24 A company is planning to buy a machine, which requires a down payment of \$150,000 and has a salvage value of \$30,000 after 10 years. The cost of maintenance is covered by the manufacturer up to the end of 3 years. For the fourth year, the maintenance cost is \$1000, paid at the end of the year. These costs increase by \$1000 each year until the end of the tenth year, when the company pays for the maintenance of the facility and sells it, as shown in Figure P6.24. The rate of interest is 10%, compounded annually. Find the PW of buying and maintaining the machine over 10 years. If the company wants to take out a fixed amount annually from its income to cover the entire expense, calculate this amount, starting at the end of the first year.

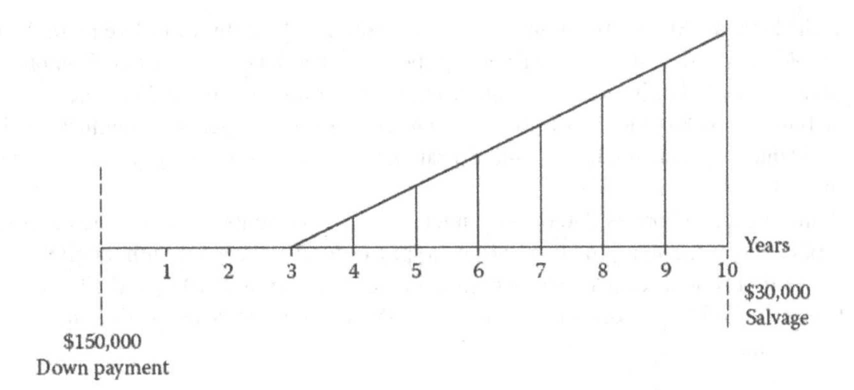

FIGURE P6.24

6.25 A manufacturing company wants to buy a welding machine, which costs $10,000. The cost of maintenance is zero in the first year, $500 in the second year, and increases by $500 each year until the eighth year when the company pays the maintenance expense and sells the facility for $2000. The maintenance expense is paid at the end of each year. The rate of interest is 9%, compounded annually. Find the PW of acquiring and maintaining this machine over 8 years.

6.26 A company is considering the purchase and operation of a manufacturing system. The initial cost of the system is $200,000 and the maintenance costs are zero at the end of the first year, $5000 at the end of the second year, $10,000 at the end of the third year, and continue to increase by $5000 each year. If the life of the system is 15 years, find the PW of buying and maintaining it over this period. Also, find the uniform annual amount that the system costs the company each year, starting after the first year. Take the interest rate as 10% compounded annually.

6.27 An industrial firm wants to acquire a laser-cutting machine. It can buy a new one by paying $150,000 now and six yearly payments of $20,000 each, starting at the end of the fifth year. It can also buy an old machine by paying $100,000 now and 10 yearly payments of $15,000, starting at the end of the first year. At the end of 10 years, the salvage value of the new machine is $80,000 and that of the old one is $60,000. Which is the better purchase for the firm, if the interest rate is 12% compounded annually? Use life-cycle savings. Repeat the calculation for a 10% interest rate.

6.28 Using the data given in Example 6.7, choose between the two machines for interest rates of 4%, 6%, and 10%. Compare the results obtained with those given in the example and discuss the implications of the observed trends.

6.29 Again using the data given in Example 6.7, study the effects of the useful lives of the machines on their economic viability. Consider useful life durations of 4, 8, and 10 years. Discuss the implications of the results obtained in making appropriate choices in the design process based on costs.

6.30 Calculate the rates of return for the two facilities given in Example 6.8 as functions of the useful lives of the facilities. Take the life as 4, 6, and 8 years, and calculate the corresponding rates of return with and without taxes at the rate of 50% of the profit taken into account. Compare these with the earlier results and comment on their significance in the design process.

6.31 A loan of $5000 is taken from a bank that charges a nominal interest rate i, compounded monthly. A monthly payment of $200, starting at the end of the first month, is needed for 36 months to pay off the loan. Write down the equation for calculating the value of i. Using root solving methods of Chapter 4, with Matlab, obtain the value of i.

6.32 If the loan is \$20,000, the monthly payment \$600, and the duration 4 years in the preceding problem, calculate the value of i, using the approach developed earlier. Compare the result obtained with that in the preceding problem and comment on the difference.

6.33 A bond of \$1000 yields 8% interest annually and has 7 years to maturity. It is sold for \$500 due to prevailing higher interest rate i. Calculate the interest rate i, using any suitable method.

6.34 A firm needs to borrow \$50,000 to undertake improvements in its existing facilities. For the repayment of the loan, the firm wishes to pay only \$1000 each month, beginning at the end of the first month. Considering possible interest rates of 8%, 10%, and 12%, determine the time required to pay off the loan for these three cases. Also, determine the amount of the last payment.

7 Problem Formulation for Optimization

7.1 INTRODUCTION

In the preceding chapters, we focused our attention on obtaining a workable, feasible, or acceptable design of a system. Such a design satisfies the requirements for the given application, without violating any imposed constraints. A system fabricated or assembled on the basis of this design is expected to perform the appropriate tasks for which the effort was undertaken. However, the design would generally not be the *best* design, where the definition of *best* is based on cost, performance, efficiency, or some other such measure. In actual practice, we are usually interested in obtaining the best quality or performance per unit cost, with acceptable environmental effects. This brings in the concept of *optimization*, which minimizes or maximizes quantities and characteristics of particular interest to a given application.

Optimization is by no means a new concept. In our daily lives, we attempt to optimize by seeking to obtain the largest amount of goods or output per unit expenditure, this being the main idea behind clearance sales and competition. In the academic world, most students try to achieve the best grades with the least amount of work, hopefully without violating the constraints imposed by ethics and regulations. The best value of various items, including consumer products such as televisions, automobiles, cameras, smart phones, vacation trips, and even education, per dollar spent, is often quoted to indicate the cost effectiveness of these items. Different measures of quality, such as durability, finish, reliability, corrosion resistance, strength, and speed, are included in these considerations, often based on actual consumer inputs, as is the case with publications such as *Consumer Reports*. Thus, buyers, who may be a student (or a parent) seeking an appropriate college for higher education, a couple looking for a cruise, or a young professional searching for his first dream car, may use information available on the best value for their money to make their choices.

7.1.1 Optimization in Design

The need to optimize is similarly very important in design and has become particularly crucial in recent times due to growing global competition. It is no longer enough to obtain a workable system that performs the desired tasks and meets the given constraints. At the very least, several workable designs should be generated and the final design, which minimizes or maximizes an appropriately chosen quantity, selected from these. In general, many parameters affect the performance and cost of a system. Therefore, if the parameters are varied, an optimum can often be obtained in quantities such as power per unit fuel input, cost, efficiency, energy consumption per unit output, and other features of the system. Different product characteristics may be of particular interest in different applications and the most important and relevant ones may be employed for optimization. For instance, weight is particularly important in aerospace and aeronautical applications, acceleration in automobiles, energy consumption in refrigerators, and flow rate in a water pumping system. Thus, these characteristics may be chosen for minimization or maximization.

Workable designs are obtained over the allowable ranges of the design variables in order to satisfy the given requirements and constraints. A unique solution is generally not obtained and different system designs may be generated for a given application. We may call the region over which acceptable designs are obtained the *domain of workable designs*, given in terms of the physical variables in the

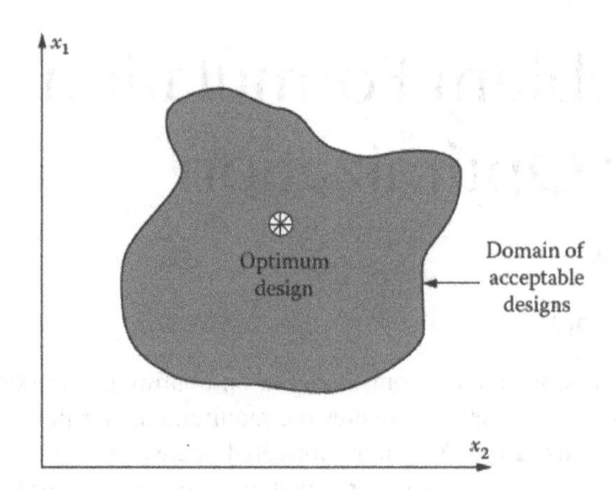

FIGURE 7.1 The optimum design in a domain of acceptable designs.

problem. Figure 7.1 shows, qualitatively, a sketch of such a domain in terms of variables x_1 and x_2, where these may be physical quantities such as the diameter and length of the shell in a shell-and-tube heat exchanger. Then, any design in this domain is an acceptable or workable design and may be selected for the problem at hand. Optimization, on the other hand, tries to find the best solution, one that minimizes or maximizes a feature or quantity of particular interest. Local extrema may be present at different points in the domain of acceptable designs. However, only one global optimal point, which yields the minimum or maximum in the entire domain, is found to arise in most applications, as indicated in the figure. It is this optimal design that is sought in the optimization process.

7.1.2 Final Optimized Design

The optimization process is expected to yield an optimal design or a subdomain in which the optimum lies. The final system design is obtained on the basis of this solution. The design variables are generally not taken as exactly equal to those obtained from the optimal solution, but are changed somewhat to use more convenient sizes, dimensions, and standard items available in the market. For instance, an optimal dimension of 4.65 m may be taken as 5.0 m, a 8.34 kW motor as a 10 kW motor, or a 1.8 kW heater as a 2.0 kW heater, because items with these specifications may be readily available, rather than having the exact ones custom made.

An important concept that is used at this stage to finalize the design variables is *sensitivity*, which indicates the effect of changing a given variable on the output or performance of the system. In addition, safety factors are employed to account for inaccuracies and uncertainties in the modeling, simulation, and design, as well as for fluctuations in operating conditions and other unforeseen circumstances. Some changes may also be made due to fabrication or material limitations. Based on all these considerations, the final system design is obtained and communicated to various interested parties, particularly those involved in fabrication and prototype development.

Generally, optimization of a system applies to its hardware, i.e., to the geometry, dimensions, materials, and components. As discussed in Chapter 2, the hardware refers to the fixed parts of the system, components that cannot be easily varied and items that determine the overall specifications of the system. However, the system performance is also dependent on operating conditions, such as temperature, pressure, flow rate, and heat input. These conditions can generally be varied quite easily, over ranges that are determined by the hardware. Therefore, the output of the system, as well as the costs incurred, may also be optimized with respect to the operating conditions. Such an optimum may be given in terms of the conditions for obtaining the highest efficiency or output. For instance, the settings for optimal output from an air conditioner or a refrigerator may be given as functions of the ambient conditions.

This chapter presents the important considerations that govern the optimization of a system. The formulation of the optimization problem and different methods that are employed to solve it are outlined, with detailed discussion of these methods taken up in subsequent chapters. It will be assumed that we have been successful in obtaining a domain of acceptable designs and are now seeking an optimal design. The modeling and simulation effort that has been used to obtain a workable design is also assumed to be available for optimization. Therefore, the optimization process is a continuation of the design process, which started with the formulation of the design problem and involved modeling, simulation, and design as presented in the preceding chapters. The conceptual design is kept unchanged during optimization because the acceptable designs were obtained with the chosen concept.

This chapter also considers special considerations that arise for thermal systems, such as the thermal efficiency, energy losses, and heat input rate, that are associated with thermal processes. Important questions regarding the implementation of the optimal solution, such as sensitivity analysis, dependence on the model, effect of quantity chosen for optimization, and selection of design variables for the final design, are considered. Many specialized books are available on optimization in design, for instance, those by Fox (1971), Vanderplaats (1984), Stoecker (1989), Rao (2009), Rhinehart (2018), Papalambros and Wilde (2017), Arora (2004), and Ravindran et al. (2006). Books are also available on the basic aspects of optimization, such as those by Beveridge and Schechter (1970), Beightler et al. (1979), and Miller (2000). These books may be consulted for further details on optimization techniques and their application to design.

7.2 BASIC CONCEPTS

We can now proceed to formulate the basic problem for the optimization of a thermal system. Because the optimal design must satisfy the given requirements and constraints, the designs considered as possible candidates must be acceptable or workable ones. This implies that the search for an optimal design is carried out in the domain of acceptable designs. The conceptual design is kept fixed so that optimization is carried out within a given concept. Generally, different concepts are considered at the early stages of the design process and a particular conceptual design is selected based on prior experience, environmental impact, material availability, etc., as discussed in Chapter 2. However, if a satisfactory design is not obtained with a particular conceptual design, the design process may be repeated, starting with a different conceptual design.

7.2.1 Objective Function

Any optimization process requires specification of a quantity or function that is to be minimized or maximized. This function is known as the *objective function*, and it represents the aspect or feature that is of particular interest in a given circumstance. Though the cost, including initial and maintenance costs, and profit are the most commonly used quantities to be optimized, many other aspects are employed for optimization, depending on the system and the application. The choice of the objective function is of critical importance in the optimization process because the results are often strongly dependent on the chosen criterion. The objective functions that are optimized for thermal systems are frequently based on the following characteristics:

1. Weight
2. Size or volume
3. Rate of energy consumption
4. Heat transfer rate
5. Efficiency
6. Overall profit

7. Costs incurred
8. Product quality
9. Environmental effects
10. Pressure head needed
11. Durability and reliability
12. Safety
13. System performance
14. Output delivered

The weight is of particular interest in transportation systems, such as airplanes and automobiles. Therefore, an electronic system designed for an airplane may be optimized in order to have the smallest weight while it meets the given requirements. Similarly, the size of the air conditioning system for environmental control of a house may be minimized in order to require the least amount of space. Energy consumption per unit output is particularly important for thermal systems and is usually indicative of the efficiency of the system. Frequently, this is given in terms of the energy rating of the system, thus specifying the power consumed for operation under given conditions. Refrigeration, heating, drying, air conditioning, and many such consumer-oriented systems are generally optimized to achieve the minimum rate of energy consumption for specified output. Costs and profits are always important considerations and efforts are made to minimize the former and maximize the latter. The output is also of particular interest in many thermal systems, such as manufacturing processes and automobiles. However, even if one wishes to maximize the thrust, torque, or power delivered by a motor vehicle, cost is still a very important consideration. Therefore, in many cases, the objective function is based on the output per unit cost. Similarly, other relevant measures of performance are considered in terms of the costs involved. Environmental effects, safety, product quality, and several other such aspects are important in various applications and may also be considered for optimization.

Let us denote the objective function that is to be optimized by U, where U is a function of the n independent variables in the problem $x_1, x_2, x_3, \ldots, x_n$. Then the objective function and the optimization process may be expressed as

$$U = U(x_1, x_2, x_3, \ldots, x_n) \rightarrow U_{\text{opt}} \tag{7.1}$$

where U_{opt} denotes the optimal value of U. The x's represent the design hardware variables as well as the operating conditions, which may be changed to obtain a workable or optimal design. Physical variables such as height, thickness, material properties, heat flux, temperature, pressure, and flow rate may be varied over allowable ranges to obtain an optimum design, if such an optimum exists. A minimum or a maximum in U may be sought, depending on the nature of the objective function.

The process of optimization involves finding the values of the different design variables for which the objective function is minimized or maximized, without violating the constraints. Figure 7.2 shows a sketch of a typical variation of the objective function U with a design variable x_1, over its acceptable range. It is seen that though there is an overall, or global, maximum in $U(x_1)$, there are several local maxima or minima. Our interest lies in obtaining this global optimum. However, the local optima can often confuse the true optimum, making the determination of the latter difficult. It is necessary to distinguish between local and global optima so that the best design is obtained over the entire domain.

7.2.2 Constraints

The constraints in a given design problem arise due to limitations on the ranges of the physical variables, and due to the basic conservation principles that must be satisfied. The restrictions on

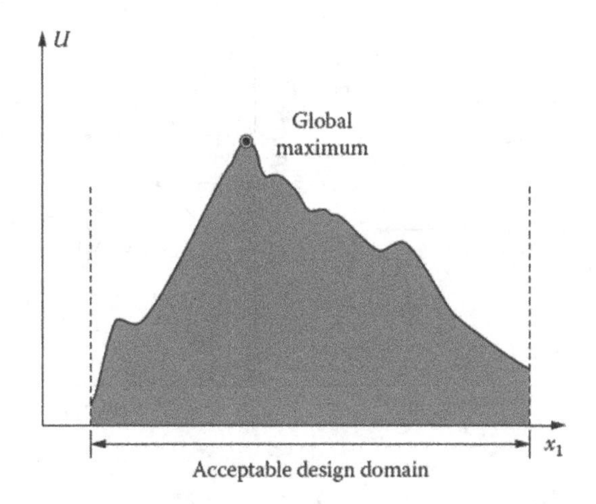

FIGURE 7.2 Global maximum of the objective function U in an acceptable design domain of the design variable x_1.

the variables may arise due to the space, equipment, and materials being employed. These may restrict the dimensions of the system, the highest temperature that the components can safely attain, allowable pressure, material flow rate, force generated, and so on. Minimum values of the temperature may be specified for thermoforming of a plastic and for ignition to occur in an engine. Thus, both minimum and maximum values of the design variables may be involved as constraints.

Many of the constraints relevant to thermal systems have been considered in earlier chapters. The constraints limit the domain in which the workable or optimal design lies. Figure 7.3 shows a few examples in which the boundaries of the design domain are determined by constraints arising from material or space limitations. For instance, in heat treatment of steel, the minimum temperature needed for the process T_{min} is given, along with the maximum allowable temperature T_{max} at which the material will be damaged. Similarly, the maximum pressure p_{max} in a metal extrusion process is fixed by strength considerations of the extruder and the minimum p_{min} is fixed by the flow stress needed for the process to occur. The limitations on the dimensions W and H define the domain in an electronic system.

Many constraints arise because of the conservation laws, particularly those related to mass, momentum, and energy in thermal systems. Thus, under steady-state conditions, the mass inflow into the system must equal the mass outflow. This condition gives rise to an equation that must be satisfied by the relevant design variables, thus restricting the values that may be employed in the search for an optimum. Similarly, energy balance considerations are very important in thermal systems and may limit the range of temperatures, heat fluxes, dimensions, etc., that may be used. Several such constraints are often satisfied during modeling and simulation because the governing equations are based on the conservation principles. Then the optimization process has already considered these constraints. In such cases, only the additional limitations that define the boundaries of the design domain are left to be considered.

There are two types of constraints, *equality* constraints and *inequality* constraints. As the name suggests, equality constraints are equations that may be written as

$$\begin{aligned} G_1(x_1, x_2, x_3, \ldots, x_n) &= 0 \\ G_2(x_1, x_2, x_3, \ldots, x_n) &= 0 \\ &\vdots \\ G_m(x_1, x_2, x_3, \ldots, x_n) &= 0 \end{aligned} \tag{7.2}$$

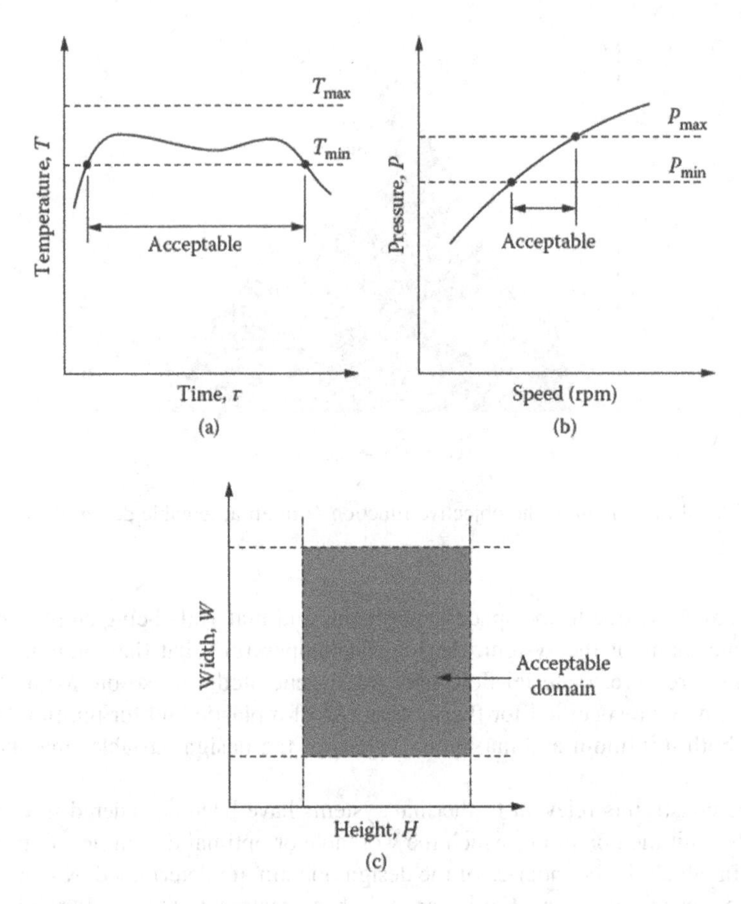

FIGURE 7.3 Boundaries of the acceptable design domain specified by limitations on the variables for (a) heat treatment, (b) metal extrusion, and (c) cooling of electronic equipment.

Similarly, inequality constraints indicate the maximum or minimum value of a function and may be written as

$$\begin{aligned}
H_1(x_1, x_2, x_3, \ldots, x_n) &\le C_1 \\
H_2(x_1, x_2, x_3, \ldots, x_n) &\le C_2 \\
H_3(x_1, x_2, x_3, \ldots, x_n) &\ge C_3 \\
&\vdots \\
H_l(x_1, x_2, x_3, \ldots, x_n) &\ge C_l
\end{aligned} \tag{7.3}$$

Therefore, either the upper or the lower limit may be given for an inequality constraint. Here, the C's are constants or known functions. The m equality and l inequality constraints are given for a general optimization problem in terms of the functions G and H, which are dependent on the n design variables or operating conditions, $x_1, x_2, \ldots, x_n$. Thus, the constraints in Figure 7.3 may be given as $T_{min} \le T \le T_{max}$, $P_{min} \le P \le P_{max}$, and so on.

The equality constraints are most commonly obtained from conservation laws; e.g., for a steady flow circumstance in a control volume, we may write

$$\sum (\text{mass flow rate})_{in} - \sum (\text{mass flow rate})_{out} = 0$$

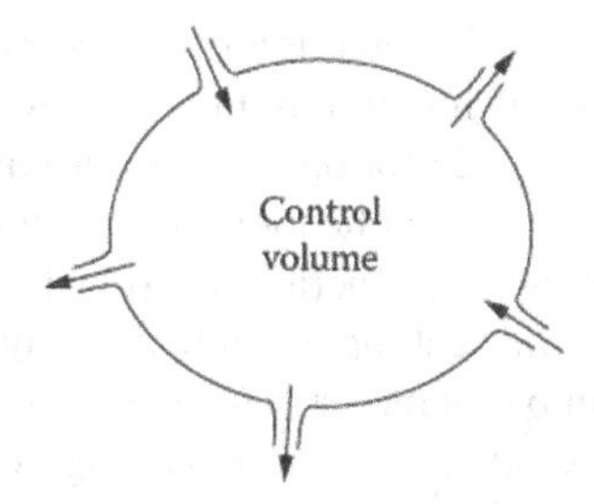

FIGURE 7.4 Inflow and outflow of material and energy in a fixed control volume.

or

$$\sum(\rho VA)_{in} - \sum(\rho VA)_{out} = 0 \tag{7.4}$$

where ρ is the mean density of the material, V is the average velocity, A is the cross-sectional area, and Σ denotes the sum of flows in and out of several channels, as sketched in Figure 7.4. Similarly, equations for energy balance and momentum-force balance may be written. The conservation equations may be employed in their differential or integral forms, depending on the detail needed in the problem.

It is generally easier to deal with equations than with inequalities because many methods are available to solve different types of equations and systems of equations, as discussed in Chapter 4, whereas no such schemes are available for inequalities. Therefore, inequalities are often converted into equations before applying optimization methods. A common approach employed to convert an inequality into an equation is to use a value larger than the constraint if a minimum is specified and a value smaller than the constraint if a maximum is given. For instance, the constraints may be changed as follows:

$$H_1(x_1, x_2, x_3, \ldots, x_n) \leq C_1 \quad \text{becomes} \quad H_1(x_1, x_2, x_3, \ldots, x_n) = C_1 - \Delta C_1 \tag{7.5a}$$

$$H_3(x_1, x_2, x_3, \ldots, x_n) \geq C_3 \quad \text{becomes} \quad H_3(x_1, x_2, x_3, \ldots, x_n) = C_3 + \Delta C_3 \tag{7.5b}$$

where ΔC_1 and ΔC_3 are chosen quantities, often known as *slack* variables, that indicate the difference from the specified limits. Though any finite values of these quantities will satisfy the given constraints, generally the values are chosen on the basis of the characteristics of the given problem and the importance of the constraint. Frequently, a fraction of the actual limiting value is used as the slack to obtain the corresponding equation. For instance, if 200°C is given as the limiting temperature for a plastic, a deviation of, say, 5% or 10°C may be taken as acceptable to convert the inequality into an equation.

7.2.3 Operating Conditions versus Hardware

It was mentioned earlier that the process of optimization might be applied to a system so that the design, given in terms of the hardware, is optimized. Much of our discussion on optimization will focus on the system so that the corresponding hardware, which includes dimensions, materials, geometry, components, etc., is varied to obtain the best design with respect to the chosen objective function. However, it is worth reiterating that once a system has been designed, its performance and characteristics are also functions of the operating conditions. Therefore, it is generally possible to obtain conditions under which the system performance is optimum. For instance, if we are interested in the minimum fuel consumption of a motor vehicle, we may be able to determine a speed, such as 88 km/h (55 miles/h), at which this condition is met. Similarly, the optimum setting for an air conditioner, at which the efficiency is maximum, may be determined as, say, 22.2°C (72°F), or the revolutions per minute of a motor as 125 for optimal performance.

The operating conditions vary from one application to another and from one system to the next. The range of variation of these conditions is generally fixed by the hardware. Therefore, if a heater is chosen for the design of a furnace, the heat input and temperature ranges are fixed by the specifications of the heater. Similarly, a pump or a motor may be used to deliver an output over the ranges for which these can be satisfactorily operated. The operating conditions in thermal systems are typically specified in terms of the following variables:

1. Heat input rate
2. Temperature
3. Pressure
4. Mass or volume flow rate
5. Speed, revolutions per minute (rpm)
6. Chemical composition

Thus, imposed temperature and pressure, as well as the rate of heat input, may be varied over the allowable ranges for a system such as a furnace or a boiler. The volume or mass flow rate is chosen, along with the speed (rpm), for a system like a screw extruder, a diesel engine, or a gas turbine. The chemical composition is important in specifying the chosen inlet conditions for a chemical reactor, such as a food extruder where the moisture content in the extruded material is an important variable.

It is useful to determine the optimum operating conditions and the corresponding system performance. The approach to optimize the output or performance in terms of the operating conditions is similar to that employed for the hardware design and optimization. The model is employed to study the dependence of the system performance on the operating conditions and an optimum is chosen using the methods discussed here.

7.2.4 Mathematical Formulation

We may now write the basic mathematical formulation for the optimization problem in terms of the objective function and the constraints. We will first consider the formulation in general terms, followed by a few examples to illustrate these ideas. The various steps involved in the formulation of the problem are

1. Determination of the design variables, x_i where $i = 1, 2, 3, \ldots, n$
2. Selection and definition of the objective function, U
3. Determination of the equality constraints, $G_i = 0$, where $i = 1, 2, 3, \ldots, m$
4. Determination of the inequality constraints, $H_i \leq$ or $\geq C_i$, where $i = 1, 2, 3, \ldots l$
5. Conversion of inequality constraints to equality constraints, if appropriate

The selection of the design variables x_i and of the objective function U is extremely important for the success of the optimization process, because these define the basic problem. The number of independent variables determines the complexity of the problem and, therefore, it is important to focus on the dominant variables rather than consider all that might affect the solution. As the number of independent variables is increased, the effort needed to solve the problem increases substantially, particularly for thermal systems, because of their generally complicated, nonlinear characteristics. Consequently, optimization of thermal systems is often carried out with a relatively small number of design variables that are of critical importance to the system under consideration. Optimization may also be done considering only one design variable at a time, with the different variables being alternated, as we advance toward the optimal solution.

Similarly, the selection of the objective function demands great care. It must represent the important characteristics and concerns of the system and of the application for which it is intended. However, it must also be sensitive to variations in the design parameters; otherwise, a clear optimal result may not emerge from the analysis. Different aspects may be combined to define the objective function, e.g., output per unit cost, efficiency per unit cost, profit per unit solid waste, and heat rejected per unit power delivered.

The constraints are obtained from the conservation laws and from limitations imposed by the materials employed; space and weight restrictions; environmental, safety, and performance considerations; and requirements of the application. If there are no constraints at all, the problem is termed *unconstrained* and is much easier to solve than the corresponding constrained problem. Efforts are usually made to reduce the number of constraints or eliminate these by substitution and algebraic manipulation to simplify the problem.

Therefore, the general mathematical formulation for the optimization of a system may be written as

$$U(x_1, x_2, x_3, \ldots, x_n) \rightarrow U_{opt} \tag{7.6a}$$

with

$$G_i(x_1, x_2, x_3, \ldots, x_n) = 0, \quad \text{for} \quad i = 1, 2, 3, \ldots, m \tag{7.6b}$$

and

$$H_i(x_1, x_2, x_3, \ldots, x_n) \leq \text{or} \geq C_i, \quad \text{for} \quad i = 1, 2, 3, \ldots, l \tag{7.6c}$$

If the number of equality constraints m is equal to the number of independent variables n, the constraint equations may simply be solved to obtain the variables and there is no optimization problem. If $m > n$, the problem is overconstrained and a unique solution is not possible because some constraints have to be discarded to obtain a solution. If $m < n$, an optimization problem is obtained. This is the case considered here and in the following chapters. The inequality constraints are generally employed to define the range of variation of the design parameters.

7.3 OPTIMIZATION METHODS

Several methods may be employed for solving the mathematical problem given by Equation (7.6) to optimize a system or a process. Each approach has its limitations and advantages over the others. Thus, for a given optimization problem, a certain method may be particularly appropriate while some of the others may not even be applicable. The choice of method largely depends on the nature of the equations representing the objective function and the constraints. It also depends on whether the mathematical formulation is expressed in terms of explicit functions or if numerical solutions or experimental data are to be obtained to determine the variation of the objective function and the constraints with the design variables. Because of the complicated nature of typical thermal systems, numerical solutions of the governing equations and experimental results are often needed to study the behavior of the objective function as the design variables are varied and to monitor the constraints. However, in several cases, detailed numerical results are generated from a mathematical model of the system or experimental data are obtained from a physical model, and these are curve fitted to obtain algebraic equations to represent the characteristics of the system. Optimization of the system may then be undertaken based on these relatively simple algebraic expressions and equations. The commonly used methods for optimization and the nature and type of equations to which these may be applied are outlined in the following sections.

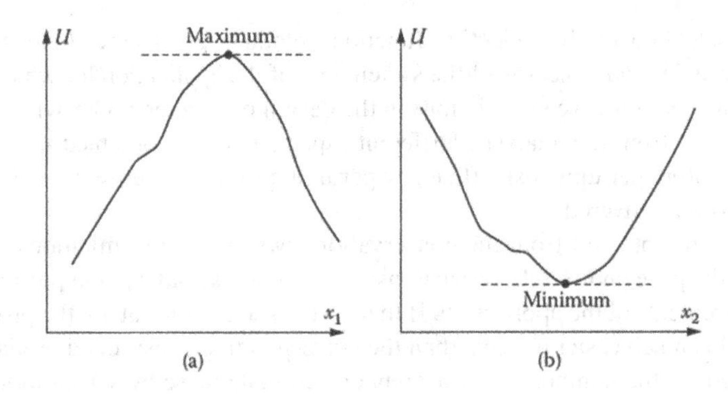

FIGURE 7.5 Maximum or minimum in the objective function U, varying with a single independent variable x_1 or x_2.

7.3.1 Calculus Methods

The use of calculus for determining the optimum is based on derivatives of the objective function and of the constraints. The derivatives are used to indicate the location of a minimum or a maximum. At a local optimum, the slope is zero, as sketched in Figure 7.5, for U varying with a single design variable x_1 or x_2. The equations and expressions that formulate the optimization problem must be continuous and well-behaved, so that these are differentiable over the design domain. An important method that employs calculus for optimization is the method of *Lagrange multipliers*. This method is discussed in detail in the next chapter. The objective function and the constraints are combined through the use of constants, known as Lagrange multipliers, to yield a system of algebraic equations. These equations are then solved analytically or numerically, using the methods presented in Chapter 4, to obtain the optimum as well as the values of the multipliers.

The range of application of calculus methods to the optimization of thermal systems is somewhat limited because of complexities that commonly arise in these systems. Numerical solutions are often needed to characterize the behavior of the system and implicit, nonlinear equations that involve variable material properties are frequently encountered. However, curve fitting may be employed in some cases to yield algebraic expressions that closely approximate the system and material characteristics. If these expressions are continuous and easily differentiable, calculus methods may be conveniently applied to yield the optimum. These methods also indicate the nature of the functions involved, their behavior in the domain, and the basic characteristics of the optimum. In addition, the method of Lagrange multipliers provides information, through the multipliers, on the sensitivity of the optimum with respect to changes in the constraints. In view of these features, it is worthwhile to apply the calculus methods whenever possible. However, curve fitting often requires extensive data that may involve detailed experimental measurements or numerical simulations of the system. Because this may demand a considerable amount of effort and time, particularly for thermal systems, it is generally preferable to use other methods of optimization that require relatively smaller numbers of simulations.

7.3.2 Search Methods

As the name suggests, these methods involve selection of the best solution from a number of workable designs. If the design variables can only take on certain fixed values, different combinations of these variables may be considered to obtain possible acceptable designs. Similarly, if these variables can be varied continuously over their allowable ranges, a finite number of acceptable designs may be generated by changing the variables. In either case, a number of workable designs are obtained, and the optimal design is selected from these. In the simplest approach, the objective function is calculated at uniformly spaced locations in the domain, selecting the design with the optimum value.

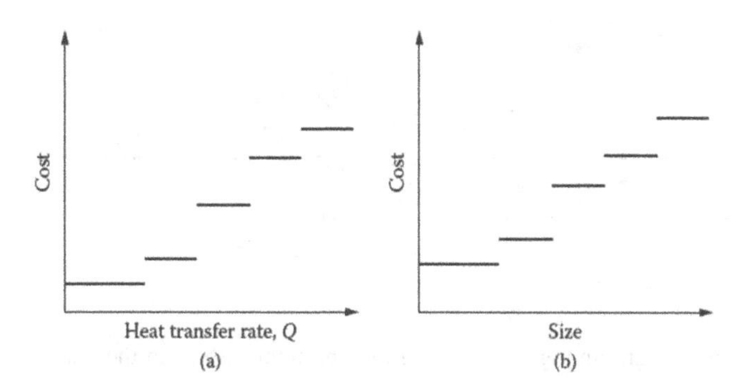

FIGURE 7.6 Variation of cost as a discrete function with (a) heat transfer rate in a heat exchanger, and (b) size of an item such as a fan or pump.

This approach, known as *exhaustive search*, is not very imaginative and is clearly an inefficient method to optimize a system. As such, it is generally not used for practical systems. However, the basic concept of selecting the best design from a set of acceptable designs is an important one and is used even if a detailed optimization of the system is not undertaken. Sometimes, an unsystematic search, based on prior knowledge of the system, is carried out instead.

Several efficient search methods have been developed for optimization and may be adopted for optimizing thermal systems. Because of the effort involved in experimentally or numerically simulating typical thermal systems, particularly large and complex systems, it is important to minimize the number of simulation runs or iterations needed to obtain the optimum. The locations in the design domain where simulations are carried out are selected in a systematic manner by considering the behavior of the objective function. Methods such as *dichotomous*, *Fibonacci*, *univariate*, and *steepest ascent* search start with an initial design and attempt to use a minimum number of iterations to reach close to the optimum, which is represented by a peak or the lowest point, as sketched in Figure 7.5.

The exact optimum is generally not obtained even for continuous functions because only a finite number of iterations are used. However, in actual engineering practice, components, materials, and even dimensions are not available as continuous quantities but as discrete steps. For instance, a heat exchanger would typically be available for discrete heat transfer rates such as 50, 100, 200 kW, etc. The cost may be assumed to be a discrete distribution rather than a continuous variation (see Figure 7.6). Similarly, the costs of items such as pumps and compressors are discrete functions of the size. Different materials involve distinct sets of properties and not continuous variations of thermal conductivity, specific heat, or other thermal properties. Search methods can easily be applied to such circumstances, whereas calculus methods demand continuous functions. Consequently, search methods are extensively used for the optimization of thermal systems. The basic strategies and their applications to thermal systems are discussed in Chapter 9.

7.3.3 Linear and Dynamic Programming

Programming as applied here simply refers to optimization. Linear programming is an important optimization method and is extensively used in industrial engineering, operations research, and many other disciplines. However, the approach can be applied only if the objective function and the constraints are all linear. Large systems of variables can be handled by this method, such as those encountered in air traffic control, transportation networks, and supply and utilization of raw materials. However, as we well know, thermal systems are typically represented by nonlinear equations. Consequently, linear programming is not particularly important in the optimization of thermal systems, though it is applicable in a few cases and some problems may be linearized to use this method. A brief outline of the method is given in Chapter 10.

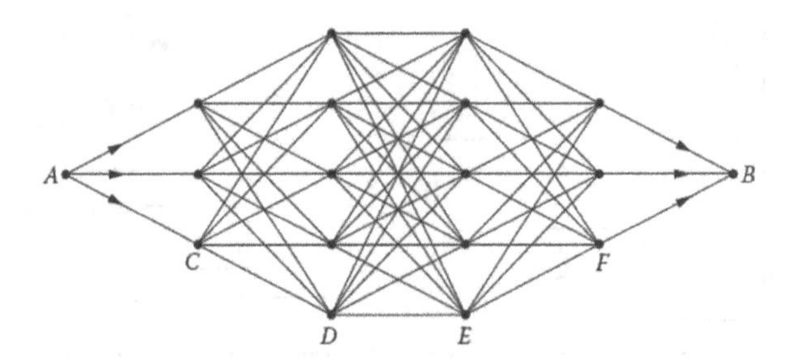

FIGURE 7.7 Dynamic programming for choosing the optimum path from the many different paths to go from point *A* to point *B*.

Dynamic programming is used to obtain the best path through a series of stages or steps to achieve a given task, for instance, the optimum configuration of an assembly line, the best path for the flow of hot water in a building, and the best layout for transport of coal in a power plant. Therefore, the result obtained from dynamic programming is not a point where the objective function is optimum but a curve or path over which the function is optimized. Figure 7.7 illustrates the basic concept by means of a sketch. Several paths can be used to connect points *A* and *B*. The optimum path is the one over which a given objective function, say, total transportation cost, is minimized. Multiple solutions are possible and additional considerations, such as safety, convenience, availability of items, etc., are used to choose the best design. Clearly, there are a few circumstances of interest in thermal systems where dynamic programming may be used to obtain the best layout to minimize losses and reduce costs. Some of these considerations are discussed in Chapter 10.

7.3.4 Geometric Programming

Geometric programming is an optimization method that can be applied if the objective function and the constraints can be written as sums of polynomials. The independent variables in these polynomials may be raised to positive or negative, integer or noninteger exponents, e.g.,

$$U = ax_1^2 + bx_2^{1.2} + cx_1x_2^{-0.5} + d \tag{7.7}$$

Here, a, b, c, and d are constants, which may also be positive or negative, and x_1 and x_2 are the independent variables. Curve fitting of experimental data and numerical results for thermal systems often leads to polynomials and power-law variations, as seen in Chapter 3. Therefore, geometric programming is useful in the optimization of thermal systems if the function to be optimized and the constraints can be represented as sums of polynomials. If the method is applicable in a particular case, the optimal solution and even the sensitivity of the solution to changes in the constraints are often obtained directly and with very little computational effort. The method is discussed in detail in Chapter 10.

However, it must be remembered that unless extensive data and numerical simulation results are available for curve fitting, and unless the required polynomial representations can be obtained to the desired accuracy level, geometric programming cannot be used for common thermal systems. In such cases, search methods provide an important approach that is widely used for large and complicated systems.

7.3.5 Other Methods

Several other optimization methods have been developed in recent years because of the strong need to optimize systems and processes. Many of these are particularly suited to specific applications and may not be easily applied to thermal systems. Among these are shape, trajectory, and structural optimization methods, which involve specialized techniques for finding the desired optimum. Frequently, finite-element solution procedures are linked with the relevant optimization strategy. Iterative shapes, trajectories, or structures are generated, starting with an initial design. For monotonically increasing or decreasing objective functions and constraints, a method known as *monotonicity analysis* has been developed for optimization. This approach focuses on the constraints and the effects these have on the optimum.

Several other methods and associated approaches have been developed and employed in recent years to facilitate the optimization of a wide variety of processes and systems. Though initially directed at linear problems, these approaches have been modified to include the optimization of nonlinear problems such as those of interest in thermal systems. Among the methods that may be mentioned are *genetic algorithms (GAs)*, *artificial neural networks (ANNs)*, *fuzzy logic,* and *response surfaces*. The first three are based on artificial intelligence approaches, as discussed later in Chapter 11. A brief discussion is included here, while the fourth method, response surfaces, is discussed in some detail in the following.

GAs are search methods used for obtaining the optimal solution and are based on evolutionary techniques that are similar to evolutionary biology, which involves inheritance, learning, selection, and mutation. The process starts with a population of candidate solutions, called individuals, and progresses through generations, with the fitness, as defined on the basis of the objective function, for each individual being evaluated. Then multiple individuals are selected from the current generation based on the fitness and modified to form a new population. This new population is used in the next iteration and the algorithm progresses toward the desired optimal point (Goldberg, 1989; Mitchell, 1998; Holland, 2002; Sheppard, 2018).

ANNs are interconnected groups of processing elements, called artificial neurons, similar to those in the central nervous system of the body and studied as *neuroscience*. The characteristics of the processing elements and the interconnections determine the processing of information and the modeling of simple and complex processes. Functions are performed in parallel and the networks have both nonadaptive and adaptive elements, which change with the input/output and the problem. Thus, nonlinear, distributed, parallel, local processing, and adaptive representations of systems are obtained (Jain and Martin, 1999).

Fuzzy logic allows one to deal with inherently imprecise concepts, such as cold, warm, very, and slight, and is useful in a wide variety of thermal systems where approximate, rather than precise, reasoning is needed (Ross, 2004). It can be used for control of systems and in problems where a sharp cutoff between two conditions does not exist.

The preceding three approaches are available in toolboxes developed by MathWorks and can thus be used easily with Matlab, along with several other optimization techniques. Many are based on the natural world such as particle swarm optimization (PSO), which uses a flock of birds searching for food as the basis for the optimization strategy.

Another approach, which has found widespread use in engineering systems, including thermal systems, is that of response surfaces. As mentioned earlier, response surface or surrogate models are approximate models that reduce the simulation effort by using the responses at intelligently selected points. The basic approach is similar to curve fitting, discussed in Chapter 3. The response surface methodology (RSM) comprises a group of statistical techniques for empirical model building, followed by the use of the model in the design and development of new products and also in the improvement of existing designs (Box and Draper, 1987, 2007). RSM is used when only a small number of computational or physical experiments can be conducted due to the high costs (monetary or computational) involved. Response surfaces are fitted to the limited data collected and are used to estimate the location of the optimum. RSM gives a fast approximation to the model, which can be used to identify important variables, visualize the relationship of the input to the output, and quantify trade-offs between multiple objectives. This approach has been found to be valuable in developing new processes and systems, optimizing their performance, and improving the design and formulation of new products (Myers and Montgomery, 2016).

Figure 7.8(a) shows graphically the relation between the response or output and two design variables x_1 and x_2. Note that for each value of x_1 and x_2, there is a corresponding value of the response. These values of the response may be perceived as a surface lying above the $x_1 - x_2$ plane, as shown in the figure. It is this graphical perspective of the problem that has led to the term *response surface methodology*. If there are two design variables, then we have a three-dimensional space in which the coordinate axes represent the response and the two design variables. When there are N design variables ($N > 2$), we have a response surface in the $N + 1$-dimensional space.

Optimization of the process is straightforward if the graphical display shown in Figure 7.8(a) could be easily constructed. However, in most practical situations, the true response function is unknown and thus the methodology consists of examining the space of design variables, empirical

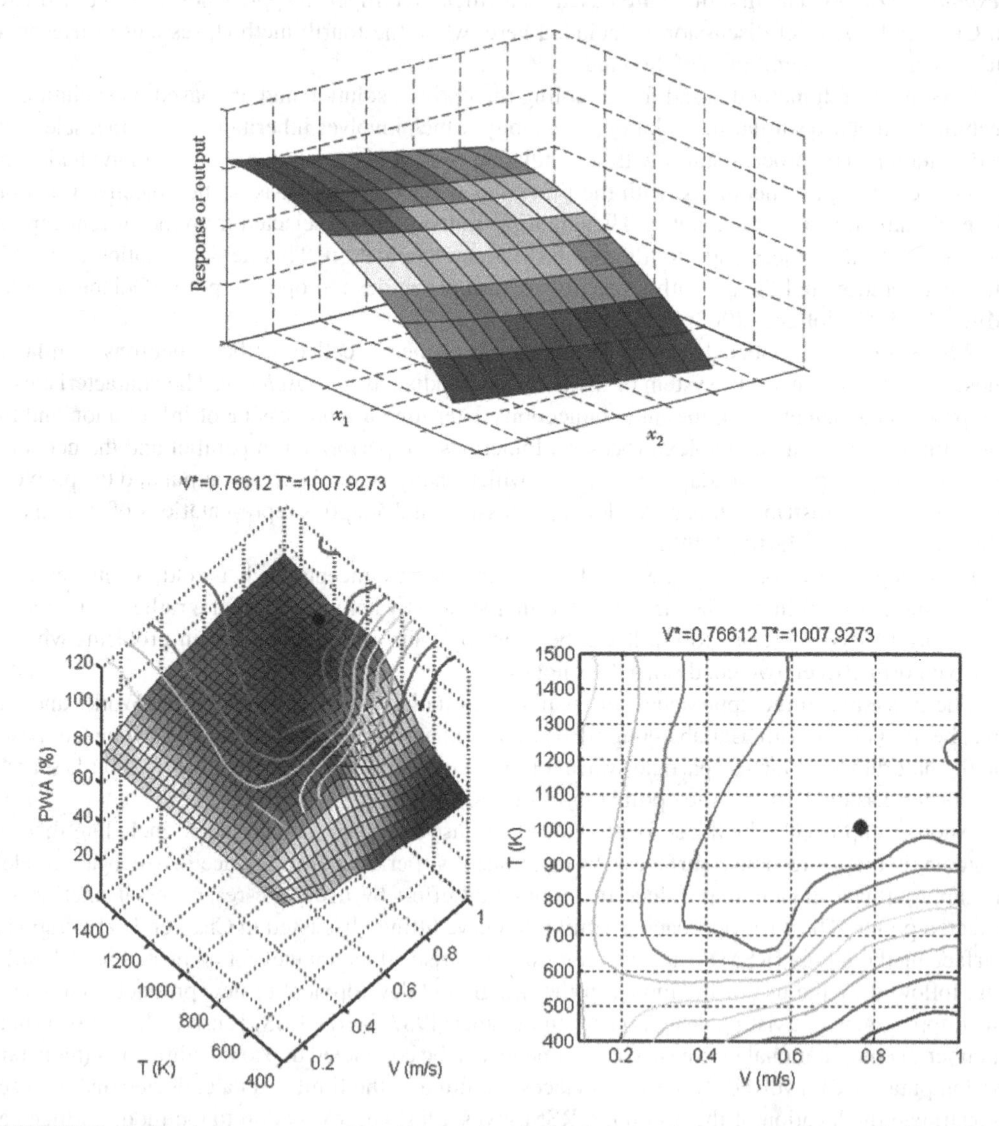

FIGURE 7.8 (a) Typical response surface showing the relation between the response or output and the design variables x_1 and x_2. (b) Results for a practical thermal system, the chemical vapor deposition process for fabrication of thin films. The response is the percentage working area (PWA), which gives the percentage area of acceptable film thickness, and the inflow velocity V and deposition surface temperature T are the two variables. The optimum is shown on the response surface and also on the constant PWA contours.

statistical modeling to develop an approximating relationship (response function) between the response and the design variables, and optimization methods for finding the values of the design variables that produce optimal values of the responses. As an example, Figure 7.8(b) shows the results for a practical thermal system. This is the chemical vapor deposition system in which a thin film is deposited on a heated surface at temperature T due to the inflow of reacting gases at velocity V into the reactor. This figure shows the percentage working area (PWA) of the surface that has a film of acceptable thickness as the response to the two variables T and V. The resulting optimum point that maximizes the percentage working area is also shown on the response curves, as well on constant response contours.

The method normally starts with a lower-order model, such as linear or second order. If the second-order model is inadequate, as judged by checking against points not used to generate the model, simulations are performed at additional design points and the data used to fit the third-order model. Then the resulting third-order model is checked against additional data points not used to generate the model. If the third-order model is found to be inadequate, then a fourth-order model is fit based on the data from additional simulations and then tested, and so on. A typical second-order model for the response, z, is

$$z=\beta_0+\beta_1 x+\beta_2 y+\beta_3 xy+\beta_4 x^2+\beta_5 y^2 \tag{7.8}$$

Similarly, a third-order model for the response, z, is

$$z=\beta_0+\beta_1 x+\beta_2 y+\beta_3 xy+\beta_4 x^2+\beta_5 y^2+\beta_6 x^2 y+\beta_7 xy^2+\beta_8 x^3+\beta_9 y^3 \tag{7.9}$$

where the β's are coefficients to be determined from the data and x, y are the two independent design variables. Once the response surface has been generated, visual inspection can be used to locate the region where the optimum is located and a closer inspection can then be used to accurately determine the location of the optimum. Calculus can also be used to identify the minimum or maximum. Both local and global optimum locations can generally be identified. However, because a limited number of data points are used in order to generate the response surface, the surface approximates the actual behavior and the results are similarly approximate, though for many practical problems this is quite adequate.

7.4 OPTIMIZATION OF THERMAL SYSTEMS

We have considered the basic formulation for optimization, as well as different methods that are available for solving these problems. Several physical problems have been mentioned as examples to illustrate the general approach. Let us now briefly consider these aspects as related to the optimization of thermal systems.

7.4.1 Important Considerations

Thermal systems are mainly concerned with energy and fluid flow. Therefore, the objective function is frequently based on energy consumption, which involves considerations of energy transport and losses, efficiency of the system and its components, energy exchange with the environment, fuel consumed, etc. A useful objective function is the rate of energy consumption per unit output, where the output may be power delivered, heat removed, products manufactured, and so on. The design that requires the least amount of energy per unit output is then the optimum. Similarly, the system that delivers the largest output per unit energy consumption is optimum. Because energy consumption can be expressed in terms of cost, this objective function can also be considered as the output per unit cost.

Similar considerations often apply to fluid flow, where again it is important to minimize the energy consumed. This frequently implies minimizing the flow rate, pressure head, and fluid leakage or loss, particularly if a closed system is needed for preserving the purity and if the fluid is expensive. A lower pressure head generally translates into lower cost of the pumping system and is desirable. Therefore, some of the physical quantities that are often *maximized* in thermal systems may be listed as

1. Efficiency
2. Output per unit energy, or fuel, consumption
3. Output per unit cost
4. Heat removal rate in electronic systems
5. Heat exchange rate

whereas the quantities that are *minimized* may be listed as

1. Energy losses
2. Energy input for cooling systems
3. Pressure head for fluid flow
4. Flow rate of fluid
5. Fluid leakage or loss
6. Rate of energy or fuel consumed per unit output

In thermal systems, the constraints arise largely from the conservation laws for mass, momentum, and energy, and from limitations of the material, space, and equipment being used, as discussed earlier. However, these usually lead to nonlinear, multiple, coupled, partial differential equations, with complicated geometries and boundary conditions in typical systems of practical interest. Other complexities may also arise due to the material characteristics, combined thermal transport mechanisms, etc., as discussed in earlier chapters.

The main problem that arises due to these complexities is that the simulation of the system for each set of conditions requires a considerable amount of time and effort. Therefore, it is usually necessary to minimize the number of simulation runs needed for optimization. For relatively simple thermal systems, numerical or experimental simulation results may be used, with curve fitting, to obtain algebraic expressions and equations to characterize the behavior of the system. Then the optimization problem becomes straightforward and many of the available methods can be used to extract the optimum. Unfortunately, this approach is possible in only a few simple, and often impractical, circumstances. For common practical systems, numerical modeling is employed to obtain the simulation results, as needed, to obtain the optimum. Experimental data are also used if a prototype is available, but again such data are limited because experimental runs are generally expensive and time consuming.

7.4.2 Different Approaches

Several different optimization methods have been mentioned earlier and will be discussed in detail in later chapters. Some of these have only limited applicability with respect to thermal systems. Search methods constitute the most important optimization strategy for thermal systems. Many different approaches have been developed and are particularly appropriate for different problems. However, the underlying idea is to generate a number of designs, which are also called *trials* or *iterations*, and to select the best among these. Effort is made to keep the number of trials small, often going to the next iteration only if necessary. This is a very desirable feature with respect to thermal systems because each trial may take a considerable amount of computational or experimental effort.

Search methods may also be combined with other methods in order to accelerate convergence or approach to the optimum. For instance, calculus methods may be used at certain stages to narrow the domain in which the optimum lies. Trials for the search method are then used to provide information for extracting the derivatives and other relevant quantities. Prior knowledge on the optimum for similar systems may also be used to develop *heuristic* rules to accelerate the search.

7.4.3 Different Types of Thermal Systems

As we have seen in the preceding chapters, thermal systems cover a very wide range of applications. Different concerns, constraints, and requirements arise in different types of systems. Therefore, the objective function and the nature of the constraints would generally vary with the application. Though costs and overall profit or return are frequently optimized, other quantities are also of interest and are used. Let us consider some of the common types of thermal systems and discuss the corresponding optimization problems.

1. *Manufacturing systems.* The objective function is typically the number of items produced per unit cost. It could also be the amount of material processed in heat treatment, casting, crystal growing, extrusion, or forming. The number of solder or welding joints made, length of material cut in gas or laser cutting, or the length of optical fiber drawn may also be used, depending on the application. Again, the output per unit cost or the cost for a given output may also be used as the objective function. The constraints are often given on the temperature and pressure due to material limitations. Conservation principles and equipment limitations restrict the flow rates, cutting speed, draw speed, etc. It is important to note that product quality is often of critical importance in this area. The quality may be defined in terms of defects, uniformity, microstructure, and other characteristics of the product.
2. *Energy systems.* The amount of power produced per unit cost is the most important measure of success in energy systems and is, therefore, an appropriate quantity to be optimized. The overall thermal efficiency is another important variable that may be optimized. Most of the constraints arise from conservation laws. However, environmental and safety considerations also lead to important limitations on items such as the water outlet temperature and flow rate from the condensers of a power plant to a cooling pond or lake. Material and space limitations will also provide some constraints on the design variables.
3. *Electronic systems.* The rate of thermal energy removed from the system as well as this quantity per unit cost are important design requirements and may, thus, be used as objective functions. The cost of the system may also be minimized while ensuring that the temperature requirements of the components are satisfied. The pressure head needed for the coolant flow is another important aspect and may be taken as an objective function to be minimized or as a constraint on the final design. The weight and volume are important considerations in portable systems and in systems used in planes and rockets. These may also be chosen for optimization. Besides the constraints due to conservation principles, space and material limitations generally restrict the temperatures, fluid flow rates, and dimensions in the system.
4. *Transportation systems.* The torque, thrust, or power delivered are important considerations in these systems. Therefore, these quantities, or these taken per unit cost, may be maximized. This feature may also be taken as the output per unit fuel consumed. The costs for a given output in thrust, acceleration, etc., may also be chosen for minimization. The thermal efficiency of the system is another important aspect that may be maximized. The constraints are largely due to material, weight, and size limitations, besides those due to conservation laws. Thus, the temperature, pressure, dimensions, and fuel consumption rate may be restricted within specified limits.
5. *Heating and cooling systems.* The amount of heat removed or provided per unit cost is a good measure of the effectiveness of these systems and may be chosen for maximization.

The system cost as well as the operating cost, which largely includes the energy costs, may be minimized while satisfying the requirements. The thermal efficiency of the system may be maximized for optimum performance. Besides those due to conservation laws, most of the constraints arise due to space limitations. Weight constraints are important in mobile systems. Fluid properties lead to constraints on the temperature and pressure in the system.

6. *Heat transfer and fluid flow equipment.* The rate of heat transfer and the total flow rate are important considerations in these systems. These quantities may be used for optimization. The heat transfer or flow rate per unit equipment, or operating, costs may also be considered. The resulting temperature of a fluid being heated or cooled, the efficiency of the equipment, energy losses, etc., may also be chosen as objective functions. Space limitations often provide the main constraints on dimensions. Constraints due to weight are also important in many cases, particularly in automobiles. Conservation laws provide constraints on temperatures and flow rates.

The foregoing discussion serves to illustrate the diversity of the objective function and the constraints in the wide range of applications that involve thermal systems. Even though costs and profit are important concerns in engineering systems, other quantities such as output, efficiency, environmental effect, etc., also provide important considerations that may be used effectively in the optimization process. Clearly, the preceding list is not exhaustive. Many other objective functions, constraints, and applications can be considered, depending on the nature and type of thermal system being optimized.

7.4.4 Examples

Example 7.1

An important problem in power generation is heat rejection. As discussed in Chapter 5, bodies of water such as lakes and ponds are frequently used for cooling condensers. The distance x between the inflow at point A into the cooling pond and outflow at point B, as shown in Figure 7.9, is an important variable that determines the performance and cost of the system. If x increases, the cost increases because of increased distance for pumping the cooling water. As x decreases, the hot water discharged into the lake can recirculate to the outflow, raising the temperature there. This effect increases the temperature of the cooling water entering the condensers of the power plant. This, in turn, raises the temperature at which heat rejection occurs and thus lowers the thermal efficiency of the plant, as is well-known from thermodynamics. Therefore, an increase in x increases the cost of the piping and pumps, while a decrease in x increases the cost of power generation by lowering the thermal efficiency. If the objective function U is taken as cost per unit of generated power, we may write

$$U(x) = F_1(x) - F_2(x) \quad \text{or} \quad U(x) = F_1(x)/F_2(x) \tag{7.10}$$

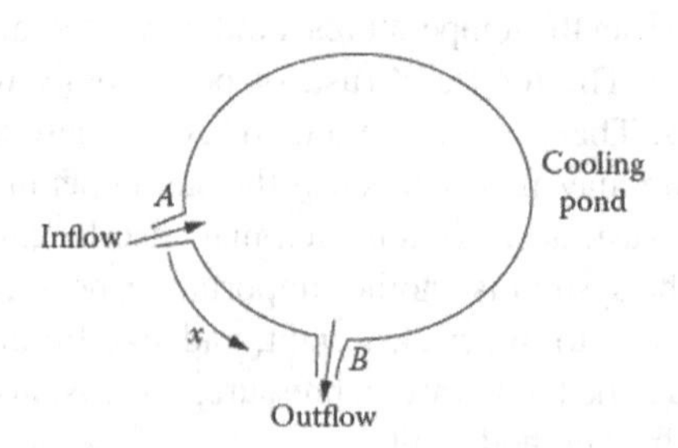

FIGURE 7.9 Heat rejection from a power plant to a cooling pond, with x as the distance between the inflow and outflow.

where $F_1(x)$ and $F_2(x)$ are costs related to piping and efficiency of the system, respectively. This implies that an optimum distance x may be obtained for minimum costs per unit output.

This is actually a very complicated problem because the model involves turbulent, multidimensional flow, complex geometries, varying ambient conditions, and several combined modes of heat transfer. The problem has to be solved numerically, with many simplifications, to obtain the desired inputs for design and optimization. Some simple problems were considered in Chapter 5. Constraints due to conservation principles are already taken into account in the numerical simulation. However, limitations on x due to the shape and size of the pond define an acceptable design domain. If the numerical simulation results are curve fitted to yield expressions of the form

$$U(x) = Ax^a + Bx^b - Cx^c \tag{7.11}$$

where A, B, C, a, b, and c are constants obtained from curve fitting, calculus methods can easily be applied to determine the optimum. However, this is a time-consuming process because adequate data points are needed and a more appropriate approach would be search methods where x is varied over the given domain and selective simulation runs are carried out at chosen locations to determine the optimum, as discussed in Chapter 9. This has been an important problem for the power industry for many years and has resulted in many different designs to obtain the highest efficiency-to-cost ratio.

Example 7.2

In an automobile, the drag force on the vehicle due to its motion in air increases with its speed V. The engine efficiency η also varies with the speed due to the higher revolutions per minute of the engine and increased fuel flow rate. The efficiency initially increases and then decreases at large V due to the effect on the combustion process. These two variations are sketched qualitatively in Figure 7.10. If the cost per mile of travel is taken as the objective function U, then we may write

$$U(V) = AF_1(V) + \frac{B}{F_2(V)} \tag{7.12}$$

where $F_1(V)$ represents the drag force and $F_2(V)$ represents the engine efficiency. An increase in drag force increases the cost, and an increase in efficiency reduces the cost. The constants A and B represent the effect of these quantities on the cost.

Again, this is a complicated numerical simulation problem because of the transient, three-dimensional problem involving turbulent flow and combustion. The constraints due to the

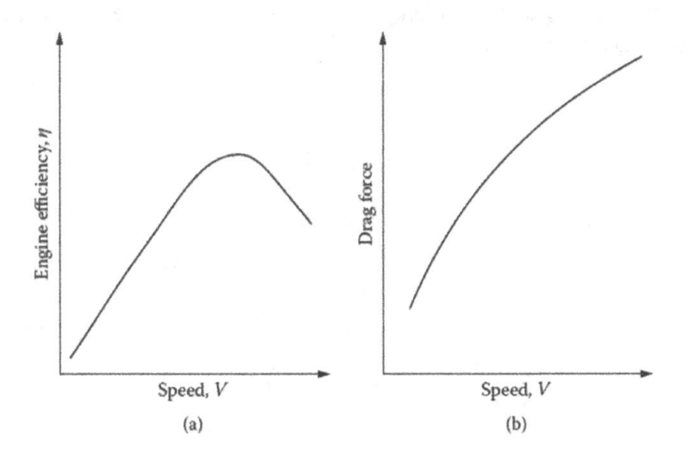

FIGURE 7.10 Dependence of engine efficiency and drag force on the speed V of an automobile.

conservation principles are already accounted for in the simulation. The physical limitations on the speed V, say, from 0 to 200 km/h for common vehicles, may be used to define the domain. If the simulation results are curve fitted with algebraic expressions, we may use calculus methods, as $dU/dV = 0$, to obtain the optimum. Search methods are more appropriate because only a limited number of simulations are needed at chosen values of V to extract the optimum.

Example 7.3

In a metal extrusion process, the total cost for a given amount of extruded material may be taken as the objective function U. This cost includes the capital or equipment cost A, the cost of the die subsystem, and the cost of the arrangement for applying the extrusion force. For the metal extrusion process sketched in Figure 7.11, the independent variables are taken as $x_1 = d/D$ and $x_2 = V_2/V_1$. Then, the objective function may be written as

$$U(x_1, x_2) = A + F_1(x_1, x_2) + F_2(x_1, x_2) = A + Bx_1^n x_2 + Cx_1 x_2^m \tag{7.13}$$

where F_1 and F_2 represent the costs for the die and for applying the extrusion force. Possible expressions from curve fitting to represent these are also given, with B, C, n, and m as constants. A constraint that arises from mass balance is given by

$$\frac{\pi D^2}{4} V_1 = \frac{\pi d^2}{4} V_2, \quad \text{or,} \quad x_1^2 x_2 = 1 \tag{7.14}$$

This constraint may be included in the analysis or may have to be brought in separately if an expression, such as Equation (7.13), is given for U. The ranges of x_1 and x_2 due to limitations on the forces exerted are used to define the design domain. This problem can be solved by calculus methods as well as by geometric programming. The effect of temperature T on the process may also be included in the optimization process.

Example 7.4

In many processes, such as optical fiber drawing, hot rolling, continuous casting, and extrusion, the material is cooled by the flow of a cooling fluid, such as inert gases in optical fiber drawing, at velocity V_1, while the material moves at velocity V_2, as shown in Figure 7.12. Numerical simulation may be used to obtain the temperature decay with distance for different values of these variables, as shown qualitatively in Figure 7.13. The temperature decay increases with increasing V_1 because of accelerated cooling, but decreases with increasing V_2 because the time available for heat removal in the cooling section of length L decreases at higher speed. The exit temperature must drop below a given value T_2.

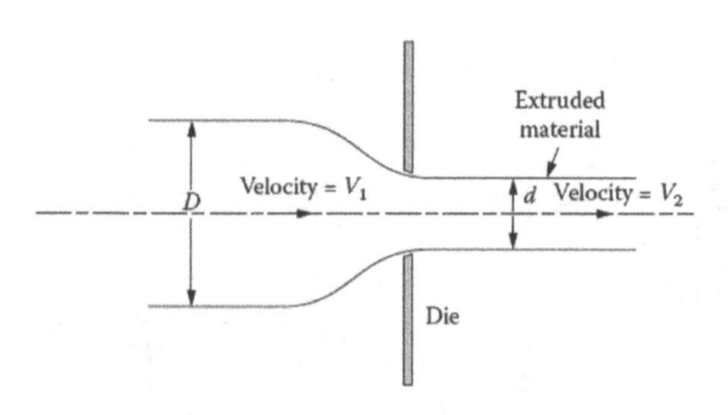

FIGURE 7.11 A metal extrusion process.

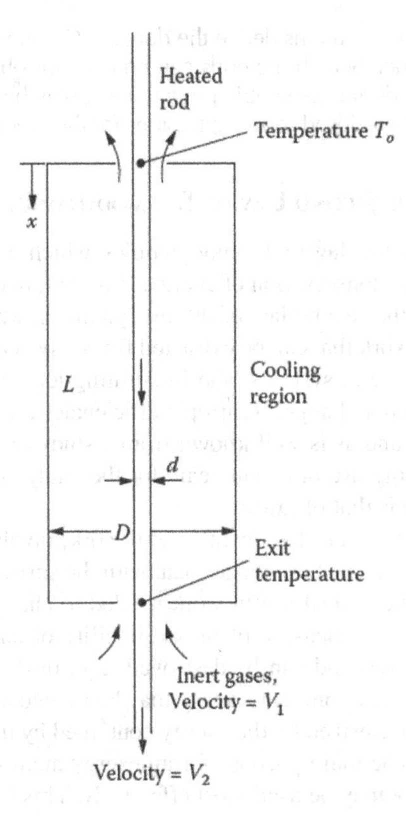

FIGURE 7.12 Cooling of a heated moving rod by the flow of inert gases.

Numerical simulation may be used to solve this combined conduction convection problem and obtain the inputs needed for design and optimization of the cooling system. If the cost per unit length of processed material is taken as the objective function U, we may write

$$U(V_1, V_2) = \frac{F_1(V_2)}{F_2(V_2)} + F_3(V_1) \quad (7.15)$$

where the function F_1 represents the costs for feeding and pickup of the material, F_2 represents the productivity, and F_3 represents the cost of the inert gas and the flow arrangement. Limitations on

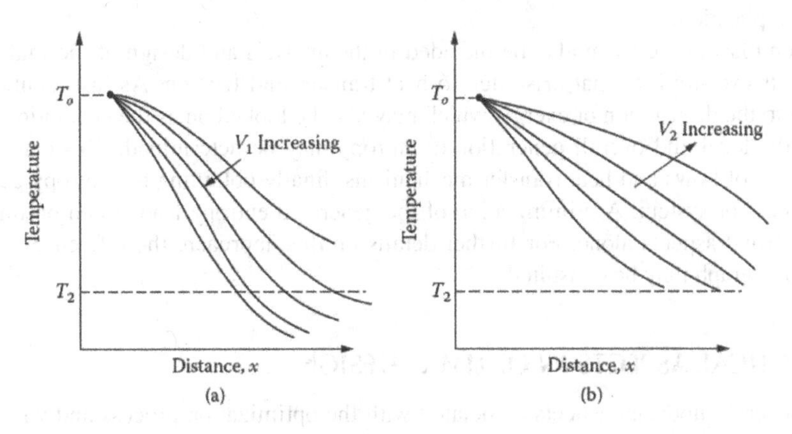

FIGURE 7.13 Dependence of temperature decay with distance x on (a) the velocity V_1 of inert gases and (b) velocity V_2 of the heated moving rod in Example 7.4.

V_1 and V_2 due to physical considerations define the domain. Constraints due to mass and energy balances are part of the model. Search methods can be used for obtaining the optimum values of V_1 and V_2. Calculus methods and geometric programming may be applicable if the simulation results are curve fitted to obtain closed-form expressions for the preceding functions.

7.4.5 Consideration of the Second Law of Thermodynamics

We have already considered the first law of thermodynamics, which states that energy cannot be created or destroyed, leading to the conservation of energy. However, in dealing with thermal systems, an important consideration is the second law of thermodynamics, which brings in the concepts of entropy and maximum useful work that can be extracted from a system. Entropy is used extensively in analyzing thermal processes and systems, and in defining ideal processes that are isentropic, i.e., in which the entropy does not change. Isentropic efficiencies are based on this ideal behavior, as has been mentioned earlier and as is well-known from a study of thermodynamics. However, a concept that is finding increasing use in recent years for the analysis, design, and optimization of thermal processes and systems is that of *exergy*.

Exergy is defined as the maximum theoretical useful work, involving shaft or electrical work, that can be obtained from a system as it exchanges heat with the surroundings to attain equilibrium. Similarly, it is the minimum theoretical useful work needed to change the state of matter, as in a refrigerator. Therefore, exergy is a measure of the availability of energy from a thermal system. Exergy is generally not conserved and can be destroyed, e.g., in the uncontrolled expansion of a pressurized gas. For a specified environment, exergy may be treated as an extensive property of the system, which can thus be characterized by the exergy contained by the system. Exergy can also be transferred between systems. The main purpose for an exergy analysis is to determine where and how losses occur so that energy may be used most effectively. This leads to an optimization of the process and thus of the system.

Several papers have focused on exergy analysis and the use of the second law of thermodynamics for the optimization of thermal systems; see, for instance, Bejan (1982, 1995), Bejan et al. (1996), and Dincer et al. (2014). Similar to the conservation of mass and energy, exergy balance equations may be written for closed systems and control volumes. The destruction of exergy due to friction and heat transfer is included in the balance. An efficiency, known as *exergetic efficiency* and based on the second law, may then be employed to give a true measure of the behavior of a thermal system. Such an efficiency can be defined for compressors, pumps, fans, turbines, heat exchangers, and other components of thermal systems. Then a maximization of this efficiency would result in the optimization of the system in order to extract the maximum amount of useful work from it. Thus, exergy may also be used as a basis for optimization and for obtaining the most cost-effective system for a given application.

The second law aspects can also be included in the analysis and design of thermal systems by considering irreversibilities that arise due to heat transfer and friction. As just mentioned, these effects lead to the destruction of exergy, which may also be looked on as the generation of entropy. Therefore, the local and overall generation of entropy may be determined. This can be done for different types of flows and heat transfer mechanisms, finally obtaining the entropy generation in a given process or system. A minimization of the generated entropy leads to an optimum system based on thermal aspects alone. For further details on this approach, the references given in the preceding paragraph may be consulted.

7.5 PRACTICAL ASPECTS IN OPTIMAL DESIGN

There are several important aspects associated with the optimization process and with the implementation of the optimal design obtained. These considerations are common to all the different approaches and address the practical issues involved in optimization. Because our interest lies in an optimum design that is both feasible and practical, it is necessary to include the following aspects in the overall design and optimization of thermal systems.

7.5.1 Choice of Variables for Optimization

Several independent variables are generally encountered in the design of a thermal system. A workable design is obtained when the design, as represented by a selection of values for these variables, satisfies the given requirements and constraints. The same variables, considered over their allowable ranges, indicate the boundaries of the domain in which the optimal design is sought. If only two design variables are considered, the objective function $U(x_1, x_2)$ may be plotted as the elevation over an $x_1 - x_2$ coordinate plane to yield a surface, as discussed earlier and as shown in Figure 7.14(a). Then, depending on the problem, the maximum or minimum value of U on this surface gives the desired optimum. Because of the difficulty of drawing such three-dimensional representations on a two-dimensional drawing surface, the variation of U with x_1 and with x_2 may be plotted separately to determine the corresponding optima, as shown in Figures 7.14(b) and (c).

Clearly, it is much easier to deal with a relatively small number of independent variables, as compared to the full set of variables. With just one or two variables, it is possible to visualize the variation of the objective function and it is easy to extract the optimum. Therefore, it is best to focus on the most important variables, as judged from a physical understanding of the system or as derived from a *sensitivity analysis*. One may start with a workable design and vary just one or two dominant design variables to obtain the optimum. For instance, after a feasible design of a power plant is obtained, the boiler pressure may be considered as the most important design variable to seek an optimum in the power output per unit cost. As the pressure is increased, the objective function increases, with local decreases resulting from the need to go to a larger boiler or one with a different

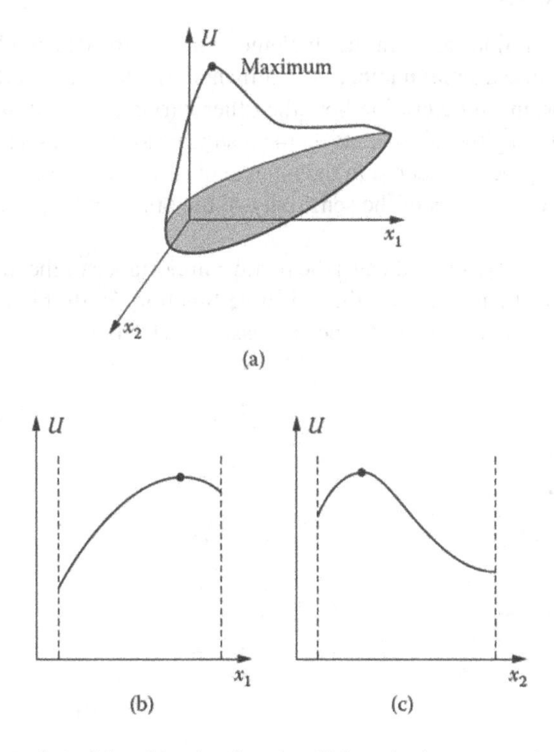

FIGURE 7.14 Optimum value of the objective function $U(x_1, x_2)$, shown on a three-dimensional elevation plot and on graphs for each of the independent variables.

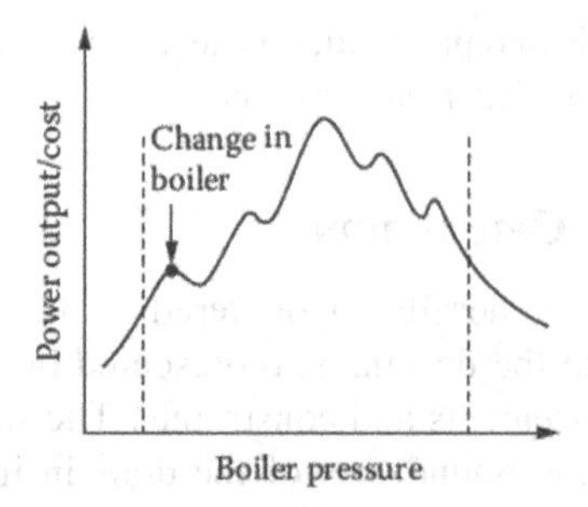

FIGURE 7.15 Variation of power output/cost ratio for a power plant as a function of the boiler pressure, showing the effect of changing the boiler size and a global maximum.

design. An overall maximum may arise, as shown in Figure 7.15, with a decrease in U beyond this value due to the higher material and construction costs at large pressures. Thus, an optimum boiler pressure may be determined. Other variables, such as condenser pressure, may also be considered to seek the optimal design.

Similarly, Figure 7.16 shows the variation of the objective function with a dominant design variable in two other cases. In the first, the objective function is the productivity per unit cost in an optical fiber drawing process and the fiber speed is the dominant variable. In the second case, the heat removal rate per unit cost for an electronic system is the objective function and the fan size or rating is the main design variable. Therefore, the optimum fiber speed and fan size may be determined by applying optimization techniques. In all such cases, effort is made to use the smallest number of variables, considering only the most crucial ones in the optimization process.

7.5.2 Sensitivity Analysis

Several important considerations arise in the implementation of the design obtained from an optimization procedure. Because a small number of dominant variables are usually employed to obtain the optimum, it is important to determine how the other variables would affect the optimum. In addition, the effect of relaxing the constraints on the results needs to be ascertained. Some changes in the design variables may be considered in the interest of convenience or reduced costs. All these aspects are best considered in terms of the sensitivity of the optimal design to the design variables and to the constraints.

A sensitivity analysis of a system indicates the relative importance of the different design parameters, as given in terms of their effect on the objective function. With this information, we could determine which parameters are crucial to the successful performance of the system. This would

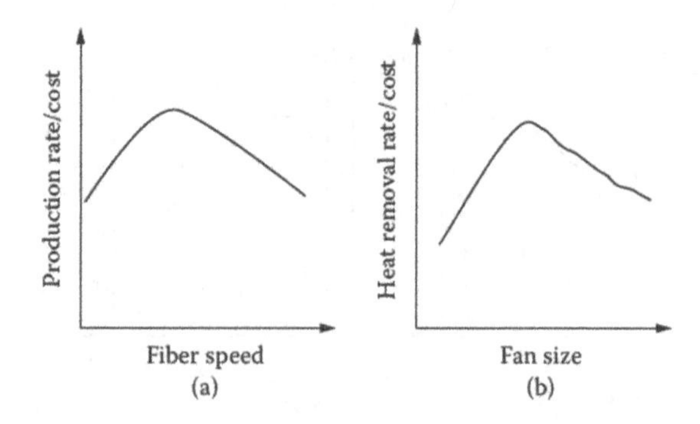

FIGURE 7.16 (a) Variation of production rate per unit cost in an optical fiber drawing system with fiber speed and (b) variation of heat removal rate per unit cost in an electronic system with the fan size.

allow us to focus on the most important parameters and their critical ranges of variation. As an example, let us take the cost per unit output for a metal forming production system as the objective function U. Let us assume that it can be expressed in terms of the pressure P, feed rate V, and heat input Q as

$$U = AP^{a} + \frac{BQ}{V^b} + CQ^d P \tag{7.16}$$

implying that the cost increases with the imposed pressure and heat input, while the output increases with the feed rate. Here, the coefficients A, B, and C, and the exponents a, b, and d are constants. Then the partial derivative of the objective function with respect to each independent variable indicates the sensitivity to that variable. These derivatives may be normalized by the values at a reference point, denoted by the subscript "ref," to give relative sensitivities, which are more useful than absolute values in determining the importance of the different variables. The reference point may be the optimum, the average value of each variable, or any other convenient value in the design domain.

Thus, the relative sensitivities S_P, S_Q, and S_V, with respect to the three variables, may be obtained analytically as

$$S_P = \frac{\partial(U/U_{ref})}{\partial(P/P_{ref})} = \left(AaP^{a-1} + CQ^d\right)\frac{P_{ref}}{U_{ref}} \tag{7.17a}$$

$$S_Q = \frac{\partial(U/U_{ref})}{\partial(Q/Q_{ref})} = \left(\frac{B}{V^b} + CdQ^{d-1}P\right)\frac{Q_{ref}}{U_{ref}} \tag{7.17b}$$

$$S_V = \frac{\partial(U/U_{ref})}{\partial(V/V_{ref})} = \left(-\frac{BbQ}{V^{b+1}}\right)\frac{V_{ref}}{U_{ref}} \tag{7.17c}$$

Numerical values of these relative sensitivities can be obtained to determine which variables are crucial and which ones are of minor importance. The optimization process is then carried out using only the dominant variables, as discussed earlier. If the optimum has already been obtained, considering a small number of variables chosen based on physical characteristics, the sensitivity analysis may be used to determine if the other variables are important and if adjustments in the optimal design with respect to these variables would significantly improve the design.

Analytical methods for sensitivity analysis, as outlined here, are generally of limited value and can be used only if closed-form expressions such as the one given in Equation (7.16) characterize the system or are available through curve fitting. If the analytical approach is not possible, numerical methods may be used. The desired partial derivatives are obtained by varying the design parameters by a small amount, say a few percent of its value at the midpoint of its range or at any other chosen location, and evaluating the derivative at this point as

$$\frac{\partial U}{\partial x} = \frac{U(x+\Delta x) - U(x)}{\Delta x} \tag{7.18}$$

where x is the independent variable under consideration. Thus, all the relevant partial derivatives may be obtained and normalized by the values at the chosen reference point to determine the dominant variables.

An important practical consideration in the implementation of the optimal design obtained from the analysis is the choice of the closest dimension, size, or rating that may be available *off the shelf*, rather than have the exact values custom made. For instance, if the optimal design yields a pipe diameter of 0.46 in. (1.17 cm), it would be desirable to use one with a diameter of 0.5 in. (1.27 cm) because of its easy availability and lower cost. Similarly, the specifications of a heater, valve, storage tank, pump, compressor, or heat exchanger may be adjusted to use readily available standard items. The sensitivity analysis is again useful in this regard because it indicates the effects of changing the design variables. Relatively large adjustments may be made if the design is not very sensitive to a given variable and small adjustments if it is.

Another important consideration is the sensitivity of the optimum to the constraints. This relates to the change in the optimum if a given constraint is relaxed in order to employ readily available items, to simplify the fabrication and assembly of the system, or to meet some other desirable goals. The relevant parameters are known as sensitivity coefficients and are obtained as part of the solution in the Lagrange multiplier method. In other approaches, the sensitivity coefficients are often derived in order to help in making adjustments in the optimal design before proceeding with its implementation.

7.5.3 Dependence on Objective Function: Trade-Offs

The optimal design of the system is obtained based on a chosen objective function that is minimized or maximized. Several examples of important objective functions relevant to thermal systems have been given earlier. However, even though several features or aspects are important in most systems, only one characteristic was chosen for optimization. Because the cost, profit, input, quality, efficiency, output, etc., are all of particular interest, these are often used separately or in combination, for example, as output/cost, quality/cost, efficiency/input, or profit/cost. Then, other important features of the system such as weight, volume, thrust, flow rate, pressure, etc., are not optimized, even though effort is often made to bring these into the optimization process through costs, profits, efficiency, and outputs. It is evident that the choice of the objective function is a very important decision and is expected to play a critical role in the determination and selection of an optimal design.

Suppose a system is optimized by maximizing the output per unit cost, but the weight is also an important consideration. If the system were then optimized by minimizing its weight, the optimal design would, in general, be different. Because both aspects are important, we need to consider *trade-offs* between the two optimum designs in order to take both of these into account. This is, by no means, an easy exercise because the behavior of the optimum with respect to the design variables may have opposite trends in the two cases. For instance, use of a different, stronger composite material may reduce the weight while increasing the cost. A smaller heating region in a glass manufacturing facility may reduce the cost, but it will also reduce the output.

One way of approaching such trade-offs is to assign a value to each important aspect, as discussed by Siddall (1982). The value represents the desirability of the given feature. For instance, a large weight is undesirable and is assigned a low value, with the value dropping to zero beyond a certain weight, as shown in Figure 7.17(a). Similarly, a high value is given to a large output/cost and a low one to a small output/cost, as shown in Figure 7.17(b). These values are obviously subjective and depend on the designer and the application. A trade-off curve may be drawn by finding the maximum output/cost for different weights. The weight then becomes a specification and the maximum output/cost is determined for each case, generating a curve such as the one shown in Figure 7.17(c). The optimum, which includes considerations of both features, is the point on this trade-off curve that has a maximum combined value for the two. This point is somewhere near the middle of the trade-off curve in the example shown.

7.5.4 Multi-Objective Optimization

It was mentioned earlier that optimal conditions are generally strongly dependent on the chosen objective function. However, as discussed in the preceding section, not one but several features or

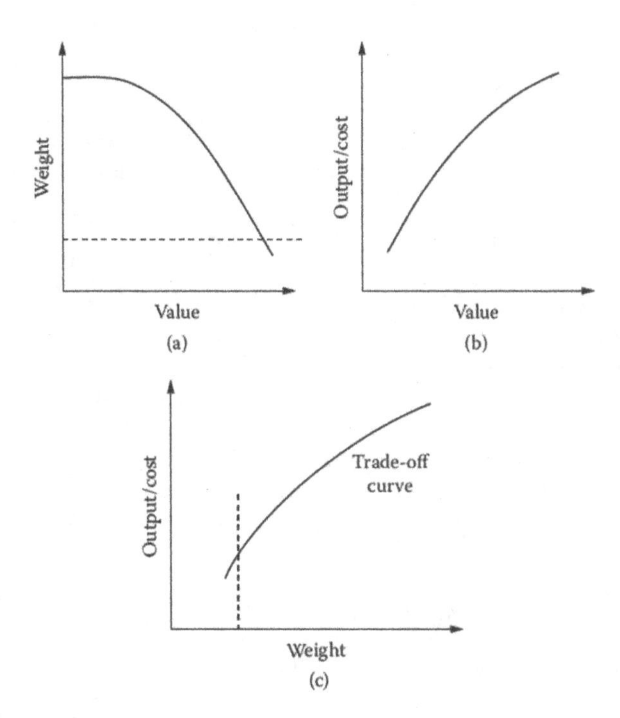

FIGURE 7.17 Typical value curves that may be used to develop a trade-off curve in optimization.

aspects are typically important in most practical applications. In thermal systems, the efficiency, production rate, output, quality, and heat transfer rate are common quantities that are to be maximized, while cost, input, environmental effect, and pressure are quantities that need to be minimized. Thus, any of these could be chosen individually as the objective function, though interest clearly lies in dealing with more than one objective function. The use of the trade-off curve was outlined in the preceding section.

A common approach to multiple objective functions is to combine them to yield a single objective function that is minimized or maximized. Examples given earlier include output/cost, quality/cost, and efficiency/input. In heat exchangers and cooling systems for electronic equipment, it is desirable to maximize the heat transfer rate. However, this comes at the cost of flow rate or pressure head. Then heat transfer rate/pressure head could be chosen as the objective function. Similarly, additional aspects could be combined to obtain a single objective function, e.g., objective function $U = \text{quality} \times \text{production rate/cost}$. However, the various quantities that compose the objective function should be scaled and weighted in order to base the system optimization on the importance of each in comparison to the others. For instance, heat transfer rate and pressure head may be scaled with the expected maximum values in a given instance so that both vary from 0 to 1. Other nondimensionalizations are also possible, as discussed earlier in Chapters 2 and 3, to ensure that equal importance is given to each of these. Weights can similarly be used to increase or decrease the importance of a given quantity compared to the others. Derived quantities like logarithm or exponential of given physical quantities may also be employed for scaling and for considering appropriate ranges of the quantities. Clearly, the objective function thus obtained is not unique and different formulations can be used to generate different functions, which could presumably yield different optimal points. A few examples on this approach are given in Chapter 9.

Another approach, which has gained interest in recent years, is that of multi-objective optimization. In this case, two or more objective functions that are of interest in a given problem are considered and a strategy is developed to trade off one objective function in comparison to the others (Miettinen, 1999; Deb, 2009). Let us consider a problem with two objective functions f_1

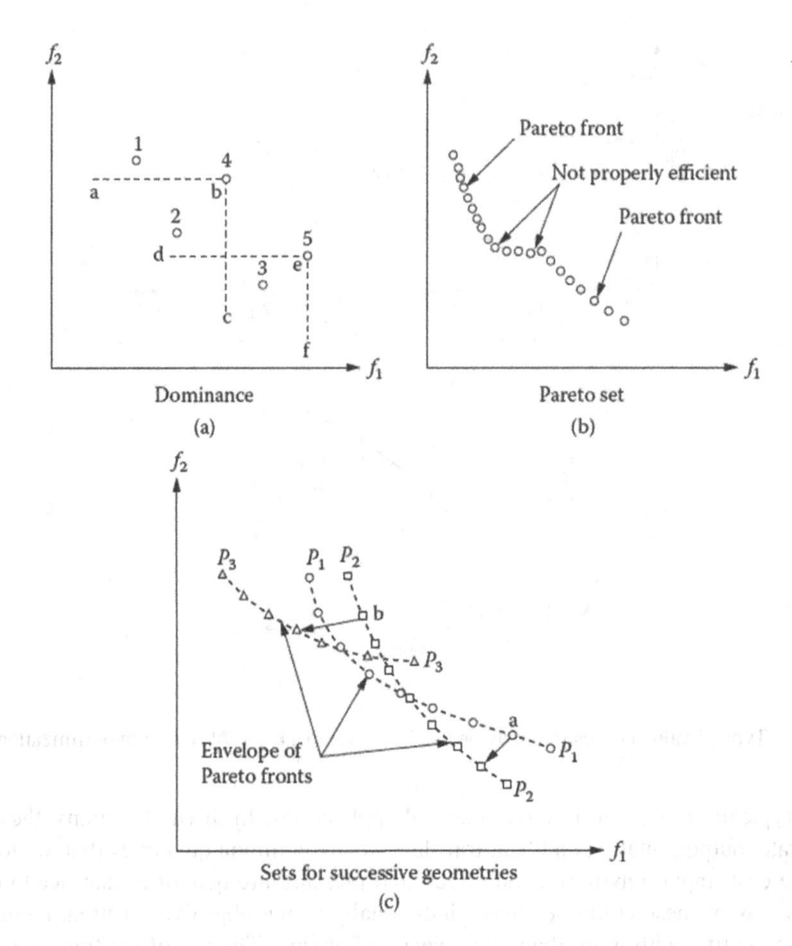

FIGURE 7.18 Multi-objective optimization with two objective functions f_1 and f_2, which are to be minimized, showing the dominant designs, the Pareto front, and the envelope of Pareto fronts for different geometric configurations.

and f_2. With no loss of generality, we can assume that each of these is to be minimized because maximization is equivalent to minimization of the negative of the function. The values of f_1 and f_2 are shown for five designs in Figure 7.18(a), each design being indicated by a point. Design 2 dominates Design 4 because both objective functions are smaller for Design 2 compared to Design 4. Similarly, Design 3 dominates Design 5. However, Designs 1, 2, and 3 are not dominated by any other design. The selection of the better design is straightforward for the dominated cases, though not so for the others. The set of nondominated designs is termed the *Pareto set*, which represents the best collection of designs. As shown in Figure 7.18(b), the near horizontal or near vertical sections are omitted to obtain proper efficiency for design selection, and a Pareto front is obtained. Then, for any design in the Pareto set, one objective function can be improved, i.e., reduced as considered here, at the expense of the other objective function. The same arguments apply for more than two objective functions. The set of designs that constitute the Pareto set represents the formal solution in the design space to the multi-objective optimization problem. The selection of a specific design from the Pareto set is left to the decision-maker or the engineer. A large literature exists on utility theory, which seeks to provide additional insight to the decision-maker to assist in selecting a specific design; see Ringuest (1992).

For different concepts, such as geometrical configurations, different Pareto fronts can be generated, with the envelope of these yielding the desired solution, as shown in Figure 7.18(c).

Many multi-objective optimization methods are available that can be used to generate Pareto solutions. Various quality metrics are often used to evaluate the "goodness" of a Pareto solution obtained and possibly improve the method as well as the optimal solution. Examples of multi-objective optimization of thermal systems are given in Chapter 9.

7.5.5 Part of Overall Design Strategy

Optimization of thermal systems, as discussed here, is treated as a step in the overall design process. Thus, the optimization process is based on the modeling, simulation, and experimental effort undertaken to obtain a feasible design. The requirements and constraints are also those that are specified for the workable design. The initial effort is directed at a workable design that satisfies the given requirements and constraints. A single design for the system or a number of acceptable designs may be generated. This completes the first phase of the overall design strategy because any one of the designs generated may be used for the intended application.

An objective function is then selected for optimization and an appropriate method is used to extract a design for which the chosen objective function is minimized or maximized. Though the exact optimal point is reached in only a few ideal cases, the optimization process generally does allow one to obtain a small region containing the optimum. The final, optimal design is then obtained using the practical considerations outlined in the preceding sections. The model or concept employed, as well as the governing equations, are the same as those used for a feasible design.

Therefore, optimization is an extension of the design process employed to generate a workable design. The main difference between the two stages concerned with workable and optimal designs of the system lies in the objective or purpose of the effort. In the first stage, we want to obtain any design that meets the given requirements and constraints so that it will perform the desired task satisfactorily. Several designs may be acceptable. It is also possible that no design satisfies the given problem, making it necessary to choose a different concept, adjust the requirements and constraints, or abandon the project. No consideration is given at this stage to finding the *best* design. In the second stage, it is assumed that we have succeeded in obtaining an acceptable design or a number of these and are now seeking a design that optimizes a chosen quantity of particular interest to the intended application. The optimal design is expected to be unique or close to it, i.e., the design lies in a small domain of the variables. The relationship between the various steps to a feasible design and the optimization process may thus be represented qualitatively by Figure 7.19, indicating optimization as a part of the overall design effort. A similar, more detailed, schematic was also shown earlier in Figure 2.14.

7.5.6 Change of Concept or Model

With optimization taken as the next step after obtaining a feasible design, it is clear that the optimal design is necessarily related to a chosen conceptual design. The model and the simulation of the system are based on the conceptual design, which forms the starting point of the design. If a feasible design is not possible with a given concept, the concept may be changed. The basic thermal process

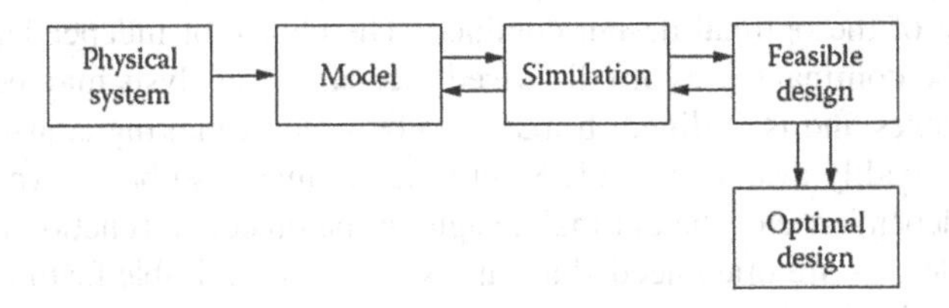

FIGURE 7.19 Optimization shown as a step in the overall design process.

may also be changed if a chosen process is not crowned with success. This is a common situation in manufacturing where one process may be replaced by another, say forming by casting, in order to satisfy the problem statement. Similarly, material substitution may be used effectively to satisfy given needs. An electric water heater may be replaced by a gas one, an evaporative cooler by an air conditioning system, a natural air drying arrangement with a forced air one, and so on, in order to satisfy the design problem.

A change in the conceptual design is undertaken at the feasible design stage, not during optimization, because one would proceed to optimize the system only after a workable design has been obtained. Therefore, optimization of the system is within the chosen concept and no variation in the conceptual design is considered. If the conceptual design is changed, the model, simulation, feasible design, and optimization all change and a design process similar to the one discussed here may be carried out for each concept considered.

7.6 SUMMARY

This chapter introduces the basic considerations in optimization and provides the general guidelines for the quantitative formulation of the problem. Starting with a discussion on the importance and need for optimizing thermal systems, the main features of the optimization process are considered. These include the objective function, which is the quantity that is to be optimized; the design variables; the operating conditions; and the constraints. Commonly used objective functions for thermal systems include energy or product output, cost, profit, output/cost, weight, volume, efficiency, energy consumption per unit output, and environmental impact. The constraints in these systems are often due to temperature and pressure limitations of materials, energy and mass conservation, ambient conditions, and practical limitations on variables such as flow rate, heat input, and dimensions. Several examples are given to illustrate the setting up of the optimization problem because the success of the optimization process is strongly dependent on an accurate and satisfactory formulation.

The chapter also outlines different optimization techniques, including calculus and search methods, and linear, dynamic, and geometric programming. The range of application of these methods to thermal systems is discussed. The calculus methods are applicable only if the objective function and the constraints are given as closed-form, differentiable expressions, severely limiting the applicability of this approach. Similarly, linear programming is applicable when only linear equations are involved in the problem, a rare circumstance in thermal processes. Dynamic programming optimizes the objective function along a path and is useful in a few problems such as flow circuits and production line design. Geometric programming requires that the problem involve sums of polynomials and, as such, is particularly useful for thermal systems in which curve fitting has been used to obtain expressions to characterize the system behavior. Search methods are clearly the most important optimization strategy for practical thermal systems because these methods search for the optimum by iterating from one design to the next, keeping the number of iterations at a minimum. Because each simulation is usually expensive and time consuming for practical systems, efficient search procedures are particularly appropriate for converging to an optimal design.

Finally, this chapter discusses several important practical issues related to optimization and to the implementation of the optimal design obtained. The choice of independent variables and the need to focus on the dominant ones are discussed. Sensitivity analysis may be used in the choice of the critical variables and is outlined. It may also be used in making adjustments to the design in order to employ readily available items. Safety factors may also be incorporated in the design at this stage. The dependence of the optimal design on the objective function is another important consideration. Trade-offs are often needed to satisfy different desirable features or multiple objective functions. Optimization follows the initial design stage, which results in a feasible design and is thus a part of the overall design process.

REFERENCES

Arora, J.S. (2004). *Introduction to optimum design* (2nd ed.). New York: Academic Press.
Beightler, C.S., Phillips, D.T., & Wilde, D.J. (1979). *Foundations of optimization* (2nd ed.). Englewood Cliffs, NJ: Prentice-Hall.
Bejan, A. (1982). *Entropy generation through heat and fluid flow*. New York: Wiley.
Bejan, A. (1995). *Entropy generation minimization*. Boca Raton, FL: CRC Press.
Bejan, A., Tsatsaronis, G., & Moran, M. (1996). *Thermal design and optimization*. New York: Wiley.
Beveridge, G.S.G., & Schechter, R.S. (1970). *Optimization: theory and practice*. New York: McGraw-Hill.
Box, G.E.P., & Draper, N.R. (1987). *Empirical model-building and response surfaces*. New York: Wiley.
Box, G.E.P., & Draper, N.R. (2007) *Response surfaces, mixtures and ridge analyses* (2nd ed.). New York: Wiley.
Deb, K. (2009). *Multi-objective optimization using evolutionary algorithms*. New York: Wiley.
Dincer, I., Midilli, A., & Kucuk, H. (2014). *Progress in exergy, energy and the environment*. Cham, Switzerland: Springer.
Fox, R.L. (1971). *Optimization methods for engineering design*. Reading, MA: Addison-Wesley.
Goldberg, D.E. (1989). *Genetic algorithms in search, optimization and machine learning*. Boston, MA: Kluwer Academic Press.
Holland, J.H. (2002). *Adaptation in natural and artificial systems*. Cambridge, MA: MIT Press.
Jain, L.C., & Martin, N.M. (Eds.). (1999). *Fusion of neural networks, fuzzy systems and genetic algorithms: industrial applications*. Boca Raton, FL: CRC Press.
Miettinen, K.M. (1999). *Nonlinear multi-objective optimization*. Boston, MA: Kluwer Academic Press.
Miller, R.E. (2000). *Optimization: foundations and applications*. New York: Wiley.
Mitchell, M. (1998). *An introduction to genetic algorithms*. Cambridge, MA: MIT Press.
Myers, R.H., & Montgomery, D.C. (2016). *Response surface methodology, process and product optimization using designed experiments* (4th ed.). New York: Wiley.
Papalambros, P.Y., & Wilde, D.J. (2017). *Principles of optimal design* (3rd ed.). New York: Cambridge University Press.
Rao, S.S. (2009). *Engineering optimization: theory and practice* (4th ed.). New York: Wiley.
Ravindran, A., Ragsdell, K.M., & Reklaitis, G.V. (2006). *Engineering optimization: methods and applications* (2nd ed.). New York: Wiley.
Rhinehart, R.R. (2018). *Engineering optimization: applications, methods and analysis*. New York: Wiley.
Ringuest, J.L. (1992). *Multiobjective optimization: behavioral and computational considerations*. Boston, MA: Kluwer Academic Press.
Ross, T.J. (2004). *Fuzzy logic with engineering applications* (2nd ed.). New York: Wiley.
Sheppard, C. (2018). *Genetic algorithms with Python*. Austin, TX: CreateSpace Independent Publishing.
Siddall, J.N. (1982). *Optimal engineering design*. New York: Marcel Dekker.
Stoecker, W.F. (1989). *Design of thermal systems* (3rd ed.). New York: McGraw-Hill.
Vanderplaats, G.N. (1984). *Numerical optimization techniques for engineering design*. New York: McGraw-Hill.

PROBLEMS

7.1 Consider plastic extrusion at temperature T and a given pressure p. The cost varies as T^a and as V^{-b}, where V is the speed of the emerging plastic billet and a, b are constants. In addition, $V(T)$ is given as a third-order polynomial. Formulate the optimization problem for this system and outline a method to obtain the solution for minimum cost at the given pressure level.

7.2 In continuous casting, the cost varies as L^c and as V^{-d}, where L is the length of the mold and V is the speed of the material. In addition, c and d are given constants. Assume that the solidification occurs entirely in the mold, with heat loss to the mold at convective heat transfer coefficient h and mold temperature T_a. Using a simple model for the process, formulate the optimization problem.

7.3 Suggest different objective functions for optimizing the thermal systems considered in Example 3.5 and Example 3.6. Choose the most appropriate one and give reasons for the choice.

7.4 You have learned in this chapter that the choice of the objective function is very important. A condenser is to be designed to condense steam to water at the same temperature, while removing thermal energy at the specified rate Q. A counter-flow heat exchanger is to be employed. Constraints on temperature rise of the colder fluid and heat exchanger dimensions are given. Suggest an objective function for optimization of the heat exchanger, giving reasons for your choice.

7.5 For the optimization of a stereo system, suggest three objective functions that can be used. Choose one and give reasons for your choice.

7.6 A refrigeration system is to be designed to provide 5 kW of cooling at –5°C, with the ambient at 25°C. If the dimensions of the region that has to be cooled are fixed, list the design variables and requirements for an acceptable design. Suggest an objective function that may be employed for optimization. Also, give the constraints, if any, in the problem.

7.7 A heat pump is being designed to supply 12 kW to a residential unit when the ambient temperature is approximately 0°C and the interior temperature is 20°C. Using any appropriate conceptual design, list the design variables, constraints, and requirements. Obtain an acceptable design to achieve the given requirements. If the energy consumption is to be minimized, formulate the optimization problem.

7.8 A condenser is being designed to condense steam at a constant temperature of 100°C, with water entering at 20°C. The total energy transfer is given as 20 kW and the UA of the heat exchanger is given as 4 kW/K, where U is the overall heat transfer coefficient and A the heat transfer area. The heat loss to the environment may be taken as negligible. Clearly, an acceptable design may be obtained for this problem over wide ranges of the governing parameters. Calculate the flow rates and give an acceptable design for this process. Suggest a few objective functions that may be used for optimizing the system and then choose one to formulate the optimization problem. What optimization technique would you use to solve this problem?

7.9 Example 5.1 presented the approach for obtaining an acceptable design. Is it possible to optimize the system in this case? If so, formulate the problem, in terms of the objective function, design variables, and constraints, and discuss the procedure that may be adopted to obtain the optimum.

7.10 Consider the condensation soldering facility discussed in detail in Chapter 2 and sketched in Figure 2.4 and Figure 2.6. The dimensions of the condensation region are fixed by the size of the electronic components submerged in this region. The fluid choice is limited by temperature needed, safety, cost, and other aspects mentioned earlier. If the fluid and the dimensions of the condensation region are taken as fixed, what are the design variables and constraints? Suggest a few objective functions that may be used to optimize the system. Choose the one that you feel is particularly appropriate for this problem, giving reasons for your choice.

7.11 An acceptable design is discussed in the coiling of plastic cords, presented in Example 5.4. If the system is to be optimized to minimize the manufacturing cost per cord, formulate the corresponding optimization problem, and give the appropriate mathematical expressions.

7.12 For the thermal systems considered in Example 5.5 and Example 5.6, suggest appropriate objective functions for optimization. Also, list the design variables and the constraints, if any. Discuss the optimization strategies you would adopt for these problems.

7.13 A circulating water loop has a heat exchanger on either side, as shown in Figure P7.13. On one side, steam condenses at a constant temperature of 90°C, and on the other side, a low-boiling fluid boils at 25°C. The total energy transfer is given as 50 kW and the overall heat transfer coefficient U is given as 25 $W/m^2 \cdot K$ for both heat exchangers. The capital cost of the heat exchangers is given as \$100 per unit area (in square meters) for heat transfer and the pumping cost over its useful life is $10^4\ \dot{m}$ in present worth, where $\dot{m}$ is the mass flow rate. If the total cost is to be minimized, formulate the optimization problem and outline a method to solve it.

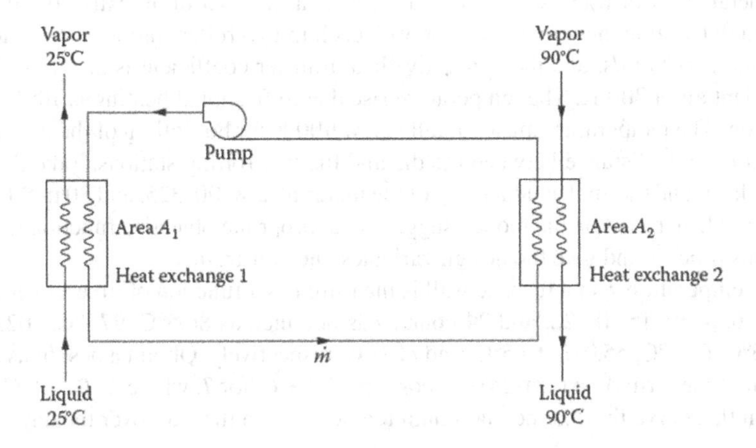

FIGURE P7.13

7.14 Water is to be taken from a purification unit to a storage tank by using two flow circuits as shown in Figure P7.14. The efficiency E of each pump, in percent, is given as

$$E = 32 + 4\dot{m} - 0.2(\dot{m})^2$$

and the pressure head P, in meters of water, versus mass flow rate $\dot{m}$, in kilograms per second, is given for the two pumps as

$$P = 20 - \dot{m} \quad \text{and} \quad P = 10 - 0.5\dot{m}$$

Either both or a single pump may be used at a given time. If the energy consumption is to be minimized, formulate the optimization problem and present the optimal method of filling the tank.

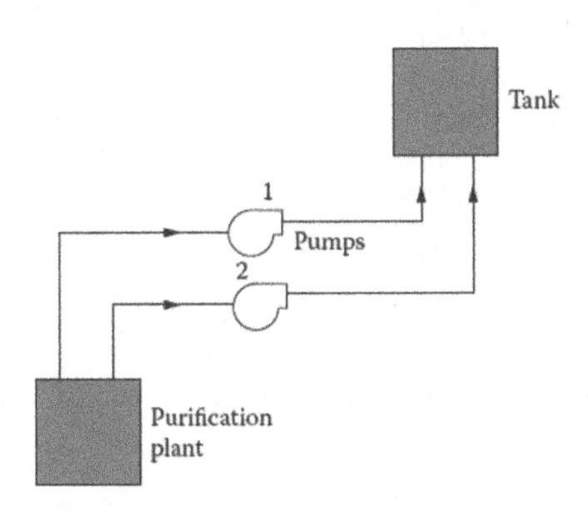

FIGURE P7.14

7.15 If the combustion efficiency of the engine of an automobile varies as V^n and the frictional force and drag on the car as V^m, where V is the speed and n and m are exponents that may be positive or negative, formulate the optimization problem to determine the speed at which the fuel consumption per unit distance traveled is minimum. What optimization technique would you use to solve this problem? Give reasons for your choice.

7.16 A metal sheet of thickness 5 cm is at 1100 K at the exit of an extrusion die. It then goes through two thickness reductions of 30% each in two roller stations. The material speed at the die is 0.25 m/s, and the convective heat transfer coefficient is given as 75 $W/m^2 \cdot K$ to ambient air at 300 K. The temperature rise due to frictional heating is 100 K at each roller station. The temperature must not fall below 900 K for hot rolling of the material. Calculate the allowable distance between the die and the two rolling stations. Take the density, specific heat, and thermal conductivity of the material as 8500, 325, and 80 in S.I. units, respectively. Then, based on the model, suggest an appropriate objective function for optimization of this process and give the design variables and constraints.

7.17 The temperature T in a furnace wall is measured as a function of time τ over a day. For τ of 2, 3, 6, 8, 10, 15, 18, 22, and 24 hours, T is obtained as 86.5°C, 97.7°C, 102.0°C, 101.7°C, 92.5°C, 62.3°C, 55.0°C, 67.5°C, and 80.0°C, respectively. Obtain a best fit assuming a variation of the form $A \sin(2\pi\tau/24) + B \cos(2\pi\tau/24) + C$, for T, where A, B, and C are constants. From this curve fit, find the maximum temperature in the wall over the day.

8 Lagrange Multipliers

8.1 INTRODUCTION TO CALCULUS METHODS

We are all quite familiar, from courses in mathematics, with the determination of the maximum or minimum of a function by the use of calculus. If the function is continuous and differentiable, its derivative becomes zero at the extremum. For a function $y(x)$, this condition is written as

$$\frac{dy}{dx} = 0 \tag{8.1}$$

where x is the independent variable. The basis for this property may be explained in terms of the extrema shown in Figure 8.1. As the maximum at point A is approached, the value of the function $y(x)$ increases and just beyond this point, it decreases, resulting in zero gradient at A. Similarly, the value of the function decreases up to the minimum at point B and increases beyond B, giving a zero slope at B.

In order to determine whether the point is a maximum or a minimum, the second derivative is calculated. Because the slope goes from positive to negative, through zero, at the maximum, the second derivative is negative. Similarly, the slope increases at a minimum and, thus, the second derivative is positive. These conditions may be written as (Keisler, 2012)

$$\text{For a maximum:} \quad \frac{d^2y}{dx^2} < 0 \tag{8.2}$$

$$\text{For a minimum:} \quad \frac{d^2y}{dx^2} > 0 \tag{8.3}$$

These conditions apply for nonlinear functions $y(x)$ and, therefore, calculus methods are useful for thermal systems, which are generally represented by nonlinear expressions. However, both the function and its derivative must be continuous for the preceding analysis to apply.

Thus, by setting the gradient equal to zero, the locations of the extrema may be obtained and the second derivative may then be used to determine the nature of each extremum. There are cases where both the first and the second derivatives are zero. This indicates an inflection point, as sketched in Figure 8.1(c), a saddle point, or a flat curve, as in a ridge or valley. It must be noted that the conditions just mentioned indicate only a local extremum. Several such local extrema may arise in the given domain. Because our interest lies in the overall maximum or minimum in the entire domain for optimizing the system, we would seek the global extremum, which is usually unique and represents the largest or smallest value of the objective function. The following simple example illustrates the use of the preceding procedure for optimization.

Example 8.1

Apply the calculus-based optimization technique just given to the minimization of cost C for hot rolling a given amount of metal. This cost is expressed in terms of the mass flow rate $\dot{m}$ of the material as

$$C = 3.5\dot{m}^{1.4} + \frac{14.8}{\dot{m}^{2.2}}$$

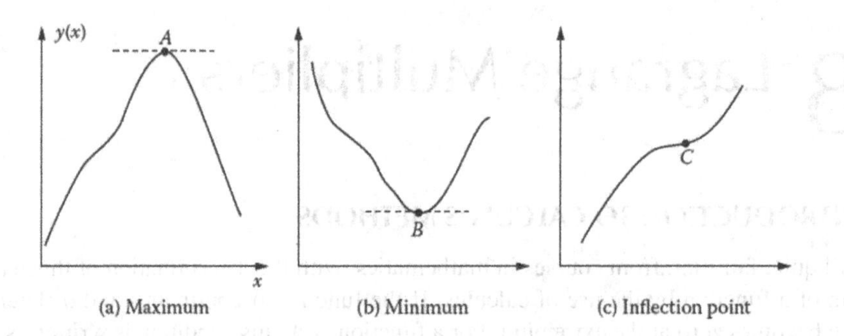

FIGURE 8.1 Sketches showing (a) a maximum, (b) a minimum, and (c) an inflection point in a function $y(x)$ plotted against the independent variable x.

where the first term on the right-hand side represents equipment costs, which increase as the flow rate increases, and the second term represents the operating costs, which go down as $\dot{m}$ increases.

SOLUTION

The extremum is given by

$$\frac{dC}{d\dot{m}} = (3.5)(1.4)\dot{m}^{0.4} - (14.8)(2.2)\dot{m}^{-3.2}$$
$$= 4.9\dot{m}^{0.4} - 32.56\dot{m}^{-3.2} = 0$$

Therefore,

$$\dot{m} = \left(\frac{32.56}{4.9}\right)^{1/3.6} = 1.692$$

The second derivative is obtained as

$$\frac{d^2C}{d\dot{m}^2} = 1.96\dot{m}^{-0.6} + 104.19\dot{m}^{-4.2}$$

which is positive because the flow rate $\dot{m}$ is positive. This implies that the optimization technique has yielded a minimum of the objective function C, as desired. Therefore, minimum cost is obtained at $\dot{m} = 1.692$, and the corresponding value of C is 11.962.

The preceding discussion and the simple example serve to illustrate the use of calculus for optimization of an unconstrained problem with a single independent variable. However, such simple problems are rarely encountered when dealing with the optimization of practical thermal systems. Usually, several independent variables are involved and constraints may have to be satisfied. This considerably complicates the application of calculus to extract the optimal solution. In addition, the use of calculus methods requires that any constraints in the problem must be equality constraints. This limitation is often circumvented by converting inequality constraints into equality ones, as outlined in Chapter 7. In many practical circumstances, the objective function is not readily available in the form of continuous and differentiable functions, such as the one given in Example 8.1. However, curve fitting of numerical and experimental data may be used in some cases to yield continuous expressions that characterize the given system and that can then be used to obtain the optimum.

Calculus methods, whenever applicable, provide a fast and convenient method to determine the optimum. They also indicate the basic considerations in optimization and the characteristics of the problem under consideration. In addition, some of the ideas and procedures used for these methods are employed in other techniques. Therefore, it is important to understand this optimization method and the basic concepts introduced by this approach. This chapter presents the Lagrange multiplier method, which is based on the differentiation of the objective function and the constraints. The physical interpretation of this approach is brought out and the method is applied to both constrained and unconstrained optimization. The sensitivity of the optimum to changes in the constraints is discussed. Finally, the application of this method to thermal systems is considered.

8.2 THE LAGRANGE MULTIPLIER METHOD

This is the most important and useful method for optimization based on calculus. It can be used to optimize functions that depend on a number of independent variables, with and without functional constraints. As such, it can be applied to a wide range of practical circumstances provided the objective function and the constraints can be expressed as continuous and differentiable functions. In addition, only equality constraints can be considered in the optimization process.

8.2.1 Basic Approach

The mathematical statement of the optimization problem was given in the preceding chapter as

$$U(x_1, x_2, x_3, \ldots, x_n) \rightarrow \text{Optimum} \tag{8.4}$$

subject to the constraints

$$\begin{aligned} G_1(x_1, x_2, x_3, \ldots, x_n) &= 0 \\ G_2(x_1, x_2, x_3, \ldots, x_n) &= 0 \\ &\vdots \\ G_m(x_1, x_2, x_3, \ldots, x_n) &= 0 \end{aligned} \tag{8.5}$$

where U is the objective function that is to be optimized and $G_i = 0$, with i varying from 1 to m, represents the m equality constraints. As mentioned earlier, if inequality constraints arise in the problem, these must be converted into equality constraints in order to apply this method. In addition, in several cases, inequality constraints simply define the acceptable domain and are not used in the optimization process. Nevertheless, the solution obtained is checked to ensure that these constraints are not violated.

The method of Lagrange multipliers basically converts the preceding problem of finding the minimum or maximum into the solution of a system of algebraic equations, thus providing a convenient scheme to determine the optimum. The objective function and the constraints are combined into a new function Y, known as the *Lagrange expression* and defined as

$$\begin{aligned} Y(x_1, x_2, \ldots, x_n) = U(x_1, x_2, \ldots, x_n) + \lambda_1 G_1(x_1, x_2, \ldots, x_n) + \lambda_2 G_2(x_1, x_2, \ldots, x_n) \\ + \ldots + \lambda_m G_m(x_1, x_2, \ldots, x_n) \end{aligned} \tag{8.6}$$

where the λ's are unknown parameters, known as *Lagrange multipliers*. Then, according to this method, the optimum occurs at the solution of the system of equations formed by the following equations:

$$\frac{\partial Y}{\partial x_1} = 0 \quad \frac{\partial Y}{\partial x_2} = 0 \quad \cdots \quad \frac{\partial Y}{\partial x_n} = 0 \tag{8.7a}$$

$$\frac{\partial Y}{\partial \lambda_1} = 0 \quad \frac{\partial Y}{\partial \lambda_2} = 0 \quad \cdots \quad \frac{\partial Y}{\partial \lambda_m} = 0 \tag{8.7b}$$

When these differentiations are applied to the Lagrange expression, we find that the optimum is obtained by solving the following system of equations:

$$\begin{gathered}
\frac{\partial U}{\partial x_1}+\lambda_1\frac{\partial G_1}{\partial x_1}+\lambda_2\frac{\partial G_2}{\partial x_1}+\ldots+\lambda_m\frac{\partial G_m}{\partial x_1}=0 \\
\frac{\partial U}{\partial x_2}+\lambda_1\frac{\partial G_1}{\partial x_2}+\lambda_2\frac{\partial G_2}{\partial x_2}+\ldots+\lambda_m\frac{\partial G_m}{\partial x_2}=0 \\
\vdots \\
\frac{\partial U}{\partial x_n}+\lambda_1\frac{\partial G_1}{\partial x_n}+\lambda_2\frac{\partial G_2}{\partial x_n}+\ldots+\lambda_m\frac{\partial G_m}{\partial x_n}=0 \\
G_1(x_1,x_2,x_3,\ldots,x_n)=0 \\
G_2(x_1,x_2,x_3,\ldots,x_n)=0 \\
\vdots \\
G_m(x_1,x_2,x_3,\ldots,x_n)=0
\end{gathered} \tag{8.8}$$

If the objective function U and the constraints G_i are continuous and differentiable, a system of algebraic equations is obtained. Because there are m equations for the constraints and n additional equations are derived from the Lagrange expression, a total of $m + n$ simultaneous equations are obtained. The unknowns are the m multipliers, corresponding to the m constraints, and the n independent variables. Therefore, this system may be solved by the methods outlined in Chapter 4 to obtain the values of the independent variables, which define the location of the optimum, as well as the multipliers. Analytical methods for solving a system of algebraic equations may be employed if linear equations are obtained and/or when the number of equations is small, typically up to around five. For nonlinear equations and for larger sets, numerical methods are generally more appropriate. Matlab is particularly well-suited for solving large sets of linear and nonlinear algebraic equations, with the latter generally requiring iteration, as discussed in Chapter 4. The optimum value of the objective function is then determined by substituting the values obtained for the independent variables into the expression for U. The optimum is often represented by asterisks, i.e., x_1^*, x_2^*, …, x_n^*, and U^*.

Thus, the preceding equations, given by Equation (8.8), represent the Lagrange multiplier method. The physical interpretation and proof of the method are given in the next section. But the solution to Equation (8.8) determines the optimum and also yields the multipliers.

8.2.2 Physical Interpretation

In order to understand the physical reasoning behind the method of Lagrange multipliers, let us consider a problem with only two independent variables x and y and a single constraint $G(x, y) = 0$. Then the optimum is obtained by solving the equations

$$\begin{gathered}
\frac{\partial U}{\partial x}+\lambda\frac{\partial G}{\partial x}=0 \\
\frac{\partial U}{\partial y}+\lambda\frac{\partial G}{\partial y}=0 \\
G(x,y)=0
\end{gathered} \tag{8.9}$$

The first two equations can be written in vector notation as

$$\nabla U+\lambda\nabla G=0 \tag{8.10}$$

where ∇ is the gradient vector. The gradient of a scalar quantity $\phi(x,y)$ is defined as

$$\nabla\phi = \frac{\partial\phi}{\partial x}\mathbf{i} + \frac{\partial\phi}{\partial y}\mathbf{j} \tag{8.11}$$

where **i** and **j** are unit vectors in the x and y directions, respectively. Therefore, $\nabla\phi$ is a vector with the two partial derivatives $\partial\phi/\partial x$ and $\partial\phi/\partial y$ as the two components in these directions. For example, if the temperature T in a region is given as a function of x and y, the rate of change of T in the two coordinate directions is given by the components of the gradient vector ∇T. This vector is used effectively in heat conduction to represent the heat flux vector **q**, which is given as $\mathbf{q} = -k\nabla T$ from Fourier's law, k being the thermal conductivity. This heat flux vector is used to determine the rate of heat transfer in different coordinate directions (Gebhart, 1971; Incropera and Dewitt, 2001).

8.2.2.1 Gradient Vector

Let us consider the gradient vector further in order to provide a graphical representation for the method of Lagrange multipliers. This discussion will also be useful in other optimization schemes that are based on the gradient vector. From the definition of $\nabla\phi$ and from calculus, the magnitude and direction of the gradient vector, as well as a unit vector **n** in its direction, may be calculated as

$$|\nabla\phi| = \sqrt{(\partial\phi/\partial x)^2 + (\partial\phi/\partial y)^2} \qquad \theta = \tan^{-1}\left[\frac{\partial\phi/\partial y}{\partial\phi/\partial x}\right],$$
$$\boldsymbol{n} = \frac{(\partial\phi/\partial x)\boldsymbol{i} + (\partial\phi/\partial y)\boldsymbol{j}}{\sqrt{(\partial\phi/\partial x)^2 + (\partial\phi/\partial y)^2}} \tag{8.12}$$

where $|\nabla\phi|$ is the magnitude of the gradient vector and θ is the inclination with the x-axis.

Let us now consider a ϕ = constant curve in the x-y plane, as shown in Figure 8.2 for three values c_1, c_2, and c_3 of this constant. Then, from the chain rule in calculus,

$$d\phi = \frac{\partial\phi}{\partial x}dx + \frac{\partial\phi}{\partial y}dy \tag{8.13}$$

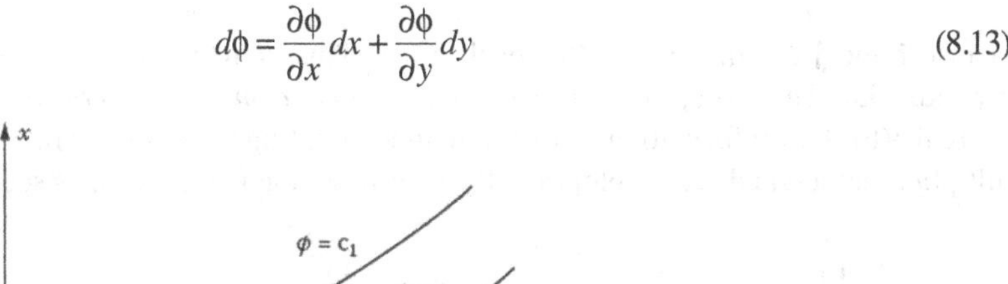

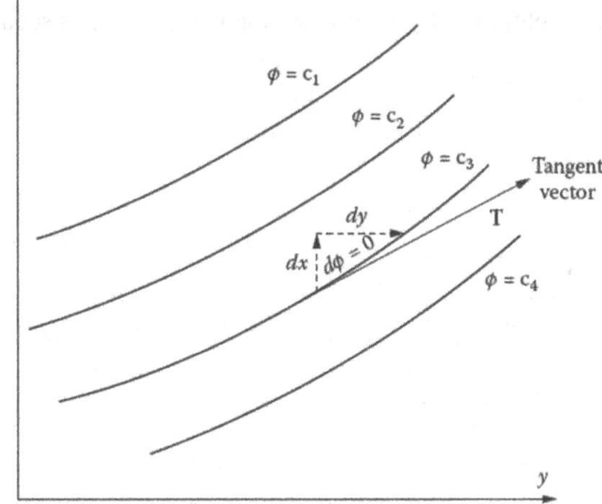

FIGURE 8.2 Contours of constant ϕ shown on an x-y plane for different values of the constant. Also shown is the tangent vector **T**, which is tangential to one such contour.

For ϕ = constant, $d\phi = 0$. If this condition is used to represent movement along the curve, we get

$$dx = -dy\left(\frac{\partial\phi/\partial y}{\partial\phi/\partial x}\right) \tag{8.14}$$

Therefore, the tangential vector **T**, shown in Figure 8.2, may be obtained by using a differential element $d\mathbf{T}$, which is given by this relationship between dx and dy as:

$$dT = dx\boldsymbol{i} + dy\boldsymbol{j} = -dy\left(\frac{\partial\phi/\partial y}{\partial\phi/\partial x}\right)\boldsymbol{i} + dy\boldsymbol{j} \tag{8.15}$$

The unit vector **t** along the tangential direction may be obtained, as done previously for the gradient vector, by dividing the vector by its magnitude. Thus,

$$\boldsymbol{t} = \frac{-dy\dfrac{\partial\phi/\partial y}{\partial\phi/\partial x}\boldsymbol{i} + dy\boldsymbol{j}}{\sqrt{\left(-dy\dfrac{\partial\phi/\partial y}{\partial\phi/\partial x}\right)^2 + (dy)^2}} = \frac{-(\partial\phi/\partial y)\boldsymbol{i} + (\partial\phi/\partial x)\boldsymbol{j}}{\sqrt{(\partial\phi/\partial x)^2 + (\partial\phi/\partial y)^2}} \tag{8.16}$$

Thus it is seen that the two vectors **n** and **t** may be represented as

$$n = c\,\mathbf{i} + d\,\mathbf{j} \quad \text{and} \quad t = -d\,\mathbf{i} + c\,\mathbf{j} \tag{8.17}$$

where c and d represent the respective components given in the preceding equations. The relationship given by Equation (8.17) applies for vectors that are normal to each other. This is shown graphically in Figure 8.3(a). Mathematically, if a dot product of two vectors that are perpendicular to each other is taken, the result should be zero. Applying the dot product to **n** and **t**, we get

$$\mathbf{n}\cdot\mathbf{t} = (c\,\mathbf{i} + d\,\mathbf{j})\cdot(-d\,\mathbf{i} + c\,\mathbf{j}) = -cd + cd = 0 \tag{8.18}$$

because **i** and **j** are independent of each other. This confirms that the two vectors **t** and **n** are perpendicular. Therefore, *the gradient vector* $\nabla\phi$ *is normal to the constant* ϕ *curve*, as shown in Figure 8.3(b). This information is useful in understanding the basic characteristics of the Lagrange multiplier method and for developing other optimization techniques, as seen in later chapters.

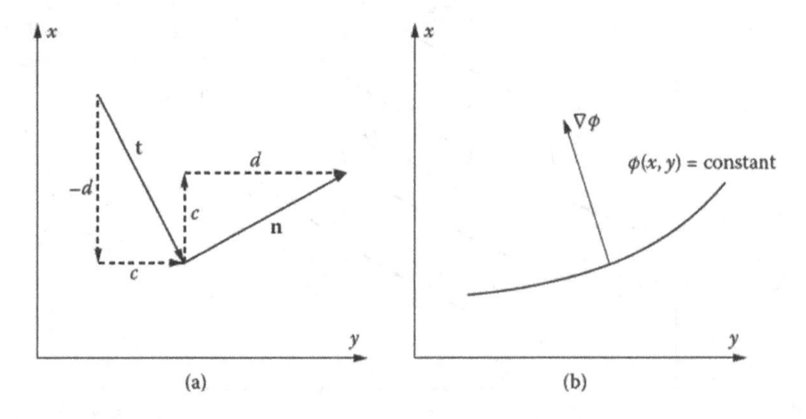

FIGURE 8.3 (a) Unit vectors **t** and **n** are perpendicular to each other; (b) gradient vector $\nabla\phi$ is normal to the ϕ = constant contour.

If three independent variables are considered, a surface is obtained for a constant value of ϕ. Then, the gradient vector $\nabla\phi$ is normal to this surface. Similar considerations apply for a larger number of independent variables. The gradient $\nabla\phi$ may be written for n independent variables as

$$\nabla\phi = \frac{\partial \phi}{\partial x_1} i_1 + \frac{\partial \phi}{\partial x_2} i_2 + \frac{\partial \phi}{\partial x_3} i_3 + \cdots + \frac{\partial \phi}{\partial x_n} i_n \tag{8.19}$$

where $\mathbf{i}_1, \mathbf{i}_2, \ldots, \mathbf{i}_n$ are unit vectors in the n directions representing the n independent variables $x_1, x_2, \ldots, x_n$, respectively. Therefore, these unit vectors are independent of each other. Though it is difficult to visualize the gradient vector for more than three independent variables, the mathematical treatment of the problem is the same as that given previously for two independent variables. Again, the **n** and **t** unit vectors may be determined and their dot product taken to show that $\mathbf{n} \cdot \mathbf{t} = 0$, indicating that $\nabla\phi$ is perpendicular to the ϕ = constant contours or surfaces. Because of this property, the gradient vector represents the direction in which the dependent variable ϕ changes at the fastest rate, this rate being given by the magnitude of the gradient. In addition, the direction in which ϕ increases is the same as the direction of the vector $\nabla\phi$. These properties are useful in many optimization strategies, particularly in gradient-based search methods.

8.2.2.2 Lagrange Multiplier Method for Unconstrained Optimization

Let us first consider the unconstrained problem for two independent variables x and y. Then the Lagrange multiplier method yields the location of the optimum as the solution to the equation

$$\nabla U = \frac{\partial U}{\partial x} i + \frac{\partial U}{\partial y} j = 0 \tag{8.20}$$

Therefore, the gradient vector, which is normal to the constant U contour, is zero, implying that the rate of change in U is zero as one moves away from the point where this equation is satisfied. This indicates a stationary point, or extremum, as shown qualitatively in Figure 8.4 for one or two independent variables. The point may be a minimum or a maximum. It may also be a saddle point, ridge, or valley (see Figure 8.1). Additional information is needed to determine the nature of the stationary point, as discussed later. Because Equation (8.20) is a vector equation, each component may be set equal to zero, giving rise to the following two equations:

$$\frac{\partial U}{\partial x} = 0 \quad \text{and} \quad \frac{\partial U}{\partial y} = 0 \tag{8.21}$$

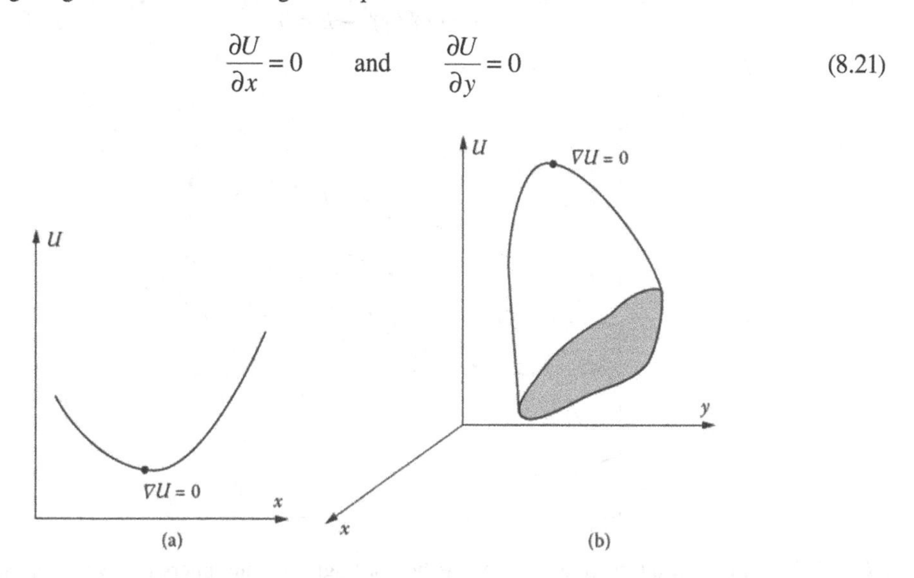

FIGURE 8.4 The minimum and maximum in an unconstrained problem, as given by $\nabla U = 0$.

which may be solved to obtain x and y at the optimum, denoted as x^* and y^*. The optimal value U^* is then calculated from the expression for U. The number of equations obtained is equal to the number of independent variables and the optimum may be determined by solving these equations.

8.2.2.3 Lagrange Multiplier Method for Constrained Optimization

The optimum for a problem with a single constraint is obtained by solving the equations

$$\nabla U + \lambda \nabla G = 0 \quad \text{and} \quad G = 0 \tag{8.22}$$

The gradient vector ∇U is normal to the constant U contours, whereas ∇G is a vector normal to the constant G contours. The Lagrange multiplier λ is simply a constant. Therefore, this equation implies that the two gradient vectors are aligned, i.e., they are both in the same straight line. The magnitudes could be different and λ can be adjusted to ensure that Equation (8.22) is satisfied. However, if the two vectors are not in the same line, the sum cannot be zero unless both vectors are zero. This result is shown graphically in Figure 8.5 for a minimum in U. As one moves along the constraint, given by $G = 0$, in order to ensure that the constraint is satisfied, the gradient ∇G varies in direction. The point where it becomes collinear with ∇U is the optimum. At this point, the two curves are tangential and thus yield the minimum value of U while satisfying the constraint. Constant U curves below the constraint curve do not satisfy the constraint and those above it give values of U larger than the optimum at the locations where they intersect with the constraint curve. Clearly, values of U smaller than that at the optimum shown in the figure could be obtained if there were no constraint, in which case the governing equations would be obtained from Equation (8.20).

As an example, consider an objective function U of the form

$$U = A(x)^a + B(y)^b \tag{8.23}$$

with a constraint of the form

$$(x)^c(y)^d = E \tag{8.24}$$

This constraint may be written as the following to put it in the form given by Equation (8.22):

$$G = (x)^c(y)^d - E = 0 \tag{8.25}$$

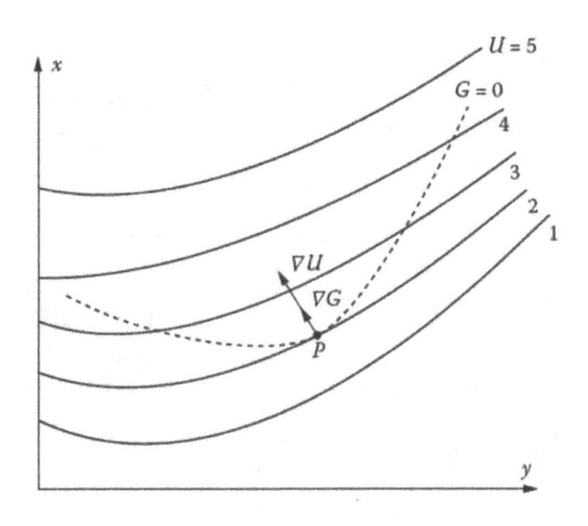

FIGURE 8.5 Physical interpretation of the method of Lagrange multipliers for two independent variables and a single constraint.

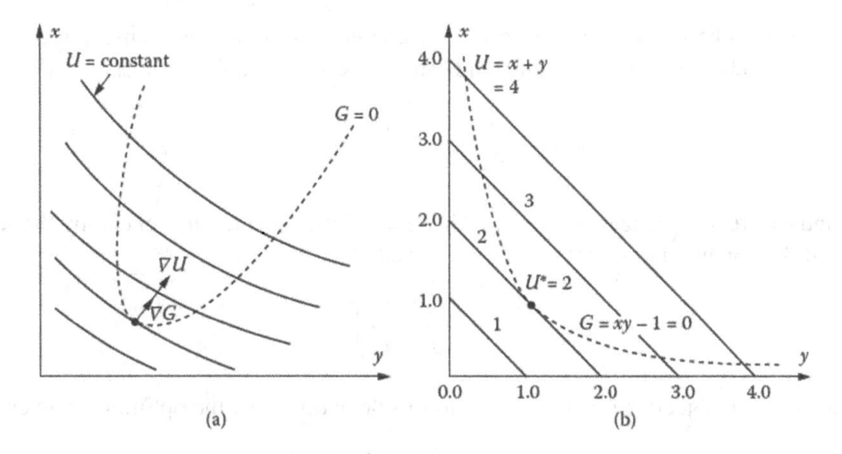

FIGURE 8.6 (a) Optimization of the simple constrained problem given by Equation (8.23) to Equation (8.25), and (b) the results when all the constants in the given expressions are unity.

Here E, the coefficients A and B, and the exponents a, b, c, and d are assumed to be known constants. Such expressions are frequently encountered in thermal systems. For example, U may be the overall cost and x and y the pump needed and pipe diameter, respectively, in a water flow system. The pressure decreases as the diameter increases, resulting in lower cost for the pump, and the cost for the pipe increases. This gives rise to a relationship such as Equation (8.24). Thus, contours of constant U may be drawn along with the constraint curve on an x - y plane, as sketched in Figure 8.6(a). Then the optimum is indicated by the location where the constant U contour becomes tangential to the constraint curve, thus aligning the ∇U and ∇G vectors. For the simple case when all the constants in these expressions are unity, i.e., $U = x + y$ and $G = xy - 1 = 0$, the constant U contours are straight lines and the constraint curve is given by $x = 1/y$, as sketched in Figure 8.6(b). The optimum is at $x^* = 1.0$ and $y^* = 1.0$, and the optimum value U^* is 2.0 for this case.

Even though only two independent variables are considered here for ease of visualization and physical understanding, the basic ideas can easily be extended to a larger number of variables. The system of equations to be solved to obtain the optimum is given by the n scalar equations derived from the vector equation

$$\nabla U + \sum_{i=1}^{m} \lambda_i \nabla G_i = 0 \tag{8.26}$$

and the m equality constraints

$$G_i = 0, \quad \text{for} \quad i = 1,2,3,\ldots,m \tag{8.27}$$

In most practical cases, these equations are solved analytically or numerically, using techniques presented in Chapter 4, to yield the multipliers and the values of the independent variables, from which the optimum value of the objective function is determined.

8.2.2.4 Proof of the Method

The proof of the method of Lagrange multipliers is available in most books on optimization. However, the mathematical analysis becomes involved as the number of independent variables and the number of constraints increase (Stoecker, 1989). Let us again consider the problem with two independent variables and a single constraint for simplicity. At the desired optimum point, the constraint must also be satisfied. If we now deviate from this point, while ensuring that the constraint continues to be satisfied, the change in G should be zero, i.e., from the chain rule,

$$dG = \left(\frac{\partial G}{\partial x}\right)dx + \left(\frac{\partial G}{\partial y}\right)dy = 0 \tag{8.28}$$

where dx and dy are the changes in x and y. Therefore, these changes are related by the following expression, if the condition of $G = 0$ is to be preserved:

$$dx = -\left(\frac{\partial G/\partial y}{\partial G/\partial x}\right)dy \tag{8.29}$$

The change in the objective function U due to this deviation from the optimum is given by

$$dU = \left(\frac{\partial U}{\partial x}\right)dx + \left(\frac{\partial U}{\partial y}\right)dy = \left[-\left(\frac{\partial G/\partial y}{\partial G/\partial x}\right)\frac{\partial U}{\partial x} + \frac{\partial U}{\partial y}\right]dy \tag{8.30}$$

If a quantity λ is defined as

$$\lambda = -\frac{\partial U/\partial x}{\partial G/\partial x} \tag{8.31}$$

then

$$dU = \left[\lambda\frac{\partial G}{\partial y} + \frac{\partial U}{\partial y}\right]dy \tag{8.32}$$

where λ is the Lagrange multiplier for this one-constraint problem. In order for the starting point to be the optimum, there should be no change in U for a differential movement from the optimum while satisfying the constraint $G = 0$, i.e., this should be a stationary point. This implies that, for the differential change in U to be zero, the quantity in the brackets in Equation (8.32) must be zero. Therefore,

$$\frac{\partial U}{\partial y} + \lambda\frac{\partial G}{\partial y} = 0 \tag{8.33}$$

Also, from Equation (8.31),

$$\frac{\partial U}{\partial x} + \lambda\frac{\partial G}{\partial x} = 0 \tag{8.34}$$

Then Equation (8.33) and Equation (8.34), along with the constraint equation $G = 0$, define the optimum for this problem. These are the same as the equations given earlier for the method of Lagrange multipliers for this single-constraint, two-independent variable problem.

8.2.3 Significance of the Multipliers

Let us again consider the optimization problem with two independent variables and one constraint, as given by Equation (8.22). Then the Lagrange multiplier may be written as

$$\lambda = -\left(\frac{\nabla U}{\nabla G}\right)^* = -\left[\frac{\frac{\partial U}{\partial x}\mathbf{i} + \frac{\partial U}{\partial y}\mathbf{j}}{\frac{\partial G}{\partial x}\mathbf{i} + \frac{\partial G}{\partial y}\mathbf{j}}\right]^* \tag{8.35}$$

Because the condition for the optimum is used here, all the derivatives are evaluated at the optimum, as indicated by the asterisks. Now, if dot products of both the numerator and the denominator are taken with $(dx\mathbf{i} + dy\mathbf{j})$, the result obtained is

$$\lambda = -\left[\frac{(\partial U/\partial x)dx + (\partial U/\partial y)dy}{(\partial G/\partial x)dx + (\partial G/\partial y)dy}\right]^* = -\left(\frac{\Delta U}{\Delta G}\right)^* = -\left(\frac{\partial U}{\partial G}\right)^* \tag{8.36}$$

where ΔU and ΔG are incremental changes in U and G from the optimum. The partial derivative is thus obtained for infinitesimal changes. This is similar to the expression given in Equation (8.31) and defines the Lagrange multiplier for this problem.

An important consideration in optimization is the effect of a change in the constraint on the objective function at the optimum. A parameter known as the sensitivity coefficient S_c is defined as the rate of change of the objective function U with the constraint G at the optimum, i.e.,

$$S_c = \left(\frac{\partial U}{\partial G}\right)^* \tag{8.37}$$

This parameter is useful in adjusting the design variables to come up with the final design, as discussed in the preceding chapter. The value of the sensitivity coefficient gives an indication of the level of adjustment needed in the constraint in order to employ standard sizes and readily available components (Dieter, 2000). It is seen from the definition of the sensitivity coefficient and Equation (8.36) for λ that the Lagrange multiplier is the negative of the sensitivity coefficient. This analysis can easily be extended to multiple constraints and larger numbers of independent variables. It can then be shown that $\lambda_1 = -(S_c)_1$, $\lambda_2 = -(S_c)_2$, …, $\lambda_m = -(S_c)_m$, where the subscripts refer to the different constraint equations. Thus, the method of Lagrange multipliers not only yields the optimum but also the sensitivity coefficients with respect to the various constraints in the problem.

Frequently, the constraint may be written as

$$G(x, y) = g(x, y) - E = 0 \tag{8.38}$$

where $g(x, y)$ is a function of the two independent variables and E is a parameter; see Equation (8.25). This parameter could be a quantity that provides the constraint for optimization such as the total volume of a tank, the total length of piping in a flow system, the total heat input in a thermal system, etc. Then, the constraint may be expressed as

$$g(x, y) = E \tag{8.39}$$

where E is an adjustable parameter. Therefore, $dg = dE$, which gives, from the preceding discussion,

$$\lambda = -\left(\frac{\partial U}{\partial E}\right)^* \tag{8.40}$$

Thus, the Lagrange multiplier λ gives the rate of change in the objective function U with the constraint parameter E, at the optimum. Therefore, the values of the Lagrange multipliers obtained in a given problem may be used to guide slight changes in the constraints in order to choose appropriate sizes, dimensions, etc., that are easily available or more convenient to fabricate.

8.3 OPTIMIZATION OF UNCONSTRAINED PROBLEMS

Most of the optimization problems encountered in the design of thermal systems are constrained due to the conservation principles and limitations imposed by the materials used, space available, safety and environmental regulations, etc. However, frequently the constraints are used to obtain the relationships between different design variables. If these are then substituted into the expression for the objective function, an unconstrained problem is obtained because the constraints have been satisfied. Sometimes, in the formulation of the optimization problem itself, the constraints are employed in deriving the appropriate expressions and additional constraints are not needed. Thus, an unconstrained problem results. Certainly, in a few cases, there are no significant constraints, and the problem is treated as unconstrained. Thus, the unconstrained optimization problem is of interest in many practical thermal systems and processes.

8.3.1 Use of Gradients for Optimization

If there are no constraints in the problem, the optimum is given by a solution to the following vector equation for $U(x_1, x_2, x_3, \ldots, x_n)$:

$$\nabla U = \frac{\partial U}{\partial x_1} \boldsymbol{i}_1 + \frac{\partial U}{\partial x_2} \boldsymbol{i}_2 + \frac{\partial U}{\partial x_3} \boldsymbol{i}_3 + \cdots + \frac{\partial U}{\partial x_n} \boldsymbol{i}_n = 0 \tag{8.41}$$

Again, it is easy to visualize the gradient vector if only two or three variables are involved, as seen in the preceding section. However, the basic concepts are the same and may be extended to any appropriate number of variables. All the components of the vector equation, Equation (8.41), must be zero in order that the vector be zero because all of the variables are taken as independent of each other. Therefore, the optimum is obtained by solving the equations

$$\frac{\partial U}{\partial x_1} = 0 \quad \frac{\partial U}{\partial x_2} = 0 \quad \frac{\partial U}{\partial x_3} = 0, \quad \ldots, \quad \frac{\partial U}{\partial x_n} = 0 \tag{8.42}$$

This is similar to the condition for a stationary point given by Equation (8.1) for a single independent variable. The objective function must be a continuous and differentiable function of the independent variables in the problem and the derivatives must be continuous.

The system of equations represented by Equation (8.42) could be linear or nonlinear, though nonlinear equations are more commonly encountered for thermal systems and processes. Mathematical analysis may be used in simple cases to obtain the solution. Otherwise, numerical techniques are needed. This method is particularly useful for systems that may be characterized by algebraic expressions that can be readily differentiated. This situation usually arises in cases where curve fitting yields such expressions for characterizing the system behavior and in small, idealized, and simple systems. The constraints are assumed to be absent or taken care of in developing the objective function. We now consider determining whether the optimum is a minimum or a maximum.

8.3.2 Determination of Minimum or Maximum

In most cases, the physical nature of the problem would indicate whether the solution obtained is a maximum, a minimum, or some other stationary point. Frequently, it is known from experience that a minimum or a maximum exists in the given domain. For instance, it may be known that a minimum in the energy consumption would arise if the pressure of the compressor were varied over the acceptable range in a refrigeration system. Similarly, a maximum thermal efficiency is expected if the speed, in revolutions per minute, of a diesel engine is varied. However, in the absence of such information, further analysis may be carried out to determine the characteristics of the optimum.

Equation (8.2) and Equation (8.3) give the conditions for a maximum and a minimum, respectively, for a single independent variable. If the second derivative is zero at the stationary point, the occurrence of a saddle point, inflexion point, ridge, or valley is indicated. Similar conditions may be derived for two or more independent variables and are given in most calculus textbooks, such as Keisler (2012) and Kaplan (2002), and in books on optimization, such as Beightler et al. (1979),

Rao (2009) and Chong and Zak (2013). For the case of two independent variables, x_1 and x_2, with $U(x_1, x_2)$ and its first two derivatives continuous, these conditions are given as

If $\frac{\partial^2 U}{\partial x_1^2} > 0$, with $S > 0$, the stationary point is a minimum

If $\frac{\partial^2 U}{\partial x_1^2} < 0$, with $S > 0$, the stationary point is a maximum

If $S < 0$, the stationary point is a saddle point

where

$$S = \frac{\partial^2 U}{\partial x_1^2}\frac{\partial^2 U}{\partial x_2^2} - \left(\frac{\partial^2 U}{\partial x_1 \partial x_2}\right)^2 \tag{8.43}$$

Therefore, for two independent variables, the optimum may be obtained by solving $\partial U/\partial x_1 = \partial U/\partial x_2 = 0$ and applying the preceding conditions. Though similar conditions may be derived for a larger number of variables, the analysis becomes quite involved. Therefore, in most practical circumstances, which involve three or more independent variables, it is more convenient and efficient to depend on the physical nature of the problem to determine if a minimum or a maximum has been obtained. In addition, the independent variables may be changed slightly near the optimum to determine if the value of the objective function increases or decreases. If the value decreases as one moves away from the optimum, a maximum is indicated, whereas if it increases, a minimum has been obtained.

Example 8.2

The cost C per unit mass of material processed in an extrusion facility is given by the expression

$$C = 2T^2V + \frac{3T}{V^2} + \frac{2}{T}$$

where T is the dimensionless temperature of the material being extruded, V is the dimensionless volume flow rate, and C includes both capital and running costs. Determine the minimum cost.

SOLUTION

Because there are no constraints, the approach given in the preceding sections may be adopted. Therefore, the location of the optimum is given by the solution of the equations

$$\frac{\partial C}{\partial V} = 2T^2 - \frac{6T}{V^3} = 0 \quad \text{and} \quad \frac{\partial C}{\partial T} = 4TV + \frac{3}{V^2} - \frac{2}{T^2} = 0$$

Because both T and V are positive quantities, we have

$$T = \frac{3}{V^3} \qquad \text{and} \qquad 4\left(\frac{3}{V^3}\right)V + \frac{3}{V^2} - \frac{2}{\left(3/V^3\right)^2} = 0$$

These equations give $V^* = 1.6930$ and $T^* = 0.6182$. When these are substituted in the expression for C, we obtain $C^* = 5.1763$. Now the second derivatives may be obtained to ascertain the nature of the critical point. Thus,

$$\frac{\partial^2 C}{\partial V^2} = \frac{18T}{V^4} \qquad \frac{\partial^2 C}{\partial T^2} = 4V + \frac{4}{T^3} \qquad \frac{\partial^2 C}{\partial V \partial T} = 4T - \frac{6}{V^3}$$

Substituting the values of V and T at the stationary point, we calculate these three second derivatives as 1.3544, 23.7023, and 1.2364, respectively. This gives $S = 30.57$. Therefore, $S > 0$ and $\partial^2 C/\partial V^2 > 0$, indicating that the minimum cost has been obtained.

8.3.3 Conversion of Constrained to Unconstrained Problem

It is evident from the preceding discussion that an unconstrained optimization problem is easier to solve, as compared to the corresponding constrained one, because the number of unknowns is smaller in the former case. Each constraint introduces a Lagrange multiplier as an unknown and an additional equation has to be satisfied. Therefore, it is desirable to convert a given constrained problem into an unconstrained one whenever possible. The constraints represent relationships between the various independent variables that must be satisfied. If these equations can be used to obtain explicit expressions for some of the variables in terms of the others, these expressions may then be substituted into the objective function to eliminate the constraints and thus convert the problem to an unconstrained one. Even if all the constraints cannot be eliminated, it is worthwhile to eliminate as many of these as possible in order to reduce the complexity of the problem.

Let us again consider the optimization problem represented by Equation (8.23) through Equation (8.25). Then, the constraint equation, Equation (8.24), may be used to express y in terms of x as

$$y = \left(\frac{E}{x^c} \right)^{1/d} \tag{8.44}$$

Substituting this value of y into Equation (8.23), we have

$$U = A(x)^a + B\left(\frac{E}{x^c} \right)^{b/d} \tag{8.45}$$

Therefore, an unconstrained problem is obtained with $U(x)$ as the objective function. The optimum is obtained by setting $\partial U/\partial x = 0$, which yields the value of x. The corresponding y at the optimum is obtained from Equation (8.44) and the optimum value of U is obtained from Equation (8.45). It is easy to see that $x^* = y^* = 1.0$ and $U^* = 2.0$ for the simple case when all the constants and exponents are unity.

Thus, it is desirable to reduce the number of constraints, which will also reduce the number of unknown variables and Lagrange multipliers, by using the constraint equations to find explicit expressions relating the variables. Obviously, it is not always possible to do so because the constraint may not yield an explicit relationship that can be used to eliminate a variable. Then, the problem has to be treated as a constrained circumstance. The following example illustrates the solution of a constrained problem by converting it into an unconstrained one. It is later solved as a constrained problem to indicate the difference between the two approaches.

Example 8.3

A cylindrical storage tank is to be designed for storing hot water from a solar energy collection system. The volume is given as 2 m^3 and the surface area is to be minimized in order to minimize the heat loss to the environment. Solve this optimization problem as an unconstrained circumstance.

SOLUTION

If r is the radius of the tank and h its height, the volume V is given by

$$V = \pi r^2 h$$

The surface area A is given by

$$A = 2\pi r^2 + 2\pi rh$$

where the first term represents the two ends and the second term represents the curved lateral surface. Then A is the objective function that is to be minimized and $V = 2$ represents the constraint. From this constraint equation, $h = V/(\pi r^2)$. Substituting this relationship into the expression for area gives

$$A = 2\pi r^2 + \frac{2\pi rV}{\pi r^2} = 2\pi r^2 + \frac{2V}{r}$$

Because the constraint has already been considered, this becomes an unconstrained problem. Differentiating A with respect to r and setting the derivative equal to zero to obtain the radius for the optimum, we get

$$\frac{dA}{dr} = 4\pi r - \frac{2V}{r^2} = 0$$

Therefore,

$$r^* = \left(\frac{V}{2\pi}\right)^{1/3} \qquad h^* = \frac{V}{\pi r^2} = \left(\frac{4V}{\pi}\right)^{1/3}$$

If V is taken as 2 m^3,

$$r^* = \pi^{-1/3} = 0.683\text{m}, \quad h^* = 1.366\text{m}, \quad A^* = 8.793\text{m}^2$$

The second derivative is calculated to determine the nature of the optimum. Thus,

$$\frac{\partial^2 A}{\partial r^2} = 4\pi + \frac{4V}{r^3}$$

Because r is positive, the second derivative is also positive, indicating that the area is a minimum, as desired in the problem. Similarly, the dimensions for other desired tank volumes may be determined for minimum surface area for heat loss.

8.4 OPTIMIZATION OF CONSTRAINED PROBLEMS

The optimization of most thermal systems is governed by constraints that arise due to the conservation laws and limitations imposed by the materials, space, cost, safety, etc. As discussed earlier, the number of equality constraints must be less than the number of independent variables for optimization to be possible. If the number of constraints equals the number of variables, the problem may simply be solved to yield the set of variables that satisfies the constraints. No flexibility is available to choose the best or optimal design. If the number of constraints is larger than the number of variables, the problem is overconstrained and some of the constraints must be discarded, resulting in arbitrariness and a lack of uniqueness in the solution. These considerations are evident from Equation (8.8), where the condition $m < n$ is needed for optimization of a system. If $m = n$, the constraint equations can be used to obtain the solution, and if $m > n$, the problem is overconstrained, and no solution is possible unless $m - n$ constraints are dropped.

Considering the optimization problem, i.e., $m < n$, the method of Lagrange multipliers may be applied to determine the optimal design. The equations that need to be solved are given by Equation (8.26) and Equation (8.27) and may be rewritten here as

$$\nabla U + \sum_{i=1}^{m} \lambda_i \nabla G_i = 0$$

$$G_i = 0, \quad \text{for} \quad i = 1,2,\ldots,m$$

where ∇U and ∇G_i are the gradient vectors, which may be expanded in terms of the n independent variables to yield Equation (8.8). Therefore, $m + n$ equations are obtained for the n independent variables and m multipliers. These equations may be linear or nonlinear, and may involve polynomials or transcendental functions such as exponential, logarithm, and hyperbolic functions. Analytical methods may be used for relatively simple cases with a small number of equations. Numerical techniques, as outlined in Chapter 4, may be used for more complicated circumstances, which commonly arise when dealing with practical thermal systems.

If there is only one constraint, $G = 0$, Equation (8.22) is obtained. Only one Lagrange multiplier λ arises and is determined from the solution of the resulting $n + 1$ equations. Consider, for instance, the simple optimization problem given by

$$U = 2x^2_1 + 5x_2 \quad \text{and} \quad G = x_1 x_2 - 12 = 0 \tag{8.46}$$

Then, the method of Lagrange multipliers yields the following equations:

$$\frac{\partial U}{\partial x_1} + \lambda \frac{\partial G}{\partial x_1} = 0, \quad \frac{\partial U}{\partial x_2} + \lambda \frac{\partial G}{\partial x_2} = 0, \quad G = 0 \tag{8.47}$$

These equations lead to

$$4x_1 + \lambda x_2 = 0, \quad 5 + \lambda x_1 = 0, \quad x_1 x_2 = 12 \tag{8.48}$$

Therefore, the solution is obtained as:

$$x_1^* = 2.466, \quad x_2^* = 4.866, \quad U^* = 36.493, \quad \lambda = -2.027 \tag{8.49}$$

It can be shown that if either x_1 or x_2 is varied from its optimum value, while ensuring that the constraint is satisfied, the objective function U increases. Therefore, the optimum obtained is a minimum in U. In this simple case, the second derivatives may also be derived to confirm that a minimum in U has been obtained. The sensitivity coefficient $S_c = -\lambda = 2.027$. This gives the effect of relaxing the constraint on the optimum value of U. For instance, if $x_1\ x_2 = 13$, instead of 12, U^* can be calculated to be 38.493, an increase of 2.0. There is a slight difference in the change in U^* obtained from solving the equations with the changed constraint and that from the calculated value of S_c. This is because nonlinear equations make S_c a function of x_1 and not a constant. The following example illustrates this treatment for the tank problem considered earlier.

Example 8.4

Solve the tank problem given in Example 8.3 as a constrained optimization problem.

SOLUTION

The objective function is the area A, which is to be minimized, and the constraint is the volume V. Thus, the optimization problem may be written as

$$U = A = 2\pi r^2 + 2\pi rh$$

and

$$G = \pi r^2 h - V = 0$$

From Equation (8.22),

$$\frac{\partial U}{\partial r} + \lambda\frac{\partial G}{\partial r} = 0, \qquad \frac{\partial U}{\partial h} + \lambda\frac{\partial G}{\partial h} = 0, \qquad G = 0$$

Therefore,

$$4\pi r + 2\pi h + \lambda(2\pi rh) = 0, \quad 2\pi r + \lambda(\pi r^2) = 0, \quad \pi r^2 h = V$$

or

$$2r + h + \lambda rh = 0, \quad 2 + \lambda r = 0, \quad \pi r^2 h = V$$

Solving these three equations yields

$$r^* = \left(\frac{V}{2\pi}\right)^{1/3} \qquad h^* = \left(\frac{4V}{\pi}\right)^{1/3} \qquad \lambda = -2\left(\frac{2\pi}{V}\right)^{1/3}$$

For $V = 2\ m^3$: $r^* = 0.683\ m$, $h^* = 1.366\ m$, $A^* = 8.793\ m^2$, $\lambda = -2.928$. These values are the same as those obtained earlier by converting the problem to an unconstrained one. Again, it can be confirmed that a minimum in the area has been obtained.

The sensitivity coefficient S_c, which is equal to $-\lambda$, is obtained as additional information. Let us assume that the constraint on the volume is relaxed from 2.0 to 2.1. Then, it can easily be shown that $r^* = 0.694$ m and $A^* = 9.078$ m^2. Therefore, $\partial A/\partial V = (9.078 - 8.793)/0.1 = 2.85$, which is close to the sensitivity coefficient S_c, that is given by $-\lambda$ and is, thus, equal to 2.928 at the optimum point. Again, the slight difference between S_c and $\partial A/\partial V$ is due to the dependence of λ on the variables.

8.5 APPLICABILITY TO THERMAL SYSTEMS

The use of calculus methods to optimize thermal systems is limited by the requirement that the objective function and constraints be continuous and differentiable. In addition, only equality constraints can be considered. Practical thermal systems seldom lead to simple analytic functions such as those considered in the examples discussed here. Only very simple systems, with a small number of components and highly idealized characteristics, can generally be represented by polynomials and other simple expressions. Governing sets of algebraic and differential equations must often be solved to determine the behavior of a given system. In addition, discrete values are frequently taken by the design variables, due to the availability of standard parts and components, making it difficult to obtain the system characteristics as continuous functions. However, the calculus methods, whenever applicable, are convenient and efficient. In addition, they form part of several other optimization strategies and are useful in understanding the nature of the optimum. Therefore, effort is often made to obtain expressions that facilitate the use of calculus methods.

8.5.1 Use of Curve Fitting

Curve fitting is certainly the most useful method of representing the results from numerical and experimental modeling in the form of simple algebraic expressions, which can easily be treated by calculus methods to optimize the system. Employing the methods outlined in Chapter 3, particularly the least squares best fit approach, the simulation results for a system may be expressed in terms of polynomials, exponentials, sinusoidal, power-law variations, and so on. The simulation itself may be very complicated, involving the numerical solution of nonlinear algebraic and differential equations and experimental data from property measurements and physical modeling. However, if continuous changes in the design variables and in the operating conditions can be considered, algebraic expressions may be employed to closely approximate the results. The choice of the type and form of the function used for curve fitting the data is based on the physical nature of the system. Polynomials may be used to curve fit the results if information is not available to choose a more specific expression.

Thus, the results from a variety of sources may often be represented by relatively simple algebraic expressions. Empirical correlations for the convective heat transfer coefficient, characteristics of a pump, and pressure-flow rate relation for an extrusion die are examples of such curve fits and may typically be written as

$$\mathrm{Nu} = \frac{hL}{k} = A_1 (\mathrm{Re})^a (\mathrm{Pr})^b \tag{8.50}$$

$$\Delta p = A_2 - A_3\, \dot{m} - A_4 \dot{m}^2 \tag{8.51}$$

$$\Delta p = A_5 + A_6\, \dot{m}^c \tag{8.52}$$

where Nu, Re, and Pr are the Nusselt, Reynolds, and Prandtl numbers, defined in Chapter 3; h is the heat transfer coefficient; L is a characteristic length; k is the thermal conductivity; Δp is the pressure rise in a pump or the pressure drop through a die; $\dot{m}$ is the mass flow rate; and the A's and the exponents a, b, c are constants obtained from curve fitting. Similar expressions can usually be derived for heat exchangers, manufacturing systems, refrigeration units, cooling towers, and other thermal systems, linking the characteristics with the design variables and operating conditions. The constraints may be satisfied during the simulation because the governing equations are based on the conservation laws, yielding an unconstrained problem, or similar algebraic equations may be derived for the constraints. Inequality constraints are also encountered in the design of thermal systems and are considered later.

Once the objective function and the constraints have been obtained in the form of algebraic equations, the method of Lagrange multipliers may be conveniently applied to obtain the location of the optimum point and the corresponding value of the objective function. As far as possible, the constrained problem should be solved, even though it is more involved, because it provides the additional information on the sensitivity coefficients, which are used to fine-tune the final design. Similarly, unless necessary, elimination should not be used to reduce the number of equations because this also removes a design variable that could be treated as an independent quantity in the design process. Numerical methods are particularly useful because large systems of nonlinear equations are frequently obtained.

8.5.2 Examples

In the preceding sections, we considered several examples to illustrate the application of calculus methods to optimization. Practical thermal systems also provide interesting examples of the use of this approach. A common circumstance encountered in the design of heat transfer equipment is that of minimizing the heat loss Q while meeting the constraints due to energy balance, strength considerations, etc. The heat transfer coefficient h may be obtained from empirical correlations available in the literature. Therefore, the optimization problem may be formulated as

$$Q = hA\Delta T \rightarrow \text{Minimize} \tag{8.53}$$

with

$$A = f_1(L_1, L_2), \quad h = f_2(L_1, L_2, \Delta T), \quad f_3(L_1, L_2, \Delta T) = C \tag{8.54}$$

where A is the surface area; ΔT is the temperature difference from the ambient; L_1 and L_2 are dimensions, such as diameter and height of the cylindrical shell of a heat exchanger; C is a constant; and f_1, f_2, and f_3 are functions. Additional dimensions may also be included if necessary. The expressions for A and h may be substituted into the objective function to obtain a single constraint problem.

The method of Lagrange multipliers may be applied to the preceding problem to obtain the dimensions at which the heat loss is a minimum. The sensitivity coefficients are also derived to determine the effect of a variation in the constraint(s) on the optimum. The following examples illustrate the solution of such a problem.

Example 8.5

In an electronic circuitry, the power source may be considered as a thin square with side dimension L in meters. It is desired to minimize the heat transfer from the surface of the power supply to the local surroundings. The heat transfer coefficient h in W/(m^2 · K) is given by the expression

$$h = \left(2 + 10L^{1/2}\right) \Delta T^{1/4} L^{-1}$$

where ΔT is the temperature difference in K from the local ambient. A constraint arises due to the strength of the bond that attaches the power supply to the electronic circuit board as $L\,\Delta T = 5.6$. Calculate the side dimension L of the square that would minimize the total heat loss, solving the problem as both an unconstrained and a constrained one.

SOLUTION

The rate of heat loss Q from the power supply is the objective function that is to be minimized and is given by the expression

$$Q = hA\Delta T = \left(2 + 10L^{1/2}\right)\Delta T^{1/4} L^{-1}\left(L^2\right)\Delta T = \left(2L + 10L^{3/2}\right)\Delta T^{5/4}$$

because the surface area A is L^2 for a square. The problem may be treated as unconstrained by substituting ΔT in terms of L from the given constraint. Thus, $\Delta T = 5.6/L$ and this is substituted in the preceding equation to yield

$$Q = \left(2L + 10L^{3/2}\right)\left(5.6L^{-1}\right)^{5/4} = 8.61\left(\frac{2}{L^{1/4}} + 10L^{1/4}\right)$$

For Q to be a minimum,

$$\frac{\partial Q}{\partial L} = 8.61\left(-\frac{2L^{-5/4}}{4} + \frac{10L^{-3/4}}{4}\right) = 0$$

which gives

$$L^* = \frac{1}{25} = 0.04\,\text{m} = 4\,\text{cm}$$

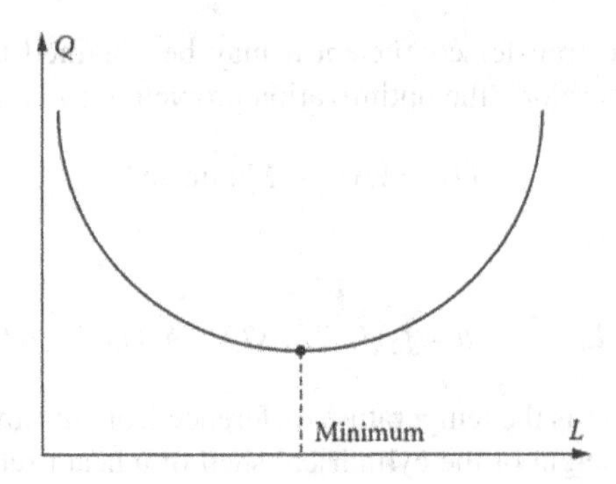

FIGURE 8.7 Variation of the heat transfer rate Q with dimension L of the power source in Example 8.5.

The second derivative is given by

$$\frac{d^2Q}{dL^2} = 8.61\left(\frac{5}{8}L^{-9/4} - \frac{15}{8}L^{-7/4}\right)$$

At $L^* = 0.04$ m, the second derivative is calculated as 3008.25, a positive quantity, indicating that Q is a minimum. Its value is obtained as $Q^* = 77.05$W. Figure 8.7 shows a sketch of the variation of Q with L, and the minimum value is indicated.

The problem can also be solved as a constrained one, with the objective function and the constraint written as

$$Q = (2L + 10L^{3/2})\Delta T^{5/4} \quad \text{and} \quad L\Delta T = 5.6$$

Therefore, from Equation (8.9), the optimum is given by the equations

$$\Delta T^{5/4}\left(2 + 15L^{1/2}\right) + \lambda\Delta T = 0$$

$$\frac{5}{4}\left(2L + 10L^{3/2}\right)\Delta T^{1/4} + \lambda L = 0$$

$$L\Delta T = 5.6$$

These equations can be solved to yield the optimum as

$$L^* = 0.04\,\text{m}; \quad \Delta T^* = 140; \quad Q^* = 77.05\,\text{W}; \quad \lambda = -17.2$$

Thus, the results obtained are identical to those obtained by solving the corresponding unconstrained problem. It can be shown that if the constraint is increased from 5.6 to 5.7, the heat transfer rate Q becomes 78.77, i.e., an increase of 1.72. This change can also be obtained from the sensitivity coefficient S_c. Here, $S_c = -\lambda = 17.2$, which gives the change in Q for a change of 1.0 in the constraint. Therefore, for a change of 0.1, Q is expected to increase by 1.72.

Example 8.6

For the solar energy system considered in Example 5.3, the cost U of the system is given by the expression

$$U = 35A + 208V$$

where A is the surface area of the collector and V is the volume of the storage tank. Find the conditions for which the cost is a minimum, and compare the solution with that obtained in Example 5.3.

SOLUTION

The objective function is $U(A, V)$, given by the preceding expression. A constraint arises from the energy balance considerations given in Example 5.3 as

$$A\left(290-\frac{100}{V}\right)=5833.3$$

Therefore, A may be obtained in terms of V from this equation and substituted in the objective function to obtain an unconstrained problem as

$$U=(35)\left(\frac{5833.3}{290-100/V}\right)+208V\rightarrow \text{Minimum}$$

or

$$U=\frac{2041.67}{2.9-1/V}+208V\rightarrow \text{Minimum}$$

Therefore, U may be differentiated with respect to V and the derivative set equal to zero to obtain the optimum. This leads to the equation

$$\frac{2041.67}{(2.9-1/V)^2}\frac{1}{V^2}=208$$

or

$$2.9V-1=(9.816)^{1/2}=3.133$$

Therefore, $V^* = 1.425\ m^3$. Then, $A^* = 26.536\ m^2$ and $U^* = 1225.16$.

It can easily be shown that if V or A is varied slightly from the optimum, the cost increases, indicating that this is a minimum. The maximum temperature T_o is obtained as 55.09°C, which lies in the acceptable range. These values may also be compared with those obtained in Example 5.3 for different values of T_o, indicating good agreement. Unique values of the area A and volume V are obtained at which the cost is a minimum, rather than the domain of acceptable designs obtained in Example 5.3. However, these values of A and V are usually adjusted for the final design in order to use standard items available at lower costs.

8.5.3 Inequality Constraints

Inequality constraints arise largely due to limitations on temperature, pressure, heat input, and other quantities that relate to material strength, process requirements, environmental aspects, and space, equipment, and material availability. For instance, the temperature T_o of cooling water at the condenser outlet of a power plant is constrained due to environmental regulations as $T_o < T_{amb} + R$, where T_{amb} is the ambient temperature and R is the regulated temperature difference. The outlet from a cooling tower has a similar constraint. Other common constraints such as

$$T\le T_{max},\quad P\le P_{max},\quad \tau\ge\tau_{min},\quad \dot{m}>\dot{m}_{min},\quad Q>Q_{min} \tag{8.55}$$

where the temperature T, pressure P, process time τ, mass flow rate $\dot{m}$, and heat input rate Q apply to a given part of the system. Such constraints, which are given in terms of the maximum or minimum values, represented respectively by subscripts max and min, have been considered earlier. The time τ_{min} represents the minimum time needed for a given thermal process, such as heat treatment.

Because only equality constraints can be considered if calculus methods are to be applied, these inequality constraints must either be converted to equality ones or handled in some other manner.

As discussed in Chapter 7, a common approach is to choose a value less than the maximum or more than the minimum for the constrained quantity. Thus, the temperature at the condenser outlet may be taken as

$$T_o = T_{amb} + R - \Delta T \tag{8.56}$$

where ΔT is an arbitrarily chosen temperature difference, which may be based on available information on the system and safety considerations. Similarly, the wall temperature may be set less than the maximum, the pressure in an enclosure less than the maximum, and so on, in order to obtain equality constraints.

In many cases, it is not possible to arbitrarily set the variable at a particular value in order to satisfy the constraint. For instance, if the temperature and pressure in an extruder are restricted by strength considerations, we cannot use this information to set the conditions at certain locations because it is not known a priori where the maxima occur. In such cases, the common approach is to solve the problem without considering the inequality constraint and then checking the solution obtained if the constraint is satisfied. If not, the design variables obtained for the optimum are adjusted to satisfy the constraint. If even this does not work, the solution obtained may be used to determine the locations where the constraint is violated, set the values at these points at less than the maximum or more than the minimum, and solve the problem again. With these efforts, the inequality constraints are often satisfied. However, if even after all these efforts the constraints are not satisfied, it is best to apply other optimization methods.

8.5.4 Some Practical Considerations

In the preceding discussion, we assumed that an optimum of the objective function exists in the design domain and methods for determining the location of this optimum were obtained. However, many different situations may and often do arise when dealing with practical thermal systems. Frequently, for unconstrained problems, several local maxima and minima are present in the domain, which is defined by the ranges of the design variables, as sketched in Figure 8.8. These optima are determined by solving the system of algebraic equations derived from the vector equation $\nabla U = 0$. Because nonlinear equations generally arise for thermal systems and processes, multiple solutions may be obtained, indicating different local optima. Because interest obviously lies in the overall or global maximum or minimum, it is necessary to consider each extremum in order

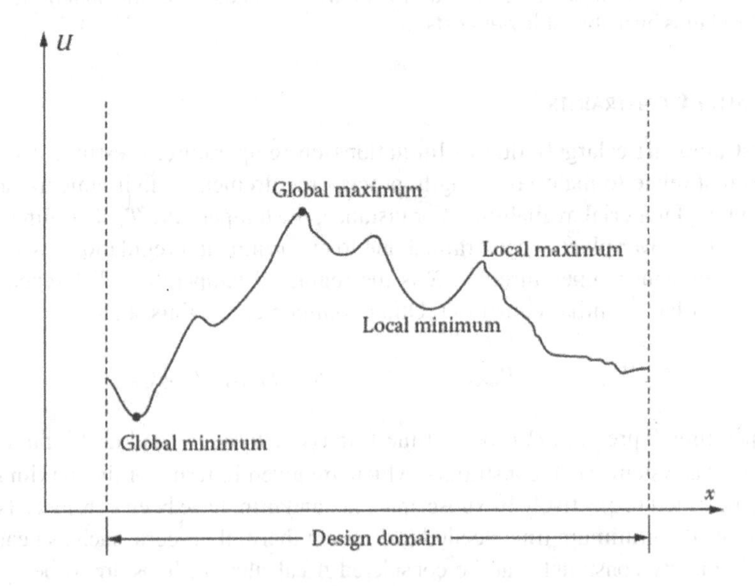

FIGURE 8.8 Local and global extrema in an allowable design domain.

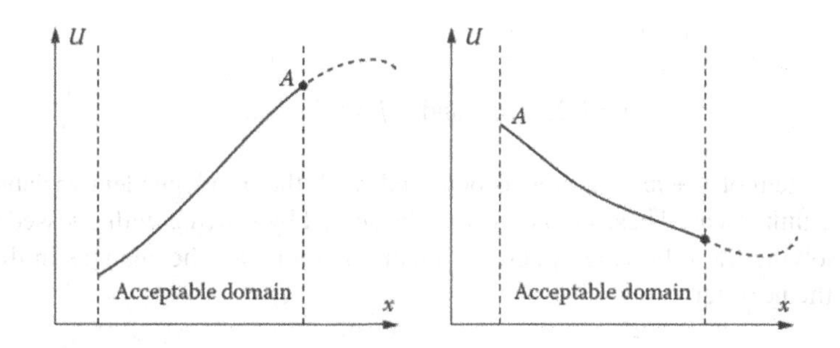

FIGURE 8.9 Monotonically varying objective functions over given acceptable domains, resulting in optimum at the boundaries of the domain.

to ensure that the global optimum has been obtained. Multiple solutions are also possible for constrained problems because of the generally nonlinear nature of the equations. Again, each optimum point must be considered and the objective function determined so that the desired best solution over the entire domain is obtained.

In many cases, the objective function varies monotonically and a maximum or minimum does not arise. In several practical systems, opposing mechanisms do give rise to optimum values, but the locations where these occur may not be within the acceptable ranges of variation of the design parameters, as shown in Figure 8.9. Therefore, the application of the method of Lagrange variables may either not yield an optimum at all or give a value outside the design domain. Both these circumstances are commonly encountered and are treated in a similar way. The desired maximum or minimum value of the objective function is obtained at the boundaries of the domain and the corresponding value of the independent variable is selected for the design, as indicated by point A in Figure 8.9 for a maximum in U.

An example of such a situation is maximization of the flow rate in a network consisting of pumps and pipes. A monatomic rise in flow rate is expected with increasing pressure head of the pump and a stationary point is not obtained. Thus, in Example 5.8, the maximum flow rate arises at the maximum allowable values of the pressure levels P_1 and P_2 (see Figure 5.38). Similarly, energy balance for materials undergoing heat treatment may yield a temperature, beyond the allowable range, at which the system heat loss is minimized. In such cases, the maximum or minimum allowable values of the design parameters that result in the desired largest or smallest value of the objective function are chosen for the design.

8.5.5 Computational Approach

Analytical methods for deriving and solving the equations for the Lagrange multiplier method are generally applicable to a relatively small number of components and simple expressions. A computational approach may be developed for problems that are more complicated. One such scheme is based on the solution of a system of nonlinear equations by the Newton-Raphson method, presented in Chapter 4. The governing equations from the method of Lagrange multipliers may be written as

$$\begin{aligned}
F_1(x_1,x_2,\ldots,x_n,\lambda_1,\lambda_2,\ldots,\lambda_m) &= \frac{\partial U}{\partial x_1}+\lambda_1\frac{\partial G_1}{\partial x_1}+\lambda_2\frac{\partial G_2}{\partial x_1}+\ldots+\lambda_m\frac{\partial G_m}{\partial x_1}=0\\
F_2(x_1,x_2,\ldots,x_n,\lambda_1,\lambda_2,\ldots,\lambda_m) &= \frac{\partial U}{\partial x_2}+\lambda_1\frac{\partial G_1}{\partial x_2}+\lambda_2\frac{\partial G_2}{\partial x_2}+\ldots+\lambda_m\frac{\partial G_m}{\partial x_2}=0\\
&\vdots\\
F_i(x_1,x_2,\ldots,x_n,\lambda_1,\lambda_2,\ldots,\lambda_m) &= \frac{\partial U}{\partial x_i}+\lambda_1\frac{\partial G_1}{\partial x_i}+\lambda_2\frac{\partial G_2}{\partial x_i}+\ldots+\lambda_m\frac{\partial G_m}{\partial x_i}=0\\
F_{n+j}(x_1,x_2,\ldots,x_n,\lambda_1,\lambda_2,\ldots,\lambda_m) &= G_j(x_1,x_2,\ldots,x_n)=0
\end{aligned} \tag{8.57}$$

where

$$i = 1,2,\ldots,n \quad \text{and} \quad j = 1,2,\ldots,m$$

Therefore, a system of $n + m$ equations is obtained, with the n independent variables and the m multipliers as the unknowns. These equations may be solved by starting with guessed values of the unknowns and solving the following system of linear equations for the changes in the unknowns, Δx_i and $\Delta\lambda_i$, for the next iteration:

$$\begin{bmatrix} \frac{\partial F_1}{\partial x_1} & \frac{\partial F_1}{\partial x_2} & \cdots & \frac{\partial F_1}{\partial \lambda_m} \\ \vdots & & & \\ \frac{\partial F_n}{\partial x_1} & \frac{\partial F_n}{\partial x_2} & \cdots & \frac{\partial F_n}{\partial \lambda_m} \\ \frac{\partial F_{n+1}}{\partial x_1} & \frac{\partial F_{n+1}}{\partial x_2} & \cdots & \frac{\partial F_{n+1}}{\partial \lambda_m} \\ \vdots & & & \\ \frac{\partial F_{n+m}}{\partial x_1} & \frac{\partial F_{n+m}}{\partial x_2} & \cdots & \frac{\partial F_{n+m}}{\partial \lambda_m} \end{bmatrix} \begin{bmatrix} \Delta x_1 \\ \vdots \\ \Delta x_n \\ \Delta\lambda_1 \\ \vdots \\ \Delta\lambda_m \end{bmatrix} = \begin{bmatrix} -F_1 \\ \vdots \\ -F_n \\ \vdots \\ -F_{n+m} \end{bmatrix} \tag{8.58}$$

Then, the values for the next iteration are given, for i varying from 1 to n and j varying from 1 to m, by

$$x_i^{l+1} = x_i^l + \Delta x_i^l \quad \text{and} \quad \lambda_j^{l+1} = \lambda_j^l + \Delta\lambda_j^l \tag{8.59}$$

where the superscripts l and $l + 1$ indicate the present and next iterations, respectively.

The initial, guessed values are based on information available on the physical system. However, values of the multipliers are not easy to estimate. Earlier analysis of the system, information on sensitivity, or estimates based on the guessed, starting values of the x's may be employed to arrive at starting values of the λ's. The partial derivatives needed for the coefficient matrix are generally obtained numerically if the expressions are not easily differentiable. Therefore, for a given function f_i, the first derivative may be obtained from

$$\frac{\partial f_i}{\partial x_j} = \frac{f_i(x_1, x_2, \ldots, x_j + \Delta x_j, \ldots, x_n) - f_i(x_1, x_2, \ldots, x_j, \ldots, x_n)}{\Delta x_j} \tag{8.60}$$

where Δx_j is a chosen small increment in x_j. Second derivatives will also be needed because the functions F_i in Equation (8.57) contain first derivatives. The second derivatives may be obtained from Gerald and Wheatley (2003), Jaluria (2012), and Chapra and Canale (2016) as

$$\frac{\partial^2 f_i}{\partial x_j^2} = \frac{\frac{\partial f}{\partial x_j}(x_1, x_2, \ldots, x_j + \Delta x_j, \ldots, x_n) - \frac{\partial f}{\partial x_j}(x_1, x_2, \ldots, x_j, \ldots, x_n)}{\Delta x_j} \tag{8.61}$$

Other finite-difference approximations can also be used, as discussed in Chapter 4.

Therefore, a numerical scheme may be developed to determine the optimum using the method of Lagrange multipliers. The guessed values are entered and the iteration process is carried out until the unknowns do not change significantly from one iteration to the next, as given by a chosen convergence criterion (see Chapter 4). However, the process is quite involved because the first and second derivatives may have to be obtained numerically and a system of linear equations is to be solved for each iteration. Such an approach is suitable for complicated expressions and for a relatively large number of independent variables and constraints, generally in the range of 5 to 10. For a still larger number of unknowns, the problem becomes very complicated and time-consuming, making it necessary to seek alternative approaches.

8.6 SUMMARY

This chapter focuses on the calculus-based methods for optimization. These methods use the derivatives of the objective function U and the constraints to determine the location where the objective function is a minimum or a maximum. For the unconstrained problem, a stationary point is indicated by the partial derivatives of the objective function U, with respect to the independent variables, going to zero. The nature of the stationary point, whether it is a maximum, a minimum, or a saddle point, is determined by obtaining the higher-order derivatives. For the constrained problem, the method of Lagrange multipliers is introduced and the system of equations, whose solution yields the optimum, is derived. Derivatives are again needed, making it a requirement for applying calculus methods that the objective function and the constraints must be continuous and differentiable. In addition, only equality constraints can be treated by this approach. The importance of this method lies not only in solving relatively simple problems, but also in providing basic concepts and strategies that can be used for other optimization methods.

The physical interpretation of the Lagrange multiplier method is discussed, using a single constraint and only two independent variables. It is seen that the gradient vector of the objective function U becomes aligned with that of the constraint G, where $G = 0$ is the constraint, at the optimum. Thus, the contours of constant U become tangential to the constraint curve at the optimum. Proof of this method is also given for the simple case of a single constraint. The characteristics and solutions of more complicated problems are discussed. The method is used for both unconstrained and constrained problems, including cases where a constrained problem may be converted into an unconstrained one by substitution. The significance of the multipliers is discussed and these are shown to be related to the sensitivity of the objective function to changes in the constraints. This is important additional information obtained by this method and forms a valuable input in deriving the final design of the system.

Finally, the application of these methods to thermal systems is considered. Because the objective function and the constraints must be continuous and differentiable, this approach is often restricted to relatively simple systems. However, curve fitting of experimental and numerical simulation results may be used to obtain algebraic expressions to characterize system behavior. Then the method of Lagrange multipliers may be employed easily to obtain the optimum. Inequality constraints may also be considered, in some cases by converting these to equality constraints and in others by checking the solution obtained, without taking these into account in the analysis, to ensure that the inequalities are satisfied. In some practical problems, an optimum may not arise in the design domain. In such cases, the largest or smallest value of the objective function is obtained at the domain boundaries and the corresponding values may be used for the best design. A few examples of thermal systems and processes are given. A computational approach for solving relatively complicated optimization problems using these methods is also presented.

REFERENCES

Beightler, C.S., Phillips, D., & Wilde, D.J. (1979). *Foundations of optimization* (2nd ed.). Englewood Cliffs, NJ: Prentice-Hall.

Chapra, S.C., & Canale, R.P. (2016). *Numerical methods for engineers* (7th ed.). New York: McGraw-Hill.

Chong, E.K.P., & Zak, S.H. (2013). *An introduction to optimization* (4th ed.). New York: Wiley-Interscience.

Dieter, G.E. (2000). *Engineering design* (3rd ed.). New York: McGraw-Hill.

Gebhart, B. (1971). *Heat transfer* (2nd ed.). New York: McGraw-Hill.
Gerald, C.F., & Wheatley, P.O. (2003). *Applied numerical analysis* (7th ed.). Reading, MA: Addison-Wesley.
Incropera, F.P., & Dewitt, D.P. (2001). *Fundamentals of heat and mass transfer* (5th ed.). New York: Wiley.
Jaluria, Y. (2012). *Computer methods for engineering with Matlab applications*. Boca Raton, FL: CRC Press.
Kaplan, W. (2002). *Advanced calculus* (5th ed.). Reading, MA: Addison-Wesley.
Keisler, H.J. (2012). *Elementary calculus* (3rd ed.). Mineola, NY: Dover Publications.
Rao, S.S. (2009). *Engineering optimization: theory and practice* (4th ed.). New York: Wiley.
Stoecker, W.F. (1989). *Design of thermal systems* (3rd ed.). New York: McGraw-Hill.

PROBLEMS

8.1 The cost C involved in the transportation of hot water through a pipeline is given by

$$C = \frac{20}{D^5} + \frac{4}{\ln\left(\frac{D+x}{D}\right)} + 2.5x + 5D$$

where the four terms represent pumping, heating, insulation, and pipe costs. Here, D is the diameter of the pipe and x is the thickness of insulation, as shown in Figure P8.1. Find the values of D and x that result in minimum cost.

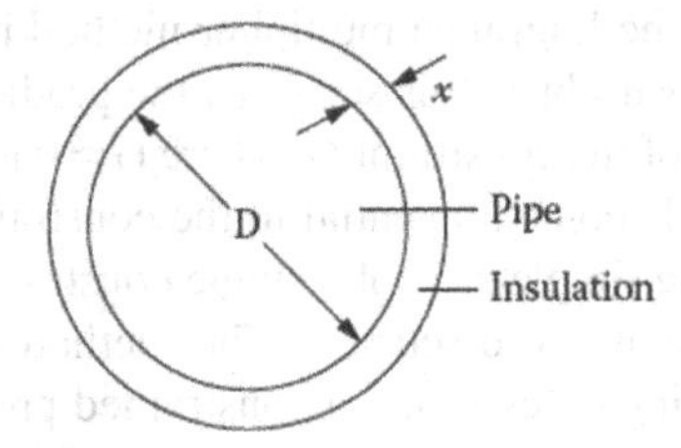

FIGURE P8.1

8.2 A manufacturer of steel cans wants to minimize costs. As a first approximation, the cost of making a can consists of the cost of the metal plus the cost of welding the longitudinal seam and the top and bottom, see Figure P8.2. The can may have any diameter D and length L, for a given volume V. The wall thickness d is 1 mm. The cost of the material is \$0.50/kg and the cost of welding is \$0.1/m of the weld. The density of the material is 10^4 kg/m^3. Using the method of Lagrange multipliers, find the dimensions of the can that will minimize cost.

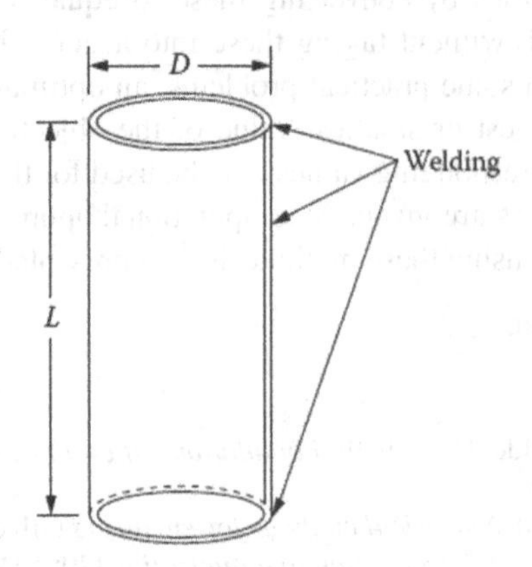

FIGURE P8.2

8.3 The cost C in a metal forming process is given in terms of the speed U of the material as

$$C = \frac{3.5KS^{1.8}}{U^{1.6}\left[2 + 17.3\left(\frac{3}{U}\right)^{2.5}\right]^2}$$

where K and S are constants. Find the speed U at which the cost is optimized. Is this a minimum or a maximum?

8.4 The cost S of a rectangular box per unit width is given in terms of its two other dimensions x and y as

$$S = 8x^2 + 3y^2$$

The volume, per unit width, is given as 12, so that $xy = 12$. Solve this problem by the Lagrange multiplier method to obtain the optimum value of S. Is it a maximum or a minimum? What is the physical significance of the multiplier l?

8.5 In a hot rolling manufacturing process, the temperature T, velocity ratio V, and thickness ratio R are the three main design variables that determine the cost C as

$$C = 65\frac{R}{V} + \frac{250}{RT} + 5T + 4RV$$

Obtain the conditions for optimal cost and determine if this is a minimum or a maximum.

8.6 A rectangular duct of length L and height H is to be placed in a triangular region, where each side is equal to 1.0 m, as shown in Figure P8.6, so that the cross-sectional area of the duct is maximized. Formulate the optimization problem as a constrained circumstance and determine the optimal dimensions.

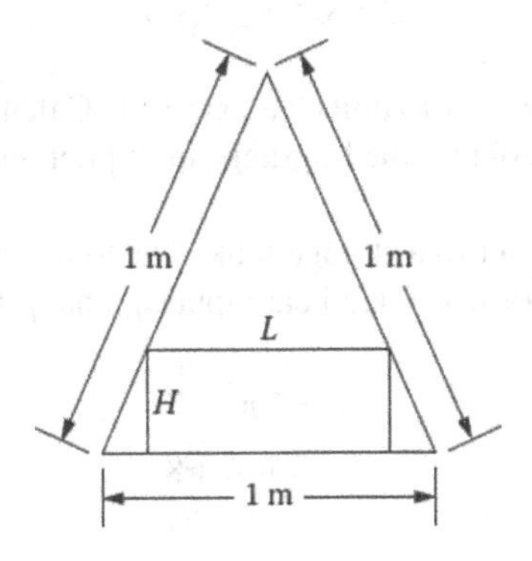

FIGURE P8.6

8.7 A rectangular box has a square base, with each side of length L, and height H. The volume of the box is to be maximized, provided the sum of the height and the four sides of the base does not exceed 100 cm, i.e., $H + 4L < 100$. Set up the optimization problem and calculate the dimensions at which maximum volume is obtained.

8.8 Consider the convective heat transfer from a spherical reactor of diameter D and temperature T_s to a fluid at temperature T_a, with a convective heat transfer coefficient h. Denoting $(T_s - T_a)$ as θ, h is given by

$$h = 2 + 0.55\theta^{0.27}D^{-1.2}$$

Also, a constraint arises from strength considerations and is given by

$$D\theta = 75$$

We wish to minimize the heat transfer from the sphere. Set up the objective function in terms of D and θ and with one constraint. Employing Lagrange multipliers for this constrained optimization, obtain the optimal values of D and θ. Also, obtain the sensitivity coefficient and explain its physical meaning in this problem. How will you use it in the final selection of the values of D and θ?

8.9 The heat lost by a thermal system is given as hL^2T, where h is the heat transfer coefficient, T is the temperature difference from the ambient, and L is a characteristic dimension. The heat transfer coefficient, in SI units, is given as

$$h = 3\left(\frac{T^{1/3}}{L^{3/2}}\right) + 8.7\left(\frac{T^{1/3}}{L^{1/2}}\right)$$

It is also given that the temperature T must not exceed $7.5\, L^{-3/4}$. Calculate the dimension L that minimizes the heat loss, treating the problem as an unconstrained one first and then as a constrained one. What information does the Lagrange multiplier yield in the latter case?

8.10 For the solar energy system considered in Example 8.6, study the effect of varying the cost per unit surface area of the reactor, given as 35 in the problem, and also of varying the cost per unit volume of the storage tank, given as 208. Vary these quantities by ±20% of the given values in turn, keeping the other coefficient unchanged, and determine, for each case, the conditions for which the cost is a minimum. Discuss the physical implications of the results obtained.

8.11 The cost C of fabricating a tank of dimensions x, y, and z is given by the expression

$$C = 8x^2 + 3y^2 + 4z^2$$

with the total volume given as 16 units, i.e., $xyz = 16$. Calculate the dimensions for which the cost is minimized. Also, obtain the Lagrange multiplier and explain its physical meaning in this problem.

8.12 Two pipes deliver hot water to a storage tank. The total flow rate in dimensionless terms is given as 10, and the nondimensional heat inputs q_1 and q_2 in the two pipes are given as

$$q_1 = 5\dot{m}_1^2 + 7$$
$$q_2 = 3\dot{m}_2^2 + 8$$

where $\dot{m}_1$ *and* $\dot{m}_2$ are the flow rates through the two pipes. If the total heat input is to be minimized, set up the optimization problem for this system. Using Lagrange multipliers for a constrained problem, obtain the optimal values of the flow rates and the sensitivity coefficient. What does it represent physically in this problem?

8.13 The mass flow rates in two pipes are denoted by $\dot{m}_1$ *and* $\dot{m}_2$. The heat inputs in these two circuits are correspondingly given as q_1 and q_2. The total mass flow rate, $\dot{m}_1 + \dot{m}_2$, is given as 14 and the following equations apply:

$$q_1 = 4\dot{m}_1^2 + 3\dot{m}_1 + 6$$
$$q_2 = 3\dot{m}_2^2 + 2\dot{m}_2 + 5$$

Obtain the values of $\dot{m}_1$ *and* $\dot{m}_2$ that optimize the total heat input, $q_1 + q_2$, using the method of Lagrange multipliers. Also, obtain the sensitivity coefficient.

8.14 The fuel consumption F of a vehicle is given in terms of two parameters x and y, which characterize the combustion process and the drag as

$$F = 10.5x^{1.5} + 6.2y^{0.7}$$

with a constraint from conservation laws as

$$x^{1.2}y^2 = 20$$

Cast this problem as an unconstrained optimization problem and solve it by the Lagrange multiplier method. Is it a maximum or a minimum?

8.15 In a water flow system, the total flow rate Y is given in terms of two variables x and y as

$$Y = 8.5x^2 + 7.1y^3 + 21$$

with a constraint due to mass balance as

$$x + y^{1.5} = 25$$

Solve this optimization problem both as a constrained problem and as an unconstrained problem, using the Lagrange multiplier method. Determine if it is a maximum or a minimum.

9 Search Methods

9.1 BASIC CONSIDERATIONS

Search methods, which are based on selecting the best design from several alternative designs, are among the most widely used methods for optimizing thermal systems. A finite number of designs that satisfy the given requirements and constraints is generated and the design that optimizes the objective function is chosen. Though particularly suited to circumstances where the design variables take on discrete values, this approach can also be used for continuous functions, such as those considered in Chapter 8. A large number of search methods have been developed to handle different kinds of problems and to provide robust, versatile, and flexible means to optimize practical systems and processes.

Comparing different alternatives and choosing the best one is not a new concept and is used extensively in our daily lives. Before purchasing a stereo system, we would generally consider different models, retailers, manufacturers, and so on, in order to procure the optimal system within our financial constraints. There is a finite number of options, with each combination of the different attributes of the system giving rise to a possible choice. The final choice is based on personal preference, finances available, reputation of the manufacturer, system features, etc. In a similar way, optimization of practical thermal systems may be based on considering a number of feasible designs and choosing the best one, as guided by the objective function.

This chapter discusses the use of search methods for the optimization of thermal systems. Because generating a feasible design is generally a time-consuming process, it is necessary to minimize the number of designs needed to reach the optimum. Therefore, efficient search methods that converge rapidly to the optimum have been developed and are extensively used for thermal systems. The efficiency of the different methods is also considered, in terms of iterative steps needed to reach the optimum.

Both constrained and unconstrained problems are considered, for single as well as multiple independent variables. As discussed in Chapter 8, a constrained problem may often be transformed into an unconstrained one by using substitution and elimination. In addition, the constraints are often included in the modeling and simulation, making the optimization problem an unconstrained one. Thus, unconstrained problems, which are often much simpler to solve than the constrained ones, arise in a wide variety of practical systems and processes. A discussion of search methods is given in this chapter, along with examples to illustrate their application to thermal systems. For further details on these methods, textbooks on optimization, such as Siddall (1982), Vanderplaats (1984), Rao (2009), Arora (2004), Ravindran et al. (2006), and Rhinehart (2018), may be consulted.

9.1.1 Importance of Search Methods

In many practical thermal systems, the design variables are not continuous functions but assume finite values over their acceptable ranges. This is largely due to the limited number of materials and components available for design. Finite numbers of components, such as pumps, blowers, fans, compressors, heat exchangers, heaters, and valves, are generally available from the manufacturers at given specifications. Even though additional, intermediate specifications can be obtained if these are custom made, it is much cheaper and more convenient to consider what is readily available and base the system design on those that are readily available. Similarly, a finite number of different materials may be considered for the system parts, leading to a finite number of discrete design choices.

In order to obtain an acceptable design, the design process, which involves modeling, simulation, and evaluation of the design, is followed. As discussed in the earlier chapters of this book, this is usually a fairly complicated and time-consuming procedure. Results from the simulation are also needed to determine the effect of the different design variables on the objective function. Because of the effort needed to simulate typical thermal systems, a systematic search strategy is necessary so that the number of simulation runs is kept at a minimum. Each run, or set of runs, must be used to move closer to the optimum. Random or unsystematic searches, where extensive simulation or experimental runs are carried out over the design domain, are inefficient and impractical.

Search methods can be used for a wide variety of problems, ranging from very simple problems with unconstrained single-variable optimization to extremely complicated systems with many constraints and variables. Because of their versatility and easy application, these methods are the most commonly used for optimizing thermal systems. In addition, these methods can be used to improve the design even if a complete optimization process is not undertaken. For instance, if an acceptable design has been obtained, the design variables may be varied from the values obtained, near the acceptable design. This allows one to search for a better solution, as given by improvement in the objective function. Similarly, several acceptable designs may be generated during the design process. Again, the best among these is selected as the optimum in the given domain.

It is obvious that search methods provide important and useful approaches for extracting the optimum design and to improve existing designs. We will focus on systematic search schemes, which may be used to determine the optimum design in a region whose boundaries are defined by the ranges of the design variables. In order to illustrate the different methods, relatively simple expressions are employed here for which search methods are really not necessary, and simpler schemes such as the calculus methods can easily be employed. However, this is only for illustration purposes and, in actual practice, each test run or simulation would generally involve considerable time and effort. Some practical systems are also considered to demonstrate the application of these methods to more complex systems.

9.1.2 Types of Approaches

Several approaches can be employed in search methods, depending on whether a constrained or an unconstrained problem is being considered and whether the problem involves a single variable or multiple variables. These approaches may be classified as follows.

9.1.2.1 Elimination Methods

In these methods, the domain in which the optimum lies is gradually reduced by eliminating regions that are determined not to contain the optimum. We start with the design domain defined by the acceptable ranges of the variables. This region is known as the initial *interval of uncertainty.* Therefore, the region of uncertainty in which the optimum lies is reduced until a desired interval is achieved. Appropriate values of the design variables are chosen from this interval to obtain the optimal design. For single-variable problems, the main search methods based on elimination are

Exhaustive search
Dichotomous search
Fibonacci search
Golden section search

All these approaches have their own characteristics, advantages, and applicability, as discussed later in detail. These methods can also be used for multivariable problems by applying the approach to one variable at a time. This technique, known as a *univariate search,* is presented later and is widely used. Exhaustive search over the domain can also be used for multivariable problems. The application of these methods to unconstrained optimization problems is discussed, along with their effectiveness in reducing the interval of uncertainty for a specified number of simulation runs.

9.1.2.2 Hill-Climbing Techniques

These methods are based on finding the shortest way to the peak of a hill, which represents the maximum of the objective function. A modification of the approach may be used to locate the bottom, or depression, which represents the minimum. The calculation proceeds so that the objective function improves with each step. Though more involved than the elimination methods, hill-climbing techniques are generally more efficient, requiring a smaller number of iterations to achieve the optimal design. These methods are applied to multivariable problems, for which some of the important hill-climbing techniques are

Lattice search
Univariate search
Steepest ascent/descent method

Though these methods are discussed in detail for relatively simple two-variable problems, they can easily be extended to a larger number of independent variables. Derivatives are needed for the steepest ascent/descent method, thus limiting its applicability to continuous and differentiable functions. The other methods mentioned in the preceding, though generally less efficient than steepest ascent, are applicable to a wider range of systems, including those that involve discrete and discontinuous values. Several other search methods have been developed in recent years because of their importance in practical systems, as outlined earlier and also presented later.

9.1.2.3 Constrained Optimization

The techniques mentioned earlier are particularly useful for unconstrained optimization problems. However, many of these can also be used, with some modifications, for constrained problems. The constraints must be satisfied while searching for the optimum. This restricts the movement toward the optimum. The constraints may also define the acceptable design domain. Two important schemes for optimizing constrained problems are

Penalty function method
Searching along a constraint

The former approach combines the objective function and the constraints into a new function that is treated as unconstrained, but which allows the effect of the constraints to be taken into account through a careful choice of weighting factors. The latter approach can be combined with the methods mentioned earlier for unconstrained optimization, particularly with the steepest ascent method. The search is carried out along the constraints so that the choices are limited and the optimal design satisfies these constraints. The procedure becomes quite involved in all but very simple cases. Therefore, effort is often directed at converting a constrained problem to an unconstrained one or the penalty function method is used.

9.1.3 Application to Thermal Systems

As discussed in preceding chapters, each simulation or experimental run is generally very involved and time-consuming for practical thermal systems. For instance, the temperature T_b of the barrel in the screw extrusion of plastics (see Figure 1.10b) is an important variable. If the optimum temperature is sought in order to maximize the mass flow rate or minimize the cost, simulation of the system must be carried out at different temperatures, over the acceptable range, to choose the best value. However, each simulation involves solving the governing partial differential equations for the flow and heat transfer of the plastic in the extruder as well as in the die. The material melts as it moves in the screw channel and the viscosity of the molten plastic varies with temperature and shear rate in the flow, the latter characteristic known as the *non-Newtonian* behavior of the fluid. Similarly, other properties are temperature-dependent. Other complexities such as the complicated geometry of the extruder, viscous dissipation effects in the flow, conjugate heat transfer in the screw, etc., must also be included. Thus, each simulation run requires substantial effort and computer time. This is typical of practical thermal systems because of the various complexities that are generally involved. Therefore, it is important to minimize the number of iterations needed to reach the optimum.

In our discussion of the various search methods, we will assume that each simulation or experimental run is complicated and time-consuming. Then the best method is the one that yields the optimum with the smallest number of runs. For illustration, we will use simple analytic expressions in many cases. These could easily be differentiated and the derivatives set equal to zero to obtain the maximum or minimum in unconstrained problems, as presented in Chapter 8. The Lagrange method of multipliers may also be used advantageously for many of these constrained or unconstrained problems. However, simple expressions are chosen only for demonstrating the use of search methods. In actual practice, such simple expressions are rarely obtained and simulation of the system, such as the extruder mentioned previously, has to be undertaken to find the optimum.

9.2 SINGLE-VARIABLE PROBLEM

Let us first consider the simplest case of an optimization problem with a single independent variable x. The mathematical statement is simply

$$U(x) \rightarrow U_{opt} \tag{9.1}$$

where U is the objective function and the optimum U_{opt} may be a maximum or a minimum. There are no constraints to be satisfied. In fact, there can be no equality constraints because only one variable is involved. If an equality constraint is given, it could be used to determine x and there would be nothing to optimize. However, inequality constraints may be given to specify an acceptable range of x over which the optimum is sought. For instance, in the plastic extrusion system considered in the preceding discussion, the barrel temperature T_b may be allowed to range from room temperature to the *charring* temperature of the plastic, which is around 250°C for typical plastics and is the temperature at which these are damaged.

The single-variable optimization problem is often of limited interest in thermal systems because several independent variables are generally important in practical circumstances. However, there are two main reasons to study the single-variable problem. First, there are systems whose performance is dominated by a single variable, even though other variables affect its performance. Examples of such a dominant single variable are heat rejected by a power plant, energy dissipated in an electronic system, temperature setting in an air conditioning or heating system, pressure or concentration in a chemical reactor, fuel flow rate in a furnace, surface area in a heat exchanger, and speed of an automobile. In such cases, the optimal design may be sought by varying only the single, dominant variable. Second, many multivariable optimization problems are solved by alternately optimizing with respect to each variable.

If $U(x)$ is a continuous, differentiable function, such as the ones shown in Figure 9.1, the maximum or the minimum can easily be found by setting the derivative $dU/dx = 0$. However, in search methods discrete runs are made at various values of x to determine the location of the optimum or the interval in which it lies, to the desired accuracy level. The objective function may be unimodal in the given domain, i.e., it has a single minimum or maximum, as sketched in Figure 9.1, or it may have several such local minima or maxima, as seen in Figure 9.2. Most of the methods discussed here assume that the objective function is unimodal. If it is not, the domain has to be approximately mapped to isolate the global optimum and apply search methods to this subdomain, as indicated in

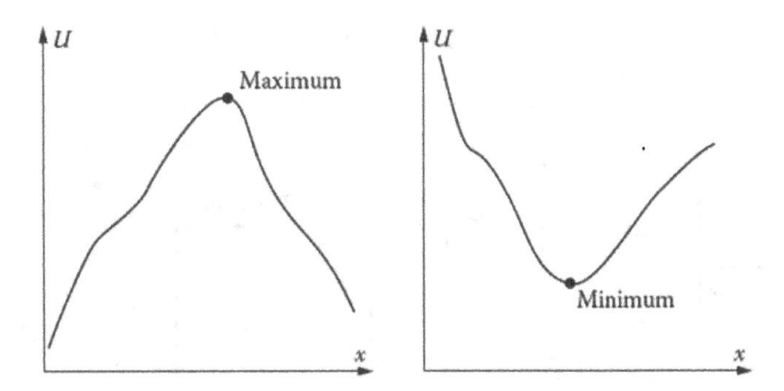

FIGURE 9.1 Unimodal objective function distributions, showing a maximum and a minimum.

Figure 9.2. Let us start with a uniform exhaustive search, which can be used effectively to determine the variation of the objective function over the entire domain and thus isolate local and global optima.

9.2.1 Uniform Exhaustive Search

As the name suggests, this method employs uniformly distributed locations over the entire design domain to determine the objective function. The number of runs n is first chosen and the initial range L_o of variable x is subdivided by placing n points uniformly over the domain. Therefore, $n + 1$ subdivisions, each of width $L_o/(n + 1)$, are obtained. At each of these n points, the objective function $U(x)$ is evaluated through simulation or experimentation of the system. The interval containing the optimum is obtained by eliminating regions where inspection indicates that it does not lie. Thus, if a maximum in the objective function is desired, the region between the location where the smaller value of $U(x)$ is obtained in two runs and the nearest boundary is eliminated, as shown in Figure 9.3 in terms of the results from three runs. In Figure 9.3(a), the region beyond C and that before A are eliminated, thus reducing the domain in which the maximum lies to the region between A and C. Similarly, in Figure 9.3(b), the region between the lower domain boundary and point B is eliminated.

Consider a chemical manufacturing plant in which the temperature T_r in the reactor determines the output M by shifting the equilibrium of the reaction. If the temperature can be varied over the

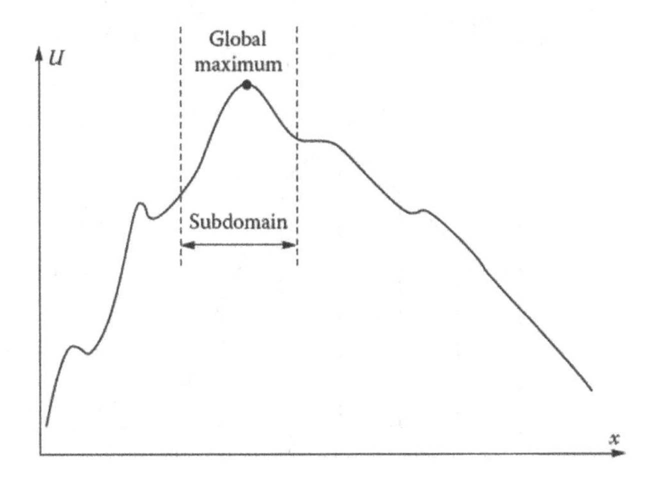

FIGURE 9.2 Variation of the objective function $U(x)$ showing local and global optima over the acceptable design domain. Also shown is a subdomain containing the global maximum.

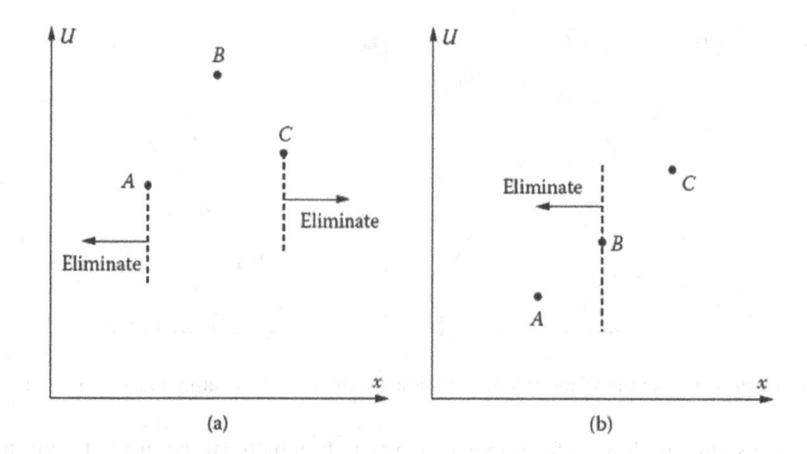

FIGURE 9.3 Elimination of regions in the search for a maximum in U.

range 300 to 600 K, the initial region of uncertainty is 300 K. The maximum output in this range is to be determined. If five trial points or runs are chosen, i.e., $n = 5$, the range is subdivided into six intervals, each of width 50 K, as shown in Figure 9.4. The output is computed from a simulation of the system at the chosen points and the results obtained are shown. From inspection, the maximum output must lie in the interval $400 < T_r < 500$. Therefore, the interval of uncertainty has been reduced from 300 to 100 as a result of five runs. The desired optimal design is then chosen from this interval. In general, the final region of uncertainty L_f is

$$L_f = \frac{2L_o}{n+1} \tag{9.2}$$

because two subintervals, out of a total of $n + 1$, contain the optimum.

The reduction of the interval of uncertainty is generally expressed in terms of the reduction ratio R, defined as

$$R = \frac{\text{Initial interval of uncertainty}}{\text{Final interval of uncertainty}} = \frac{L_o}{L_f} \tag{9.3}$$

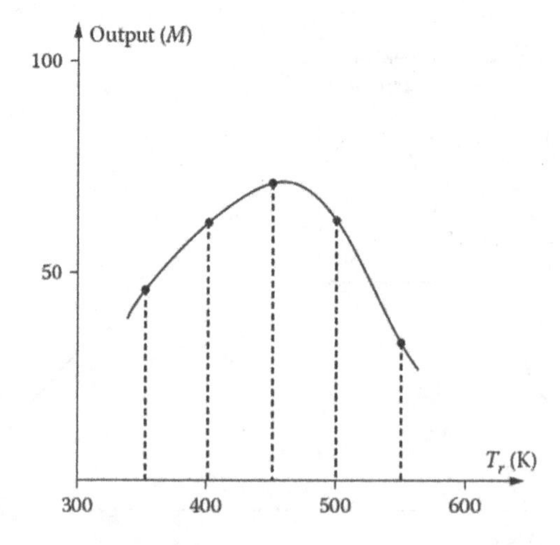

FIGURE 9.4 Uniform exhaustive search for the maximum in the output M in a chemical reactor, with the temperature T_r as the independent variable.

For the uniform exhaustive search method, the reduction ratio is

$$R = \frac{n+1}{2} \tag{9.4}$$

Therefore, the number of experiments or trial runs n needed for obtaining a desired interval of uncertainty may be determined from this equation. For instance, if, in the preceding example, the region containing the optimum is to be reduced to 30 K, then the reduction ratio is 10 and the number n of trial runs needed to accomplish this is 19.

The exhaustive search method is not a very efficient strategy to determine the optimum because it covers the entire domain uniformly. However, it does reveal the general characteristics of the objective function being optimized, particularly whether it is unimodal, whether there is indeed an optimum, and whether it is a maximum or a minimum. Therefore, though inefficient, this approach is useful for circumstances where the basic trends of the objective function are not known because of the complexity of the problem or because it is a new problem with little prior information. It is not unusual to encounter thermal systems with unfamiliar characteristics of the chosen objective function. Even in the case of the plastic screw extruder, considered earlier, the effect of the barrel temperature is not an easy one to predict because of the dependence of system behavior on the material, whose properties vary strongly with temperature and thus affect the flow and heat transfer characteristics. The exhaustive search helps in defining the optimization problem more sharply than the original formulation. Only a small number of runs may be made initially to determine the behavior of the function. Using the information thus obtained, one of the more efficient approaches, presented in the following, may then be selected for optimization.

9.2.2 Dichotomous Search

In a dichotomous search, trial runs are carried out in pairs, separated by a relatively small amount, in order to determine whether the objective function is increasing or decreasing. Therefore, the total number of runs must be even. Again, the function is assumed to be unimodal in the design domain, and regions are eliminated using the values obtained in order to reduce the region of uncertainty that contains the maximum or the minimum. The dichotomous search method may be implemented in the following two ways.

9.2.2.1 Uniform Dichotomous Search

In this case, the pairs of runs are spread evenly over the entire design domain. Therefore, the approach is similar to the exhaustive search method, except that pairs of runs are used in each case. Each pair is separated by a small amount ε in the independent variable. Considering the example shown earlier in Figure 9.4, the total design domain stretches from 300 to 600 K. We may decide to use four runs, placing one pair at 400 K and the other at 500 K, with a separation ε of 10 K in each case. As seen in Figure 9.5, the left pair allows us to eliminate the region from the left boundary to point A and the right pair the region beyond point b. Here, the pairs A, a and B, b are located at equal distance on either side of the chosen values of 400 K and 500 K, with a difference of ±5 K from these. The separation ε must be larger than the error in fixing the value of the variable in order to obtain accurate and repeatable results.

For n runs or simulations, the initial range L_o is divided into $(n/2) + 1$ subintervals, neglecting the region between a single pair. Because the final interval of uncertainty L_f has the width of a single subdivision, the reduction ratio R is obtained, neglecting the separation ε, as

$$R = \frac{n}{2} + 1 \tag{9.5}$$

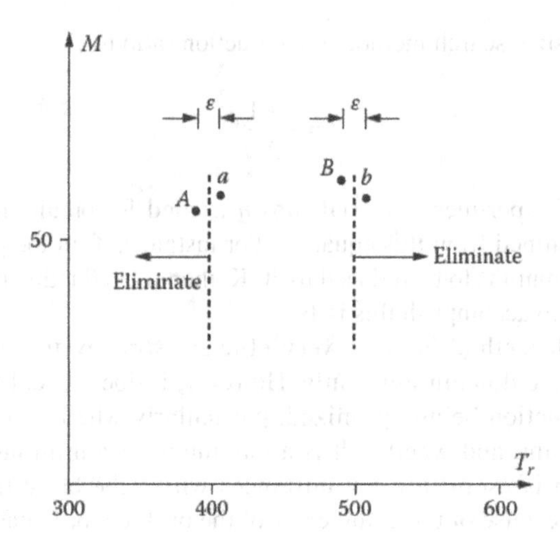

FIGURE 9.5 Uniform dichotomous search for a maximum in *M*.

Therefore, the initial interval of uncertainty is reduced to one-third, or $400 < T_r < 500$ K, after four runs. With exhaustive search, 40% of the domain is left after four runs, as seen from Equation (9.4). Therefore, the uniform dichotomous search has slightly faster convergence than the uniform exhaustive search. However, the sequential dichotomous search, discussed next, is a considerable improvement over both of these.

9.2.2.2 Sequential Dichotomous Search

As before, this method uses pairs of experiments or simulations to ascertain whether the function is increasing or decreasing and thus reduce the interval containing the optimum. However, it also uses the information gained from one pair of runs to choose the next pair. The first pair is located near the middle of the given range and about half the domain is eliminated. The next pair is then located near the middle of the remaining domain and the process repeated. This process is continued until the desired interval of uncertainty is obtained. Because pairs of runs are used, the total number of runs is even.

Considering, again, the example used earlier, let us locate the first pair of points, *A* and *a*, on either side of $T_r = 450$ K with a separation ε of 10 K. Because we are seeking a maximum in the output and because $M_a > M_A$, where M_a and M_A are the values of the objective function at these two points, the region to the left is eliminated, and the new interval of uncertainty is $450 < T_r < 600$ K if ε is neglected. The next pair, *B* and *b*, is then placed at the middle of this domain, i.e., at $T_r = 525$ K, as shown in Figure 9.6. Again, by inspection, because $M_B > M_b$, the region to the right of the pair is eliminated, leaving the interval $450 < T_r < 525$ K. Therefore, the interval of uncertainty is reduced to 25%, or one-fourth, of its initial value. With each pair, the region of uncertainty is halved. Therefore, neglecting the separation ε, the interval is halved *n*/2 times, where *n* is the total number of runs and is an even number. Therefore, the reduction ratio is obtained as $R = 2^{n/2}$. This implies that an even number of runs may be chosen *a priori* to reduce the region of uncertainty to obtain the desired accuracy in the selection of the independent variable for optimal design.

9.2.3 Fibonacci Search

The Fibonacci search is an efficient technique to narrow the domain in which the optimum value of the design variable lies. It uses a sequential approach based on the Fibonacci series, which is a series of numbers derived by Fibonacci, a mathematician in the thirteenth century. The series is given by the expression

$$F_n = F_{n-2} + F_{n-1} \tag{9.6}$$

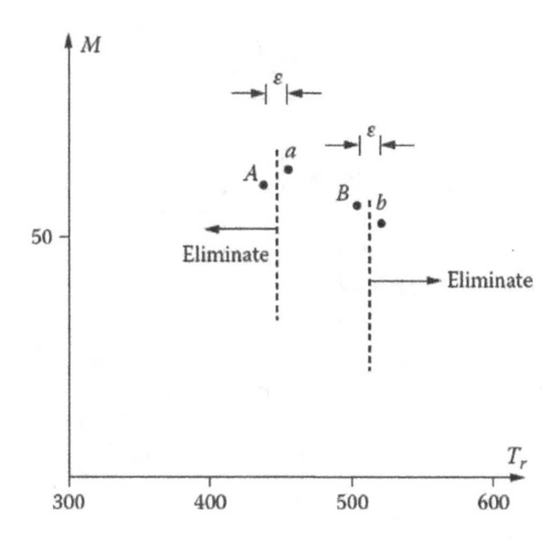

FIGURE 9.6 Sequential dichotomous search for a maximum in M.

where $F_0 = F_1 = 1$.

Therefore, the first two numbers in the series are unity and the nth number is the sum of the preceding two numbers. The Fibonacci series may thus be written as

n:	0	1	2	3	4	5	6	7	8	9	10...
F_n:	1	1	2	3	5	8	13	21	34	55	89...

It can be seen from this series that the numbers increase rapidly as n increases. The fact that, for $n \geq 2$, each number is a sum of the last two numbers is used advantageously to distribute the trial runs or experiments.

The method starts by choosing the total number of runs n. This choice is based on the reduction ratio, as discussed later. The initial range of values L_o is assumed to be given. Then the Fibonacci search places the first two runs at a distance $d_1 = (F_{n-2}/F_n)L_o$ from either end of the initial interval. For $n = 5$, this implies placing the runs at $d_1 = F_3/F_5 = (3/8)L_o$ from the two ends of the range. The simulation of the system is carried out at these two values of the design variable and the corresponding objective function determined. The values obtained are used to eliminate regions from further consideration, as discussed earlier and shown in Figure 9.3. The remaining interval of width L is now considered and runs are carried out at a distance of d_2 from each end of this interval, where $d_2 = (F_{n-3}/F_{n-1})L$. The location of one of the runs coincides with that for one of the previous runs, due to the nature of the series, and only one additional simulation is needed for the second set of points. Again, regions are eliminated from further consideration and points for the next iteration are placed at distance d_3 from the two ends of the new interval, where $d_3 = (F_{n-4}/F_{n-2})L$, L being the width of this interval. Thus, the region of uncertainty is reduced. This process is continued until the nth run is reached. This run is placed to the right of an earlier simulation near the middle of the interval left, and thus the region is further halved to yield the final interval of uncertainty L_f. The following simple example illustrates this procedure.

Example 9.1

For a heating system, the objective function $U(x)$ is the heat delivered per unit energy consumed. The independent variable x represents the temperature setting and has an initial range of 0 to 8. A maximum in U is desired to operate the system most efficiently. The objective function is given as $U(x) = 7 + 17x - 2x^2$. Obtain the optimum using the Fibonacci search method.

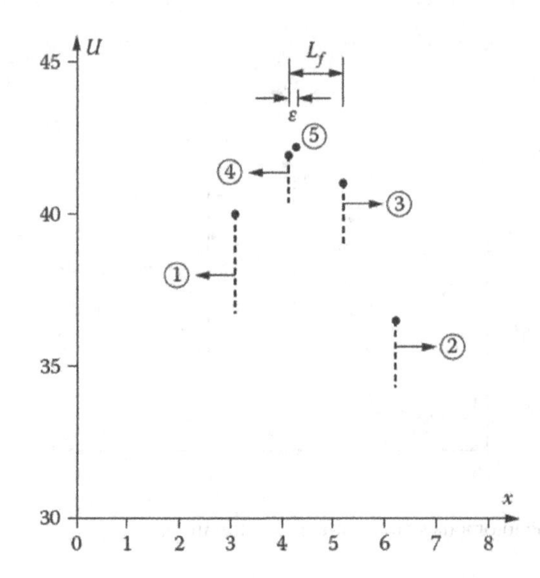

FIGURE 9.7 Use of Fibonacci search method to reduce the interval of uncertainty in Example 9.1.

SOLUTION

Let us choose the total number of runs as five. Then the first two runs are made at $d_1 = (F_3/F_5)L_o = (3/8)8 = 3$ from either end, i.e., at $x = 3$ and $x = 5$. The value at $x = 5$ is found to be larger than that at $x = 3$. Therefore, for a maximum in $U(x)$, the region $0 < x < 3$ is eliminated, leaving the domain from 3 to 8. The next two points are located at $d_2 = (F_2/F_4)L = (2/5)5 = 2$ from either end of the new interval of width $L = 5$. Thus, the two points are located at $x = 6$ and at $x = 5$. This latter location has already been simulated. The results from the run at $x = 6$ indicate that the objective function is smaller than that at $x = 5$. Therefore, the region beyond $x = 6$ is eliminated, leaving the domain from $x = 3$ to $x = 6$ for future consideration.

The next two points are located at $d_3 = (F_1/F_3)L = (1/3)3 = 1$ from the two ends of the domain, i.e., at $x = 5$ (which is already available) and at $x = 4$. Thus, simulation is carried out at $x = 4$, and the objective function is found to be greater than that at $x = 5$. The region beyond $x = 5$ is eliminated, leaving the domain $3 < x < 5$. The fifth and final run is now made at a point just to the right of $x = 4$ to determine if the function is increasing or decreasing. The value of the function is found to be higher at this point, indicating an increasing function with increasing x. Therefore, the region $3 < x < 4$ is eliminated, giving $4 < x < 5$ as the final region of uncertainty. If $x = 4.5$ is chosen as the setting for optimal U, the maximum heat delivered per unit energy consumed is obtained as 43.125. The value of U at $x = 0$ is 7 and that at $x = 8$ it is 15. Therefore, substantial savings are obtained by optimizing. Figure 9.7 shows the various steps in the determination of the final interval of uncertainty.

The initial range is reduced to one-eighth of its value in just five runs. Because $F_n = 8$ for $n = 5$, this also indicates that the reduction ratio is F_n, a statement that can be proved more rigorously by taking additional examples as well as by mathematics. Thus, this search method converges very rapidly to the optimum and only a few runs are often adequate for obtaining the desired accuracy level. For this simple case, calculus may be used as a check on the results. Calculus yields the optimum at $x = 4.25$, which is in the domain of uncertainty obtained by the search method and close to the optimum selected.

9.2.4 Golden Section and Other Search Methods

The golden section search method is derived from the Fibonacci search and, though not as efficient, is often more convenient to use. It is based on the fact that the ratio of two successive Fibonacci numbers is approximately 0.618 for $n > 8$, i.e., $F_{n-1}/F_n = 0.618$. This ratio has been known for a long time and was of interest to the ancient Greeks as an aesthetic and desirable ratio of lengths in their constructions. The ratio of the height to the base of the Great Pyramid is also 0.618. The reciprocal of this ratio is 1.618, which has also been used as a number with magical properties. The term for the method itself comes from Euclid, who called the ratio the *golden mean* and pointed out that a length divided in this ratio results in the same ratio between the smaller and larger segments (Vanderplaats, 1984; Dieter, 2000).

The golden section search uses the ratio 0.618 to locate the trial runs or experiments in the search for the optimum. The first two runs are located at 0.618 L_o from the two ends of the initial range. As before, an interval is eliminated by inspection of the values of the objective function obtained at these points. The new interval of length L is then considered and the next two runs are located at 0.618 L from the two ends of this interval. The result for one of the points is known from the previous calculations, and only one more simulation is needed. Again, an interval is eliminated and the domain in which the optimum lies is reduced. This procedure is continued until the optimum is located within an interval of desired uncertainty. The final run may be made at a location close to the middle of the interval, in order to reduce the uncertainty by approximately half, as done earlier for the Fibonacci search.

Therefore, the total number of runs n need not be decided *a priori* in this method. This allows us to employ additional runs near the optimum if the curve is very steep there, or to use fewer points if the curve is flat. In the Fibonacci search, we are committed to the total number of runs and cannot change it based on the characteristics of the optimum. In the golden section search, the trial runs are always located at 0.618 L from the two ends of the interval of width L at a given search step. This makes it somewhat less efficient than the Fibonacci search, particularly for small values of n.

Similarly, other search strategies have been developed to extract the optimum design. Several of these are combinations of the various methods presented here. For instance, an exhaustive search may be used to determine if the function is unimodal and to determine the subinterval in which the global optimum lies. This may be followed by more efficient methods such as the Fibonacci search. An unsystematic search, though generally very inefficient, is nevertheless used in some cases because of the inherent simplicity and because the physical nature of the problem may guide the user to the narrow domain in which the optimum lies. In general, information available on the system is very valuable in the search for the optimum because it can be used to narrow the range, determine the acceptable level of uncertainty in the variables, and choose the most appropriate strategy.

9.2.5 Comparison of Different Elimination Methods

The reduction ratio R, defined in Equation (9.3), gives the ratio of the initial interval of uncertainty to the interval obtained after n runs. Therefore, it is a measure of the efficiency of the method. It can also be used to select the number of runs needed to obtain a desired uncertainty in locating the optimum. The reduction ratios for the various methods presented here for the optimization of a single-variable problem are given in Table 9.1. Here, the effect of the separation ε between pairs of runs on the reduction ratio is neglected.

TABLE 9.1
Reduction Ratios for Single-Variable Search Methods

	Reduction Ratio		
Search Method	**General Formula**	$n = 5$	$n = 12$
Uniform exhaustive	$(n + 1)/2$	3	6.5
Uniform dichotomous	$(n + 2)/2$	3.5	7.0
Sequential dichotomous	$2^{n/2}$	5.66	64
Fibonacci	F_n	8	233
Golden section		6.86	199

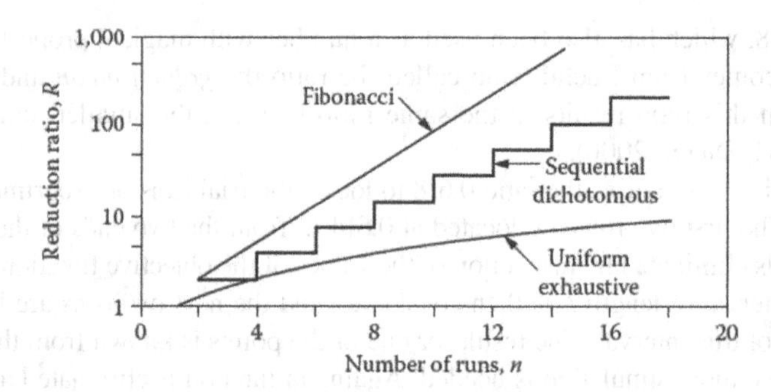

FIGURE 9.8 Reduction ratio R as a function of the number of runs n for different elimination search methods.

If ε is retained, the final interval can be shown to be

$$L_f = \frac{L_o}{F_n} + \varepsilon, \quad \text{for Fibonacci search} \tag{9.7}$$

$$L_f = \frac{L_o}{2^{n/2}} + \varepsilon\left(1 - \frac{1}{2^{n/2}}\right), \quad \text{for sequential dichotomous search} \tag{9.8}$$

when the second point of the pair is always located to the right of the first point at a separation of ε (Stoecker, 1989). Thus, the reduction ratios given in Table 9.1 are obtained when ε is neglected. The corresponding results are also shown graphically in Figure 9.8.

It is clearly seen that the Fibonacci search is an extremely efficient method and is, therefore, widely used. It is particularly valuable in multivariable optimization problems, which are based on alternating single-variable searches, and in the optimization of large and complicated systems that require substantial computing time and effort for each simulation run. For small and relatively simple systems, the exhaustive search provides a convenient, though not very efficient, approach to optimization.

Example 9.2

Formulate the optimization problem given in Example 8.6 and Example 5.3 in terms of the maximum temperature T_o as the independent variable and solve it by the uniform exhaustive search and Fibonacci search methods to reduce the interval of uncertainty to 0.1 of its initial value.

SOLUTION

The initial interval of uncertainty in T_o is from 40°C to 100°C, or 60°C. This is to be reduced to an interval of 6°C by the use of two elimination methods. Using the reduction ratios given in Table 9.1, we have

$$\frac{n+1}{2} = 10 \quad \text{or} \quad n = 19 \text{ for the uniform exhaustive search method}$$

and

$$F_n = 10 \quad \text{or} \quad n = 6 \text{ for the Fibonacci method}$$

The objective function is given by the equation

$$U = 35A + 208V = f(T_o)$$

and the dependence of A and V on T_o is given by the equations

$$A = \frac{5833.3}{[290 - 2(T_o - 20)]}$$

$$V = \frac{50}{T_o - 20}$$

Therefore, T_o may be varied over the given domain of 40°C to 100°C and the objective function determined using these equations. This problem thus illustrates the use of results from the model as one proceeds with the optimization. For complicated thermal systems, the results will generally require numerical simulation to obtain the desired results.

From the preceding calculation of the required value of n, we may choose n as 20 for a uniform exhaustive search, for convenience, and to ensure that at least a tenfold reduction in the interval of uncertainty is achieved. The value for n is taken as 6 for the Fibonacci search because this gives a reduction ratio of 13. For the uniform exhaustive search, the width of each subinterval is 60/21, and 20 computations are carried out at uniformly distributed points. The point where the minimum value of U occurs, as well as the two points on either side of this point, yields the following results:

T_o	A	V	U
51.43	25.68	1.59	1229.75
54.29	26.34	1.46	1225.37
57.14	27.04	1.35	1226.46

Therefore, the minimum lies in the interval 51.43 to 57.14. If the value at the midpoint, $T_o = 54.29$°C, is chosen, the cost is 1225.37. These values are close to those obtained in Example 8.6 by using the Lagrange multiplier method.

The Fibonacci search method is more involved because decisions on eliminating regions have to be made. Six runs are made, with 5/13, 3/8, 2/5, and 1/3 of the interval of uncertainty taken at successive steps to locate two points at equal distances from the boundaries. The first step requires two calculations and the next three require only one calculation each because points are repeated. The final calculation is taken at a point to the right of a point near the middle of the interval of uncertainty after five runs. The results obtained are summarized as

T_o	A	V	U	Action Taken
63.08	28.62	1.16	1243.01	
76.92	33.11	0.88	1341.72	Eliminate region beyond 76.92
53.85	26.24	1.48	1225.67	Eliminate region beyond 63.08
49.23	25.19	1.71	1237.56	Eliminate region 40 to 49.23
58.46	27.38	1.30	1228.58	Eliminate region beyond 58.46
53.90	26.25	1.47	1225.62	Eliminate region 49.23 to 53.85

The last point is just to the right of 53.85, which is close to the middle of the region 49.23 to 58.46 left after five runs. Therefore, the final region of uncertainty is from 53.85 to 58.46, which has a width of 4.61°C. The optimum design may be taken as a point in this region. The results agree with the earlier results from the Lagrange multiplier and the uniform exhaustive search methods. Therefore, only six runs are needed to reduce the interval of uncertainty to less than one-tenth of its initial value. The Fibonacci method is very efficient and is extensively used, though the programming is more involved than for the exhaustive search method. It must also be noted that, in some cases, the optimum may lie at the boundaries. Then the final region of uncertainty would lie adjacent to the boundary and the value at the boundary may be calculated to narrow the region further.

9.3 UNCONSTRAINED SEARCH WITH MULTIPLE VARIABLES

Let us now consider the search for an optimal design when the system is governed by two or more independent variables. For ease of visualization and discussion, we will largely consider only two variables, later extending the techniques to a larger number of variables that arise in more complicated systems. However, the complexity of the problem rises sharply as the number of variables increases and, therefore, attention is generally directed at the most important variables, usually restricting these to two or three. In addition, many practical thermal systems can be well-characterized in terms of two or three predominant variables. Examples of this include the length and diameter of a heat exchanger, fluid flow rate and evaporator temperature in a refrigeration system, height of a cooling tower and the energy rejected by it, volume of a combustion chamber and the fuel flow rate, and so on.

In order to graphically depict the iterative approach to the optimum design, a convenient method is the use of contours or lines of constant values of the objective function. Figure 9.9 shows a typical contour plot where each contour represents a particular value of the objective function and the maximum or minimum is indicated by the innermost contour. This plot is similar to the ones used in topology to represent different heights or elevations in mountainous regions. The peak represents a maximum and the depression or bottom represents a minimum. Increasing height on the mountain is thus similar to advancing toward the center of the contour plot. Such a graphical representation works well for a two-variable problem because the plane of the figure is adequate to show the movement toward the peak or the bottom. However, a three-dimensional representation is needed for three variables, with each contour replaced by a surface. This becomes quite involved for visualization and the complexity increases with increasing number of variables. However, the extension of the mathematical treatment to a larger number of variables is straightforward and can be employed for more complicated problems.

The methods presented here for multivariable, unconstrained optimization are based on moving the calculation in the direction of increasing objective function for a maximum and in the direction of decreasing objective function for a minimum. Therefore, the procedure for determining a maximum is similar to climbing toward the peak of a mountain or hill, so these methods are known as *hill-climbing techniques*. The three methods discussed in detail here are *lattice search*, *univariate search*, and *steepest ascent*. Elimination methods, which reduce the interval of uncertainty by eliminating regions, may also be combined with these techniques, particularly with an univariate search, to obtain the optimum.

9.3.1 LATTICE SEARCH

This search method is based on calculating the objective function U in the neighborhood of a chosen starting point and then moving this point to the location that has the largest value of U, if

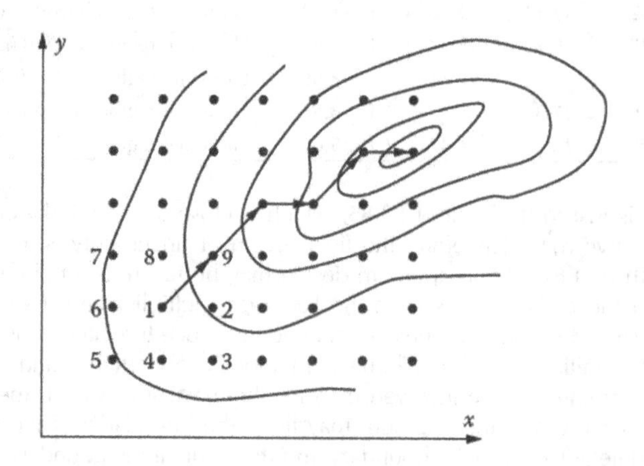

FIGURE 9.9 Lattice search method in a two-variable space.

the search is for a maximum. Thus, the calculation moves in the direction of increasing value of the objective function for locating a maximum. The maximum is reached when the value at the central point is higher than the values at its neighboring points. Though the search for a maximum in U is considered here, a similar procedure may be followed for a minimum, moving the calculation in the direction of decreasing value of the objective function.

A grid lattice is superimposed on the design domain, as shown in Figure 9.9 in terms of the contour plots on a two-dimensional space. The starting point may be chosen based on available information on the system characteristics or on the location of the maximum from previous efforts; otherwise, a point away from the boundaries of the region may be selected, such as point 1 in the figure. The objective function is evaluated at all the neighboring points, 2–9. If the maximum value of the objective function turns out to be at point 9, then this point becomes the central point for the next set of calculations. Because the values at points 1, 2, 8, and 9 are known, only the values at the remaining five points, 10 through 14, are needed. Again, the trial point is moved to the location where the objective function is the largest. This process is continued until the maximum value appears at the central point itself.

Clearly, this is not a very efficient approach and involves exhaustive search in the neighborhood of a central point, which is gradually moved toward the optimum. However, it is more efficient than using an exhaustive search over the entire region because only a portion of the region is involved in a lattice search and the previously calculated values are used at each step. The efficiency of a lattice search, compared to an exhaustive search, is expected to be even higher for a larger number of variables and finer grids. It is also obvious that the convergence to the optimum depends on the grid. It is best to start with a coarse grid, employing only a few grid points across the region. Once the maximum is found with this grid, the grid may be refined and the previous maximum taken as the starting point. Further grid refinement may be used as the calculations approach the optimum. The method is fairly robust and versatile. It can even be used for discontinuous functions and for discrete values, as long as the objective function can be evaluated. The approach can be extended easily to a problem with more than two variables. However, the number of points in the neighborhood of the central point, including this point, rises sharply as the number of variables increases, being 3^2 for two, 3^3 for three, 3^4 for four variables, and so on.

9.3.2 Univariate Search

A univariate search involves optimizing the objective function with respect to one variable at a time. Therefore, the multivariable problem is reduced to a series of single-variable optimization problems, with the process converging to the optimum as the variables are alternated. This procedure is shown graphically in Figure 9.10. A starting point is chosen based on available information on the system or at a point away from the boundaries of the region. First, one of the variables, say x, is held constant and the function is optimized with respect to the other variable y. Point A represents the optimum thus obtained. Then y is held constant at the value at point A and the function is optimized with respect to x to obtain the optimum given by point B. Again, x is held constant at the value at point B and y is varied to obtain the optimum, given by point C. This process is continued, alternating the variable, which is changed while the others are held constant, until the optimum is attained. This is indicated by the change in the objective function, from one step to the next, becoming less than a chosen convergence criterion or tolerance.

Therefore, the two-variable problem is reduced to two single-variable problems applied alternately. The basic procedure can easily be extended to three or more independent variables. In solving the single-variable problem, the search methods presented earlier, such as Fibonacci and golden section searches, may be used. This provides a very useful method for optimizing thermal systems, particularly those that have discrete values for the design variables and those that have to be simulated for each trial run. Efficient search methods, rather than exhaustive searches, are of interest in such cases. Calculus methods may also be used if continuous, differentiable functions

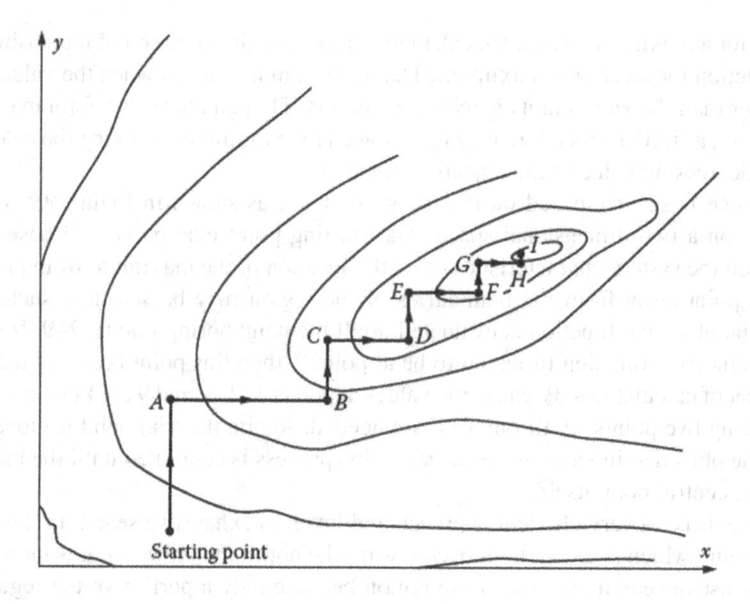

FIGURE 9.10 Various steps in the univariate search method.

are involved, as illustrated in the following example. There are certain circumstances where an univariate search may fail, such as those where ridges and very sharp changes occur in the objective function (Stoecker, 1989). However, by varying the starting point, interval of search, and method for single-variable search, such difficulties can often be overcome.

Example 9.3

The objective function U, which represents the cost of a fan and duct system, is given in terms of the design variables x and y, where x represents the fan capacity and y the duct length, as

$$U = \frac{x^2}{6} + \frac{4}{xy} + 3y$$

Both x and y are real and positive. Using the univariate search, obtain the optimum value of U and the corresponding values of x and y. Is this optimum a minimum or a maximum?

SOLUTION

Calculus methods may be used for the two single-variable optimization problems that are obtained in the univariate search. If y is kept constant, the value of x at the optimum is given by

$$\frac{\partial U}{\partial x} = \frac{2x}{6} - \frac{4}{x^2 y} = 0 \qquad \text{i.e., } x = \left(\frac{12}{y}\right)^{1/3}$$

Similarly, if x is held constant, the value of y at the optimum is given by

$$\frac{\partial U}{\partial y} = -\frac{4}{xy^2} + 3 = 0 \qquad \text{i.e., } y = \left(\frac{4}{3x}\right)^{1/2}$$

Because the only information available on x and y is that these are real and greater than 0, let us choose $x = y = 0.5$ as the starting point. If a solution is not obtained, the starting point may be varied.

First x is held constant and y is varied to obtain an optimum value of U. Then y is held constant and x is varied to obtain an optimum value of U. In both cases, the preceding equations are used.

The results obtained are tabulated as

x	y	U
0.5	1.633	9.840
1.944	1.633	6.788
1.944	0.828	5.598
2.438	0.828	5.456
2.438	0.740	5.428
2.532	0.740	5.423
2.532	0.726	5.423
2.548	0.726	5.422
2.548	0.723	5.422
2.550	0.723	5.422
2.550	0.723	5.422

For each step, one of the variables is held constant, as indicated, and the optimum is obtained in terms of the other variable. The procedure is repeated until the overall optimum, which is a minimum in U, is attained. The iteration is terminated when x and y stop changing. A convergence criterion can also be used to stop the iterative process. The procedure is quite straightforward and converges quite rapidly for this simple problem. Even for substantially different starting points, the method converges to the optimum. The optimum can also be obtained by calculus methods, as discussed in Chapter 8. The results are identical to those obtained here by the univariate search, providing validation for this scheme. If U is not calculated at each step, it can be confirmed that a minimum in cost is achieved by varying x or y from the values obtained at the optimum. The value of U increases if either of these is varied, indicating that indeed a minimum is obtained.

9.3.3 Steepest Ascent/Descent Method

The steepest ascent/descent method is an efficient search method for multivariable optimization and is widely used for a variety of applications, including thermal systems. It is a hill-climbing technique in that it attempts to move toward the peak, for maximizing the objective function, or toward the bottom, for minimizing the objective function, over the shortest possible path. The method is termed *steepest ascent* in the former case and *steepest descent* in the latter. At each step, starting with the initial trial point, the direction in which the objective function changes at the greatest rate is chosen for moving the location of the point, which represents the design on the multivariable space. Figure 9.11 shows

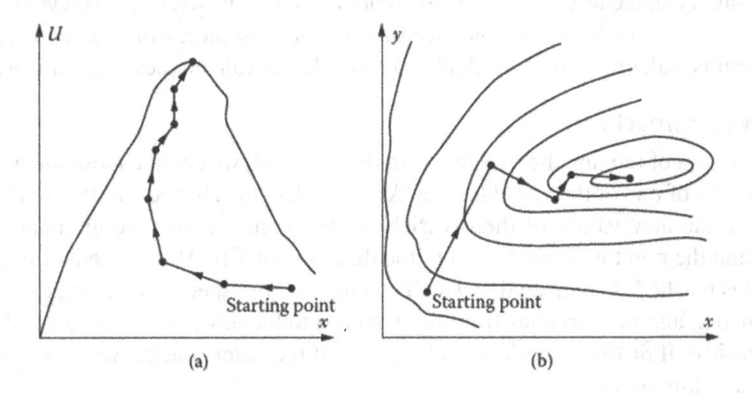

FIGURE 9.11 Steepest ascent method, shown in terms of (a) the climb toward the peak of a hill and (b) in terms of constant U contours.

this movement schematically on a hill as well as on a two-variable contour plot. Because the search always moves in the direction of the greatest rate of change of U, the number of trial runs needed to reach the optimum is expected to be relatively small and the method to be very efficient. However, it does require the evaluation of gradients in order to determine the appropriate direction of motion, limiting the application of the method to problems where the gradients can be obtained accurately and easily. Numerical differentiation may be used if an algebraic expression is not available for the objective function, which is often the case for thermal systems.

It was seen in Section 8.2 that the gradient vector ∇U is normal to the constant U contour line in a two-variable space, to the constant U surface in a three-variable space, and so on. Because the normal direction represents the shortest distance between two contour lines, the direction of the gradient vector ∇U is the direction in which U changes at the greatest rate. For a multivariable problem, the gradient vector may be written as

$$\nabla U = \frac{\partial U}{\partial x_1}\mathbf{i}_1 + \frac{\partial U}{\partial x_2}\mathbf{i}_2 + \frac{\partial U}{\partial x_3}\mathbf{i}_3 + \cdots + \frac{\partial U}{\partial x_n}\mathbf{i}_n \tag{9.9}$$

where $\mathbf{i}_1, \mathbf{i}_2, \ldots, \mathbf{i}_n$ are the unit vectors in the $x_1, x_2, \ldots, x_n$ directions, respectively. At each trial point, the gradient vector is determined and the search is moved along this vector, the direction being chosen so that U increases if a maximum is sought, or U decreases if a minimum is of interest.

The direction represented by the gradient vector is given by the relationship between the changes in the independent variables. Denoting these by $\Delta x_1, \Delta x_2, \ldots, \Delta x_n$, we have from vector analysis

$$\frac{\Delta x_1}{\partial U/\partial x_1} = \frac{\Delta x_2}{\partial U/\partial x_2} = \frac{\Delta x_3}{\partial U/\partial x_3} = \cdots = \frac{\Delta x_n}{\partial U/\partial x_n} \tag{9.10}$$

Therefore, if Δx_1 is chosen, the changes in the other variables must be calculated from these equations. In addition, Δx_1 is taken as positive or negative, depending on whether U increases or decreases with x_1 and whether a maximum or a minimum is sought. For a maximum in U, Δx_1 is chosen so that U increases, i.e., Δx_1 is positive if $\partial U/\partial x_1$ is positive and negative if $\partial U/\partial x_1$ is negative. The partial derivatives, such as $\partial U/\partial x_1$, are generally obtained numerically by using expressions such as

$$\frac{\partial U}{\partial x_1} = \frac{U(x_1 + h, x_2, \ldots, x_n) - U(x_1, x_2, \ldots, x_n)}{h} \tag{9.11}$$

where h is a small change in x_1. Similarly, the other partial derivatives may be evaluated. If an algebraic expression is available for the objective function, for instance, from curve fitting of numerical simulation results, calculus can be used advantageously to evaluate these derivatives.

9.3.3.1 Two Approaches

There are two ways of moving the trial point. In the first case, we could choose the magnitude of the step size in terms of one of the variables, say Δx_1, calculate the changes in the remaining variables, and determine the new values of these variables. At the new point, the gradient vector is again determined and the point is again moved in the direction of ∇U. This procedure is continued until the optimum is reached, as indicated by small changes, within specified convergence criteria, in the objective function and the variables from one trial run to the next. Again, for a maximum in U, Δx_1 is taken as positive if $\partial U/\partial x_1$ is positive and negative if the latter is negative, these conditions being reversed for a minimum in U.

The second approach is to move the trial point along the direction of the gradient vector until an optimum is reached. This becomes the new trial point. The gradient vector is evaluated, the new direction of movement determined, and the trial point moved in this direction until, again, an optimum is reached. This procedure is continued until the overall optimum is attained. This approach is the one shown in Figure 9.11(b). Because the calculation of the gradients may be time-consuming, the second approach is often preferred because fewer calculations of the gradient are needed. In addition, the first approach could run into a problem if the objective function U varies very slowly or rapidly with the variable, say x_1, whose step size is chosen.

These two approaches for applying the steepest ascent/descent method may be summarized for a two-variable (x and y) problem as follows.

First approach. Choose a starting point. Select Δx. Calculate the derivatives. Decide the direction of movement, i.e., whether Δx is positive or negative. Calculate Δy. Obtain the new values of x, y, and U. Calculate the derivatives again at this point. Repeat previous steps to attain new point. This procedure is continued until the change in the variables between two consecutive iterations is within a desired convergence criterion.

Second approach. Choose a starting point. Calculate the derivatives. Decide the direction of movement, i.e., whether x must increase or decrease. Vary x, using a chosen step size Δx and calculating the corresponding Δy. Continue to vary x until the optimum in U is reached. Obtain the new values of x, y, and U. Calculate the derivatives again at this point and move in the direction given by the derivatives. This procedure is continued until the change in the variables from one trial point to the next is within a desired tolerance.

The two approaches are, therefore, similar, except that the second approach involves much fewer calculations of the derivatives. Similarly, other schemes may be developed for applying the steepest ascent/descent method. The application of these two approaches is illustrated in the following example.

Example 9.4

Consider the simple problem of Example 9.3 and apply the two approaches just discussed for the steepest ascent/descent method to obtain the minimum cost U.

SOLUTION

The objective function U for this unconstrained optimization problem is given by

$$U = \frac{x^2}{6} + \frac{4}{xy} + 3y$$

The partial derivatives in terms of the independent variables x and y are

$$\frac{\partial U}{\partial x} = \frac{2x}{6} - \frac{4}{x^2 y}$$

$$\frac{\partial U}{\partial y} = -\frac{4}{xy^2} + 3$$

To move the trial point in the direction of ∇U, the following relationship applies:

$$\frac{\Delta x}{\partial U / \partial x} = \frac{\Delta y}{\partial U / \partial y}$$

Therefore, Δx may be chosen and Δy calculated from this equation. If $\partial U/\partial x$ is positive, Δx is taken as positive for search for a maximum in U. In the present case, we want a minimum in U. Therefore, Δx is taken as positive if $\partial U/\partial x$ is negative.

For the first approach, the derivatives are calculated at each point obtained by changing x by Δx and y by Δy, where Δy is obtained from the preceding relationship between Δx and Δy. The starting point is taken as $x = y = 0.5$. The results obtained for different values of Δx are

Δx	No. of Iterations	x	y	U
0.5	3	2.0	0.699	5.625
0.1	20	2.5	0.731	5.423
0.05	40	2.5	0.731	5.423
0.01	205	2.55	0.723	5.422
0.005	410	2.55	0.722	5.422

Clearly, only a few iterations are needed to reach close to the optimum, but a much larger number is needed to obtain it with a high level of accuracy, as achieved for very small Δx. Because the final design is generally not the exact optimum, but close to it, so that standard available items may be used for the system, there is no reason to insist on very high accuracy for the optimum.

In the second approach, the derivatives are calculated at a trial point which is then moved in the direction of ∇U until an optimum is obtained. This optimum point is obtained by monitoring U and stopping at the minimum value. This becomes the new trial point and the process is repeated. The results obtained in terms of trial points, with the same starting point as the first approach, are

x	y	U
0.5	0.5	17.542
0.995	0.951	7.245
1.490	1.340	6.139
1.985	0.721	5.615
2.09	0.844	5.528
2.245	0.718	5.475
2.295	0.782	5.453
2.385	0.717	5.438
2.41	0.752	5.431
2.47	0.716	5.427
2.48	0.733	5.424
2.54	0.733	5.423
2.54	0.733	5.423

Again, convergence near the optimum is quite slow. It is also interesting to note that the values of y fluctuate and are not monotonic as in the first approach. This is because the derivatives are not calculated after each increase in x but are kept constant until an optimum is reached. These results are obtained with a step size Δx of 0.005. The overall convergence is slower than that in the first approach, because several calculations are needed to obtain the trial points shown in the table. However, if the calculation of derivatives is involved and time-consuming, this approach could be more efficient than the first one.

The preceding examples demonstrate the use of univariate search and steepest ascent/descent in obtaining the optimum. Simple examples are taken to show the various steps involved and the convergence to the optimum. However, these methods can easily be implemented on the computer in order to deal with practical thermal systems, which are much more complex and which may involve simulations at the trial points to obtain the desired results. Then the optimization scheme is coupled with the numerical model and simulation results are used in reaching the optimum.

9.4 MULTIVARIABLE CONSTRAINED OPTIMIZATION

We now come to the problem of constrained optimization, which is much more involved than the various unconstrained optimization cases considered thus far in this chapter. The number of independent variables must be larger than the number of equality constraints; otherwise, these constraints may simply be used to determine the variables and no optimization is possible. Inequality constraints often indicate the feasible domain of the variables. There is no restriction on the number of inequality constraints that may be used to define the region in which the optimum must lie.

Constrained problems are quite common in the design of thermal systems. The inequality constraints are often due to various limitations imposed on the system by practical considerations, such as temperature and pressure limitations on the materials to maintain the structural integrity of a containment. The equality constraints are largely due to the basic conservation principles for mass, momentum, and energy. For instance, the speed of material emerging from the rollers in hot rolling may be obtained in terms of the speed before the rollers and the dimensions on the two sides by using mass conservation. However, in most practical cases, the numerical simulation of the system includes the conservation equations and other restrictions on the variables. Then the results obtained have already taken care of the constraints and the problem may be treated as unconstrained. Similarly, in several cases, the constraints are used to eliminate some of the variables from the problem and thus make it unconstrained, as seen in Chapter 8. All such attempts are made to convert constrained problems into unconstrained ones because of the complexity introduced by the constraints.

Despite various efforts to remove the constraints from the optimization problem, many problems still cannot be simplified and need to be solved as constrained problems. In addition, the elimination of an equality constraint results in the elimination of an independent variable. The constraint itself may be an important consideration and its retention desirable for the system being considered. As discussed in Chapter 8, the sensitivity coefficient, which indicates the effect of relaxing the constraint on the optimum, is an important feature that is useful in the final design of the system. Therefore, the constrained problem is of interest in a variety of applications. Several techniques are available for solving constrained optimization problems (Haug and Arora, 1979; Rao, 2009; Arora, 2004). We shall consider two approaches that are of particular interest to thermal systems.

9.4.1 PENALTY FUNCTION METHOD

The basic approach of this method is to convert the constrained problem into an unconstrained one by constructing a composite function using the objective function and the constraints. The method uses certain parameters, known as *penalty parameters*, that penalize the optimization of the composite function for violation of the constraints. The penalty is larger if the violation is greater. The composite function is then optimized using any of the techniques applicable for unconstrained problems. The penalty parameters are varied and the resulting composite functions are optimized. The process is continued until there is no significant change in the optimum when the penalty parameters are varied.

Let us consider the optimization problem given by the equations

$$U(x_1, x_2, x_3, \ldots, x_n) \rightarrow \text{Minimum/Maximum} \quad (9.12)$$

$$G_i(x_1, x_2, x_3, \ldots, x_n) = 0, \quad \text{where } i = 1, 2, 3, \ldots, m \quad (9.13)$$

where only equality constraints are considered and, therefore, $n > m$. The composite function, also known as the *penalty function*, may be formulated in many different ways. A commonly used formulation is given here. If a maximum in U is being sought, a new objective function V is defined as

$$V = U - \left[r_1(G_1)^2 + r_2(G_2)^2 + r_3(G_3)^2 + \ldots + r_m(G_m)^2\right] \quad (9.14)$$

and if a minimum in U is desired, the new objective function is defined as

$$V = U + \left[r_1(G_1)^2 + r_2(G_2)^2 + r_3(G_3)^2 + \ldots + r_m(G_m)^2 \right] \tag{9.15}$$

Therefore, the squares of the constraints are included in the new objective function V. The use of the squares ensures that the magnitude of the violation of a constraint is considered, and not its positive or negative value that may cancel out with the violation in other constraints. Here the r's are scalar quantities that vary the importance given to the various constraints and are known as penalty parameters. They may all be taken as equal or different. Higher values may be taken for the constraints that are critical and smaller values for those that are not as important.

If the penalty parameters are all taken as zero, the constraints have no effect on the solution and, therefore, the constraints are not satisfied. On the other hand, if these parameters are taken as large, the constraints are satisfied but the convergence to the optimum is slow. Therefore, by varying the penalty parameters we can vary the rate of convergence and the effect of the different constraints on the solution. The general approach is to start with small values of the penalty parameters and gradually increase these as the G's, which represent the constraints, become small. This implies going gradually and systematically from an unconstrained problem to a constrained one. The values of the G's at a point in the iteration may also be used to choose the penalty parameters, using larger values for larger G's so that these are driven more rapidly toward zero. Figure 9.12 shows schematically the effect of the penalty parameter r on the penalty function and on the minimum obtained for a single constraint. Clearly, the unconstrained minimum is obtained at $r = 0$ and at small values of r. The constrained minimum is attained at larger values of r.

The preceding formulation is one of the many that can be developed to use unconstrained optimization techniques for constrained problems. Several other formulations are given in the literature (Vanderplaats, 1984; Arora, 2004). Such techniques are often known as *sequential unconstrained minimization techniques* (SUMTs). The method can be used for both equality and inequality constraints. For instance, consider the following optimization problem with one variable x:

$$U = \frac{(x+3)^2}{12} \rightarrow \text{Minimum} \tag{9.16}$$

$$G_1 = \frac{2-x}{5} \leq 0 \tag{9.17}$$

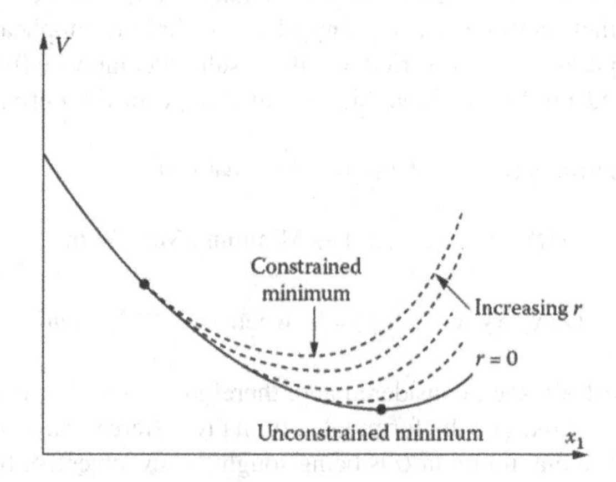

FIGURE 9.12 Penalty function method for the combined objective function V and different values of the penalty parameter r.

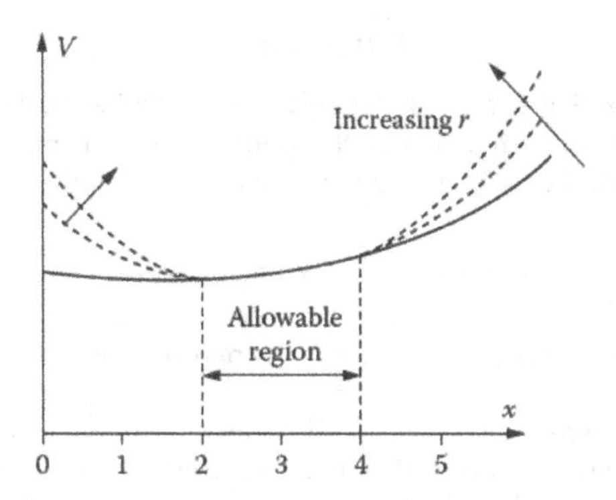

FIGURE 9.13 The penalty function method for an acceptable domain defined by inequality constraints.

$$G_2 = \frac{x-4}{6} \leq 0 \tag{9.18}$$

where U is to be minimized. The inequality constraints give the feasible domain as $2 \leq x \leq 4$. Without the constraints, the optimum is at $x = -3$, where U is zero. With the constraints, the minimum is at $x = 2$, where $U = 25/12 = 2.08$. The penalty function may be written as

$$V = \frac{(x+3)^2}{12} + r\left\{\left[\max\left(0, \frac{2-x}{5}\right)\right]^2 + \left[\max\left(0, \frac{x-4}{6}\right)\right]^2\right\} \tag{9.19}$$

where the maximum values in the ranges are used to satisfy the given inequalities. Figure 9.13 shows the penalty function for different values of the penalty parameter r. The feasible domain and the minimum for this constrained problem are also shown. Similarly, problems with a larger number of variables and constraints may be considered.

Generally, the problem is well-behaved and easy to optimize at small values of r. However, the solution may not be in the feasible region, and the optimum derived is not the desired one. As r increases, the nonlinearity in the function increases, making convergence difficult. The solutions at smaller r-values may then be used to provide the initial estimate to the optimum. As r increases, the desired optimum for the constrained problem is approached. Thus, at large values of r, $r \to \infty$, the optimum in the feasible region is obtained. The following example illustrates the use of the penalty function method for a single-equality constraint with two independent variables.

Example 9.5

In a two-component system, the cost is the objective function given by the expression

$$U(x,y) = 2x^2 + 5y$$

where x and y represent the specifications of the two components. These variables are also linked by mass conservation to yield the constraint

$$G(x,y) = xy - 12 = 0$$

Solve this problem by the penalty function method to obtain minimum cost.

SOLUTION

This is a simple problem, which can easily be solved by calculus methods. However, it can be used to illustrate some important features of the penalty function method. The new objective function $V(x, y)$, consisting of the objective function and the constraint, is defined as

$$V(x,y) = 2x^2 + 5y + r(xy - 12)^2$$

where r is a penalty parameter and the form used is the one given in Equation (9.15) for a minimum in the objective function.

We can now choose different values of r and minimize the unconstrained function $V(x, y)$. Any method for unconstrained optimization may be used for obtaining the optimum. Let us use the univariate search method, with an exhaustive search for each variable because of the simplicity of the method and of the given functions. If r is taken as zero, the constraints are not satisfied, and if r is taken as large, the constraints are satisfied, but the convergence is slow. We start with small values of r and then increase it until the results do not vary significantly with a further increase. Some typical results, obtained for different values of r, are given in the following table:

r	x	y	xy	U
0.3	2.15	3.86	8.30	28.55
0.5	2.33	4.20	9.79	31.86
1.0	2.39	4.58	10.96	34.32
10.0	2.46	4.84	11.90	36.29
100.0	2.48	4.83	11.99	36.48

Different subinterval sizes were used in the exhaustive search to obtain the desired accuracy in the results. It is seen that at small values of r, the constraint $xy = 12$ is not satisfied, and the optimum value is not the correct one. As r increases, the constraint is approximately satisfied, and the optimum value becomes independent of r. However, if r is increased to still higher values, the constraint is closely satisfied, but convergence is very slow and requires a large number of runs to obtain accurate results. Therefore, the optimum may be taken as $x^* = 2.48$, $y^* = 4.83$, and $U^* = 36.48$.

Calculus methods may also be used for this simple problem, yielding $x^* = 2.47$, $y^* = 4.87$, and $U^* = 36.49$. We may also derive x and y in terms of the penalty parameter r, by differentiating V with respect to x and y, and equating the resulting expressions to zero, as

$$x = \frac{24ry}{4 + 2ry^2} \quad \text{and} \quad y = \frac{24rx - 5}{2rx^2}$$

These equations may also be used instead of exhaustive search. It can easily be seen that as $r \to \infty$, the constraint $G = xy - 12 = 0$ is satisfied. However, as $r \to 0$, the constraint is not satisfied because x approaches zero and y approaches ∞. Therefore, the correct optimum for this constrained problem is obtained at large r.

9.4.2 Search along a Constraint

Several methods for optimization of constrained problems are based on reaching the constraint and then moving along the constraint in order to search for the optimum. These include the *gradient projection* method, the *generalized reduced gradient* method, and the *hemstitching* method (Arora, 2004; Stoecker, 1989). All these methods are quite similar, in that they search for the optimum while staying on or close to the constraint, though there are differences in their implementation. Inequality constraints generally determine the feasible region in which the optimum is sought and the search is carried out along the equality constraints so that the optimum satisfies all the constraints. In addition, inequality constraints can be converted to equalities by the use of slack variables, which ensure that the given limits are not violated, as outlined in Chapter 7. However, these methods are best suited to problems for which the gradients of the objective function and the constraints are defined and easy to determine, analytically or numerically.

Let us first consider the *hemstitching* method. The main steps involved in this method are

1. Start with a trial point.
2. Move toward and reach the constraint(s).
3. Move tangentially along the constraint(s).
4. Bring point back to the constraint(s).

The direction of the tangential move is chosen so that the objective function increases if a maximum is being sought and decreases if a minimum is of interest. The application of this method depends on the number of variables and the number of equality constraints. It is useful in a variety of thermal problems that can be represented by continuous functions.

For the simplest case of a single constraint in a two-variable space, the basic approach involves choosing an initial guess or starting point in the feasible domain. We then move to the constraint and obtain a point on the constraint. From this point, we move tangentially to the constraint. This takes the trial point off the constraint in nonlinear optimization problems and the next step is used to bring the point back to the constraint. This process is repeated, moving along the constraint, until the optimum value of the objective function is obtained. If two constraints are involved in a three-variable space, the movement of the trial point is along the tangent to both the constraints. This approach is applicable for all cases in which the number of variables n is greater than the number of constraints m by one. If this difference is greater than one, the move may be made in a direction that yields the greatest change in the objective function.

Figure 9.14 shows the hemstitching method for a two-variable, single-constraint problem. The first step involves reaching the constraint by keeping one of the two variables, x_1 or x_2, fixed and varying the other until the constraint is satisfied. For example, if the constraint is

$$x_1^2 x_2^4 = M \tag{9.20}$$

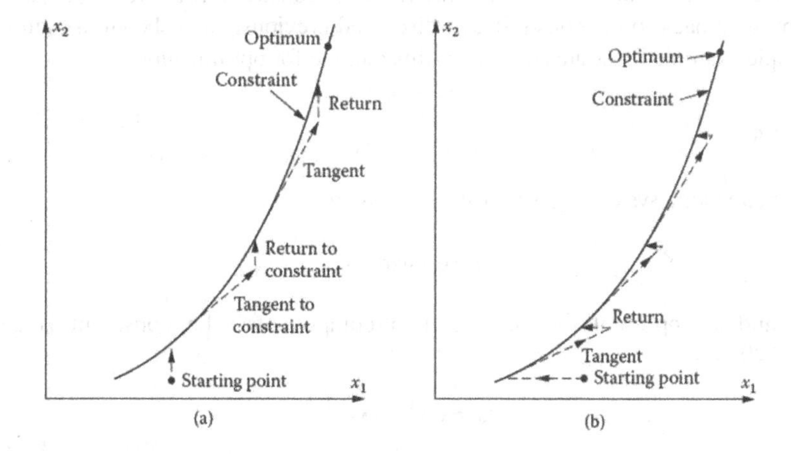

FIGURE 9.14 The hemstitching method with return to the constraint obtained by keeping (a) x_1 fixed, and (b) x_2 fixed.

where M is a constant, we can keep either x_1 or x_2 fixed to obtain the value of the other variable at the constraint as

$$x_1 = \left(\frac{M}{x_2^4}\right)^{1/2} \quad \text{or} \quad x_2 = \left(\frac{M}{x_1^2}\right)^{1/4} \tag{9.21}$$

Therefore, a point on the constraint can be located. These equations can also be used to return to the constraint if a move tangential to the constraint takes the point away from the constraint, as shown in Figure 9.14 for the two schemes of keeping x_1 or x_2 fixed.

To optimize the objective function $U(x_1, x_2)$, the trial point is moved tangentially to the constraint, which implies that $\Delta G = 0$, where ΔG is the change in the constraint. From the chain rule

$$\Delta G = \frac{\partial G}{\partial x_1}\Delta x_1 + \frac{\partial G}{\partial x_2}\Delta x_2 \tag{9.22}$$

Therefore, if ΔG is set equal to zero in this equation, Δx_1 and Δx_2 must satisfy the equation

$$\frac{\Delta x_1}{\Delta x_2} = -\frac{\partial G/\partial x_2}{\partial G/\partial x_1} \tag{9.23}$$

The change in the objective function ΔU is given by

$$\Delta U = \frac{\partial U}{\partial x_1}\Delta x_1 + \frac{\partial U}{\partial x_2}\Delta x_2 \tag{9.24}$$

Therefore, the change in $U(x_1, x_2)$ due to a move tangential to the constraint is given by

$$\Delta U = \left(\frac{\partial U}{\partial x_2} - \frac{\partial U}{\partial x_1}\frac{\partial G/\partial x_2}{\partial G/\partial x_1}\right)\Delta x_2 = S\Delta x_2 \tag{9.25}$$

If a maximum in U is being sought, Δx_2 should be positive if S is positive so that the value of U increases because of the move. This means that x_2 should be increased and the corresponding changes in the values of x_1 and U determined from Equation (9.23) and Equation (9.25), respectively. Because a tangential move takes the point away from the constraint if the functions are nonlinear, the point is brought back to the constraint, as discussed previously and shown in Figure 9.14. The following simple example illustrates the use of this method for optimization.

Example 9.6

The cost function for a system is given by the expression

$$U = 6 + 4x_1^2 + 5x_2$$

where x_1 and x_2 represent the sizes of two components. The constraint is given by Equation (9.20) as

$$G = x_1^2 x_2^3 - 35 = 0$$

Using the hemstitching method, obtain the minimum cost. Take $x_1 = 2.0$ as the starting point in the region and keep x_1 constant to return to the constraint. Take 0.2 as the step size in x_2.

SOLUTION

Because the objective function and the constraint are simple analytic expressions, calculus may be used to calculate the derivatives needed for the method. Therefore,

$$\frac{\partial U}{\partial x_1} = 8x_1 \qquad \frac{\partial U}{\partial x_2} = 5 \qquad \frac{\partial G}{\partial x_1} = 2x_1x_2^3 \qquad \frac{\partial G}{\partial x_2} = 3x_1^2x_2^2$$

This gives

$$S = 5 - \frac{12x_1^2}{x_2} \quad \text{and} \quad \Delta x_1 = -\Delta x_2\left(\frac{3x_1}{2x_2}\right)$$

The starting point is taken as $x_1 = 2$, so that $x_2 = (35/4)^{1/3} = 2.061$ satisfies the constraint. Then S is calculated and x_2 is varied, with a chosen step size of 0.2. If $S < 0$, x_2 is increased by this amount, because a minimum in U is to be obtained. Then Δx_1 is calculated from the relationship just shown between Δx_2 and Δx_1. From this result, the new x_1 is calculated as $x_1 + \Delta x_1$. The new objective function is determined, the point is brought back to the constraint, and the process is repeated. The results obtained are shown in the following table.

x_1	x_2	U	G	Next Move
2.0	2.061	32.303	0	Increment x_2
1.735	2.261	29.338	−18.294	Return to constraint
1.735	2.266	29.364	0	Increment
1.524	2.466	27.614	−10.935	Return
1.524	2.470	27.637	0	Increment
1.352	2.670	26.668	−6.275	Return
1.352	2.675	26.690	0	Increment
1.211	2.875	26.243	−3.205	Return
1.211	2.879	26.262	0	Increment
1.093	3.079	26.174	−0.128	Return
1.093	3.083	26.192	0	Increment

The problem may also be solved easily by calculus methods from Chapter 8 to yield $x_1^* = 1.127$, $x_2^* = 3.02$, and $U^* = 26.181$ for the location and value of the desired optimum. Therefore, these results are close to those obtained here by the hemstitching method. As we approach the optimum, the change in U from one iteration to the next becomes small. A zero change, i.e., $S = 0$, indicates that the optimum has been attained. Oscillations may arise near the optimum and the step size must be reduced if a closer approximation to the analytical result is desired. However, such an accurate determination of the optimum is rarely needed in practical problems because the variables are generally adjusted for the final design on the basis of convenience and available standard system parts. In the preceding example, x_1^* may be taken as 1.1 and x_2^* as 3.1 for defining the optimum. This example illustrates the hemstitching procedure for finding the optimum of a constrained problem. The evaluation of the derivatives is the major limitation on the use of this approach. Numerical differentiation is needed in most practical problems. The procedure could get fairly involved as the number of variables increases and would fail if the functions are not continuous and well-behaved.

If the optimization problem involves *two constraints* and *three variables*, the first two steps are the same as before, i.e., a trial point is chosen and moved until it reaches the constraints, which are now surfaces in a three-dimensional space. One of the variables is held constant and the two constraint equations are solved to determine the other two variables. Once on the constraints, the move is taken as tangential to both constraints. Therefore, the increments in the three variables are linked by the equations

$$\Delta G_1 = \frac{\partial G_1}{\partial x_1}\Delta x_1 + \frac{\partial G_1}{\partial x_2}\Delta x_2 + \frac{\partial G_1}{\partial x_3}\Delta x_3 = 0 \tag{9.26}$$

$$\Delta G_2 = \frac{\partial G_2}{\partial x_1}\Delta x_1 + \frac{\partial G_2}{\partial x_2}\Delta x_2 + \frac{\partial G_2}{\partial x_3}\Delta x_3 = 0 \tag{9.27}$$

where $G_1(x_1, x_2, x_3) = 0$ and $G_2(x_1, x_2, x_3) = 0$ are the two equality constraints. Therefore, if the increment in one of the variables, say Δx_1, is chosen, the other two, Δx_2 and Δx_3, may be calculated from the preceding equations. The change in the objective function $U(x_1, x_2, x_3)$ is given by the equation

$$\Delta U = \frac{\partial U}{\partial x_1}\Delta x_1 + \frac{\partial U}{\partial x_2}\Delta x_2 + \frac{\partial U}{\partial x_3}\Delta x_3 \tag{9.28}$$

The step size Δx_1 is chosen, increments Δx_2 and Δx_3 are calculated from Equation (9.26) and Equation (9.27), and the change in U is obtained from Equation (9.28). This determines whether Δx_1 should be positive or negative for a desired change in U. After a move, which is tangential to both constraints, the point is brought back to the constraints by keeping one of the variables fixed. The process is repeated until a negligible change in the objective function is obtained from one step to the next. This procedure can be extended to problems with a larger number of independent variables as long as the number of equality constraints m is one less than the number of variables n.

For circumstances where an arbitrary number of independent variables and constraints is involved, the move is made tangential to the constraints such that the change in U is the largest for a fixed distance d of movement. For three variables, this distance d is given by the equation

$$d^2 = (\Delta x_1)^2 + (\Delta x_2)^2 + (\Delta x_3)^2 \tag{9.29}$$

For maximum ΔU, given by Equation (9.28), and subjected to constraints due to tangential direction and fixed distance, such as Equation (9.26), Equation (9.27), and Equation (9.29), the Lagrange multipliers method may be employed to determine the increments Δx_1, Δx_2, etc. (Stoecker, 1989). With these increments, the new point may be obtained for the desired favorable change in U. The point is brought back to the constraint and the process is repeated until convergence is achieved, as indicated by a small change in the objective function from one iteration to the next.

Several methods have been developed with this general approach to solve constrained optimization problems. These include the constrained steepest descent (CSD) method, the method of feasible directions, the gradient projection method, and the generalized reduced gradient (GRG) method. Many efficient algorithms have been developed to obtain the optimum with the least number of trials or iterations. Some of these are available in the public domain, while others are available commercially. The difference between all these methods lies in deciding on the direction of the move and the scheme used to return to the constraint. The major problem remains the calculation of the gradients. Linearization of the nonlinear optimization problem is also carried out in some cases, and linear programming techniques can then be used for the solution. For details on these and other methods, see Arora (2004), Bertsekas (2016), and the various other references mentioned earlier.

9.5 EXAMPLES OF THERMAL SYSTEMS

We have discussed a wide range of search methods and their application to thermal systems in Chapter 7 and in this chapter. A few examples are given here for illustration of the application of these methods to practical thermal systems.

Optimization of the optical fiber drawing furnace, as shown in Figure 1.10(c), can be carried out on the basis of the numerical simulation of the process. Because of the dominant interest in fiber quality, the objective function can be based on the tension, defect concentration, and velocity difference across the fiber, all these being the main contributors to lack of quality. These are then scaled by the maximum values obtained over the design domain to obtain similar ranges of variation. The objective function U is taken as the square root of the sum of the squares of these three quantities and is minimized. The two main process variables are taken as the furnace temperature, representing the maximum in a parabolic distribution, and the draw speed. The univariate search method is applied, using the golden section search for each variable and alternating from one variable to the other. Figure 9.15 shows typical results from this search strategy for the optimal draw temperature and draw speed. The results from the first search are used in the second search, following the univariate search strategy, to obtain optimal design in terms of these two variables. The optimization process can be continued though additional iterations to narrow the domain further. However, each iteration is time-consuming and expensive. Several other results

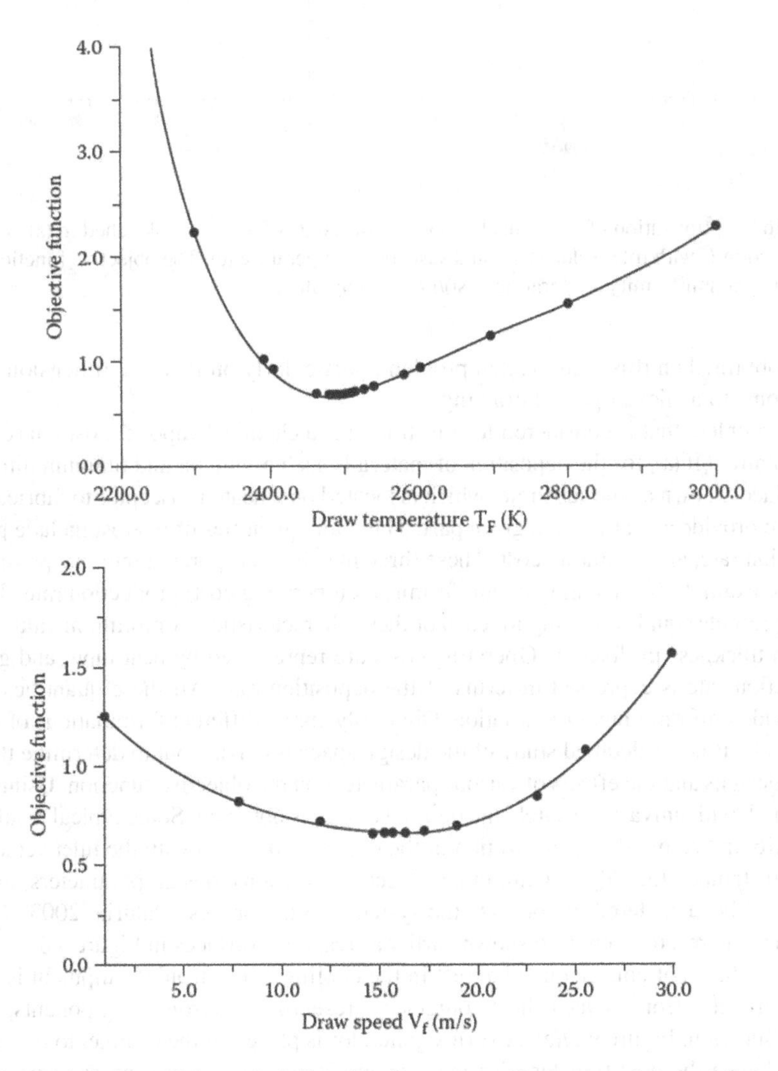

FIGURE 9.15 Optimization of the optical fiber drawing process: Evaluation of optimal furnace draw temperature at a draw speed of 15 m/s and the optimal draw speed at a draw temperature of 2489.78 K by using the golden section search method. The objective function U is chosen to represent defects in the fiber and is to be minimized.

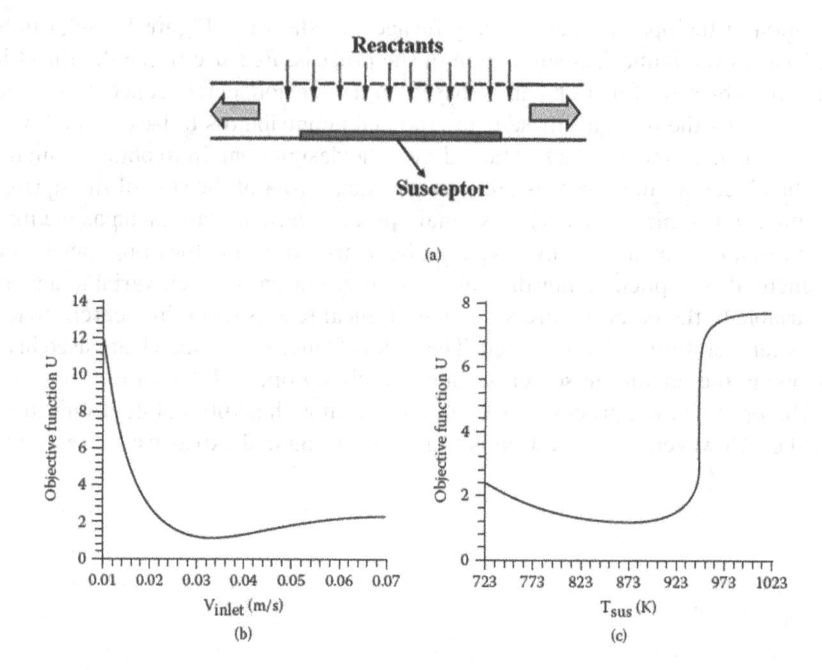

FIGURE 9.16 Optimization of a chemical vapor deposition (CVD) system, sketched in (a). Variation of the objective function U with inlet velocity (b) and susceptor temperature (c). The objective function U is defined as U = (coating nonuniformity × operating cost)/production rate.

have been obtained on this complicated problem, particularly on furnace dimensions and operating conditions, to achieve optimal drawing.

Another problem that is considered for illustration is a chemical vapor deposition (CVD) system, shown in Figure 9.16(a), for the deposition of materials such as silicon and titanium nitride (TiN) on a given surface, known as the substrate, which is located on a heated susceptor to fabricate electronic devices or to provide a coating on a given part. The main quantities of interest include product quality, production rate, and operating cost. These three may be incorporated into one possible objective function, for example U = (coating nonuniformity) × (operating cost)/production rate. The objective function represents equal weighting for each of these characteristics. A minimum value in U implies greater film thickness uniformity. Operating costs are represented by heat input and gas flow rate. The production rate is expressed in terms of the deposition rate. All these quantities are normalized to provide uniform ranges of variation. Obviously, many different formulations of the objective function can be used. A detailed study of the design space is carried out to determine the domain of acceptable designs and the effects of various parameters on the objective function. Using the steepest ascent method, with univariate search, the optimal design is obtained. Some typical results are shown in Figure 9.16, indicating the minimization of the objective function with the inlet velocity V_{inlet} and the susceptor temperature T_{sus}. Again, other objective functions, design parameters, and operating conditions can be considered to optimize the system and the process (Jaluria, 2003). Some typical results for a CVD reactor were also shown earlier as response surfaces in Figure 7.8.

A problem that is of considerable interest in the cooling of electronic equipment is one pertaining to heat transfer from isolated heat sources, representing electronic components, located in a channel, as shown in Figure 9.17(a). A vortex generator is placed in the channel to oscillate the flow and thus enhance the heat transfer. The main quantities of interest are the pressure head ΔP and the heat transfer rates Q_1 and Q_2 from the two sources. It is desirable to maximize the heat transfer rates from the two sources, to accommodate more electronic components in a given space, and to minimize the pressure head, which affects the cost of the cooling system. These three quantities

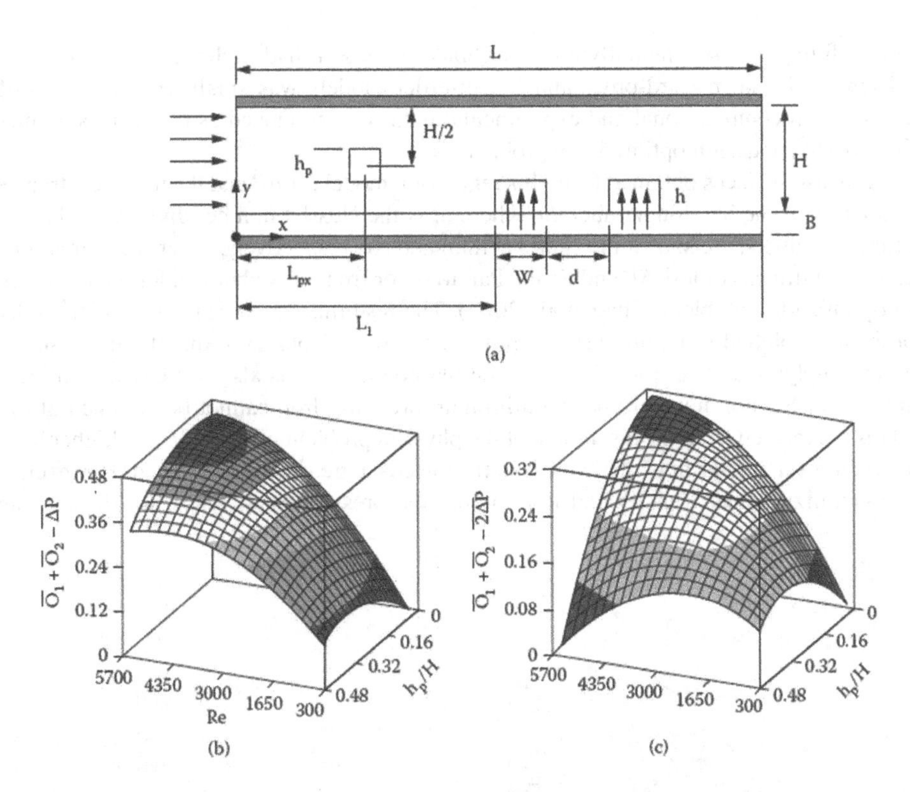

FIGURE 9.17 (a) A simple system for cooling of electronic equipment, consisting of two heat sources representing electronic components and a vortex promoter. (b) Response surface for the objective function $F = W_1Q_1 + W_2Q_2 - W_3\Delta P$ for $W_1 = W_2 = W_3$. (c) Response surface for the objective function F for $W_1 = W_2 = W_3/2$.

can be considered separately as a multi-objective function problem, or they can be combined, in their normalized forms, to form a single-objective function, F, which is then maximized. One such objective function, with the normalized quantities indicated by overbars is

$$F = W_1\overline{Q_1} + W_2\overline{Q_2} - W_3\Delta\overline{P} \tag{9.30}$$

where the W's are the weights of the three individual objective functions. The choice of the weights strongly depends on the design priorities. The responses for two different objective functions, as obtained from different values of the W's, are presented in Figure 9.17. Here, both experimental and numerical results are employed for the data points. For the first case, the optimal Reynolds number is obtained as 5600 and the height of the vortex promoter h_p/H as 0.12. For the second case, the Reynolds number is obtained as 5460 and the vortex promoter height as 0. Thus, a greater emphasis on pressure in the second case leads to a better solution without a promoter. If the weight W_3 for pressure is made half of the weights for the heat transfer rates, the optimal promoter height is obtained as 0.26. Similarly, other weights and promoter geometries can be considered (Icoz and Jaluria, 2006).

The preceding electronic cooling system can also be considered without the vortex promoter. The total heat transfer rate and the pressure head are taken as the two main objective functions. Response surfaces can be drawn from these to investigate the optimum. Multi-objective function optimization can also be used, as discussed in Chapter 7. As was done for the preceding problem, both experimental and computational data are used to build the database for the response surfaces in terms of length dimensions L_1 and L_2. Second-order, third-order, and higher-order regression models are considered. Comparing the second order with the third order, it was observed that the third-order fitting was substantially a better choice because it had higher correlation coefficients. The difference between third-order and fourth-order models was small. Hence, the third-order model, based on computational and experimental data, was employed as the regression model for the multi-objective design optimization problem.

The response surfaces obtained from this regression model for ΔP and the total heat transfer rate, given in terms of the Stanton number, *St*, where *St* is the Nusselt number divided by the Reynolds and Prandtl numbers, are shown in Figures 9.18(a) and (b), respectively. After the regression model is obtained for dimensionless ΔP and *St*, the Pareto set or frontier is obtained for the multi-objective design optimization problem (Zhao et al., 2007). The resulting Pareto set, which is the solution to this problem, is plotted in Figure 9.18(c). From the figure, it is observed that if the pressure drop is decreased, implying a lower pumping cost, the Stanton number is also decreased, and vice versa. The maximum Stanton number and the minimum pressure drop cannot be obtained at the same time. This is expected from the discussion of the physical problem given earlier. A higher heat transfer rate requires a greater flow rate, which in turn needs a greater pressure head. However, interest lies in maximizing heat transfer and minimizing the pressure head. Thus, for decision-making,

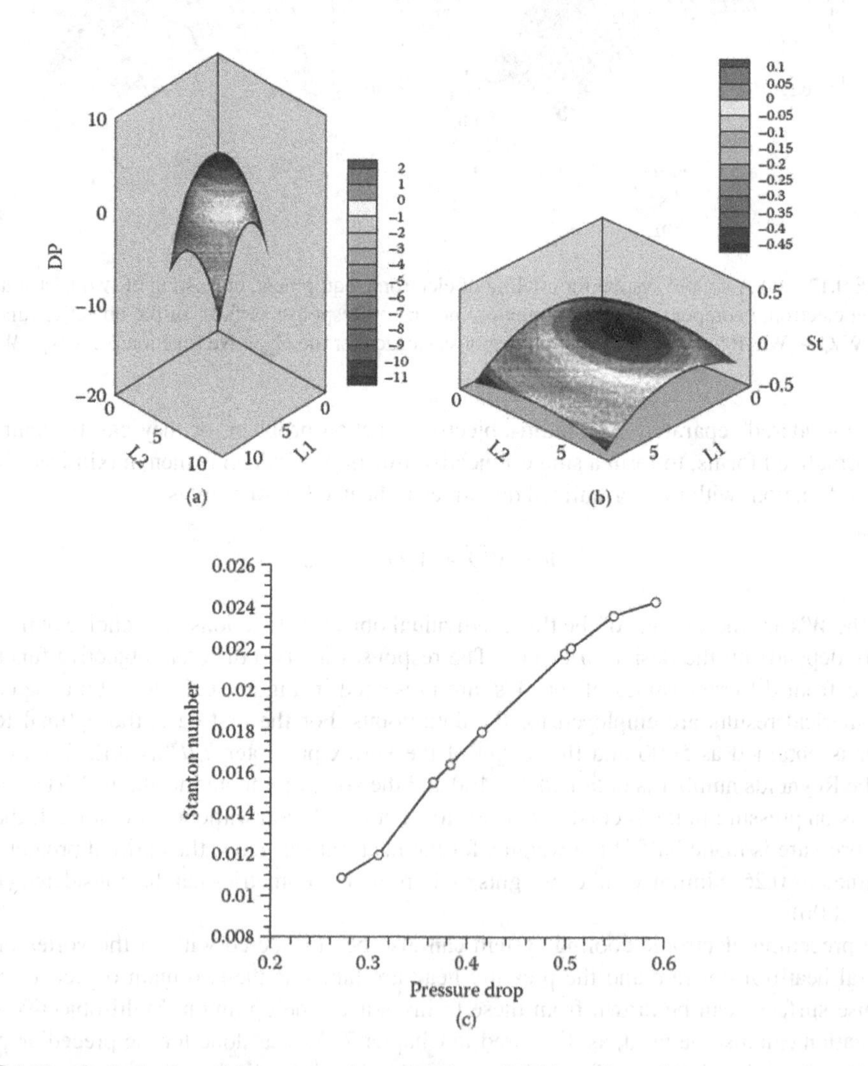

FIGURE 9.18 Optimization of the system shown in Figure 9.17(a), without the vortex promoter. (a) Third-order response surface for ΔP. (b) Third-order response surface for Stanton number, *St*. (c) Pareto front for multi-objective design optimization.

other considerations have to be added to select the proper solution from the Pareto frontier, as outlined earlier in Chapter 7. A trade-off between the two objective functions is needed and is based on the requirements and preferences for a given circumstance.

9.6 SUMMARY

This chapter presents search methods, which constitute one of the most important, versatile, and widely used approaches for optimizing thermal systems. Search methods can be used if the objective function and constraints are continuous functions as well as if these take on discrete values. In many circumstances, combinations of the components and other design variables yield a finite number of feasible designs. Search methods are ideally suited for such problems to determine the best or optimum design. Both constrained and unconstrained optimizations can be carried out using search methods.

The simplest problem of single-variable unconstrained optimization is considered first. Such circumstances are of limited practical interest, but are illustrative of the optimization techniques for more complicated problems. In addition, multivariable problems are often broken down into simpler single-variable problems for which these methods can be used. An exhaustive search for the optimum in the feasible design domain is sometimes used because of its simplicity and to determine subdomains containing the optimum, even though it is not a very efficient approach. In addition, experience and information available on the system can sometimes be used with an unsystematic search to focus on a particular subdomain to extract the optimum. Efficient elimination methods, such as Fibonacci and dichotomous schemes, are presented next. The efficiency of these methods in reducing the interval of uncertainty for a given number of iterative designs is discussed. These schemes are quite commonly used for the optimization of thermal systems.

Multivariable unconstrained problems are discussed next. A lattice search, which is relatively easy to use but is an inefficient method, is considered, followed by a univariate search strategy, which breaks the problem down into alternating searches with a single variable. This is an important approach because it allows the use of efficient methods, such as Fibonacci and calculus methods, to solve the problem as a series of single-variable problems. Hill-climbing techniques, such as steepest ascent, are very efficient for multivariable unconstrained problems. However, this approach requires the determination of the derivatives of the objective function. These derivatives are obtained analytically in relatively simple cases and numerically in cases that are more complicated. However, this does limit the use of the method to problems that can be represented by continuous functions and expressions.

Constrained multivariable problems are the most complicated ones encountered in the optimization of thermal systems. Because of their complexity, efforts are made to include the constraints in the objective function, thus obtaining an unconstrained problem. The inequality constraints generally define the feasible domain and the equality constraints often arise from conservation principles. In the simulation of most thermal systems, the equations stemming from conservation laws are generally part of the solution and do not result in equality constraints. However, there are problems that have to be solved as constrained problems. Two main approaches are presented in this chapter. The first is the penalty function method, which defines a new objective function, with the constraints included, and imposes a penalty if the constraints are not satisfied. The second approach is based on searching along the constraint. Derivatives are needed for the implementation of this method, thus restricting its applicability to continuous and differentiable functions. Examples of the application of search methods to practical thermal systems are finally outlined.

REFERENCES

Arora, J.S. (2004). *Introduction to optimum design* (2nd ed.). New York: Academic Press.

Bertsekas, D. (2016). *Nonlinear programming* (3rd ed.). Nashua, NH: Athena Scientific.

Dieter, G.E. (2000). *Engineering design: a materials and processing approach* (3rd ed.). New York: McGraw-Hill.

Haug, E.J., & Arora, J.S. (1979). *Applied optimal design*. New York: Wiley.
Icoz, T., & Jaluria, Y. (2006). Design optimization of size and geometry of vortex promoter in a two-dimensional channel, *ASME Journal of Heat Transfer*, 128, 1081–1092.
Jaluria, Y. (2003) Thermal processing of materials: from basic research to engineering, *Journal of Heat Transfer*, 125, 957–979.
Rao, S.S. (2009) *Engineering optimization: theory and practice* (4th ed.). New York: Wiley.
Ravindran, A., Ragsdell, K.M., & Reklaitis, G.V. (2006). *Engineering optimization*. New York: Wiley.
Rhinehart, R.R. (2018). *Engineering optimization: applications, methods and analyses*. New York: Wiley.
Siddall, J.N. (1982). *Optimal engineering design*. New York: Marcel Dekker.
Stoecker, W.F. (1989). *Design of thermal systems* (3rd ed.). New York: McGraw-Hill.
Vanderplaats, G.N. (1984). *Numerical optimization techniques for engineering design*. New York: McGraw-Hill.
Zhao, H., Icoz, T., Jaluria, Y., & Knight, D. (2007). Application of data driven design optimization methodology to a multi-objective design optimization problem. *Journal of Engineering Design*, 18, 343–359.

PROBLEMS

9.1 Use Fibonacci search to find the minimum of the function $U(x)$, where

$$U(x)=\frac{[\ln(x)]\sin\left(x^2/5\right)}{2x+3}$$

Obtain a final interval of uncertainty in x of 0.1 or less.

9.2 Reduce the cylindrical storage tank problem considered in Example 8.3 to its unconstrained form and determine the optimal dimensions using the following search methods:
a. Uniform exhaustive search
b. Dichotomous search
c. Fibonacci search

Compare the number of trial runs needed in the three cases and the final solution obtained. Take the desired final interval of uncertainty for the radius as 5 cm.

9.3 The amount of ammonia produced in the chemical reactor considered in Example 4.6 is to be optimized by varying the bleed over the range of 0 to 40 moles/s. Using any search method, with the numerical model given earlier, determine if an optimum in ammonia production exists in this range and obtain the applicable bleed rate if it does.

9.4 An optimum flow rate is to be achieved in the fan and duct system considered in Example 4.7 by varying the constants 15 and 80, which represent the zero pressure and the zero flow parameters. Use any suitable search method to determine if the flow can be optimized by varying these two parameters over the range ±30% of the given base values.

9.5 Use univariate search to find the optimum of the unconstrained objective function $U(x,y)$ given by

$$U(x,y)=2x+\frac{20}{xy}+\frac{y}{3}$$

Show that you have obtained a minimum. Also, use calculus methods to obtain the minimum and compare the results from the two approaches.

9.6 The cost C of a storage chamber is given in terms of its three dimensions as

$$C=12x^2+2y^2+5z^2$$

with the volume given as 10 units, i.e., $xyz = 10$. Recast this problem as an unconstrained optimization problem with two independent variables. Applying univariate search, determine the dimensions that minimize the cost.

9.7 We wish to minimize the cost U of a system, where U is given in terms of the three independent variables x, y, and z as

$$U = xy + \frac{1}{xz} - 16y^2 + z$$

Starting with the initial point (1, 0.5, 0.5), in x, y, and z, respectively, obtain the optimum by the univariate search method as well as by the steepest descent method with $|\Delta x| = 1.0$. Compare the results and number of trial runs in the two cases. Is the given value of $|\Delta x|$ satisfactory?

9.8 Apply any search method to solve the optimization problem for the solar energy system considered in Example 8.6. Employ the area A and the volume V as the two independent variables. Compare the results obtained with those presented in the example and the computational effort needed to obtain the solution.

9.9 In Example 5.1, an acceptable design of a refrigeration system was obtained to achieve the desired cooling. As seen earlier, an acceptable design may be selected from a wide domain. Considering the evaporator and condenser temperatures as the only design variables, formulate the optimization problem for maximizing the coefficient of performance. Using any suitable search method, determine the optimal design of the system.

9.10 The heat transfer Q from a spherical reactor of diameter D is given by the equation $Q = h \cdot T \cdot A$, where h is the heat transfer coefficient, T is the temperature difference from the ambient, and $A(=\pi D^2)$ the surface area of the sphere. Here, h is given by the expression

$$h = 2 + 0.5T^{0.2} D^{-1}$$

A constraint also arises from material limitations as

$$DT = 20$$

Set up the optimization problem for minimizing the total heat transfer Q. Using the method of steepest descent, obtain the optimum, starting at the initial point $D = 0.1$ and $T = 50$, with step size in T equal to 10. Also, obtain the minimum by simple differentiation of the unconstrained objective function and compare the results from the two approaches.

9.11 In a water flow system, the total flow rate Y is given in terms of two variables x and y as

$$Y = 8.5x^2 + 7.1y^3 + 25$$

with a constraint due to mass balance as

$$x + y^{1.75} = 32$$

Solve this optimization problem both as a constrained problem and as an unconstrained problem, using any appropriate search method for the purpose.

9.12 The heat loss Q from a furnace depends on the temperature a and the wall thickness b as

$$Q = \frac{a^2}{2} + \frac{4}{ab} + 2b$$

Starting with the initial point (1,1), use the univariate search method to obtain the optimum value of Q. Also, apply the method of steepest ascent to obtain the optimum. Is it a maximum or a minimum?

9.13 The cost of a thermal system is given by the expression

$$C=(3.3x^2+4y^2)+\left(\frac{1400}{x}+\frac{1500}{y}\right)$$

where x and y are the sizes of two components. The terms within the first parentheses represent the capital costs and the terms within the second parentheses quantify the maintenance costs. Using the method of steepest ascent, calculate the values of x and y that optimize the cost.

9.14 In an extrusion process, the diameter ratio x, the velocity ratio y, and the temperature z are the main design variables. The cost function is obtained after including the constraints as

$$C=58\frac{x}{y}+\frac{305}{xz}+3xy+4z$$

Using any suitable optimization technique, obtain the optimal cost or show a few steps toward the minimum.

9.15 Apply the method of searching along a constraint to solve the constrained two-variable problem given in Example 9.5. Compare the results obtained with those from the penalty function method given in the example. Also, present the trial runs needed to obtain the solution.

9.16 Solve the constrained optimization problem considered in Example 9.6 by the penalty function method. Compare the results obtained with those given in the example. Also, compare the computational effort needed by the two methods.

9.17 Solve the constrained optimization problem given in Problem 8.12 by any appropriate search method given in this chapter.

9.18 Solve Problem 8.6 as a constrained optimization problem by the hemstitching method.

10 Geometric, Linear, and Dynamic Programming and Other Methods for Optimization

Several optimization methods are applicable only for certain types of functions or for specific problems. In the former category are methods such as *geometric* and *linear programming*. As mentioned in Chapter 7, geometric programming can be employed for problems in which the objective function and the constraints can be represented as sums of polynomials. Linear programming is applicable when these can be represented as linear combinations of the independent variables. In the second category are techniques such as *dynamic programming* and those for optimizing form, shape, and structure. Dynamic programming is applicable to continuous processes that can be represented by a sequence of stages or steps so that the optimum path may be determined. Shape and structural optimization focus on varying the geometrical form or configuration of an item to obtain the optimum characteristics for a given application.

All these optimization methods have seen a considerable increase in interest and activity in recent years. This has been mainly due to growing global competition, the advent of new and diverse technological fields, new demands being placed on technology, and increasing computational power. However, except for geometric programming, many of these methods are generally not easily applicable to practical thermal systems and have been employed mainly in other fields such as communications, transportation, aircraft structures, and construction. Therefore, this chapter presents a detailed discussion of geometric programming and a brief outline of the other techniques for the sake of completeness and for indicating recent trends in optimization. In addition, a few specific problems as well as modifications in the problem formulation for some thermal systems may allow the use of these specialized techniques.

10.1 GEOMETRIC PROGRAMMING

Geometric programming can be used to solve problems that are characterized by nonlinear functions and it is, therefore, a nonlinear optimization technique. Because thermal systems are often nonlinear in character, geometric programming is a useful method for optimizing these systems. The method, as presented here, is very easy to apply, because it involves the solution of linear equations, rather than nonlinear equations that have to be solved for the calculus methods of optimization. It first yields the optimum objective function, which is then used to determine the independent variables at the optimum. The relative contributions of the various terms in the objective function are also obtained, indicating dominant as well as negligible aspects. However, the method is convenient and easy to use only if certain conditions are met, as outlined in the next section. If these conditions are not satisfied, it is best to go to some other optimization technique, though in some cases it may be possible to modify the problem formulation in order to satisfy these conditions.

10.1.1 Applicability

Geometric programming is applicable to problems in which both the objective function and the constraints can be expressed as sums of polynomials of the independent variables. The exponents of the variables can be integer or noninteger, positive or negative, quantities. A few examples of the objective function, $U(x_1, x_2, \ldots, x_n)$, in unconstrained problems, which can be treated by geometric programming are

$$U = 2x_2^{3/2} + x_1^2 + 7x_1x_2^{1/3} \tag{10.1}$$

$$U = 4 + 3x_1 - 2x_1^{1.6} \tag{10.2}$$

$$U = 550 + 105x_1 + \frac{120,000}{x_1^3} - 10x_1^{1.4} \tag{10.3}$$

$$U = x_1x_2^2 + 3\frac{x_1}{x_3^{1.3}} - 3.7x_2x_3^2 + 8x_2^{2.2} \tag{10.4}$$

Similarly, for constrained problems, the following form is suitable for geometric programming:

$$U = 4x_1^2x_2^3 + 8x_1^{-1/2}x_2^{-1/3} - 6x_1^{0.6} + 3x_2 \tag{10.5}$$

with the constraint

$$x_1x_2^{1.2} = 20 \tag{10.6}$$

Therefore, fractions or integers, positive or negative, exponents and coefficients may be considered. Because such polynomial or power-law representations are quite common in thermal systems, particularly from curve fitting of experimental or simulation results, as seen in Chapter 3, geometric programming is a useful technique for optimizing these systems.

10.1.1.1 Degree of Difficulty

An important consideration that determines how the method is to be applied and whether it can be used to yield the optimum directly and without detailed analysis is the *degree of difficulty D*, which is defined as

$$D = N - (n+1) \tag{10.7}$$

where N is the total number of terms in the objective function and in the constraints and n is the number of independent variables. Because the addition of a constant to the objective function does not affect the location of the optimum, only the terms containing polynomials are counted. For instance, the degree of difficulty for the problem given by Equation (10.1) is $D = 3 - (2+1) = 0$, because there are three polynomial-containing terms and two independent variables x_1 and x_2. Similarly, the degree of difficulty D for Equation (10.2), Equation (10.3), and Equation (10.4) are obtained as, respectively, 0, 1, and 0. For the problem given by Equation (10.5) and Equation (10.6), there are five polynomial terms in the objective function and the constraint and only two variables, resulting in a degree of difficulty D of 2.

Geometric programming is particularly useful when the degree of difficulty is zero. In this case, the optimum value of the objective function can be written right away, without resorting to any analysis, and the independent variables can be derived from this result. It is then a fairly simple method to use. In some cases, if D is not zero, terms may be combined to reduce the degree of difficulty to zero. We shall consider only the circumstance of $D = 0$ here because if D is not zero and cannot be reduced to zero, the application of geometric programming involves the solution of nonlinear equations and becomes quite complicated. Other techniques such as search and calculus methods may be easier to use in this case. For further details, see Duffin et al. (1967), Zener (1971), Beightler and Phillips (1976), Wilde (1978), Stoecker (1989), Chiang (2005), Boyd et al. (2007), and Islam and Mandal (2019).

10.1.2 UNCONSTRAINED OPTIMIZATION

Let us first consider the application of geometric programming to unconstrained optimization problems. Because we are interested in problems with degree of difficulty zero, the number of terms must be greater than the number of variables by one.

10.1.2.1 Single Independent Variable

The objective function U may be written in terms of the independent variable x as

$$U = Ax^a + Bx^b \tag{10.8}$$

The two terms may be denoted as u_1 and u_2, where $u_1 = Ax^a$ and $u_2 = Bx^b$. Therefore, these terms represent the individual contributions to the overall objective function. For instance, the cost of producing an item using a manufacturing system may be taken as the sum of the contributions due to initial and operating costs, both of which are functions of the capacity or size x of the system. As the system size increases, the initial costs increase, but the operating and maintenance costs per item decrease, as shown in Figure 10.1.

According to geometric programming, the optimum value of the objective function U^* in the preceding equation is given by the expression

$$U^* = \left(\frac{Ax^a}{w_1}\right)^{w_1}\left(\frac{Bx^b}{w_2}\right)^{w_2} \tag{10.9}$$

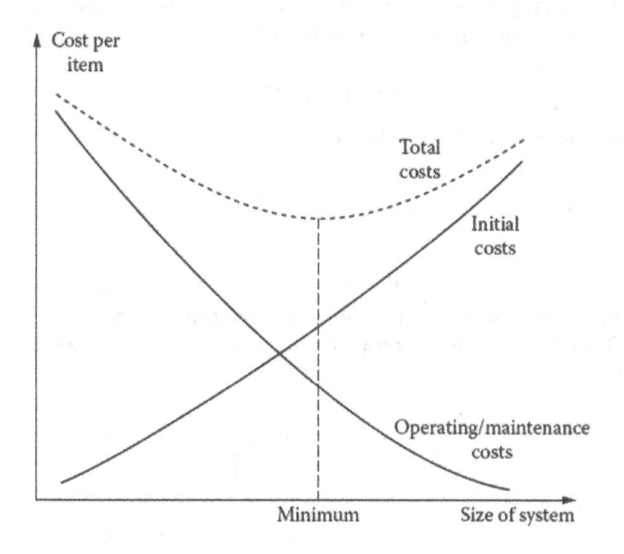

FIGURE 10.1 Variation of initial and operating costs involved in producing an item with the size of the system, indicating a minimum cost per item.

with

$$w_1 + w_2 = 1 \tag{10.10}$$

and

$$aw_1 + bw_2 = 0 \tag{10.11}$$

where w_1 and w_2 are parameters to be determined from Equation (10.10) and Equation (10.11). The latter equation results in the elimination of the independent variable x in Equation (10.9), yielding

$$U^* = \left(\frac{A}{w_1}\right)^{w_1}\left(\frac{B}{w_2}\right)^{w_2} \tag{10.12}$$

Also, it will be shown later, in Section 10.1.3, that

$$w_1 = \frac{u_1^*}{u_1^* + u_2^*} = \frac{u_1^*}{U^*} \tag{10.13}$$

$$w_2 = \frac{u_2^*}{u_1^* + u_2^*} = \frac{u_2^*}{U^*} \tag{10.14}$$

Therefore, w_1 and w_2 represent the weights or relative contributions of the two terms to the total objective function at the optimum. This is an important piece of information obtained in geometric programming and may be used for further improvements in system design. From the preceding equations for w_1 and w_2, the independent variable x at the optimum may be determined from $u_1 = Ax^a$ or $u_2 = Bx^b$. Let us consider a simple problem to illustrate this procedure.

Example 10.1

For the cost function C given in Example 8.1 for a metal rolling process, determine the minimum cost and the corresponding mass flow rate $\dot{m}$ using geometric programming. Compare the results with those obtained earlier using the calculus method.

SOLUTION

The objective function is the cost C given by the expression

$$C = 3.5\dot{m}^{1.4} + \frac{14.8}{\dot{m}^{2.2}}$$

with $\dot{m}$ as the mass flow rate, which is the independent variable. The two terms in the objective function for this unconstrained problem are polynomials and the degree of difficulty $D = 2 - (1 + 1) = 0$. Therefore, geometric programming may be employed easily to write down the optimum cost C^* as

$$C^* = \left(\frac{3.5}{w_1}\right)^{w_1}\left(\frac{14.8}{w_2}\right)^{w_2}$$

with

$$w_1 + w_2 = 1$$

and

$$1.4w_1 - 2.2w_2 = 0$$

From these equations,

$$w_1 = 0.611 \quad \text{and} \quad w_2 = 0.389.$$

Therefore,

$$C^* = (3.5/0.611)^{0.611}(14.8/0.389)^{0.389} = 11.965$$

Then, from Equation (10.13) and Equation (10.14),

$$3.5\dot{m}^{1.4} = 0.611C^*$$

which gives

$$\dot{m} = (0.611C^*/3.5)^{1/1.4} = 1.692$$

These values are close to those obtained earlier in Example 8.1 by using a calculus-based optimization technique. The minimum cost is written directly, without any mathematical analysis; the factors w_1 and w_2 are obtained by solving two linear equations; and the mass flow rate $\dot{m}$ is calculated from Equation (10.13). The values of w_1 and w_2 indicate the contributions of the two terms to the objective function at the optimum. Therefore, the first term contributes 61.1% and the second term contributes 38.9%, indicating the dominance of the first term, which represents the equipment costs. The second term represents operating costs. Therefore, if further reduction in cost is desired, it will be best to focus on equipment costs.

Example 10.2

In a system for providing hot water for industrial use, the heating unit has a power input of 150 kW and a thermal efficiency of $100(0.2 + 0.045T^{0.5})$, in percent, where T is the operating temperature difference from the ambient temperature in degrees centigrade. The rate of heat loss to the environment, in kW, is represented by the expression $0.12T^{1.25}$. Formulate the optimization problem to maximize the rate of energy supplied to the industry and obtain the optimum by using geometric programming. Also, solve the problem by minimizing the energy loss and show that the results obtained are the same as before.

SOLUTION

The objective function E is the rate of energy supplied to the company and is obtained by subtracting the rate of heat loss from the net energy input rate, which is itself a product of the input power and the efficiency. Therefore,

$$E = 150(0.2 + 0.045T^{0.5}) - 0.12T^{1.25} = 30 + 6.75T^{0.5} - 0.12T^{1.25}$$

Because the constant does not affect the location of the optimum, we can optimize the function

$$U = 6.75T^{0.5} - 0.12T^{1.25}$$

As seen in Figure 10.2, the first term increases as T increases. The second term, which is negative, also increases in magnitude as T increases. The sum of the two terms indicates a maximum in the domain of interest. Such problems are commonly encountered in thermal systems.

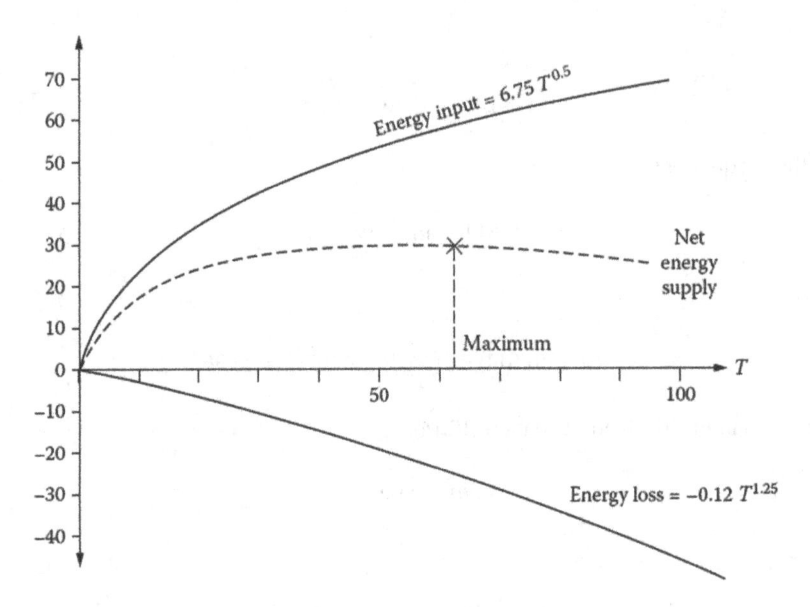

FIGURE 10.2 Variation of energy input and heat loss with temperature in Example 10.2, along with the rate of net energy supply.

The degree of difficulty $D = 2 - (1 + 1) = 0$, because there are two terms and one independent variable. Therefore, the optimum is given by

$$U^* = \left(\frac{6.75}{w_1}\right)^{w_1}\left(\frac{-0.12}{w_2}\right)^{w_2}$$

with

$$w_1 + w_2 = 1 \quad \text{and} \quad 0.5w_1 + 1.25w_2 = 0$$

From these equations, $w_1 = 1.667$ and $w_2 = -0.667$. Therefore,

$$U^* = \left(\frac{6.75}{1.667}\right)^{1.667}\left(\frac{-0.12}{-0.667}\right)^{-0.667} = 32.31$$

The operating temperature T is obtained from the equation for w_1, i.e.,

$$w_1 = \frac{6.75T^{0.5}}{U^*} = 1.667$$

which gives

$$T = \left(\frac{1.667 \times 32.31}{6.75}\right)^2 = 63.68$$

We can now add the constant dropped from the original objective function to give the maximum rate of energy supplied as $30 + 32.31 = 62.31$ kW.

Similarly, the rate of energy loss E_l can be written as

$$E_l = \text{Power input} - \text{Rate of energy supplied to the industry}$$
$$= 150 - (30 + 6.75T^{0.5} - 0.12T^{1.25}) = 120 + 0.12T^{1.25} - 6.75T^{0.5}$$

Then the objective function to be minimized may be taken as

$$U_l = 0.12T^{1.25} - 6.75T^{0.5}$$

Following the procedure just given, the optimum objective function is obtained as

$$U_l^* = \left(\frac{0.12}{w_1}\right)^{w_1}\left(\frac{-6.75}{w_2}\right)^{w_2}$$

where calculations yield $w_1 = -0.667$ and $w_2 = 1.667$, in this case. Therefore,

$$U_l^* = \left(\frac{0.12}{-0.667}\right)^{-0.667}\left(\frac{-6.75}{1.667}\right)^{1.667}$$

Even though the quantities within the parentheses are negative, resulting in complex numbers, the negative sign from both the terms may be combined to yield

$$U_l^* = (-1)\left(\frac{0.12}{0.667}\right)^{-0.667}\left(\frac{6.75}{1.667}\right)^{1.667} = -32.31$$

This gives the minimum rate of energy loss as 120 – 32.31 = 87.69 kW, and thus the maximum rate of energy supply as 150 – 87.69 = 62.31 kW, as before.

This example shows that the w's can be positive or negative and may also be larger than 1.0 in order to satisfy the governing equations. If negative terms appear within the parentheses, leading to complex numbers, the negative sign is extracted to yield –1.0 as a coefficient. Therefore, different forms of the objective function may be optimized by this approach. This problem can also be solved easily by calculus methods to obtain values very close to those given here. However, geometric programming yields the contributions of the two terms to the objective function at the optimum. Thus, the heat input along with the thermal efficiency of the heating unit contribute 1.667, whereas the heat losses contribute –0.667, indicating the greater effect of the thermal efficiency as well as the expected negative effect of heat losses on energy supply.

10.1.2.2 Multiple Independent Variables

The preceding procedure for optimizing an unconstrained single-variable problem can easily be extended to unconstrained multiple-variable problems. However, as before, the objective function must consist of terms that are polynomials or power-law variations and the degree of difficulty must be zero. The weighting factors w_1, w_2, w_3, etc., are introduced with each term and equations are written to ensure that the sum of all the w's is unity and that the independent variables are eliminated from the expression for the optimum value of the objective function. Therefore, the objective function is written as

$$U(x,y) = u_1 + u_2 + u_3 \tag{10.15}$$

where u_1, u_2, and u_3 are polynomials in terms of two independent variables x and y. Because there are three terms in the objective function and two independent variables, the degree of difficulty is $D = 3 - (2 + 1) = 0$. Then the optimum value of the objective function may be written as

$$U^* = \left(\frac{u_1}{w_1}\right)^{w_1}\left(\frac{u_2}{w_2}\right)^{w_2}\left(\frac{u_3}{w_2}\right)^{w_3} \tag{10.16}$$

with

$$w_1 + w_2 + w_3 = 1 \tag{10.17}$$

Two other equations for w_1 and w_2 are obtained from Equation (10.16) by equating the sum of the exponents of each variable to zero in order to eliminate these variables from the optimum value of U. Thus, we have three linear equations that may easily be solved for w_1, w_2, and w_3. Then U^* is computed and the independent variables obtained from the equations

$$w_1 = \frac{u_1}{U^*} \quad w_2 = \frac{u_2}{U^*} \quad w_3 = \frac{u_3}{U^*} \tag{10.18}$$

The mechanics of the procedure outlined here are best illustrated by means of examples, which follow.

Example 10.3

In Example 10.2, if the height H of the system is also included as an additional independent variable, the thermal efficiency, in percent, is represented by the expression $100(0.2 + 0.07H\,T^{0.5} - 0.08H^2)$ and the rate of energy loss by $0.15H\,T^{1.25}$. If the power input is still 150 kW, formulate the optimization problem to maximize the rate of energy supply and solve it by geometric programming.

SOLUTION

The rate of energy supply E is given by the expression

$$E = 150(0.2 + 0.07H\,T^{0.5} - 0.08H^2) - 0.15H\,T^{1.25}$$
$$= 30 + 10.5H\,T^{0.5} - 12H^2 - 0.15H\,T^{1.25}$$

Therefore, the objective function U may be taken as

$$U = 10.5H\,T^{0.5} - 12H^2 - 0.15H\,T^{1.25}$$

All three terms are polynomials and the degree of difficulty $D = 3 - (2 + 1) = 0$. Therefore, from geometric programming, the optimum value of the objective function is given by

$$U^* = \left(\frac{10.5HT^{0.5}}{w_1}\right)^{w_1} \left(\frac{-12H^2}{w_2}\right)^{w_2} \left(\frac{-0.15HT^{1.25}}{w_3}\right)^{w_3}$$

with

$$w_1 + w_2 + w_3 = 1$$
$$w_1 + 2w_2 + w_3 = 0$$
$$0.5w_1 + 1.25w_3 = 0$$

The last two equations are obtained by setting the sum of the exponents of H and T equal to zero, respectively, to eliminate these variables from the expression for U^*. This gives $w_1 = 10/3$, $w_2 = -1$, and $w_3 = -4/3$. Therefore,

$$U^* = \left(\frac{10.5}{10/3}\right)^{10/3} \left(\frac{-12}{-1}\right)^{-1} \left(\frac{-0.15}{-4/3}\right)^{-4/3} = 70.30$$

The independent variables H and T may be obtained from the expressions for the weighting factors, given by Equation (10.18). Thus,

$$10.5HT^{0.5} = w_1U^* \quad \text{and,} \quad -12H^2 = w_2U^*$$

From these equations, we obtain $H = 2.42$ m and $T = 85.02$°C at the optimum. The corresponding equation for the last term may also be used as a check on these values. Therefore, the maximum rate of energy supply is 30 + 70.30 = 100.30 kW. It is also seen that the first term in the objective function is the dominant one, with the contributions of the remaining two terms being of similar magnitude and less than half that of the first term. This information can be used in adjusting the design variables to improve the energy supply.

Example 10.4

In a manufacturing system, rectangular boxes of length x, height y, and width z (in meters) and open at the top, as shown in Figure 10.3, are used for storing and conveying material. The cost of material and fabrication is \$150 per unit area in square meters, and the cost of storage varies inversely as the volume xyz, being 10^3 per unit volume in cubic meters. Formulate the optimization problem for minimizing the cost and obtain the optimum by using geometric programming.

SOLUTION

The objective function U is the total cost of material/fabrication and storage/conveying. Therefore, it may be written as

$$U(x,y,z) = 150(xz + 2xy + 2yz) + \frac{1{,}000}{xyz}$$

because the top of area xz is open. All the terms are polynomials and the degree of difficulty $D = 4 - (3 + 1) = 0$. Therefore, the optimum value of the cost is given by

$$U^* = \left(\frac{150}{w_1}\right)^{w_1}\left(\frac{300}{w_2}\right)^{w_2}\left(\frac{300}{w_3}\right)^{w_3}\left(\frac{1{,}000}{w_4}\right)^{w_4}$$

with

$$\begin{aligned} w_1 + w_2 + w_3 + w_4 &= 1 \\ w_1 + w_2 - w_4 &= 0 \\ w_2 + w_3 - w_4 &= 0 \\ w_1 + w_3 - w_4 &= 0 \end{aligned}$$

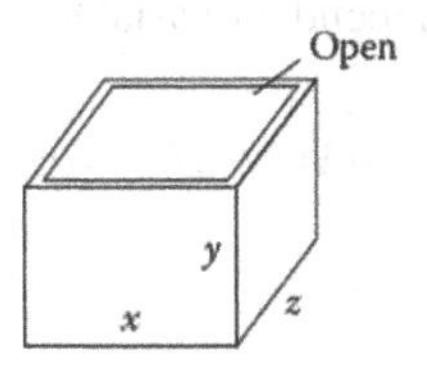

FIGURE 10.3 Rectangular box considered in Example 10.4.

The last three equations ensure that the independent variables x, y, and z are eliminated from the objective function at the optimum, resulting in the preceding expression for U^*. The first equation ensures that the sum of the weighting factors is unity.

This system of linear equations may be solved easily to yield $w_1 = w_2 = w_3 = 1/5$, and $w_4 = 2/5$. This implies that the last term is twice as important as each of the other terms. Therefore, the optimum value of the objective function is

$$U^* = \left(\frac{150}{1/5}\right)^{1/5}\left(\frac{300}{1/5}\right)^{1/5}\left(\frac{300}{1/5}\right)^{1/5}\left(\frac{1{,}000}{2/5}\right)^{2/5} = \$1601.86$$

The independent variables x, y, and z may be obtained from the equations

$$150xz = w_1U^* \quad 300xy = w_2U^* \quad 300yz = w_3U^*$$

which give $x = 1.46$ m, $y = 0.73$ m, and $z = 1.46$ m at the optimum. It can easily be confirmed that the cost obtained is a minimum by changing the variables slightly, away from the optimum.

This problem can also be solved by the use of calculus methods for optimization. However, as mentioned earlier, geometric programming involves the solution of a set of linear equations, whereas calculus methods may require solving nonlinear equations. This is a substantial advantage in most practical problems. In addition, the weighting factors obtained as part of the solution indicate the relative importance of the various terms and can be used to improve the design by focusing on the dominant terms.

10.1.3 Mathematical Proof

The procedure for the application of geometric programming to unconstrained nonlinear optimization problems has been outlined and a few examples have been given to illustrate the use of the method. Let us now consider why this procedure yields the optimum and show that the weighting factors represent the relative contributions of the individual terms in the objective function.

10.1.3.1 Single Variable

The objective function for the simple case of a single independent variable may be written as

$$U = u_1 + u_2 = Ax^a + Bx^b \tag{10.19}$$

Then, using calculus methods, the location of the optimum is found by differentiating this equation and setting the derivative equal to zero, i.e.,

$$\frac{dU}{dx} = aAx^{a-1} + bBx^{b-1} = 0 \tag{10.20}$$

Multiplying this equation by the independent variable x, we get

$$aAx^a + bBx^b = 0$$

or

$$au_1^* + bu_2^* = 0 \tag{10.21}$$

where the asterisks indicate values at the optimum. The preceding simplification is possible only because the two terms are polynomials.

We now define a function F as

$$F = \left(\frac{u_1}{w_1}\right)^{w_1} \left(\frac{u_2}{w_2}\right)^{w_2} \quad (10.22)$$

with

$$w_1 + w_2 = 1 \quad (10.23)$$

To determine the values of w_1 and w_2 at which an optimum in F is obtained, we may apply the Lagrange multiplier method to the function F with a constraint given by Equation (10.23), i.e.,

$$\text{Maximize} \quad \ln F = w_1(\ln u_1 - \ln w_1) + w_2(\ln u_2 - \ln w_2)$$
$$\text{with} \quad G = w_1 + w_2 - 1 = 0$$

Therefore, from the Lagrange multiplier method

$$\nabla(\ln F) + \lambda \nabla G = 0$$
$$G = 0$$

with w_1 and w_2 as the independent variables. Therefore, the equations for the variables w_1, w_2, and λ are obtained as

$$\ln u_1 - \ln w_1 - 1 + \lambda = 0$$
$$\ln u_2 - \ln w_2 - 1 + \lambda = 0$$
$$w_1 + w_2 - 1 = 0$$

These equations may be solved to obtain w_1 and w_2 as

$$w_1 = \frac{u_1}{u_1 + u_2} \qquad w_2 = \frac{u_2}{u_1 + u_2} \quad (10.24)$$

Substituting these expressions in Equation (10.22), we get

$$F = (u_1 + u_2)^{u_1/(u_1+u_2)} (u_1 + u_2)^{u_2/(u_1+u_2)} = u_1 + u_2 \quad (10.25)$$

Therefore, F is made equal to the objective function U by a proper choice of w_1 and w_2, as given by Equation (10.24), so that the optimum value of U is equal to that for F.

Using Equation (10.21) and Equation (10.24), we have

$$w_1 = \frac{-(b/a)u_2^*}{-(b/a)u_2^* + u_2^*} = -\frac{b}{a-b}$$

$$w_2 = \frac{u_2^*}{-(b/a)u_2^* + u_2^*} = \frac{a}{a-b}$$

The optimum value of the objective function is thus obtained as

$$\begin{aligned} U^* = F^* &= \left(\frac{Ax^a}{w_1}\right)^{-b/(a-b)} \left(\frac{Bx^b}{w_2}\right)^{a/(a-b)} \\ &= \left(\frac{A}{w_1}\right)^{-b/(a-b)} \left(\frac{B}{w_2}\right)^{a/(a-b)} \\ &= \left(\frac{A}{w_1}\right)^{w_1} \left(\frac{B}{w_2}\right)^{w_2} \end{aligned} \tag{10.26}$$

It is seen that the independent variable x is eliminated from the optimum value of the objective function. In addition, the weighting factors w_1 and w_2 are shown to indicate the relative contributions of the two terms u_1 and u_2 at the optimum.

10.1.3.2 Multiple Variables

The proof just given can be extended to unconstrained multiple-variable optimizations as long as the number of terms is greater than the number of variables by one (degree of difficulty is zero) and all the terms are polynomials. The optimum of the objective function is obtained by differentiating it with respect to each of the independent variables x_i, in turn, and setting the derivative equal to zero. If each of these equations is multiplied by the corresponding x_i, the resulting system of equations is of the form

$$\begin{aligned} a_{11}u_1^* + a_{12}u_2^* + a_{13}u_{31}^* + \ldots + a_{1,n+1}u_{n+1}^* &= 0 \\ a_{21}u_1^* + a_{22}u_2^* + a_{23}u_{31}^* + \ldots + a_{2,n+1}u_{n+1}^* &= 0 \\ &\vdots \\ a_{n1}u_1^* + a_{n2}u_2^* + a_{n3}u_{31}^* + \ldots + a_{n,n+1}u_{n+1}^* &= 0 \end{aligned} \tag{10.27}$$

Thus, there are n independent variables and $n + 1$ terms. The coefficients a_{ij} are the exponents, which appear as coefficients due to differentiation. By forming a function $F(x_1, x_2, \ldots, x_n)$ as done in Equation (10.22) for a single variable and optimizing ln F, subject to

$$w_1 + w_2 + \ldots + w_{n+1} = 1 \tag{10.28}$$

we get

$$w_i = \frac{u_i}{U} = \frac{u_i}{\Sigma u_i} \quad \text{for} \quad i = 1, 2, \ldots, n+1 \tag{10.29}$$

When these equations are employed with Equation (10.27), the independent variables x_i are eliminated from the optimum value of the objective function U. Therefore, the optimum and the weighting factors are obtained by the geometric programming procedure outlined and applied earlier.

It is seen that the weighting factors depend only on the exponents, not on the coefficients in the various terms. This means that the relative importance of each term remains unchanged as long as the exponents are the same. However, the optimum value and its location will change if the coefficients vary, for instance, because of changes in cost per unit item, energy consumption, etc. The exponents represent the dependence of the objective function on the different variables and are often fixed for a given system or process.

10.1.4 Constrained Optimization

Geometric programming can also be used for optimizing systems with equality constraints. The degree of difficulty is again taken as zero, so that the total number of polynomial terms in the objective function and the constraints is greater than the number of independent variables by one. Let us consider the constrained optimization problem given by the objective function

$$U = u_1 + u_2 + u_3 \tag{10.30}$$

subject to the constraint

$$u_4 + u_5 = 1 \tag{10.31}$$

with x_1, x_2, x_3, and x_4 as the four independent variables. The unity on the right-hand side of Equation (10.31) can be obtained by normalizing the equation if a quantity other than unity appears in the equation, which is often the case. Following the treatment given in the preceding section, the objective function and the constraint may be written as

$$U = \left(\frac{u_1}{w_1}\right)^{w_1}\left(\frac{u_2}{w_2}\right)^{w_2}\left(\frac{u_3}{w_3}\right)^{w_3} \tag{10.32}$$

with

$$w_1 + w_2 + w_3 = 1 \quad \text{and} \quad w_i = \frac{u_i}{U}$$

$$1 = \left(\frac{u_4}{w_4}\right)^{w_4}\left(\frac{u_5}{w_5}\right)^{w_5} \tag{10.33}$$

as well as

$$w_4 + w_5 = 1 \quad \text{and} \quad w_4 = \frac{u_4}{1},\, w_5 = \frac{u_5}{1}$$

Equation (10.33) may be raised to the power of an arbitrary constant p, and the objective function may be written as

$$U = \left(\frac{u_1}{w_1}\right)^{w_1}\left(\frac{u_2}{w_2}\right)^{w_2}\left(\frac{u_3}{w_3}\right)^{w_3}\left(\frac{u_4}{w_4}\right)^{pw_4}\left(\frac{u_5}{w_5}\right)^{pw_5} \tag{10.34}$$

Now, we may apply the method of Lagrange multipliers to obtain the optimum. The corresponding equations are

$$\nabla(u_1 + u_2 + u_3) + \lambda\nabla(u_4 + u_5) = 0$$
$$u_4 + u_5 = 1$$

Again, as was done in the preceding section, the derivatives are taken with respect to the independent variables x_i, one at a time, and the resulting equations multiplied by x_i.

The constant p is arbitrary and can be taken as λ/U^*. Then the equations for the w's are obtained as

$$
\begin{aligned}
a_{11}w_1 + a_{12}w_2 + a_{13}w_3 + pa_{14}w_4 + pa_{15}w_5 &= 0 \\
a_{21}w_1 + a_{22}w_2 + a_{23}w_3 + pa_{24}w_4 + pa_{25}w_5 &= 0 \\
&\vdots \\
a_{41}w_1 + a_{42}w_2 + a_{43}w_3 + pa_{44}w_4 + pa_{45}w_5 &= 0
\end{aligned}
\tag{10.35}
$$

with

$$w_1 + w_2 + w_3 = 1 \quad \text{and} \quad p(w_4 + w_5) = p$$

These linear equations may be solved for w_1, w_2, w_3, w_4, w_5, and p. Therefore, Equation (10.34) gives the optimum value of the objective function and the independent variables are obtained from the expressions for the weighting factors, as was done before. The sensitivity coefficient $S_c = -\lambda = -pU^*$ and has the same physical interpretation as discussed in Chapter 8 for the Lagrange multiplier method, i.e., it is the negative of the rate of change in the optimum with respect to a change in the adjustable parameter E in the constraint $G = g - E = 0$. The preceding approach may be extended easily to more than one constraint as long as the degree of difficulty is zero. The following examples illustrate the use of the method for constrained optimization.

Example 10.5

For the problem considered in Example 10.4, minimize the cost of material and fabrication of the box for a given total volume of 5 m^3, using geometric programming.

SOLUTION

The costs of the material and fabrication vary directly as the total surface area of the rectangular container, which is open at the top. Therefore, the objective function U may be taken as the area, given by

$$U(x, y, z) = xz + 2xy + 2yz$$

with the constraint due to the total volume given as

$$xyz = 5$$

In order to apply geometric programming, the constraint is written as

$$0.2(xyz) = 1$$

All the four relevant terms in the objective function and in the constraint are polynomials and the number of independent variables is three. Therefore, the degree of difficulty $D = 4 - (3 + 1) = 0$.

From geometric programming for constrained optimization, the optimum value of the objective function may be written as

$$U^* = \left(\frac{xz}{w_1}\right)^{w_1} \left(\frac{2xy}{w_2}\right)^{w_2} \left(\frac{2yz}{w_3}\right)^{w_3} \left(\frac{0.2xyz}{w_4}\right)^{pw_4}$$

In order to eliminate the independent variables x, y, and z from the preceding equation for the objective function at the optimum, we have, respectively,

$$w_1 + w_2 + pw_4 = 0$$
$$w_2 + w_3 + pw_4 = 0$$
$$w_1 + w_3 + pw_4 = 0$$

Also,

$$w_1 + w_2 + w_3 = 1$$

and

$$pw_4 = p$$

This system of linear equations may be solved easily to yield

$$w_1 = w_2 = w_3 = \frac{1}{3} \qquad w_4 = 1 \qquad \text{and} \qquad pw_4 = -\frac{2}{3}$$

Therefore, the optimum value of the objective function is obtained as

$$\begin{aligned} U^* &= \left(\frac{1}{w_1}\right)^{w_1}\left(\frac{2}{w_2}\right)^{w_2}\left(\frac{2}{w_3}\right)^{w_3}\left(\frac{0.2}{w_4}\right)^{pw_4} \\ &= \left(\frac{1}{1/3}\right)^{1/3}\left(\frac{2}{1/3}\right)^{1/3}\left(\frac{2}{1/3}\right)^{1/3}\left(\frac{0.2}{1}\right)^{-2/3} \\ &= 13.92 \text{ m}^2 \end{aligned}$$

The independent variables are obtained from the equations

$$xz = w_1U^* = \frac{1}{3}U^* \quad 2xy = w_2U^* = \frac{1}{3}U^* \quad 2yz = w_3U^* = \frac{1}{3}U^*$$

Therefore, these equations are solved to obtain $x = 2.15$ m, $y = 1.08$ m, and $z = 2.15$ m at the optimum. Again, it can be confirmed that the area obtained at the optimum is a minimum by calculating U for small changes in x, y, and z from the optimum values. This simple example illustrates the use of geometric programming for constrained nonlinear optimization. Even though the requirements of polynomial expressions and zero degree of difficulty limit the applicability of this approach, the method is useful in a variety of problems, particularly in thermal systems, where polynomials are frequently used to represent the characteristics.

Example 10.6

In a hot-rolling process, the cost C of the system is a function of the dimensionless temperature T, the thickness ratio x, and the velocity ratio y, before and after the rolls, and is given by the expression

$$C = 1.5 + 5x^2y + \frac{10}{T^2}$$

subject to the constraints due to mass and energy balance given, respectively, as

$$xy = 1 \quad \text{and} \quad T = \frac{5x}{y}$$

Formulate this optimization problem and apply geometric programming to determine the optimum.

SOLUTION

The constant in the objective function does not affect the optimum and the second constraint must be written in a form suitable for applying geometric programming. Therefore, the optimization problem may be written as

$$U = 5x^2y + \frac{10}{T^2}$$

subject to

$$xy = 1 \quad \text{and} \quad \frac{Ty}{5x} = 1$$

All the terms are polynomials and the degree of difficulty is zero because the total number of polynomial terms is four and the number of variables is three. Therefore, the optimum value of the objective function is given by

$$U^* = \left(\frac{5x^2y}{w_1}\right)^{w_1}\left(\frac{10}{T^2w_2}\right)^{w_2}\left(\frac{xy}{w_3}\right)^{p_1w_3}\left(\frac{Ty}{5xw_4}\right)^{p_2w_4}$$

with the following equations for the unknowns w_1, w_2, w_3, w_4, p_1, and p_2:

$$w_1 + w_2 = 1$$
$$p_1w_3 = p_1$$
$$p_2w_4 = p_2$$
$$2w_1 + p_1w_3 - p_2w_4 = 0$$
$$w_1 + p_1w_3 + p_2w_4 = 0$$
$$-2w_2 + p_2w_4 = 0$$

where the last three equations ensure that x, y, and T, respectively, are eliminated from the expression for U^*. These equations are solved to yield $w_1 = 0.8$, $w_2 = 0.2$, $w_3 = w_4 = 1$, $p_1 = -1.2$, and $p_2 = 0.4$. This gives

$$U^* = \left(\frac{5}{0.8}\right)^{0.8}\left(\frac{10}{0.2}\right)^{0.2}\left(\frac{1}{1}\right)^{-1.2}\left(\frac{1}{5}\right)^{0.4} = 4.976$$

Therefore, the optimum cost $C^* = 1.5 + 4.976 = 6.476$. Employing the equations $5x^2y = w_1U^*$ and $10/T^2 = w_2U^*$, along with the constraints, we obtain $x = 0.796$, $y = 1.256$, and $T = 3.170$. The two Lagrange multipliers are $\lambda_1 = p_1U^* = -5.971$ and $\lambda_2 = p_2 U^* = 1.99$, yielding corresponding sensitivity coefficients as $(S_c)_1 = -\lambda_1$ and $(S_c)_2 = -\lambda_2$. Therefore, the first constraint is more important and an increase of 0.1 in the constant, which is unity, in the constraint will increase the dimensionless cost by 0.5971. Similarly, an increase of 0.1 in the constant in the second constraint decreases the cost by 0.199. This information can be used to adjust the design variables for convenience and to use readily available items for the final design.

10.1.5 Nonzero Degree of Difficulty

For the application of geometric programming to the optimization of systems, we have considered only those cases where the degree of difficulty D is zero. For this particular circumstance, the method requires the solution of linear equations and, consequently, provides a simple approach for optimization. However, there are obviously many problems for which the degree of difficulty is not zero, as can be seen from the examples discussed in preceding chapters. If the degree of difficulty is higher than zero, geometric programming can be used, but it involves solving a system of nonlinear equations. This considerably complicates the solution and it is then probably best to use some other optimization technique. Efficient computational algorithms may also be developed for solving such nonlinear systems, as discussed earlier in Chapter 4. Then geometric programming may be employed for a broader range of problems than if we are constrained to problems with zero degree of difficulty. Inequality constraints can also be converted into equality constraints, as discussed in earlier chapters, for applying this method of optimization.

Despite the possibility of solving problems with degree of difficulty greater than zero, geometric programming is clearly best suited to cases where it is zero. Therefore, effort is often directed at reducing the problem with a nonzero degree of difficulty to one with zero degree of difficulty. One technique of achieving this is *condensation,* in which terms of similar characteristics may be combined to reduce the number of terms. For instance, in the rectangular container problem of Example 10.4, if an additional term $200z$ arises due to side supports to the box, the objective function becomes

$$U(x,y,z) = 150(xz + 2xy + 2yz) + \frac{1{,}000}{xyz} + 200z \qquad (10.36)$$

The degree of difficulty is one in this case. However, we may combine two terms, say the first and last, to reduce the degree of difficulty to zero. Writing these terms according to the geometric programming approach, we have

$$\left(\frac{150xz}{1/2}\right)^{1/2}\left(\frac{200z}{1/2}\right)^{1/2} = \left(120{,}000xz^2\right)^{1/2} = 346.41x^{0.5}z \qquad (10.37)$$

where the two terms have been taken to be of equal importance. With this combined term, the degree of difficulty becomes zero and the approach given in this chapter may be applied. Similarly, other terms may be combined to make the degree of difficulty zero. In some cases, information on the physical characteristics of the system may be used to eliminate relatively unimportant terms. The number of independent variables may also be reduced by holding one or more constant for the optimization in order to bring the degree of difficulty to zero. All such techniques and procedures expand the application of geometric programming. For additional information on geometric programming, the references given earlier may be consulted.

10.2 LINEAR PROGRAMMING

Linear programming is an important optimization technique that has been applied to a wide range of problems, particularly to those in mathematics, economics, industrial engineering, power transmission, and material flow. This method is applicable if the objective functions as well as the constraints are linear functions of the independent variables. The constraints may be equalities or inequalities. Since its first appearance about 70 years ago, linear programming has found increasing use due to the need to model, manage, and optimize large systems such as those concerned with manufacturing, transportation, energy, and telecommunications (Hadley, 1962; Murtagh, 1981; Dantzig, 1998; Dantzig and Thapa, 2003; Gass, 2010; Karloff, 2006; Vanderbei, 2014). A large number of efficient optimization algorithms for linear programming have been developed and are available commercially as well as in the public domain. For instance, Matlab toolboxes have software that can be easily employed to solve linear programming problems for system or process optimization.

The applicability of linear programming to thermal systems is somewhat limited because of the generally nonlinear equations that represent these systems. However, several problems are concerned with the distribution and allocation of resources in various industries, such as manufacturing and the petroleum industry, which may be solved by linear programming techniques.

In addition, because of the availability of efficient linear programming software, nonlinear optimization problems are solved, in certain cases, by using the following approaches:

1. Using transformations to convert nonlinear terms into linear ones, as discussed in Chapter 3 for curve fitting of exponential, power-law, and other nonlinear variations.
2. Focusing on local regions so that the variations are not large and the terms may be approximated as linear without significant error. Thus, a piecewise linear programming is carried out.
3. Converting the nonlinear problem into a sequence of linear problems, as discussed in Chapter 4 for nonlinear algebraic systems. Iteration is then used, starting with an initial guessed solution, to converge to the optimum.

With the use of these techniques, linear programming may be used effectively for many thermal processes and systems.

A common method of linearization is to use the known values from the previous iteration for the nonlinear terms. For instance, an objective function of the form $U = 3x_1x_2^2 + 4x_2 + x_1^3$ may be linearized as

$$U = 3x_1\left(x_2^2\right)^l + 4x_2 + \left(x_1^3\right)^l$$

where the superscript l indicates values from the previous iteration, the others being from the current iteration. Therefore, the function becomes linear because the quantities within the parentheses are known from the previous iteration. Then, linear programming may be used, with iteration, to obtain the solution. However, despite these efforts, linear programming finds its greatest use in the various areas mentioned previously, rather than in thermal engineering, for which nonlinear optimization techniques are often necessary. Therefore, only the essential features of this optimization technique and a few representative examples are given here. For further details, various references already given may be consulted.

10.2.1 Formulation and Graphical Method

The problem statement for linear programming is given in terms of the objective function and the constraints, which must both be linear functions of the independent variables. Therefore, the objective function that is to be minimized or maximized is written as

$$U\left(x_1, x_2, x_3, \ldots, x_n\right) = b_1x_1 + b_2x_2 + b_3x_3 + \cdots + b_nx_n = \sum_i b_ix_i \tag{10.38}$$

subject to the constraints

$$G_i = \sum_i a_{ij}x_j \quad <, =, or > C_i \tag{10.39}$$

where b_i, a_{ij}, and C_i are constants. The constraints may be equalities or inequalities, with G_i greater or smaller than the constants C_i. There are n variables and m linear equations and/or inequalities that involve these variables. In linear programming, because of inequality constraints, n may be greater than, equal to, or smaller than m, unlike the method of Lagrange multipliers, which is applicable only for equality constraints and for n larger than m. We are interested in finding the values of these variables that satisfy the given equations and inequalities and also maximize or minimize the linear objective function U.

Let us illustrate the application of linear programming with the following problem involving two variables x and y:

$$U(x,y) = 5x + 2y \tag{10.40a}$$

$$4x + 3y \leq 16 \tag{10.40b}$$

$$y - 2x \geq -4 \tag{10.40c}$$

This simple problem can be solved graphically, as sketched in Figure 10.4. The inequality constraints define the feasible region in which the solution must lie. Therefore, the shaded area in the figure represents the feasible domain. The objective function is defined by a family of parallel straight lines intersecting the two axes, with the value of U increasing as one moves away from the origin. For instance, 5x + 2y = 5 gives a straight line joining x = 1, y = 0 and x = 0, y = 2.5. Similarly, 5x + 2y = 10 gives a straight line joining x = 2, y = 0 and x = 0, y = 5, and so on. Therefore, the maximum value of U is obtained by the line that touches point A, which is at the intersection of the two constraints. At this point, $x = 2.8$, $y = 1.6$, and $U = 17.2$. Therefore, the optimum occurs on the boundary of the feasible domain. This is a particular feature of linear programming and most efficient algorithms seek to move rapidly along the boundary, including the axes, to obtain the optimum.

Similarly, the optimum value of U may be obtained for a different set of constraints. For instance, let us replace Equation (10.40c) by

$$x \leq 2, \text{ or } 4x + y \leq 8 \tag{10.41}$$

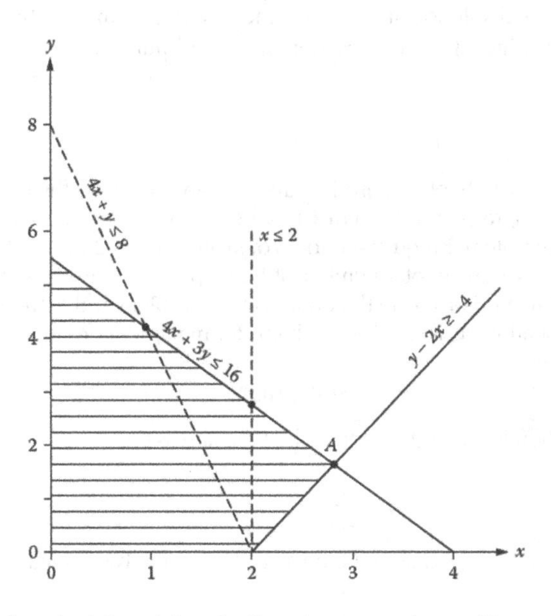

FIGURE 10.4 Graphical method for solving the linear programming problems given by Equation (10.40) and Equation (10.41).

In the first case, the optimum is obtained at $x = 2$ and $y = 8/3$, yielding $U = 46/3$. Again, the optimum is given by the line of constant U passing through the point given by the intersection of the two constraints. In the second case, the optimum is at $x = 1$ and $y = 4$, giving $U = 13.0$. As expected, the optima occur at the boundary of the feasible domain.

10.2.2 Slack Variables

The preceding linear programming problems may also be solved by algebra by converting the inequalities into equalities. As mentioned in Chapter 7, additional constants, known as *slack variables*, may be included in order to ensure that the inequalities are not violated. Thus, by adding a constant s_1, where $s_1 > 0$, to the left-hand side of Equation (10.40b), we can write an equation of the form

$$4x + 3y + s_1 = 16 \tag{10.42}$$

Similarly, Equation (10.40c) may be written as

$$y - 2x - s_2 = -4 \tag{10.43}$$

with $s_2 > 0$. The other inequalities just considered may also be written as equalities by using slack variables. Therefore,

$$x + s_3 = 2, \text{ and, } 4x + y + s_4 = 8 \tag{10.44}$$

The slack variables indicate the difference from the constraint. Therefore, the optimization problem considered here now involves the four variables, x_1, x_2, s_1, and s_2, and there are only two equations, i.e., $n = 4$ and $m = 2$. To find the optimum value of U, two variables, in turn, are set equal to zero and the remaining variables are obtained from a solution of the two equations. For this problem, there are six such combinations, this number being in general $[n!]/[m!(n - m)!]$, where $n!$ is *n factorial.* The optimum is the maximum obtained from these combinations. It can easily be shown that the results given earlier from a graphical solution are also obtained by employing algebra, as outlined here. The following example further illustrates the extraction of the optimum by linear programming.

Example 10.7

A company produces x quantity of one product and y of another, with the profit on the two items being four and three units, respectively. Item 1 requires 1 hour of facility A and 3 hours of facility B for fabrication, whereas item 2 requires 3 hours of facility A and 2 hours of facility B, as shown schematically in Figure 10.5. The total number of hours per week available for the two facilities is 200 and 300, respectively. Formulate the optimization problem and solve it by linear programming to obtain the allocation between the two items for maximum profit.

SOLUTION

The objective function is the profit U, given by the expression

$$U = 4x + 3y$$

The constraints arise due to the maximum time available for the two facilities. Therefore,

$$x + 3y \le 200$$
$$3x + 2y \le 300$$

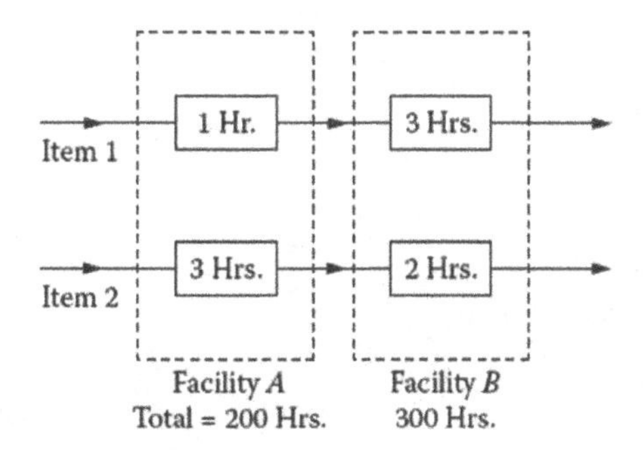

FIGURE 10.5 Schematic showing the utilization of facilities *A* and *B* to manufacture items 1 and 2 in Example 10.7.

Introducing slack variables s_1 and s_2, we have the following two equations:

$$x + 3y + s_1 = 200$$
$$3x + 2y + s_2 = 300$$

Two variables are set equal to zero, in turn, and these equations are solved for the other two variables. The value of the objective function is determined in each case. If s_1 or s_2 is negative, the solution is not allowed, because these were assumed to be positive to satisfy the constraints. The six combinations yield the following results:

x	y	s_1	s_2	U
0	0	200	300	0
0	66.67	0	166.66	200.01
0	150	−250	0	Negative s_1, not allowed
200	0	0	−300	Negative s_2, not allowed
100	0	100	0	400
71.43	42.86	0	0	414.3

Therefore, a maximum value of 414.3 is obtained for U at $x = 71.43$ and $y = 42.86$. This problem can also be solved graphically, as shown in Figure 10.6, to confirm that the optimum arises at the intersection of the two constraints and is thus on the boundary of the feasible region.

10.2.3 Simplex Algorithm

Extensive literature is available on linear programming and many efficient algorithms have been developed over the years for solving large optimization problems consisting of many terms. Most computer systems include software packages for linear programming. Among these is the *simplex* algorithm, which searches through the many possible combinations for the optimum value of the objective function. It is based on the Gauss-Jordan elimination procedure, outlined in Chapter 4, for solving a set of simultaneous linear equations. Therefore, the normalization of the pivot element to obtain unity, as well as eliminating the other coefficients in the column containing the pivot element, are used during the solution procedure.

The method uses a tabular form of data presentation known as a *programming tableau.* The initial tableau is formed by using slack variables s_1 and s_2, as described earlier. For instance, let us consider the linear programming problem given by Equation (10.40a,b,c). Equation (10.42) and Equation (10.43) give the corresponding slack variables. These equations form the initial tableau. The simplex algorithm focuses on the variable among x, y, s_1, and s_2 that affects the solution the

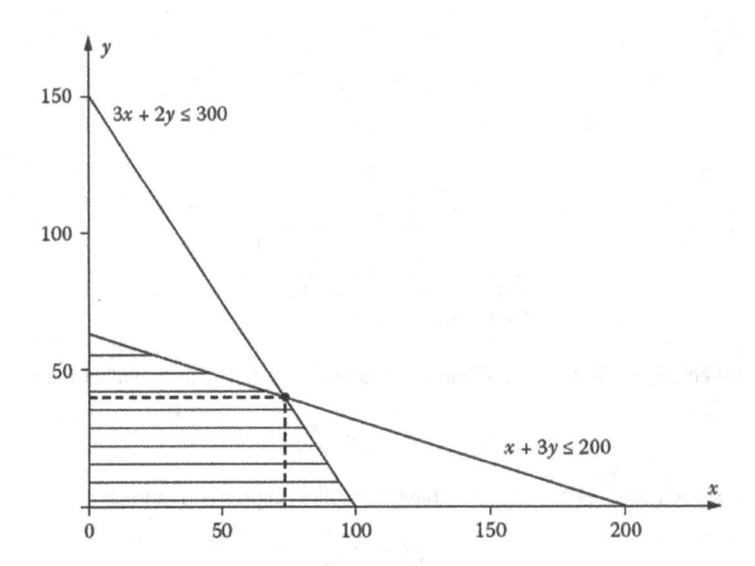

FIGURE 10.6 Graphical solution to the linear programming problem posed in Example 10.7.

most by determining the benefit obtained by adding a unit of each quantity. The one that shows the greatest positive change is used to replace one of the active variables in the previous iteration, the first iteration being initiated with s_1 and s_2 as the first set of active variables. This process is continued, with interchange of variables, until the benefit to the optimum by a change in the variables is small. The optimum solution is then obtained by putting the slack variables equal to zero, i.e., $s_1 = s_2 = 0$. Because the method involves matrix manipulations, it is well-suited for solution on a digital computer. If a minimum of the objective function is sought instead of a maximum, the signs of the coefficients in the given expression for the objective function U are changed. Similarly, the signs of the coefficients of the slack variables in the constraint equations are changed if the constraint is greater than a given quantity instead of being less than it, as discussed earlier. The following example illustrates the use of the simplex algorithm.

Example 10.8

The objective function for an optimization problem is taken as the total income, which involves an income of five units on item *A* and seven units on item *B*. Item *A* requires 2.5 hours of cutting and 1.5 hours of polishing, whereas item *B* requires 4 hours of cutting and 1 hour of polishing. If the total labor hours available for cutting are 4000 and for polishing 2000, formulate the optimization problem and solve it by the simplex algorithm to obtain the optimum.

SOLUTION

The optimization problem reduces to the objective function

$$U = 5x_1 + 7x_2$$

where x_1 and x_2 represent the amounts of items *A* and *B*, respectively, that are produced. The constraint equations are

$$2.5x_1 + 4x_2 \leq 4000$$
$$1.5x_1 + x_2 \leq 2000$$

Because the objective function and the constraints are all linear expressions, linear programming may be used to obtain the solution. In order to use the simplex algorithm, the problem is written in terms of the slack variables as

$$U = 5x_1 + 7x_2 + 0s_1 + 0s_2$$
$$2.5x_1 + 4x_2 + s_1 + 0s_2 = 4000$$
$$1.5x_1 + x_2 + 0s_1 + s_2 = 2000$$

We start by choosing the slack variables s_1 and s_2 as the active variables, with the coefficients k_i for these equal to zero. The coefficients for x_1 and x_2 are 5.0 and 7.0, respectively. The column represented by b, in the following table, contains the constants from the constraint equations. The objective function U_j is evaluated by the equation $U_j = \Sigma k_i a_{ij}$, where a_{ij} represents the coefficients in the matrix. The row C_j (= $k_j - U_j$) gives the improvement in the objective function due to the addition of a unit of each variable. This yields the initial tableau, as follows.

Initial Tableau

	k_j	**5.0**	**7.0**	**0.0**	**0.0**		
Active	k_i	x_1	x_2	s_1	s_2	b	g
s_1	0	2.5	<u>4</u>	1	0	4000	1000
s_2	0	1.5	1	0	1	2000	2000
U_j		0	0	0	0	0	
C_j		5	7	0	0		

The largest positive value for C_j is obtained for x_2. Therefore, x_2 is made an active variable in the next iteration. The column headed by g, in the preceding table, is obtained by dividing the value of b for each row by the value in the pivot column for that row. This indicates the contribution of each row, and the one with the smallest contribution is removed. Therefore, s_1 is dropped from the active variables. The underlined number is the pivot element given by the intersection of the row being removed and the column containing the new active variable.

Now, the Gauss-Jordan elimination procedure, presented in Chapter 4, is used. The pivot element is made 1.0 by dividing all the elements in the pivot row by the value of the pivot element. Elimination is used to make all other elements in the pivot column go to zero. Therefore, the second tableau is obtained as follows.

Second Tableau

	k_j	**5.0**	**7.0**	**0.0**	**0.0**		
Active	k_i	x_1	x_2	s_1	s_2	b	g
x_2	7	5/8	1	1/4	0	1000	1600
s_2	0	<u>7/8</u>	0	−1/4	1	1000	1142.9
U_j		35/8	7	7/4	0	7000	
C_j		5/8	0	−7/4	0		

Following the procedure outlined previously for the third tableau, x_1 is made an active variable, and s_2 is dropped using the results given for the second tableau. The pivot element is underlined. Again, Gauss-Jordan elimination is used to make the pivot element unity and all the other elements in the pivot column zero. This gives the third tableau as follows.

Third Tableau

	k_j	**5.0**	**7.0**	**0.0**	**0.0**	
Active	k_i	x_1	x_2	s_1	s_2	b
x_1	5	1	0	−2/7	8/7	1142.9
x_2	7	0	1	3/7	−5/7	285.7
U_j		5	7	11/7	5/7	7714.4
C_j		0	0	−11/7	−5/7	

This iteration is carried out until C_j values are less than or equal to zero. Because this is the case for the third tableau, this is the optimum condition and we stop at this stage. The resulting equations are

$$x_1 = 1142.9 + \frac{2}{7}s_1 - \frac{8}{7}s_2$$

$$x_2 = 285.7 - \frac{3}{7}s_1 + \frac{5}{7}s_2$$

$$U = 7714.4 - \frac{11}{7}s_1 - \frac{5}{7}s_2$$

Then the optimum solution is obtained by putting $s_1 = s_2 = 0$. This gives $x_1 = 1142.9$, $x_2 = 285.7$, and $U = 7714.4$. It can easily be shown that the results obtained by the graphical method are close to these values. However, the application of the simplex algorithm, as outlined here, is well-suited for digital computation and can be used effectively for large systems.

Efficient procedures are also available for several important practical problems in linear programming. A few examples are indicated in terms of the following problems:

1. *The transportation problem.* This problem concerns the optimum way to distribute an item or product from a number of production plants to a number of warehouses. If a company has several plants with different outputs and several warehouses with different requirements, a shipping pattern may be established that minimizes transportation and manufacturing costs.
2. *The allocation problem.* As the name suggests, this problem deals with the allocation of resources such as machines and workers in order to maximize the output or minimize the costs. If an industrial establishment manufactures several items and each has its own requirements of time on various machines such as milling, grinding, and heat treatment facilities, the time allocated to different products may be varied for optimization. Example 10.7 considered this application to illustrate the use of linear programming.
3. *Critical-path problems.* These problems focus on finding the most efficient path through many tasks that must be carried out to complete a given operation such as the fabrication of a heat exchanger. The tasks that cause excessive delay are determined and other tasks arranged around these to minimize the total time.
4. *The blending problem.* In this case, the inflow of different raw materials and the production of finished products are blended or mixed in such a way that the profit is maximized. An example of this problem arises in oil companies that buy crude oil of different quality from different sources and produce different finished products such as diesel, gasoline, and polymeric materials.

Problems such as the ones just outlined give rise to large linear systems for which linear programming may be applied to minimize costs, maximize profits, or seek the optimum of some other objective function. The simplex algorithm, mentioned earlier, is one efficient scheme that searches along the boundary of the feasible domain from one *vertex*, given by the intersection of constraints, to the next until the optimum is obtained. For large-scale problems, such as telecommunication networks and electric power grids, other specialized and more efficient schemes have been developed. One such scheme is by Karmarkar (1984), which searches for efficient directions while working in the interior of the feasible domain, rather than searching only at the outer boundary. It can thus converge very rapidly to the desired optimum.

10.3 DYNAMIC PROGRAMMING

Several engineering processes consist of a sequence of stages or continuous operations that can be approximated as a series of interconnected steps. In thermal systems, discrete stages such as pumps, compressors, evaporators, and condensers are frequently encountered. The output from one stage is the input to another stage, thus coupling all the stages. In addition, in many cases, such as chemical reactors, the process may be broken down into a series of smaller steps. Figure 10.7 shows a schematic of the various stages or steps in the manufacture of an insulated wire, going from the raw material to a spool of plastic-insulated wire. *Dynamic programming* is an optimization technique that is applicable to such processes and seeks to find the path through these stages or steps that would minimize cost, maximize output, or optimize some other chosen objective function. The word *dynamic* refers to iterative changes in the path and not to the usual connotation of changes with time.

Dynamic programming is quite different from the other optimization techniques such as Lagrange multiplier and geometric programming approaches because it yields an optimal function rather than an optimal point. It is similar to the use of calculus of variations to determine the path that would minimize, for instance, the distance traveled or energy consumed by an automobile in going from one point to another under given constraints. In dynamic programming, the total path is divided into a finite number of smaller steps, and variations in the location and sequence of the steps are used to obtain the optimal path. The method is well-suited to problems that involve a number of activities that can be treated as stages and can be varied in their relative positions in the process. By dividing a large complicated optimization problem into a series of smaller steps, the overall effort to obtain the optimum is reduced (Nemhauser, 1967; Stoecker, 1989; Bellman, 2003; Denardo, 2003; Lew and Mauch, 2006; Bertsekas, 2012) .

In going from one point to another through a sequence of steps, dynamic programming starts with one stage, analyzes it, and determines the optimum corresponding to it. It then proceeds to the next stage, obtains its optimum, and combines it with the previous stage. Optimal plans are established for subsections of the problem. Thus, it does not consider all possible combinations, but uses the optimal plans for the subsections, ignoring other nonoptimal plans. Once an optimum is determined for a particular subsection, it is not repeated for future calculations of other subsections and the final optimum. Figure 10.8 shows a sketch of a typical problem amenable to a solution by dynamic programming. Three possible locations exist for each of the three stations on the transport of material from *A* to *B*. The costs between different points are given and dynamic programming seeks a path through the three stations that would minimize the overall cost. Therefore, if c_i represents the cost in each stage of the transportation of the material from one location to another, dynamic programming seeks to minimize the total cost *C* for *n* stages given by

$$C = \sum_{i=1}^{n} C_i \tag{10.45}$$

Dynamic programming is a useful technique for a variety of engineering and management problems, such as those encountered in plant layouts, transportation networks, pipeline for oil and water distribution, and manufacturing systems. Chemical engineers have extensively used dynamic programming for the design, optimization, and control of chemical reactors and processes. However, the use of this optimization technique for thermal processes and systems is rather limited because it

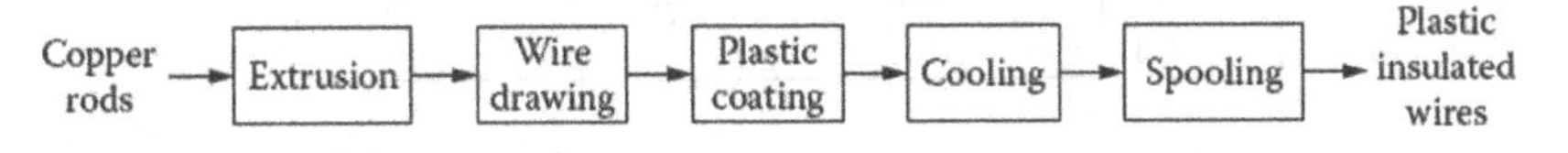

FIGURE 10.7 Discrete stages in the manufacture of a plastic insulated electrical wire.

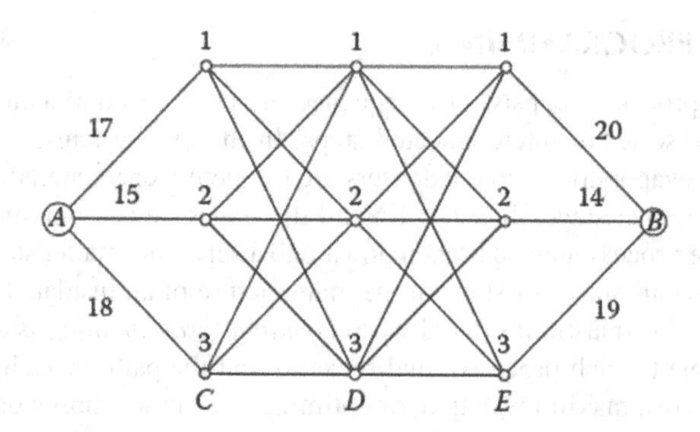

FIGURE 10.8 Use of dynamic programming to minimize the cost of going from point A to point B.

is often difficult to divide continuous processes into steps. When stages do arise, as in heating and cooling systems, the number of stages is often small and is generally not interchangeable or movable within the process.

Example 10.9

Use dynamic programming to find the path for minimum cost of transportation from point A to B in Figure 10.8 while passing through one of the three stations of locations C, D, and E. Employ the costs given in the figure and the following costs for going between the other locations:

1–1	1–2	1–3	2–1	2–2	2–3	3–1	3–2	3–3
10	14	20	14	15	16	20	16	15

SOLUTION

Starting with point A, there are three ways to reach C. However, we cannot eliminate the non-optimal paths at this stage because the combination with the next step may change the optimum. In order to reach $D1$, we can go through $C1$, $C2$, or $C3$, resulting in three different subsections. Similarly, $D2$ and $D3$ can be reached by going through these three stations of C. Therefore, the total cost in going from A to D is given by

Through	Cost to $D1$	Cost to $D2$	Cost to $D3$
$C1$	<u>27</u>	31	37
$C2$	29	<u>30</u>	<u>31</u>
$C3$	38	34	33

The optimal path for each subsection is underlined. We can similarly consider reaching $E1$, $E2$, and $E3$ through $D1$, $D2$, and $D3$. In each case, only the optimal solution for the subsection up to D is used and the others are ignored. This means that the cost up to $D1$ is taken as 27, up to $D2$ as 30, and up to $D3$ as 31. Therefore, the total cost in going from A to E is given by

Through	Cost to $E1$	Cost to $E2$	Cost to $E3$
$D1$	<u>37</u>	<u>41</u>	47
$D2$	44	45	<u>46</u>
$D3$	51	47	<u>46</u>

Again, the optimal solutions for the path up to E are underlined, indicating two choices with same cost for reaching $E3$. The costs from E to B are given in Figure 10.8. When these are added to the optimum costs for reaching the three stations for E, it is seen that the cheapest is through $E2$ and has a total cost of 55. Therefore, the optimal path is given as $A - C1 - D1 - E2 - B$. By using the optimal solutions obtained earlier for the subsections of the overall path, the number of computations is reduced. The total number of combinations is $3 \times 3 \times 3 \times 1 = 27$. The number of calculations needed here is $9 + 9 + 3 = 21$. Clearly, the benefit of dynamic programming in reducing the computational effort will increase as the number of steps is increased.

10.4 OTHER METHODS

Several other optimization techniques have been developed in recent years to meet the growing demand for optimization in many specialized and emerging areas. Some of these methods are new while others are based on many of the techniques discussed in this and previous chapters. A common circumstance encountered in design of systems is that of the objective function and the constraints increasing or decreasing monotonically with respect to a given design variable. *Monotonicity analysis* is a technique that may be used for such problems to determine which constraints directly affect the optimum and thus locate it. It also indicates how the feasible domain may be modified to improve the design in terms of the objective function (Papalambros and Wilde, 2000).

Another area of optimization that has received a lot of attention lately is that of *shape optimization*. In this optimization problem, the geometry or topology of the item is a variable, rather than just its dimensions. Thus, the shape of the part may be varied to minimize cost, maximize heat transfer rate, minimize weight, and so on, while the given constraints are satisfied. Generally, an initial geometry or shape is chosen and a numerical computation, usually based on the finite element method because of its versatility, is carried out to determine the objective function. The shape is changed iteratively within the feasible domain to optimize the objective function subject to the constraints. The design variables include those that define the boundary or shape of the item under consideration. Such an iterative procedure, with changes in the shape, is possible mainly because of the availability of efficient computational schemes for analysis and fast computers. These ideas have been extended to the optimization of topology, profile, trajectory, and configuration in different types of systems and applications. Though an active area for research in the design of structures, shape optimization has not been used much in thermal systems and processes. Several other methods were outlined in Chapter 7 and some examples were given in Chapter 9. These include response surfaces and multi-objective design optimization. Genetic algorithms, artificial neural networks, and fuzzy logic are other approaches that are used in the optimization process, as discussed in Chapter 7.

Another approach to the optimization of processes and phenomena was proposed by Bejan (2000) in terms of the *constructal law*, which was stated as, "For a finite-size system to persist in time, it must evolve in such a way that it provides easier access to the imposed currents that flow through it." This concept has been used to determine optimum shapes, patterns, and configurations for a wide range of applications. These include heat exchangers, electronic equipment, fluid flow systems, and other thermal processes and systems (Bejan and Lorente, 2008).

10.5 SUMMARY

This chapter presents several optimization methods that are of interest in engineering systems. However, some of these are not very useful for thermal systems, though they are important in other applications. Geometric programming is a nonlinear optimization technique that requires the objective function and the constraints to be sums of polynomials. Because many thermal systems can be represented by polynomials and power-law expressions with exponents that are positive, negative, fractions, or whole numbers, this technique is of particular value in the optimization of these systems. However, the results from analysis or curve fitting of numerical/experimental data must be expressed in the form of polynomials. If the number of polynomial terms in the objective function and the constraints is greater than the number of variables by one, the degree of difficulty is said to be zero, and geometric programming is probably the simplest method to obtain the optimum. If the degree of difficulty is not zero, it is sometimes possible to reduce the problem to one with zero degree of difficulty. Otherwise, it may be best to use some other method.

Linear programming, which is extensively used in industrial engineering, economics, traffic flow, telecommunications, and many other important applications, requires that the objective function and the constraints be linear functions of the variables. Because thermal processes are generally nonlinear, linear programming is not very useful for thermal systems. However, some problems may be linear or the equations may be linearized in some cases, allowing the use of linear programming. The basic approach for solving linear programming problems is discussed using graphical methods and algebra with slack variables. The frequently used simplex algorithm is also presented. The occurrence of the optimum at the domain boundaries is an important feature of these problems, and efficient methods are employed to move rapidly along the boundary or to go from one point on the boundary to the other through the interior region of the domain.

Dynamic programming leads to an optimal function rather than an optimal point. It is applicable to processes that involve several discrete stages or that can be approximated by a series of steps. Thus, it seeks to optimize the path through the various steps. It is a useful technique for plant layout and production planning. In thermal systems, discrete steps, such as compressors, pumps, and turbines, are involved in many cases. However, there is generally little freedom to choose the sequence because the process determines this. In addition, only a few stages are often encountered. Thus, dynamic programming, though important for a variety of problems, is of limited interest in thermal systems. Similarly, other specialized techniques such as structural, shape, and trajectory optimization are outlined. Again, these approaches are of considerable interest in many engineering problems but are of limited use in the optimization of thermal systems. Search methods remain the most important optimization technique for thermal systems and various strategies have been developed to facilitate the use of these methods.

REFERENCES

Arora, J.S. (2004). *Introduction to optimum design* (2nd ed.). New York: Academic Press.

Beightler, C.S., & Phillips, D.T. (1976). *Applied geometric programming.* New York: Wiley.

Bejan, A. (2000). *Shape and structure, from engineering to nature.* Cambridge, UK: Cambridge University Press.

Bejan, A., & Lorente, S. (2008). *Design with constructal theory.* New York: Wiley.

Bellman, R.E. (2003). *Dynamic programming.* Mineola, NY: Dover Publications.

Bertsekas, D.P. (2012). *Dynamic programming and optimal control* (4th ed.). Nashua, NH: Athena Scientific.

Boyd, S., Kim, S.J., Vandenberghe, L., & Hassibi, A. (2007). A tutorial on geometric programming, *Optimization and Engineering, 8*, 67–127.

Chiang, M. (2005). *Geometric programming for communication systems.* Hanover, MA: Now Publishers.

Dantzig, G.B. (1998). *Linear programming and extension.* Princeton, NJ: Princeton University Press.

Dantzig, G.B., & Thapa, M.N. (2003). *Linear programming 2: theory and extensions.* New York: Springer.

Denardo, E.V. (2003). *Dynamic programming: models and applications.* Mineola, NY: Dover Publications.

Duffin, R.J., Peterson, E.L., & Zener, C.M. (1967). *Geometric programming.* New York: Wiley.

Gass, S.I. (2010). *Linear programming: methods and applications* (5th ed.). Mineola, NY: Dover Publications.

Hadley, G.H. (1962). *Linear programming.* Reading, MA: Addison-Wesley.

Islam, S., & Mandal, W.A. (2019). *Fuzzy geometric programming techniques and applications.* Singapore: Springer.

Karloff, H. (2006). *Linear programming.* Heidelberg, Germany: Springer.

Karmarkar, N. (1984). A new polynomial-time algorithm for linear programming. *Combinatorica, 4*(4), 373–395.

Lew, A., & Mauch, H. (2006). *Dynamic programming: a computational tool.* Heidelberg, Germany: Springer-Verlag.

Murtagh, B.A. (1981). *Advanced linear programming.* New York: McGraw-Hill.

Nemhauser, G.L. (1967). *Introduction to dynamic programming*. New York: Wiley.
Papalambros, P.Y., & Wilde, D.J. (2000). *Principles of optimal design* (2nd ed.). New York: Cambridge University Press.
Stoecker, W.F. (1989). *Design of thermal systems* (3rd ed.). New York: McGraw-Hill.
Vanderbei, R.J. (2014). *Linear programmimg: foundations and extensions* (4th ed.). New York: Springer.
Wilde, D.J. (1978). *Globally optimum design*. New York: Wiley-Interscience.
Zener, C.M. (1971). *Engineering design by geometric programming*. New York: Wiley-Interscience.

PROBLEMS

10.1 Solve the unconstrained optimization problem given in Example 8.2 using the geometric programming method. Compare the results obtained with those presented earlier and discuss the advantages and disadvantages of this method over calculus methods.

10.2 Hot water is delivered by pipe systems with flow rates $\dot{m}_1$ and $\dot{m}_2$. The total heat input Q, which is to be minimized, is given as

$$Q = 3(\dot{m}_1)^2 + 4(\dot{m}_2)^2 + 15$$

with the constraint

$$\dot{m}_1\,\dot{m}_2 = 20$$

Use geometric programming to obtain the optimal flow rates. First, recast the problem into an unconstrained circumstance and solve it. Then, solve it as the given constrained problem and compare the two approaches.

10.3 Minimize the cost U of a system, where U is given in terms of the three independent variables x, y, and z as

$$U = 2xy + \frac{2}{xz} - 12y^2 + 3z$$

using geometric programming. Compare the results obtained with those from calculus methods. Comment on the differences between the two methods.

10.4 The cost C in an extrusion process is given in terms of the diameter ratio x, the velocity ratio y, and the temperature z as

$$C = 50\frac{x}{y} + \frac{300}{xz} + 4xy + 5z$$

Using geometric programming for this unconstrained problem, obtain the minimum cost and the values of the independent variables at the optimum.

10.5 The cost C in a metal processing system is given in terms of the speed V of the material as

$$C = \frac{\pi K S^{4/3}}{V^{5/4}\left[2 + 17.5(3/V)^{7/3}\right]^2}$$

where K and S are constants. Using geometric programming, find the speed V at which the cost C is optimum. Is this point a maximum or a minimum?

10.6 Solve the cylindrical storage tank problem given in Example 8.3 by geometric programming as a constrained problem to determine the optimal values of the design variables. Also, calculate the sensitivity coefficient.

10.7 The cost C of a storage chamber varies with the three dimensions x, y, and z as

$$C = 12x^2 + 2y^2 + 5z^2$$

and the volume is given as 10 m^3 so that

$$xyz = 10$$

Using geometric programming, calculate the dimensions that yield the minimum cost. Also, calculate the sensitivity coefficient. What does this quantity mean physically in this problem?

10.8 The heat transfer rate Q from a spherical reactor of diameter D is given by the equation $Q = h\Delta TA$, where h is the heat transfer coefficient, ΔT is the temperature difference from the ambient, and A is the surface area, i.e., $A = \pi D^2$. Here, h is given by the expression

$$h = 4.5D^{-1} + 2.0\ \Delta T^{0.25} D^{-2}$$

A constraint arises from the energy input and is given as

$$\Delta T D^{0.5} = 40$$

Set up the optimization problem for the total heat transfer rate Q. Using geometric programming, find the optimum value of Q and the corresponding diameter. Also, find the sensitivity coefficient.

10.9 The fuel consumption F of a vehicle is given in terms of two parameters x and y, which characterize the combustion process and the drag, as

$$F = 10.5x^{1.5} + 6.2y^{0.7}$$

with a constraint from conservation laws as

$$x^{1.2}y^2 = 20$$

Cast this problem as an unconstrained optimization problem and solve it by the Lagrange multiplier method and by geometric programming. Is it a maximum or a minimum?

10.10 For the problem given in Example 10.5, if the total volume is given as 10 m^3, obtain the resulting optimal conditions and compare them with those presented in the example. Also, calculate the sensitivity coefficient and discuss your results in terms of the value obtained.

10.11 Simplify the problem given in Example 10.6 by reducing the number of constraints to one by elimination. Then apply geometric programming to obtain the optimum and compare the results obtained with those given earlier.

10.12 The cost C of a system consisting of two components is given by the linear expression

$$C = 2x_1 + 6x_2$$

where x_1 and x_2 are independent variables that characterize the two components and must satisfy the constraints

$$x_1 \geq 0$$
$$x_2 \geq 0$$
$$2x_1 + 4x_2 \geq 4$$
$$1.5x_1 + 3x_2 \leq 4.5$$

Solve this problem by linear programming using slack variables, as well as the graphical method, to obtain the optimal value of C and the corresponding values of x_1 and x_2.

10.13 Obtain the solution to the preceding optimization problem by using the simplex algorithm and compare the results with those obtained from the graphical method.

10.14 Use the simplex algorithm to obtain the optimum for the constraints given in Problem 10.12 if the objective function is given, instead, as $C = 3.5x_1 + 4.0x_2$. Comment on the difference in the results from those obtained in the earlier problem.

10.15 Using the simplex method, derive the optimum for the problem posed in Problem 10.12 if the last two constraints are replaced by the inequalities

$$2x_1 + 4x_2 \geq 8.0$$
$$1.5x_1 + x_2 \leq 3.0$$

while the remaining problem remains unchanged.

10.16 Using the graphical linear programming method, determine the variables x_1 and x_2 that yield an optimal value of the objective function

$$U = x_1 + 2.5x_2$$

subjected to the constraints

$$x_1 + 3x_2 \geq 35$$
$$x_1 + x_2 \geq 18$$
$$5.5x_1 + 2.5x_2 \leq 110$$
$$x_1 \geq 0$$
$$x_2 \geq 0$$

10.17 The number of components produced by a company in two different categories are x and y. The objective function is the overall income U, given by

$$U = 1.25x + 1.75y$$

subjected to the constraints

$$x + y \leq 450$$
$$2x + 6y \leq 750$$
$$4x + 7y \leq 1480$$

Solve this problem by any linear programming method to obtain the optimum number of components produced.

10.18 Determine whether your conclusions will be affected if, in Example 10.9, all the costs given for going between the following locations are all increased by a fixed amount of 2; i.e.,

1–1	1–2	1–3	2–1	2-2	2–3	3–1	3–2	3–3
12	16	22	16	17	18	22	18	17

while the remaining costs are unchanged. What would happen if these were increased by 4 instead? Discuss your findings.

10.19 Solve the dynamic programming problem shown in Figure 10.8, if the costs involved in going between various locations are

A–1	A–2	A–3	1–B	2–B	3–B
19	20	17	18	16	21

and for the others

1–1	1–2	1–3	2–1	2–2	2–3	3–1	3–2	3–3
12	15	19	15	13	17	18	18	16

10.20 Solve the dynamic programming problem shown in Figure P10.20, with the time (in minutes) between the various locations given as

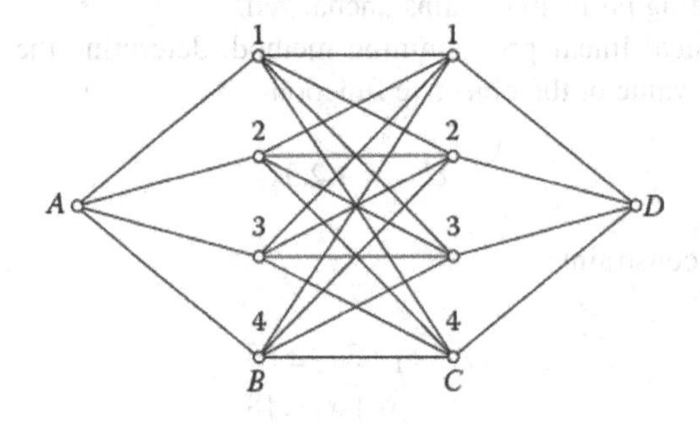

FIGURE P10.20

A–1	A–2	A–3	A–4	1–D	2–D	3–D	4–D
10	14	12	15	16	12	10	8

and

1–1	1–2	1–3	1–4	2–1	2–2	2–3	2–4
16	20	14	17	12	14	15	6
3–1	3–2	3–3	3–4	4–1	4–2	4–3	4–4
10	11	13	18	8	11	14	22

Determine the optimum path that leads to the minimum amount of time in going from *A* to *D*.

11 Knowledge-Based Design and Additional Considerations

The basic approach to the design and optimization of thermal systems has been presented in the preceding chapters. Different steps in the design process, starting with the formulation and the concept, have been presented to obtain acceptable designs, followed by optimization. Many additional aspects that must be included in practice for a successful design have also been outlined. These include economic, safety, environmental, regulatory, legal, and other issues that may be technical or nontechnical in nature. These considerations are of critical importance, because a design that is technically feasible may not be acceptable because of the excessive cost or because it violates regulations regarding safety or the environmental impact.

In this chapter, we will consider *knowledge-based design*, which is a non-traditional design methodology based on experience, informal approaches or heuristics, information on existing systems, and current practice. It includes the knowledge base pertaining to the process or system under consideration, as well as reasoning or inference that can be used to acquire new knowledge and suggest solutions. The main elements of this method and the overall scheme are outlined, followed by examples of a few thermal systems to demonstrate the power and usefulness of this approach.

Also considered in this chapter are some additional important issues with respect to the design of thermal systems, such as professional ethics and other constraints. The sources of information that may be employed to provide inputs for design are also outlined. Some of the important sources for information on material property data, characteristics of components, economic variables, optimization techniques, computer software, etc., are given. An overview of the design and optimization of thermal systems is also presented. This overview serves to put the entire design and optimization process in perspective. Several design projects are included as problems at the end of the chapter to cover the entire process for common thermal systems. Groups of students may use these as projects in design courses that involve design and optimization undertaken over the period of a semester.

11.1 KNOWLEDGE-BASED SYSTEMS

With the extensive growth in computer-based design, considerable effort has been directed at streamlining the design process, improving the design methodology, automating the use of existing information, and developing strategies for rapid convergence to the final design. Many of these techniques are discussed in the literature (Suh, 1990; Rosenman et al., 1990; Sobieszczanski-Sobieski et al., 2015). A particularly important approach that is finding increasing use in the design process is that of *knowledge-based design*. The development and use of this tool is based on the premise that the more the machine or computer knows or learns as it proceeds, the more effective and efficient this process will be. Therefore, an attempt is made to include relevant information on the system, process, and current practice, adding to this information with time and employing the information base to guide the design. The experience gained by a designer over time and various so-called rules-of-thumb or heuristics are also included. Recent advancements in computer science in areas such as information storage and retrieval, artificial intelligence, and symbolic languages are used in developing knowledge-based design aids, which are then used to provide inputs to the design process.

11.1.1 Introduction

Storing the knowledge and experience of an *expert* in a particular area and using these to make logical decisions for selection, diagnostics, and design is the basic concept behind knowledge-based systems. Therefore, knowledge-based systems are also sometimes known as *expert systems* and involve artificial intelligence (AI) features such as

1. Stored expert knowledge and experience
2. Reasoning
3. Decision making and logic
4. Learning

AI is used in many areas such as natural languages, database systems, expert consulting systems, theorem proving, manufacturing, scheduling, pattern recognition, image processing, model development, and design. Recent years have seen considerable growth in AI and its application to different problems, and the basic approaches, scope, and definitions in the field have undergone many changes over the last decade. Examples of software in use include MYSIN, which diagnoses diseases; PROSPECTOR, which evaluates potential ore deposits; MACSYMA, which solves problems in calculus by using symbolic manipulation; and DENDRAL, which finds structures of complex organic compounds. Several expert systems have been developed for the design of different types of systems, including thermal systems, and are discussed later in this chapter.

The knowledge-based methodology requires efficient storage of the knowledge base so that repetition is avoided, minimum space is taken, and rapid retrieval of information is possible. A common arrangement used for the storage of information is a *tree structure*, in which objects are organized in a hierarchical scheme with certain objects taken as subclasses of other objects. These *subclasses* inherit the common features from the objects above it. Therefore, a relationship similar to that of a parent and child is established with respect to inheritance of characteristics and properties. Figure 11.1 shows a tree structure for storing information on animals, with only two choices at each step. So if we consider a cat, its hierarchy indicates that it is a nonvegetarian, nonflying land animal. Only information specific to cats needs to be placed at the particular location, with more

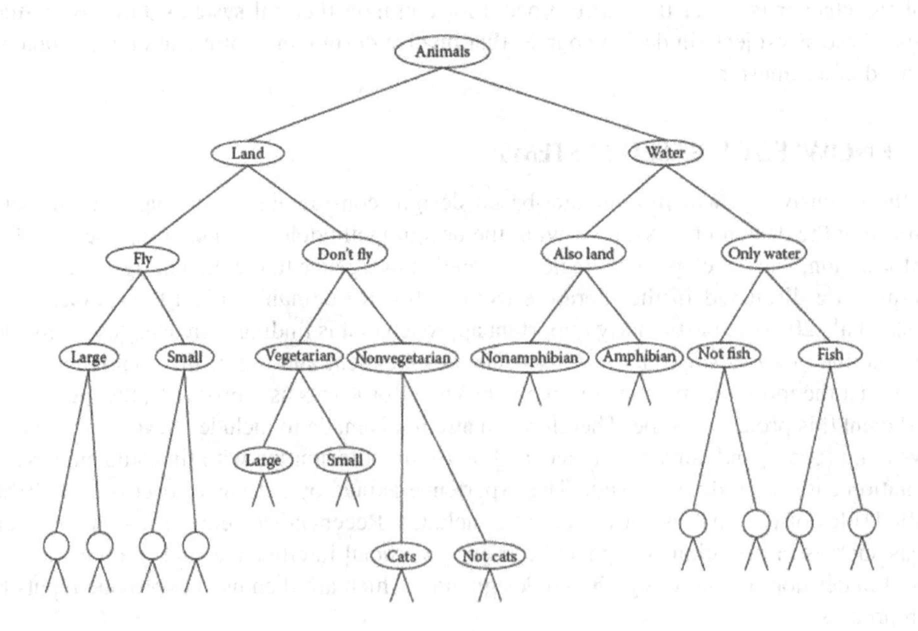

FIGURE 11.1 Tree structure for storing data on animals.

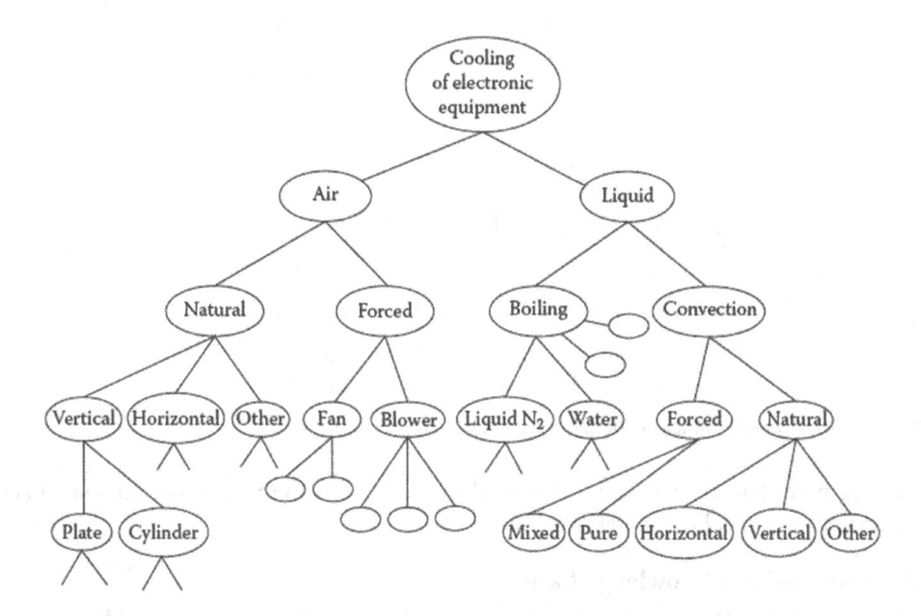

FIGURE 11.2 An example of a tree structure for storing data on cooling of electronic equipment.

general features being derived from its hierarchy. Genetic algorithms, discussed earlier, employ many of these concepts for optimization.

Similar tree structures can be developed for thermal processes, as given in Figure 11.2 for cooling systems for electronic equipment. Different types of cooling arrangements, fluids, and transport mechanisms are included. In earlier chapters, Figure 2.7 gave a similar tree structure for forced convection cooling, considering different types of systems. Similarly, Figure 2.32 gives a tree structure that can be used to store information on different types of materials. Again, the use of subclasses helps in information storage and retrieval. The types of information that may be stored are knowledge and experience available with an expert or design engineer, material characteristics, design rules, empirical data, and other inputs that may be used for design. In many practical cases, intuitive ideas, heuristics, and general features are used to guide the design. These may also be built into the system to obtain an acceptable or optimal design.

11.1.2 Basic Components

The main components of a knowledge-based design system, shown in Figure 11.3, are

1. Front end
2. Computational modules
3. Material databases
4. Graphics output
5. Knowledge base

The user interacts with the *front end*, which interfaces with the other components of the system. Numerical, symbolic, or graphical inputs are provided by the user to the front end. The geometry, configuration, dimensions, materials, and operating conditions are entered. This is quite similar to a wide range of commercially available software in different areas. The front end then obtains material property data and supplies these to the computational modules to obtain the simulation results needed for design. Empirical data, correlations, component characteristics, etc., may be included in the computational modules to complete the modeling and the simulation. These are linked with

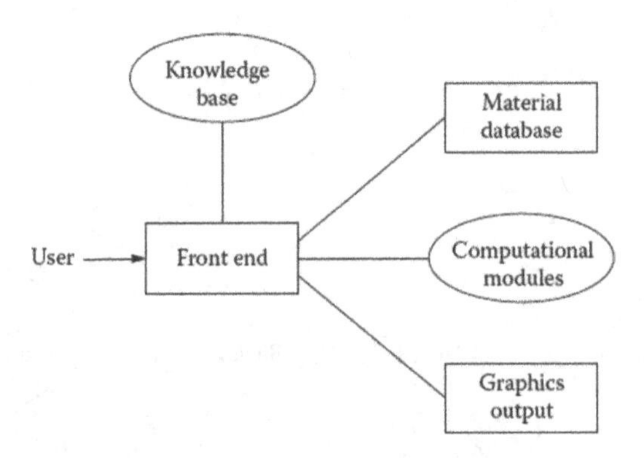

FIGURE 11.3 Components of a knowledge-based system for design.

the knowledge base. The given design rules are then used to obtain the final design, which is then communicated as graphical or tabulated results.

11.1.2.1 Front End and Knowledge Base

The front end contains the design rules, constraints, requirements, design variables, and other aspects pertaining to the given system. Some of these, particularly constraints due to material limitations, are obtained from the databases associated with the system. The knowledge base contains the relevant knowledge, which includes

1. Information from previous designs
2. Rules of thumb or common known features of the system
3. Heuristics based on informal methods
4. Safety and environmental regulations
5. Information on existing and similar systems
6. Current engineering practice
7. Other information that constitutes the experience of a designer

All this knowledge may be used in the development of a realistic and successful design. Therefore, the knowledge base is an important component of this design methodology. It helps a designer avoid mistakes made in the past and use earlier design efforts for accelerating the iterative design process. It is worth noting that many of these aspects are typically employed in the design process even if the systematic approach given here is not followed.

11.1.2.2 Computational Modules

The computational modules house algorithms for numerical simulation of the system. This component is particularly important for the design and optimization of thermal systems, because computer simulation results usually form the basis for design. Even if items such as pumps, heat exchangers, and compressors are only to be selected for the thermal system, computational effort is needed for analyzing the system to ensure that the given requirements and constraints are satisfied. Various computational techniques are stored in the form of subroutines, which can be called from the front end to provide computational results. Examples of methods that may be included for thermal systems are

1. Gauss-Jordan method for matrix inversion
2. Least squares method for best fit
3. Numerical differentiation and integration
4. Successive over relaxation (SOR) method for linear algebraic equations

5. Runge-Kutta method for ordinary differential equations
6. Finite-difference and finite-element methods for partial differential equations

Separate modules may be developed for a given problem, such as a glass furnace, air conditioning system, diesel engine, etc. Information on the discretization methodology, convergence criteria, data storage for graphics, etc., is included to enable accurate results to be obtained and linked with the other parts of the system. Programming languages such as Fortran and C or software like Matlab and Mathcad are used for carrying out rapid computations. Parallel computing, with a large number of processors, is commonly employed for faster response from these modules. Empirical data, usually in the form of correlations, are also included here.

11.1.2.3 Material Databases

The material databases contain information on various materials that are of interest for the types of systems under consideration. Important items that may be included are

1. Thermal properties
2. Allowable ranges of temperature and temperature gradient
3. Strength data, hardness, malleability, and other physical characteristics
4. Cost per unit mass or volume
5. Availability, including import considerations
6. Manufacturability or ease of fabrication
7. Environmental effect

Thermal properties, such as thermal conductivity, diffusivity, specific heat, density, and latent heat, are stored for thermal systems, usually at different temperatures or as functions of temperature. In order to avoid damaging them, constraints on temperature and temperature gradient are given for the various materials. Damage may occur, for instance, due to the melting or charring of the material, thermal stresses, and deformation at high temperatures. Cost, availability, manufacturability, strength, and other relevant properties are important in material selection and should also be included. The information stored is usually a strong function of the application. Again, the information is stored in terms of classes and subclasses of materials, as shown in Figure 2.32, to facilitate inclusion of additional property data and information retrieval.

11.1.2.4 Graphical Input/Output

The graphical output is important for convenience and proper use of the system for design. Impressive advancements have been made in graphics software, and it is quite easy to obtain the outputs in different forms suitable for a wide variety of applications. Some of the important features available in current systems are

1. Line graphs and contour plots
2. Menu-driven software
3. Real-time output
4. Three-dimensional plots
5. Choice of scales
6. Different viewing angles
7. Color graphics

Therefore, the outputs can be fine-tuned to a given application. For example, if a plastic screw extruder is being designed, the pressure and temperature rise in the extruder as the material flows from the hopper to the die may be displayed. As an example, Figure 11.4 shows the temperature distribution in the channel of a plastic extruder in terms of isotherms. The temperature distributions

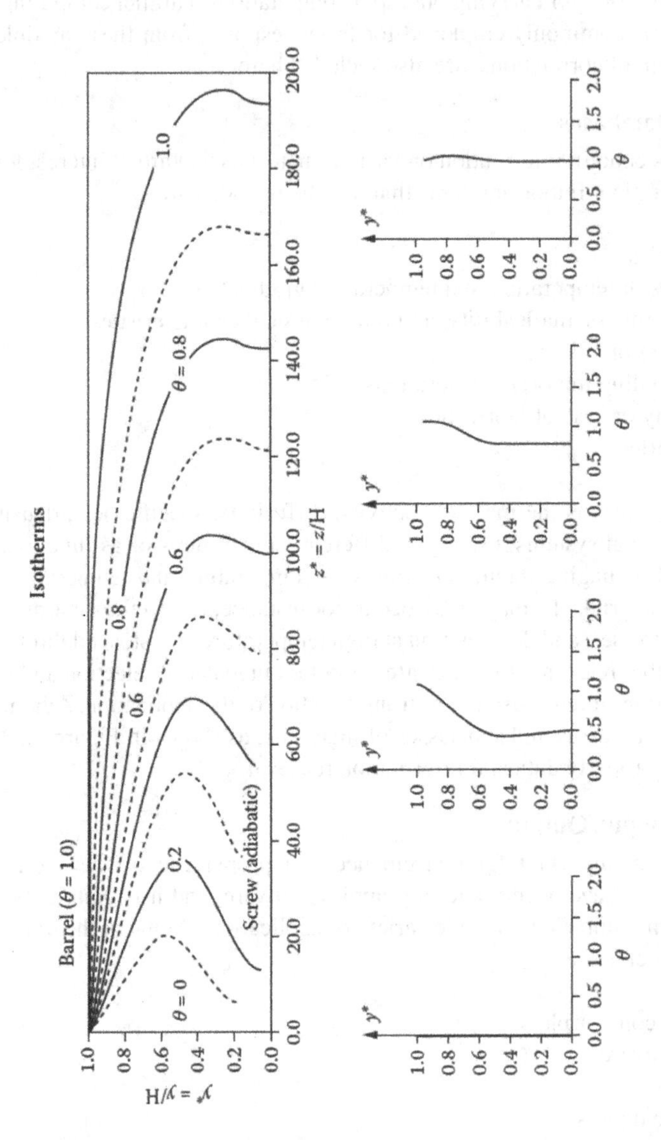

FIGURE 11.4 Isotherms and temperature distributions in the channel of a plastic screw extruder. Here θ, y^*, and z^* are dimensionless temperature, cross-channel coordinate distance, and down-channel distance, respectively. H is the channel height. (Adapted from Karwe and Jaluria, 1990.)

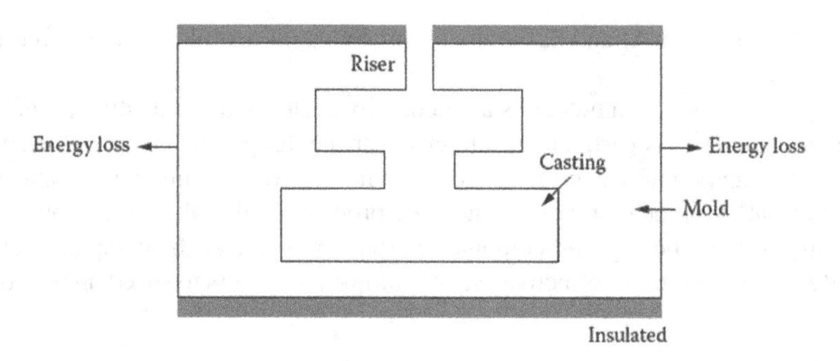

FIGURE 11.5 Geometrical configuration of an ingot casting process.

across the channel at four down-channel locations are also shown. This figure shows how the plastic heats up as it moves from the hopper at $z^* = 0$ to the die at the other end of the channel. In iterative design, the results may be displayed after each iteration, allowing the user to observe the convergence to the final design and to intervene if the iterative process appears to be diverging or if the design emerging from the design process is not satisfactory. Color graphics may also be used to indicate hot and cold regions in the flow.

Graphical inputs to the front end are important in many applications because the geometry, boundary conditions, and dimensions are most conveniently entered on a graphical display. An example of casting is seen in Figure 11.5, where the mold, cast cavity, runner, and thermal conditions at the outer surfaces are shown. Such a schematic may be displayed and the user may interactively enter the appropriate quantities and parameters such as dimensions, heat transfer coefficients, and materials. Many available programming languages, such as Visual Basic, are particularly suited for such graphical inputs.

11.1.2.5 Languages

The programming language employed in the knowledge-based design system forms another important consideration. Languages, such as LISP, PROLOG, SMALLTALK, and PYTHON, which allow the use of symbols rather than just numbers for manipulation and control of the software, are particularly useful for the front end and the knowledge base. For instance, descriptions of a surface as "smooth," viscosity as "high," and disturbances as "small" are all symbolic in form and digital values may or may not be associated with these. This is similar to the concept of fuzzy logic discussed in Chapter 7. In the storage and use of knowledge, we need symbolic representations for

1. Symbolic manipulation of objects
2. Rules of thumb and heuristic arguments in symbolic form
3. Inputs/outputs given in symbolic form
4. Use of symbolic notation for storage of data
5. Symbolic representation of design rules

A symbolic environment allows the versatility and flexibility needed for specifying design rules, constraints, expert knowledge, and other pertinent information. LISP was a commonly used language and variations of LISP are often used to develop expert system shells in which the rules and expert knowledge for a given application can be easily entered (Winston and Horn, 1989). PROLOG and its various versions, such as Sigma PROLOG, have a variety of other features that may be appropriate for certain types of applications (Clark and McCabe, 1984; Bratko, 2011). One of these features is easy link with computational modules, which are often based on languages such as FORTRAN and C, and this makes PROLOG attractive for the design of thermal systems. SMALLTALK is a more powerful language, but it is more complicated and more difficult to implement (Brauer, 2015).

PYTHON is widely used for AI and has many useful packages that may be employed for different applications (Lutz, 2013).

As previously mentioned, computations are generally performed in scientific programming languages. Parallel computing is particularly attractive for the design and control of thermal systems because very fast computations can be achieved, allowing real-time simulation of many systems and providing appropriate graphical outputs. Multi-core processors, distributive computing, massively parallel systems, grid computing, and computer clusters are some of the computing environments that employ algorithms that make effective use of multiple processors to speed up the computations.

11.1.3 Expert Knowledge

Expert systems are based on the knowledge, or *expertise*, of an expert, so that the logical decisions made by an expert in a given area can be made by the computer itself. Empirical data, heuristic arguments, and rules for making decisions are all part of this knowledge-based methodology. The expert knowledge is obviously specific to a given application and represents the knowledge and experience acquired by the expert over a long period of work in the area of interest (Jackson, 1999; Giarratano and Riley, 2005; Ryan, 2018).

For instance, if an expert system is to be developed for the solution of differential equations, an expert will tell us that the solution depends on the nature of the equation, some of the important characteristics being

1. Ordinary or partial differential equation (ODE or PDE)
2. Linear or nonlinear
3. Order of the equation
4. Initial-value or boundary-value ODE
5. Characteristics of PDE: elliptic, parabolic, hyperbolic

Using mathematics, we can develop rules to determine the nature of the equations. A database of analytical solutions can be built and expert knowledge can be used to determine if a particular equation can be solved analytically. Then a search in the database may yield the desired solution. If a numerical solution is necessary, the knowledge base may again be used to choose the following:

1. Numerical scheme
2. Grid size and discretization
3. Appropriate time step so that numerical stability is ensured
4. Convergence criterion to terminate iteration
5. Initial guessed values, if needed

A computational expert uses his or her experience and knowledge in deciding many issues such as the method, grid for desired accuracy, termination of scheme, obtaining an analytical solution if possible, accuracy of the numerical results, and so on. Therefore, one who is presumably not an expert in this area can use this expertise effectively to guide the solution. A database of analytical solutions, different numerical methods, stability criteria, and other constraints and rules is built into the expert system (Russo et al., 1987). The knowledge base will be useful for an inexperienced user as well as users who may have a stronger background in the area, because the information available can guide and confirm the decisions to be made. We all use such knowledge in a variety of actions and decisions in our daily lives.

An important element in the development and use of knowledge-based methodology is *object-oriented programming*, developed using a programming language such as C++, Python, or Java. In a non-object-oriented, or *procedural*, environment, a programmer begins with an initial state and, through a set of prescribed procedures, arrives at the goal. For example, to invert a matrix in procedural programming, one must prescribe every step of the method. In an object-oriented environment, a *message* would be sent to an *object* called *matrix inversion*, which is like a subroutine and which would invert the matrix. The object would already have all the necessary information on the procedures for inverting a matrix. Thus, an object is the housing used to store information. This information comes in the form of procedures known as *methods*, which can act on the given data, being a matrix in this example, upon receiving a message to do so. Objects are organized in a hierarchical scheme with certain objects taken as *subclasses* of other objects, as discussed earlier. These subclasses, therefore, *inherit* the common features from the methods of the object above it. The three components of the object-oriented system described here are *encapsulation, message passing,* and *inheritance* (Cox and Novobilski, 1991; Budd, 2002).

All the information needed to use an object is stored in the object itself. The methods embedded or encapsulated in each of the objects are unique to that object. Therefore, encapsulation makes different objects reusable and reduces duplication within a program. Message passing is independent of the methods and is like a call statement, making it possible to execute the methods stored in the object. Inheritance gives each object access to the features of the methods from the class above it on a tree structure. This aspect allows any system to be easily modified by adding an object, or subroutine, without affecting any other object. With the use of these features of object-oriented programming, it is easy to store and retrieve expert knowledge. Many of the features outlined here have been built into many common software platforms used in engineering simulation and design.

11.1.4 Design Methodology

In the traditional design process, the designer uses the results from modeling and simulation of a system to vary the design variables and to choose appropriate values that would lead to a design that satisfies the given requirements and constraints. In this process, the designer often uses additional information on environmental and safety regulations, material properties, empirical data on the characteristics of some components, economic issues, and so on, to obtain an acceptable or optimal design. All these aspects have been considered in the preceding chapters. However, the designer also relies on experience, current engineering practice, and existing information on the design of similar systems to choose between different alternatives at various stages of the design process. The inclusion of this expert knowledge is often crucial to the development of a successful design. For instance, the designer may be aware of the types of materials that have been used in previous designs and may be able to narrow the search rapidly by this expert knowledge. Similarly, the choice of the type and size of the compressor to be employed for an air conditioning system will benefit from the knowledge of an expert who has worked on such systems and knows the typical components used.

A typical design process involves an initial concept, analysis, choice of design variables, evaluation, and redesign, until acceptable or optimal design is obtained. All of these features may be built into the design procedure to accelerate convergence to the final design as well as to ensure that realistic and practical designs are considered. Therefore, a knowledge-based system for design includes

1. *Design rules.* Requirements, constraints, heuristics, priority of different considerations, and rules of thumb.
2. *Knowledge base.* Expert knowledge, accepted practice, information on existing systems and previous designs, material properties, and federal and state regulations.
3. *Simulation.* Mathematical, numerical, and other models, graphical and numerical input/outputs, numerical methods, design variables, off-design conditions, etc.
4. *Design.* Use of computed results with design rules and knowledge base to obtain acceptable designs; evaluation of designs.
5. *Optimization.* Design variables are adjusted to optimize chosen objective function, knowledge base is used to guide the process and select the domain.

The knowledge base is used at various stages of the design process, starting with the selection of an initial design and varying dimensions, geometry, materials, and other design variables to obtain the final design. The model used for simulation is also varied, as needed, for accurate results. The most important contribution of this methodology is that it includes the experience and expertise of a knowledgeable designer. This helps in avoiding unrealistic and impractical designs, thus focusing rapidly on a domain of acceptable designs. This methodology is particularly valuable in optimization because a lot of unnecessary expense is avoided by considering the iterative improvement in the design in the context of the relevant knowledge base, stopping if a particular approach is unacceptable.

Finally, it must be pointed out that even though knowledge-based design aids are often considered nontraditional, the experience and knowledge of a designer are routinely used in design without actually developing a system to bring this information into the process. Therefore, the ideas presented here can be effectively used in the various design procedures presented in this book by including a knowledge base that is linked with the simulation and design evaluation processes.

11.1.5 Application to Thermal Systems

A lot of work has been done on the use of knowledge-based design aids in areas such as electronic and mechanical systems, particularly in the selection of components like resistors, capacitors, gears, bearings, cams, and dampers. As discussed in Chapter 1, selection is a much simpler process than design, though it may form part of the overall design process. Information on available components is stored in a hierarchical manner, as outlined earlier, and the given requirements are matched with the available items using various selection rules and logic. Expert systems have found use in many other areas in a similar way for selecting items and making logical decisions based on expert knowledge. However, the use of knowledge-based methodology for design is a relatively recent phenomenon because the problem is much more involved as compared to selection (Dixon, 1986; Sriram and Fenves, 1988; Rychener, 1988; Luger and Stubblefield, 1989; Sriram, 1997; Tong and Sriram, 2007).

The application of knowledge-based methodology to the design of thermal systems has received even less attention because of the complexity of these systems and the need to couple numerical simulation with design rules for typical systems. Nevertheless, there is growing interest in knowledge-based design aids for thermal systems because the work done thus far has indicated the advantages and the power of this approach. Even if an entire knowledge-based system is not developed, the basic ideas contained in this methodology can be incorporated into the design process in order to accelerate convergence to the final design and to ensure that the design is realistic and in line with current engineering practice and knowledge.

Most of the examples on design discussed in this book would benefit from the inclusion of knowledge-based engineering in the process. To some extent, this knowledge is available to the designer who uses it at various stages of the design process. Some of the activities where expert knowledge is particularly useful are

1. Obtaining a conceptual or initial design
2. Developing a suitable model
3. Choosing domain and conditions for simulation
4. Choosing appropriate materials
5. Evaluating different designs
6. Formulating the optimization problem
7. Developing the final design, considering cost, safety, environmental, and other issues

At all these stages, the experience of the designer can play a significant role. We have already discussed creativity and use of information available from existing and previous designs for initial concept, model development, and optimization. Similarly, knowledge of the types of materials available, manufacturing processes, applicable regulations, decisions made in the past, success and failure in previous designs, and current trends will help in keeping the design process contained within a realistic domain so that the chances of a successful design are high. The following examples illustrate the use of knowledge-based design methodology for the design and optimization of thermal systems.

Example 11.1: Casting

Let us consider the casting of a material in an enclosed region, as sketched in Figure 1.3. This is an important manufacturing process and is used extensively for metals, alloys, and other materials. The need for design and optimization of the system arises because of the desire to reduce the solidification time to enhance production rate and to improve the product quality in order to meet desired requirements. A large number of design parameters and operating conditions arise in this problem, such as materials, geometry, initial melt pour temperature, cooling fluid and its flow rate, and dimensions. The quality of the casting is determined by grain size, composition, directional strength, concentration of defects, voids, thermal stresses, etc. It is necessary to carry out a thermal analysis of the solidification process, using modeling and simulation, to obtain inputs for design and to evaluate the nature of the casting. Therefore, the decision-making module must be coupled with the computational module. Figure 11.6 shows the algorithm for the design of the thermal system for ingot casting, using inputs from expert knowledge on this process. This includes constraints on the time rate of temperature change $\partial T/\partial \tau$. Heuristics are useful in specifying the characteristics of the casting such as the shape of liquid-solid interface, grain structure, and smoothness of surface.

A PROLOG-based decision-making front end was interfaced by Viswanath and Jaluria (1991) with a FORTRAN-based computational engine, using the various other system components discussed earlier in this chapter, for rapid design. Several different analytical and numerical models,

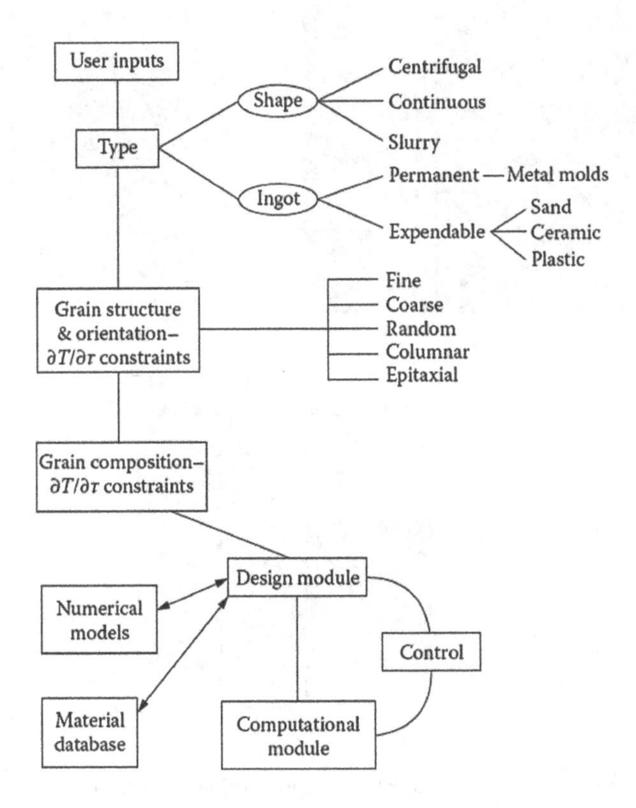

FIGURE 11.6 Algorithm for the design of ingot casting process.

with different levels of accuracy and complexity, are available for this problem (Ghosh and Mallik, 1986; Jaluria, 2018). These include the following:

1. *Steady conduction in solid model:* Melt is taken at freezing temperature, mold at fixed temperature, and steady conduction in the solidified region is assumed.
2. *Chvorinov model:* Entire thermal resistance is assumed to be due to the mold, and energy balance is used.
3. *Lumped mold model:* Temperature in the mold is assumed to be uniform and time-dependent, melt is taken at freezing temperature, and steady conduction in the solid is assumed (see Example 4.8).
4. *Semi-infinite model:* Semi-infinite approximations are used for the mold and the solidified region.
5. *One-dimensional conduction model:* Transient one-dimensional temperature distribution is assumed in the mold, solidified region, and the melt (see Section 5.3.2).
6. *Two- and three-dimensional models:* Natural convection flow in the melt is included along with conduction in the mold and the solidified region.
7. *More sophisticated models:* Needed for alloys, generation of voids and defects, complicated geometries, etc.

Three of the simpler models are shown in Figure 11.7. The governing equations and analytical/numerical techniques for solving them were discussed previously. Therefore, different models may be chosen, depending on the application and materials involved. Expert knowledge plays a major role here. For instance, if an insulating material such as ceramic or sand is used for the mold, the Chvorinov model may yield good results, because most of the thermal resistance is in the mold.

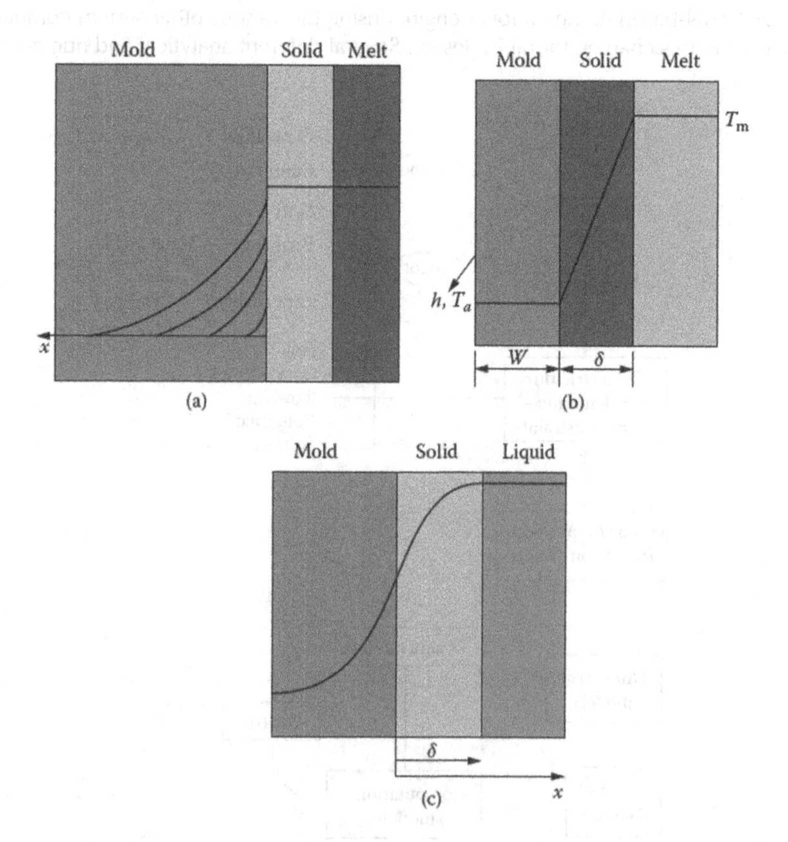

FIGURE 11.7 Different mathematical models for ingot casting. (a) Chvorinov model, (b) lumped mold model, and (c) semi-infinite model.

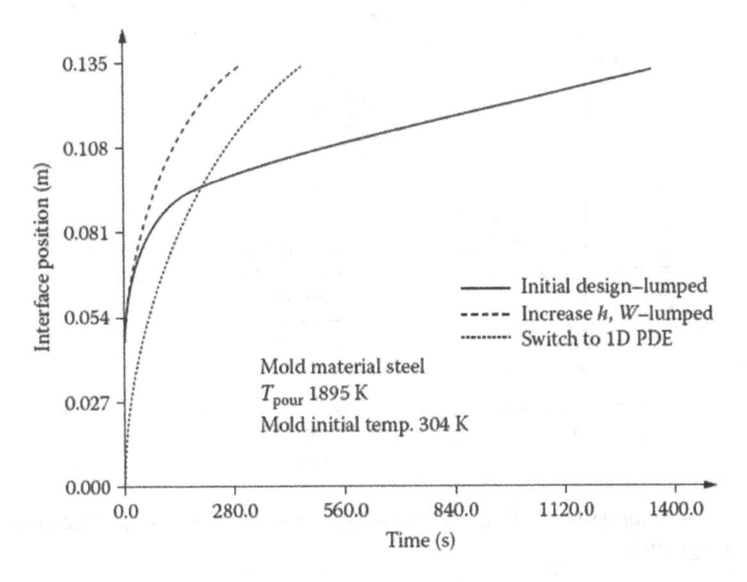

FIGURE 11.8 Results for design of an ingot casting system, showing solid-liquid interface movement with time, and switching to a more complex model after many design trials.

One-dimensional models are adequate for solidification near the boundaries. Sophisticated models are needed for alloy solidification and for considering the microstructure in the casting.

The optimal design may be obtained with solidification time for a given casting being chosen as the objective function and employing constraints from the expert knowledge to avoid unacceptable thermal stresses and defects in the casting. We can start with the simplest model and keep on moving to models with greater complexity until the desired details are obtained or the results remain essentially unchanged from one model to the next. Thus, models may be automatically selected using decision-making based on accuracy considerations.

In a typical design session, the cooling parameters are first varied to reduce the solidification time. If the solidification time τ_s does not reach the desired value, the pour temperature of the melt may be varied. If even this does not satisfy the requirements, the thickness of the mold wall may be changed. The material of the wall may also be varied, if needed. Thus, by first varying the operating conditions and then the dimensions and materials, the solidification time may be minimized or brought below a desired value. Figure 11.8 and Figure 11.9 show some typical runs for the design of the given system. Each successful design may be stored for help in future designs. This is the process of improving the system through learning from past experiences. Figure 11.8 also shows the switch to a more accurate model to improve the simulation results.

Example 11.2: Die Design

Die design is an important consideration in plastic extrusion because the operation of the extruder system and the quality of the final product are strongly influenced by the die. Even though a die is often treated as a component of the extruder, practical dies are usually subsystems consisting of different parts, such as entrance, flow channel, and exit regions, which are attached to each other through couplings and screws, as well as the heating/cooling arrangement at the outer wall to maintain a desired temperature. Pressure and temperature monitoring/control arrangements are also included.

Figure 11.10 gives a sketch of a simple extrusion die with a circular cross-section. The inlet and outlet diameters are given, the former from the design of the screw extruder to which the die is coupled and the latter from the desired outflow or extrudate diameter for the given application. The materials to be extruded and the allowable range of mass flow rates are also given. The design variables are the geometry or shape of the flow channel, entrance and land lengths, material

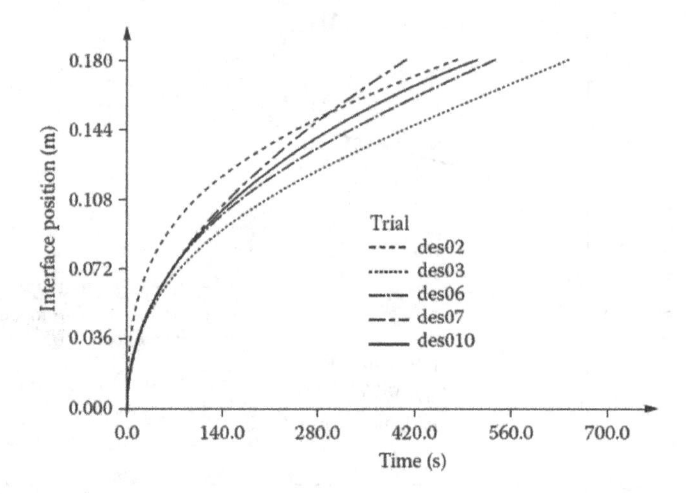

FIGURE 11.9 Results for design of an ingot casting system, showing solid-liquid interface movement with time for many design trials.

and thickness of the walls, and temperature or heat transfer conditions at the outer surface of the wall. Constraints arise due to the material limitations and determine the acceptable pressure and temperature ranges. We also want to avoid stagnation and recirculating flow in the die in order to avoid overheating or damage to the extrudate and thus obtain products of high quality. An acceptable design is one that gives the desired flow rates without violating the constraints. An optimal design may also be obtained so that the pressure needed at the die entrance for a given flow rate is minimum. These considerations define the design and optimization problem. Other objective functions are also possible and may be used, depending on the given application.

The analysis of the flow and heat transfer in the die may be undertaken by means of available information on characterizing equations and numerical techniques (Michaeli, 2017;

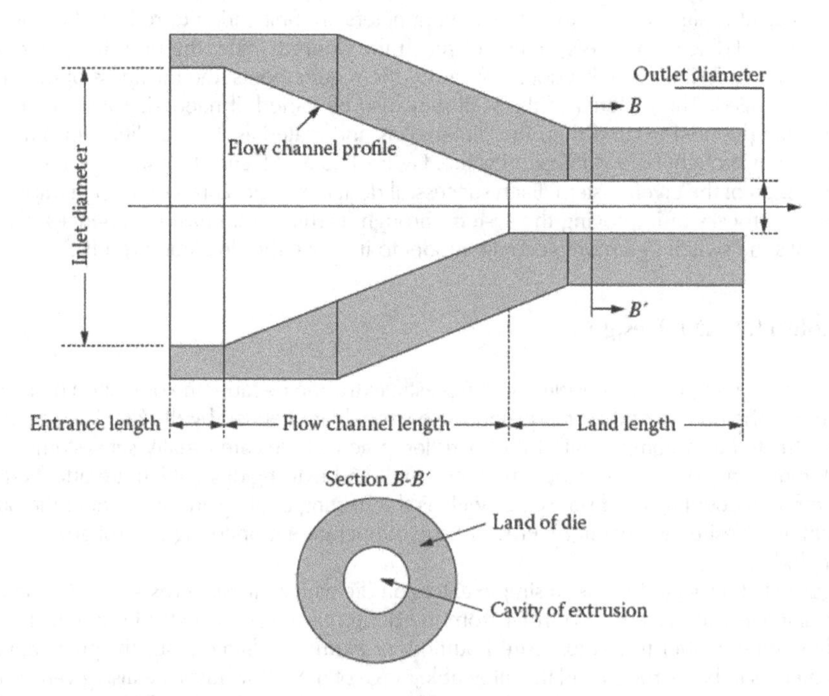

FIGURE 11.10 A cross-sectional view of an extrusion die and its circular outlet.

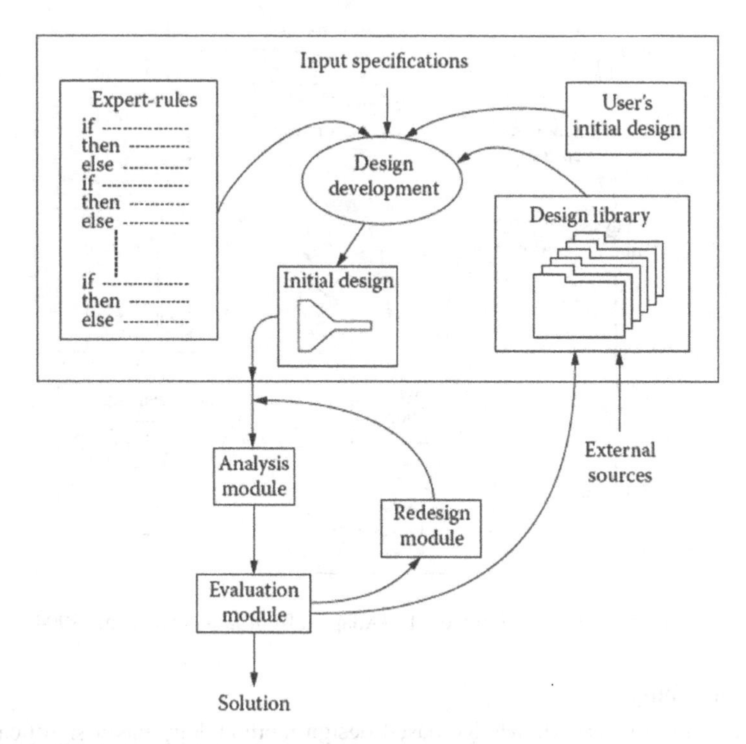

FIGURE 11.11 Initial design module. (Adapted from Jamalabad et al., 1994.)

Rauwendaal, 2014) for different designs. An initial design is taken, analyzed, and evaluated by using the simulation results. If the design does not satisfy the given requirements, a new design is generated and the procedure is repeated. This iterative process is carried out until an acceptable design is obtained or the objective function does not vary significantly from one iteration to the next (Jamalabad et al., 1994).

Knowledge-based design methodology is used effectively for selecting an initial design. Two strategies may be used for this purpose. The first is based on a library of designs built using information from earlier design efforts and from existing systems. The design closest to the given problem may be selected by comparing the designs in the library with the desired specifications. The second approach is based on expert rules for die design. Employing knowledge and experience used by an expert, rules may be set down to generate a design for the given requirements and constraints. Preliminary evaluations and estimates are used to develop a possible design that is used as an initial design. Of course, the user can always enter his or her own initial design if the output from the library or the expert rules is not satisfactory. Figure 11.11 shows the initial design module giving these different strategies.

The knowledge base is also used in the redesign module to evaluate a given design and, if this is not satisfactory, to generate a new design. Expert rules are written based on earlier experience and knowledge from an expert. These establish the relationship between a design variable and the objective function, which may be a performance indicator based on several items such as the pressure needed for a given flow rate, flow characteristics, temperature gradients, etc. Several efficient strategies can be developed for selecting the design variables to go from one design to the next. A single variable may be considered at a time or all the important ones may be varied to obtain new designs, known as *spawns* of the old design, or *parent*. Figure 11.12 shows a schematic of the redesign module based on these concepts. The selection of design variables for the new design is guided by expert rules as well as the results of the design process up to the given instant. Thus, the efficiency of the iterative process is substantially improved. Figure 11.13 shows the advantage of selective search, which avoids local infeasible minima, as compared to an exhaustive search, in the drive toward the optimum given by a minimum value of a numeric heuristic performance indicator.

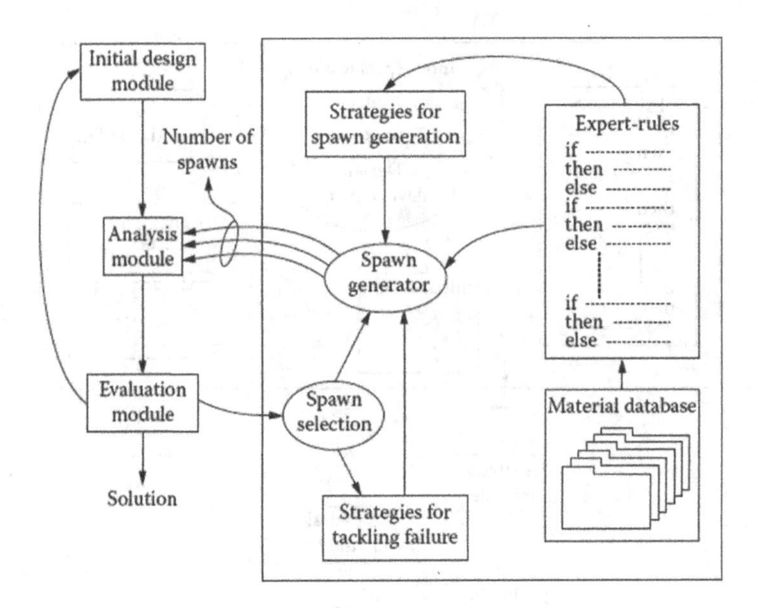

FIGURE 11.12 Schematic for the redesign module. (Adapted from Jamalabad et al., 1994.)

11.1.5.1 Other Examples

Several other examples where knowledge-based design methodology has a significant advantage over the traditional approach may be given. The computer-aided design system considered in Example 2.6 and the cooling system for electronics considered in Example 5.5 are two design problems that were solved earlier by traditional methods and that will benefit substantially by the use of

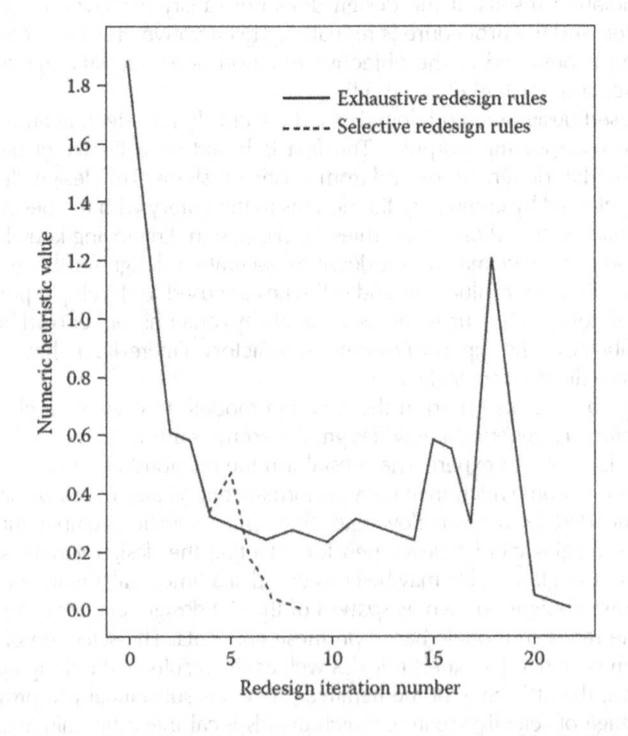

FIGURE 11.13 Comparison of redesign strategies for a 1 cm outlet circular die for extruding low-density polyethylene (LDPE) at 400 kg/h. (Adapted from Jamalabad et al., 1994.)

a knowledge base. Both of these problems are complex, even though simple mathematical models were employed for simulating the systems for idealized and simplified conditions. More sophisticated and complex models would generally be needed for practical problems, resulting in a large number of design variables and additional constraints. There is a considerable amount of expert knowledge available on these systems that may be used to narrow the domain of feasible designs and to facilitate the convergence to the acceptable or optimal design.

Considering first the forced-air oven of Example 2.6, the formulation of the design problem, in terms of given quantities, requirements, design variables, and constraints, was given earlier. A knowledge-based system will enhance the design process by including variation in geometry, model used, materials employed, and even the conceptual design, which is often kept fixed in traditional design. By using symbolic notation and expert knowledge, these variables may be changed easily if the design does not satisfy the given requirements and constraints over the ranges of the chosen design variables.

Some of the important items contained in the expert knowledge base used for this problem are

1. Design rules: Problem formulation, material limitations
2. Different models: Lumped, one-dimensional, two-dimensional, etc.
3. Choice of heat transfer correlations: Laminar/turbulent flow, geometry
4. Priority for changing variables: First heat input Q, then flow rate, followed by dimensions, materials, and geometry
5. Different computational approaches: Runge-Kutta, finite-difference, Matlab
6. Information on existing designs: Provides starting values for Q and heat transfer coefficient h, and initial geometry and dimensions
7. Redesign within acceptable domain: From knowledge base on cost, availability, size, limits on Q, flow rate, etc.
8. Material database
9. Storage/retrieval of collected information: Storage of simulation and design results for use in future designs
10. Different design concepts: Used to vary conceptual design

All these features make the design strategy presented here very powerful because a wide spectrum of changes can be made and the available material database, stored computational methods, different models, and knowledge base can be used to obtain a feasible or optimal design. The priority for changing design variables is based on the expert knowledge and is used to make decisions that would generate new designs. If the changes in Q, flow rate, and dimensions do not lead to a feasible design, the materials are changed, using inputs from the database for properties, cost, and limitations. If even this does not yield an acceptable design, the geometry may be varied, using the expert knowledge to consider different geometries and to obtain the relevant model and numerical scheme. Finally, if a feasible design is still not obtained, other conceptual designs may be considered. A similar approach may be used for obtaining an optimal design. Clearly, a versatile, efficient, and powerful design strategy can be developed using knowledge-based methodology to design thermal systems in a given area or knowledge domain. Jaluria and Lombardi (1991) and Lombardi et al. (1992) have presented this approach in detail and some of the results obtained on this problem.

Similarly, the design of the cooling system for electronic equipment, considered in Example 5.5, may be enhanced by the use of knowledge-based design methodology. The main requirement is that the allowable temperature of the electronic components must not exceed a given value for proper functioning of the system. The expert knowledge on these systems may be used to expand the items that may be varied for a feasible design. For instance, we could consider

1. Different modes of cooling: Air cooling, liquid immersion, boiling
2. Different types of flow systems: Fans, blowers, natural convection

3. Different geometries: Arrangement of boards, sources, number of boards
4. Different materials: Choice of materials from list of allowable ones
5. Dimensions: Height, width, and thickness of boards; spacing between these

Several of these variables were considered in Example 5.5 to obtain a feasible design. However, variation of materials is simplified by the use of the knowledge base because only acceptable materials, with appropriate properties and cost, are considered, and a materials database is available. Variations in geometry, mode of cooling, and flow equipment are more involved and a knowledge base helps in staying within realistic and practical choices. Again, if changes in dimensions and spacing of boards do not lead to a feasible design, other items may be varied according to a chosen priority to obtain a feasible or optimal design.

Knowledge-based design methodology is an important and valuable tool in the design process. It allows the experience and knowledge of an expert to be used in making design decisions, selecting materials, choosing variables to change, obtaining an initial design, and accelerating convergence to the final design. Expert systems may also be used to help in the development of a model (Ling et al., 1993). The information available from earlier design efforts, from existing systems, and from engineering practice is also used to stay within a domain of realistic designs. The methodology substantially increases the versatility and flexibility of the design procedure. Many of these ideas are brought into the design process by the user anyway, but the process is facilitated by a systematic incorporation of the relevant features in the computer system used for design.

It must be noted that only a brief discussion on knowledge-based methodology has been given in this chapter to present its essential features and application to thermal systems. There has been considerable interest in design and optimization methods based on AI approaches, similar to the various aspects outlined here. Some of these were discussed in Chapter 7, particularly with respect to genetic algorithms, artificial neural networks, and fuzzy logic. Artificial neural networks incorporate important features like distributed, or parallel, information processing and object-oriented programming. The different neurons can be programmed to perform certain simple tasks and the neurons are linked in the network to provide fast and efficient processing that is needed for design and optimization. Similarly, fuzzy logic allows one to consider heuristics and imprecise characteristics to consider practical situations and to ensure that a realistic design has been obtained. Genetic algorithms use the concepts of inheritance, evolution, selection, mutation, and recombination to generate global search heuristics. The ideas are quite similar to those discussed here for the knowledge base in terms of parent, child, individual, population, and selection. All such methods are considered as AI-based strategies. Over the last decade, there has been considerable work done on improving these methods for design and optimization and on applying them to a wide variety of practical and complex problems.

11.2 ADDITIONAL CONSTRAINTS

Throughout this book, a wide range of constraints imposed on the design has been considered. Many of these were due to limitations on the materials used, space available, weight, heat input, etc. Such limitations result in constraints on the temperature, pressure, flow rate, and other variables in the problem and are generally used to define the boundaries of the domain of acceptable designs. Constraints also arise from conservation principles, such as mass and energy conservation, that lead to equations that link different variables. However, all such constraints are technical in nature and affect the design by restricting the ranges of the design variables and operating conditions.

There are also many nontechnical constraints that are important in the design of thermal systems and, in some cases, can have a dominant influence on the acceptability of a design. These constraints arise due to the following considerations:

1. Economic aspects
2. Safety issues

3. Environmental effects
4. Legal aspects
5. Social and national issues

We have discussed the importance of economic considerations in the viability of a project and also at various stages in the design process. The constraints on the design generally arise in terms of the costs involved. If the cost per unit mass or volume of a material exceeds a certain value, it may be unacceptable. Similarly, the price of a component such as a fan, compressor, or heat exchanger may be constrained to be below specified values. The component may be changed or the design altered to meet such constraints. Constraints may also be placed on the maintenance and service costs demanded by the system. If these exceed the constraints, the design is unacceptable, even though it may be technically feasible. Many such considerations arise in practice and must be taken care of in a successful design.

Safety issues are obviously critical in the acceptability of a design. Constraints that are more conservative than the limitations on the pressure, temperature, stresses, etc., because of technical reasons are generally used to ensure the safe operation of the system. Safety factors that are based on possible variations in the operating conditions, the user, ambient conditions, raw materials, and other unpredictable parameters are used to yield an acceptable margin of safety. For instance, a dimension obtained as 2 cm from the feasible or optimal design may be increased to 3 cm for safety, giving a factor of safety of 1.5, particularly if this dimension refers to the outer wall of a system and may fail. Higher safety factors are used if damage to an operator may result from leakage of radiation, hot fluid, high-pressure gases, etc. This is particularly true in the design of nuclear reactors, boilers, furnaces, and many other such systems where high temperatures and pressures arise. Therefore, the design obtained on technical grounds is adjusted to account for the safety of the user as well as of the materials and the system.

Similarly, environmental concerns can affect the design. Constraints are placed on the types of materials that can be used, on temperature and concentration of discharges into the environment, on maximum flow rates that may be used, on types and amount of solid waste, and so on. For instance, chlorofluorocarbons (CFCs) are not allowed because of their effect on the ozone layer. The temperature of the cooling water (from the condensers of a power plant) that is discharged into a lake or a river is constrained to, say, within 10°C of the ambient water temperature by federal regulations. This imposes an important constraint on the design of the system. Similarly, the concentration of pollutants and the flow rate of the fluid discharged into air are constrained by regulations to ensure a clean environment. Due to concerns with climate change, the amount of greenhouse gases released may be restricted or a carbon penalty imposed. Again, these aspects arise as additional constraints on the design and must be considered.

The legal issues are often related to federal and state regulations on waste disposal, zoning, procurement, transportation, security, permits, etc., and can generally be considered in terms of the cost incurred, because these involve additional expenditures by a company. Similarly, social issues can often be translated into costs for the company. For instance, providing housing, health care, child care, and other benefits to the employees results in expenses for the firm. Therefore, these aspects lead to financial constraints that may be quantified. However, many aspects are not easy to account for, such as the social effects of layoffs resulting from a particular industrial development. There may be substantial effect at the local level. However, these aspects are largely considered at higher levels of management and affect the design only as far as the viability of the overall project is concerned. National considerations, such as those pertaining to defense, sometimes override the technical and economic aspects. Items and materials that have to be imported may be substituted by those that are available in the domestic market. Thus, constraints may be placed on the choice of materials, components, and subsystems to take these concerns into account.

These additional constraints are usually considered after a feasible design has been obtained. At this stage, the constraints due to the costs, safety, environment, and other aspects are considered to ensure that these are not violated by the design. However, some of these constraints may also be more effectively used at earlier stages of the design process. The temperature and pressure limitations arising from safety concerns, for instance, may be built into the design and optimization process. The final design that is communicated to the management and fabrication facilities must satisfy all such additional constraints.

11.3 UNCERTAINTY AND RELIABILITY-BASED DESIGN

As discussed in Chapter 5, an important consideration in the design of thermal systems is the presence of uncertainties that could arise in the design parameters and operating conditions. Even if an acceptable design is obtained from deterministic models, the uncertainties can make the design unsatisfactory. Due to the existence of uncertainties, the traditional deterministic formulation is no longer reliable to generate safe and acceptable designs because it may lead to a design with high risk of system failure. In order to achieve high reliability in the final design, it is necessary to develop reliability-based design, which results in failure rate lower than an accepted level. A systematic strategy for the modeling and optimization of a thermal system including the effects of uncertainties was presented by Lin et al. (2010), as mentioned earlier. An impingement type chemical vapor deposition reactor for thin film deposition, on which some results have been presented earlier in Figure 7.8(b) and Figure 9.16, was taken as an example. Some of the major uncertainties that arise in this process were also mentioned earlier.

Thus, in practical problems, the design and optimization with uncertainties is based on reliability, which refers to the percentage failure rate of the design. Reliability-based design optimization (RBDO) algorithms are developed, with chosen distributions of uncertainty in the different parameters (Tu et al., 1999; Youn and Choi, 2004). Much of the uncertainty arises in the operating conditions, such as heat input, flow rates and concentration levels. Again, objectives such as product quality and production rate, may be considered for optimization. Normal probability distributions are frequently used, though other distributions may also be employed for different applications. Then the optimal solutions are obtained, using techniques discussed earlier, subject to the allowable level of failure probability.

Figure 11.14 shows a sampling of results for the chemical vapor deposition reactor for different constraints, with normal distribution of the variables. The failure is brought down to less than 0.13%, which is usually the accepted level in RBDO. The optimal point moves away from the one obtained for deterministic conditions due to uncertainties in order to satisfy this condition. For the two cases shown in this figure, the failure rate was over 40% if uncertainties are not considered. But by including uncertainties, a more realistic and practical optimal design is obtained. The importance of uncertainties in design and optimization has only recently been considered for thermal systems.

Uncertainties are particularly critical in the manufacture of microscale and nanoscale devices. These include processes like the fabrication of optical fibers and the fabrication of microchannels for use in microfluidics, in heat removal, or in other applications. The microchannel is typically in the range 20–100 μm in diameter, width, or height. Then uncertainties in the fabrication process can substantially affect the final shape, quality, and dimensions obtained.

Sensitivity analysis is also an important tool that is generally used to study the importance of various parameters and focus on the dominant ones for design and optimization. In the absence of the information from a sensitivity study, the work would consider essentially all the parameters, making it a fairly prohibitive task. Clearly, much work is needed for thermal systems to obtain optimal designs that are more reliable and realistic.

11.4 PROFESSIONAL ETHICS

A topic that has come under considerable scrutiny in recent years is that of *professional ethics*. This is due to the large number of cases that have been uncovered indicating lack of proper behavior, and resulting in damage to life and property. *Ethics* refers to the principles that govern

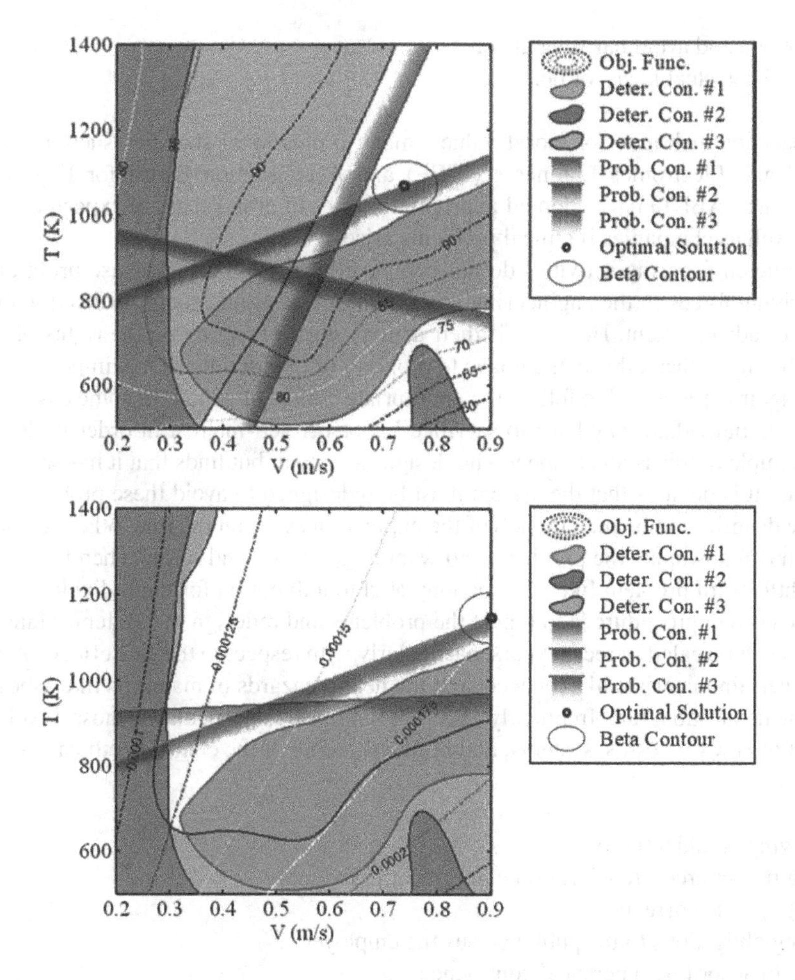

FIGURE 11.14 Results for impingement type chemical vapor deposition used for the fabrication of thin films. Different constraints are considered, with some given as probabilistic with normal distributions, as indicated by the spread in the constraint. The optimum point moves away from the deterministic results due to uncertainties.

the conduct of an individual or groups of people involved in a profession. Every profession sets down its own code of ethics to provide rules of behavior that are proper, fair, and morally correct. It is important for individuals involved in a profession that is self-regulated, such as engineering, to know the ethical behavior that is expected of them and to preserve their integrity under various circumstances. Though many recent cases have involved the legal and medical professions, mainly because of their direct impact on people, engineers are also concerned with many decisions that involve professional ethics. Certainly, design is an area that has far-reaching implications for the profession and must, therefore, be carried out with strict adherence to ethical standards.

Ethics has its roots in moral philosophy, and the basic features are simply the moral values that govern personal behavior. Some of the important aspects of proper conduct can be listed as follows:

1. Be fair to others.
2. Respect the rights of others.
3. Do not do anything illegal.
4. Do not break contracts.

5. Help others and avoid harming anyone.
6. Do not cheat, steal ideas, or lie.

On the basis of such a set of moral values, many professional societies such as the Institute of Electrical and Electronics Engineers (IEEE) and Accreditation Board for Engineering and Technology, Inc. (ABET) have adopted appropriate codes of ethics that are expected to guide the members as well as companies if ethical problems arise.

Most routine engineering activities do not involve ethical conflicts. Because profit and growth are major driving forces in the engineering profession, most engineers are involved with pursuing these goals for advancement. However, if their actions start infringing on the rights of others and may lead to harm to others, through damage to property or to individuals, it is important to balance self-interest against ethics and to follow the appropriate course of action. In some cases, the choice is clear, but the individual may have to sacrifice his or her self-interest in order to do the proper thing. An example of this is an engineer who designs a system, but finds that it has safety problems during testing. It is obvious that the system must be redesigned to avoid these problems. However, there may be deadlines to be met, the job of the engineer may be on the line, others involved in the design may try to downplay the problems and want to go ahead, and so on. Therefore, even a relatively straightforward problem like this one may lead to a dilemma for the individuals concerned. However, the correct procedure is to report the problems and redesign the system. Many unethical actions have been revealed in recent years, particularly with respect to the side effects of medicines, addiction to certain materials like tobacco, and the health hazards of materials like asbestos.

Such ethical questions are frequently faced by engineers, particularly those involved in the development of new processes, systems, and products. Some of the common ethical situations that arise are

1. Preserving confidentiality
2. Giving proper credit to appropriate groups or individuals
3. Reporting data correctly
4. Meeting obligations to the public versus the employer
5. Acting in accord with personal conscience
6. Responding to inappropriate behavior by others in the company

Many readers may have already experienced some of the preceding situations. Giving proper credit to different people involved in a project is always a touchy issue, but it is important to be fair to all. Similarly, proprietary information must be maintained as confidential because access to such information or data is allowed only to a few individuals under the conditions of confidentiality. A subcontractor who is designing a component for a system manufactured by a different organization is often provided with essential details on the system. These details are for use in the design process and are not to be revealed to other parties who are not involved. Correct reporting of data has been in the news a lot in the last few years, as some researchers have been found to falsify experimental results to downplay the damaging effects of the product and to claim important advances. When uncovered, these actions have led to large financial settlements and public ridicule.

The issue of inappropriate behavior by others is a very difficult one and requires considerable care. Suppose an engineer finds out that his or her company is violating federal regulations on dumping of chemical hazardous wastes. Does he or she simply ignore the problem? According to the code of ethics, this matter has to be reported to the proper authorities. This is known as *whistle blowing* and it cannot be taken lightly. It is necessary to be sure that the activity is illegal and is adversely affecting public safety or welfare. Proper documentation is needed to support the accusation, and it must also be confirmed that management is aware of the activities. If the problem lies within the company and is a consequence of oversight, a simpler solution may be possible. Otherwise, appropriate regulatory bodies may be contacted. Whistle blowing obviously requires great moral courage, and the present position as well as the future advancement of anyone who undertakes this effort are at stake. Such instances are rare, but they have been increasing and they test the professional code of ethics.

In most cases where ethical conflicts need to be resolved, internal appeal processes are adequate. These start with the immediate supervisor, or with someone outside the region of conflict, and follow the internal chain of command. Satisfactory documentation is crucial in such appeals. Conflicts arise, for instance, due to unfair treatment, improper credit being given, falsification of data, improper use of funds, etc. Only if the internal appeal process is unsuccessful does one need to resort to external options. These include contacting professional societies, the media, regulatory agencies, personal legal counsel, and other external sources for settling the conflict. Many case studies are given by Ertas and Jones (1997) and the IEEE code of ethics is given by Dieter (2000) to indicate the basic issues involved and the types of ethical conflicts encountered in practice.

11.5 SOURCES OF INFORMATION

An important ingredient in design is availability of relevant information on a variety of topics that are needed for developing a successful design. Information is needed to provide accurate data for the modeling and simulation of the system and to use past experience and results for help with the design process. Without adequate information on previous design efforts and existing systems, we could repeat past mistakes, spend time on obtaining information that is readily available, or generate designs that do not meet appropriate regulations and standards. Therefore, it is crucial that we spend extensive effort on gathering the most accurate and up-to-date information available on all facets of the given design problem.

The types of information needed for design are obviously functions of the system under consideration. However, the information sought for the design of thermal systems is generally in the following main areas:

1. Material property data, including cost and availability
2. Design and operation of existing or similar systems and processes
3. Availability and cost of different types of components
4. Manufacturing processes available and the costs involved
5. Available computer software
6. Available empirical results, including heat transfer correlations, relevant technical data, and characteristics of equipment
7. Federal, state, and local regulations on safety and environment
8. Standards and specifications set by appropriate professional bodies
9. Current financial parameters, including rates of interest and inflation, different costs, and market trends

We have seen in earlier chapters that all this information is needed at various stages of the design process to obtain the desired inputs as well as to evaluate the design. Property data are particularly important because the accuracy of the simulation results, which are crucial to the design process, is determined by the data provided to the model. Economic data are needed to evaluate costs, make economic decisions during design, and determine if a project is financially viable. The design must meet the standards and regulations specified for the given process or application and, therefore, it is necessary to obtain accurate information on these. Finally, information on existing systems, software, manufacturability, and technical results helps in minimizing the effort and avoiding duplication of work.

Throughout this book, references have been made to books, papers, and other publications that deal in depth with a particular topic or area. Certainly, textbooks in the areas of heat transfer, thermodynamics, and fluid mechanics are a good starting point for the design of thermal systems.

Similarly, text and reference books on numerical methods, engineering economics, optimization, and design are important sources of relevant information. These can be used to provide the basic technical background needed for the various aspects that are involved in design. Different methods for analyzing various processes and systems, including mathematical, numerical, and experimental techniques, are given in detail in such books, along with characteristic results. Some information is also available on common materials; components of interest in thermal systems such as pumps, compressors, and heat exchangers; and general features of standard systems, such as air conditioners and diesel engines. The references quoted in these books and papers may be further used to expand the source base. However, for detailed information on material properties, applicable regulations, current economic parameters, existing systems, characteristics of available equipment, and current trends in industry, other sources of information are needed.

There are two main types of sources of information for design. These are

1. *Public sources:* Libraries, universities, research organizations, departments and agencies of the federal, state, and local government
2. *Private sources:* Manufacturing and supplying companies, banks, professional societies, consultants, individuals, computer software companies, and membership and trade associations

Public sources are extensively used because the information obtained is free or relatively inexpensive. Government reports and publications from departments such as commerce, defense, energy, and labor are very valuable because these provide detailed information on results obtained, requirements and regulations, methods used, material property data, and various other guidelines. Patents issued also provide an excellent source of information on existing systems. These are available from public sources, through libraries and the Internet.

Private sources are usually expensive, but a lot of information can be obtained through company brochures and websites used for advertising their products. The specifications, cost, maintenance, servicing, and performance of available equipment can be obtained from the supplier. In many cases, additional details on the materials used, tests performed, basic design of the component, and even samples can be obtained if one is interested in a particular item. Though complete information on the component is generally confidential, enough information can be obtained to decide if a given item is appropriate for the system being designed. Similar considerations apply to commercially available software. Extensive catalogs of manufacturers and suppliers are available from various listings and the Internet, which is an important and expanding source of such information.

Most engineering companies maintain their own libraries that contain books, technical magazines, journals, reports, information on their own products and systems, listing of suppliers, and so on. In the present information age, they also have access to information available in the public domain, through the Internet, library exchanges, and agreements with other research or professional establishments. Such a collection may be large or small, depending on the size of the company itself, but the available methods of literature search and procurement make it an important source of information. In addition, results from earlier efforts, information on existing systems, properties of materials used in the past, information on relevant regulatory and legal aspects, and so on, as applicable to the given industry, are probably best stored here.

Therefore, there are many important sources of information on materials, detailed technical data, existing processes and systems, items available in the market, economic data, and on other inputs needed for design. These may be listed as follows:

1. Handbooks
2. Encyclopedias
3. Monographs and books
4. Journals: technical and professional
5. Catalogs

6. Indexes and abstracts
7. Technical reports
8. Internet
9. Patents

Handbooks and encyclopedias are very useful in obtaining relevant and detailed technical information for a given process or system. Encyclopedias are available on physics, materials science, chemistry, fluid mechanics, and so on. McGraw-Hill's *Encyclopedia of Science and Technology* is an example of such reference books. Similarly, handbooks are available on pumps, air conditioning, heat exchangers, material properties, industrial engineering, manufacturing, etc. *Marks' Standard Handbook for Mechanical Engineering*, also published by McGraw-Hill, is an example. A substantial amount of relevant and focused information is generally available in such sources. Listings of indexes and abstracts allow one to search for the appropriate source rapidly, particularly if the information exists in journals, translations, and published reports. Technical and professional journals are also good sources, though these are often either too detailed or too sketchy. Nevertheless, these can be used effectively to narrow the search to specific and relevant sources of information. The Internet is certainly one of the most important sources of information today.

11.6 AN OVERVIEW OF DESIGN OF THERMAL SYSTEMS

Basic aspects. In this book, we have considered the design and optimization of systems in which thermal transport, which involves heat and mass transfer, fluid flow, and thermodynamics, play a dominant role. Many different types of thermal systems, ranging from refrigeration, heating, transportation, and power systems to manufacturing and electronic equipment cooling systems, are employed as examples. The complicated nature of these systems, particularly their typically nonlinear, time-dependent, three-dimensional, geometrically complex, and combined-mode characteristics, leads to coupled systems of partial differential equations that describe the system. These equations are simplified through modeling to obtain algebraic equations and ordinary differential equations in many cases. Such models are combined with experimental results, material property data, and other available information, often using curve fitting, to obtain a complete model for the system. This model is then used to simulate the system and obtain detailed results, which can be used for the design and optimization of the system. Thus, modeling and simulation form the core of the design effort for thermal systems, and the successful completion of a project is closely linked with the accuracy and validity of the model. This aspect of design is stressed throughout the book.

The design of a thermal system starts with a close look at the problem. This involves determining what is given or fixed in the problem, what can be varied to obtain an acceptable design, what the main requirements are, and what constraints or limitations must be satisfied by the design. This consideration leads to the problem formulation in terms of given quantities, design variables, operating conditions, requirements, and constraints. It also defines the feasible domain for the design. The next step is to obtain a conceptual design to achieve the desired goals. Different ideas for the system are considered based on what is presently in use and the design of other similar systems. Innovative concepts are employed as well. A particular conceptual design is chosen by employing simple estimates and contrasting different ideas.

With the design problem formulated and a conceptual design chosen, we can now proceed to a detailed design process, the main steps of which are

1. Characterization of the physical system
2. Modeling
3. Simulation
4. Evaluation of different designs
5. Iteration to obtain an acceptable design

6. Optimization
7. Automation and control
8. Communication of the final design

Workable or acceptable design. Though all the preceding steps are involved in the development of a successful design, modeling and simulation are particularly crucial because most of the relevant inputs for design and optimization are obtained from a simulation of the system using analytical, numerical, and experimental approaches to study the model developed for the system. Different types of models and simulation strategies are available to generate the inputs needed for design. Even though the basic approach to modeling can be established and applied to common thermal processes and systems, modeling remains a very elusive and difficult element in the design process. It involves simplification, approximation, and idealization to obtain a model that may be used to study the behavior of the actual physical system. Creativity and experience are important ingredients in the development of a model. A good understanding of the physical characteristics of the system and the basic processes that govern its behavior is essential in deciding what to neglect or how to approximate. The governing equations are obtained based on the conservation principles for mass, momentum, and energy, taking these simplifications into account.

A very important question that must be answered for a model is if it accurately and correctly predicts the characteristics of the actual system. This requires a detailed validation of the model and estimation of the accuracy of the results obtained. In many cases, simple models are first obtained by neglecting many effects that complicate the analysis. These models are then improved by including additional effects and features to bring them closer to the real system. This fine-tuning of the model is generally based on the simulation, experimental, and prototype testing results. The model is used to study the system response and behavior under a wide variety of conditions, including those that go beyond the expected region of operation to establish safety limits and ensure satisfactory operation in real life. Simulation results are also used to determine if a particular design, specified in terms of the design variables, satisfies the requirements and the constraints.

All these ideas concerning problem formulation, conceptual design, modeling, simulation, and design evaluation may then be put together for obtaining acceptable designs. Different areas such as manufacturing, energy, transportation, and air conditioning, where thermal systems are of interest, are considered to apply the various steps in the design process. It can again be seen that modeling and simulation are at the very core of the design effort since the results obtained are used to choose the appropriate design variables, evaluate different designs, and obtain a domain of acceptable designs. Many additional aspects, such as safety and environmental issues, may also be considered at this stage. A unique solution is generally not obtained and several acceptable designs are often generated. Thus, the systematic progression from problem formulation and conceptual design to an acceptable design is highlighted in the first five chapters. Though relatively simple thermal systems are considered in many cases to present the methodology, much more complicated problems that typically arise in actual practice can be treated in a similar way. Some actual industrial systems are considered to demonstrate the use of this methodology.

Optimization. Economic considerations and optimization need to be considered to complete the design and optimization process. Economic aspects are obviously very important in most problems of practical interest and are of particular significance in optimization because minimization of costs and maximization of profit are important criteria for an optimal design. Economic considerations, such as cost, return, payment, investment, depreciation, inflation, and time value of money, in the design process are discussed. Examples are given to show how economic considerations can affect decisions on design in areas such as material selection, choice of components, energy source, and so on. The importance of financial aspects in design cannot be exaggerated because the viability and success of the project itself is usually determined by the profit, return, stock price, etc. Again, several relatively simple situations are discussed to illustrate the approach for considering economic factors. However, these ideas can easily be extended to more complicated circumstances where several different considerations may arise and interact with each other.

Optimization is discussed in detail to stress the crucial need to optimize thermal systems in today's competitive international market. The design process generally leads to a domain of acceptable designs with no unique solution. Optimization with respect to a chosen criterion or objective function narrows this domain substantially, as desired for a given application, so that the final design is chosen from a small range of design variables, making it close to a unique solution. Optimization in terms of the design variables as well as in terms of the operating conditions is considered.

Many different strategies are available for optimization. Search methods are particularly useful for thermal systems because discrete values of the variables are often encountered and simulation is complicated and time-consuming, making it necessary to limit the number of runs. Methods such as Fibonacci and univariate searches are efficient and easy to use. Hill climbing techniques can also be used with numerically determined derivatives. Calculus methods and geometric programming are useful if the simulation results are curve fitted to obtain continuous expressions to represent system characteristics. Other methods such as linear and dynamic programming have limited use for thermal systems. Efficient computer programs are available for most of the techniques presented here as well as for new and emerging techniques, and may be used when dealing with the large and complicated systems encountered in actual practice.

The choice of the optimization method is guided by the form in which the simulation results are available and how involved each simulation run is. Several simple examples are employed here to illustrate the basic ideas and the techniques. These may be extended easily to many of the more complicated problems discussed in earlier chapters and examples of practical thermal systems given in the book. Typical large systems are considered first in terms of the components and subsystems, with the overall model and simulation scheme obtained by coupling the simpler sub-models. The complete model is simulated to obtain the characteristics and behavior of the system. These results are then used for developing acceptable designs, followed by an optimal design. There are obviously cases where an acceptable design is not obtained, with the given requirements and constraints, or where acceptable designs lie within a narrow range of variables, making it unnecessary to optimize the system.

Concluding remarks. The design and optimization of thermal systems is an important, though complicated, field. Many different situations can arise in actual practice and may require specialized treatment. However, the basic approach given in this book provides the general framework under which the design and optimization of a thermal system may be undertaken. Some modifications may be necessary in a few cases and additional information pertaining to materials, economic parameters, regulations, available components, and so on, may have to be obtained for specific applications. Creativity and originality are also important ingredients in design, particularly in the development of the concept and the model. In engineering practice, commercially available software packages are often used for system simulation and optimization. However, some of the software may be developed or programs in the public domain may be employed to provide the flexibility and versatility needed in many cases. Additional aspects, such as safety and environmental issues, are of concern in most problems and are built into the decision making. The final design is communicated to the appropriate groups such as the management and fabrication facilities. Depending on management decisions, this could lead to prototype development, testing, marketing, and sales. All these aspects are expected to be considered in the design projects included at the end of this chapter.

11.7 SUMMARY

This chapter concludes the presentation on the design and optimization of thermal systems by considering some recent trends, particularly knowledge-based design aids, and some additional considerations. Knowledge-based, or expert, systems, which use the expertise of people who are proficient in given areas, have been used effectively in several fields, such as medicine, chemical analysis, and mining. In traditional design, such expertise is traditionally brought in by the designer, who uses his or her knowledge of the process, materials, and system to make decisions throughout the design process in order to obtain a realistic and practical design. Expert systems use computer software based on AI techniques to bring such considerations into the design process. The basic components, including expert knowledge and databases, in knowledge-based design methodology are discussed. The advantages of this approach over traditional design methods are indicated. Several examples of thermal systems are taken to illustrate the use of this methodology. There is growing interest in these methods and many relevant techniques have been developed in recent years to help convergence to a realistic acceptable or optimal design.

This chapter also discusses the important topics of uncertainty, reliability-based design, professional ethics, sources of information, and additional constraints. These aspects arise in most engineering endeavors, but these are particularly significant for design because of the innovative and creative ideas that are often involved. Uncertainties are encountered in most engineering processes and systems. A brief discussion of uncertainty and reliability-based design is given. Different sources of information are discussed, indicating methods to locate these and extract the relevant information from them. Professional ethics can play a significant role in the development of the design, in the use of available information, in the implementation of the design, and in the progress of the entire project. A brief discussion of the important issues is given. Additional constraints that could affect the feasibility of a design are presented and discussed in the context of the material covered in the book. These could be of crucial importance in many cases and could determine whether it is worthwhile to proceed with the implementation of a given design. Finally, the chapter integrates all the ideas presented in this book as an overview of the design of thermal systems and includes a few design projects as problems.

REFERENCES

Bratko, I. (2011). *Prolog programming for artificial intelligence* (4th ed.). Toronto, Canada: Pearson Education.

Brauer, J. (2015). *Programming Smalltalk—object-orientation from the beginning: an introduction to the principles of programming.* Wiesbaden, Germany: Springer Vieweg.

Budd, T. (2002). *An introduction to object-oriented programming.* Boston, MA: Addison-Wesley.

Clark, K.L., & McCabe, F.G. (1984). *Micro-prolog: programming in logic.* Englewood Cliffs, NJ: Prentice-Hall.

Cox, B.J., & Novobilski, A.J. (1991). *Object oriented programming. An evolutionary approach.* Reading, MA: Addison-Wesley.

Dieter, G.E. (2000). *Engineering design: a materials and processing approach* (3rd ed.). New York: McGraw-Hill.

Dixon, J.R. (1986). Artificial intelligence and design: a mechanical engineering view. *AAAI, 86,* 872–877.

Ertas, A., & Jones, J.C. (1997). *The engineering design process* (2nd ed.). New York: Wiley.

Ghosh, A., & Mallik, A.K. (1986). *Manufacturing science.* New York: Wiley.

Giarratano, J.C., & Riley, G.D. (2005). *Expert systems: principles and programming* (4th ed.). Boston, MA: Thompson Course Technology.

Jackson, P. (1999). *Introduction to expert systems* (3rd ed.). Boston, MA: Addison Wesley.

Jaluria, Y. (2018). *Advanced materials processing and manufacturing.* Cham, Switzerland: Springer.

Jaluria, Y., & Lombardi, D. (1991). Use of expert systems in the design of thermal equipment and processes. *Research in Engineering Design,* 2(4), 239–253.

Jamalabad, V.R., Langrana, N.A., & Jaluria, Y. (1994). Rule-based design of a materials processing component. *Engineering with Computers, 10*(2), 81–94.

Karwe, M.V., & Jaluria, Y. (1990). Numerical simulation of fluid flow and heat transfer in a single screw extruder for non-Newtonian fluids. *Numerical Heat Transfer, 17,* 167–190.

Lin, P.T., Gea, H.C., & Jaluria, Y. (2010). Systematic strategy for modeling and optimization of thermal systems with design uncertainties. *Frontiers Heat Mass Transfer,* 1, 013003-1-20.

Ling, S.K.R., Steinberg, L., & Jaluria, Y. (1993). MSG: a computer system for automated modeling of heat transfer. *AI EDAM,* 7, 287–300.

Lombardi, D., Jaluria, Y., & Viswanath, R. (1992). Simulation of the transport processes in a thermal manufacturing system using symbolic computation. *Engineering Applications of Artificial Intelligence,* 5(2), 155–166.

Luger, G.F., & Stubblefield, W.A. (1989). *Artificial intelligence and the design of expert systems*. New York: Benjamin/Cummings.
Lutz, M. (2013). *Learning Python* (5th ed.). Sebastopol, CA: O'Reilly Media.
Michaeli, W. (2017). *Extrusion dies for plastics and rubber: design and engineering computations* (4th ed.). Munich, Germany: Carl Hanser Verlag.
Rauwendaal, C. (2014). *Polymer extrusion* (5th ed.). Munich, Germany: Carl Hanser Verlag.
Rosenman, M.A., Radford, A.D., Balachandran, M., & Gero, J.S. (1990). *Knowledge-based design systems*. Reading, MA: Addison-Wesley.
Russo, M.F., Peskin, R.L., & Kowalski, A.D. (1987). A Prolog-based expert system for modeling with partial differential equations. *Simulation*, 49, 150–157.
Ryan, D. (Ed.). (2018). *Expert systems: design, applications and technology*. Hauppauge, NY: Nova Science Pub.
Rychener, M.D. (Ed.). (1988). *Expert systems for engineering design*. Boston, MA: Academic Press.
Sobieszczanski-Sobieski, J., Morris, A., & van Tooren, M. (2015). *Multidisciplinary design optimization supported by knowledge based engineering*. New York: Wiley.
Sriram, R.D. (1997). *Intelligent systems for engineering: a knowledge-based approach*. London: Springer-Verlag.
Sriram, D., & Fenves, S.J. (Eds.). (1988). *Knowledge-based expert systems for engineering*. Ashurst Lodge, Southampton, UK: Computational Mechanics Publications.
Suh, N.P. (1990). *The principles of design*. New York: Oxford University Press.
Tong, C., & Sriram, D. (2007). *Artificial intelligence in engineering design*. Boston, MA: Academic Press.
Tu, J., Cho, J., & Park, Y.H. (1999). A new study on reliability based design optimization. *Journal of Mechanical Design*, 121, 557–564.
Viswanath, R., & Jaluria, Y. (1991). Knowledge-based system for the computer aided design of ingot casting processes. *Engineering with Computers*, 7, 109–120.
Winston, P.H., & Horn, B.K.P. (1989). *LISP* (3rd ed.). Reading, MA: Addison-Wesley.
Youn, B.D., & Choi, K.K. (2004). An investigation of nonlinearity of reliability-based design optimization approaches. *Journal of Mechanical Design*, 126, 403–411.

PROBLEMS

11.1 For any thermal system mentioned in Chapter 1, draw a simple sketch of the system considered. Give the main assumptions you would make to obtain an appropriate mathematical model. Justify these and discuss changes that you would make in order to improve the accuracy of your model. In addition, considering yourself an expert, what expert knowledge would be worth including in the design process?

11.2 An expert system is to be developed for the solution of algebraic equations. Using the information presented in Chapter 4, give a suitable tree structure for the storage and retrieval of the relevant expert knowledge. Consider single and multiple equations, as well as linear and nonlinear systems. Also, include the various methods available for solution.

11.3 List the important knowledge that may be used for developing an expert system for the design of a building air conditioning system. The relevant information may be obtained from reference books, as outlined in this chapter. Of particular interest are current practice, available equipment, and costs involved. Be brief.

11.4 For the design of a solar thermal energy storage system, using water for storage, choose a conceptual design and give the design variables. Then discuss the knowledge base that could be employed, in conjunction with the basic design process, to obtain an optimal, practical design of the system.

11.5 Present the knowledge base for the design of a furnace for heat treatment of steel rods as a tree structure. Consider different types of furnaces, different heat sources, and different configurations.

11.6 What would the material knowledge base for the design of an electronic cooling system consist of? Use the information given in earlier chapters on such systems and a few relevant references.

11.7 An expert system is to be developed for solving ordinary differential equations (ODEs). List the expert knowledge needed to solve an ODE, analytically or numerically. Also, present the tree structure for storing this knowledge. Outline the scheme employed by the expert system to solve a given ODE.

11.8 You are working as an engineer in the testing division of a large company. The division is involved in testing and evaluating equipment from various subcontractors and recommending the best ones to the production department. You are asked to test two different heaters, *A* and *B*, for durability and performance. You find that heater *B* is superior and inform your boss of your recommendation. Then you find out that he has already recommended heater *A* because of time constraints and he just wanted you to confirm his cursory evaluation of the heaters. Now he wants you to manipulate the data so that heater *A* is shown to be better. What are the ethical questions involved and how will you handle the situation? Your job is probably on the line here.

11.9 After working as a design engineer in Company *A* for 4 years, you are hired by Company *B*, which is a competitor of Company *A*. You are asked to reveal various items concerning your previous employer, such as the new products being developed and new facilities being acquired. Discuss the ethical issues involved here, for you as well as for the new employer. How would you handle this problem?

11.10 You and a colleague discussed the solution to a problem found to arise in a manufacturing process. During the discussions, an interesting solution is proposed. You do not remember who suggested it or how it was brought in. Nevertheless, it is a promising idea and it is decided that both of you should look into the proposed concept in detail. However, you have to travel for a couple of days and when you come back you find that your colleague has already passed on the idea to management as his own, with no acknowledgment of your contribution. Initial tests have shown that the concept will work and this would mean a patent for your colleague and possible advancement. Obviously, you are hurt at this betrayal of trust. But, you have no written proof of your discussions with your colleague. How will you proceed with this situation?

11.11 For the design of the thermal systems corresponding to the following applications, list the sources that may be consulted and the type of information being sought:
  a. Residential air conditioning and heating
  b. Hot water storage and transportation in a large industrial building
  c. Manufacture of molded plastic parts
  d. Annealing of steel rods
  e. Air conditioning of aircraft cabins
  f. Cooling of a data center

11.12 Choose any thermal system on the basis of your experience. Discuss its basic features and formulate the design and optimization problem for the system. Develop a simple mathematical model and the corresponding simulation scheme. Outline how you would proceed to obtain an acceptable design and then an optimal one. Do you expect to find any use for the knowledge base available on similar systems?

DESIGN PROJECTS

Give the following information on the various design projects:

1. *Problem statement:* Fixed quantities, design variables, requirements, and constraints.
2. *Conceptual design:* Sketch various possible systems, evaluate each, and select one for design.
3. *Mathematical modeling of the chosen design:* Obtain the equations that represent the system, constraints, and requirements.

4. *Computer simulation of the system:* Off-design conditions, effect of various parameters.
5. *Economic analysis and optimization of the system:* Discuss the initial cost and the operating costs. Also, outline the control and safety of your design.
6. *Final report:* Description of design and operation of the system, numerical results, drawings, references, specifications of components. The design report should be concise, but complete and detailed.

Complete the design and optimization of the following problems, making appropriate assumptions and obtaining the relevant inputs:

1. A plastic reclamation system employs the extrusion of heated plastic to obtain cylindrical rods from scrap plastic. Design a system to heat and extrude the plastic at the rate of 10 kg/h, assuming the maximum allowable temperature to be 250°C and the minimum temperature for extrusion to be 100°C.
2. In a solar energy collection and storage system, hot water is available at 90°C, when the ambient temperature is 20°C. If a total flow rate of 500 kg/s is possible, design a system for power generation, employing a low-boiling fluid like R134a for the turbines.
3. A storage room 3 m × 3 m × 3 m is to be maintained at −5°C, when the ambient temperature is 30°C in Arizona, for food storage. Design a vapor refrigeration system for the purpose.
4. Steel is to be heat-treated in a furnace, which contains an inert nitrogen gas atmosphere and which is heated electrically. The process involves steel rods of length 1 m and diameter 20 cm. These are placed on a conveyor, which takes the rods through the furnace. The rods must be heated up to 700°C, with a maximum allowable temperature of 850°C, followed by water-spray cooling after the furnace. Design the furnace and the conveyor for the system.
5. A three-bedroom house in Boston is to be maintained at 25°C during windy winter days when the outside temperature is −15°C. Design a heat pump for heating, assuming a one-story house that has no other heat input.
6. Design a flat-plate solar collector system for heating applications. It is to be used in Denver, Colorado, in winter and we wish to obtain water at 60°C at the rate of 200 kg/h. Assume solar flux incident on a normal surface to be 200 W/m^2.
7. In a manufacturing process, only the surface of spherical steel balls of diameter 10 cm is to be heat-treated. Interest lies in heating the material to a depth of at least 1 mm, with a maximum of 2 mm, and then cooling it in air. The desired temperature for heat treatment is 550°C, with a maximum of 650°C. Design a system to accomplish this at the rate of one ball per second.
8. Design the heat exchanger system to remove the heat rejected by a power plant of 500 MW capacity and 33% efficiency. Assume that the heat exchangers are cooled by cold water at 10°C from a neighboring lake.
9. An industrial heat rejection system supplies 20 MW of thermal energy in the form of hot water at 60 K above the ambient temperature. Design a system to recover as much as possible of this waste energy and to provide it as electricity.
10. Design an internal combustion engine, using the Otto cycle, to obtain a power of 150 kW. The maximum temperature and pressure in the system must not exceed 2000 K and 2.5 MPa, respectively. A heat loss of 10% to 15% of the total energy input may be assumed to the surroundings.
11. A piece of electronic equipment dissipates a total of 400 W. Its base dimensions must not exceed 40 cm × 30 cm and the height must be less than 15 cm. A maximum of six boards can be employed to mount the components. Design a system to restrict the temperature anywhere in the system to less than 120°C, using an appropriate cooling method.
12. Design a hot water storage and supply system to fit in a cubic region of side 1 m. An electric heater located in the water tank is to be used for the energy input, and the inflow of water

is at 20°C. The hot water must be supplied at 70°C ± 5°C, at a flow rate of 0 to 5 kg/s. The system must be capable of handling both steady-flow and transient situations.

13. Design a continuous thermal annealing system in which spherical aluminum pieces of diameter 2 cm are heat treated by moving them on a conveyor belt and heating them by overhead radiation lamps. The pieces must undergo fast heating to a given temperature of 350°C, held at this temperature for a given time period of 30 seconds, and then cooled slowly for annealing at a rate of less than 1.5°C/min. The system is to be designed to achieve the given temperature cycle, within an allowable tolerance of ±5%. The conveyor, the heating configuration, and the enclosure have to be designed and fabricated.
14. Design an experimental system for the visualization of flow over 2 to 5 isolated heat sources, representing electronic components, located on a flat horizontal plate in a wind tunnel. The system must be able to move, rotate, and position the sources, as well as the board, and then visualize the flow by means of smoke and a schlieren optical system. The heat input by each source is 100 W and the airflow velocity ranges from 4 to 15 m/s. The sources may be taken to be 1 cm wide and 0.5 cm high, being large in the transverse direction so that a two-dimensional flow may be assumed.
15. Design an applicator for polymer coating of optical fibers. The manufacture of optical fibers involves a process where a thin coat of polymer coating is applied on the fiber. To maintain strength and integrity, coatings must be concentric, continuous, and free of bubbles. Many experimental coating applicators have limitations of performance because of their large size and difficulty of alignment. The fiber diameter is 200 μm and the speed is 5 to 15 m/s. The clearance between the fiber and applicator surfaces should not exceed 50 μm. Design an applicator that will meet these requirements.
16. Model and design a carbon furnace for use in a vacuum chamber. A cylindrical carbon furnace is used in the materials processing of special reactive materials, and the heating elements, which are made of carbon, break. The heating capacity of the furnace depends upon these element sizes, and their arrangement. The design problem would be to choose the right size and number of elements for uniform heating. Ease of repair is also a design criterion. The maximum temperature should be 2000 K ± 5%, and the height and length of the furnace are given as 0.4 m and 0.2 m, respectively.
17. Design a warm air medication dispensing system. The employment of aerosol as a method for drug dispensing is used widely with both liquid and dry powder. What all devices that produce aerosolized drugs have in common is that the inhaled aerosol is at, or below, room temperature. Preliminary testing with liquid aerosol has shown that a warm aerosol is more comfortable to inhale and thus more effective than a cold aerosol. The scope of this project is to design and build the handheld portion of an aerosol delivery system that delivers warm aerosol at the mouthpiece. Ideal temperatures are around body temperature. Use typical breathing air volume for calculating the flow rate. Assume fluid properties to be those of air.
18. An impingement oven is to be designed for cooking. Hot air at temperatures in the range 80°C–120°C impinges on food items that are placed on a conveyor. If the total load is 50 kg/hour, obtain a conceptual design, formulate the design problem, and obtain an acceptable design.
19. Hollow hemispherical glass bowls, each weighing 4 kg and 20 cm in diameter, are to be annealed after forming. The temperature gradient must be less than 10°C/min for the annealing process. Design a system to anneal bowls that are at temperature 750°C after the forming process. Annealing is complete at a temperature of 300°C and the system should be able to anneal 10 bowls per minute.

Appendix A: Computer Programs

MATLAB

A.M.1: Common MATLAB commands for matrices
A.M.2: Common MATLAB commands for polynomials
A.M.3: Programs for root solving
A.M.4: Solution of a system of linear or nonlinear equations
A.M.5: Interpolation and curve fitting
A.M.6: Solution of ordinary differential equations
A.M.7: Solution of partial differential equations

FORTRAN

A.F.1: Gaussian elimination method for a tridiagonal system of linear equations
A.F.2: Gauss-Seidel iterative method for solving a system of linear equations
A.F.3: Secant method for finding the roots of an algebraic equation
A.F.4: Newton-Raphson method for finding the real roots of an algebraic equation
A.F.5: Successive over-relaxation method for the Laplace equation
A.F.6: Successive substitution method for solving the system of nonlinear algebraic equations in Example 4.6
A.F.7: Runge-Kutta method for second-order ODE

A NOTE ON THE COMPUTER PROGRAMS

This appendix presents several computer programs that are commonly used for the numerical modeling and simulation of thermal systems and processes. The numerical methods covered here include those for curve fitting, solution of a system of linear equations, root solving, solution of ordinary and partial differential equations, and solution of a system of nonlinear equations. Details on the methods used are given in Chapter 3 and Chapter 4, as well as in references cited therein.

The programs are written for MATLAB, which has become the most frequently used numerical computing environment and software for solving mathematical equations that arise in scientific and engineering problems, and in Fortran, which was probably the most common programming language used in engineering applications in the past and which continues to be important even today. The Fortran programs have been compiled and executed on a Unix-based computer system and may easily be modified for other versions of Fortran, programming languages and computer systems. A few MATLAB commands were discussed in Chapter 3 and Chapter 4, and several references for numerical analysis and computer solutions were given. Additional MATLAB commands and programs are given here. The results obtained from some of the programs given in this appendix are discussed in the text. The main purpose of presenting these computer programs is to provide ready access to a few important and simple programs that may be used for obtaining the inputs necessary for the design and optimization of thermal systems. In addition, the programs present the basic algorithm and the logic employed for code development. This information may be used for developing a desired numerical model or for linking with available codes in the public or commercial domain to obtain the results needed for design.

A.M.1

For matrices a and b:

`a.*b  a./b  a.\b`	element by element arithmetic; a and b must have identical rows and columns
`a*b  a/b  a\b`	matrix algebra; a and b must have appropriate rows and columns to perform these operations
`rand(n)`	generates random numbers between 0 and 1 for a n x n matrix
`b=25*rand(3)-10`	generates 3 x 3 matrix of random numbers between -10 and 15
`max(a)`	gives maximum element in one-dimensional array a
`min(a)`	gives minimum element in array a
`max(max(a))`	gives maximum element in matrix a
`min(min(a))`	gives minimum element in matrix a
`[i,j]=find(a== max(max(a)))`	gives row and column where maximum element is located

For system of equations ax = b

`inv(a)`	gives inverse A^{-1} of the matrix A
$aa^{-1} = I$	identity matrix
$x = a^{-1}b$	yields the solution x; b is column vector
Therefore,	
`x = inv(a)*b`	yields the solution vector x
`x = a\b`	backslash operator; also gives the solution x
`[l,u,p] = lu(a)`	L U matrix decomposition; p is permutation matrix
`Y = l/(p*b); x = u/y`	yields the solution x

Output:

```
>> a = 2.0;
>> b = 4.5;
>> s = ['The number that is obtained is ', num2str(a)]
```

Yields

```
The number that is obtained is 2.0
```

```
>> s = sprintf('The number %.5g is modified to %.5g.',a,b)
```

yields

```
The number 2.0 is modified to 4.5.
```

Similarly, try other formats. %.0g gives integers. Try %8.3f for floating point numbers, and so on. Use of disp(s) suppresses the printing of s =

Input:

```
>> x = input ('Enter the value of x, x = ' ) ;
>> x = input ('Initial guess  =' ) ;
```

Functions may be defined by using the function file as:

```
function z = fn1 (x,y)
```

```
z = 2*y + 3*x
end

function [s, p] = addmult (x, y)
% Compute sum and product of two matrices
s = x + y;
p = x*y;
end
```

Thus

```
>>[s, p] = addmult (x,y)
```

Will yield the values of the sum and product, respectively

These functions are stored as fn1.m and addmult.m, respectively, and can be used in the computations as functions.

Functions may also be defined within the program by the use of *inline* function as:

```
f = inline ('x.^3 + 3 * x.^2 - 4*x + 2')
f1 = inline ('(2/(pi^0.5)*exp(-x.*x)', 'x' ) ;
f2 = inline ('exp(x) +x^2', 'x' ) ;
```

Anonymous functions may also be used to define functions as:

```
g=@(x) sin(x)/x
to give g(2) for x = 2
g=@(x,r) sin(r*pi*x)/x
to give g(3, 2) for x = 3 and r = 2
```

A.M.2

Polynomials

Roots

```
>> p = [1 -4 7 -6 2]
```

represents

$x^4 - 4x^3 + 7x^2 - 6x + 2$

```
>> r = roots(p)
```

gives the roots as:

```
1.00 + 1.00 i
1.00 - 1.00 i
1.00
1.00

>> pp = poly(r)
pp =
```

```
1.00   -4.00   7.00 +0.00 i   -6.00 - 0.00 i   2.00 + 0.00 i
```

gives the polynomial with the array r as the roots.

Algebra

```
>> a = [1 2 3 4];
>> b = [1 4 9 16];
>> c =  conv (a,b)                  % convolution (multiplication) of
                                      polynomials
c =
1 6 20 50 75 84 64
>> d = a + b                        % addition
d =
2 6 12 20
  >> d = b - a                      % subtraction
d =
0 2 6 12
  >> [q,r] =deconv(c,b)             % division
q =
  1 2 3 4                           % quotient polynomial
r =
  0 0 0 0                           % remainder polynomial
>> g = [1 6 20 48 69 72 44];
>> h = polyder(g)
h =
6 30 80 144 138 72
>> x = [0 .1 .2 .3 .4 .5 .6 .7 .8 .9 1.0];
>> y = [-.45 1.98 3.28 6.16 7.08 7.34 7.66 9.56 9.48 9.30 11.2];
```

Curve Fitting, Plotting

```
>> n = 2;                           order of the polynomial for best fit
>> p = polyfit(x,y,n)               gives best fit with nth order polynomial
p =
-9.81 20.13 -0.03                   arranged in descending powers of x

>> xi = linspace(0, 1, 100);        100 evenly spaced points between 0 and 1
>> yi = polyval(p, xi);             values of polynomial at given x values
>> plot (x,y,'g*',xi,yi,'b-')       plotting with given symbols, color, labels
                                    and title
>> xlabel('x'), ylabel('y = f(x)')
>> title ('Second Order Curve Fitting')
```

A.M.3 ROOT SOLVING

A.M.3.1

```
% Bisection Method for Finding Roots of Equation f(x) = 0
%
format short g
%
%
f=inline('0.6*5.67*10^-8*(850^4-x^4)-40*(x-350)');
```

```
%
% Enter limits of the domain
%
a = input ('Enter lowest value of interval, a =');
b = input ('Enter highest value of interval, b =');
%
% Apply Bisection method
%
for i = 1:20
fa = f(a);
fb = f(b);
c(i) = (a+b)/2;
fc = f(c(i));
%
% Check for convergence
%
if(abs(fc) <= 0.02)
disp (sprintf('Iteration converged'))
break
end
%
% Next iteration
%
if(fa*fc < 0)
b = c(i);
else
a = c(i);
end
end
c=c'
```

A.M.3.2

```
% Secant Method for Root Solving
%
function [p1,err,k] = secant(f,p0,p1,delta,max1)
%
% Apply Secant method
%
for k=1:max1
p2=p1 - feval(f,p1)*(p1-p0)/(feval(f,p1)-feval(f,p0));
%
% Calculate error
%
err=abs(p2-p1);
%
% Update values
%
p0=p1;
p1=p2;
%
```

```
% Apply convergence condition
%
if (k>2)&(err<delta)
break
end
end
%
% Indicate convergence not achieved
%
if(k==max1)
disp('Max number of iterations reached')
end
sprintf('The root is = %7.3f',p1)
```

A.M.3.3

```
%      Newton-Raphson Method for Root Solving
%
%      Given equation: f1(x) = exp(x) - x^2 = 0
%
format short g
%
f1=@(x) exp(x)-x^2;
%
% Enter starting value of the root
%
x(1) = input('Enter the initial guess, x(1) = ');
%
% Apply Newton-Raphson method
%
for i=1:20
x(i+1) = x(i) - f1(x(i))/(exp(x(i))-2*x(i));
%
% Check for convergence
%
if(abs(x(i+1)-x(i)) <= 0.01)
disp(sprintf('Iteration converged'))
break
end
end
x=x'
```

A.M.3.4

```
% Successive Substitution Method for Root Solving
%
% Given equation: f(x) = 0, rewritten as x = z = g(x)
%
% The equation considered is the one obtained by
% eliminating P from the two equations in Example 4.7 to obtain a
% single nonlinear equation for R
```

```
%
% conv is the convergence parameter, x is the
% current approximation to the root and z the next approximation
%
% Enter initial guess for the root and convergence % parameter
%
x = input('Enter the value of x, x =  ');
conv=input('Enter the value of Convergence Parameter, conv=');
fprintf('X=  %.2f   CONV=  %.4f\n',x,conv);
%
% Apply successive substitution
%
for i=1:20
z = (((((15-x)/(7.5*10^-5))^ .5)-80)/10.5)^ .6;
fprintf('X=%.4f  Z=%.4f\n',x,z);
%
% Check for divergence of scheme
%
if abs(z-x)>1/conv
disp('Convergence not achieved');
break
end
%
% Check for convergence
%
if abs(z-x)<conv
%
% Print results
%
fprintf('THE REQUIRED ROOT IS X=%.4f\n',x);
break
elseif abs(z-x)>=conv
x=z;
end
end
```

A.M.4 LINEAR OR NONLINEAR EQUATIONS

A.M.4.1

```
% Direct Solution of a System of Linear Equations
%
% Enter given coefficients and constants
%
% Solving the problem given in Example 4.1
%
a = [2 1 0 6; 5 2 0 0; 0 7 2 2; 0 0 8 9];
b = [64;37;66;104];
%
% Obtain solution by matrix inversion
```

```
%
x = inv(a)*b;
x1 = a\b;
%
% Obtain solution by LU decomposition
%
[l,u,p] = lu(a);
y = l\(p*b);
x3 = u\y;
x
x1
x3
```

A.M.4.2

```
% Gauss-Seidel Method for Linear Equations
%
% Solving the problem given in Example 4.1
%
% Enter Initial Guess
%
x=[0 0 0 0];
%
% Gauss Seidel Iteration
%
for k=1:100
%
% Store Old Values
%
xold=x;
%
% Calculate New Values
%
x(1)=(37-2*x(2))/5;
x(2)=(66-2*x(3)-2*x(4))/7;
x(3)=(104-9*x(4))/8;
x(4)=(64-2*x(1)-x(2))/6;
%
% Check for Convergence
%
if abs(x-xold)<=0.001
disp(sprintf('No. of iterations = %g x(1)= %5.3f x(2)= %5.3f x(3)...
= %5.3f,x(4)=%5.3f',k,x))
break
end
end
```

A.M.4.3

```
% Gauss-Seidel Method for Finite Difference Solution of ODE
%
```

```
% Problem in Example 4.4
%
% Enter initial guess
%
for i=1:31
x(i)=0.0;
xold(i)=0.0;
end
S=50.41*(0.01^2)+2
%
% Enter boundary values
%
x(1)=100;
x(31)=100;
%
% Apply Gauss-Seidel method
%
for k=1:1000
xold(i)=x(i);
for i=2:30
x(i)=(x(i+1)+x(i-1))/S;
end
%
% Check for convergence
%
if abs(x(i)-xold(i))<=0.001
sprintf('The Solution is:')
%
% Print the results obtained
%
x
break
end
end
%
% Plot the results
%
j=1:31;
plot(j,x)
```

A.M.4.4

```
% Gaussian Elimination for a Tridiagonal Coefficient Matrix (TDMA)
%
% Problem in Example 4.4
%
% n is the number of unknowns, s is a parameter from the problem
% being solved, a, b and c are coefficients in the tridiagonal
% matrix, f is the constant vector and tp is the physical
% temperature
%
```

```
% Enter input data
%
s=0.071^2;
n=29;
f(1)=100;f(29)=100;
f(2:28)=0;
a(2:n)=-1;b(1:n)=2+s;c(1:n-1)=-1;
%
% Apply tridiagonal matrix algorithm
%
for i=2:n;
d=a(i)./b(i-1);
b(i)=b(i)-c(i-1).*d;
f(i)=f(i)-f(i-1).*d;
end
%
% Apply back-substitution
%
t(n)=f(n)./b(n);
for i=1:n-1;
j=n-i;
t(j)=(f(j)-c(j).*t(j+1))./b(j);
end
%
% Plot the results obtained
%
tp(2:30)=t(1:29)+20;
tp(1)=120;tp(31)=120;
x=linspace(0,30,31);
plot(x,tp,'k')
xlabel('Distance x (cm)', 'Fontsize' ,14)
ylabel( 'Physical Temperature Tp (Degrees C)', 'Fontsize' ,14)
```

A.M.4.5

```
% Successive Substitution Method for Solving a System of Nonlinear
% Algebraic Equations
%
% Problem in Example 4.6
%
% ep is the convergence parameter, b, p, f1, f2 are parameters
% in the problem, and c is the total flow rate of the mixture
% entering the plant
%
ep=0.0000001;
b=0.1;
c=180.0;
bo=b;
                disp('ARGON     TOTAL FLOW  AMMONIA')
for i=1:50
    f1=0.9/(1.0-b);
```

```
    p=1.0-0.57*exp(-0.0155*f1);
    f2=90.0/(1.0-b*p);
    b=1.0-23.5/(4.0*f2*p+f1);
    c=f1+4.0*f2;
    d=0.57*exp(-0.0155*f1)*2.0*f2;
        fprintf('%.4f    %.4f    %.4f\n',f1,c,d)
    if (abs(b-bo))<ep
        disp('Iteration has converged')
        disp('Converged results are')
        fprintf('ARGON=%.4f   TOTAL FLOW= %.4f   AMMONIA=. . .
        %.4f\n',f1,c,d)
        break
    end
    bo=b;
end
```

A.M.4.6

```
% Newton's Method for Solving a System of Nonlinear Algebraic
% Equations
%
% Problem in Example 4.7
%
% r and p are parameters in the problem, ep is the convergence
% parameter and dr, dp are the increments in r and p, respectively
%
% Enter starting values
%
r = input('Enter the value of parameter r, r =');
p = input('Enter the value of parameter p, p =');
ep = input('Enter the value of convergence parameter ep, ep =');
for i=1:10
    r1=((p-80)/10.5)^0.6-r;
    p1=((15-r)*(10^6)/75)^0.5-p;
    b=r1^2+p1^2;
%
% Check for convergence
%
            if b<ep
                disp('THE REQUIRED SOLUTION IS:')
    fprintf('The flow rate R = %.4f   The pressure P = %.4f\n',r,p)
    break
            end
%
% Calculate partial derivatives
%
    rr=-1;
    rp=3/(5*(10.5^0.6)*((p-80)^0.4));
    pr=-1/(2*((7.5*10^-5)^0.5)*((15-r)^0.5));
    pp=-1;
    d=rr*pp-rp*pr;
```

```
%
% Determine increments for the next iteration
%
   dr=(-r1*pp+p1*rp)/d;
    dp=(-p1*rr+r1*pr)/d;
%
% Calculate values of r and p for the next iteration
%
   r=r+dr;
    p=p+dp;
%
%       Print results
%
    fprintf('R =%.4f    P =%.4f\n',r,p)
end
```

A.M.5

A.M.5.1

```
% Interpolation
%
% Exact fit with general form of the polynomial given in Eq. (3.29)
%
% Enter data
%
x=[1 2 3 4 5 6];
y=[106.4 57.79 32.9 19.52 12.03 7.67]';
%
% Fifth order polynomial for exact fit of 6 data points
% Form Matrix for exact fit
%
c=[x.^0;x;x.^2;x.^3;x.^4;x.^5]';% Find coefficients of polynomial
%
disp('Coefficients of the polynomial are:')
a=c\y
plot(x,y,'*')
hold
%
% Find value at x = 3.4
%
p=a(6:-1:1);
x1=3.4;
y1=polyval(p,x1);
fprintf('Interpolated value from exact fit y = %.4f\n',y1)
```

A.M.5.2

```
% Use of Matlab functions
%
y2=interp1(x,y',x1,'linear');
y3=interp1(x,y',x1,'spline');
```

```
fprintf('Value from linear interpolation y = %.4f\n',y2)
fprintf('Value from spline interpolation y = %.4f\n',y3)
x=linspace(1,6,20);
y=polyval(p,x);
plot(x,y,'-g')
xlabel('x','Fontsize',14);ylabel('y','Fontsize',14)
```

A.M.5.3

```
% Curve Fitting
%
% Enter data
%
x=[1 2 3 4 5 6];
y=[106.4 57.79 32.9 19.52 12.03 7.67];
%
% Use MATLAB function "polyfit" for curve fitting
%
p1=polyfit(x,y,1)
p2=polyfit(x,y,2)
%
% Plot results
%
xi=linspace(1,6,100);
z1=polyval(p1,xi);
z2=polyval(p2,xi);
plot(x,y,'*',xi,z1,'g',xi,z2,'b')
```

A.M.5.4

```
% Exponential Expressions
%
% Solving problem given in Example 3.8
%
% Enter data
x=[0.2  0.6  1.0  1.8  2.0  3.0  5.0  6.0  8.0];
y=[146.0  129.5  114.8  90.3  85.1  63.0  34.6  25.6  14.1];
%
% Define New Variables for Linearization
%
z=log(y);
%
% Use MATLAB function "polyfit" for curve fitting
%
p=polyfit(x,z,1)
%
% Obtain Constants A and a
%
  a=p(1)
  A=exp(p(2))
```

A.M.5.5

```
% Plotting Polynomials
 x=0:0.01:2.5;
 p1=[1 -4 7 -6 2];
 p2=[1 -10 35 -50 24];
 p3=[1 -7 17 -17 6];
 y1=polyval(p1,x);
 y2=polyval(p2,x);
 y3=polyval(p3,x);
 plot(x,y1,'g',x,y2,'b',x,y3,'m')
 hold on
 y = zeros(size(x));
 plot(x,y,'r')
```

A.M.6

A.M.6.1

```
% Ordinary Differential Equations
%
% Given ODE: dh/dt = [6x10^-4 - 3x10^-4 x h^0.5]/0.03
%
% Enter given ODE
%
dhdt=inline('(6*10^(-4) - 3*10^(-4)*(h^(0.5)))/0.03','t','h');
%
% Choose step size and total time
%
dt=10;
tn=200*dt;
h0=0;
t=(0:dt:tn)';
n=length(t);
h=h0*ones(n,1);
%
% Euler's Method
%
for j=2:n;
  h(j)=h(j-1)+dt*dhdt(t(j-1),h(j-1));
end
  plot(t,h,'-b')
%
% Heun's Method
%
t=(0:dt:tn)';
for j=2:n
  h1=h(j-1)+dt*dhdt(t(j-1),h(j-1));
  h(j)=h(j-1)+(dt/2)*(dhdt(t(j),h1) + dhdt(t(j-1),h(j-1)));
end
  hold on
```

```
  plot(t,h,'-g')
%
% Using ode23
%
[t,h] = ode23(dhdt,2000,0);
Plot(t,h,'-r')
```

A.M.6.2

```
% Fourth Order Runge-Kutta Method
%
% Similar to problem in Example 4.3
%
% Enter the function f for the ODE dv/dt=f(t,v)
%
f=inline('-9.8-(0.01*v+0.001*v^2)');
%
% Choose time step and enter initial conditions
%
dt=input('Step size dt =');
t=0;
x=0;
v=100.0;
i=1;
%
while v>=0
%
% Initialize variables
%
    q=x;z=v;
    tp(i)=t;xp(i)=x;vp(i)=v;
%
% Apply 4th order Runge-Kutta formulas
%
rk1x=dt*z;
rk1v=dt*f(z);
rk2x=dt*(z+rk1v/2);
rk2v=dt*f(z+rk1v/2);
rk3x=dt*(z+rk2v/2);
rk3v=dt*f(z+rk2v/2);
rk4x=dt*(z+rk3v);
rk4v=dt*f(z+rk3v);
x=q+(rk1x+2*rk2x+2*rk3x+rk4x)/6;
v=z+(rk1v+2*rk2v+2*rk3v+rk4v)/6;
%
% Advance to next time step
%
t=t+dt;
i=i+1;
end
%
```

```
% Plot results
%
plot(tp,xp,'-',tp,vp,'--')
```

A.M.6.3

```
% Higher order ODE
%
% Second order ODE: d²x/dt² = 9.8 - 0.05 (dx/dt)
% Replaced by two first order ODEs: dx/dt = y; dy/dt = 9.8 - 0.05 y
%
% Enter two first-order equations
%
dxdt=inline('y','t','x','y');
dydt=inline('9.8-0.05*y','t','x','y');
%
% Give step size, end point and starting conditions
%
dt=0.5;
tn=40*dt;
x0=0;
y0=0;
t=(0:dt:tn)';
n=length(t);
x=x0*ones(n,1);
y=y0*ones(n,1);
%
% Euler's Method
%
for j=2:n;
  x(j)=x(j-1)+dt*dxdt(t(j-1),x(j-1),y(j-1));
  y(j)=y(j-1)+dt*dydt(t(j-1),x(j-1),y(j-1));
end
 plot(t,x,'-b',t,y,'-g')
```

A.M.6.4

```
% Using ode45
%
% Define function
%
function dydt=rhs(t,y)
dydt=[y(2);9.8-0.05*y(2)];
end

% Solve ODE given by function 'rhs'
%
y0=[0;0];
[t,y] = ode45('rhs',20,y0)
%
```

```
% Then y(1) gives t and y(2) gives y
Plot(t,y,'-b')
```

A.M.6.5

```
% Finite Difference Method for Solving Second-Order ODE
%
% s, p are parameters in the problem, nt is the total number of
% grid points, a, b and c are coefficients in the tridiagonal
% matrix, f is the constant vector, t is the dimensionless
% temperature and tp is the physical temperature
%
% Enter input data
%
p=input('Parameter P = ');
nt=input('Total number of grid points = ');
n=nt-2;
s=2+(p^2)*((1.0/(nt-1))^2);
%
% Enter boundary conditions and form tridiagonal matrix
%
f(1)=1;f(n)=0.5;
f(2:n-1)=0;
a(2:n)=-1;b(1:n)=s;c(1:n-1)=-1;
%
% Apply tridiagonal matrix algorithm
%
for i=2:n;
d=a(i)./b(i-1);
b(i)=b(i)-c(i-1).*d;
f(i)=f(i)-f(i-1).*d;
end
%
% Apply back-substitution
%
t(n)=f(n)./b(n);
for i=1:n-1;
j=n-i;
t(j)=(f(j)-c(j).*t(j+1))./b(j);
end
%
% Calculate resulting temperature distribution
%
tp(2:nt-1)=t(1:n);
tp(1)=1;tp(nt)=0.5;
%
% Plot the results obtained
%
x=linspace(0,1,51);
plot(x,tp,'k')
```

```
xlabel('Distance X', 'Fontsize' ,14)
ylabel( 'Temperature T', 'Fontsize' ,14)
```

A.M.7

A.M.7.1

```
% Forward Time Central Space (FTCS) Method
%
% Solution of 1D transient conduction problem by the FTCS method,
% see Ch. 4, Section 4.2.4
%
% th is the unknown theta, or dimensionless temperature, tint is the
% initial value of th taken as uniform, kmax is the maximum number of
% time steps, kprint the steps after which results are printed or
% plotted,dx is the grid size, dt the time step, n the number of grid
% points, k represents the time step and i the spatial grid point
%
% Enter starting values
%
tint=input('Enter the initial condition tint = ');
n=input('Enter number of grid points n = ');
kmax=input('Enter maximum number of time steps kmax = ');
kprint=input('Time steps after which results are plotted kprint = ');
%
% Specify boundary conditions
%
th(1,2:n)=tint;
th(1:kmax,1)=1.0;
dx=1/(n-1);
%
% Calculate maximum time step to avoid numerical instability
%
dt=(dx^2)/2;
    for k=2:kmax;
%
% Apply FTCS method
%
        for i=2:n-1;
        th(k,i)=th(k-1,i)+dt*(th(k-1,i+1)-2*th(k-1,i) . . .
        +th(k-1,i-1))/(dx^2);
        end
    end
%
% Store results for plotting
%
    for j=1:10;
        m=kprint*j+1;
        time=(m-1)*dt;
        fprintf('Time = %.4f\n',time)
        tp(j,1:n)=th(m,1:n);
```

```
    end
%
%     Plot results
%
x=linspace(0,1,n);
plot(x,tp)
xlabel('X');ylabel('Dimensionless temperature, \theta');
title('Temperature Versus Distance at Different Times')
```

A.M.7.2

```
% Crank-Nicolson Method
%
% See Example 4.5
%
% t is the unknown dimensionless temperature, tint is the initial
% value of t taken as uniform,kmax is the maximum number of time
% steps, kprint the steps after which results are printed or plotted,
% dx is the grid size, dt the time step, n the number of grid points,
% k represents the time step and i the spatial grid point, and
% a, b, c and f are the parameters of the tridiagonal system
%
% Enter starting values
%
tint=input('Enter the initial condition tint = ');
n=input('Enter number of grid points n = ');
dt=input('Enter the time step dt = ');
kmax=input('Enter maximum number of time steps kmax = ');
kprint=input('Time steps after which results are plotted . . .
kprint = ');
%
% Specify boundary conditions
%
t(1,2:n)=tint;
t(1:kmax,1)=1.0;
dx=1/(n-1);
%
% Calculate the parameters of the tridiagonal system
%
for k=2:kmax;
    a(1:n-2)=-dt/(2*dx^2);
    b(1:n-2)=1+dt/(dx^2);
    c(1:n-2)=-dt/(2*dx^2);
    for i=2:n-1;
        f(i-1)=t(k-1,i)+dt*(t(k-1,i+1)-2*t(k-1,i)+t(k-1,i-1))/ . . .
        (2*dx^2);
    end
    f(1)=f(1)-a(1)*t(k,1);
    a(n-2)=a(n-2)-c(n-2)/3;
    b(n-2)=b(n-2)+4*c(n-2)/3;
%
```

```
% Use the TDMA function file to obtain temperatures at the next
% time step
%
    t(k,2:n-1)=tdma(a,b,c,f,n-2);
%
% Apply boundary condition at the right boundary
%
    t(k,n)=4*t(k,n-1)/3-t(k,n-2)/3;
end
%
% Store results for plotting
%
    for j=1:10;
        m=kprint*j+1;
        time=(m-1)*dt;
        fprintf('Time = %.4f\n',time)
        tp(j,1:n)=t(m,1:n);
  end
%
% Plot results
%
x=linspace(0,1,n);
plot(x,tp)
xlabel('X');ylabel('Dimensionless Temperature, \theta');
title('Temperature Versus Distance at Different Times')

% Function tdma
function t = tdma( a,b,c,f,n)
for i=2:n;
d=a(i)./b(i-1);
b(i)=b(i)-c(i-1).*d;
f(i)=f(i)-f(i-1).*d;
end
%
%   Apply back-substitution
%
t(n)=f(n)./b(n);
for i=1:n-1;
j=n-i;
t(j)=(f(j)-c(j).*t(j+1))./b(j);
end
```

A.M.7.3

```
% Gauss-Seidel Method for an Elliptic PDE
%
% Laplace Equation
%
% See Ch. 4, Section 4.2.4
%
% m and n are grid points in x and y directions, imax is maximum
```

```
% number of iterations, phi is the unknown dependent variable,
% phiol the value of phi at the previous iteration, and ep the
% convergence parameter
%
% Input given data
%
m=input('Enter number of grid points in x direction m = ');
n=input('Enter number of grid points in y direction n = ');
phint=input('Enter initial guess for phi taken as uniform phint = ');
imax=input('Enter maximum number of iterations imax = ');
ep=input('Enter convergence parameter ep = ');
%
% Calculate grid or mesh lengths
%
dx=1/(m-1);
dy=1/(n-1);
%
% Apply boundary conditions
%
phi(2:m-1,2:n-1)=phint;
phi(1,1:n)=0;
phi(m,1:n)=0;
phi(1:m,1)=0;
phi(1:m,n)=1;
%
% Apply Gauss-Seidel iterative scheme
%
for i=1:imax;
    phiol(1:m,1:n)=phi(1:m,1:n);
    for j=2:m-1;
        for k=2:n-1;
            phi(j,k)=((phi(j+1,k)+phi(j-1,k))/ . . .
            (dx^2)+(phi(j,k+1)+ phi(j,k-1))/(dy^2))/. . .
            (2/(dx^2)+2/(dy^2));
        end
    end
%
% Check for convergence
%
    if abs(phi-phiol)<ep
        break
    end
end
%
% Plot results
%
xp=linspace(0,1,m);
nn=(n+1)/2;
plot(xp,phi(1:m,nn-4),xp,phi(1:m,nn-3),xp,phi(1:m,nn-2), . . .
xp,phi(1:m,nn-1),xp,phi(1:m,nn))
```

A.M.7.4

```
% SOR Method for an Elliptic PDE
%
% Laplace Equation
%
% See Ch.4, Section 4.2.4
%
% m and n are grid points in x and y directions, respectively,
% imax is maximum number of iterations, phi is the unknown
% dependent variable, phiol the value of phi at the previous
% iteration, and ep the convergence parameter
%
% Input given data
%
m=input('Enter number of grid points in x direction m = ');
n=input('Enter number of grid points in y direction n = ');
phint=input('Enter initial guess for phi taken as uniform phint = ');
imax=input('Enter maximum number of iterations imax = ');
ep=input('Enter convergence parameter ep = ');
%
% Calculate grid or mesh lengths
%
dx=1/(m-1);
dy=1/(n-1);
%
% Specify relaxation parameter w
%
w=0.5;
%
% Apply boundary conditions
%
phi(2:m-1,2:n-1)=phint;
phi(1,1:n)=0;
phi(m,1:n)=0;
phi(1:m,1)=0;
phi(1:m,n)=1;
%
% Apply SOR iterative scheme
%
for i=1:imax;
    phiol(1:m,1:n)=phi(1:m,1:n);
    for j=2:m-1;
        for k=2:n-1;
            phi(j,k)=w*((phi(j+1,k)+phi(j-1,k))/(dx^2)+(phi . . .
            (j,k+1)+ phi(j,k-1))/(dy^2))/(2/(dx^2)+2/. . .
            (dy^2))+(1-w) *phiol(j,k);
        end
    end
%
% Check for convergence
```

```
%
    if abs(phi-phiol)<ep
        break
    end
end
%
%   Plot results
%
xp=linspace(0,1,m);
nn=(n+1)/2;
plot(xp,phi(1:m,nn-4),xp,phi(1:m,nn-3),xp,phi(1:m,nn-2), . . .
xp,phi(1:m,nn-1), xp,phi(1:m,nn))
```

A.F.1

```
C     GAUSSIAN ELIMINATION FOR A TRIDIAGONAL SYSTEM
C
C     A(I), B(I) AND C(I) ARE THE THREE ELEMENTS IN EACH ROW OF
C     THE GIVEN SYSTEM OF EQUATIONS, F(I) REPRESENTS THE CONSTANTS
C     ON THE RIGHT-HAND SIDE OF THE EQUATIONS, T(I) ARE THE
C     TEMPERATURE DIFFERENCES TO BE COMPUTED, G IS A PARAMETER
C     DEFINED IN THE GIVEN PROBLEM, N IS THE NUMBER OF EQUATIONS
C     AND TP REPRESENTS THE PHYSICAL TEMPERATURE, WHERE
C     TP = T + 20. THE SYSTEM OF EQUATIONS TO BE SOLVED IS THE
C     ONE GIVEN IN EXAMPLE 4.4.
C
      PARAMETER (IN=30)
      DIMENSION A(IN),B(IN),C(IN),T(IN),F(IN)
C
C     SPECIFY INITIAL PARAMETERS
C
      CALL INPUT(A,B,C,F,N)
      CALL TDMA(A,B,C,F,N,T)
C
C     COMPUTE ACTUAL TEMPERATURES FROM THE TEMPERATURE
C     DIFFERENCES T(I)
C
      WRITE (6,7)
 7    FORMAT(2X,'THE REQUIRED PHYSICAL TEMPERATURES IN CELSIUS ARE'/)
      DO 8 I=1,N
            TP=T(I)+20.0
            WRITE (6,9)I,TP
 8    CONTINUE
 9    FORMAT(2X,'TP(',I2,')=',F10.4)
      STOP
      END
C******************************************************************
C     GET THE INPUT DATA
C******************************************************************
      SUBROUTINE INPUT(A,B,C,F,N)
      PARAMETER (IN=30)
```

```
      DIMENSION A(IN),B(IN),C(IN),F(IN)
      PRINT *, 'GIVE THE VALUE OF N'
      READ *, N
      G=50.41*(0.01**2)
C
C     'FMTDM' FORMS THE TRIDIAGONAL MATRIX AND THE RIGHT HAND SIDE
C     COLUMN MATRIX
C
      CALL FMTDM(G,N,A,B,C,F)
      RETURN
      END
C***************************************************************
C     THE FOLLOWING SUBROUTINE FORMS THE TRIDIAGONAL MATRIX OF
C     THE FORM
C
C     A*T(I-1) + B*T(I) + C*T(I+1) = R
C***************************************************************
      SUBROUTINE FMTDM(G,N,A,B,C,R)
      DIMENSION A(N),B(N),C(N),R(N)
C
C     ENTER THE CONSTANTS ON THE RIGHT-HAND SIDE OF THE EQUATIONS
C
      R(1)=100.0
      R(N)=100.0
      NN=N-1
      DO 1 I=2,NN
      R(I)=0.0
 1    CONTINUE
C
C     ENTER THE MATRIX COEFFICIENTS
C
      DO 2 I=1,N
      B(I)=2.0+G
 2    CONTINUE
      DO 3 I=1,NN
      C(I)=-1.0
 3    CONTINUE
      DO 4 I=2,N
            A(I)=-1.0
 4    CONTINUE
      RETURN
      END
C***************************************************************
C     TRIDIAGONAL MATRIX ALGORITHM
C***************************************************************
      SUBROUTINE TDMA(A,B,C,F,N,T)
C
C     N IS THE ORDER OF THE TRIDIAGONAL MATRIX
C     A IS THE SUBDIAGONAL OF THE TRIDIAGONAL MATRIX
C     B IS THE DIAGONAL OF THE TRIDIAGONAL MATRIX
C     C IS THE SUPERDIAGONAL OF THE TRIDIAGONAL MATRIX
```

```
C     F IS THE RIGHT HAND SIDE VECTOR
C     T IS THE SOLUTION VECTOR
C
      DIMENSION A(N),B(N),C(N),F(N),T(N)
      NN=N-1
      DO 5 I=2,N
            D=A(I)/B(I-1)
            B(I)=B(I)-C(I-1)*D
            F(I)=F(I)-F(I-1)*D
 5    CONTINUE
C
C     APPLY BACK SUBSTITUTION
C
      T(N)=F(N)/B(N)
      DO 6 I=1,NN
            J=N-I
            T(J)=(F(J)-C(J)*T(J+1))/B(J)
 6    CONTINUE
      RETURN
      END
```

A.F.2

```
C     GAUSS-SEIDEL METHOD FOR SOLVING A SYSTEM OF LINEAR
      EQUATIONS
C
C
C     T(I) REPRESENTS THE TEMPERATURE DIFFERENCES FROM THE AMBIENT
C     TEMPERATURE, TO(I) DENOTES THE TEMPERATURE DIFFERENCES AFTER
C     THE PREVIOUS ITERATION, TP IS THE ACTUAL TEMPERATURE, S IS A
C     CONSTANT DEFINED IN THE PROBLEM AND N IS THE NUMBER OF
C     EQUATIONS. THE PROBLEM CONSIDERED IS THE ONE GIVEN IN
C     EXAMPLE 4.4.
C
C
C     ENTER VALUES OF RELEVANT PARAMETERS
C
      PARAMETER (IN=30)
      DIMENSION T(IN),TO(IN)
      S=50.41*(0.01**2)+2.0
      PRINT *, 'GIVE THE NUMBER OF EQUATIONS : '
      READ (5,*) N
      NN=N-1
      EPS=0.1
C
C     DIFFERENT CONVERGENCE PARAMETER EPS
C
      DO 10 K=1,5
C
C     INPUT STARTING VALUES
C
```

```
      J=0
      DO 1 I=1,N
            T(I)=0.0
 1          CONTINUE
C
C     STORE COMPUTED VALUES AFTER EACH ITERATION
C
 2          DO 3 I=1,N
                  TO(I)=T(I)
 3          CONTINUE
C
C     COMPUTE THE END VALUES T(1) AND T(N)
C
            T(1)=(T(2)+100.0)/S
            T(N)=(100.0+T(N-1))/S
C
C     COMPUTE INTERMEDIATE VALUES
C
            DO 4 I=2,NN
                  T(I)=(T(I+1)+T(I-1))/S
 4          CONTINUE
C
C     CHECK FOR CONVERGENCE
C
            J=J+1
            DO 5 I=1,N
                  IF(ABS(TO(I)-T(I)) .GT. EPS) GO TO 2
 5          CONTINUE
            WRITE(6,6)EPS
 6          FORMAT(//2X,'EPS=',F10.5)
            WRITE(6,7)J
 7          FORMAT(/2X,'NUMBER OF ITERATIONS=',I4/)
C
C     COMPUTE ACTUAL TEMPERATURES
C
            DO 8 I=1,N
                  TP=T(I)+20.0
                  WRITE(6,9)I,TP
 8          CONTINUE
 9          FORMAT(2X,'TP(',I2,')=',F12.4)
            EPS=EPS/10.0
 10   CONTINUE
      STOP
      END
```

A.F.3

```
C     ROOT SOLVING WITH THE SECANT METHOD

C     X IS THE INDEPENDENT VARIABLE, FUN(X) IS THE GIVEN FUNCTION,
C     X1 AND X2 ARE THE X VALUES FROM THE TWO PREVIOUS ITERATIONS,
```

```
C      STARTING WITH THE TWO POINTS BOUNDING THE REGION, X3 IS THE
C      APPROXIMATION TO THE ROOT, F1, F2 AND F3 ARE THE CORRESPONDING
C      VALUES OF THE FUNCTION, AND EPS IS THE CONVERGENCE CRITERION
C      THE FUNCTION USED IS THE ONE IN EXAMPLE 4.2.
C
       EXTERNAL FUN
       PRINT *, 'ENTER THE TWO STARTING VALUES OF X'
       READ (5,*) X1,X2
C
C      STORE STARTING VALUES

       X1I=X1
       X2I=X2
       XOLD=X1
       WRITE(6,12) X1,X2
 12    FORMAT(/10X,'INITIAL X1=',F7.2,10X,'INITIAL X2=',F7.2//)
       EPS=0.01
       DO 2 I=1,4
 1           F1=FUN(X1)
             F2=FUN(X2)
C
C      COMPUTE THE APPROXIMATION TO THE ROOT
C
             X3=(X1*F2-X2*F1)/(F2-F1)
             F3=FUN(X3)
             XNEW=X3
C
C      CHECK FOR CONVERGENCE

             IF (ABS(XNEW-XOLD) .GT. EPS) THEN
                   X1=X2
                   X2=X3
                   XOLD=X3
                   WRITE(6,10)X3,F3
 10                FORMAT(2X,'TEMPERATURE T =',F10.4,4X,'FUNCTION
      $            F(T) =', F12.6)
                   GO TO 1
             ELSE
 11                WRITE(6,13)EPS,X3,F3
 13          FORMAT(//2X,'EPS=',F9.6,4X,'TEMPERATURE T =',F10.4,4X,
      $                  'FUNCTION F(T)=',F12.6//)
             END IF
C
C      VARY CONVERGENCE CRITERION
C
             EPS=EPS/10
             X1=X1I
             X2=X2I
             XOLD=X1
 2     CONTINUE
       STOP
```

```
      END
C
C     DEFINE THE FUNCTION
C
      FUNCTION FUN(X) FUN=(0.6*5.67*((850.0**4.0)-(X**4.0))/
      $      (10.0**8.0))-40.0*(X-350.0)
      RETURN
      END
```

A.F.4

```
C     THIS PROGRAM FINDS THE REAL ROOTS OF AN EQUATION F(X)=0
C     BY THE NEWTON-RAPHSON METHOD
C
C
C
C     HERE X IS THE INDEPENDENT VARIABLE, Y1 THE VALUE OF THE
C     FUNCTION AT X, Y2 THE FUNCTION AT X+0.001, YD THE
C     DERIVATIVE, DX THE INCREMENT IN X FOR THE NEXT ITERATION,
C     EPS THE CONVERGENCE CRITERION ON THE FUNCTION AND XMAX
C     THE MAXIMUM VALUE OF X. THE FUNCTION USED IS THE ONE IN
C     EXAMPLE 4.2.
C
C
C     DEFINE FUNCTION AND SPECIFY INPUT PARAMETERS
C
      EXTERNAL Y
      EPS=0.001
      WRITE(6,15)EPS
 15   FORMAT(2X,'EPS=',F8.4/)
      PRINT *, ' ENTER AN INITIAL GUESS FOR X'
      READ (5,*) X
      XMAX=850.0
 1    Y1=Y(X)
      WRITE(6,10) X,Y1
C
C     CHECK FOR CONVERGENCE
C
      IF (ABS(Y1) .GT. EPS) THEN
            XN=X+0.001
            Y2=Y(XN)
            YD=(Y2-Y1)/0.001
C
C     CHECK IF RESULTS DIVERGE
C
            IF (YD .GE. (1.0/EPS)) GO TO 20
C
C     COMPUTE NEW APPROXIMATION TO THE ROOT
C
            DX=-Y1/YD
            X=X+DX
```

```
            IF (X .GE. XMAX) GO TO 20
            GO TO 1
      ELSE
 5          WRITE(6,12) X,Y1
 12         FORMAT(/2X,'TEMPERATURE T =',F8.4,4X,'FUNCTION
 $          F(T)=',F12.6)
 10         FORMAT(2X,'TEMPERATURE T =',F8.4,4X,'FUNCTION
 $          F(T)=',F12.6)
      END IF
 20   STOP
      END
C
C     DEFINE THE FUNCTION
C
      FUNCTION Y(X)
      Y=(0.6*5.67*((850.0**4.0)-(X**4.0)))/(10.0**8.0)
      $      - 40.0*(X-350.0)
      RETURN
      END
```

A.F.5

```
C     THIS PROGRAM SOLVES THE LAPLACE EQUATION BY EMPLOYING
C     THE SUCCESSIVE OVER RELAXATION (SOR) ITERATION METHOD.
C
C     WHEN THE PROGRAM IS RUN IT PROMPTS FOR THE INPUT VALUES
C     REQUIRED.
C     ENTER THE INPUT VALUES AND YOUR OUTPUT WILL BE IN A FILE CALLED
C     'SOR.DAT'
C
C
C     DESCRIPTION OF INPUT PARAMETERS:
C
C     IL           IS THE NUMBER OF GRID POINTS IN THE X DIRECTION.
C     JL           IS THE NUMBER OF GRID POINTS IN THE Y DIRECTION.
C     DX       IS THE GRID SIZE IN X DIRECTION.
C     DY       IS THE GRID SIZE IN Y DIRECTION.
C     OMEGA    IS THE RELAXATION PARAMETER
C     PHIINT   IS THE INITIAL GUESS FOR PHI TAKEN UNIFORM OVER THE
C              WHOLE DOMAIN.
C     ITMAX    IS THE NUMBER OF MAXIMUM ITERATIONS BEFORE STOPPING.
C     EPSI     IS THE CONVERGENCE CRITERION.
C
C
C     DESCRIPTION OF OTHER VARIABLES:
C
C     PHI      IS THE SOLUTION VARIABLE AT NTH TIME STEP.
C     PHIOL    IS THE SOLUTION VARIABLE AT N-1TH TIME STEP.
C
C
      CHARACTER*2 XFILE(5)
```

```
      CHARACTER*2 YFILE(5)
      DIMENSION PHI(11,11),PHIOL(11,11)
      PRINT*,'ENTER INITIAL GUESS FOR PHI TAKEN UNIFORM OVER THE'
      PRINT*,'WHOLE DOMAIN'
      READ(5,*)PHIINT
      PRINT*,'ENTER GRID SIZE DX=, DY='
      READ(5,*)DX,DY
      PRINT *,'ENTER NO. OF GRID POINTS IL= , JL= '
      PRINT*,' MAXIMUM POSSIBLE IS 11 FOR BOTH IL AND JL,'
      PRINT*,'UNLESS DIMENSION STATEMENTS ARE CHANGED.'
      READ(5,*)IL,JL
      PRINT *,'ENTER THE RELAXATION PARAMETER'
      READ(5,*)OMEGA
      PRINT*,'ENTER MAXIMUM NO. OF ITERATIONS ALLOWED BEFORE
     $       STOPPING'
      READ(5,*)ITMAX
      PRINT *,'ENTER CONVERGENCE CRITERION'
      READ(5,*)EPSI
      PRINT*,'THE INPUT VALUES ARE:'
      PRINT*,'INITIAL GUESS FOR PHI=',PHIINT
      PRINT*,'DX=',DX,'DY=',DY
      PRINT*,'IL=',IL,'JL=',JL
      PRINT*,'MAX NO. OF ITERATIONS=',ITMAX
      PRINT*,'CONVERGENCE CRITERION=',EPSI
      ITERATION=0
C
C     OPEN THE DATA FILES FOR GRAPHING
C
      XFILE(1)='X1'
      XFILE(2)='X2'
      XFILE(3)='X3'
      XFILE(4)='X4'
      XFILE(5)='X5'
      YFILE(1)='Y1'
      YFILE(2)='Y2'
      YFILE(3)='Y3'
      YFILE(4)='Y4'
      YFILE(5)='Y5'
C
C     SET INITIAL DISTRIBUTION OF PHI
C
      DO 51 I=1,IL
            DO 5 J=1,JL
                  PHI(I,J)=PHIINT
 5          CONTINUE
 51   CONTINUE
C
C     START SOLVING FOR PHI.
C
 15   ITERATION=ITERATION+1
      IF(ITERATION.GE.ITMAX)GO TO 40
```

```
C
C     SAVE THE FIELD AT PREVIOUS TIME STEP.
C
      DO 101 I=1,IL
            DO 10 J=1,JL
                  PHIOL(I,J)=PHI(I,J)
 10         CONTINUE
 101  CONTINUE
C
C     DO SOR ITERATIONS ON PHI ON INTERIOR POINTS.
C
      DO 201 J=2,JL-1
            DO 20 I=2,IL-1
                  PHIGS=(PHI(I+1,J)+PHI(I-1,J))/DX**2+
   $                    (PHI(I,J+1)+PHI(I,J-1))/DY**2
                  PHIGS=PHIGS/(2./DX**2+2./DY**2)
                  PHI(I,J)=OMEGA*PHIGS+(1.-OMEGA)*PHIOL(I,J)
 20         CONTINUE
 201  CONTINUE
C
C     IMPOSE THE BOUNDARY CONDITIONS
C
      CALL BCOND(PHI,IL,JL)
C
C     CHECK FOR CONVERGENCE
C
      DO 351 I=1,IL
            DO 35 J=1,JL
                  IF(ABS(PHI(I,J)-PHIOL(I,J)).GE.EPSI)GO TO 15
 35         CONTINUE
 351  CONTINUE
      GO TO 50
 40   PRINT*,'SOLN. DOES NOT CONVERGE IN',ITMAX,'ITERATIONS'
 50   OPEN(UNIT=10,FILE='SOR.DAT')
      WRITE(10,110)EPSI
 110  FORMAT(1X,'CONVERGENCE CRITERION ='1X,E9.1)
      WRITE(10,115)OMEGA
 115  FORMAT(//,1X,'W=',F5.2)
      WRITE(10,120)ITERATION
 120  FORMAT(//,1X,'NO. OF ITERATIONS TO CONVERGE=',1X,I4,//)
      WRITE(10,130)
 130  FORMAT(1X,'PHI DISTRIBUTION IS:',//)
      WRITE(10,140)(I,I=1,IL)
 140  FORMAT(1X,'I=',8X,11(I2,8X))
      DO 60 J=1,JL
            WRITE(10,100)J,(PHI(I,J),I=1,IL)
 60   CONTINUE
 100  FORMAT(1X,'J=',I2,3X,11(F8.5,2X))
C
C     OUTPUT FOR GRAPHICS
C
```

```
      II=1
      DO 70 I=1,5
            II=II+1
            OPEN (UNIT=12,FILE=XFILE(I))
            DO 66 J=1,JL
                  WRITE(12,*)PHI(II,J)
 66         CONTINUE
            CLOSE(UNIT=12)
 70   CONTINUE
      JJ=1
      DO 71 J=1,4
            JJ=JJ+2
            OPEN(UNIT=12,FILE=YFILE(J))
            DO 72 I=1,IL
                  WRITE(12,*)PHI(I,JJ)
 72         CONTINUE
            CLOSE(UNIT=12)
 71   CONTINUE
      OPEN(UNIT=12,FILE='XX')
      DO 73 I=1,IL
            XX=FLOAT(I-1)*DX
            WRITE(12,*)XX
 73   CONTINUE
      CLOSE(UNIT=12)
      OPEN(UNIT=12,FILE='YY')
      DO 74 J=1,JL
            YY=FLOAT(J-1)*DY
            WRITE(12,*)YY
 74   CONTINUE
      CLOSE(UNIT=12)
      STOP
      END
C*****************************************************
      SUBROUTINE BCOND(PHI,IL,JL)
C
C     THIS SUBROUTINE IMPLEMENTS APPROPRIATE BOUNDARY CONDITIONS.
C
      DIMENSION PHI(11,11)
C     SET THE CONDITIONS ON I=1 AND I=IL SURFACES.
C
      DO 25 J=1,JL
            PHI(1,J)=0.
            PHI(IL,J)=0.
 25   CONTINUE
C
C     SET THE CONDITIONS ON J=1 AND J=JL SURFACES
C
      DO 30 I=1,IL
            PHI(I,1)=0.
            PHI(I,JL)=1.
 30   CONTINUE
```

```
      RETURN
      END
```

A.F.6

```
C     THE SUCCESSIVE SUBSTITUTION METHOD FOR NONLINEAR EQUATIONS
C
C
C     HERE F1 IS THE FLOW RATE OF ARGON IN MOLES/S, F2 IS THE FLOW
C     RATE OF NITROGEN, C IS THE TOTAL FLOW RATE, B AND P ARE THE
C     PARAMETERS DEFINED IN THE PROBLEM (EXAMPLE 4.6), D IS THE
C     AMOUNT OF AMMONIA COLLECTED IN MOLES/S, CO IS THE VALUE OF C
C     AFTER THE PREVIOUS ITERATION AND EPS IS THE CONVERGENCE
C     CRITERION APPLIED TO THE TOTAL FLOW RATE C
C
C
C     INPUT OF STARTING VALUES
C
      EPS=0.0001
      B=0.1
      C=180.0
 1    CO=C
C
C     COMPUTATION OF UNKNOWN QUANTITIES
C
            F1=0.9/(1.0-B)
            P=1.0-0.57*EXP(-0.0155*F1)
            F2=90.0/(1.0-B*P)
            B=1.0-23.5/(4.0*F2*P+F1)
            C=F1+4.0*F2
            D=0.57*EXP(-0.0155*F1)*2.0*F2
      WRITE (6,2)F1,C,D
 2    FORMAT(2X,'ARGON:',F12.5,4X,'FLOW:',F12.5,4X,'NH3:', F12.5)
C
C     CONVERGENCE CHECK
C
      IF (ABS(C-CO) .LE. EPS) THEN
            PRINT*,'THE SOLUTION HAS CONVERGED'
            PRINT*,'THE SOLUTION IS:'
            WRITE (6,3)F1,C,D
 3          FORMAT(/2X,'ARGON:',F12.5,4X,'TOTAL FLOW:',F12.5,4X,
   $        'AMMONIA:',F12.5)
            ELSE
                  GO TO 1
      END IF
      STOP
      END
```

A.F.7

```
C     RUNGE-KUTTA METHOD
```

```
C
C      THIS PROGRAM NUMERICALLY SOLVES A SECOND ORDER DIFFERENTIAL
C      EQUATION USING THE 4TH ORDER RUNGE-KUTTA METHOD
C
C      IN THE FOLLOWING PROGRAM
C
C            T STANDS FOR TIME         DT STANDS FOR STEP SIZE IN T
C
C            X STANDS FOR DISPLACEMENT      V STANDS FOR VELOCITY
C
C            A AND B ARE THE CONSTANTS APPEARING IN THE
C            DIFFERENTIAL EQN.
C
C            G IS THE ACCELERATION DUE TO GRAVITY = 9.8 M/(SEC**2)
C
            IMPLICIT REAL (A-H,O-Z)
            OPEN(UNIT=14,FILE='RT')
            OPEN(UNIT=15,FILE='RX')
            OPEN(UNIT=16,FILE='RV')
C
C      VALUES OF T ARE WRITTEN IN FILE RT
C      VALUES OF X ARE WRITTEN IN FILE RX
C      VALUES OF V ARE WRITTEN IN FILE RV
C
C
C      INPUT PARAMETERS
            PRINT*,'INPUT PARAMETERS'
            PRINT*,'A=   ','B=   '
             READ*,A,B
            PRINT*,'DT=   '
            READ*,DT
            G=9.8
C
C      SET INITIAL CONDITIONS
C
            T=0.
            X=0.
            V=100.
            WRITE(14,*)T
            WRITE(15,*)X
            WRITE(16,*)V
C
C      NEXT TIME STEP
C
 11         Q=X
            Z=V
C
C      Q AND Z ARE VALUES OF X AND V, AT PREVIOUS TIME STEP
C
C      CALCULATIONS FOR NEXT STEP USING 4TH ORDER RK METHOD
C
```

```
          RK1X = DT*Z
          RK1V = DT*(-G -A*Z -B*(Z**2))
          RK2X = DT*(Z + RK1V/2.)
          RK2V = DT*(-G -A*(Z + RK1V/2.) -B*(Z + RK1V/2.)**2)
          RK3X = DT*(Z + RK2V/2.)
          RK3V = DT*(-G -A*(Z + RK2V/2.) -B*(Z + RK2V/2.)**2)
          RK4X = DT*(Z + RK3V)
          RK4V = DT*(-G -A*(Z + RK3V) -B*(Z +RK3V)**2)
          X = Q +(RK1X +2.*RK2X + 2.*RK3X + RK4X)/6.
          V = Z +(RK1V + 2.*RK2V + 2.*RK3V + RK4V)/6.
          T = T + DT
C
C     CALCULATIONS ARE STOPPED WHEN V BECOMES ZERO.
C
          IF(V.GT.0.) THEN
          WRITE(14,*)T
          WRITE(15,*)X
          WRITE(16,*)V
          GO TO 11
          END IF
C
C         OUTPUT RESULTS
C
          PRINT*,'THE VELOCITY HAS BECOME ZERO OR NEGATIVE'
          PRINT*,'TOTAL TIME TAKEN TO REACH MAXIMUM HEIGHT
     $    =',T,'SEC'
          PRINT*,'TOTAL HEIGHT REACHED BY THE PROJECTILE =
     $    ',X,'METERS'
          CLOSE(UNIT=14)
          CLOSE(UNIT=15)
          CLOSE(UNIT=16)
          STOP
          END
```

Appendix B: Material Properties

A NOTE ON MATERIAL PROPERTIES

These tables on the properties of common materials are provided for quick reference and convenience. However, for detailed design and optimization of practical systems, the various handbooks, encyclopedias, and references cited in the text should be used instead, for the most appropriate and accurate property data.

TABLE B.1
Properties of Dry Air at Atmospheric Pressure—SI Units

Temperature			Properties							
K	°C	°F	ρ	c_p	c_p/c_v	μ	k	Pr	h	a
100	−173.15	−280	3.598	1.028		6.929	9.248	0.770	98.42	198.4
110	−163.15	−262	3.256	1.022	1.4202	7.633	10.15	0.768	108.7	208.7
120	−153.15	−244	2.975	1.017	1.4166	8.319	11.05	0.766	118.8	218.4
130	−143.15	−226	2.740	1.014	1.4139	8.990	11.94	0.763	129.0	227.6
140	−133.15	−208	2.540	1.012	1.4119	9.646	12.84	0.761	139.1	236.4
150	−123.15	−190	2.367	1.010	1.4102	10.28	13.73	0.758	149.2	245.0
160	−113.15	−172	2.217	1.009	1.4089	10.91	14.61	0.754	159.4	253.2
170	−103.15	−154	2.085	1.008	1.4079	11.52	15.49	0.750	169.4	261.0
180	−93.15	−136	1.968	1.007	1.4071	12.12	16.37	0.746	179.5	268.7
190	−83.15	−118	1.863	1.007	1.4064	12.71	17.23	0.743	189.6	276.2
200	−73.15	−100	1.769	1.006	1.4057	13.28	18.09	0.739	199.7	283.4
205	−68.15	−91	1.726	1.006	1.4055	13.56	18.52	0.738	204.7	286.9
210	−63.15	−82	1.684	1.006	1.4053	13.85	18.94	0.736	209.7	290.5

(Continued)

TABLE B.1 (CONTINUED)
Properties of Dry Air at Atmospheric Pressure—SI Units

Temperature			Properties							
K	°C	°F	ρ	c_p	c_p/c_v	μ	k	Pr	h	a
215	−58.15	−73	1.646	1.006	1.4050	14.12	19.36	0.734	214.8	293.9
220	−53.15	−64	1.607	1.006	1.4048	14.40	19.78	0.732	219.8	297.4
225	−48.15	−55	1.572	1.006	1.4046	14.67	20.20	0.731	224.8	300.8
230	−43.15	−46	1.537	1.006	1.4044	14.94	20.62	0.729	229.8	304.1
235	−38.15	−37	1.505	1.006	1.4042	15.20	21.04	0.727	234.9	307.4
240	−33.15	−28	1.473	1.005	1.4040	15.47	21.45	0.725	239.9	310.6
245	−28.15	−19	1.443	1.005	1.4038	15.73	21.86	0.724	244.9	313.8
250	−23.15	−10	1.413	1.005	1.4036	15.99	22.27	0.722	250.0	317.1
255	−18.15	−1	1.386	1.005	1.4034	16.25	22.68	0.721	255.0	320.2
260	−13.15	8	1.359	1.005	1.4032	16.50	23.08	0.719	260.0	323.4
265	−8.15	17	1.333	1.005	1.4030	16.75	23.48	0.717	265.0	326.5
270	−3.15	26	1.308	1.006	1.4029	17.00	23.88	0.716	270.1	329.6
275	−1.85	35	1.235	1.006	1.4026	17.26	24.28	0.715	275.1	332.6
280	6.85	44	1.261	1.006	1.4024	17.50	24.67	0.713	280.1	335.6
285	11.85	53	1.240	1.006	1.4022	17.74	25.06	0.711	285.1	338.5
290	16.85	62	1.218	1.006	1.4020	17.98	25.47	0.710	290.2	341.5
295	21.85	71	1.197	1.006	1.4018	18.22	25.85	0.709	295.2	344.4
300	26.85	80	1.177	1.006	1.4017	18.46	26.24	0.708	300.2	347.3
305	31.85	89	1.158	1.006	1.4015	18.70	26.63	0.707	305.3	350.2
310	36.85	98	1.139	1.007	1.4013	18.93	27.01	0.705	310.3	353.1
315	41.85	107	1.121	1.007	1.4010	19.15	27.40	0.704	315.3	355.8
320	46.85	116	1.103	1.007	1.4008	19.39	27.78	0.703	320.4	358.7
325	51.85	125	1.086	1.008	1.4006	19.63	28.15	0.702	325.4	361.4
330	56.85	134	1.070	1.008	1.4004	19.85	28.53	0.701	330.4	364.2
335	61.85	143	1.054	1.008	1.4001	20.08	28.90	0.700	335.5	366.9
340	66.85	152	1.038	1.008	1.3999	20.30	29.28	0.699	340.5	369.6
345	71.85	161	1.023	1.009	1.3996	20.52	29.64	0.698	345.6	372.3
350	76.85	170	1.008	1.009	1.3993	20.75	30.03	0.697	350.6	375.0
355	81.85	179	0.9945	1.010	1.3990	20.97	30.39	0.696	355.7	377.6
360	86.85	188	0.9805	1.010	1.3987	21.18	30.78	0.695	360.7	380.2
365	91.85	197	0.9672	1.010	1.3984	21.38	31.14	0.694	365.8	382.8
370	96.85	206	0.9539	1.011	1.3981	21.60	31.50	0.693	370.8	385.4
375	101.85	215	0.9413	1.011	1.3978	21.81	31.86	0.692	375.9	388.0
380	106.85	224	0.9288	1.012	1.3975	22.02	32.23	0.691	380.9	390.5
385	111.85	233	0.9169	1.012	1.3971	22.24	32.59	0.690	386.0	393.0
390	116.85	242	0.9050	1.013	1.3968	22.44	32.95	0.690	391.0	395.5
395	121.85	251	0.8936	1.014	1.3964	22.65	33.31	0.689	396.1	398.0
400	126.85	260	0.8822	1.014	1.3961	22.86	33.65	0.689	401.2	400.4
410	136.85	278	0.8608	1.015	1.3953	23.27	34.35	0.688	411.3	405.3
420	146.85	296	0.8402	1.017	1.3946	23.66	35.05	0.687	421.5	410.2

TABLE B.1 (CONTINUED)
Properties of Dry Air at Atmospheric Pressure—SI Units

Temperature			Properties							
K	°C	°F	ρ	c_p	c_p/c_v	μ	k	Pr	h	a
430	156.85	314	0.8207	1.018	1.3938	24.06	35.75	0.686	431.7	414.9
440	166.85	332	0.8021	1.020	1.3929	24.45	36.43	0.684	441.9	419.6
450	176.85	350	0.7342	1.021	1.3920	24.85	37.10	0.684	452.1	424.2
460	186.85	368	0.7677	1.023	1.3911	25.22	37.78	0.683	462.3	428.7
470	196.85	386	0.7509	1.024	1.3901	25.58	38.46	0.682	472.5	433.2
480	206.85	404	0.7351	1.026	1.3892	25.96	39.11	0.680	482.8	437.6
490	216.85	422	0.7201	1.028	1.3881	26.32	39.76	0.680	493.0	442.0
500	226.85	440	0.7057	1.030	1.3871	26.70	40.41	0.680	503.3	446.4
510	236.85	458	0.6919	1.032	1.3861	27.06	41.06	0.680	513.6	450.6
520	246.85	476	0.6786	1.034	1.3851	27.42	41.69	0.680	524.0	454.9
530	256.85	494	0.6658	1.036	1.3840	27.78	42.32	0.680	534.3	459.0
540	266.85	512	0.6535	1.038	1.3829	28.14	42.94	0.680	544.7	463.2
550	276.85	530	0.6416	1.040	1.3818	28.48	43.57	0.680	555.1	467.3
560	286.85	548	0.6301	1.042	1.3806	28.83	44.20	0.680	565.5	471.3
570	296.85	566	0.6190	1.044	1.3795	29.17	44.80	0.680	575.9	475.3
580	306.85	584	0.6084	1.047	1.3783	29.52	45.41	0.680	586.4	479.2
590	316.85	602	0.5980	1.049	1.3772	29.84	46.01	0.680	596.9	483.2
600	326.85	620	0.5881	1.051	1.3760	30.17	46.61	0.680	607.4	486.9
620	346.85	656	0.5691	1.056	1.3737	30.82	47.80	0.681	628.4	494.5
640	366.85	692	0.5514	1.061	1.3714	31.47	48.69	0.682	649.6	502.1
660	386.85	728	0.5347	1.065	1.3691	32.09	50.12	0.682	670.9	509.4
680	406.85	764	0.5189	1.070	1.3668	32.71	51.25	0.683	692.2	516.7
700	426.85	800	0.5040	1.075	1.3646	33.32	52.36	0.684	713.7	523.7
720	446.85	836	0.4901	1.080	1.3623	33.92	53.45	0.685	735.2	531.0
740	466.85	872	0.4769	1.085	1.3601	34.52	54.53	0.686	756.9	537.6
760	486.85	903	0.4643	1.089	1.3580	35.11	55.62	0.687	778.6	544.6
780	506.85	944	0.4524	1.094	1.3559	35.69	56.68	0.688	800.5	551.2
800	526.85	950	0.4410	1.099	1.354	36.24	57.74	0.689	822.4	557.8
850	576.85	1070	0.4152	1.110	1.349	37.63	60.30	0.693	877.5	574.1
900	626.85	1160	0.3920	1.121	1.345	38.97	62.76	0.696	933.4	589.6
950	676.85	1250	0.3714	1.132	1.340	40.26	65.20	0.699	989.7	604.9
1000	726.85	1340	0.3529	1.142	1.336	41.53	67.54	0.702	1046	619.5
1100	826.85	1520	0.3208	1.161	1.329	43.96			1162	648.0
1200	926.85	1700	0.2941	1.179	1.322	46.26			1279	675.2
1300	1026.85	1580	0.2714	1.197	1.316	48.46			1398	701.0
1400	1126.85	2060	0.2521	1.214	1.310	50.57			1518	725.9
1500	1226.85	2240	0.2353	1.231	1.304	52.61			1640	749.4
1600	1326.85	2420	0.2206	1.249	1.299	54.57			1764	772.6
1800	1526.85	2780	0.1960	1.288	1.288	58.29			2018	815.7

(Continued)

TABLE B.1 (CONTINUED)
Properties of Dry Air at Atmospheric Pressure—SI Units

Temperature			Properties							
K	°C	°F	ρ	c_p	c_p/c_v	μ	*k*	Pr	*h*	*a*
2000	1726.85	3140	0.1764	1.338	1.274				2280	855.5
2400	2126.85	3860	0.1467	1.574	1.238				2853	924.4
2800	2526.85	4580	0.1245	2.259	1.196				3599	983.1

Symbols and units: K, absolute temperature, kelvins; °C, temperature, degrees Celsius; °F, temperature, degree Fahrenheit; ρ, density, kg/m³; c_p, specific heat capacity, kJ/kg·K; c_p/c_v, specific heat capacity ratio, dimensionless; μ, viscosity [for N·s/m² (= kg/m·s) multiply tabulated values by 10^{-6}]; *k*, thermal conductivity, mW/m·K; Pr, Prandtl number, dimensionless; *h*, enthalpy,. KJ/kg; *a*, sound velocity, m/s.

Source: Reprinted with permission from R. C. Weast, ed., *Handbook of Tables for Applied Engineering Scioence*. Copyright © 1970, CRC Press, Inc., Boca Raton, FL.

TABLE B.2
Property Values of Gases at Atmospheric Pressure

T, K	ρ, kg/m³	c_p, Ws/kg · K	μ, kg/ms	ν, m²/s	k, W/m · K	α, m²/s	Pr
Helium							
3		5.200×10^3	8.42×10^{-7}		0.0106		
33	1.4657	5.200	50.2	3.42×10^{-6}	0.0353	0.04625×10^{-4}	0.74
144	3.3799	5.200	125.5	37.11	0.0928	0.5275	0.70
200	0.2435	5.200	156.6	64.38	0.1177	0.9288	0.694
255	0.1906	5.200	181.7	95.50	0.1357	1.3675	0.70
366	0.13280	5.200	230.5	173.6	0.1691	2.449	0.71
477	0.10204	5.200	275.0	269.3	0.197	3.716	0.72
589	0.08282	5.200	311.3	375.8	0.225	5.125	0.72
700	0.07032	5.200	347.5	494.2	0.251	6.661	0.72
800	0.06023	5.200	381.7	634.1	0.275	8.774	0.72
900	0.05286	5.200	413.6	781.3	0.298	10.834	0.72
Hydrogen							
30	0.84722	10.840×10^3	1.606×10^{-6}	1.895×10^{-6}	0.0228	0.02493×10^{-4}	0.759
50	0.50955	10.501	2.516	4.880	0.0362	0.0676	0.721
100	0.24572	11.229	4.212	17.14	0.0665	0.2408	0.712
150	0.16371	12.602	5.595	34.18	0.0981	0.475	0.718
200	0.12270	13.540	6.813	55.53	0.1282	0.772	0.719
250	0.09819	14.059	7.919	80.64	0.1561	1.130	0.713
300	0.08185	14.314	8.963	109.5	0.182	1.554	0.706
350	0.07016	14.436	9.954	141.9	0.206	2.031	0.697
400	0.06135	14.491	10.864	177.1	0.228	2.568	0.690
450	0.05462	14.499	11.779	215.6	0.251	3.164	0.682
500	0.04918	14.507	12.636	257.0	0.272	3.817	0.675
550	0.04469	14.532	13.475	301.6	0.292	4.516	0.668

TABLE B.2 (CONTINUED)
Property Values of Gases at Atmospheric Pressure

T, K	ρ, kg/m^3	c_p, Ws/kg · K	μ, kg/ms	ν, m^2/s	k, W/m · K	α, m^2/s	Pr
600	0.04085	14.537	14.285	349.7	0.315	5.306	0.664
700	0.03492	14.574	15.89	455.1	0.351	6.903	0.659
800	0.03060	14.675	17.40	569	0.384	8.563	0.664
900	0.02723	14.821	18.78	690	0.412	10.217	0.676
1000	0.02451	14.968	20.16	822	0.440	11.997	0.686
1100	0.02227	15.165	21.46	965	0.464	13.726	0.703
1200	0.02050	15.366	22.75	1107	0.488	15.484	0.715
1300	0.01890	15.575	24.08	1273	0.512	17.394	0.733
1333	0.01842	15.638	24.44	1328	0.519	18.013	0.736
Oxygen							
100	3.9918	0.9479×10^3	7.768×10^{-6}	1.946×10^{-6}	0.00903	0.023876×10^{-4}	0.815
150	2.6190	0.9178	11.490	4.387	0.01367	0.05688	0.773
200	1.9559	0.9131	14.850	7.593	0.01824	0.10214	0.745
250	1.5618	0.9157	17.87	11.45	0.02259	0.15794	0.725
300	1.3007	0.9203	20.63	15.86	0.02676	0.22353	0.709
350	1.1133	0.9291	23.16	20.80	0.03070	0.2968	0.702
400	0.9755	0.9420	25.54	26.18	0.03461	0.3768	0.695
450	0.8682	0.9567	27.77	31.99	0.03828	0.4609	0.694
500	0.7801	0.9722	29.91	38.34	0.04173	0.5502	0.697
550	0.7096	0.9881	31.97	45.05	0.04517	0.6441	0.700
600	0.6504	1.0044	33.92	52.15	0.04832	0.7399	0.704
Nitrogen							
100	3.4808	1.0722×10^3	6.862×10^{-6}	1.971×10^{-6}	0.009450	0.025319×10^{-4}	0.786
200	1.7108	1.0429	12.947	7.568	0.01824	0.10224	0.747
300	1.1421	1.0408	17.84	15.63	0.02620	0.22044	0.713
400	0.8538	1.0459	21.98	25.74	0.03335	0.3734	0.619
500	0.6824	1.0555	25.70	37.66	0.03984	0.5530	0.684
600	0.5687	1.0756	29.11	51.19	0.04580	0.7486	0.686
700	0.4934	1.0969	32.13	65.13	0.05123	0.9466	0.691
800	0.4277	1.1225	34.84	81.46	0.05609	1.1685	0.700
900	0.3796	1.1464	37.49	91.06	0.06070	1.3946	0.711
1000	0.3412	1.1677	40.00	117.2	0.06475	1.6250	0.724
1100	0.3108	1.1857	42.28	136.0	0.06850	1.8591	0.736
1200	0.2851	1.2037	44.50	156.1	0.07184	2.0932	0.748
Carbon Dioxide							
220	2.4733	0.783×10^3	11.105×10^{-6}	4.490×10^{-6}	0.010805	0.05920×10^{-4}	0.818
250	2.1657	0.804	12.590	5.813	0.012884	0.07401	0.793
300	1.7973	0.871	14.958	8.321	0.016572	0.10588	0.770
350	1.5362	0.900	17.205	11.19	0.02047	0.14808	0.755
400	1.3424	0.942	19.32	14.39	0.02461	0.19463	0.738
450	1.1918	0.980	21.34	17.90	0.02897	0.24813	0.721
500	1.0732	1.013	23.26	21.67	0.03352	0.3084	0.702
550	0.9739	1.047	25.08	25.74	0.03821	0.3750	0.685
600	0.8938	1.076	26.83	30.02	0.04311	0.4483	0.668

(Continued)

TABLE B.2 (CONTINUED)
Property Values of Gases at Atmospheric Pressure

T, K	ρ, kg/m³	c_p, Ws/kg · K	μ, kg/ms	ν, m²/s	k, W/m · K	α, m²/s	Pr
			Carbon Monoxide				
220	1.55363	1.0429×10^3	13.832×10^{-6}	8.903×10^{-6}	0.01906	0.11760×10^{-4}	0.758
250	0.8410	1.0425	15.40	11.28	0.02144	0.15063	0.750
300	1.13876	1.0421	17.843	15.67	0.02525	0.21280	0.737
350	0.97425	1.0434	20.09	20.62	0.02883	0.2836	0.728
400	0.85363	1.0484	22.19	25.99	0.03226	0.3605	0.722
450	0.75848	1.0551	24.18	31.88	0.0436	0.4439	0.718
500	0.68223	1.0635	26.06	38.19	0.03863	0.5324	0.718
550	0.62024	1.0756	27.89	44.97	0.04162	0.6240	0.721
600	0.56850	1.0877	29.60	52.06	0.04446	0.7190	0.724
			Ammonia, NH_3				
220	0.3828	2.198×10^3	7.255×10^{-6}	1.90×10^{-5}	0.0171	0.2054×10^{-4}	0.93
273	0.7929	2.177	9.353	1.18	0.0220	0.1308	0.90
323	0.6487	2.177	11.035	1.70	0.0270	0.1920	0.88
373	0.5590	2.236	12.886	2.30	0.0327	0.2619	0.87
423	0.4934	2.315	14.672	2.97	0.0391	0.3432	0.87
473	0.4405	2.395	16.49	3.74	0.0467	0.4421	0.84
			Steam (H_2O vapor)				
380	0.5863	2.060×10^3	12.71×10^{-6}	2.16×10^{-5}	0.0246	0.2036×10^{-4}	1.060
400	0.5542	2.014	13.44	2.42	0.0261	0.2338	1.040
450	0.4902	1.980	15.25	3.11	0.0299	0.307	1.010
500	0.4405	1.985	17.04	3.86	0.0339	0.387	0.996
550	0.4005	1.997	18.84	4.70	0.0379	0.475	0.991
600	0.3652	2.026	20.67	5.66	0.0422	0.573	0.986
650	0.3380	2.056	22.47	6.64	0.0464	0.666	0.995
700	0.3140	2.085	24.26	7.72	0.0505	0.772	1.000
750	0.2931	2.119	26.04	8.88	0.0549	0.883	1.005
800	0.2739	2.152	27.86	10.20	0.0592	1.001	1.010
850	0.2579	2.186	29.69	11.52	0.0637	1.130	1.019

Source: Eckert, E.R.G., & Drake, R.M. (1972). *Analysis of heat and mass transfer.* New York: McGraw-Hill.

TABLE B.3
Properties of Saturated Water

T (°C)	c_p (kJ/kg · °C)	ρ (kg/m³)	μ × 10³ (kg/m · s)	ν × 10⁶ (m²/s)	k (W/m · °C)	α × 10⁷ (m²/s)	β × 10³ (1/K)	Pr
0	4.218	99.8	1.791	1.792	0.5619	1.332	–0.0853	13.45
5	4.203	1000.0	1.520	1.520	0.5723	1.362	0.0052	11.16
10	4.193	999.8	1.308	1.308	0.5820	1.389	0.0821	9.42
15	4.187	999.2	1.139	1.140	0.5911	1.413	0.148	8.07
20	4.182	998.3	1.003	1.004	0.5996	1.436	0.207	6.99
25	4.180	997.1	0.8908	0.8933	0.6076	1.458	0.259	6.13
30	4.180	995.7	0.7978	0.8012	0.6150	1.478	0.306	5.42
35	4.179	994.1	0.7196	0.7238	0.6221	1.497	0.349	4.83
40	4.179	992.3	0.6531	0.6582	0.6286	1.516	0.389	4.34
45	4.182	990.2	0.5962	0.6021	0.6347	1.533	0.427	3.93
50	4.182	998.0	0.5471	0.5537	0.6405	1.550	0.462	3.57
55	4.184	985.7	0.5043	0.5116	0.6458	1.566	0.496	3.27
60	4.186	983.1	0.4668	0.4748	0.6507	1.581	0.529	3.00
65	4.187	980.5	0.4338	0.4424	0.6553	1.596	0.560	2.77
70	4.191	977.7	0.4044	0.4137	0.6594	1.609	0.590	2.57
75	4.191	974.7	0.3783	0.3881	0.6633	1.624	0.619	2.39
80	4.195	971.6	0.3550	0.3653	0.6668	1.636	0.647	2.23
85	4.201	968.4	0.3339	0.3448	0.6699	1.647	0.675	2.09
90	4.203	965.1	0.3150	0.3264	0.6727	1.659	0.702	1.97
95	4.210	961.7	0.2978	0.3097	0.6753	1.668	0.728	1.86
100	4.215	958.1	0.2822	0.2945	0.6775	1.677	0.755	1.76
120	4.246	942.8	0.2321	0.2461	0.6833	1.707	0.859	1.44
140	4.282	925.9	0.1961	0.2118	0.6845	1.727	0.966	1.23
160	4.339	907.3	0.1695	0.1869	0.6815	1.731	1.084	1.08
180	4.411	886.9	0.1494	0.1684	0.6745	1.724	1.216	0.98
200	4.498	864.7	0.1336	0.1545	0.6634	1.706	1.372	0.91
220	4.608	840.4	0.1210	0.1439	0.6483	1.674	1.563	0.86
240	4.770	813.6	0.1105	0.1358	0.6292	1.622	1.806	0.84
260	4.991	783.9	0.1015	0.1295	0.6059	1.549	2.130	0.84
280	5.294	750.5	0.0934	0.1245	0.5780	1.455	2.589	0.86
300	5.758	712.2	0.0858	0.1205	0.5450	1.329	3.293	0.91
320	6.566	666.9	0.0783	0.1174	0.5063	1.156	4.511	1.02
340	8.234	610.2	0.0702	0.1151	0.4611	0.918	7.170	1.25
360	16.138	526.2	0.0600	0.1139	0.4115	0.485	21.28	2.35

Source: A.J. Chapman. (1984). *Heat transfer* (4th ed.). New York: Macmillan. Reprinted with permission of Simon & Schuster, copyright © 1984.

TABLE B.4
Properties of Common Liquids—SI Units[a]

Common Name	Density, kg/m³	Specific Heat, kJ/kg · K	Viscosity, N · s/m²	Thermal Conductivity, W/m · K	Freezing Point, K	Latent Heat of Fusion, kJ/kg	Boiling Point, K	Latent Heat of Evaporation, kJ/kg	Coefficient of Cubical Expansion, K⁻¹
Acetic acid	1049	2.18	0.001155	0.171	290	181	391	402	0.0011
Acetone	784.6	2.15	0.000316	0.161	179.0	98.3	329	518	0.0015
Alcohol, ethyl	785.1	2.44	0.001095	0.171	158.6	108	351.46	846	0.0011
Alcohol, methyl	786.5	2.54	0.00056	0.202	175.5	98.8	337.8	1100	0.0014
Alcohol, propyl	800.0	2.37	0.00192	0.161	146	86.5	371	779	
Ammonia (aqua)	823.5	4.38		0.353					
Benzene	873.8	1.73	0.000601	0.144	278.68	126	353.3	390	0.0013
Bromine		0.473	0.00095		245.84	66.7	331.6	193	0.0012
Carbon disulfide	1261	0.992	0.00036	0.161	161.2	57.6	319.40	351	0.0013
Carbon tetrachloride	1584	0.866	0.00091	0.104	250.35	174	349.6	194	0.0013
Castor oil	956.1	1.97	0.650	0.180	263.2				
Chloroform	1465	1.05	0.00053	0.118	209.6	77.0	334.4	247	0.0013
Decane	726.3	2.21	0.000859	0.147	243.5	201	447.2	263	
Dodecane	754.6	2.21	0.001374	0.140	247.18	216	489.4	256	
Ether	713.5	2.21	0.000223	0.130	157	96.2	307.7	372	0.0016
Ethylene glycol	1097	2.36	0.0162	0.258	260.2	181	470	800	
Fluorine refrigerant R-11	1476	0.870[b]	0.00042	0.093[b]	162		297.0	180[c]	
Fluorine refrigerant R-12	1311	0.971[b]		0.071[b]	115	34.4	243.4	165[c]	

Fluorine refrigerant									
R-22	1194	1.26[b]		0.086[b]	113	183	232.4	232[c]	
Glycerine	1259	2.62	0.950	0.287	264.8	200	563.4	974	0.00054
Heptane	679.5	2.24	0.000376	0.128	182.54	140	371.5	318	
Hexane	654.8	2.26	0.000297	0.124	178.0	152	341.84	365	
Iodine		2.15			386.6	62.2	457.5	164	
Kerosene	820.1	2.09	0.00164	0.145				251	
Linseed oil	929.1	1.84	0.0331		253		560		
Mercury		0.139	0.00153		234.3	11.6	630	295	0.00018
Octane	698.6	2.15	0.00051	0.131	216.4	181	398	298	0.00072
Phenol	1072	1.43	0.0080	0.190	316.2	121	455		0.00090
Propane	493.5	2.41[b]	0.00011		85.5	79.9	231.08	428[c]	
Propylene	514.4	2.85	0.00009		87.9	71.4	225.45	342	
Propylene glycol	965.3	2.50	0.042		213		460	914	
Sea water	1025	3.76–4.10			270.6				
Toluene	862.3	1.72	0.000550	0.133	178	71.8	383.6	363	
Turpentine	868.2	1.78	0.001375	0.121	214		433	293	0.00099
Water	997.1	4.18	0.00089	0.609	273	333	373	2260	0.00020

[a] At 1.0 atm pressure (0.101325 MN/m^2), 300 K, except as noted.

[b] At 297 K, liquid.

[c] At 0.101325 MN/m^2, saturation temperature.

Source: Reprinted with permission from R. C. Weast, ed., *Handbook of Tables for Applied Engineering science*. Copyright © 1970, CRC Press, Inc., Boca Raton, FL.

(*Continued*)

TABLE B.5
Thermal Properties of Metals and Alloys

Metal	Properties at 20°C				Thermal Conductivity, k (W/m · °C)									
	ρ (kg/m^3)	c_p (kJ/kg · °C)	k (W/m · °C)	$\alpha \times 10^5$ (m^2/s)	−100°C	0°C	100°C	200°C	300°C	400°C	600°C	800°C	1000°C	1200°C
Aluminum														
Pure	2,707	0.896	204	8.418	215	202	206	215	228	249				
Al–Cu (Duralumin) 94–96 AL 3–5 Cu, trace Mg	2,787	0.883	164	6.676	126	159	182	194						
Al–Mg (Hydronalium) 91–95 Al, 5–9 Mg	2,611	0.904	112	4.764	93	109	125	142						
Al–Si (Silumin) 87 Al, 13 Si	2,659	0.871	164	7.099	149	163	175	185						
Al–Si (Silumin copper bearing) 86.5 Al, 12.5 Si, 1 Cu	2,659	0.867	137	5.933	119	137	144	152	161					
Al–Si (Alusil) 78–80 Al, 20–22 Si	2,627	0.854	161	7.172	144	157	168	175	178					
Al–Mg–Si, 97 Al, 1 Mg, 1 Si, 1 Mn	2,707	0.892	177	7.311		175	189	204						

Lead	11,373	0.130	35	2.343	36.9	35.1	33.4	31.5	29.8					
Iron														
Pure	7,897	0.452	73	2.026	87	73	67	62	55	48	40	36	35	36
Wrought iron (C < 0.5%)	7,849	0.46	59	1.626		59	57	52	48	45	36	33	33	33
Cast iron (C ≈ 4%)	7,272	0.42	52	1.703										
Steel (Cmax. ≈ 1.5%)														
Carbon Steel, C ≈ 0.5%	7,833	0.465	54	1.474		55	52	48	45	42	35	31	29	31
1.0%	7,801	0.473	43	1.172		43	43	42	40	36	33	29	28	29
1.5%	7,753	0.486	36	0.970		36	36	36	35	33	31	28	28	29
Iron														
Steel														
Nickel steel, Ni ≈ 0%	7.897	0.452	73	2.026	87	73	67	62	55	48	40	36	35	36
10%	7,945	0.46	26	0.720										
20%	7,993	0.46	19	0.526										
30%	8,073	0.46	12	0.325										
40%	8,169	0.46	10	0.279										
50%	8,266	0.46	14	0.361										
60%	8,378	0.46	19	0.493										
70%	8,506	0.46	26	0.666										
80%	8,618	0.46	35	0.872										

(Continued)

TABLE B.5 (CONTINUED)
Thermal Properties of Metals and Alloys

	Properties at 20°C				Thermal Conductivity, k (W/m · °C)									
Metal	ρ (kg/m³)	c_p (kJ/kg · °C)	k (W/m · °C)	$\alpha \times 10^5$ (m²/s)	−100°C	0°C	100°C	200°C	300°C	400°C	600°C	800°C	1000°C	1200°C
90%	8,762	0.46	47	1.156										
100%	8,906	0.448	90	2.276										
Invar, Ni = 36%	8,137	0.46	10.7	0.286										
Chrome steel, Cr = 0%	7,897	0.452	73	2.026	87	73	67	62	55	48	40	36	35	36
1%	7,865	0.46	61	1.665		62	55	52	47	42	36	33	33	
2%	7,865	0.46	52	1.443		54	48	45	42	38	33	31	31	
5%	7,833	0.46	40	1.110		40	38	36	36	33	29	29	29	
10%	7,785	0.46	31	0.867		31	31	31	29	29	28	28	29	
20%	7,689	0.46	22	0.635		22	22	22	22	24	24	26	29	
30%	7,625	0.46	19	0.542										
Cr–Ni (chrome-nickel);														
15 Cr, 10 Ni	7,865	0.46	19	0.526										
18 Cr, 8 Ni (V2A)	7,817	0.46	16.3	0.444		16.3	17	17	19	19	22	26	31	
20 Cr, 15 Ni	7,833	0.46	15.1	0.415										
25 Cr, 20 Ni	7,865	0.46	12.8	0.361										
Ni–Cr (nickel-chrome);														
80 Ni, 15 Cr	8,522	0.46	17	0.444										
60 Ni, 15 Cr	8,266	0.46	12.8	0.333										
40 Ni, 15 Cr	8,073	0.46	11.6	0.305										
20 Ni, 15 Cr	7,865	0.46	14.0	0.390		14.0	15.1	15.1	16.3	17	19	22		

Cr–Ni–Al; 6 Cr, 1.5 Al, 0.55 Si (Sicromal 8)	7,721	0.490	22	0.594										
24 Cr, 2.5 Al, 0.55 Si (Sicromal 12)	7,673	0.494	19	0.501										
Manganese steel, Mn = 0%	7,897	0.452	73	2.026	87	73	67	62	55	48	40	36	35	36
1%	7,865	0.46	50	1.388										
2%	7,865	0.46	38	1.050		36	36	36	36	35	33			
5%	7,849	0.46	22	0.637										
10%	7,801	0.46	17	0.483										
Tungsten steel, W = 0%	7,897	0.452	73	2.026	87	73	67	62	55	48	40	36	35	36
1%	7,913	0.448	66	1.858										
2%	7,961	0.444	62	1.763		62	59	54	48	45	36			
5%	8,073	0.435	54	1.525										
10%	8,314	0.419	48	1.391										
20%	8,826	0.389	43	1.249										
Silicon steel, Si = 0%	7,897	0.452	73	2.026	87	73	67	62	55	48	40	36	35	36
1%	7,769	0.46	42											
2%	7,673	0.46	31											
5%	7,417	0.46	19											
Copper														

(Continued)

TABLE B.5 (CONTINUED)
Thermal Properties of Metals and Alloys

Metal	Properties at 20°C				Thermal Conductivity, k (W/m · °C)									
	ρ (kg/m³)	c_p (kJ/kg · °C)	k (W/m · °C)	$\alpha \times 10^5$ (m²/s)	−100°C	0°C	100°C	200°C	300°C	400°C	600°C	800°C	1000°C	1200°C
Pure	8,954	0.3831	386	11.234	407	386	379	374	369	363	353			
Aluminum bronze:														
95 Cu,														
5 Al	8,666	0.410	83	2.330										
Bronze: 75 Cu, 25 Sn	8,666	0.343	26	0.859										
Red brass: 85 Cu, 9 Sn, 6 Zn	8,714	0.385	61	1.804		59	71							
Brass: 70 Cu, 30 Zn	8,522	0.385	111	3.412	88		128	144	147					
German silver: 62 Cu, 15 Ni,														
22 Zn	8,618	0.394	24.9	0.733	19.2		31	40	45	48				
Constantan: 60 Cu, 40 Ni	8,922	0.410	22.7	0.612	21		22.2	26						
Magnesium														
Pure	1,746	1.013	171	9.708	178	171	168	163	157					
Mg–Al (electrolytic):														
6–8 Al, 1–2 Zn	1,810	1.00	66	3.605		52	62	74	83					
Mg–Mn: 2% Mn	1,778	1.00	114	6.382	93	111	125	130						
Molybdenum	10,220	0.251	123	4.790	138	125	118	114	111	109	106	102	99	92
Nickel														
Pure (99.9%)	8,906	0.4459	90	2.266	104	93	83	73	64	59				
Impure (99.2%)	8,906	0.444	69	1.747		69	64	59	55	52	55	62	67	69
Ni–Cr: 90 Ni 10 Cr	8,666	0.444	17	0.444		17.1	18.9	20.9	22.8	24.6				
80 Ni, 20 Cr	8.314	0.444	12.6	0.343		12.3	13.8	15.6	17.1	18.9	22.5			

Silver												
Purest	10.524	0.2340	419	17.004	419	417	415	412				
Pure (99.9%)	10.524	0.2340	407	16.563	419	410	415	374	362	360		
Tin, pure	7,304	0.2265	64	3.884	74	65.9	59	57				
Tungsten	19,350	0.1344	163	6.271		166	151	142	133	126	112	76
Zinc, pure	7,144	0.3843	112.2	4.106	114	112	109	406	100	93		

Source: Data collected by R. Koch and R. M. Drake, as given in Eckert, E. R. G., and Drake, R. M. (1972) *Analysis of Heat and Mass Transfer*, McGraw-Hill, New York.

TABLE B.6
Properties of Other Materials

Description/ Composition	Temperature, K	Density, ρ, kg/m^3	Thermal Conductivity, k, W/m · K	Specific Heat, c_p, J/kg · K
Asphalt	300	2115	0.062	920
Bakelite	300	1300	1.4	1465
Brick, refractory				
Carborundum	872	—	18.5	—
	1672	—	11.0	—
Chrome brick	473	3010	2.3	835
	823		2.5	
	1173		2.0	
Diatomaceous silica, fired	478	—	0.25	—
	1145	—	0.30	
Fire clay, burnt 1600 K	773	2050	1.0	960
	1073	—	1.1	
	1373	—	1.1	
Fire clay, burnt 1725 K	773	2325	1.3	960
	1073		1.4	
	1373		1.4	
Fire clay brick	478	2645	1.0	960
	922		1.5	
	1478		1.8	
Magnesite	478	—	3.8	1130
	922	—	2.8	
	1478		1.9	
Clay	300	1460	1.3	880
Coal, anthracite	300	1350	0.26	1260
Concrete (stone mix)	300	2300	1.4	880
Cotton	300	80	0.06	1300
Foodstuffs				
Banana (75.7% water content)	300	980	0.481	3350
Apple, red (75% water content)	300	840	0.513	3600
Cake, batter	300	720	0.223	—
Cake, fully baked	300	280	0.121	—
Chicken meat, white	198	—	1.60	—
(74.4% water content)	233	—	1.49	
	253		1.35	
	263		1.20	
	273		0.476	
	283		0.480	
	293		0.489	
Glass				
Plate (soda lime)	300	2500	1.4	750
Pyrex	300	2225	1.4	835
Ice	273	920	1.88	2040
	253	—	2.03	1945
Leather (sole)	300	998	0.159	—

TABLE B.6 (CONTINUED)
Properties of Other Materials

Description/ Composition	Temperature, K	Density, ρ, kg/m³	Thermal Conductivity, k, W/m · K	Specific Heat, c_p, J/kg · K
Paper	300	930	0.180	1340
Paraffin	300	900	0.240	2890
Rock				
Granite, Barre	300	2630	2.79	775
Limestone, Salem	300	2320	2.15	810
Marble, Halston	300	2680	2.80	830
Quartzite, Sioux	300	2640	5.38	1105
Sandstone, Berea	300	2150	2.90	745
Rubber, vulcanized				
Soft	300	1100	0.13	2010
Hard	300	1190	0.16	—
Sand	300	1515	0.27	800
Soil	300	2050	0.52	1840
Snow	273	110	0.049	—
		500	0.190	—
Teflon	300	2200	0.35	—
	400		0.45	—
Tissue, human				
Skin	300	—	0.37	—
Fat layer (adipose)	300	—	0.2	—
Muscle	300	—	0.41	—
Wood, cross gain				
Balsa	300	140	0.055	—
Cypress	300	465	0.097	—
Fir	300	415	0.11	2720
Oak	300	545	0.17	2385
Yellow pine	300	640	0.15	2805
White pine	300	435	0.11	—
Wood, radial				
Oak	300	545	0.19	2385
Fir	300	420	0.14	2720

Source: Incropera, F.P., & Dewitt, D. P. (1990). *Fundamentals of heat and mass transfer* (3rd ed.). New York: Wiley. Copyright © 1990. Used with permission of John Wiley & Sons, Inc.

TABLE B.7
Emissivities ε_n of the Radiation in the Direction of the Normal to the Surface and ε of the Total Hemispherical Radiation for Various Materials for the Temperature $T^{\dagger\ddagger}$

Surface	T, °C	ε_n	ε
Gold, polished	130	0.018	
	400	0.022	
Silver	20	0.020	
Copper, polished	20	0.030	
Lightly oxidized	20	0.037	
Scraped	20	0.070	
Black oxidized	20	0.78	
Oxidized	131	0.76	0.725
Aluminum, bright rolled	170	0.039	0.049
	500	0.050	
Aluminum paint	100	0.20–0.40	
Silumin, cast polished	150	0.186	
Nickel, bright matte	100	0.041	0.046
Polished	100	0.045	0.053
Manganin, bright rolled	118	0.048	0.057
Chrome, polished	150	0.058	0.071
Iron, bright etched	150	0.128	0.158
Bright abrased	20	0.24	
Red rusted	20	0.61	
Hot rolled	20	0.77	
	130	0.60	
Hot cast	100	0.80	
Heavily rusted	20	0.85	
Heat-resistant oxidized	80	0.613	
	200	0.639	
Zinc, gray oxidized	20	0.23–0.28	
Lead, gray oxidized	20	0.28	
Bismuth, bright	80	0.340	0.366
Corundum, emery rough	80	0.855	0.84
Clay, fired	70	0.91	0.86
Lacquer, white	100	0.925	
Red lead	100	0.93	
Enamel, lacquer	20	0.85–0.95	
Lacquer, black matte	80	0.970	
Bakelite lacquer	80	0.935	
Brick, mortar, plaster	20	0.93	
Porcelain	20	0.92–0.94	
Glass	90	0.940	0.876

TABLE B.7 (CONTINUED)
Emissivities ε_n of the Radiation in the Direction of the Normal to the Surface and ε of the Total Hemispherical Radiation for Various Materials for the Temperature T[†‡]

Surface	T, °C	ε_n	ε
Ice, smooth, water	0	0.966	0.918
Rough crystals	0	0.985	
Waterglass	20	0.96	
Paper	95	0.92	0.89
Wood, beech	70	0.935	0.91
Tarpaper	20	0.93	

†: From measurements by E. Schmidt and E. Eckert.

‡: For metals, the emissivities rise with rising temperature, but for nonmetallic substances (metal oxides, organic substances) this rule is sometimes not correct. Where the exact measurements are not given, take for bright metal surfaces an average ratio $\varepsilon/\varepsilon_n = 1.2$ and for other substances with smooth surfaces $\varepsilon/\varepsilon_n = 0.95$; for rough surfaces use $\varepsilon/\varepsilon_n = 0.98$.

Source: Eckert, E.R.G., & Drake, R.M. (1972). *Analysis of heat and mass transfer.* New York: McGraw-Hill.

Appendix C: Interest Tables

C.1: 4 percent compound interest rate
C.2: 10 percent compound interest rate
C.3: 16 percent compound interest rate

A NOTE ON INTEREST TABLES

These tables are provided as a convenient reference for checking calculations on economic analysis. Some additional factors are also introduced. For other interest rates and compounding frequencies (these tables are for annual compounding), the formulas given in Chapter 6 may be used and other references cited in the chapter may be consulted.

TABLE C.1
4 Percent Compound Interest Rate

	Single Payment		Uniform Annual Series	
	Compound-Amount Factor	Present-Worth Factor	Series Present-Worth Factor	Capital-Recovery Factor
	$(1+i)^n$	$1/(1+i)^n$	$[(1+i)^n-1]/[i(1+i)^n]$	$[i(1+i)^n]/[(1+i)^n-1]$
	P to *F*	*F* to *P*	*S* to *P*	*P* to *S*
n	*F/P*	*P/F*	*P/S*	*S/P*
1	1.0400E 00	9.6154E-01	9.6154E-01	1.0400E 00
2	1.0816E 00	9.2456E-01	1.8861E 00	5.3020E-01
3	1.1249E 00	8.8900E-01	2.7751E 00	3.6035E-01
4	1.1699E 00	8.5480E-01	3.6299E 00	2.7549E-01
5	1.2167E 00	8.2193E-01	4.4518E 00	2.2463E-01
6	1.2653E 00	7.9031E-01	5.2421E 00	1.9076E-01
7	1.3159E 00	7.5992E-01	6.0021E 00	1.6661E-01
8	1.3686E 00	7.3069E-01	6.7327E 00	1.4853E-01
9	1.4233E 00	7.0259E-01	7.4353E 00	1.3449E-01
10	1.4802E 00	6.7556E-01	8.1109E 00	1.2329E-01
11	1.5395E 00	6.4958E-01	8.7605E 00	1.1415E-01
12	1.6010E 00	6.2460E-01	9.3851E 00	1.0655E-01

(Continued)

TABLE C.1 (CONTINUED)
4 Percent Compound Interest Rate

	Single Payment		Uniform Annual Series	
	Compound-Amount Factor	Present-Worth Factor	Series Present-Worth Factor	Capital-Recovery Factor
	$(1+i)^n$	$1/(1+i)^n$	$[(1+i)^n-1]/[i(1+i)^n]$	$[i(1+i)^n]/[(1+i)^n-1]$
	P to *F*	*F* to *P*	*S* to *P*	*P* to *S*
n	*F/P*	*P/F*	*P/S*	*S/P*
13	1.6651E 00	6.0057E-01	9.9856E 00	1.0014E-01
14	1.7317E 00	5.7748E-01	1.0563E 01	9.4669E-02
15	1.8009E 00	5.5526E-01	1.1118E 01	8.9941E-02
16	1.8730E 00	5.3391E-01	1.1652E 01	8.5820E-02
18	2.0258E 00	4.9363E-01	1.2659E 01	7.8993E-02
20	2.1911E 00	4.5639E-01	1.3590E 01	7.3582E-02
25	2.6658E 00	3.7512E-01	1.5622E 01	6.4012E-02
30	3.2434E 00	3.0832E-01	1.7292E 01	5.7830E-02
35	3.9461E 00	2.5342E-01	1.8665E 01	5.3577E-02
40	4.8010E 00	2.0829E-01	1.9793E 01	5.0523E-02
45	5.8412E 00	1.7120E-01	2.0720E 01	4.8262E-02
50	7.1067E 00	1.4071E-01	2.1482E 01	4.6550E-02

Reproduced with permission from Jelen, F. C., ed. (1970) *Cost and Optimization Engineering*, McGraw-Hill, New York.

		Uniform Gradient Series	Depreciation Series	
	Capitalized-Cost Factor	Present-Worth Factor	Sum-of-Digits Present-Worth Factor	Straight-Line Present-Worth Factor
	$[(1+i)^n]/[(1+i)^n-1]$	$[(P/S)-(nP/F)]/i$	$[n-(P/S)]/[0.5n(n+1)i]$	$1/[ni(K/P)]$
	P to *K*	*C* to *P*		
n	*K/P*	*P/C*		
1	2.6000E 01	0.0000E 00	9.6154E-01	9.6154E-01
2	1.3255E 01	9.2456E-01	9.4921E-01	9.4305E-01
3	9.0087E 00	2.7025E 00	9.3712E-01	9.2503E-01
4	6.8873E 00	5.2670E 00	9.2526E-01	9.0747E-01
5	5.6157E 00	8.5547E 00	9.1363E-01	8.9036E-01
6	4.7690E 00	1.2506E 01	9.0222E-01	8.7369E-01
7	4.1652E 00	1.7066E 01	8.9102E-01	8.5744E-01
8	3.7132E 00	2.2181E 01	8.8004E-01	8.4159E-01
9	3.3623E 00	2.7801E 01	8.6926E-01	8.2615E-01

		Uniform Gradient Series	Depreciation Series	
	Capitalized-Cost Factor	Present-Worth Factor	Sum-of-Digits Present-Worth Factor	Straight-Line Present-Worth Factor
	$[(1+i)^n]/[(1+i)^n-1]$	$[(P/S)-(nP/F)]/i$	$[n-(P/S)]/[0.5n(n+1)i]$	$1/[ni(K/P)]$
	P to *K*	*C* to *P*		
n	*K/P*	*P/C*		
10	3.0823E 00	3.3881E 01	8.5868E-01	8.1109E-01
11	2.8537E 00	4.0377E 01	8.4830E-01	7.9641E-01
12	2.6683E 00	4.7248E 01	8.3812E-01	7.8209E-01
13	2.5036E 00	5.4455E 01	8.2812E-01	7.6813E-01
14	2.3667E 00	6.1962E 01	8.1830E-01	7.5451E-01
15	2.2485E 00	6.9735E 01	8.0867E-01	7.4123E-01
16	2.1455E 00	7.7744E 01	7.9921E-01	7.2827E-01
18	1.9748E 00	9.4350E 01	7.8080E-01	7.0329E-01
20	1.8395E 00	1.1156E 02	7.6306E-01	6.7952E-01
25	1.6003E 00	1.5610E 02	7.2138E-01	6.2488E-01
30	1.4458E 00	2.0106E 02	6.8322E-01	5.7640E-01
35	1.3394E 00	2.4488E 02	6.4823E-01	5.3327E-01
40	1.2631E 00	2.8653E 02	6.1607E-01	4.9482E-01
45	1.2066E 00	3.2540E 02	5.8647E-01	4.6045E-01
50	1.1638E 00	3.6116E 02	5.5917E-01	4.2964E-01

TABLE C.2
10 Percent Compound Interest Rate

	Single Payment		Uniform Annual Series	
	Compound-Amount Factor	Present-Worth Factor	Series Present-Worth Factor	Capital-Recovery Factor
	$(1+i)^n$	$1/(1+i)^n$	$[(1+i)^n-1]/[i(1+i)^n]$	$[i(1+i)^n]/[(1+i)^n-1]$
	P to *F*	*F* to *P*	*S* to *P*	*P* to *S*
n	*F/P*	*P/F*	*P/S*	*S/P*
1	1.1000E 00	9.0909E-01	9.0909E-01	1.1000E 00
2	1.2100E 00	8.2645E-01	1.7355E 00	5.7619E-01
3	1.3310E 00	7.5131E-01	2.4869E 00	4.0211E-01
4	1.4641E 00	6.8301E-01	3.1699E 00	3.1547E-01
5	1.6105E 00	6.2092E-01	3.7908E 00	2.6380E-01
6	1.7716E 00	5.6447E-01	4.3553E 00	2.2961E-01
7	1.9487E 00	5.1316E-01	4.8684E 00	2.0541E-01
8	2.1436E 00	4.6651E-01	5.3349E 00	1.8744E-01

(*Continued*)

TABLE C.2 (CONTINUED)
10 Percent Compound Interest Rate

	Single payment		Uniform annual series	
	Compound-Amount Factor	Present-Worth Factor	Series Present-Worth Factor	Capital-Recovery Factor
	$(1+i)^n$	$1/(1+i)^n$	$[(1+i)^n - 1]/[i(1+i)^n]$	$[i(1+i)^n]/[(1+i)^n - 1]$
	P to *F*	*F* to *P*	*S* to *P*	*P* to *S*
n	*F/P*	*P/F*	*P/S*	*S/P*
9	2.3579E 00	4.2410E-01	5.7590E 00	1.7364E-01
10	2.5937E 00	3.8554E-01	6.1446E 00	1.6275E-01
11	2.8531E 00	3.5049E-01	6.4951E 00	1.5396E-01
12	3.1384E 00	3.1863E-01	6.8137E 00	1.4676E-01
13	3.4523E 00	2.8966E-01	7.1034E 00	1.4078E-01
14	3.7975E 00	2.6333E-01	7.3667E 00	1.3575E-01
15	4.1772E 00	2.3939E-01	7.6061E 00	1.3147E-01
16	4.5950E 00	2.1763E-01	7.8237E 00	1.2782E-01
18	5.5599E 00	1.7986E-01	8.2014E 00	1.2193E-01
20	6.7275E 00	1.4864E-01	8.5136E 00	1.1746E-01
25	1.0835E 01	9.2296E-02	9.0770E 00	1.1017E-01
30	1.7449E 01	5.7309E-02	9.4269E 00	1.0608E-01
35	2.8102E 01	3.5584E-02	9.6442E 00	1.0369E-01
40	4.5259E 01	2.2095E-02	9.7791E 00	1.0226E-01
45	7.2890E 01	1.3719E-02	9.8628E 00	1.0139E-01
50	1.1739E 02	8.5186E-03	9.9148E 00	1.0086E-01

Reproduced with permission from Jelen, F. C., ed. (1970) *Cost and Optimization Engineering*, McGraw-Hill, New York.

		Uniform Gradient Series	Depreciation Series	
	Capitalized-Cost Factor	Present-Worth Factor	Sum-of-Digits Present-Worth Factor	Straight-Line Present-Worth Factor
	$[(1+i)^n]/[(1+i)^n - 1]$	$[(P/S) - (nP/F)]/i$	$[n - (P/S)]/[0.5n(n+1)i]$	$1/[ni(K/P)]$
	P to *K*	*C* to *P*		
n	*K/P*	*P/C*		
1	1.1000E 01	0.0000E 00	9.0909E-01	9.0909E-01
2	5.7619E 00	8.2645E-01	8.8154E-01	8.6777E-01
3	4.0211E 00	2.3291E 00	8.5525E-01	8.2895E-01
4	3.1547E 00	4.3781E 00	8.3013E-01	7.9247E-01
5	2.6380E 00	6.8618E 00	8.0614E-01	7.5816E-01
6	2.2961E 00	9.6842E 00	7.8321E-01	7.2588E-01
7	2.0541E 00	1.2763E 01	7.6128E-01	6.9549E-01

n	Capitalized-Cost Factor	Uniform Gradient Series Present-Worth Factor	Depreciation Series Sum-of-Digits Present-Worth Factor	Straight-Line Present-Worth Factor
	$[(1+i)^n]/[(1+i)^n-1]$	$[(P/S)-(nP/F)]/i$	$[n-(P/S)]/[0.5n(n+1)i]$	$1/[ni(K/P)]$
	P to *K*	*C* to *P*		
	K/P	*P/C*		
8	1.8744E 00	1.6029E 01	7.4030E-01	6.6687E-01
9	1.7364E 00	1.9421E 01	7.2022E-01	6.3989E-01
10	1.6275E 00	2.2891E 01	7.0099E-01	6.1446E-01
11	1.5396E 00	2.6396E 01	6.8257E-01	5.9046E-00
12	1.4676E 00	2.9901E 01	6.6491E-01	5.6781E-00
13	1.4078E 00	3.3377E 01	6.4798E-01	5.4641E-01
14	1.3575E 00	3.6800E 01	6.3174E-01	5.2619E-01
15	1.3147E 00	4.0152E 01	6.1616E-01	5.0707E-01
16	1.2782E 00	4.3416E 01	6.0120E-01	4.8898E-01
18	1.2193E 00	4.9640E 01	5.7302E-01	4.5563E-01
20	1.1746E 00	5.5407E 01	5.4697E-01	4.2568E-01
25	1.1017E 00	6.7696E 01	4.8994E-01	3.6308E-01
30	1.0608E 00	7.7077E 01	4.4243E-01	3.1423E-01
35	1.0369E 00	8.3987E 01	4.0247E-01	2.7555E-01
40	1.0226E 00	8.8953E 01	3.6855E-01	2.4448E-01
45	1.0139E 00	9.2454E 01	3.3949E-01	2.1917E-01
50	1.0086E 00	9.4889E 01	3.1439E-01	1.9830E-01

TABLE C.3
16 Percent Compound Interest Rate

n	Single Payment Compound-Amount Factor	Present-Worth Factor	Uniform Annual Series Series Present-Worth Factor	Capital-Recovery Factor
	$(1+i)^n$	$1/(1+i)^n$	$[(1+i)^n-1]/[i(1+i)^n]$	$[i(1+i)^n]/[(1+i)^n-1]$
	P to *F*	*F* to *P*	*S* to *P*	*P* to *S*
	F/P	*P/F*	*P/S*	*S/P*
1	1.1600E 00	8.6207E-01	8.6207E-01	1.1600E 00
2	1.3456E 00	7.4316E-01	1.6052E 00	6.2296E-01
3	1.5609E 00	6.4066E-01	2.2459E 00	4.4526E-01
4	1.8106E 00	5.5229E-01	2.7982E 00	3.5738E-01
5	2.1003E 00	4.7611E-01	3.2743E 00	3.0541E-01
6	2.4364E 00	4.1044E-01	3.6847E 00	2.7139E-01
7	2.8262E 00	3.5383E-01	4.0386E 00	2.4761E-01
8	3.2784E 00	3.0503E-01	4.3436E 00	2.3022E-01
9	3.8030E 00	2.6295E-01	4.6065E 00	2.1708E-01
10	4.4114E 00	2.2668E-01	4.8332E 00	2.0690E-01
11	5.1173E 00	1.9542E-01	5.0286E 00	1.9886E-01
12	5.9360E 00	1.6846E-01	5.1971E 00	1.9241E-01

(*Continued*)

TABLE C.3 (CONTINUED)
16 Percent Compound Interest Rate

	Single Payment		Uniform Annual Series	
	Compound-Amount Factor	Present-Worth Factor	Series Present-Worth Factor	Capital-Recovery Factor
	$(1 + i)^n$	$1/(1 + i)^n$	$[(1 + i)^n - 1]/[i(1 + i)^n]$	$[i(1 + i)^n]/[(1 + i)^n - 1]$
	P to *F*	*F* to *P*	*S* to *P*	*P* to *S*
n	*F/P*	*P/F*	*P/S*	*S/P*
13	6.8858E 00	1.4523E-01	5.3423E 00	1.8718E-01
14	7.9875E 00	1.2520E-01	5.4675E 00	1.8290E-01
15	9.2655E 00	1.0793E-01	5.5755E 00	1.7936E-01
16	1.0748E 01	9.3041E-02	5.6685E 00	1.7641E-01
18	1.4463E 01	6.9144E-02	5.8178E 00	1.7188E-01
20	1.9461E 01	5.1385E-02	5.9288E 00	1.6867E-01
25	4.0874E 01	2.4465E-02	6.0971E 00	1.6401E-01
30	8.5850E 01	1.1648E-02	6.1772E 00	1.6189E-01
35	1.8031E 02	5.5459E-03	6.2153E 00	1.6089E-01
40	3.7872E 02	2.6405E-03	6.2335E 00	1.6042E-01
45	7.9544E 02	1.2572E-03	6.2421E 00	1.6020E-01
50	1.6707E 03	5.9855E-04	6.2463E 00	1.6010E-01

Reproduced with permission from Jelen, F.C. (Ed.). (1970). *Cost and optimization engineering*. New York: McGraw-Hill.

		Uniform Gradient Series	Depreciation Series	
	Capitalized-Cost Factor	Present-Worth Factor	Sum-of-Digits Present-Worth Factor	Straight-Line Present-Worth Factor
	$[(1 + i)^n]/[(1 + i)^n - 1]$	$[(P/S) - (nP/F)]/i$	$[n - (P/S)]/[0.5n(n + 1)i]$	$1/[ni(K/P)]$
	P to *K*	*C* to *P*		
n	*K/P*	*P/C*		
1	7.2500E 00	0.0000E 00	8.6207E-01	8.6207E-01
2	3.8935E 00	7.4316E-01	8.2243E-01	8.0262E-01
3	2.7829E 00	2.0245E 00	7.8553E-01	7.4863E-01
4	2.2336E 00	3.6814E 00	7.5114E-01	6.9955E-01
5	1.9088E 00	5.5858E 00	7.1904E-01	6.5486E-01
6	1.6962E 00	7.6380E 00	6.8907E-01	6.1412E-01
7	1.5476E 00	9.7610E 00	6.6103E-01	5.7694E-01
8	1.4389E 00	1.1896E 01	6.3479E-01	5.4295E-01
9	1.3568E 00	1.4000E 01	6.1020E-01	5.1184E-01
10	1.2931E 00	1.6040E 01	5.8713E-01	4.8332E-01
11	1.2429E 00	1.7994E 01	5.6547E-01	4.5715E-01
12	1.2026E 00	1.9847E 01	5.4510E-01	4.3309E-01

	Capitalized-Cost Factor	Uniform Gradient Series: Present-Worth Factor	Depreciation Series: Sum-of-Digits Present-Worth Factor	Depreciation Series: Straight-Line Present-Worth Factor
	$[(1+i)^n]/[(1+i)^n-1]$	$[(P/S)-(nP/F)]/i$	$[n-(P/S)]/[0.5n(n+1)i]$	$1/[ni(K/P)]$
	P to *K*	*C* to *P*		
n	*K/P*	*P/C*		
13	1.1699E 00	2.1590E 01	5.2594E-01	4.1095E-01
14	1.1431E 00	2.3217E 01	5.0789E-01	3.9054E-01
15	1.1210E 00	2.4728E 01	4.9086E-01	3.7170E-01
16	1.1026E 00	2.6124E 01	4.7479E-01	3.5428E-01
18	1.0743E 00	2.8583E 01	4.4525E-01	3.2321E-01
20	1.0542E 00	3.0632E 01	4.1878E-01	2.9644E-01
25	1.0251E 00	3.4284E 01	3.6352E-01	2.4388E-01
30	1.0118E 00	3.6423E 01	3.2020E-01	2.0591E-01
35	1.0056E 00	3.7633E 01	2.8556E-01	1.7758E-01
40	1.0026E 00	3.8299E 01	2.5737E-01	1.5584E-01
45	1.0013E 00	3.8660E 01	2.3405E-01	1.3871E-01
50	1.0006E 00	3.8852E 01	2.1448E-01	1.2493E-01

Appendix D: Heat Transfer Correlations

A NOTE ON HEAT TRANSFER CORRELATIONS

These tables present some of the most frequently used correlations for the convective heat transfer coefficients for a few common geometries and conditions. For other circumstances, various handbooks, encyclopedias, and other reference books mentioned in the text may be consulted.

TABLE D.1
Summary of Natural Convection Correlations for External Flows over Isothermal Surfaces

Geometry	Recommended Correlation[a]	Range
1. Vertical flat surfaces	$\mathrm{Nu} = \left\{0.825 + \dfrac{0.387\ \mathrm{Ra}^{1/6}}{[1+(0.492/\mathrm{Pr})^{9/16}]^{8/27}}\right\}^2$	$10^{-1} < \mathrm{Ra} < 10^{12}$
2. Inclined flat surfaces	Above equation with g replaced with $g\cos\gamma$	$\gamma \lesssim 60°$
3. Horizontal flat surfaces	$\mathrm{Nu} = 0.54\ \mathrm{Ra}^{1/4}$	$10^5 \lesssim \mathrm{Ra} \lesssim 10^7$
(a) Heated, facing upward	$\mathrm{Nu} = 0.15\ \mathrm{Ra}^{1/3}$	$10^7 \lesssim \mathrm{Ra} \lesssim 10^{10}$
(b) Heated, facing downward	$\mathrm{Nu} = 0.27\ \mathrm{Ra}^{1/4}$	$3 \times 10^5 \lesssim \mathrm{Ra} \lesssim 3 \times 10^{10}$
4. Horizontal cylinders	$\mathrm{Nu} = \left\{0.6 + \dfrac{0.387\ \mathrm{Ra}^{1/6}}{[1+(0.559/\mathrm{Pr})^{9/16}]^{8/27}}\right\}^2$	$10^{-5} \lesssim \mathrm{Ra} \lesssim 10^{12}$
5. Spheres	$\mathrm{Nu} = 2 + 0.43\ \mathrm{Ra}^{1/4}$	$\mathrm{Pr} \approx 1$ $1 \lesssim \mathrm{Ra} \lesssim 10^5$

[a] The average Nusselt number Nu and the Rayleigh number Ra are based on height L for the vertical plate, length L for inclined and horizontal surfaces, and diameter D for horizontal cylinders and spheres. All fluid properties are evaluated at the film temperature T_f, which is the average of surface and ambient temperatures.

Source: Kakac, S., Shah, R. K., Aung, W., eds. (1987) *Handbook of Single-Phase Convective Heat Transfer,* Wiley, New York. Copyright ©1987. Used with permission of John Wiley & Sons, Inc.

TABLE D.2
Correlation Equations for Natural Convection in Vertical Two-Dimensional Rectangular Enclosures

Aspect Ratio, $\alpha^* = H/W$	Laminar	Turbulent
Large	$\mathrm{Nu}_H = \dfrac{0.364[\mathrm{Ra}_H f_1(\mathrm{Pr})]^{1/4}}{G(\mathrm{Ra}_H, \alpha^*)}$	$\mathrm{Nu}_H = 0.05[\mathrm{Ra}_H f_1(\mathrm{Pr})]^{1/3}$
	$f_1(\mathrm{Pr}) = \left[1 + \left(\dfrac{0.5}{\mathrm{Pr}}\right)^{9/16}\right]^{16/9}$	$\mathrm{Ra}_H \alpha^{*-4} > 10^6$
	$G(\mathrm{Ra}_H, \alpha^*) = \left[1 + \dfrac{0.231}{(\mathrm{Ra}_H \alpha^{*4})^{1/4}}\right]^2$	
	$\alpha^* > 5,\ 10^2 < \mathrm{Ra}_H \alpha^{*-4} < 10^6$	
≈ 1	$\mathrm{Nu}_H = \alpha^* \left\{\left[\left(1 + \dfrac{\mathrm{Ra}_H^2 \alpha^{*-7}}{20{,}000}\right)^{-8} + \left(\dfrac{\mathrm{Ra}_H \alpha^{*-4} f_1(\mathrm{Pr})}{57G}\right)^{-2}\right]^{-3/8} + \dfrac{\mathrm{Ra}_H \alpha^{*-3} f_1(\mathrm{Pr})}{8000}\right\}^{-1/3}$	
	$\mathrm{Ra}_H < 10^9,\ \mathrm{Pr} \approx 0.7,\ \alpha^* \approx 1$	

$$\mathrm{Nu}_H = a\mathrm{Ra}_H^b$$

Pr	a	b
0.01	0.1344	0.259
0.03	0.1521	0.266
0.06	0.1613	0.271
0.10	0.1605	0.277

$\mathrm{Ra}_H < 10^6$, $\alpha^* = 1$

$$\mathrm{Nu}_H = \begin{cases} 0.082\ \mathrm{Ra}_H^{0.329} & \text{for} \quad 10^6 < \mathrm{Ra}_H \le 10^{12} \\ 1.325\ \mathrm{Ra}_H^{0.245} & \text{for} \quad 10^{12} < \mathrm{Ra}_H < 10^{16} \end{cases}$$

Small

$$\mathrm{Nu}_H = 1 + \left\{ \left(\frac{\mathrm{Ra}_H \alpha^*}{602.4} \right)^{-0.8} + \left[\left(\frac{\mathrm{Ra}_H f_1(\mathrm{Pr})}{10.66} \right)^{1/5} \alpha^{*-1} \right]^{-0.4} \right\}^{-2.5}$$

$\alpha^* < 0.6$, $\mathrm{Pr} \ge 0.7$, $\mathrm{Ra}_H < 10^8$

Source: Kakac, S., Shah, R. K., Aung, W., eds. (1987) *Handbook of Single-Phase Convective Heat Transfer*, Wiley, New York. Copyright ©1987. Used with permission of John Wiley & Sons, Inc.

TABLE D.3
Summary of Forced Convection Heat Transfer Correlations for External Flow

Correlation	Geometry	Conditions
$\delta = 5x\text{Re}_x^{-1/2}$	Flat plate	Laminar, T_f
$C_{f,x} = 0.664\text{Re}_x^{-1/2}$	Flat plate	Laminar, local, T_f
$\text{Nu}_x = 0.332\text{Re}_x^{1/2}\text{Pr}^{1/3}$	Flat plate	Laminar, local, T_f, $0.6 \lesssim \text{Pr} \lesssim 50$
$\delta_t = \delta\,\text{Pr}^{-1/3}$	Flat plate	Laminar, T_f
$\bar{C}_{f,x} = 1.328\text{Re}_x^{-1/2}$	Flat plate	Laminar, average, T_f
$\overline{\text{Nu}}_x = 0.664\text{Re}_x^{1/2}\text{Pr}^{1/3}$	Flat plate	Laminar, average, T_f, $0.6 \lesssim \text{Pr} \lesssim 50$
$\text{Nu}_x = 0.565\text{Pe}_x^{1/2}$	Flat plate	Laminar, local, T_f, $\text{Pr} \lesssim 0.05$
$C_{f,x} = 0.0592\text{Re}_x^{-1/5}$	Flat plate	Turbulent, local, T_f, $\text{Re}_x \lesssim 10^8$
$\delta = 0.37x\text{Re}_x^{-1/5}$	Flat plate	Turbulent, local, T_f, $\text{Re}_x \lesssim 10^8$
$\text{Nu}_x = 0.0296\text{Re}_x^{4/5}\text{Pr}^{1/3}$	Flat plate	Turbulent, local, T_f, $\text{Re}_x \lesssim 10^8$, $0.6 \lesssim \text{Pr} \lesssim 60$
$\bar{C}_{f,L} = 0.074\text{Re}_L^{-1/5} - 1742\text{Re}_L^{-1}$	Flat plate	Mixed, average, T_f, $\text{Re}_{x,c} = 5 \times 10^5$, $\text{Re}_L \lesssim 10^8$
$\overline{\text{Nu}}_L = (0.037\text{Re}_L^{4/5} - 871)\text{Pr}^{1/3}$	Flat plate	Mixed, average, T_f, $\text{Re}_{x,c} = 5 \times 10^5$, $\text{Re}_L \lesssim 10^8$, $0.6 < \text{Pr} < 60$
$\overline{\text{Nu}}_D = C\text{Re}_D^m\text{Pr}^{1/3}$ (Table D3.1)	Cylinder and noncircular cylinder	Average, T_f, $0.4 < \text{Re}_D < 4 \times 10^5$, $\text{Pr} \gtrsim 0.7$
$\overline{Nu}_D = C\text{Re}_D^m\text{Pr}^n(\text{Pr}/\text{Pr}_s)^{1/4}$ (Table D3.2)	Cylinder	Average, T_∞, $1 < \text{Re}_D < 10^6$, $0.7 < \text{Pr} < 500$
$\overline{\text{Nu}}_D = 0.3 + [0.62\text{Re}_D^{1/2}\text{Pr}^{1/3} \times [1 + (0.4/\text{Pr})^{2/3}]^{-1/4}] \times [1 + (\text{Re}_D/282{,}000)^{5/8}]^{4/5}$	Cylinder	Average, T_f, $\text{Re}_D\,\text{Pr} > 0.2$
$\overline{\text{Nu}}_D = 2 + (0.4\text{Re}_D^{1/2} + 0.06\text{Re}_D^{2/3})\text{Pr}^{0.4} \times (\mu/\mu_s)^{1/4}$	Sphere	Average, T_∞, $3.5 < \text{Re}_D < 7.6 \times 10^4$, $0.71 < \text{Pr} < 380$, $1.0 < (\mu/\mu_s) < 3.2$
$\overline{\text{Nu}}_D = 2 + 0.6\text{Re}_D^{1/2}\text{Pr}^{1/3}$	Falling drop	Average, T_∞
$\overline{\text{Nu}}_D = 1.13C_1\text{Re}_{D,\max}^m\text{Pr}^{1/3}$ (Table D3.3)	Tube bank	Average, $\bar{T}_f$, $2000 < \text{Re}_{D,\max} < 4 \times 10^4$, $\text{Pr} \geq 0.7$

TABLE D.3 (CONTINUED)
Summary of Forced Convection Heat Transfer Correlations for External Flow

Correlation	Geometry	Conditions
$\overline{\mathrm{Nu}}_D = C\mathrm{Re}_{D,\max}^{m}\,\mathrm{Pr}^{0.36}(\mathrm{Pr}/\mathrm{Pr}_s)^{1/4}$ (Table D3.4)	Tube bank	Average, $\overline{T}$, $1000 < \mathrm{Re}_D < 2 \times 10^6$, $0.7 < \mathrm{Pr} < 500$
Single round nozzle	Impinging jet	Average T_f, $2000 < \mathrm{Re} < 4 \times 10^5$, $2 < (H/D) < 12$, $2.5 < (r/D) < 7.5$
Single slot nozzle	Impinging jet	Average, T_f, $3000 < \mathrm{Re} < 9 \times 10^4$, $2 < (H/W) < 10$, $4 < (x/W) < 20$

Here δ, C_f, and Nu are the boundary layer thickness, friction coefficient, and Nusselt number; H for a nozzle is distance from surface, W the slot width, and D the nozzle diameter.

Source: Incropera, F. P., and Dewitt, D. P. (1996) *Fundamentals of Heat and Mass Transfer*, 6th ed., Wiley, New York, Copyright ©1996. Used with permission of John Wiley & Sons, Inc.

TABLE D.3.1

Re_D	C	m
0.4–4	0.989	0.330
4–40	0.911	0.385
40–4000	0.683	0.466
4000–40,000	0.193	0.618
40,000–400,000	0.027	0.805

Geometry	Re_D	C	m
Square			
V → (diamond) D	5×10^3–10^5	0.246	0.588
V → (square) D	5×10^3–10^5	0.102	0.675
Hexagon			
V → (hexagon) D	5×10^3–1.95×10^4	0.160	0.638
	1.95×10^4–10^5	0.0385	0.782
V → (hexagon) D	5×10^3–10^5	0.153	0.638
Vertical plate			
V → (plate) D	4×10^3–1.5×10^4	0.228	0.731

TABLE D.3.2

RE_D	C	m
1–40	0.75	0.4
40–1000	0.51	0.5
10^3–2×10^5	0.26	0.6
2×10^5–10^6	0.076	0.7

Source: Incropera, F. P., and Dewitt, D. P. (1996) *Fundamentals of Heat and Mass Transfer*, 6th ed., Wiley, New York. Copyright ©1996. Used with permission of John Wiley & Sons, Inc.

TABLE D.3.3

	S_T/D							
	1.25		1.5		2.0		3.0	
S_L/D	C_1	m	C_1	m	C_1	m	C_1	m
Aligned								
1.25	0.348	0.592	0.275	0.608	0.100	0.704	0.0633	0.752
1.50	0.367	0.586	0.250	0.620	0.101	0.702	0.0678	0.744
2.00	0.418	0.570	0.299	0.602	0.229	0.632	0.198	0.648
3.00	0.290	0.601	0.357	0.584	0.374	0.581	0.286	0.608
Staggered								
0.600	—	—	—	—	—	—	0.213	0.636
0.900	—	—	—	—	0.446	0.571	0.401	0.581
1.000	—	—	0.497	0.558	—	—	—	—
1.125	—	—	—	—	0.478	0.565	0.518	0.560
1.250	0.518	0.556	0.505	0.554	0.519	0.556	0.522	0.562
1.500	0.451	0.568	0.460	0.562	0.452	0.568	0.488	0.568
2.000	0.404	0.572	0.416	0.568	0.482	0.556	0.449	0.570
3.000	0.310	0.592	0.356	0.580	0.440	0.562	0.428	0.574

C_2: $\overline{Nu_D}|_{N_L<10} = C_2 \overline{Nu_D}|_{N_L \geq 1}$

N_L	1	2	3	4	5	6	7	8	9
Aligned	0.64	0.80	0.87	0.90	0.92	0.94	0.96	0.98	0.99
Staggered	0.68	0.75	0.83	0.89	0.92	0.95	0.97	0.98	0.99

S_L and S_T are tube spacings along and transverse to the flow direction, respectively; N_L is the number of rows in the flow direction.

Source: Incropera, F. P., and Dewitt, D. P. (1996) *Fundamentals of Heat and Mass Transfer*, 6th ed., Wiley, New York. Copyright ©1996. Used with permission of John Wiley & Sons, Inc.

TABLE D.3.4

Configuration	$Re_{D,max}$	C	m
Aligned	$10–10^2$	0.80	0.40
Staggered	$10–10^2$	0.90	0.40
Aligned	$10^2 – 10^3$	Approximate as a single (isolated) cylinder	
Staggered	$10^2 – 10^3$		
Aligned ($S_T/S_L > 0.7$)[a]	$10^3–2 \times 10^5$	0.27	0.63
Staggered ($S_T/S_L < 2$)	$10^3–2 \times 10^5$	$0.35(S_T/S_L)^{1/5}$	0.60
Staggered ($S_T/S_L > 2$)	$10^3–2 \times 10^5$	0.40	0.60
Aligned	$2 \times 10^5–2 \times 10^6$	0.021	0.84
Staggered	$2 \times 10^5–2 \times 10^6$	0.022	0.84

C_2: $\overline{Nu_D}|_{N_L<20} = C_2 \overline{Nu_D}|_{N_L \geq 20}$

N_L	1	2	3	4	5	7	10	13	16
Aligned	0.70	0.80	0.86	0.90	0.92	0.95	0.97	0.98	0.99
Staggered	0.64	0.76	0.84	0.89	0.92	0.95	0.97	0.98	0.99

Source: Incropera, F. P., and Dewitt, D. P. (1996) *Fundamentals of Heat and Mass Transfer*, 6th ed., Wiley, New York. Copyright ©1996. Used with permission of John Wiley & Sons, Inc.

TABLE D.4
Summary of Forced Convection Correlations for Flow in a Circular Tube

Correlation	Conditions
$f = 64/Re_D$	Laminar, fully developed
$Nu_D = 4.36$	Laminar, fully developed, uniform q_s, $Pr \gtrsim 0.6$
$Nu_D = 3.66$	Laminar, fully developed, uniform T_s, $Pr \gtrsim 0.6$
$\overline{Nu}_D = 3.66 + \dfrac{0.0668(D/L)Re_D Pr}{1+0.04[(D/L)Re_D Pr]^{2/3}}$	Laminar, thermal entry length ($Pr >> 1$ or an unheated starting length), uniform T_s
or	
$\overline{Nu}_D = 1.86\left(\dfrac{Re_D Pr}{L/D}\right)^{1/3}\left(\dfrac{\mu}{\mu_s}\right)^{0.14}$	Laminar, combined entry length $\{[Re_D Pr/(L/D)]^{1/3}(\mu/\mu_s)^{0.14}\} \gtrsim 2$, uniform T_s, $0.48 < Pr < 16{,}700$, $0.0044 < (\mu/\mu_s) < 9.75$
$f = 0.316 Re_D^{-1/4}$	Turbulent, fully developed, $Re_D \lesssim 2 \times 10^4$
$f = 0.184 Re_D^{-1/5}$	Turbulent, fully developed, $Re_D \gtrsim 2 \times 10^4$

(Continued)

TABLE D.4 (CONTINUED)
Summary of Forced Convection Correlations for Flow in a Circular Tube

Correlation	Conditions
or	
$f = (0.790 \ln Re_D - 1.64)^{-2}$	Turbulent, fully developed, $3000 \lesssim Re_D \lesssim 5 \times 10^6$
$Nu_D = 0.023 Re_D^{4/5} Pr^n$	Turbulent, fully developed, $0.6 \leq Pr \leq 160$, $Re_D \geq 10{,}000$, $(L/D) \gtrsim 10$, $n = 0.4$ for heating and $n = 0.3$ for cooling
or	
$Nu_D = 0.027 Re_D^{4/5} Pr^{1/3} \left(\frac{\mu}{\mu_s}\right)^{0.14}$	Turbulent, fully developed, $0.7 \leq Pr \leq 16{,}700$, $Re_D \gtrsim 10{,}000$, $L/D \gtrsim 10$
or	
$Nu_D = \frac{(f/8)(Re_D - 1000) Pr}{1 + 12.7 (f/8)^{1/2} (Pr^{2/3} - 1)}$	Turbulent, fully developed, $0.5 < Pr < 2000$, $3000 \lesssim Re_D \lesssim 5 \times 10^6$, $(L/D) \gtrsim 10$
$Nu_D = 4.82 + 0.0185 (Re_D Pr)^{0.827}$	Liquid metals, turbulent, fully developed, uniform q_s, $3.6 \times 10^3 < Re_D < 9.05 \times 10^5$, $10^2 < Pe_D < 10^4$
$Nu_D = 5.0 + 0.025 (Re_D Pr)^{0.8}$	Liquid metals, tubulent, fully developed, uniform T_s, $Pe_D > 100$

Here f is friction factor, L is length, and D is diameter.

Source: Incropera, F. P., and Dewitt, D. P. (1996), *Fundamentals of Heat and Mass Transfer*, 6th ed., Wiley, New York.

Index

Note: Page numbers in **bold** indicate tables; those in *italics* indicate figures.

D

E

N

Q

R

S

U